LONGMAN DICTIONARY OF

Environmental Science

LONGMAN DICTIONARY OF

Environmental Science

Eleanor Lawrence
Andrew R. W. Jackson
Julie M. Jackson

LONGMAN

Addison Wesley Longman Limited
Edinburgh Gate, Harlow
Essex CM20 2JE
England

and Associated Companies throughout the World

First published 1998

ISBN 0 582–25356–X

British Library Cataloguing in Publication Data
A catalogue record for this book is
available from the British Library.

Set by 35 in 8/9pt Times
Produced by Addison Wesley Longman Singapore (Pte) Ltd.,
Printed in Singapore

CONTENTS

PREFACE

Environmental science is the systematic study of the natural and man-made world. Interest in the environment and the impact of human activity upon it has grown steadily in recent years. Now a major discipline in its own right, environmental science draws on expertise from many traditional subject areas such as biology, geography, geology, chemistry and agriculture. The aim of this wide-ranging dictionary is to provide clear, concise definitions across this broad spectrum of subject areas, with a special emphasis on the scientific terms that are likely to be encountered in a study of environmental science.

The classification of the living world followed in this edition is that of the 'five kingdoms' scheme. Entries in the body of the dictionary are given for all the main phyla, divisions and classes of plants, fungi, animals, protists and prokaryotes, with some orders being included for groups such as insects, birds, mammals and flowering plants. Entries for many common names of organisms are also included. The appendices at the back of the book include basic information on the chemical elements, the geological timescale, and an outline of the various kingdoms of living organisms.

Terms are arranged in alphabetical order, with abbreviations and acronyms included in their appropriate place within the body of the dictionary. Numbers, Greek letters, and configurational letters at the beginning of chemical names are ignored for alphabetization purposes. Within an entry, different meanings of a term are numbered and separated by semicolons. The abbreviations (*bot.*), (*zool.*), (*chem.*), (*geol.*) etc. have been used in some cases to indicate more clearly which subject area a definition refers to. We aim to have defined most of the technical terms that may be used as part of a definition. To avoid complicating the text with excessive cross-referencing therefore, words that are defined elsewhere in the dictionary are not always indicated by (*q.v.*).

Common suffixes and prefixes derived from Latin and Greek are entered in the body of the dictionary along with their usual meanings, and Appendix 11 gives etymological origins of some common word elements.

We should like to thank the staff of Addison Wesley Longman for their help and encouragement throughout the project. Comments concerning errors or omissions in this edition will be greatly appreciated, so that they may be rectified in future reprints and editions.

Eleanor Lawrence
Andrew R. W. Jackson
Julie M. Jackson

ABBREVIATIONS

a. adjective
adv. adverb
agric. agriculture
alt. alternative (synonym)
appl. applies or applied to
bact. bacteriology
behav. behavioural science
biochem. biochemistry
biol. biology
bot. botany
ca. circa (approximately)
cf. compare
chem. chemistry
clim. climatology
ecol. ecology
e.g. for example
esp. especially
et al. and others
etc. and so forth
genet. genetics
geog. geography
geol. geology
Gk Greek
hort. horticulture

i.e. that is
L. Latin
microbiol. microbiology
mol. biol. molecular biology
mycol. mycology
n. noun
n haploid number of chromosomes
oceanogr. oceanography
pert. pertaining to
phys. physics
physiol. physiology
plant ecol. plant ecology
plu. plural
p.p.m. parts per million
q.v. see
SI Système International
sing. singular
soil. sci. soil science
sp. species (*sing.*)
spp. species (*plu.*)
v. verb
virol. virology
zool. zoology

UNITS AND CONVERSIONS
(see dictionary for further definition)

Basic SI units
ampere, A (electric current)
candela, cd (luminous intensity)
kelvin, K (temperature)
kilogram, kg (mass)
metre, m (length)

Units and conversions

mole, mol (amount of substance)
second, s (time)

Some common derived SI units

joule, J (energy)	$kg\ m^2\ s^{-2}$
molar, M (concentration)	$mol\ dm^{-3}\ (mol\ l^{-1})$
newton, N (force)	$kg\ m\ s^{-2}$
pascal, Pa (pressure)	$kg\ m^{-1}\ s^{-2}$
volt, V (electrical potential)	
watt, W (power)	$kg\ m^2\ s^{-3}$

Conversions from SI or metric

centimetre, cm (10^{-2} m)	0.394 inches (in.)
degree Celsius (centigrade), °C*	(9/5) °F
gram, g (10^{-3} kg)	0.035 ounces (oz.)
hectare, ha (10^4 m^2)	2.471 acres
joule, J (kg m^2 s^{-2})	0.239 calories (cal.)
kilogram, kg	2.20 pounds (lb)
litre, l (dm^3)	1.76 pints (pt)
metre	39.37 in.
millimetre, mm (10^{-3} m)	0.039 in.
tonne (metric ton), t (1 Mg)	0.984 tons

Conversions to SI or metric

acre (4840 sq. yd)	4046.86 m^2
ångström unit, Å	10^{-10} m
atmosphere, standard, atm (14.72 p.s.i.)	101 325 Pa
bar	10^5 Pa
British thermal unit, Btu	1.055 kJ
British thermal unit/hour, Btu h^{-1}	0.293 W
bushel, bu	0.0364 m^3
bushel (US), bu	0.0352 m^3
calorie, thermochemical	4.184 J
cubic foot, cu. ft, ft^3	0.0283 m^3
cubic inch, cu. in., in^3	16.387 cm^3
cubic yard, cu. yd, yd^3	0.7645 m^3
degree Fahrenheit, °F*	(5/9) °C
dram (avoirdupois), dr	1.772 g
fathom (6 ft)	1.829 m
fluid ounce, fl. oz	28.413 cm^3
foot, ft	0.3048 m
gallon, gal	4.546 dm^3
gallon (US), US gal	3.785 dm^3
grain, gr	64.799 mg
hundredweight, cwt	50.802 kg
inch, in.	25.4 mm

kilocalorie/h, kcal h^{-1}	1.163 W
mile	1.6093 km
millibar, mbar	100 Pa
millimetre of mercury, mm Hg	13.332 Pa
millimetre of water	9.807 Pa
ounce (avoirdupois), oz	0.0283 kg
pint, pt	0.568 dm^3
pound (avoirdupois), lb	0.4536 kg
square foot, sq. ft, ft^2	0.0929 m^2
square inch, sq. in., in^2	645.16 m^2
square mile, sq. mile (640 acres)	2.590 km^2
square yd, sq. yd, yd^2	0.836 m^2
ton (long) (2240 lb)	1016.05 kg
yard, yd	0.9144 m

* To convert temperature in °C to °F, multiply by 9/5 and add 32; to convert °F to °C subtract 32, then multiply by 5/9.

SI PREFIXES

The following prefixes may be used to construct decimal multiples of units.

Multiple	Prefix	Symbol	Multiple	Prefix	Symbol
10^{-1}	deci	d	10	deca	da
10^{-2}	centi	c	10^2	hecto	h
10^{-3}	milli	m	10^3	kilo	k
10^{-6}	micro	μ	10^6	mega	M
10^{-9}	nano	n	10^9	giga	G
10^{-12}	pico	p	10^{12}	tera	T
10^{-15}	femto	f			
10^{-18}	atto	a			

GREEK ALPHABET

Name	Greek letter		English equivalent	Name	Greek letter		English equivalent
alpha	A	α	a	theta	Θ	θ	th
beta	B	β	b	iota	I	ι	i
gamma	Γ	γ	g	kappa	K	κ	k
delta	Δ	δ	d	lambda	Λ	λ	l
epsilon	E	ε	e	mu	M	μ	m
zeta	Z	ζ	z	nu	N	ν	n
eta	H	η	e	xi	Ξ	ξ	x

Name	Greek letter		English equivalent	Name	Greek letter		English equivalent
omicron	O	*o*	o	upsilon	Y	*υ*	u
pi	Π	*π*	p	phi	Φ	*φ*	ph
rho	P	*ρ*	r	chi	X	*χ*	ch
sigma	Σ	*σ*	s	psi	Ψ	*ψ*	ps
tau	T	*τ*	t	omega	Ω	*ω*	o

COMMON LATIN AND GREEK NOUN ENDINGS*

sing.	*plu.*
-a	-ae (L.)
-a	-ata (Gk)
-is	-es (L.)
-on	-a (Gk)
-um	-a (L.)
-us	-i (L.)

* Some familiar exceptions to these rules include genus (*sing.*), genera (*plu.*) and opus (*sing.*), opera (*plu.*).

A

α alpha, (1) symbol used to represent the alpha particle (*q.v.*); (2) for headwords with prefix α- refer to alpha-.

A (1) absorbance (*q.v.*); (2) adenine (*q.v.*); (3) alanine (*q.v.*); (4) ampere *q.v.*

Å Ångström *q.v.*

A mass number *q.v.*

aa *n.* a viscous, slow-moving lava flow that hardens to give a rough surface composed of jagged clinker-like fragments. *cf.* blocky lava.

abacá Manila hemp, *see* hemp.

abattoir *n.* slaughter-house for cattle and other domestic livestock.

abaxial *a. pert.* the surface of a structure that is furthest from, or is turned away from, the axis.

Aberfan disaster the collapse and subsequent slide of the Aberfan colliery tip (Merthyr Valley, South Wales, UK) on 21 October 1966, which engulfed the village school and neighbouring houses, claiming the lives of 144 people, 116 of them children.

aberrant *a.* with characteristics not in accordance with type, *appl.* species, etc.

abiocoen *n.* the non-living parts of the environment in total.

abiogenesis *n.* the production of living from non-living matter, as in the origin of life.

abiology *n.* the study of non-living things in a biological context.

abiosis *n.* apparent suspension of life.

abiotic *a.* non-living.

abiotic environment that part of an organism's environment consisting of non-biological factors, such as topography, geology, climate and inorganic nutrients.

abiotic phase *see* biogeochemical cycle.

A-bomb *see* nuclear bomb.

above-ground *a.* in an ecosystem, *appl.* the material, structures and organisms that occur above the soil surface, e.g. stems and leaves of plants. *cf.* below-ground.

abrasion *n.* erosion of a rock surface by material carried across it by wind, ice, water or mass movement. *alt.* corrasion.

abrupt speciation the formation of a species as a result of a sudden change in chromosome number or constitution.

abscisic acid (ABA) sesquiterpene plant hormone first discovered through its promotion of leaf senescence and leaf fall, but whose major functions are in adaptation of plants to environmental stresses such as drought, where it induces stomatal closure and synthesis of proteins that protect against the effects of desiccation. It is also involved in the maintenance of dormancy in seeds and buds.

absenteeism *n.* the practice of certain animals of nesting away from their offspring and visiting them from time to time to provide them with food and a minimum of care.

absolute age the age of a rock, fossil or archaeological specimen in years before present (BP), which can be determined within the bounds of experimental error by radioisotope dating methods, or for wooden artefacts sometimes determined by dendrochronology.

absolute dating techniques *see* dating techniques.

absolute error (*chem.*) the difference between the measured value and the true, or accepted, value. Where the measured value is greater than the true value, its absolute error is shown as a positive number.

absolute humidity mass of water vapour present in a given volume of air, expressed as grams per cubic metre (g m^{-3}).

1

absolute resource scarcity a lack of a particular resource so that present or future demand cannot be met. *cf.* relative resource scarcity.

absolute temperature a quantitative characteristic of a system, at thermal equilibrium, that is a correlate of the thermal energy embodied in the random motion of its constituent particles. In the SI system, absolute temperature is recorded in kelvins (K). In terms of magnitude, 1 K is equal to 1 °C. To convert from degrees Celsius to kelvins, add 273.15. *alt.* thermodynamic temperature.

absolute zero the temperature at which a perfect crystal has zero entropy. This is the lowest temperature theoretically achievable and equals zero kelvin (i.e. 0 K).

absorbance (A) *n.* a spectrophotometric measurement of the absorption of light at a particular wavelength by a substance in solution, which can be used, e.g., to determine the concentration of the substance in solution and to follow conversion of substrate to product in enzyme reactions. *alt.* extinction, optical density.

absorption *n.* (1) the uptake of one substance by another, as the uptake of a gas by a solid or liquid; (2) uptake of fluid and solutes by living cells and tissues, *alt.* absorptive nutrition; (3) of light, when neither reflected nor transmitted.

absorption spectrum the characteristic plot of wavelength versus intensity of electromagnetic radiation that has been absorbed by a given substance.

absorptive capacity measure of the amount of waste that can be discharged into a given environment without causing any adverse ecological or landscape changes. *alt.* assimilative capacity.

abstraction river capture *q.v.*

abundance *n.* general term for the total number of individuals of a species, type, etc., present in a given area.

abyssal *a. appl.* (1) or *pert.* the ocean depths, below 2000 m, the depths of the ocean beyond the continental shelf, where light does not penetrate, and to the organisms and material usually found there; (2) the depths of a lake where light does not penetrate.

abyssal benthic zone that part of the sea bottom that is at great depths (variously defined as being >1000 m or >2000 m deep).

abyssal plain flat, extensive area of seafloor, found at an average depth of around 4000 m. Such plains are not generally featureless but traversed by deep-sea trenches and dotted with seamounts (*q.v.*). *alt.* deepsea plain.

abyssal rocks plutonic rocks *q.v.*

abyssobenthic *a. pert.* or found on the ocean floor at the depths of the ocean, in the abyssal zone.

abyssopelagic *a. pert.* or inhabiting the ocean depths of the abyssal zone, but floating, not on the ocean floor.

Ac symbol for the chemical element actinium *q.v.*

acacias *n.plu.* trees and shrubs of the tropical and subtropical leguminous genus *Acacia*, which may form the dominant vegetation in arid areas. Many commercial products are derived from acacias and some species can be exploited for human and animal food and as a fast-growing source of fuelwood. *see also* brigalow forest.

Acanthaster *see* crown of thorns starfish.

Acanthocephala *n.* a phylum of pseudocoelomate animals, commonly called thorny-headed worms, that as adults are intestinal parasites of vertebrates and as larvae have an arthropod host.

Acanthodii *n.* extinct group of fishes, first present in the Silurian, which are the first known jawed vertebrates.

Acanthopterygii, acanthopterygians *n.*, *n.plu.* large group of teleost fishes, having spiny fins and spiny scales, and including perch, mackerel and plaice.

Acari, Acarina *n.* a very large and varied order of arachnids, commonly called mites and ticks, an individual usually having a rounded body carrying the four pairs of legs. Ticks are relatively large and parasitic, living as ectoparasites on mammals and sucking their blood, and are the vectors of several serious diseases including Rocky Mountain spotted fever and tick-borne encephalitis. Mites are smaller, inhabiting both plants and animals and are common in soil. They include both parasites and non-parasites, and are the vectors of scrub typhus in humans.

acaricide *n.* any chemical that kills ticks and mites (Acarina).

acarology *n.* the study of mites and ticks.

accelerated eutrophication the speeding up of the natural process of eutrophication (*q.v.*) in waterbodies, as a consequence of human activity, effected through the addition of those nutrients that are normally limiting to plant growth. In freshwater lakes and rivers, the limiting nutrient element is usually phosphorus, whereas in marine waters nitrogen is often in short supply. Currently, in most locations, anthropogenic phosphorus (in the form of the phosphate anion, PO_4^{3-}) comes largely from domestic sewage effluent, while the largest source of nitrogen (in various forms, including the nitrate anion, NO_3^-, and ammonia, NH_3) is probably nitrogen-containing fertilizers. *alt.* artificial eutrophication, cultural eutrophication.

accelerated soil erosion *see* soil erosion.

acceptable daily intake (ADI) daily consumption of a substance, such as a vitamin, nutrient or toxic material, deemed, on the basis of available knowledge, to be without risk when taken at that level during the lifespan of the organism. Expressed as milligrams of chemical per kilogram of organism body weight.

access agreement agreement allowing the public access to privately owned open country suitable for open-air leisure activities.

access order in the UK, a planning order granting the public access to privately owned land where an access agreement cannot be made or is proving ineffective.

accessory mineral (*geol.*) a minor constituent of a rock that is not used in its classification.

accessory species (*plant ecol.*) a species that is found in one-quarter to one-half of the area of a stand. *cf.* accidental species.

accidental species (*plant ecol.*) a species that is found in less than one-quarter of a stand. *cf.* accessory species.

accipiters *n.plu.* the hawks, medium-sized birds of prey with rounded wings and long tails, part of the family Accipitridae.

acclimation *n.* the physiological habituation of an organism to a change in a particular environmental factor, generally used in reference to laboratory-based experiments. *cf.* acclimatization, adaptation.

acclimatization *n.* the physiological and/or behavioural habituation of an organism to a different climate or environment, generally used in reference to organisms in the wild. *cf.* acclimation, adaptation.

accrescence *n.* (1) growth through addition of similar tissues; (2) continued growth after flowering. *a.* accrescent.

accretion *n.* growth by external addition of new matter.

accumulators *n.plu.* plants that accumulate relatively high concentrations of certain chemical elements, such as heavy metals, in their tissues.

accuracy *n.* (*chem.*) the degree of agreement between the outcome of a quantitative analytical procedure and the true, or accepted, value. High accuracy equates with close agreement. Accuracy is often difficult to quantify as, frequently, the true value is unknown, because if it were there would be no need for the analysis to be performed! *cf.* precision.

acellular *a.* not divided into cells.

acellular slime moulds a group of simple eukaryotic soil microorganisms (Myxomycetes) whose vegetative phase consists of a plasmodium – a naked, creeping, multinucleate mass of protoplasm sometimes covering up to several square metres.

Acerales Sapindales *q.v.*

acetic acid bacteria bacteria (e.g. *Acetobacter*, *Acetomonas*) that partially oxidize ethyl alcohol to produce acetic acid, a reaction used in manufacturing vinegar. *alt.* acetobacters.

acetyl group chemical group, $-COCH_3$, formed by removal of OH from acetic acid.

ACGIH American Conference of Government Industrial Hygienists, *see* threshold value limit.

achene *n.* a one-seeded, dry, indehiscent fruit formed from one carpel, usually with one seed that is not fused to the fruit wall.

aciculilignosa *n.* evergreen forest and bush made up of needle-leaved coniferous trees and shrubs.

acid *a.* (*geol.*) *appl.* igneous rocks that contain >66% silica (SiO_2).

acid *n.* unless otherwise stated, this is, in most cases, synonymous with a Brønsted–Lowry acid (*q.v.*). *see also* Arrhenius acid, Lewis acid.

acid deposition rain (acid rain) or other form of precipitation, or dry deposition, that contains acids and acid-forming compounds and has a pH of less than 5.6. It can cause acidification of lakes, with harmful effects on the aquatic flora and fauna, and damage to terrestrial vegetation. Acid deposition is caused mainly by atmospheric sulphur dioxide (SO_2) produced by the burning of coal and other fossil fuels, which is precipitated as sulphuric acid and sulphates. It is also caused by nitrogen oxides emitted from fossil fuel burning and vehicle exhausts, which form nitric acid and nitrogen dioxide (NO_2). *see also* acidification of natural waters, NO_x, sulphur dioxide.

acid dew-point the temperature at and below which an acidic condensate appears from a gas, *appl.* esp. stack gases that contain both water vapour ($H_2O_{(g)}$) and sulphur trioxide ($SO_{3(g)}$). In this context, it is the temperature at and below which these species will combine and condense to form liquid dilute sulphuric acid ($H_2SO_{4(aq)}$). The acid dew-point, in this case, is dependent on the $SO_{3(g)}$ and $H_2O_{(g)}$ content of the stack gas. If the waste gases are allowed to cool to the acid dew-point before they exit the stack, excessive corrosion of the installation will take place.

acid hydrolysis hydrolysis reactions (*q.v.*) involving acid solutions.

acidic *a.* (1) having the properties of an acid; (2) *appl.* stains such as eosin that react with basic components of protoplasm, such as cytoplasm and collagen.

acidification of natural waters the increase in acidity of natural waters esp. as a result of acid deposition (*q.v.*), acid mine drainage (*q.v.*) or the planting of extensive upland tracts of coniferous forests. The direct effect of decreased pH on the aquatic flora and fauna is one of overall impoverishment, with a sharp decline in species richness. In addition, the acidification of natural waters causes an increase in the aqueous concentration of toxic metal ions, e.g. aluminium ions, that are particularly harmful to fish.

acid ionization constant (K_a) an equilibrium constant that relates to the dissociation of weak Brønsted bases (a process that may be represented by the general equation $HA + H_2O \rightleftharpoons H_3O^+ + A^-$). The acid ionization constant is expressed as $K_a = \{H_3O^+\}\{A^-\}/\{HA\}$, where the curly brackets, $\{ \}$, represent activity based on molar concentration. In dilute solutions, the activity of a solute approximates to its molar concentration.

acid lava a highly viscous lava formed from magma with a high silica content. Acid lavas (also known as rhyolitic lavas) flow for only short distances and form local features. This type of lava is found at convergent plate boundaries where oceanic crust is subducted under continental crust. *cf.* basic lava.

acid mine drainage the acid effluent from mines, principally coal mines, which can cause acidification of natural waters.

acidophil(e) *n.* organism, usually a microorganism, that thrives in acidic (<pH 5) conditions, and can be isolated on acidic media. *a.* acidophilic, acidophilous.

acid precipitation *see* acid deposition.

acid rain *see* acid deposition.

acid shock the biological disruption due to rapid acidification of aquatic ecosystems.

aciduric *a.* tolerating acid conditions, *see* acidophile.

acoelomate *a. appl.* animals not having a true coelom, i.e. sponges, sea anemones and corals, nematodes, rotifers, platyhelminths and nemertean worms. *alt.* acelomate, acoelous.

acorn worms Enteropneusta *q.v.*

acquired behaviour behaviour brought on by conditioning or learning.

acquired character modification or permanent structural change brought on during the lifetime of an individual by use or disuse of a particular organ, or by disease, trauma, or other environmental influence, and which is not heritable.

acquired immune deficiency syndrome (AIDS) disease caused by infection with the human immunodeficiency virus (HIV) in which a severe deficiency of a particular class of lymphocyte develops, often a number of years after the initial infection. The immunodeficient patient is susceptible to infection by opportunistic pathogens and to the development of unusual cancers. HIV is transmitted sexually or by infected blood products and needles, and can also be transmitted from mother to child at birth or by breast feeding.

Acrania *n.* a group of chordates, sometimes considered as a subphylum, which includes the urochordates and the cephalochordates, and excludes the craniates.

Acraniata *n.* invertebrates *q.v.*

Acrasiales, Acrasiomycota, Acrasiomy-cotina, Acrasiomycetes cellular slime moulds *q.v.*

acre *n.* British unit of measurement of area, applied to land. 1 acre = 0.4047 hectare.

acre-foot volume of any substance that will cover 1 acre of a level surface to the depth of 1 foot (in SI units, 1232.75 m^3).

Acrididae *n.* a family of grasshoppers (order Orthoptera) with antennae shorter than their bodies and that includes the locust, *Locusta migratoria*. *alt.* short-horned grasshoppers.

acridines *n.plu.* a group of dyes (e.g. acridine orange, acriflavin, ethidium bromide and proflavin) that are mutagenic because they intercalate into DNA causing addition or deletion of single bases during DNA replication.

actinides *n.plu.* series of radioactive metallic elements with atomic numbers from 90 (thorium) to 103 (lawrencium) inclusive, although actinium (at. no. 89) is considered by some to be an actinide. They include thorium, proactinium, uranium (at. no. 92), neptunium, plutonium, americium, curium, berkelium, californium, einsteinium, fermium, mendelevium, nobelium and lawrencium. Those with atomic number >92 do not occur naturally but may be formed by bombardment of heavy nuclei, e.g. iron, with light nuclei. *see also entries for individual elements. alt.* actinide series, actinoids, actinons.

actinium (Ac) *n.* a radioactive element of the actinide series [at. no. 89, r.a.m. (of its most stable known isotope) 227]. It has two naturally occurring isotopes, ^{227}Ac and ^{228}Ac, and *ca.* 20 artificial ones.

Actinobacteria Actinomycetes *q.v.*

actinolite *n.* $Ca_2(Mg,Fe)_5Si_8O_{22}(OH)_2$, the iron-containing members of the tremolite–actinolite series *q.v.*

actinomorphic *a.* (1) radially symmetrical; (2) regular.

Actinomycetes, actinomycetes *n., n.plu.* a group of Gram-positive prokaryotic microorganisms found in soil, river muds and lake bottoms that grow as slender branched filaments (hyphae) and that include many species (e.g. *Streptomyces*) producing antibiotics.

actinomycins *n.plu.* antibiotics produced by species of the actinomycete *Streptomyces*, which block transcription in both bacterial and eukaryotic cells by binding to DNA. Actinomycin C (cactinomycin) and actinomycin D (dactinomycin) have been used as anticancer drugs.

Actinopoda, actinopods *n., n.plu.* phylum of non-photosynthetic protists characterized by long slender cytoplasmic projections (axopodia) that are stiffened by a bundle of microtubules. They include the marine radiolarians and the mainly freshwater heliozoans. The radiolarians are divided into three groups: acantharians, with a radially symmetrical skeleton of rods of strontium sulphate, and polycystids and phaeodarians, both with silica spicules.

Actinopodea, actinopods in older classifications, the class of protozoans including the radiolarians and heliozoans. *see* Actinopoda.

Actinopterygii, actinopterygians *n., n.plu.* a subclass of bony fishes (Osteichthyes), the ray-finned fishes, including all extant bony fishes except the lungfishes (Dipnoi) and the coelacanth. *see also* Acanthopterygii.

actinorrhiza *n.* nitrogen-fixing nodule-like structure formed on the roots of some non-legumes (e.g. alders, *Alnus* spp.) by infection with nitrogen-fixing actinomycetes of the genus *Frankia*.

Actinozoa Anthozoa *q.v.*

actinula *n.* larval stage in some jellyfish, which develops into the medusa.

action area urban area given special priority within a wider regeneration plan, selected because of pressing social need.

action spectrum the range of wavelengths of light or other electromagnetic radiation within which a certain photochemical process takes place, often *pert.* to photosynthetic processes.

activated carbon a porous form of carbon with an extremely high surface area to mass ratio (up to 10^3 m^2 g^{-1}). It is made by heating animal or vegetable remains in the absence of air. It is used to adsorb a multitude of organic and inorganic compounds from gaseous and aqueous waste streams. Available in granular (granular activated carbon, GAC) and powdered forms. *alt.* activated charcoal, active charcoal.

activated complex a term used in chemical kinetics to denote a transitory combination of two potentially reacting chemical species, formed by the collision of these species. On disintegration, an activated complex will yield either the products of the reaction or the reactant species unchanged.

activated sludge process a sewage treatment process in which a mixture of protozoa and bacteria (activated sludge) is added to aerated sewage to break down the organic matter. As the micoorganisms use the organic matter for food they multiply, producing more activated sludge.

activated sludge tank a type of reactor commonly used in the secondary treatment of sewage. The settled sewage is oxidized by a suspension of microorganisms and the oxygen (O_2) for this biological process is supplied by active aeration. The secondary sludge produced is removed downstream in final settlement tanks, and most of this is returned to the aerated tank in order to maintain its biological community. *see also* sewage treatment.

activation energy the free energy of activation, the minimum amount of energy required to initiate a spontaneous chemical reaction. In a chemical reaction A → B, it is the difference in free energy between A and the intermediate transition state of higher energy on which the rate of the reaction depends. Enzymes act as catalysts by lowering the free energy of activation.

active dispersal any process by which seeds are actively scattered by force, such as by explosive opening of pods, rather than by purely passive means, such as wind or water.

active layer (*geog.*) thin surface layer of ground overlying permafrost (*q.v.*), which is subject to seasonal freezing and thawing.

active solar heating systems *see* solar heating systems.

active space of a pheromone or other chemical signal, the space within which the chemical is above the threshold concentration for its detection by another individual.

active volcano a volcano that has been seen to erupt.

activity *n.* (1) (*ecol.*) the total flow of energy through a system in unit time; (2) of enzymes, the rate at which they catalyse a chemical reaction; (3) (*chem.*) the effective concentration of a solute or the effective partial pressure of a gas. The relationship between activity, a, and concentration or partial pressure, c, takes the form $a = fc/s$, where f is the activity coefficient (obtained from experimental measurement) and s is the standard concentration or partial pressure (1 mol dm^{-3} for a solute, 1 atm (i.e. 101 325 pascals) for a gas). As the units of c are mol dm^{-3} for a solute and atmospheres for a gas, this means that activities are dimensionless, and therefore have no units. By convention, chemically pure solids and liquids are given unit activity.

activity coefficient *see* activity.

activity node any location where people assemble for a particular purpose, e.g. shopping.

actual evapotranspiration (AET) the total water loss over a given area with time by the removal of water from soil and other surfaces by evaporation and from plants by transpiration.

actual vegetation (*ecol.*) the vegetation existing at the time of observation.

acute *a.* (1) ending in a sharp point; (2) temporarily severe, not chronic; *appl.* (3) toxicity: causing severe poisoning, leading to severe illness, damage or death, within 24–96 h of exposure to the substance; (4) medical conditions or diseases that progress rapidly to a crisis. *cf.* chronic.

adaptation *n.* (1) evolutionary process involving genetic change by which a population becomes fitted to its prevailing environment; (2) a structure or habit fitted for some special environment or activity;

(3) process by which an organ or organism becomes habituated to a particular level of stimulus and ceases to respond to it, a more intense stimulus then being needed to produce a response.

adaptive *a*. (1) capable of fitting different conditions; (2) adjustable; (3) *appl.* any trait that confers some advantage on an organism and thus is maintained in a population by natural selection. Traits can only be defined as adaptive with reference to the environment pertaining at the time, as a change in environment can render a previously adaptive trait non-adaptive, and vice versa.

adaptive radiation evolutionary process in which numerous species descended from a common ancestor multiply and diverge to occupy different ecological niches.

adaxial *a. pert.* the surface of a structure (e.g. a leaf) that is nearest to, or turned towards, the axis.

additive *n*. any substance that is deliberately added to products to bring about a specific desired effect. *see* antiknock agent, feed additive, food additive.

additive mortality total mortality due to all factors (e.g. predation, disease, accident) over a given period.

additive variance in quantitative genetics, that part of the genotypic variance that can be attributed to the additive effects of genes.

adeciduate *a*. not falling or coming away.

adenine (A) *n*. a purine base, one of the four nitrogenous bases in DNA and RNA, in which it pairs with thymine (T) and uracil (U), respectively. It is also a component of the nucleoside adenosine, the nucleotides AMP, ADP and ATP, the nicotinamide cofactors NAD/NADH and NADP/NADPH, and the flavin nucleotide cofactors FAD and FMN.

adenosine *n*. a nucleoside made up of the purine base adenine linked to the sugar ribose.

adenosine monophosphate, adenosine diphosphate *see* adenosine triphosphate.

adenosine triphosphatase (ATPase) enzyme activity found in many different proteins that converts ATP to ADP and inorganic phosphate with the release of energy.

adenosine triphosphate (ATP) nucleotide made of the purine base adenine linked to the pentose sugar ribose that carries three phosphate groups linked in series. It is important as a source of energy and phosphate groups for metabolic reactions in all living cells through its conversion to ADP (adenosine diphosphate) with release of energy and phosphate (P_i) or to AMP (adenosine monophosphate) with release of energy and pyrophosphate (PP_i). In aerobic organisms, it is chiefly regenerated from ADP by the process of oxidative phosphorylation. It is also produced from ADP during photosynthesis.

Adenoviridae, adenoviruses *n., n.plu.* family of double-stranded DNA viruses including adenovirus type 2, which causes a mild respiratory infection in various mammals, including humans, but which is tumorigenic in newborn hamsters.

ADI (1) acceptable daily intake (*q.v.*); (2) aerobic dependence index *q.v.*

adiabatic *a*. without gain or loss of heat, *appl.* any change in or to a system (*q.v.*) that does not involve the transfer of heat across its boundary (*q.v.*). For example, rising air will expand as the air pressure drops. Under adiabatic conditions, this requires the conversion of some of its internal energy into work, thereby causing its temperature to fall. Such adiabatic cooling of water-saturated air will cause condensation to occur. This may ultimately lead to cloud formation and precipitation.

adiabatic cooling of air *see* adiabatic.

adiabatic warming of air the warming observed in a portion of air that is descending under adiabatic conditions. Warming occurs because the surrounding atmosphere does pressure–volume work (*q.v.*) on the air, increasing its temperature.

adiabatic wind air movement brought about by adiabatic warming or adiabatic cooling, e.g. local winds such as the Chinook, föhn and Samun.

adipose tissue the 'fat' in animals, a type of connective tissue that is made of cells (adipocytes) filled almost completely with fat droplets. *see* brown adipose tissue, white adipose tissue.

adit *n*. horizontal tunnel excavated into a hillside that connects with a mine, for access or drainage purposes.

ADMR average daily metabolic rate *q.v.*

ADP adenosine diphosphate.

adret slope sunny (south-facing) side of an Alpine valley. *alt.* sonnenseite slope. *cf.* ubac slope.

adsere *n.* (*plant ecol.*) that stage in a plant succession that precedes its development into another at any time before the climax stage is reached.

adsorbate *see* adsorption.

adsorbent *see* adsorption.

adsorption *n.* (1) the adhesion of chemical species to the surface of a solid or liquid, by virtue of weak interspecies forces (physisorption) or covalent bonds (chemisorption). The substance adhering to the surface is called the adsorbate, the substance to which it adheres is the adsorbent; (2) formation of a thin layer of a substance at the surface of a solid or liquid.

adsorption complexes clay–humus complexes *q.v.*

adtidal *a. appl.* organisms living just below the low-tide mark.

adularia *n.* $KAlSi_3O_8$, one of the feldspar minerals *q.v.*

adult *n.* an organism that has reached full sexual maturity.

advanced *a.* of more recent evolutionary origin. *cf.* primitive.

advanced gas-cooled reactor (AGR) a type of fission nuclear reactor that utilizes stainless steel-clad fuel pellets of enriched uranium oxide. It has carbon dioxide as the coolant and a graphite moderator.

advanced sewage treatment tertiary sewage treatment, *see* sewage treatment.

advection cooling cooling of warm, moist air as a result of its transit over cooler land or ocean surfaces. This type of cooling can lead to the formation of advection fog.

advection fog type of fog formed when warm, moist air moves over a colder land or sea surface. The cooling of the air mass leads to condensation of atmospheric water vapour.

adventitious *a.* (1) accidental; (2) found in an unusual place; (3) *appl.* tissues and organs arising in unusual positions, such as the prop roots of mangrove and maize that arise from the stem above ground.

adventive *a.* (1) not native; (2) *appl.* organism in a new habitat but not completely established there.

AEC anion exchange capacity *q.v.*

aeolian *a.* (1) involving the action of wind, *appl.* processes and landforms; (2) windborne, *appl.* deposits. *alt.* eolian.

Aepyornithiformes *n.plu.* order of very large Pleistocene birds of the subclass Neornithes from Madagascar, known as elephant birds.

aerenchyma *n.* (1) relatively unspecialized plant tissue containing large intercellular spaces; (2) air-storing tissue in the cortex of some aquatic plants.

aerial *a.* above ground *appl.* (1) roots growing above ground; (2) small bulbs appearing in leaf axils.

aerial plankton spores, pollen, bacteria and other microorganisms floating in the air. *alt.* aeroplankton.

aeroallergen *n.* any substance present in the air, such as pollen, that causes an allergy in susceptible individuals when inhaled.

aerobe *n.* any organism capable of living in the presence of atmospheric oxygen, obligate aerobes being unable to live without oxygen. All animals and higher plants are obligate aerobes, many bacteria and some fungi are anaerobes or facultative anaerobes. *a.* aerobic. *cf.* anaerobe.

aerobic *a.* using or requiring oxygen.

aerobic decomposition the oxygen-requiring breakdown of organic material by aerobic microorganisms. *alt.* aerobic digestion. *cf.* anaerobic decomposition, fermentation.

aerobic dependence index (ADI) a measure of the relative use of aerobic and anaerobic metabolism in muscle during exercise in animals, which is calculated using the lactate content of resting muscles after exercise as a measure of anaerobic metabolism and maximum oxygen consumption as a measure of aerobic metabolism. Higher ADIs tend to indicate more active animals.

aerobic digestion aerobic decomposition *q.v.*

aerobic endospore-forming rods group of aerobic Gram-positive bacteria whose members produce highly resistant internal spores (endospores) that may persist for many years. Genera include *Bacillus* and *Clostridium*.

aerobic respiration respiration occurring in the presence of molecular oxygen, producing large amounts of ATP, and producing carbon dioxide and water as waste products. *cf.* anaerobic respiration, fermentation.

aerobiology *n.* study of airborne organisms and their distribution.

aerobiotic *a.* living mainly in the air.

aerolites *see* meteorites.

aeroplankton aerial plankton *q.v.*

aerosol *n.* (1) a suspension of solid or liquid particles within a gas, e.g. wind-blown pollen, *see also* particulates; (2) colloquially, a pressurized canister from which a true aerosol is sprayed. These canisters contain a propellant that forces the desired contents of the canister to exit through a nozzle, out of which the propellant also escapes. At one time chlorofluorocarbons (*q.v.*) were widely used as propellants but, because of concern over their deleterious effects on stratospheric ozone levels, their use is now restricted.

aerotaxis *n.* movement towards or away from a source of oxygen. *a.* aerotactic.

aerotropism *n.* reaction to gases, generally to oxygen, particularly the growth curvature of roots and other parts of plants in response to changes in oxygen tension. *a.* aerotropic.

aestidurilignosa *n.* mixed evergreen and deciduous broad-leaf forest.

aestilignosa *n.* broad-leaf deciduous forest that loses its leaves in winter.

aestival *a.* (1) produced in, or *pert.* summer; (2) *pert.* early summer.

aestivation *n.* (1) (*bot.*) the mode in which different parts of a flower are arranged in the bud; (2) (*zool.*) torpor during heat and drought during summer in some animals.

AET actual evapotranspiration *q.v.*

aetiology *n.* cause, usually *pert.* a disease. *alt.* etiology.

AFDW ash-free dry weight.

affinity chromatography technique for purifying specific proteins (or other macromolecules) by their affinity for particular chemical groups, by passage through a column filled with an inert granular matrix to which are attached those groups to which the protein binds.

afforestation *n.* the production of forest over an area of open land, either by planting or by allowing natural regeneration or colonization.

aflatoxins *n.plu.* toxic compounds formed by the mould *Aspergillus flavus*, which may be found on cereals and esp. groundnuts (peanuts). Ingestion of foods contaminated with these toxins may lead to cancer and liver damage.

African subkingdom subdivision of the Palaeotropical Floral Realm, comprising all of the African continent, Madagascar, Ascension Island and St Helena.

afro-alpine *n.* term used in East Africa for the vegetation of high tropical mountains.

after-burner *n.* waste gas treatment device that reduces the levels of certain components via incineration. Most commonly used in furnaces or motor vehicles, in which it is placed downstream of the main combustion zone. It can also be used to treat the gaseous wastes from non-combustion processes.

after-ripening period after dispersal in which a seed cannot yet germinate, even if conditions are favourable.

Ag symbol for the chemical element silver *q.v.*

agamic *a.* (1) asexual; (2) parthenogenetic.

agamic complex group of asexually reproducing plants that are usually allopolyploids and consist of many different biotypes, forming a taxonomically difficult group.

agamodeme *n.* small assemblage of closely related individuals consisting predominantly of apomictic plants or asexual organisms.

agamogenesis *n.* (1) any reproduction without participation of a male gamete; (2) asexual reproduction. *a.* agamogenetic.

agar *n.* gelatinous substrate for bacterial cultures and constituent of some gels used for electrophoresis, prepared from agar-agar, a gelatinous polysaccharide extracted from red algae.

agarics *n.plu.* common name for a large group of basidiomycete fungi, the order Agaricales, typified by mushrooms and toadstools, with conspicuous fruiting bodies, usually composed of a stalk and cap, with the spores borne on the surface of gills on the underside of the cap. In some classifications the order Agaricales covers all gilled fungi, in others, it includes only certain families, but the term agarics is often used for the gilled fungi as a whole.

agate *n.* SiO_2, varieties of the mineral chalcedony (*q.v.*) that contain curved zones of different colours. *cf.* onyx.

age distribution *see* age structure.

Agenda 21 a blueprint produced by the United Nations Conference on Environment and Development (1992) for pursuing sustainable development in the twenty-first century.

Agent Orange defoliant used extensively by the US army in the Vietnam War (1963–1973). This herbicide consisted of the butyl esters of 2,4,5-trichloro- and 2,4-dichlorophenoxyacetic acids, in a 1 : 1 mixture. The highly toxic dioxin TCDD is also present in small quantities as a contaminant.

age polyethism in social animals, the changing of roles by members of a society as they age.

age ratio of a population, the relative frequencies of different age groups within it.

age-specific death rate mortality rate for different age groups within a population.

age-specific natality fecundity *q.v.*

age-specific reproductive value an index of the extent to which the members of a given age group contribute to the next generation between now and the time they die.

age structure of a population, the percentage of the population at each age level, or the number of individuals of each sex at each age level. *alt.* age distribution.

agglomerate *n.* rock consisting of sub-rounded to angular volcanic fragments (>6.4 cm diameter) embedded in a finer-grained matrix.

aggradation *n.* (*geog.*) process whereby a land surface becomes built up by the deposition of water-borne or wind-blown material. *cf.* degradation.

aggregate *a.* (1) formed in a cluster; *appl.* (2) fruit formed from a gynoecium having separate or partially separate carpels, such as a raspberry; (3) fruit formed from several flowers, such as pineapple.

aggregate materials mineral materials, such as sands and gravels, used (with little preprocessing) in the construction of buildings and roads.

aggregation *n.* a group of individuals of the same species gathered in the same place but not socially organized or engaged in cooperative behaviour.

aggression *n.* an act or threat of action by one individual that limits the freedom of action of another individual, often shown by animals to each other in ritualized trials of strength, e.g. to gain mates, territory, etc.

aggressive mimicry mimicking a harmless species in order to suprise and attack an unsuspecting prey.

Agnatha, agnathans *n.*, *n.plu.* subphylum, class or superclass of primitive jawless vertebrates, including the lampreys (subclass Monorhina), hagfishes (subclass Diplorhina) and their extinct relatives. The internal skeleton is made of cartilage. The agnathans were the first vertebrates to evolve, and many extinct forms typically had bodies heavily armoured with bony plates and scales. Extinct agnathans include the anaspids (the probable ancestors of the modern lampreys), the heterostracans, the osteostracans and the thelodontids.

agonistic behaviour any activity related to fighting, whether aggressive or conciliatory.

AGR advanced gas-cooled reactor *q.v.*

agrarian *a.* (1) relating to cultivated land; (2) *appl.* societies in which most people are still involved in agriculture.

Agreement on Conservation of Polar Bears international agreement (1976) between the former USSR, Norway, Canada and Denmark, drafted by the International Union for the Conservation of Nature and Natural Resources, which forbids the killing or taking of polar bears except for scientific purposes or by local native peoples and requires governments to protect polar bear habitats.

agrestal *a.* *appl.* uncultivated plants growing on arable land.

agricultural cooperative association of several independent farmers who act jointly to purchase farm equipment, market produce, provide credit, etc.

agricultural hearth geographical site identified as being a centre of origin for the development of agriculture, found, e.g., in Asia and the Mediterranean Basin.

agricultural industrialization transformation of traditional small-scale family farms into highly mechanized, large-scale farming units geared to high productivity and profit margins.

Agricultural Revolution the gradual shift from a hunter–gatherer way of life to settled agriculture, in which animals and wild plants were domesticated, which began between 10 000 and 12 000 years ago.

agriculture *n*. the cultivation of land for the purposes of crop production and/or the rearing of livestock, primarily for food but also to provide materials for, e.g., fuel, clothing and shelter.

Agrobacterium tumefaciens a plant pathogenic bacterium, the cause of crown gall in numerous dicotyledonous plants. It carries a plasmid, the Ti plasmid, that becomes integrated into the chromosomes of infected tissue. *Agrobacterium* and its plasmid have been extensively modified by genetic engineering and are widely used to introduce novel genes into plant cells to produce transgenic plants. *see* Fig. 14, p. 172.

agrochemicals *n.plu*. chemical substances, either produced industrially or mined from natural deposits, used in agriculture as, e.g., fertilizers or pesticides.

agroclimatology *n*. branch of climatology relating to how the weather affects crops.

agroecosystem *n*. an ecosystem that develops on farmed land and that includes the indigenous microorganisms, plants and animals, and the crop species. *alt*. agrocoenosis.

agroforestry n. mixed cultivation of trees and herbaceous crops, traditionally practised in some areas of the humid tropics, such as Sri Lanka. *alt*. farm forestry, social forestry.

agroinfection, agroinoculation *n*. the infection or inoculation of plants with genetically engineered *Agrobacterium* as a means of introducing novel genes.

agronomy *n*. the scientifically based management of land for field-crop production.

agrostology *n*. that part of botany dealing with grasses.

A-horizon the upper dark-coloured layer of a soil, commonly called the topsoil, which includes partially decomposed organic

matter (humus), plant roots, soil organisms and some minerals. In some classifications the A-horizon also includes the E-horizon (present in some soils only), a lighter-coloured deeper zone of low organic content through which dissolved or suspended materials move downward. *see* Fig. 28, p. 385. *alt*. eluvial layer, zone of leaching. *see also* Ap-horizon, soil horizons, surface soil.

AI, AID artificial insemination by donor, widely used in livestock breeding.

aid *n*. assistance, esp. *appl*. that in the form of resources (food, medicine, etc.), finance and technology, channelled towards developing countries from developed countries and international organizations. This type of aid is also called overseas development assistance (ODA).

AIDS acquired immune deficiency syndome *q.v.*

air classifier air separator *q.v.*

air mass large body of air that shows relatively little horizontal variation in temperature or humidity. Air masses may be classified on the basis of characteristics dictated by their source regions. *see* polar continental, polar maritime, tropical continental, tropical maritime.

air pollution any gaseous or particulate matter in the air that is not a normal constituent of air or not normally present in such high concentrations. It may be the result of human activity, as sulphur dioxide from burning of coal, and carbon monoxide and nitrogen oxides from exhaust emissions, or can result from natural causes, as desert dust, or methane and hydrogen sulphide from microbial activity in bogs, or volcanic debris in the atmosphere.

air pressure atmospheric pressure *q.v.*

air quality usually means the concentration of specific pollutant(s) in air, often averaged over a specified time period. It can also be used to mean the approximate opposite of air pollution, i.e. air quality is high when pollution is low. In this latter context, air quality may be judged by a variety of criteria, e.g. chemical or physical analysis, or damage to plants, buildings or animals.

air quality standards recommendations for, or legally enforceable, maximum concentration(s) of specific pollutant(s) in air (often averaged over a specified time period, e.g. 1 h mean, 8 h mean), in a given geographical location.

air separator device used to separate shredded domestic refuse into different fractions on the basis of density. *alt.* air classifier.

aitionastic *a. appl.* curvature of part of a plant induced by a diffuse stimulus.

Aitken nuclei solid particles or liquid droplets each $0.005–0.1$ μm in diameter and suspended in the atmosphere, mainly produced by the condensation of hot vapours.

aklé *n.* a dune network composed of parallel cresent-shaped sand ridges that interlock. This pattern, which resembles fish scales, is formed transverse to the prevailing wind direction.

Al symbol for the chemical element aluminium *q.v.*

alabaster *n.* a fine-grained granular variety of gypsum *q.v.*

alanine (Ala, A) *n.* alpha-amino propionic acid, a simple amino acid with a methyl side chain, a constituent of protein, nonessential in the human diet as it can be synthesized in the body by the reaction of pyruvate and glutamate to give alanine and alpha-ketoglutarate.

alarm behaviour diverse types of behaviour shown by animals when disturbed and that are intended to distract predators or to hide the animal from view.

alarm pheromone chemical substance released in minute amounts by an animal that induces a fright response in other members of the species.

ALARP acronym for 'as low as reasonably practicable', a principle of risk management that aims to reduce risks as low as reasonably possible, accepting the concept that zero risk is unattainable.

alas *n.* a large thermokarst (*q.v.*) depression, up to 40 m deep and 15 km across, which may contain a lake. Several of these periglacial features may coalesce to form an alas valley, stretching for tens of kilometres.

Alaskan oil spill *see Exxon Valdez* incident.

alas valley *see* alas.

albedo *n.* (1) the ratio of the amount of light reflected by a surface to the amount of incident light; (2) (*bot.*) mesocarp, the white tissue of rind of orange, lemon, etc.

albic horizon *see* spodosol.

albinism *n.* genetically determined or environmentally induced absence of pigmentation in structures normally pigmented, leading, e.g., to lack of pigmentation in hair, skin and eyes in animals or of green colour in plants. Organisms showing albinism are known as albinos.

albite *n.* $NaAlSi_3O_8$, one of the feldspar (*q.v.*) minerals.

alboll *n.* a suborder of the order mollisol of the USDA Soil Taxonomy system comprising mollisols with a bleached eluvial horizon from which iron oxides and clay have been removed.

albumins *n.plu.* general name for a class of small globular proteins found in blood (serum albumin), synovial fluid, milk and other mammalian secretions, and as storage proteins in plant seeds, characterized originally as proteins that are soluble in water and dilute buffers at neutral pH.

alburnum *n.* sapwood, splintwood, the young wood of dicotyledons.

alcids *n.plu.* common name for seabirds of the family Alcidae of the order Charadriiformes, e.g., the razorbills, guillemots, auks and puffins. They often suffer heavily in oil spillages.

alcohol *n.* any organic compound that has a structure equivalent to that of a hydrocarbon with one or more of its hydrogen atoms each replaced by a hydroxyl (-OH) group. In non-technical usage usually refers to ethanol (ethyl alcohol) or, less commonly, methanol (methyl alcohol).

alcoholic fermentation the production in anaerobic conditions of ethyl alcohol from carbohydrates by yeasts and other microorganisms.

Alcyonaria *n.* soft corals, sea pens and sea fans, a class of colonial coelenterates in which the lower parts of the polyps fuse to form a soft mass.

aldehyde *n.* an organic compound that contains a carbonyl group (C=O), the carbon atom of which is bonded to both a hydrogen atom and an aryl or alkyl radical.

alder flies *see* Neuroptera.

aldicarb *n.* systemic carbamate pesticide used to control nematodes and aphids on sugar beet, potatoes and onions.

aldose *n.* any monosaccharide containing an aldehyde group. *cf.* ketose.

aldrin *n.* contact organochlorine insecticide used to control soil pests, such as wireworms and leatherjackets, on potatoes. Its use is restricted due to its persistence in the environment.

aleurone grains storage granules in endosperm and cotyledon of seeds, consisting of proteins, phytin and hydrolytic enzymes, forming an aleurone layer in seeds such as cereals.

alfalfa lucerne *q.v.*

alfalfa mosaic virus group plant virus group containing a single member, alfalfa mosaic virus, which is a rod-shaped single-stranded RNA virus causing mosaic symptoms (mottling of leaves). It is a multicomponent virus in which the four separate genomic RNAs are encapsidated in different virus particles.

alfisol *n.* an order of the USDA Soil Taxonomy system consisting of moist mineral soils that lack the dark surface layer of the mollisols, but which exhibit clay accumulation in the illuvial (B) horizon. They develop mainly in humid regions under native deciduous forest. Five suborders are recognized, *see* aqualf, boralf, udalf, ustalf, xeralf.

algae *n.plu.* a general term, with no taxonomic status, for a heterogeneous group of unicellular, colonial and multicellular eukaryotic photosynthetic organisms of simple structure. Traditionally included in the plant kingdom (*see* Appendix 5), the different groups of algae are now often classified as divisions of the kingdom Protista (*see* Appendix 8). They are aquatic or live in damp habitats on land and include unicellular organisms such as *Chlamydomonas* and diatoms, colonial forms such as *Volvox*, the multicellular green, red and brown seaweeds, and freshwater multicellular green algae such as *Spirogyra*. In multicellular forms the algal body is known as a thallus and is generally filamentous or flattened into a thin sheet or ribbon. The so-called blue-green algae are not true algae but prokaryotes, the cyanobacteria (*q.v.*). Planktonic algae and cyanobacteria form the basis of the food chain in marine and freshwater habitats. *sing.*

alga. *see* Bacillariophyta, Charophyta, Chlorophyta, Chrysophyta, Cryptophyta, Dinoflagellata, Euglenophyta, Eustigmatophyta, Gamophyta, Glaucophyta, Haptophyta, Phaeophyta, Pyrrophyta, Rhodophyta, Xanthophyta, zooanthellae, zoochlorellae, zooxanthellae.

algal bloom an exceptional growth of algae or cyanobacteria in lakes, rivers or oceans, which may occur in particular climatic conditions or as a result of excess nutrients in the water. In some cases, the microorganisms produce toxic compounds.

algicide *n.* any substance that kills algae.

alginate *n.* carbohydrate polymer derived from algin, a gel-like polysaccharide found in the cell walls of brown algae, which is used as a food-stabilizing and texturing agent and in dental moulding materials.

Algonkian *a. pert.* the late Proterozoic era.

alien *n.* plant thought to have been introduced by man but now more or less naturalized.

Alismales, Alismatales *n.* order of herbaceous monocots, aquatic or partly aquatic, and including the families Alismaceae (water plantain), Butomaceae (flowering rush) and Limnocharitaceae.

alkali *n.* (1) any substance that neutralizes acids and turns red litmus (a vegetable dye) blue; (2) any water-soluble metal hydroxide, esp. a hydroxide of any one of the Group 1 elements of the periodic table, *see* Appendix 2; (3) a solution of a base in water.

alkali feldspars *see* feldspar minerals.

alkali lake playa lake *q.v.*

alkali metals the elements lithium (Li), sodium (Na), potassium (K), rubidium (Rb), caesium (cesium, Cs) and francium (Fr), i.e. Group 1 of the periodic table, *see* Appendix 2.

alkali soils soils with a very high surface content of mineral salts, such as sodium chloride, sodium sulphate, sodium carbonate and borax, which are formed in dry regions where evaporation is much greater than rainfall. *see* solonchaks, solonetz. *cf.* alkaline soils.

alkaline *a. appl.* any substance that exhibits the properties of an alkali *q.v.*

alkaline earth metals the elements beryllium (Be), magnesium (Mg), calcium (Ca), strontium (Sr), barium (Ba) and radium (Ra), i.e. Group 2 of the periodic table, *see* Appendix 2.

alkaline soils (*agric./hort.*) temperate-region soils rich in calcium compounds, with a pH >7.5 and up to 8 or 9, which develop over chalk or limestone, *alt.* calcareous soils. *cf.* alkali soils.

alkalization *n.* process whereby sodium salts accumulate in a soil and its pH value rises to >7. This process occurs in arid and semi-arid regions. *alt.* sodification.

alkaloid *n.* any of a group of nitrogenous organic bases found in plants, often with toxic or medicinal properties, such as caffeine, morphine, nicotine, strychnine.

alkane *n.* any member of the homologous series of saturated hydrocarbons, each of which has a stoichiometry that accords with the general formula C_nH_{2n+2}. When found in soil and rocks they are considered to indicate present or past activities of living organisms.

alkene *n.* any member of the homologous series of unsaturated hydrocarbons, each of which has a stoichiometry that accords with the general formula C_nH_{2n} and contains a C=C double bond.

alkyl *n.* any chemical group that conforms to the general formula $-C_nH_{2n+1}$ and that results when a hydrogen atom is removed from a saturated hydrocarbon. Examples are methyl $-CH_3$ (derived from methane, CH_4), ethyl $-C_2H_5$ (derived from ethane, C_2H_6).

alkyl group a chain of CH_2 groups of any length, i.e. $(CH_2)_n$ where $n = 0$ to infinity, terminating in a CH_3 group.

alkyl mercury highly toxic mercury compounds of general formula RHg^+, where R is an alkyl group, which have been used as seed dressings and fungicides.

alkyl sulphonates surface-active agents used in detergents, with the general formula $RSO_3^-M^+$, where R is an alkyl group and M is either a hydrogen atom, an alkali metal cation or an organic cation. Those with a branched structure, e.g. alkyl benzene sulphonate, are not readily biodegradable and cause foaming in water into which they are discharged.

alkylating agents highly reactive organic chemicals that attach to bases in DNA causing errors in replication and thus are potential mutagens, as well as causing widespread damage to tissues, e.g. mustard gas.

alkyne *n.* any member of the homologous series of unsaturated hydrocarbons, each of which has a stoichiometry that accords with the general formula C_nH_{2n-2} and contains a carbon–carbon triple bond.

allanite *n.* a mineral of the epidote (*q.v.*) group that is found as an accessory in many granitic rocks, syenites, gneisses and skarns, *alt.* orthite.

allele *n.* one of a number of alternative forms of a gene that can occupy a given genetic locus on a chromosome.

allelic *a.* (1) *pert.*, or the state of being, alleles; (2) *appl.* two or more mutations mapping to the same area of the chromosome, which do not complement each other in the heterozygous state, showing that they are affecting the same genetic locus.

allelochemical *n.* chemical produced by one plant that limits the growth of another.

allelomimetic *a. appl.* animal behaviour involving imitation of another animal, usually of the same species.

allelopathic *a. pert.* the influence or effects (sometimes inhibitory or harmful) of a living plant on other nearby plants or microorganisms. *n.* allelopathy.

allelotype *n.* the frequency of different alleles in a population.

Allen's rule rule that in a widely distributed species of endothermic animal (animals that generate their own body heat), the extremities (e.g. ears, feet, tail) tend to be smaller in the colder regions of the species range than in the warmer regions.

allergen *n.* substance to which an individual is hypersensitive and that causes an immune response usually characterized by local inflammatory reactions (causing, e.g., runny nose and eyes, difficulty in breathing), but sometimes by severe shock symptoms (anaphylactic shock).

allergic rhinitis hayfever *q.v.*

alley-cropping *n.* a type of agroforestry where arable crops are cultivated between rows of trees. The tree species are usually leguminous, providing the arable crops with an enhanced nitrogen supply.

alliaceous *a. pert.* or like garlic or onion, *appl.* to any member of the genus *Allium.*

allo- prefix from Gk *allos,* meaning different or other.

allochroic *a.* (1) able to change colour; (2) with colour variation.

allochronic *a.* (1) not contemporary, *appl.* species in evolutionary time; (2) *appl.* species or populations that have non-overlapping breeding seasons or flowering periods.

allochthonous *a. appl.* material or species that has originated elsewhere.

alloenzyme allozyme *q.v.*

allogenic *a. appl.* plant successions caused by external factors, such as fire or grazing.

allogenous *a. appl.* flora persisting from an earlier environment.

allogrooming *n.* grooming directed at another individual of the same species.

allometric *a.* (1) differing in growth rate; (2) *pert.* allometry.

allometry *n.* (1) study of relative growth; (2) change of proportions with increase in size; (3) growth rate of a part differing from standard growth rate or from growth rate as a whole.

allomixis *n.* cross-fertilization (*q.v.*). *a.* allomictic, *appl.* to plants that reproduce by or have arisen from cross-fertilization.

allomone *n.* chemical secreted by one individual that causes an individual of another species to react favourably to it, such as scent given out by flowers to attract pollinating insects.

alloparental care assistance in the care of the young by individuals other than the parents.

allopatric *a.* having separate and mutually exclusive areas of geographical distribution.

allopatric speciation evolution of a new species from a geographically isolated population of the ancestral species. *cf.* sympatric speciation.

allophane *n.* approx. chemical composition $Al_2O_3.2SiO_2.H_2O$, a microcrystalline or amorphous material that is found in the clay fraction of many soils, particularly those formed on recently deposited volcanic ash.

allopolyploid *n.* a polyploid containing two or more different haploid sets of chromosomes derived from two or more different species.

allosematic *a.* having markings or coloration imitating warning signs in other, usually dangerous, species. *cf.* aposematic.

allotetraploid *n.* a tetraploid produced by hybridization between two different species followed by doubling of chromosome number. Allotetraploids are generally fertile. *alt.* amphidiploid.

allotments *n.plu.* small plots of land rented by amateur gardeners to grow their own fruit, flowers and vegetables.

allotopic *a. appl.* sympatric populations occupying different habitats within the same geographical range of distribution.

allotrope *n.* form of an element that is in a similar state to another, chemically identical, form of the element, but that has different physical properties. Diamond and graphite are two allotropes of carbon.

allozyme *n.* one of a number of forms of the same enzyme having different electrophoretic mobilities. *alt.* alloenzyme, isoenzyme, isozyme.

alluvial *a. pert.* soils composed of sediment transported and deposited by flowing stream or river water.

alluvial apron bajada *q.v.*

alluvial fan a fan-shaped deposit of alluvial material that may occur, e.g., at the edge of a glacier or where a stream emerges along a mountain front.

alluvial mining recovery of metals, e.g. gold, from river deposits.

alluvium *n.* unconsolidated material transported and then deposited by flowing streams or river water.

almandine *n.* one of the garnet minerals *q.v.*

alpestrine *a.* growing high on mountains but not above the tree line.

alpha *n.* (*biol.*) the highest-ranking individual within a dominance hierarchy.

alpha decay a type of radioactive decay in which the parent nucleus emits a fast-moving helium nucleus, called an alpha particle *q.v.*

alpha diversity ecological diversity within a habitat due to the coexistence of many subtly different ecological niches (*q.v.*), each occupied by a different species. This type of diversity results from competition between species that reduces the variation within particular species as they become more finely adapted to their ecological niches. *alt.* niche diversification. *cf.* beta diversity.

alpha-mesosaprobic category in the saprobic classification of river organisms comprising those, e.g. the water-louse (*Asellus*), that can live in polluted water in which decomposition is partly aerobic and partly anaerobic. *cf.* beta-mesosaprobic, oligosaprobic, polysaprobic.

alpha particle the fast-moving helium nucleus (symbolized variously as α, $^4_2α$ or ^4_2He) emitted during alpha decay *q.v.*

alphaviruses *n.plu.* group of viruses in the Togaviridae (*q.v.*), including Semliki Forest virus.

alpine *a.* (1) *appl.* the part of a mountain above the tree line and below the permanent snow line, *appl.* species mainly restricted to this zone; (2) loosely *appl.* the flora and fauna of high mountains.

alpine farming type of extensive agriculture in which livestock is moved to graze on the high mountain pastures during the short summer growing season experienced at such altitudes.

alpine glaciers valley glaciers *q.v.*

alpine grassland grassland found above the tree line on high mountains.

alpine permafrost sporadic permafrost *q.v.*

alpine tundra the zone of tundra-like vegetation found on high mountains above the alpine grasslands and below the permanent snow line.

alps *n.plu.* benches of land above the steep sides of U-shaped glaciated valleys, used for the summer grazing of livestock.

alternation of generations the alternation of haploid and diploid stages that occurs in the life cycle of sexually reproducing eukaryotic organisms. In some organisms, e.g. mosses, the haploid phase is predominant. In others, e.g. flowering plants and many animals, the diploid phase is dominant and the haploid phase is represented only by the gametes. In others, e.g. ferns and hydrozoan coelenterates, independent diploid and haploid organisms alternate.

alternative energy energy derived from sources other than fossil or nuclear fuels.

alterne *n.* vegetation exhibiting disturbed zonation due to abrupt change in environ-ment, or to interference with normal plant succession.

altimeter *n.* instrument that measures height above a surface, e.g. by radar.

altiplanation terraces cryoplanation terraces *q.v.*

altocumulus *n.* white-grey, medium-altitude cloud composed of individual spheres or waves separated by patches of blue sky, indicative of fine weather.

altostratus *n.* dark-grey, medium-altitude cloud that forms continuous cover.

altricial *a.* requiring care or nursing after hatching or birth.

altruism *n.* any act or behaviour that results in an individual increasing the genetic fitness of another at the expense of its own, e.g. by devoting large amounts of time and resources to caring for another individual's offspring at the expense of producing its own. *a.* altruistic. *see also* reciprocal altruism.

alum *n.* a mineral that approximates to $K_2SO_4.Al_2(SO_4)_3.24H_2O$ as it is subject to isomorphous replacement, *see* alums. It is used, e.g., in the dyeing, leather, paper and water industries. *alt.* for pure $K_2SO_4.Al_2(SO_4)_3.24H_2O$ is potash alum.

aluminium (Al) *n.* metallic element (at. no. 13, r.a.m. 26.98), forming a silvery metal of low density. Present as a constituent of a wide range of minerals, including beryl (emerald), corundum (ruby and sapphire), topaz, tourmaline, feldspars and other minerals. It is extracted chiefly from the ore bauxite (mixtures of aluminium oxides). The metal is widely used in ships and aircraft, in saucepans and in kitchen foil. *alt.* aluminum (US spelling).

aluminium sulphate white soluble salt used as a coagulant in water treatment. Available in either its anhydrous form, $Al_2(SO_4)_3$, or as the hydrate, $Al_2(SO_4)_3.18H_2O$.

aluminosilicates *n.plu.* a large group of minerals in which aluminium atoms have been substituted for some of the silicon atoms in the oxygen–silicon framework.

aluminum *alt.* spelling of aluminium *q.v.*

alums *n.plu.* a series of double salts that share the general formula $M_2^ISO_4.M_2^{III}(SO_4)_3.24H_2O$, where M^I is a unipositive cation (e.g. Na^+, K^+ or NH_4^+) and M^{III} is a tripositive cation (e.g. Al^{3+}, Cr^{3+} or Fe^{3+}). *see also* alum.

alum-stone, alumstone alunite *q.v.*

alunite *n.* $K_2SO_4.Al_2(SO_4)_3.4Al(OH)_3$, a mineral used as a source of alum (*q.v.*). *alt.* alumstone, alum-stone.

Am symbol for the chemical element americium *q.v.*

amaranths *n.plu.* plants of the genus *Amaranthus*, several species of which are cultivated as crop plants in South America. They produce seed-heads resembling those of sorghum that contain protein-rich seeds with a high lysine content. The leaves are also edible.

Amastigomycota *n.* in some classifications, a major division of the fungi including the subdivisions Zygomycotina, Ascomycotina, Basidiomycotina, and the form subdivision Deuteromycotina (Fungi Imperfecti). They are terrestrial fungi with a usually well developed mycelium and do not have motile flagellate zoospores or gametes. They include the familiar moulds (e.g. *Mucor*, *Penicillium*) and the mushrooms and toadstools.

ambari hemp *see* hemp.

amber *n.* hard translucent yellow or brown material, which is the fossilized resin of coniferous trees. It sometimes contains insects or other material trapped and preserved within it.

ambergris *n.* waxy secretion of the sperm whale, formerly used as a musk fragrance in perfumery, now superseded by synthetic compounds.

ambient *a.* surrounding.

amblygonite *n.* $(Li,Na)Al(PO_4)(F,OH)$, a rare mineral that is used as a source of lithium.

amenities *n.plu.* in relation to the environment, often means services, such as carparks, restaurants, etc., provided for the convenience of those visiting an area. *cf.* amenity.

amenity *n.* those properties of an environment that have no quantifiable monetary value, such as its beauty, tranquility, wildlife, etc., but from which people derive pleasure, education and enrichment.

amenity clause term sometimes used for legislation that requires a company or agency to safeguard natural beauty, i.e. landscape, wildlife and habitat, as it carries out its functions.

amensalism *n.* a form of competition between two species in which one is inhibited and the other is not. *alt.* antagonism.

americium (Am) *n.* a transuranic element [at. no. 95, r.a.m. (of its most stable known isotope) 243], a white metal when in elemental form, used as a source of alpha particles.

Ames test simple *in vitro* test devised by the American biochemist Bruce Ames to screen compounds for potential mutagens and carcinogens by their ability to cause mutations in bacteria.

ametabolous *a.* not changing form, *appl.* the orders of primitive wingless insects (the Ametabola) in which the young hatch from the egg resembling young adults, and comprising the Diplura (two-pronged bristletails), Thysanura (three-pronged bristletails), Collembola (springtails) and the Protura (bark-lice).

amethyst *n.* SiO_2, purple quartz.

ametoecious *a.* parasitic on one host during one life cycle. *alt.* autoecious.

amictic *a. appl.* (1) eggs that cannot be fertilized and that develop parthenogenetically into females; (2) females producing such eggs.

amide *n.* organic compound that contains the group $-CO.NH_2$, biological amides being derived from carboxylic acids and amino acids by replacement of the -OH of the carboxyl group with $-NH_2$.

amidine *n.* a compound that contains the group $-CNH.NH_2$.

amines *n.plu.* organic analogues of ammonia (NH_3) in which one or more of its hydrogen atoms has been replaced by a univalent hydrocarbon radical (R). Primary amines take the form NH_2R, whereas secondary amines and tertiary amines are NHR_2 and NR_3 respectively.

amino acid any carboxylic acid that contains an amino group. In nature, nearly all amino acids have the general formula $RCH(NH_2)COOH$ (alpha-amino acids), where R is a distinctive side chain, and which can occur as optically active D- and L-isomers, of which only L-isomers are found in proteins. Around 20 different amino acids are present in proteins, all of which can be synthesized by autotrophs but which in heterotrophs are chiefly obtained by breakdown of dietary protein. Amino acids are also biosynthetic precursors of many important molecules, such as purines, pyrimidines, histamine, thyroxine, adrenaline, melanin, serotonin, the nicotinamide ring and porphyrins, among others. D-amino acids are found in peptides of bacterial cell walls and in a few other instances.

amino acid analysis determination of the amino acid composition of a protein.

amino acid racemization the conversion of L-amino acids to D-amino acids, which occurs at a very slow rate in nature. It can be used to date certain fossils by measuring the amount of D-amino acids in the sample and calculating the time taken for them to form from the original L-amino acids (only L-amino acids are found in living tissue to any appreciable extent), this method being applicable to fossils between 15 000 and 100 000 years old.

amino acid sequence the order of amino acids in a polypeptide or protein.

aminoacyl-tRNA a transfer RNA with an amino acid attached.

amino group the univalent radical -NH₂, derived from ammonia (NH₃) by the removal of a hydrogen atom.

aminotriazole *n.* translocated herbicide used, e.g., to control the growth of broad-leaved weeds and grasses prior to sowing or planting crops.

amitraz *n.* contact pesticide used to control ticks on cattle and sheep and red spider mites on pears and apples.

amixis *n.* absence of fertilization. *a.* amictic.

ammonia (NH₃) *n.* chemical compound that either directly or in the form of the ammonium ion (NH₄⁺) is a source of nitrogen for plants and microorganisms. In the environment it is produced by biological fixation of atmospheric nitrogen, by the action of soil microorganisms that break down protein in dead organic matter, by the decomposition of animal urine, and as an excretion product of, e.g., teleost fishes. Industrially it can be produced by the Haber process (*q.v.*). In concentrated form it is irritant and highly toxic to humans.

ammonification *n.* the production of ammonia (as ammonium ion) by the decomposition of organic nitrogenous compounds, such as proteins, carried out by a variety of heterotrophic microorganisms (ammonifiers). *see* nitrogen cycle.

ammonifier *n.* any of a diverse group of heterotrophic bacteria and fungi that can produce ammonia and/or ammonium ion from organic matter breakdown, esp. in soil.

ammonites *n.plu.* an extinct group of abundant cephalopods familiar from their coiled fossil shells, ranging from the Devonian to Upper Cretaceous. They were similar to the nautiloids but probably had a calcareous larval shell.

ammonium ion *see* ammonia.

ammonium nitrate NH₄NO₃ (*agric.*) inorganic nitrogen fertilizer containing 35% by weight N, supplied as granules or prills (spherical granules).

ammonium sulphate (NH₄)₂SO₄ (*agric.*) inorganic nitrogen fertilizer containing 21% by weight N, supplied in crystalline form, but now rarely used.

ammonotelic *a.* excreting nitrogen mainly as ammonia, as most aquatic invertebrates, tadpoles, and some teleost fish.

Ammophila genus of grasses including marram grass *q.v.*

amniotes *n.plu.* the land-living vertebrates, which have an amnion around the developing embryo, such as reptiles, birds and mammals.

Amoco Cadiz **incident** a major oil spill that occurred when the supertanker *Amoco Cadiz* ran aground on rocks off the Brittany coast on 16 March 1978. At 223 000 tonnes, the oil spill was four times greater than that from the *Torrey Canyon* and nearly 200 miles of the French coastline suffered serious pollution as a result.

amoeba *n.* unicellular non-photosynthetic wall-less protozoan, whose shape is subject to constant change due to formation and retraction of pseudopodia. The amoebas are classified in the protist phylum Rhizopoda. *plu.* amoebae, amoebas. *alt.* ameba.

amoebiasis amoebic dysentery *q.v.*

amoebic *a. pert.*, or caused by, amoebae. *alt.* amebic.

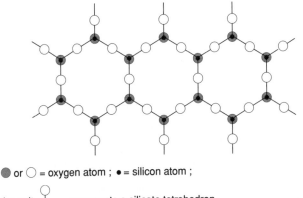

● or ○ = oxygen atom ; • = silicon atom ;

the unit ⬦ represents a silicate tetrahedron
(these are linked via shared oxygen atoms)

Note, those oxygen atoms represented thus ● reside above the silicon atoms

Fig. 1 Schematic representation of the structure of the double chains of silicate tetrahedra, as found in amphibole minerals.

amoebic dysentery *n.* a disease, endemic in the subtropics and tropics, caused by the parasitic protozoan *Entamoeba histolytica* and transmitted by faecally contaminated drinking water. *alt.* amoebiasis.

amorphous *a.* (1) of indeterminate or irregular form; (2) with no visible differentiation in structure.

amosite *n.* a highly fibrous variety of the mineral cummingtonite. *alt.* brown asbestos. *see also* asbestos.

AMP adenine monophosphate.

ampere (A) *n.* the basic SI unit of electric current, defined as the constant current that, when maintained in two parallel conductors of infinite length and negligible cross-section positioned 1 m apart, will produce a force equal to 2×10^{-7} N m^{-1} between them.

amphi- Gk prefix denoting both, on both sides.

Amphibia, amphibians *n., n.plu.* vertebrate class including the extant subclass Lissamphibia, comprising the frogs and toads (order Anura), newts and salamanders (order Urodela) and the worm-like caecilians (order Apoda). There are also a number of extinct subclasses dating from the Devonian onwards, of which the ichthyostegalians comprise the earliest amphibian fossils found. Amphibians are ectothermic anamniote tetrapod vertebrates that typically return to the water for reproduction and pass through an aquatic larval stage with gills. Adults generally have lungs, are carnivorous, may be at least partly land-living, and modern amphibians have a moist skin without scales, which is permeable to water and gases. *see also* anthracosaurs, ichthyostegalians, lepospondyls, temnospondyls.

amphibian, amphibious *a.* adapted for life either in water or on land.

amphibole minerals a large group of structurally related rock-forming silicate minerals formed in both igneous and metamorphic rocks. They all contain double chains of silicate tetrahedra (*see* Fig. 1). Each chain may be pictured as being made up of two of the chains found in pyroxene minerals (*q.v.*) linked via shared 'basal' oxygens. This generates the repeating pattern of six-membered rings that can be discerned within the double chains. Approximately in the centre of each of these rings lies an oxygen atom of a hydroxyl (OH) group. Within any one double chain there are two approximately planar arrays of oxygen atoms. The first is made up of the 'bases' of the tetrahedra, while the second contains both their apices and the hydroxyl oxygens. These planes are parallel to the axis of the chain and to one another. The double chains are held to each other by cationic counterions. Amphiboles share the general formula $A_{0-1}B_2C_5D_8O_{22}E_2$, where, in common amphiboles, A = Na or K, B = Na, Ca, Mg or divalent Fe, C = Al, trivalent or divalent Fe or Mg, D = Si or Al, E = OH, partially replaced with F, Cl or O. Extensive atomic substitution and some solid-solution series occur within the amphibole group.

amphidiploid allotetraploid *q.v.*

amphidromous *a.* going in both directions, *appl.* animal migration.

amphimict *n.* (1) a group of individuals resulting from sexual reproduction; (2) an obligate sexual organism.

amphimixis *n.* reproducing by fusion of gametes from two organisms in sexual reproduction. *a.* amphimictic. *cf.* apomixis.

Amphineura *n.* class of marine molluscs, commonly called chitons, having an elongated body and a mantle bearing calcareous plates.

amphioxus *n.* the lancelet, a cephalochordate of the genus *Branchiostoma*. They are small eel-like animals, living in sand in shallow water.

amphiploid allopolyploid *q.v.*

amphipod *n.* member of the Amphipoda, an order of terrestrial, marine and freshwater malacostracan crustaceans, having a laterally compressed body, elongated abdomen and no carapace, e.g. sand-hopper.

amphiprotic *a. appl.* any chemical species that is capable of either donating or accepting a proton(s) (H^+) depending on the nature of the other reactants present.

Amphisbaenia, amphisbaenians, amphisbaenids *n.*, *n.plu.*, *n.plu.* group of worm-like, burrowing, generally limbless reptiles with inconspicuous eyes and rounded tails.

amphitoky *n.* parthenogenetic reproduction of both males and females.

amphoteric *a.* (1) *appl.* chemical species that are capable of acting as either an acid or a base depending on the nature of the other reactants present. Examples are amino acids; (2) with opposite characters.

amplexus *n.* mating embrace in frogs and toads when eggs are shed into the water and fertilized.

AMTS aquatic macrophyte treatment system *q.v.*

a.m.u. atomic mass unit *q.v.*

amygdales *n.plu.* (*geol.*) term *appl.* to those vesicles (*q.v.*, (6)) which are occupied by low-temperature minerals (e.g. quartz, zeolites or calcite).

amygdaloidal *a.* (*geol.*) *appl.* structure of those igneous rocks that contain amygdales *q.v.*

amylases *n.plu.* enzymes that digest glucose polysaccharides, such as starch and glycogen, to give dextrin, maltose or glucose.

amylolytic a. starch-digesting.

anabatic wind wind that blows up the sides of mountain valleys during the daytime. This localized airflow develops during calm, clear weather conditions when air above the valley top is warmed, consequently rises and is replaced by cooler air flowing upslope. *alt.* valley wind. *cf.* katabatic wind.

anabiosis *n.* a condition of apparent death or suspended animation produced in certain organisms by, e.g., desiccation, and from which they can be revived to normal metabolism.

anabolism *n.* the constructive biochemical processes in living organisms involving the formation of complex molecules from simpler ones and the storage of energy. *a.* anabolic.

anabolite *n.* any substance involved in anabolism.

anachoresis *n.* the phenomenon of living in holes or crevices.

anadromous *a. appl.* fishes that migrate from salt to fresh water to spawn. *cf.* catadromous.

anaerobe *n.* any organism that can live in the absence of oxygen, obligate anaerobes being unable to live in even low oxygen concentrations, facultative anaerobes being able to live in low or normal oxygen concentrations as well. *a.* anaerobic. *cf.* aerobe.

anaerobic *a. appl.* (1) organisms that live in the absence of gaseous oxygen (O_2); (2) environments or conditions lacking in oxygen.

anaerobic decomposition, anaerobic digestion the breakdown of organic matter by microorganisms in the absence of gaseous oxygen (O_2).

anaerobic respiration the oxidative breakdown of foodstuffs when it occurs in the absence of gaseous oxygen. *alt.* glycolysis, fermentation.

anaerobiotic *a.* lacking or depleted in oxygen, *appl.* habitats.

anagenetic *a. appl.* evolution occurring by the gradual change of one type into another. *n.* anagenesis.

analcime, analcite *n.* $NaAlSi_2O_6.H_2O$, one of the zeolite minerals *q.v.*

analogous *a. appl.* structures that are similar in function but not in structure and developmental origin, e.g. the wings of insects and birds. *cf.* homologous.

analogue *n.* (1) any organ or part similar in function to one in a different plant or animal, but of unlike origin; (2) any compound chemically related to but not identical with another and that in the case of analogues of natural metabolites compete with them for binding sites on enzymes, receptors, etc., often blocking the normal reaction. *alt.* analog.

analogy *n.* resemblance in function though not in structure or origin. *a.* analogous.

analysis of variance (ANOVAR) a statistical method by which the variance of a set of data can be apportioned to different causes.

analyte *n.* (*chem.*) that part of a sample that is to be identified or, more usually, quantified.

anamniotes *n.plu.* fishes, amphibians and the Agnatha (lampreys and hagfishes), characterized by the absence of an amnion around the embryo.

anamorphosis *n.* evolution from one type to another through a series of gradual changes.

anapsids *n.plu.* tortoises and turtles and extinct members of the reptilian subclass Anapsida, characterized by a sprawling gait and a skull with no temporal opening (anapsid skull).

anatomical *a. pert.* the structure of a plant or animal.

anatomy *n.* study of the structure of plants and animals as determined by dissection.

ancient forest, ancient woodland native forest or woodland, which may be virgin forest or old secondary forest developed by secondary succession, that has been continuously present on a site, often with management but without extensive clear-felling, for hundreds of years and that can be recognized by its characteristic flora. *cf.* old-growth forest, secondary forest.

andalusite *n.* Al_2SiO_5, a nesosilicate mineral of metamorphic rocks (formed under low pressure) and some pegmatites. Green transparent andalusite is prized as a gemstone.

Andean Floral Region Pacific South American Floral Region *q.v.*

andesine *n.* one of the plagioclase (*q.v.*) minerals.

andesite *n.* a fine-grained, frequently porphyritic, igneous rock. Phenocrysts are plagioclase feldspars together with biotite, pyroxene (typically augite) and/or hornblende. The groundmass has essentially the same mineralogy as the phenocrysts.

andesitic lava acid lava *q.v.*

andisol *n.* an order of the USDA Soil Taxonomy system consisting of weakly developed, fairly fertile soils formed on volcanic ash and located mainly within the Pacific Ring of Fire. Seven suborders are recognized: aquand (wet areas), cryand (cold areas), torrand (warm, arid climates), ustand (monsoon climates), udand (humid climates), xerand (mediterranean climates) and vitrand (associated with volcanic glass).

andradite *n.* one of the garnet minerals *q.v.*

Andreaeidae granite mosses *q.v.*

androconia *n.plu.* modified wing scales producing a sexually attractive scent in certain male butterflies.

androecium *n.* (1) male reproductive organs of a plant; (2) stamens collectively.

androgen *n.* any of various male steroid sex hormones concerned with development of male reproductive system and production and maintenance of secondary sexual characteristics, and secreted chiefly by testis, e.g. androsterone and testosterone.

androgynous *a.* (1) hermaphrodite; (2) (*bot.*) bearing both staminate and pistillate flowers in the same inflorescence; (3) (*mycol.*) with antheridium and oogonium on the same hypha.

androtype *n.* type specimen of the male of a species.

anemo- prefix derived from Gk *anemos*, wind.

anemochory *n.* dispersal of seeds by the wind. *a.* anemochorous.

anemometer *n.* instrument used for measuring wind speed.

anemophily *n.* wind pollination or any other type of fertilization brought about by wind. *a.* anemophilous.

anemotaxis *n.* movement in response to air currents.

aneroid barometer instrument used to measure air pressure, which consists of a metal box enclosing a vacuum. Changes in atmospheric pressure cause the expansion or contraction of an attached spring, which moves a pointer over a scale marked in millibars.

aneuploid *a.* (1) having more or less than the normal number of chromosomes; (2) *appl.* chromosomal abnormalities that disrupt relative gene dosage, such as deletions, *alt.* unbalanced. *n.* aneuploidy.

Angiospermae, angiosperms Anthophyta *q.v.*

anglesite *n.* $PbSO_4$, a secondary lead mineral.

Ångström (Å) *n.* unit of ultramicroscopic measurement, 10^{-10} m or 0.1 nm.

anhydrite *n.* a mineral of formula $CaSO_4$.

anhydrous *a.* without water, frequently *appl.* salts that do not contain water of crystallization.

anhydrous ammonia NH_3, (*agric.*) inorganic nitrogen fertilizer containing 82% by weight N, supplied as a liquid, delivered by injection into the soil.

animal unit unit of livestock, based on food needs and used in calculating the stocking density a piece of land can sustain. The unit is equal to the needs of one cow (a mature cow of about 454 kg), which is considered equivalent to one horse, five sheep, five pigs or six goats.

Animalia, animals *n.*, *n.plu.* a kingdom of multicellular heterotrophic eukaryotic organisms with wall-less, non-photosynthetic cells, which feed by taking in solid organic material (holozoic nutrition). All multicellular animals except the sponges possess some form of nervous system and contractile muscle or muscle-like cells, and most can move about. In some classifications, non-photosynthetic unicellular Protista, e.g. protozoans such as *Amoeba*, are also included in the animal kingdom. *see* Appendix 7.

animal manures dung and urine of animals that can be used as a manure. *cf.* green manures.

anion *n.* negatively charged ion, e.g. Cl^-, that moves towards the anode, the positive electrode. *cf.* cation.

anion exchange the replacement of anions, i.e. negatively charged ions, held on the surface of a solid that has a positive charge with anions in a solution in contact with that solid.

anion exchange capacity (AEC) quantification of the ability of a positively charged solid to exchange anions held on its surface with anions in the surrounding solution. It is commonly expressed in milliequivalents per 100 g of solid (meq per 100 g, meq $100 g^{-1}$), where in this context, an equivalent is a mole of charge. It may also be expressed as centimoles of charge per kilogram of solid (cmol kg^{-1}).

aniso- prefix from Gk *anisos*, unequal.

anisopterans *n.plu.* dragonflies, members of the suborder Anisoptera of the order Odonata.

anisotropic *a. appl.* solids, e.g. crystals and rocks, that exhibit differences in physical properties in different directions, e.g. certain rocks (e.g. slate) that tend to split more readily in some directions than in others.

Annelida, annelids *n.*, *n.plu.* phylum of segmented coelomate worms, commonly called ringed worms, having a soft elongated body with a muscular body wall, divided into many similar segments, usually separated by septa, and the body covered with a thin, flexible collagenous cuticle. They possess a blood system, nephridia and a central nervous system. The Annelida contains three main classes, Polychaeta (ragworms, lugworms), Oligochaeta (e.g. earthworms) and Hirudinea (leeches).

annidation *n.* situation in which a mutant organism survives in a population because an ecological niche exists that the normal individual cannot use.

Annonales Magnoliales *q.v.*

annotinous *a. appl.* growth during the previous year.

annual *a.* (1) occurring or happening once every year; (2) *appl.* structures or growth features that are marked off or completed yearly; (3) living for a year only; (4) completing life cycle in a year from germination.

annual *n.* plant that completes its life cycle in a year.

annual allowable cut the established permissible harvest of some resource.

annual increment the increase in biomass of an organism or community in a year.

annual ring growth ring *q.v.*

annular drainage *see* drainage pattern.

Annulata, annulates *n.*, *n.plu.* a group of invertebrates including the annelid worms, arthropods and some related forms, individuals having bilateral symmetry and true metameric segmentation.

anoestrus *n.* (1) the non-breeding period; (2) period of absence of sexual receptiveness in females.

anomaly *n.* any departure from type characteristics.

anopheline *a. appl.* mosquitoes of the genus *Anopheles*, vectors of malaria and some other diseases.

Anoplura *n.* order of insects, the sucking or body lice, that are ectoparasites of mammals. The human body louse *Pediculus humanus* is the vector of typhus. *alt.* Siphunculata.

anorthite *n.* CaAl$_2$Si$_2$O$_8$, one of the feldspar minerals *q.v.*

anorthoclase *n.* (Na,K)AlSi$_3$O$_8$, one of the feldspar minerals (*q.v.*). An alkali feldspar.

anorthogenesis *n.* evolution showing changes in direction of adaptations.

anorthosite *n.* a white–grey medium- or coarse-grained igneous rock with a colour index <10 that is *c.* 90% plagioclase feldspar (bytownite, labradorite, andesine, oligoclase).

ANOVAR analysis of variance *q.v.*

anoxia *n.* lack of oxygen.

anoxic *a.* devoid of molecular oxygen, *appl.* habitats.

Anoxyphotobacteria *n.* class of bacteria including photosynthetic bacteria other than the cyanobacteria, i.e. those that do not produce oxygen as a byproduct of photosynthesis, and including the green and purple photosynthetic sulphur bacteria.

Anseriformes *n.* a large order of birds, the waterfowl, including ducks, geese and swans. *a.* anseriform.

anserine *a. pert.* a goose.

antagonism *n.* the effect of a chemical or biological species that counteracts, opposes or decreases the presence or effects of one or more other chemical or biological species. Examples are a hormone that counteracts the effects of another, a plant that excretes substance(s) from its roots that inhibit other plants nearby, a toxicant or pollutant that ameliorates the effect of another toxicant or pollutant. *a.* antagonistic.

Antarctic Circle the circle of latitude in the southern hemisphere located at 66° 32′ S. South of the Antarctic Circle, the Earth's surface receives sunlight for a period of 24 h at least once a year.

Antarctic kingdom phytogeographical area comprising the Antarctic, New Zealand and the southern tip of South America.

ante- prefix from L. *ante*, before, in front of.

antecedent drainage a pre-existing drainage pattern that is maintained during tectonic uplift. This occurs when the rate of downward erosion by the stream matches or exceeds that of uplift.

antenna *n.* (*zool.*) (1) one of a pair of jointed feelers on head of various arthropods; (2) feeler of rotifers; (3) in some fish, a modified flap on dorsal fin that attracts prey; (4) (*bot.*) group of chlorophyll and other molecules involved in light capture during photosynthesis. *plu.* antennae. *a.* antennal, antennary.

antennation *n.* touching with the antenna, serving as tactile communication signal or an exploratory probing in insects and other arthropods.

anterior *a.* (1) nearer head end; (2) ventral in human anatomy; (3) facing outwards from axis; (4) previous. *cf.* posterior.

anthelmintic drugs *n.plu.* drugs used in the treatment and control of parasitic flatworms (flukes and tapeworms) and roundworms.

anther *n.* terminal part of stamen, which produces the pollen. *see* flower, Fig. 12, p. 157.

antheridium *n.* sex organ in which male gametes are produced in many cryptogams (ferns, mosses, etc.) and algae. *plu.* antheridia.

antherozoid *n.* motile male gamete of algae, mosses and ferns and other lower plants. *alt.* sperm, spermatozoid.

anthesis *n.* (1) stage or period at which flower bud opens; (2) flowering.

Anthocerophyta, Anthoceratopsida, Anthocerotae, Anthocerotales *n.* small spore-bearing non-vascular green plants with a thalloid gametophyte and rosette-like habit of growth, commonly called hornworts. The cells typically contain a single large chloroplast. Hornworts often carry symbiotic photosynthetic cyanobacteria in the intercellular spaces. The sporophyte is typically an upright elongated sporangium on a stalk (the foot) growing from the gametophyte. *see also* bryophytes.

anthocyanin *n.* any of a group of important plant pigments found in flowers, fruits, leaves and stems, which are sap-soluble flavonoid glycosides giving scarlet, purple and blue colours. Also found in some insects, having been absorbed from the plant on which the insects feed.

anthophilous *a.* (1) attracted by flowers; (2) feeding on flowers.

anthophyllite *n.* $(Mg,Fe)_7Si_8O_{22}(OH)_2$, one of the amphibole minerals *q.v.*

Anthophyta *n.* the flowering plants, one of the five main divisions of extant seed-bearing plants. Reproductive organs (stamens and ovary) are carried in flowers (*q.v.*, *see* Fig. 12, p. 157), in which they are typically surrounded by sterile leaves (petals and sepals). After pollination and fertilization the closed ovary containing the seeds develops into a fruit. The gametophyte generation is much reduced, being restricted to the male and female gametes. *alt.* angiosperms, Magnoliophyta.

anthoxanthins *n.plu.* sap-soluble flavone flower pigments giving colours from ivory to deep yellow, also found in insects, having been absorbed from the plant on which the insect feeds.

Anthozoa, anthozoans *n.*, *n.plu.* class of coelenterates of the phylum Cnidaria, comprising the soft corals, sea pens and sea fans (subclass Alcyonaria), and the sea anemones and stony corals (subclass Zoantharia). The soft corals, sea pens, sea fans and stony corals are generally colonial, with individual polyps connected by living tissue and, in the case of the stony corals, embedded in a calcium carbonate matrix. Sea anemones are generally solitary.

anthracite *n.* hard, black, high-rank coal with a high carbon content and a very low moisture and volatile substance content. *alt.* carbonaceous coal.

anthracnose *n.* fungal disease of plants typically causing black spots on leaves.

Anthracosauria, anthracosaurs *n.*, *n.plu.* order of Carboniferous–Permian labyrinthodont amphibians, among whose members were the ancestors of the reptiles.

anthraquinone *n.* any of a class of orange or red pigments found in lichens, fungi, higher plants and insects, such as the cochineal beetle and lac insect.

anthropic zone (1) that area of the Earth's surface that is under the influence of humans; (2) a stratigraphic zone in which human remains are found.

anthropocentrism *n.* an exclusively human-centred view that human activities are paramount and need take no account of non-human species. *cf.* biocentrism. *a.* anthropocentric.

anthropochory *n.* accidental dispersal by man (via spores, pollen, etc.).

anthropogenic *a.* produced or caused by man.

anthropoid *a.* resembling or related to humans, as the anthropoid apes: orang utan, chimpanzee and gorilla. Gibbons are sometimes included.

Anthropoidea *n.* the suborder of primates consisting of monkeys, apes and humans.

anthropology *n.* the scientific study of human beings and human societies, especially differences in social organization, racial differences, physiological differences and social and religious development.

anthropometry *n.* study of proportional measurements of parts of the human body.

anthropomorphism *n.* ascribing human emotions to animals.

anti- prefix from Gk *anti*, against or L. *ante*, before.

Anticline   Recumbent fold   Overturned fold

Syncline

Fig. 2 Schematic vertical cross-section of sedimentary rock strata showing types of folds.

antibiosis *n.* antagonistic association of organisms in which one produces compounds, antibiotics, harmful to the other(s).

antibiotic *n.* any of a diverse group of organic compounds produced by microorganisms that selectively inhibits the growth of or kills other microorganisms, many antibiotics being used therapeutically against bacterial and fungal infections in man and animals. *a.* killing or inhibiting growth.

antibiotic resistance *see* drug-resistance factors, R factor.

antiboreal *a. pert.* cool or temperate regions of the southern hemisphere.

anticlines *n.plu.* arch-like upfolds in rock strata, produced by folding. When folding is very intense, the folds become distorted to form recumbent folds or even overturned folds. *see* Fig. 2.

anticryptic *a. appl.* protective coloration of an animal facilitating its ability to attack another.

anticyclogenesis *n.* development of anticyclones (high-pressure cells). *cf.* cyclogenesis.

anticyclones *n.plu.* areas of high pressure, frequently associated with clear skies and fine weather. *alt.* highs.

antidromic *a.* contrary to normal direction.

antifouling paints special paints containing biocides used to coat the marine surfaces of ships and docks to prevent colonization by marine organisms. Tributyltins (*q.v.*) were used extensively in these paints. However, increasing environmental concern over the leaching of these toxic substances into the marine environment has led to their use being restricted or banned in some countries. This has prompted the development of effective alternatives, e.g. slippery epoxy urethane coatings.

antifreeze compounds (1) compounds such as glycerol, sorbitol and mannitol, which lower the freezing point of body fluids and protect against freezing, found in the haemolymph of some insects; (2) glycoproteins found in the blood of some polar fish, which depress the freezing point of the blood by enveloping small ice crystals that would otherwise form ice nuclei and cause the blood to freeze.

antigiberellin *n.* any compound (e.g. phosphon, maleic hydrazide) with an action on plant growth opposite to that of giberellins. They cause plants to grow with short thick stems.

antigorite *n.* $Mg_3Si_2O_5(OH)_4$, a variety of the mineral serpentine *q.v.*

antiknock agents compounds that are added to petrol to increase its octane number, thereby decreasing the propensity of an engine to knock. Tetraalkyl lead compounds, including tetraethyl lead $[Pb(CH_2CH_3)_4]$, are widely used antiknock agents.

antimonite stibnite *q.v.*

antimony (Sb) *n.* silvery crystalline metalloid element (at. no. 51, r.a.m. 121.75), found in nature as stibnite (antimony sulphide) and in other minerals. Its compounds are used in dyes and lead–acid batteries and in the production of semiconductors. Prolonged exposure can lead to heart disease.

antimony glance stibnite *q.v.*

antinatalist *a. appl.* policies, or persons supporting them, designed to reduce the birth rate.

antireductionism *see* reductionism.

antiseptic *n.* substance that destroys microorganisms, esp. bacteria. *a.* preventing growth of microorganisms.

antitype *n*. a specimen of the same type as that chosen for designation of a species, and gathered at the same time and place.

antiviral *a. appl.* antibodies, drugs etc. that destroy or neutralize a virus.

antlers *n.plu.* paired bony growths, projections from the skull, on the heads of members of the deer family, which are often branched, are shed annually and are usually confined to males. *cf.* horns.

ant lions a group of insects in the order Neuroptera *q.v.*

ants *n.plu.* social insects of the superfamily Formicoidea of the order Hymenoptera, which live in colonies composed of a queen, with male, worker and, in some cases, soldier, castes.

Anura, anurans *n., n.plu.* one of the three orders of extant amphibians, comprising the frogs and toads. In some classifications called the Salientia.

anvil cloud cumulonimbus *q.v.*

AONB Area of Outstanding Natural Beauty (*q.v.*), *see also* National Park and Access to the Countryside Act.

apatetic coloration protective coloration that misleads the predator, such as markings that mask the body outline, or markings that produce the appearance of eyes where there are none.

apatite *n*. $Ca_5(PO_4)_3(F,Cl,OH)$, a mineral used to produce phosphate fertilizers. Varieties of apatite include fluorapatite [$Ca_5(PO_4)_3F$, *alt.* fluor-apatite], hydroxylapatite [$Ca_5(PO_4)_3OH$, *alt.* hydroxy-apatite] and chlorapatite [$Ca_5(PO_4)_3Cl$]. Massive cryptocrystalline apatite is sometimes called collophane.

apes *see* Primates.

apetalous *a*. without petals.

apex *n*. tip or summit, as of wing, heart, lung, root, shoot. *plu.* apices.

Aphaniptera Siphonaptera *q.v.*

Aphasmidia *n*. class of nematode worms with no phasmids, and whose amphids open on to the posterior part of the head capsule.

aphicide *n*. chemical that kills aphids.

aphids *n.plu.* insects of the family Aphididae (Aphidae) of the order Hemiptera with mouth-parts adapted for piercing and sucking plants, which are of economic importance as plant pests and vectors of virus

diseases and that have a parthenogenetic and a sexual reproductive phase.

Ap-horizon ploughed layer of cultivated soils. *alt.* plough layer.

aphotic *a. pert.* absence of light.

aphotic zone that part of the water column below the photic zone (*q.v.*), usually below *ca.* 100 m. In this zone, lack of sunlight for photosynthesis becomes a limiting factor and there is no net growth of phytoplankton as a result.

aphototropism *n*. tendency to turn away from light.

Aphragmabacteria *n*. in some classifications the name for the group of prokaryotes comprising the mycoplasmas (*q.v.*) and similar organisms, small prokaryotes lacking the typical bacterial cell wall, and bounded by a triple-layered lipid membrane. They live mostly as obligate intracellular parasites of plants and animals.

aphthoviruses *n.plu.* group of picornaviruses including foot-and-mouth disease virus.

aphyllous *a*. without foliage leaves. *n*. aphylly.

aphytic *a*. without plant life, *appl.* zone of coastal waters below *ca.* 100 m, the bottoms of deep lakes, etc.

Apiales Cornales *q.v.*

apical *a*. (1) at the tip of any cell, structure or organ; (2) *pert.* distal end; (3) *appl.* cell at tip of growing point.

apical–basal axis the shoot tip to root tip axis of a plant.

apical dominance the situation where the apical bud suppresses the outgrowth of the lateral buds below it on the stem. It is due to the production of auxin by the apical bud.

apical meristems dividing tissue at tip of developing shoot and young root of plants, at which growth occurs.

apices *plu.* of apex.

Apicomplexa *n*. phylum of non-photosynthetic heterotrophic protists parasitic in animals, comprising the sporozoan protozoans, e.g. gregarines, coccidians, *Plasmodium* and piroplasms. They are transmitted from host to host in the form of 'spores', small infective bodies produced by schizogony. *alt.* Sporozoa.

apiculture *n*. bee-keeping.

apivorous *a*. feeding on bees.

Aplysia genus of opisthobranch molluscs known as sea hares.

apocentric *a.* diverging or differing from the original type.

apocratic *a.* opportunistic, *appl.* species.

Apoda *n.* (1) an order of limbless burrowing amphibians, commonly known as caecilians, an individual having a reduced or absent larval stage and minute calcified scales in the skin. In some classifications called the Gymnophiona; (2) the name has also been given to orders of parasitic barnacles (crustaceans) and burrowing sea cucumbers (echinoderms).

Apodiformes *n.* an order of birds including the swifts and hummingbirds.

apolegamic *a. appl.* mating associated with sexual selection.

apomict *n.* an organism reproducing by apomixis. *a.* apomictic.

apomixis *n.* an asexual reproductive process without fertilization in plants, akin to parthenogenesis but including development from cells other than ovules. *see also* vegetative apomixis. *cf.* amphimixis.

apomorphous *a.* in cladistic phylogenetics, *appl.* novel character evolved from a pre-existing character. The two form a homologous pair of characters, termed an evolutionary transformation series. *see also* synapomorphy.

apophysis *n.* (1) (*geol.*) a protuberance or vein extending from an igneous intrusion into the country rock. May be a mineral vein or a small-scale sill or dyke; (2) (*zool.*) a projecting process on bone or other skeletal material, usually for muscle attachment. *plu.* apophyses.

apoplast *n.* the cell walls collectively of a tissue or a complete plant. *a.* apoplastic.

aposematic *a. appl.* warning coloration or markings that signal to a predator that an organism is toxic, dangerous or distasteful. *cf.* epidemic, arasematic.

apostatic *a.* differing markedly from the normal.

apostatic selection type of frequency-dependent selection in which a predator selects the most common morph in the population, thus tending to increase the proportions of the other morphs in the population.

apothecium *n.* (1) cup-shaped fruiting body of Discomycetes (cup fungi, morels and truffles) bearing asci on the inner surface; (2) fruiting body of some lichens. *plu.* apothecia.

apparent dip dip, apparent *q.v.*

apparent mortality a measure of mortality in a population at a given developmental stage, e.g. age groups, expressed as the percentage of the number alive at the beginning of the stage. *alt.* percentage successive mortality.

appeasement *n. appl.* behaviour that ends the attack of one animal on another of the same species by the loser adopting a submissive posture or gesture.

appendage *n.* organ or part attached to a trunk, as limb, branch, etc.

appetitive *a. appl.* behaviour at the beginning of a fixed behaviour pattern, which can be very variable, from unoriented wanderings to apparently purposeful behaviour.

application factor a factor used to determine the maximum safe concentration of a substance for an organism. It is the ratio of the concentration of the substance that produces a certain long-term response in the organism to the concentration causing death in 50% of the population within a given time period.

apposition *n.* laying down of material on a preformed surface, as in growth of cell wall, bone, etc.

appressed *a.* pressed together without being united.

appressorium *n.* (1) adhesive disc, as of sucker or haustorium; (2) modified hyphal tip that may form haustorium or penetrate substrate, as of parasitic fungi.

Apterygiformes *n.* an order of flightless birds including the kiwis.

apterygote, apterygotous *a. appl.* a group of insects, the subclass Apterygota, that have no wings, little or no metamorphosis, and abdominal appendages in the adult, and that comprises the orders Thysanura, Diplura, Protura and Collembola (*see* individual entries).

aquaculture *n.* raising of algae, fish and shellfish for human use in artificial or natural freshwater ponds, lakes, irrigated fields and irrigation ditches, and, for marine organisms, in enclosures in coastal inlets and estuaries.

aqualf *n.* a suborder of the order alfisol of the USDA Soil Taxonomy system, consisting of alfisols seasonally saturated with water.

aquand *see* andisol.

aquatic *a.* living in or near water.

aquatic ecosystems any ecosystem of which the principal component is water, such as ponds, lakes, rivers, streams and oceans. *cf.* wetland.

aquatic macrophyte treatment system (AMTS) any wastewater treatment system, e.g. reed-bed system, in which aquatic macrophytes are used to purify effluent by facilitating the breakdown of contaminants and/or their direct uptake and accumulation.

aquent *n.* a suborder of the order entisol of the USDA Soil Taxonomy system consisting of entisols that show evidence of seasonal saturation.

aqueous ammonia *n.* $NH_{3(aq)}$, (*agric.*) inorganic nitrogen fertilizer containing 21–29% N, supplied as a liquid.

aquept *n.* a suborder of the order inceptisol of the USDA Soil Taxonomy system comprising wet, poorly drained inceptisols.

aquiclude *see* aquifer.

aquifer *n.* a layer of water-bearing rock located underground. Unconfined aquifers are found where water from precipitation percolates through the ground until an impermeable rock layer is met. Confined (artesian) aquifers form when a permeable rock layer is sandwiched between two layers of impermeable rock (known as aquicludes) and these strata are tilted, thus allowing water to seep into the central permeable rock layer. *see also* perched aquifer.

aquiherbosa *n.* herbaceous vegetation growing submerged in water.

aquitard *n.* semi-porous rock bed that allows some water to seep through it.

aquod *n.* a suborder of the order spodosol of the USDA Soil Taxonomy system, consisting of spodosols seasonally saturated with water.

aquoll *n.* a suborder of the order mollisol of the USDA Soil Taxonomy system consisting of mollisols that are seasonally saturated with water.

aquox *n.* a suborder of the order oxisol of the USDA Soil Taxonomy system, consisting of oxisols formed under the influence of water.

aquult *n.* a suborder of the order ultisol of the USDA Soil Taxonomy system, consisting of ultisols seasonally saturated with water.

Ar symbol for the chemical element argon *q.v.*

arable *a.* *appl.* land that is cultivated, and to crops grown on cultivated land, except trees. *cf.* pasture, rangeland.

Arachnida, arachnids *n.*, *n.plu.* class of mainly terrestrial, carnivorous arthropods, in some classifications part of the phylum Chelicerata, comprising spiders (order Araneae), scorpions (order Scorpiones), mites and ticks (order Acari), false scorpions (order Pseudoscorpiones), palpigrades (order Palpigrada), the solifugids (order Solifugae) and harvestmen (order Opiliones). In arachnids, the body is usually divided into a prosoma of eight fused segments and a posterior opisthosoma of 13 fused segments. The prosoma is not differentiated into a head and thorax and bears the clawed and prehensile chelicerae (poison claws), the pedipalps and four pairs of walking legs.

arachnoid *a.* (1) *pert.* or resembling a spider; (2) like a cobweb; (3) consisting of fine entangled hairs.

aragonite *n.* $CaCO_3$, a major carbonate mineral, a crystalline form of calcium carbonate, that is not as abundant as the alternative form, calcite (*q.v.*). Aragonite is one of the constituents of mollusc shells.

Arales *n.* an order of herbaceous monocots comprising the families Araceae (arum) and Lemnaceae (duckweed).

Araliales Cornales *q.v.*

Aral Sea, decline of a major ecological disaster in which the Aral Sea in Central Asia, once the fourth largest lake in the world, has lost around two-thirds of its water during the last 30 years. This has been caused by the heavy withdrawal of water from the Syr and Amu tributaries, for use in intensive, irrigated cotton cultivation. The contraction of the Aral Sea has resulted in the destruction of its once-thriving fishing industry, and also the loss of its navigation routes. In addition, surrounding areas, often of productive agricultural land, have been adversely affected by huge salt dust storms that originate from exposed areas of the former sea bed.

Araneida spiders *q.v.*

araneose, araneous *a.* covered with or consisting of fine entangled filaments.

arboreal *a.* (1) living in trees; (2) *pert.* trees.

arboretum *n.* a collection of species of trees.

arboriculture *n.* the cultivation of trees.

arboroid *a.* tree-like.

arbovirus *n.* arthropod-borne virus that replicates in an arthropod as its intermediate host and in a vertebrate as its definitive host, e.g. yellow fever, transmitted by mosquitoes, and the tick-borne encephalitis complex. Previously known as arbor viruses, arborviruses.

arbuscular *a.* (1) shrub-like; (2) *appl.* complex branched hyphal systems of vesicular-arbuscular mycorrhizae.

arbuscule *n.* small tree-like shrub or dwarf tree.

arbutoid *a. appl.* endomycorrhizas formed on members of the tribe Arbutoideae (family Ericaceae), with a well-defined fungal sheath and Hartig net and extensive penetration of the cells of the root cortex.

arch-, arche- prefix from Gk *archē*, beginning.

archae- prefix from Gk *archaios*, primitive.

Archaea *n.* superkingdom of prokaryotic microorganisms also known as the archaebacteria. It includes extreme thermophiles, extreme halophiles and methanogens (methane producers), typically found in extreme environments, e.g. hot springs, salt lakes, etc. They differ in many ways from other prokaryotes, e.g. in the structure of their cell walls and membrane lipids and possession of introns in some genes, and are thought to represent a quite separate group of prokaryotes from the Bacteria (eubacteria). In some classifications they form the class Mendosicutes of the kingdom Prokaryotae. *see* Appendix 9.

Archaean *a. appl.* the earlier eon of the Precambrian, ending at around 2500 million years ago. *alt.* Archean.

archaebacteria *n.plu. see* Archaea.

Archaeopteryx and ***Archaeornis*** genera of fossil birds from the Jurassic, which show many reptilian features.

Archaeornithes *n.* a subclass of extinct primitive birds containing the fossils *Archaeopteryx* and *Archaeornis*.

Archaeozoic *a. pert.* earliest geological era, the lower division of the Precambrian, the time of Archaean rocks and solely unicellular life. *alt.* Archaean.

Archean Archaean *q.v.*

archecentric *a.* conforming more or less with the original type.

archegonium *n.* female sex organ on gametophyte of liverworts, mosses, ferns and related plants, consisting of a multicellular, flask-shaped structure containing one ovum (oosphere) in the base. *plu.* archegonia.

archi- prefix from Gk *archi*, first.

archibenthic *a. pert.* sea bottom from edge of continental shelf to upper limit of abyssal zone, at depths of *ca.* 200–1000 m.

archibenthic zone area of the sea bottom that extends from the edge of the continental shelf to a depth *ca.* 1000 m. *cf.* abyssal benthic zone.

archipelago *n.* (1) a group or string of islands; (2) a sea with many islands. *plu.* archipelagos, archipelagoes.

architype *n.* an original type from which others may be derived. *alt.* archetype.

Archosauria, archosaurs *n., n.plu.* subclass of reptiles, the 'ruling reptiles', that included the dinosaurs. Mostly extinct except for the living crocodilians.

Arctic Circle the circle of latitude in the northern hemisphere located at 66° 32′ N. North of the Arctic Circle, the Earth's surface receives sunlight for a period of 24 h at least once a year.

Arctic Floral Region the region of the Holarctic Realm that extends from the far north, south to central Alaska, Labrador, central Scandinavia and northern Siberia.

Arctogaea, Arctogea *n.* zoogeographical area comprising Holarctic, Ethiopian and Oriental regions.

arcuate delta *see* delta.

Area of Outstanding Natural Beauty (AONB) in England and Wales (UK), a planning designation used to protect special areas of landscape outside the National Parks. *see* National Park and Access to the Countryside Act.

area strip mining mining of mineral materials (e.g. coal and phosphates) found fairly near the surface by cutting deep trenches. *cf.* contour strip mining, opencast mining.

Arecales *n.* order of tree-like or climbing monocots with feather- or fan-like leaves, comprising the family Arecaceae (Palmae) (palms, yuccas, etc.).

arena *n.* (*zool.*) area used for communal courtship displays.

arenaceous *a.* having properties or appearance of sand.

arenaceous rocks *see* clastic rocks.

Arenaviridae, arenaviruses *n., n.plu.* family of animal viruses with single-stranded RNA genomes in two parts and that include Lassa fever and lymphocytic choriomeningitis virus.

arene *n.* any aromatic hydrocarbon, e.g. benzene.

arenicolous *a.* living or growing in sand.

arent *n.* a suborder of the order entisol of the USDA Soil Taxonomy system consisting of entisols found where human activity, such as ploughing, has prevented the development of soil horizons.

arescent *a.* becoming dry.

arêtes *n.plu.* narrow, steep-sided ridges found between adjacent cirques (*q.v.*). *alt.* coombe ridges, grats.

Argentinian Floral Region part of the Austral Realm that includes Argentina, Paraguay, southern Chile and the offshore islands, including the Falkands (Malvinas).

argids *see* aridisol.

argillaceous *a.* (1) having clay-sized particles, *appl.* soil; (2) having the properties of clay.

argillaceous rocks *see* clastic rocks.

argillic horizon *see* ultisol.

arginine (Arg, R) *n.* basic amino acid, positively charged at physiological pH, essential in human diet, constituent of protein, hydrolysed to ornithine and urea in urea cycle.

argon (Ar) *n.* colourless inert gas (at. no. 18, r.a.m. 39.95), one of the noble gases, which forms around 1% by volume of the atmosphere. It has a radioactive isotope, ^{41}Ar, with a half-life of 30 days. *see also* potassium–argon dating.

ariboflavinosis *n.* condition of skin cracking and lesions caused by a deficiency of the vitamin riboflavin.

arid *a. appl.* climate or habitat with less than 250 mm annual rainfall, very high evaporation and sparse vegetation.

aridisol *n.* an order of the USDA Soil Taxonomy system consisting of mineral soils that are dry for most of the year, found in arid and semi-arid regions. Aridisols usually have a thin, light-coloured surface horizon, low in organic matter (ochric epipedon) and often contain subsurface horizons rich in gypsum, calcium carbonate or more soluble salt minerals. Two suborders are recognized: argids (with horizon of clay accumulation) and orthids (without horizon of clay accumulation).

aridity index a measure of the aridity of an area that takes into account rainfall and evaporation.

arid zone region within latitudes 15–30° on both sides of the Equator in which rainfall is very low and either evaporates in the high daytime temperatures or drains away rapidly so that it is unavailable to vegetation. This zone contains most of the world's deserts. Parts of the zone support vegetation and some cultivation but are subject to overgrazing and overcultivation, which can, in times of drought, lead to desertification.

aril *n.* additional covering formed on some seeds after fertilization, which may be spongy, fleshy (as the red aril of yew berries) or a tuft of hairs.

Aristolochiales *n.* an order of dicots of the subclass Magnoliidae, containing one family, the birthworts (Aristolochiaceae), which comprise herbs and climbing plants.

arithmetic growth linear growth *q.v.*

arithmetic mean ($\bar{x}$) the statistical measure commonly known as the average, which is obtained by adding up all the numbers in a set and dividing the sum by the number of elements in the set. *see also* mean.

ark *n.* a portable structure, often triangle-shaped, used to provide shelter for poultry or pigs.

arkose *n.* a sedimentary rock derived from granite or granite gneisses that is composed mainly of quartz and $\geqslant 25\%$ feldspar and that is generally cemented with iron oxides or calcite.

armature *n.* any structure that serves as a defence, e.g. hairs, prickles, thorns, spines, stings.

Armero natural disaster one of the worst natural disasters of the 1980s. On 13 November 1985, the eruption of the Nevado del Ruiz volcano in the Columbian Andes caused melting of its snowcap and the production of a massive mudflow (known as a lahar). More than 20 000 people perished when the lahar engulfed the nearby town of Armero.

arms race (1) (*biol.*) the sequence of evolutionary changes seen in, e.g., a predator and its prey, as each advantageous adaptation in one organism is countered by a further adaptation in the other; (2) in human society, the reciprocal build-up of weapons between antagonistic geopolitical groupings.

Aroclor a trade name of the company Monsanto for a series of polychlorinated biphenyl (PCB) formulations.

aromatic amino acids amino acids with an aromatic side chain: phenylalanine, tryptophan, tyrosine.

aromatic compounds (1) benzene and its derivatives that contain the benzene ring; (2) unsaturated cyclic organic compounds and ions that share characteristic chemical properties with benzene.

arousal *n*. level of responsiveness to a stimulus in an animal.

array *n*. arrangement in order of magnitude.

Arrhenius acid a hydrogen-containing compound that releases $H^+_{(aq)}$ (i.e. aqueous hydrogen ions) when in water, e.g. nitric acid (HNO_3) or trichloroacetic acid (CCl_3COOH).

Arrhenius base a compound that, when dissolved in water, produces aqueous hydroxide ions ($OH^-_{(aq)}$), e.g. sodium hydroxide (NaOH), which dissociates in water into $Na^+_{(aq)}$ and $OH^-_{(aq)}$, or ammonia (NH_3), which reacts with water to liberate OH^-, i.e. $NH_{3(aq)} + H_2O_{(l)} \rightleftharpoons NH_{4(aq)}^+ + OH^-_{(aq)}$.

arrhenogenic *a*. producing offspring preponderantly or entirely male. *n*. arrhenogeny.

arrhenotoky *n*. type of parthenogenesis where males are formed from unfertilized eggs and are haploid.

arrow-worms Chaetognatha *q.v.*

arroyo *n*. term used in the southwestern USA for a steep-sided, flat-floored, ephemeral stream channel. This is an equivalent feature to the wadi of the Middle East and North Africa.

arsenic (As) *n*. metalloid element (at. no. 33, r.a.m. 74.92), which can form three different allotropes: a silvery-grey crystal-

line form, a black form and a yellow form. It is present in nature as an impurity in other metal ores and also as arsenopyrite. Arsenic compounds are highly toxic and have been used as insecticides, rodenticides and herbicides.

artefact, artifact *n*. (1) an apparent structure or experimental result obtained due to method of preparing specimen or experimental conditions; (2) a man-made object.

artenkreis *n*. complex of species that replace one another geographically.

artesian aquifer *see* aquifer.

artesian basin bowl-shaped area of land that overlies impermeable rock strata enclosing an aquifer.

artesian well well sunk into a confined (artesian) aquifer. The release of hydrostatic pressure forces the tapped groundwater to the surface without the aid of pumping.

Arthropoda, arthropods *n., n.plu.* very large phylum of segmented invertebrate animals with heads, jointed appendages (feelers, mouthparts and legs) and thickened chitinous cuticles forming exoskeletons. The main body cavity is a haemocoel. The phylum is sometimes divided into several subphyla, most commonly the Chelicerata, Uniramia, Crustacea and the extinct Trilobita. In this classification, the Chelicerata includes the classes Arachnida (spiders, ticks, mites, scorpions, pycnogonids) and Merostomata (horseshoe crabs and the extinct eurypterids), the Uniramia includes the insects (class Insecta) and the class Myriapoda (centipedes and millipedes), and the Crustacea includes the crustaceans (e.g. crabs, shrimps, barnacles). The velvet worms (class Onychophora) are sometimes placed in a separate phylum.

arthrous articulate *q.v.*

Articulata *n*. class of brachiopods with shells joined by a hinge joint in which two teeth on one shell move in sockets on the other. *cf.* Inarticulata.

articulate *a*. (1) jointed; (2) separating easily at certain points.

articulation *n*. joint between bones or between segments of a stem or fruit.

artifact *alt.* spelling of artefact *q.v.*

artificial classification or key a classification that groups organisms or objects together on the basis of a few convenient characteristics rather than on the basis of evolutionary relationships. *cf.* natural classification.

artificial eutrophication accelerated eutrophication *q.v.*

artificial fertilizers inorganic fertilizers *q.v.*

artificial formation a pattern of vegetation caused by human activity.

artificial insemination (AI, AID) the artificial introduction of sperm collected from a male into the female reproductive tract in mammals.

artificial levees *n.plu.* ridges built along the sides of river channels in order to prevent flooding during periods of high river discharge.

artificial reservoir *see* reservoir (2).

artificial selection the selection of particular forms of an organism as a result of environmental pressures deliberately imposed.

Artiodactyla, artiodactyls *n., n.plu.* even-toed ungulates, including pigs, sheep, cattle and camels, an individual having a complex stomach for dealing with plant food and the third and fourth digit forming a cloven hoof.

aryl *n.* the chemical group that results when a hydrogen atom is removed from an arene.

As symbol for the chemical element arsenic *q.v.*

asbestiform *a.* extremely fibrous, *appl.* asbestos minerals.

asbestos *n.* any of a group of minerals that share the common characteristic of a very fibrous habit. They are able to withstand heat, fire and chemical attack and are used in a wide variety of applications. The fibres can be separated and made into a felt-like material that has been widely used as an insulating material. Exposure to large amounts of asbestos dust is highly damaging to health and can cause the respiratory disease asbestosis (*q.v.*) and the otherwise rare lung cancer mesothelioma (*q.v.*). Consequently, asbestos use has been severely curtailed in recent years. The most common asbestos minerals are crocidolite (*q.v.*) (the most harmful), chrysotile (*q.v.*) and amosite (*q.v.*).

asbestosis *n.* a type of incapacitating lung disease caused by the prolonged inhalation of asbestos fibres, which accumulate in the lungs and reduce lung function.

Aschelminthes, aschelminths *n., n.plu.* a group of pseudocoelomate, mainly worm-like animals including the phyla Gastrotricha, Kinorhyncha, Nematoda, Nematomorpha and Rotifera.

Ascidiacea, ascidians *n., n.plu.* class of marine tunicates (urochordates), commonly called sea squirts, in which the adults are often colonial and fixed to a substrate.

Ascolichenes *n.* lichens in which the fungal partner is an ascomycete.

Ascomycota, Ascomycotina, ascomycetes *n., n., n.plu.* large group (considered as a division or a class) of terrestrial fungi, commonly called sac fungi, which have a septate mycelium and develop their spores in sac-like structures called asci and that includes the yeasts, leaf-curl fungi, black and green moulds, powdery mildews, cup fungi, morels and truffles.

ascorbic acid *see* vitamin C.

-ase suffix denoting an enzyme, usually joined to a root denoting the substance acted on or the type of reaction, e.g. proteinase, lipase, glucosidase, asparaginase, hydrolase, oxidase.

asemic *a.* without markings.

asepsis *n.* sterile conditions. *see* aseptic.

aseptic *a.* (1) sterile; *appl.* (2) cultures in which growth of bacteria is prevented; (3) certain infectious diseases in which no bacterial agent can be isolated, such as aseptic meningitis, and that may be due to viruses or other infectious agents.

asexual *a.* (1) having no gender; (2) lacking functional sexual organs.

asexual reproduction reproduction that does not involve formation and fusion of gametes and that may be by binary fission, budding, asexual spore formation or vegetative propagation, resulting in progeny with an identical genetic constitution to the parent and to each other.

ash, volcanic *see* pyroclasts.

Asiatic subregion subdivision of the Boreal phytogeographical kingdom, comprising Central Asia between latitudes 50° and 70° N.

asparagine (Asn, N) *n.* amino acid, uncharged derivative of aspartic acid, constituent of proteins, first discovered in asparagus, important in nitrogen metabolism in plants. Required in human diet.

aspartate, aspartic acid (Asp, D) *n.* amino succinic acid. One of the amino acid constituents of proteins, and important in metabolic transamination reactions. Required in human diet.

aspect *n.* (1) direction in which a surface faces; (2) appearance or look; (3) seasonal appearance.

asperate *a.* having a rough surface.

asphyxiator *n.* agent capable of depriving the body of oxygen. This can be done by either physically stopping inhalation, e.g. hydrogen sulphide can paralyse the respiratory system, or by decreasing the oxygen-carrying capacity of the blood, e.g. carbon monoxide can bind strongly to haemoglobin, displacing the oxygen that it should carry.

asporogenous *a.* (1) not originating from spores; (2) not producing spores.

ASPT average score per taxon *see* Biological Monitoring Working Party (BMWP) score.

assay *n.* a procedure for measurement or identification.

assembly *n.* the smallest community unit of plants or animals, e.g. a colony of aphids on a stem.

assimilate *n.* any of the first organic compounds produced during assimilation in autotrophs. *v.* to absorb and incorporate. *see also* assimilation.

assimilated energy in an animal's energy budget, assimilated energy (A) = consumption (C) − faeces (F) and is equivalent to production (P) + energy lost as heat during respiration (R) + energy lost by small metabolites voided in urine (U).

assimilate stream the movement of sugars out of the leaves where they are manufactured during photosynthesis to other parts of the plant via the phloem.

assimilation *n.* (1) in autotrophic organisms, the uptake of elements and simple inorganic chemical species, such as CO_2, NO_3^-, H_2O, from the environment and their incorporation into complex organic compounds; (2) in heterotrophs, the conversion of digested food material into complex biomolecules; (3) (*geol.*) the incorporation of wall rock into igneous melt. *v.* assimilate.

assimilation efficiency in animal physiology and ecophysiology, a measure of the efficiency of utilization of food. It is expressed as a ratio of assimilated energy (A) divided by energy consumed (C) and is mainly influenced by the nature of the food consumed, carnivores having a greater assimilation efficiency than herbivores.

assimilative *a.* (1) *pert.* or used for assimilation; (2) *appl.* growth preceding reproduction.

assimilative capacity absorptive capacity *q.v.*

association *n.* (1) plant community forming a division of a larger unit of vegetation and characterized by a dominant species; (2) form of learning in which repetition of a stimulus followed by a reward (or disincentive) produces an increased positive (or negative) response.

associative nitrogen fixation non-symbiotic nitrogen fixation by bacteria (e.g. *Azospirillum*, *Azotobacter*) associated with the rhizosphere of certain grasses and cereals.

associes *n.* an association representing a stage in the process of succession.

assortative mating non-random mating within a population where individuals tend to mate with individuals resembling themselves. In human populations, for example, mating tends to be random for certain characteristics such as blood groups and assortative for others such as height, ethnic group, etc.

astatine (At) *n.* radioactive element (at. no. 85, r.a.m. 210).

Asteraceae in some classifications the name for the Compositae *q.v.*

Asterales *n.* an order of herbaceous dicots, rarely trees or woody climbers, with flowers usually crowded into closely packed heads and comprising the family Asteraceae (Compositae) (daisy, etc.).

asteroid *a.* (*biol.*) (1) star-shaped; (2) *pert.* starfish.

Asteroidea, asteroids *n.*, *n.plu.* class of echinoderms, commonly called starfish or sea stars, an individual having a star-shaped body with five radiating arms not sharply marked off from the central disc.

asthenosphere *n.* the soft, partly molten layer of the Earth's mantle that underlies the rigid lithosphere (*q.v.*). The hot, plastic nature of the asthenosphere allows the lithosphere to move over it. *see* plate tectonics.

asthma *n*. respiratory disease, often caused by exposure to one or more airborne allergens, e.g. pollen, house dust or PVC. Symptoms include wheezing, persistent cough, laboured breathing and cyanosis.

asulam *n*. soil-acting and translocated carbamate herbicide used, e.g., as an aquatic herbicide and to control bracken in forests.

asymmetrical *a*. (1) *pert*. lack of symmetry; (2) having two sides unlike or disproportionate; (3) *appl*. structures that cannot be divided into similar halves by any plane.

At symbol for the chemical element astatine *q.v.*

atacamite *n*. $Cu_2Cl(OH)_3$, a secondary copper mineral.

atavism *n*. presence of an ancestral characteristic not observed in more recent progenitors.

ateleosis *n*. dwarfism where individual is a miniature adult.

Atherinomorpha *n*. small group of advanced teleost fish including the tooth-carps (guppies and swordtails).

athermopause *n*. dormancy of animals due to lack of water or food.

Atlantic North American Floral Region part of the Holarctic Floral Realm comprising North America east of the Rockies, south to the Gulf of Mexico and north to the Arctic Circle.

atmosphere *n*. (1) envelope of gas that surrounds the Earth (*see* Fig. 16, p. 236), the bulk of which (99% by mass) is found within 50 km of the Earth's surface, *see* lower atmosphere; (2) unit of pressure, equal to 101 325 Pa or 760 mm Hg.

atmospheric pressure force per unit area exerted in all directions on any body within the atmosphere by the weight of the atmosphere. *alt*. air pressure.

at. no. atomic number *q.v.*

atoll *n*. a ring-like coral reef, enclosing a central lagoon, which develops in association with a subsiding volcanic island.

atom *n*. the smallest unit of matter that is capable of entering into a chemical reaction.

atom bomb *see* nuclear bomb.

atomic mass the mass of an atom in daltons *q.v.*

atomic mass unit (a.m.u.) unit of atomic mass that is defined as one-twelfth the mass of an atom of ^{12}C. *alt*. dalton.

atomic number (Z) the total number of protons within each atom of the element concerned.

atomic weight of an element, the mass of an atom of an element expressed in terms of the mass of an atom of ^{12}C, which is defined as 12 atomic mass units (a.m.u.). The different isotopes of an element have different atomic weights and if a particular isotope is not specified, the atomic weight given for an element will represent a weighted average of the atomic weights of the naturally occurring isotopes, taking into account their relative frequency. The preferred term for atomic weight is now relative atomic mass *q.v.*

ATP adenosine triphosphate *q.v.*

ATPase adenosine triphosphatase *q.v.*

atrazine *n*. soil-acting herbicide used to control germinating weeds in crops such as maize.

atrophy *n*. diminution in size and function.

atropine *n*. alkaloid obtained from the deadly nightshade, *Atropa belladonna* and other plants of the Solanaceae, used medically as a muscle relaxant.

attenuate *v*. (1) to make thinner; (2) to reduce concentration, intensity, magnitude, strength or pathogenicity; (3) to taper to a point.

attenuated *a*. *appl*. vaccines against bacteria or viruses that are made from strains of the disease agent that have become non-pathogenic after growth in culture.

atto- (a) SI prefix denoting 10^{-18}.

attribute *a*. a qualitative property or characteristic of an individual, e.g. whether it is male or female.

Au symbol for the chemical element gold *q.v.*

auditory *a*. (1) *pert*. sense of hearing; (2) *pert*. hearing apparatus, as auditory organ, auditory canal, auditory meatus, auditory ossicle, auditory capsule, *see* ear; (3) *appl*. nerve: eighth cranial nerve, connecting inner ear with hindbrain, carrying sensations of sound and pitch from cochlea for relay to auditory area of cerebral cortex, and postural information from semicircular canals to cerebellum.

auger *n*. boring device used for taking soil samples.

augite *n.* the most common of the pyroxene minerals (*q.v.*). It is a clinopyroxene and its composition is variously reported as $(Ca,Na)(Mg,Fe,Al)(Si,Al)_2O_6$, $(Ca,Mg,Fe,Al)_2(Si,Al)_2O_6$ or $(Ca,Mg,Fe,Ti,Al)(Si,Al)_2O_6$.

aural *a. pert.* ear or hearing.

aureole *see* contact metamorphic aureole.

auriferous *a.* gold-bearing.

aurora australis *see* thermosphere.

aurora borealis *see* thermosphere.

austral *a. appl.* or *pert.* southern biogeographical region, or restricted to North America between transitional and tropical zones.

Austral Realm the floristic area that includes the southern part of South America, Australasia, the southern tip of Africa and the southern oceanic islands. It comprises five floral regions: Argentinian, Australian, New Zealand, South African and South Oceanic (*see* individual entries). *alt.* Southern Realm.

Australasian Region zoogeographical region including Australia, Tasmania, New Zealand, Papua-New Guinea, Sulawesi and other islands to the south and east of Wallace's line, which runs between Borneo and Sulawesi and between the Indonesian islands of Bali and Lombok.

Australian Floral Region part of the Austral Realm that includes Australia and Tasmania. *alt.* Australian kingdom.

Australian kingdom *see* Australian Floral Region.

Australian Region *see* Australasian Region.

australopithecines, *Australopithecus* *n.plu.,* *n.* genus of fossil hominids, found in southern and eastern Africa and believed to have lived from at least 4 million until 1 million years ago. They include the 'gracile' australopithecines (*Australopithecus africanus*) at around 2.5 million years, the 'robust' australopithecines (*A. robustus*) at around 1.5 million years, and an older species, *A. afarensis* at around 3.5 million years, found in Kenya and Ethiopia. They had an upright posture and a relatively small brain compared with *Homo*. Fossils now accepted as australopithecine were also formerly known as *Paranthropus* and *Zinjanthropus*. The evolutionary position of the australopithecines in relation to *Homo* is still unclear, but *A. afarensis* is thought by some palaeoanthropologists to be directly ancestral to the human line. The later *A. africanus* and *A. robustus* are offshoots from the direct line of human evolution and *A. robustus* is contemporaneous with early species of *Homo*.

autecious, autecism *alt.* spelling of autoecious, autoecism *q.v.*

autecology *n.* (1) the biological relations between a single species and its environment; (2) the ecology of a single organism. *cf.* synecology.

authigenesis *n.* the formation of minerals within a host sediment. *a.* authigenic.

authigenisis *n.* (*oceanogr.*) chemical formation of new mineral phases either on the surface of the seabed or in the water column above it.

auto- prefix from Gk *autos*, self.

autobahn motorway *q.v.*

autochory *n.* self-dispersal of spores or seeds by an explosive mechanism.

autochthonous *a.* (1) indigenous; (2) *appl.* indigenous soil microflora that is normally active; (3) formed where found.

autoclave *n.* equipment for sterilization of glassware and media by steam heat.

autocoprophagy refection *q.v.*

autodeliquescent *a.* becoming liquid as the result of self-digestion, as the cap and gills of fungi of the genus *Coprinus* (ink caps).

autoecious *a.* parasitic on one host only during a complete life cycle. *n.* autoecism.

autogenic *a.* caused by interactions between the members of the community itself, *appl.* plant successions.

autogenous *a.* (1) produced in same organism, *cf.* exogenous; (2) *appl.* adult female insect that does not need to feed for her eggs to mature.

autolith *n.* (*geol.*) a piece of genetically related material embedded in an igneous rock. Such inclusions are concentrations of crystals formed from the magma in an early stage in its solidification.

automimicry *n.* (1) intraspecific mimicry as when some members of a species are unpalatable, and palatable members of the same species mimic them; (2) imitation by a member of the opposite sex or another age-group of a communication signal used by a particular sex or age-group in a species.

autoparasite *n.* parasite subsisting on another parasite.

autoparthenogenesis *n.* development from unfertilized eggs activated by a chemical or physical stimulus.

autophagous *a. appl.* birds capable of running about and securing food for themselves when newly hatched.

autophilous *a.* self-pollinating.

autopolyploid, autoploid *n.* (1) organism having more than two sets of homologous chromosomes; (2) polyploid in which chromosome sets are all derived from a single species. *a. pert.* to such an organism.

autosomal dominant inheritance pattern of inheritance characteristic of a dominant allele carried on an autosome. The associated phenotypic trait is displayed by individuals who carry only one copy of the allele so that each offspring of a heterozygous individual has a one in two chance of inheriting the allele and showing the trait. *cf.* autosomal recessive inheritance, X-linked inheritance.

autosomal recessive inheritance typical pattern of inheritance of a simple recessive Mendelian genetic trait carried on an autosome. The offspring of parents both heterozygous for the recessive allele will each have a one in four chance of being homozygous for the allele and displaying the trait.

autosome *n.* any chromosome other than a sex chromosome.

autostrada motorway *q.v.*

autotomy *n.* shedding of a part of the body, as in some worms, arthropods and lizards, which can be used to escape capture by a predator.

autotroph *n.* any organism able to utilize inorganic sources of carbon (as carbon dioxide), nitrogen (as nitrates, ammonium salts), and other elements as starting materials for biosynthesis, using either sunlight (photoautotroph) or inorganic chemicals (chemoautotroph) as energy sources. *a.* autotrophic.

autotrophic succession succession in which the rate of primary production due to photosynthesis is greater than the rate of respiration, thus leading to the accumulation of organic matter.

autumn *n.* the season between summer and winter; (1) in the northern hemisphere, it is commonly taken to be September, October and November; (2) astronomically, the season that commences at the autumnal equinox and terminates at the winter solstice, i.e. in the northern hemisphere, between 23 September and 22 December. *alt.* fall.

autumnal equinox *see* equinox.

autunite *n.* $Ca(UO_2)_2(PO_4)_2.10–12H_2O$, a uranium ore mineral.

auxetic *a. appl.* growth due to increase in cell size rather than proliferation of cells. *n.* auxesis.

auxiliaries *n.plu.* female social insects that associate with other females of the same generation and become workers.

auxins *n.plu.* various related plant growth hormones, of which indoleacetic acid is most common in nature. They are involved in cellular elongation and differentiation, in root growth, the development of vascular tissue, phototropism, the development of fruits and the normal suppression of the growth of lateral buds. They are responsible for the curvature of plant shoots towards the light by causing a differential elongation of cells on the side away from light, where a greater concentration of auxin accumulates. Auxin produced by the tip of the shoot also normally suppresses the development of lateral buds.

auxotrophic *a. appl.* (1) any organism, esp. a microorganism, that has a nutritional requirement for some specific substance (e.g. vitamins, amino acids, nucleotides, etc.) because it is unable to synthesize it; (2) esp. to mutant bacteria or fungi that have lost the ability to synthesize enzymes present in the parental strain and that therefore require a specific nutritional supplement. *n.* auxotroph.

available *a. appl.* elements or other materials (e.g. nitrogen, water) in soil that are in a form that can be utilized by a plant.

available water capacity the difference between the moisture content of a soil at field capacity (*q.v.*) and at wilting point *q.v.*

avalanche *see* flow movement.

average *n.* a value that summarizes or represents a set of values and which is usually the arithmetic mean or, more rarely, the geometric mean.

average daily metabolic rate (ADMR) a measure of the daily energy requirements of an animal engaging in its normal activities and that is calculated from the energy content of the food eaten in a day.

average life expectancy the number of years a newborn individual can be expected to live, averaged over the population.

Average Score Per Taxon (ASPT) *see* Biological Monitoring Working Party (BMWP) score.

average transit time residence time *q.v.*

Aves *n.* the birds, a class of bipedal homiothermic vertebrates having bodies clothed with feathers, and front limbs modified as wings, and the skin of the jaw forming horny bills (beaks). They are thought to be descended from the extinct archosaurian reptiles. The earliest known fossil bird is *Archaeopteryx* from the Upper Jurassic. Modern birds belong to the subclass Neornithes. *see* Appendix 7.

avian *a. pert.* birds.

avicide *n.* chemical that kills birds.

avifauna *n.* all the bird species of a region or period.

avirulence gene gene found in some bacterial and plant pathogens that determines their ability to cause disease on a host plant containing a corresponding resistance gene, *see* gene-for-gene resistance.

avirulent *a. appl.* a strain of bacterium, virus or other potential pathogen that does not cause disease.

Avogadro's number the number of particles in a mole of those particles. It equals $6.022\ 52 \times 10^{23}\ \text{mol}^{-1}$. For example, exactly 12 g of ^{12}C will contain $6.022\ 52 \times 10^{23}$ carbon atoms. Symbol L or N_A. *alt.* Avogadro constant.

avoidance behaviour a wide range of defensive behaviour in animals (e.g. freez-

ing posture, running for cover, giving warning signals), which may be innate or learned, by which they minimize exposure to apparently harmful situations.

avoidance reaction movement away from stimulus.

axenic *a. appl.* pure cultures of microorganisms *in vitro* in which the organism is grown in the absence of any other contaminating microorganism or, in the case of parasites or symbionts, its plant or animal host.

axial *a.* (1) *pert.* axis or stem; *appl.* (2) structures arranged along an axis; (3) filaments and other structures running longitudinally along stem, axon of nerve cell, etc.

axial plane (*geol.*) an imaginary plane at the apex of an anticline, or the base of a syncline, that bisects the angle made where the two sloping limbs of the fold meet.

axil *n.* the angle between leaf or branch and the axis from which it springs.

axillary *a.* (1) (*bot.*) *pert.* axil, growing in axil, as buds; (2) (*zool.*) *pert.* armpit.

axis *n.* (1) the central line of a structure or body; (2) any of several directions along which body structure is organized, e.g. apical–basal axis, antero–posterior axis, dorso–ventral axis, medio–lateral axis; (3) the main stem of plant or central cylinder of plant stem; (4) the second cervical vertebra.

azinphos-methyl *n.* contact organophosphate insecticide and acaricide used for fruit and vegetable crops.

aziprotryne *n.* soil-acting herbicide used to control annual weeds in, e.g., onion and brassica crops.

azoic *a.* (1) uninhabited; (2) without remains of organisms or their products.

azonal *a. appl.* soils without definite horizons.

azurite *n.* $Cu_3(CO_3)_2(OH)_2$, an intensely blue copper mineral that is used as a decorative stone. *alt.* chessylite.

azygote *n.* organism resulting from haploid parthenogenesis.

B

β beta (1) symbol used to represent the beta particle (*q.v.*); (2) for headwords with prefix β- refer to beta-.

B (1) symbol for the chemical element boron (*q.v.*); (2) single-letter code for either asparagine or aspartic acid; (3) occasionally used as a symbol for billion (10^9).

Ba symbol for the chemical element barium *q.v.*

bacca *n.* berry, esp. if formed from an inferior ovary.

Bacillariophyta *n.* the diatoms, a division of unicellular algae in some classifications; an individual has a silicified wall in two halves, and chlorophyll and carotenoid pigments, and stores oils and leucosin instead of starch. *alt.* Bacillariophyceae.

bacillus *n.* (1) formerly much used, esp. in medical bacteriology, for any rod-shaped bacterium; (2) more specifically any member of the genus *Bacillus*, aerobic spore-forming rods, widely distributed in the soil. *plu.* bacilli.

Bacillus thuringiensis bacterium that produces endotoxins active against the larvae of many insect pests and that is used as a biological insecticide.

back cross a cross between the heterozygous F_1 generation and the homozygous recessive parent, which allows the different genotypes present in the F_1 to be distinguished.

background *a. appl.* (1) existing conditions that might affect the study of a phenomenon or process; (2) mutation, the rate at which spontaneous mutations occur in any particular organism; (3) concentration of pollutants, or level of radiation, expected in the absence of a given source(s) of contamination.

backing *a. appl.* winds, a change in direction anticlockwise in the northern hemisphere, and clockwise in the southern hemisphere. *cf.* veering.

back mutation a mutation that reverses the effects of a previous mutation in the same gene, either by an exact reversal of the original mutation or by a compensatory mutation elsewhere in the gene. *alt.* reverse mutation, reversion.

backshore *n.* the zone of beach that lies between the high-water mark and the dune or cliff line, i.e. landward of the foreshore.

backswamp zone a waterlogged area of floodplain. These zones are commonly found where rivers are bordered by natural levees. Backswamp zones are often drained to provide fertile agricultural land.

backwash *see* swash.

backwashing *n.* deliberate flow reversal within a waste treatment plant with the objective of cleaning filters or reconversion of an ion-exchanger to a useful form.

Bacteria, bacteria *n., n.plu.* a very large group, now sometimes considered as a superkingdom, of extremely metabolically diverse, prokaryotic, unicellular microorganisms usually possessing cell walls, sometimes forming filaments, which are found in soil, water, or parasitic or saprophytic on plants and animals, the parasitic forms causing many familiar diseases. Bacteria reproduce by binary fission or asexual spores and also transfer genetic material by sexual processes (conjugation) and by virus(bacteriophage)-mediated transfer (transduction). They comprise the so-called true bacteria (eubacteria), the actinomycetes and related organisms, mycoplasmas, rickettsias and rickettsia-like organisms, and spirochaetes (*see under individual entries*). The archaebacteria are commonly also called bacteria, but are now often considered as a separate prokaryotic super-kingdom, the Archaea. *sing.* bacterium. *a.* bacterial. *see* Appendix 9.

bactericidal, bacteriocidal *a.* causing death of bacteria.

bacteriochlorophyll *n.* various photosynthetic pigments related to chlorophyll, which act as light-collecting pigments in photosynthetic bacteria.

Bacteriological Code international code of nomenclature of bacteria.

bacteriology *n.* the study of bacteria. *a.* bacteriological.

bacteriophage *n.* a virus infecting bacteria, such as lambda, T2, T4 (infecting *E. coli*). *alt.* phage.

bacteriophagous, bacteriverous *a.* feeding on bacteria.

bacteriostatic *a.* inhibiting the growth of, but not killing, bacteria.

bacterium *sing.* of bacteria *q.v.*

bacteroid *n.* irregularly shaped rhizobial cell, the form in which rhizobia are found in root nodules of legumes.

badland *n.* area of rugged terrain composed of deep gullies and ravines separated by ridges, buttes and mesas, created as a result of severe soil erosion.

baffles *n.plu.* devices placed within a flue system in order to slow down and change the direction of flow. They can be used to facilitate the removal of larger-sized particulates from gaseous waste streams.

bagasse *n.* the fibrous waste material produced during the processing of sugar-cane after the sugar has been extracted.

bag filter *see* fabric filter.

bag house a chamber within a flue system that contains fabric filters (*q.v.*) for the removal of particulates from the gaseous waste stream.

bajada *n.* alluvial plain in lowlands or at the foot of mountains that is formed by the coalition of individual alluvial fans. *alt.* alluvial apron.

balaenoid whales group of toothless whales with large heads, which filter feed by means of baleen plates. They include the rorquals, right whales and the blue whale.

balanced polymorphism the stable co-existence of two or more distinct types of individual, forms of a character or different alleles of a gene in a population, the proportions of each being maintained by selection.

balance of nature ecological balance *q.v.*

balance trials a methodology for determining quantitative nutrional requirements of animals, in which total intake and total loss of a particular nutrient are measured over a long period.

balance year *pert.* glaciers, the time interval between the point of maximum ablation (melting) in two successive summers. The balance year may not correspond exactly to a calendar year.

balancing selection selection that maintains a balanced polymorphism.

Balanopales *n.* an order of dicot trees and shrubs comprising the family Balanopaceae, sometimes placed in the Fagales.

baleen *n.* whalebone, horny plates attached to upper jaw in some whales, used to filter plankton, etc., from water.

ballast *n.* (1) material carried by a ship to confer stability, esp. that placed within its hold on an empty (return) voyage (e.g. usually seawater in the case of oil tankers); (2) coarsely crushed rock, or similar material, used in the foundation of railways or roads; (3) coarse aggregate used in concrete; (4) (*bot.*) elements present in plants that are not apparently essential for growth, such as aluminium or silicon.

ballistic *a. appl.* fruits that explode when ripe, forcibly discharging seeds.

balsam *n.* any of various complex fragrant substances found in some plants, consisting of resin acids, esters and terpenes, often exuded from wounds.

bamboos *n.plu.* members of the family Bambusaceae, woody plants of the monocot order Gramineae that are important species in some tropical and subtropical ecosystems (e.g. bamboo forest) and that have many uses.

banana *see Musa.*

banded (*geol.*) layered *q.v.*

band screen a perforated band, e.g., of wire mesh, in perpetual motion, employed to extract solid material from liquid effluent.

bank calving the fall movement (*q.v.*) of soil that occurs along river banks that have been undercut by streams.

bar *n.* unit of pressure, equal to 10^5 Pa (*ca.* 0.986 atm.). In meteorology, atmospheric pressure is frequently cited in millibars (10^{-3} bar).

baraesthesia, baresthesia *n.* sensation of pressure.

barban *n.* translocated carbamate herbicide used to control blackgrass and wild oats in, e.g., wheat, bean and pea crops.

Barbeyales *n.* order of dicot trees with one family, Barbeyaceae, including one genus *Barbeya*.

barchan *n.* a cresent-shaped dune with a gently sloping windward (stoss) slope and a steep slip-face on the leeward side bounded by two horns. Barchans form transverse to the prevailing wind direction and are common in many desert and coastal regions. Individual barchans may coalesce to form a depositional feature known as a barchanoid ridge.

barchanoid ridge *see* barchan.

barite baryte *q.v.*

barium (Ba) *n.* reactive metallic element (at. no. 56, r.a.m. 137.33), forming a soft silvery metal in elemental form. In nature, it is found chiefly as the ore baryte ($BaSO_4$) and barium carbonate. It is chemically similar to calcium and some barium compounds are toxic.

bark *n.* in strict botanical terms, the layer of tissue external to the vascular cambium in woody plants, comprising the secondary phloem, cortex and periderm, the periderm being the layer commonly known as the bark.

bark lice common name for some members of the Psocoptera (*q.v.*), small wingless insects with globular abdomens and incomplete metamorphosis, inhabiting the bark of trees.

barley *n. Hordeum sativa*, a small-grained temperate cereal crop used mainly in animal feed stuffs and also for malting *q.v.*

barley yellow dwarf virus a disease of barley, wheat and oats that is caused by a virus spread by cereal aphids. Infected plants are severely stunted and yields are drastically reduced.

barley yellow mosaic virus a soil-borne viral disease of barley. Infected plants exhibit stunted growth and late maturity.

barnacle *n.* common name for a member of the Cirripedia *q.v.*

barognosis *n.* capacity to detect changes in pressure.

barometer *n.* instrument used to measure atmospheric pressure. *see* mercury barometer, aneroid barometer.

barometric gradient pressure gradient *q.v.*

barotaxis *n.* directed movement or orientation in response to a pressure stimulus.

barrage *n.* barrier built across moving water in order to control its flow and/or increase its depth. *see also* tidal barrage.

barrel *n.* volumetric measure, equivalent to 42 US gallons or 159 l, used to quantify crude oil production.

barrens *n.plu.* flat areas with poor soil of low productivity, usually sandy or serpentine, that are unable to support a normal vegetation and often have a flora of specialized endemic species.

barrier bar barrier island *q.v.*

barrier beach barrier island *q.v.*

barrier forest forest in the mountains that holds back snow from the lower slopes.

barrier island a large, permanent offshore island that runs parallel to the mainland coast. These islands are found in association with 10–14% of the world's coastline. *alt.* barrier beach, barrier bar, offshore bar.

barrier reef a coral reef that is separated from the mainland or island by a shallow lagoon. The Great Barrier Reef, north-east Australia, is the world's largest example, extending for 2500 km.

bars *n.plu.* accumulations of sediment (sands or gravels) that, in contrast to islands, lack vegetation and are relatively unstable. Bars may form in a number of different aquatic situations. *see also* braided stream.

bar screen screen consisting of equidistant bars, used to remove the larger objects, such as lumps of wood, toilet paper, etc., from raw sewage.

baryte, barytes *n.* $BaSO_4$, the most common barium mineral. *alt.* heavy spar. *alt.* barite.

basal *a.* pert., at, or near the base.

basal area measure of the area of a forest covered by the trunks of the trees. It is measured as the sum of the cross-sectional areas of all the trees within the area, and expressed as square metres per hectare.

basal ice the lowest ice layer of a glacier.

basal metabolic rate minimum metabolic rate required for survival, measured in humans at complete rest in a thermally neutral environment after fasting for 12 h. Expressed as percentage of normal heat production per hour per square metre surface area.

basal metabolism normal state of metabolic activity of organism at rest.

basal sliding a mechanism of ice movement that accounts for most of the movement (up to 90%) of warm-based glaciers (*q.v.*) but is largely inactive in cold-based glaciers (*q.v.*). In basal sliding, a thin layer of water between the basal ice and the underlying bedrock acts as a lubricant. This enables the glacier to slide forwards, eroding the rocks beneath and picking up material in the process. *alt.* glacial sliding.

basalt *n.* a fine-grained basic igneous rock with a mineralogy similar to that of gabbro (*q.v.*). The essential minerals of basalt are anorthite-rich plagioclase feldspar and augite (a pyroxene). It is typically found as a lava flow.

basaltic lava basic lava *q.v.*

base *n.* (1) unless otherwise stated, this is, in most cases, synonymous with a Brønsted–Lowry base (*q.v.*), *see also* Arrhenius base, Lewis base; (2) in biochemistry, often refers to the nitrogenous bases, i.e. the purine and pyrimidine constituents of nucleotides.

base analogue a substance chemically similar to one of the normal nucleotide bases and that is incorporated into DNA, often causing mutations.

base exchange capacity cation exchange capacity *q.v.*

base hydrolysis (*chem.*) hydrolysis reactions (*q.v.*) involving alkaline solutions.

base ionization constant (K_b) an equilibrium constant that relates to the ionization of weak Brønsted bases (a process that may be represented by the general equation $B + H_2O \rightleftharpoons BH^+ + OH^-$). The base ionization constant is expressed as $K_b = \{BH^+\}\{OH^-\}/\{B\}$, where curly brackets, $\{\}$, represent activity based on molar concentration. In dilute solutions, the activity of a solute approximates to its molar concentration.

base level the level in the landscape below which erosion by running water cannot occur. In the case of rivers, this is ultimately represented by sea-level.

base pair (bp) a single pair of complementary nucleotides from opposite strands of the DNA double helix. The number of base pairs is used as a measure of length of a double-stranded DNA.

base-rich *a. appl.* soils containing a relatively large amount of free basic ions, such as magnesium or calcium.

base saturation of soils, the situation in which the cation exchange capacity of a soil is saturated with the exchangeable cations. Expressed as a percentage of the total cation exchange capacity.

base sequence nucleotide sequence *q.v.*

base subsistence density of a human population, the density above which continued survival is impossible.

base substitution replacement of one nucleotide with another in DNA.

basic *a.* (1) having the properties of a base (*q.v.*); *appl.* (2) (*biol.*) stains that act in general on the nuclear contents of the cell; (3) soils, rich in alkaline minerals; (4) (*geol.*) igneous rocks that contain 45–52 wt % silica (SiO_2).

basic lava a low-viscosity lava derived from magma with a low silica content. Basic lavas (also known as basaltic lavas) flow for considerable distances and produce extensive, gently sloping landforms. This type of lava is found at divergent plate boundaries where magma rises from the mantle, e.g. from fissures along the mid-oceanic ridges. *cf.* acid lava.

basic slag inorganic phosphorus fertilizer, of variable chemical composition, with a phosphorus content equivalent to 8–22% P_2O_5, produced as a byproduct of steel-making. Basic slags have a liming effect and are therefore particularly effective on acid soils.

Basidiolichenes *n.* group of lichens in which the fungal partner is a basidiomycete.

Basidiomycota, Basidiomycotina, basidiomycetes *n., n., n.plu.* large group of fungi (usually considered as a division or a class) that have septate hyphae and bear their spores on the outside of spore-producing bodies (basidia) that are often borne on or in conspicuous fruiting structures. Basidiomycetes include the rusts, smuts, jelly fungi, mushrooms and toadstools, puffballs, stinkhorns, bracket fungi and bird's nest fungi.

basifuge *n.* a plant unable to tolerate basic soils. *alt.* calcifuge.

basin peat fen peat *q.v.*

basinym *n.* the name on which new names of species, etc., have been based. *alt.* basionym.

basket-of-eggs landscape *see* drumlin.

basolateral *a. pert.* sides and base of any cell, structure or organ.

bass (1) (*bot.*) bast (*q.v.*); (2) (*zool.*) type of fish.

bast *n.* (1) an inner fibrous layer of some trees; (2) phloem fibres.

bat *n.* common name for a member of the mammalian order Chiroptera *q.v.*

BAT (1) acronym for best available technique, a term used in risk analysis; (2) brown adipose tissue, brown fat *q.v.*

Batesian mimicry resemblance of one animal (the mimic) to another (the model) to the benefit of the mimic, as when the model is dangerous or unpalatable, described by the English naturalist H. W. Bates. *cf.* aposematic, Müllerian mimicry.

batholith *n.* large, dome-shaped discordant intrusion of plutonic rock with a surface area greater than 100 km^2.

bathyal *a. appl.* or *pert.* zone of seabed between the edge of the continental shelf and the abyssal zone at a water depth of 2000 m.

bathylimnetic *a.* living or growing in the depths of lakes or marshes.

bathymetric *a. pert.* vertical distribution of organisms in water.

bathypelagic *a.* inhabiting the deep sea (1000–3000 m).

bathyplankton *n.* plankton that undergo a daily migration, moving up towards the surface at dusk, and down to lower depths at dawn.

bathysmal *a. pert.* deepest depths of the sea.

BATNEEC acronym for best available technology not entailing excessive costs, a term used in cost–benefit–risk analysis, esp. in the UK, to set regulatory limits on, e.g., emission of pollutants.

batrachians *n.plu.* frogs and toads.

batrachosaurs *n.plu.* a group of labyrinthodonts of the Carboniferous–Permian, which may include the ancestors of reptiles.

battery production the most intensive method of egg production in which hens spend their egg-laying lives in tiny cages, indoors. *cf.* free-range system.

bauxite *n.* a soft rock that is the principal aluminium ore. It is almost entirely hydrated aluminium oxide ($Al_2O_3.nH_2O$) and contains the minerals gibbsite [$Al(OH)_3$], boehmite [$AlO(OH)$] and diaspore [$AlO(OH)$].

baymouth bar an elongated accumulation of beach sediment (spit) that protrudes into a bay and may eventually seal off its mouth entirely.

BCF bioconcentration factor *q.v.*

BCG Bacille Calmette–Guerin, a modified variant of a bovine strain of *Mycobacterium tuberculosis* used as a vaccine against human tuberculosis.

Be symbol for the chemical element beryllium *q.v.*

beach *n.* an accumulation of sediment, often sand, which is deposited on the shore of a lake or the sea by water currents and waves. *see* backshore, foreshore, nearshore, seashore.

beach cusps cresent-shaped surface features that develop parallel to the shore in the upper swash zone of a beach. These form a regular, repetitious pattern and are best developed in poorly sorted beach sediment.

beachrock *n.* a type of rock, common on tropical beaches, formed from beach sediments lithified by the precipitation of calcium carbonate.

beak *n.* (1) bill (*q.v.*) of birds; (2) elongated jaws or mandibles of other animals, as the elongated jaw of a dolphin; (3) (*bot.*) long angled projections on certain fruits, as those of cranesbills (Geraniales).

beard worms Pogonophora *q.v.*

Beaufort scale internationally recognized numerical scale of wind force ranging from zero (calm, average wind speed <1 km h^{-1}, smoke rises vertically) to 12 (hurricane, average wind speed >119 km h^{-1}, devastation).

becquerel (Bq) *n.* the derived SI unit for expressing the activity of a radionuclide, which is equal to 2.7×10^{-11} Ci (curies).

bed *n.* (*geol.*) a layer of rock (usually sedimentary) laid down at the Earth's surface, which differs in mineralogy and/or texture from neighbouring layers.

bedding *n.* (*geol.*) a structural feature, almost ubiquitous in sedimentary rocks, consisting of layers of rock differentiated from one another in terms of texture and/ or mineralogy.

bedding planes boundaries of individual sedimentary rock layers that represent periods when the process of sedimentation either ceased or changed, or when a period of erosion intervened.

bedeguar *n.* moss-like outgrowth produced on rose bushes by gall wasps.

bedload *n.* solid material that is moved along the bed of a river by rolling or sliding (traction) or in a series of hops (saltation). *alt.* bottom load, traction load.

bedrock *n.* unweathered rock found beneath the regolith *q.v.*

bees *n.plu.* insects of the superfamily Apoidea of the order Hymenoptera, some of which are social and some solitary, and that include the honey bees (*Apis*), bumble-bees (*Bombus*) and flower bees (*Anthophora*). They feed themselves and their young on pollen and nectar gathered from flowers, and are important plant pollinators. The social bees form colonies with a single queen, males (drones) and workers.

beetles *n.plu.* common name for the insect order Coleoptera *q.v.*

beets *n.plu.* various cultivated forms of *Beta vulgaris*, including beetroot, sugar-beet, chard, mangolds.

Begoniales *n.* an order of mostly succulent dicot herbs, but also shrub-like herbs and some large trees, comprising the families Begoniaceae (begonia) and Datiscaceae.

behavioural ecology the relationship of animals to their environment and to other animals extended to take in the effects of their behaviour and the way it may be modified by environmental factors and by interactions with members of their own species.

behavioural genetics the study of the genetic basis of behaviour, which includes the study of the contribution of environment and nurture and heritable traits to behaviour.

beheaded stream *see* river capture.

beheading river capture *q.v.*

beidellite *n.* $(Ca,Na)_{0.3}Al_2(OH)_2(Al,Si)_4O_{10} \cdot 4H_2O$, one of the clay minerals (*q.v.*) of the smectite group *q.v.*

belemnites *n.plu.* order of extinct gastropod molluscs present from the Triassic to the end of the Cretaceous, with conical uncoiled shells.

below-ground *a.* in an ecosystem, *appl.* the material, structures and organisms that occur below the soil surface, e.g. roots of plants and soil microorganisms.

belt transect *see* transect, transect sampling.

Beltian bodies, Belt's bodies small nutritive organs containing oils and proteins, borne at the tips of leaves of swollen-thorn acacias and that provide food for ants that live on the plant and help it to survive by protecting against attack by other insect pests and by damaging neighbouring seedlings.

benazolin *n.* translocated herbicide used, mainly in mixtures, to control many broadleaved weeds.

bendiocarb *n.* carbamate insecticide used, e.g., to control frit fly in maize, and nuisance pests, e.g., cockroaches, in buildings.

beneficiation *n.* separation of an ore mineral from the waste mineral material.

benefit *n.* in animal behaviour, the quantity that is maximized by the behavioural choices made. *alt.* negative cost. *cf.* cost.

benefit–cost analysis cost–benefit analysis *q.v.*

benign tumour an abnormal growth of cells (a tumour) that is, however, not invasive and does not metastasize. *cf.* cancer.

benodanil *n.* systemic fungicide used to control rust fungi on cereal crops.

benomyl *n.* systemic carbamate fungicide used to control a wide range of fungus diseases on many crops.

Benson–Calvin–Bassham cycle Calvin cycle *q.v.*

benthic *a. pert.*, or living on, the bottom of sea, lake, river, *etc. alt.* benthal.

benthos *n.* flora and fauna of sea or lake bottom from high water mark down to the deepest levels.

benthophyte *n.* a bottom-living plant.

bentonite *n.* a rock composed principally of the clay mineral montmorillonite *q.v.*

benzene *n.* C_6H_6, a volatile liquid hydrocarbon with a cyclic structure. A known carcinogen.

benzo(a)pyrene, benzpyrene *n.* a carcinogenic polycyclic aromatic hydrocarbon (*q.v.*) produced by burning fossil fuels. Also present in coal tar. *see* Fig. 22, p. 323.

benzoyl prop-ethyl *n.* translocated herbicide used to control wild oats in, e.g., wheat and oilseed rape crops.

benzyladenine *n.* a synthetic plant growth hormone.

Bergeron–Findeisen process ice-crystal process *q.v.*

Bergmann's rule the idea that geographically variable warm-blooded animal species have smaller body sizes in the warmer parts of their range than in the colder.

bergschrund *n.* a deep chasm that forms between the steep headwall of a cirque (*q.v.*) and the glacier it contains.

beri-beri *n.* vitamin deficiency disease caused by a lack of thiamine (vitamin B_1) in the diet.

berkelium (Bk) *n.* a transuranic element [at. no. 97, r.a.m. (of its most stable known isotope) 247].

berm *n.* (1) a depositional feature often present in the backshore zone of a sandy beach. It consists of a slope that rises gently seawards and terminates in a ridge, known as the berm crest; (2) narrow, man-made ledge created during opencast mining or quarrying and often used for transport purposes.

berm crest *see* berm.

berry *n.* (1) a several-seeded indehiscent fruit with a fleshy covering and without a stony layer surrounding the seeds; (2) the dark knoblike structure on bill of swan.

beryl *n.* $Be_3Al_2(Si_6O_{18})$, a cyclosilicate mineral found in granites (usually in cavities), granite pegmatites and some metamorphic rocks. It is exploited as a gemstone (some varieties only) and as a source of beryllium.

beryllium (Be) *n.* metallic element (at. no. 4, r.a.m. 9.01), forming a grey metal in elemental form. It is used to make very hard alloys and as a moderator in the nuclear industry. It is a constituent of beryl, and its coloured forms, emerald and aquamarine. Beryllium is chemically similar to magnesium and it and its compounds are toxic. Prolonged inhalation of particles of beryllium can cause lung cancer.

best available technology (BAT), best available technology not entailing excessive costs (BATNEEC) concepts used in cost–benefit–risk analysis to set regulatory limits on, e.g., emission of pollutants.

best practicable environmental option (BPEO) a concept used in cost–benefit–risk analysis, esp. in UK, to set regulatory limits on, e.g., emission of pollutants.

best practicable means (BPM) concept applied in pollution control, esp. in UK, whereby the best possible level of control is attained within constraints imposed by the technological and economic capabilities of the polluter.

betacyanins *n.plu.* complex flavonoid pigments that give a reddish colour to some flowers.

beta decay a type of radioactive decay in which the parent nucleus emits a fast-moving electron, called a beta particle (*q.v.*). During this process a neutron is converted into a proton, resulting in an increase in the atomic number by one.

beta diversity ecological diversity resulting from competition between species that produces a finer adaptation of a given species to the complete habitat, thus narrowing its range of tolerance to other environmental factors. Beta diversity is represented by the rate and extent of change in species composition along an environmental gradient, e.g. of altitude, from one habitat to another – the greater the change, the higher the beta diversity. *alt.* habitat diversification. *cf.* alpha diversity.

beta-lactam antibiotic any of a large group of antibiotics, including the penicillins and cephalosporins, that contain a beta-lactam group.

beta-mesosaprobic category in the saprobic classification of river organisms comprising those that can live in water mildly polluted with organic pollutants, in which organic decomposition is mainly aerobic, e.g. the three-spined stickleback (*Gasterosteus aculeata*) and Canadian pondweed (*Elodea canadensis*). *cf.* alpha-mesosaprobic, oligosaprobic, polysaprobic.

beta-oxidation pathway metabolic pathway by which fatty acids are degraded in the mitochondria to yield acetyl CoA, two-carbon units being removed at each round of reaction.

beta particle the fast-moving electron (symbolized variously as β, β^- or $_{-1}^{0}e$) emitted during beta decay *q.v.*

Beta vulgaris sugar-beet *q.v.*

Betulales *n.* in some classifications an order of dicot trees including the families Betulaceae (birch) and Corylaceae (hazel).

Bhopal disaster the world's worst industrial disaster in terms of fatalities. It occurred in Bhopal, India, on 3 December 1984. The escape into the atmosphere of vaporized methylisocyanate (*q.v.*) from a local pesticide plant claimed the lives of over 2500 people and injured *ca.* 200 000 more.

B-horizon layer of deposition and accumulation of minerals below the topmost layer [the A-horizon and (if present) the E-horizon] in soils, colloquially known as the subsoil. *see* Fig. 28, p. 385. *alt.* illuvial layer. *see also* soil horizons.

Bi symbol for the chemical element bismuth *q.v.*

bi- prefix from L. *bis*, twice, often indicating having two of.

bicentric *a. pert.* two centres, *appl.* discontinuous distribution of species, etc.

Bicornes Ericales *q.v.*

bidentate *a.* (1) with two teeth or tooth-like indentations; (2) (*chem.*) *appl.* ligands with two donor atoms *q.v.*

biennial *n.* plant living for two years and flowering and fruiting only in the second. *a.* biennial.

big cats common name for the lion, tiger, leopard, jaguar and other large members of the cat family (Felidae).

bigeneric *a. appl.* hybrids between two different genera.

bilateral *a. pert.* or having two sides.

bilateral symmetry having two sides symmetrical only about one median axis. *cf.* radial symmetry.

bilayer *see* lipid bilayer.

bilharzia schistosomiasis *q.v.*

bill *n.* the beak of a bird, formed from outgrowths of cornified skin at the corners of the jaws.

binapacryl *n.* contact acaricide and fungicide used on apples to control red spider mite and mildew.

binary *a.* (1) composed of two units; (2) (*chem.*) *appl.* compounds of only two chemical elements.

binary fission in prokaryotic organisms, the chief mode of division, in which a cell divides into two equal daughter cells, each containing a copy of the DNA.

binding energy, nuclear *see* nuclear binding energy.

binocular *a.* (1) *pert.* both eyes; (2) stereoscopic, *appl.* vision.

binomial *a.* consisting of two names, *appl.* nomenclature or classification, the system of double Latin names given to plants and animals, consisting of a generic name followed by a specific name, e.g. *Felis* (genus) *tigris* (species).

-bio- word element derived from Gk *bios*, life, indicating living, *pert.* living organisms, etc.

bioaccumulation *n.* the build-up of a pollutant within the body of an aquatic organism as a consequence of uptake from food and direct uptake from the surrounding water. Often used as a synonym for bioconcentration but, technically, bioconcentration refers to uptake from ambient water only. *cf.* biomagnification.

bioaccumulator *n.* plant or animal species that accumulates heavy metals or other environmental contaminants (e.g. fat-soluble pesticides) in its tissues, and can be used as an indicator of the presence of chronic pollution by these compounds, especially where amounts of pollutant in the environment are too low to be easily detectable.

bioassay *n.* (1) any biological assay; (2) use of a living organism or tissue for assay purposes; (3) a quantitative biological analysis.

bioavailability *n.* the availability of nutrients and/or pollutants to living organisms in general, or to a specific organism in particular.

biocenosis biocoenosis *q.v.*

biocentrism *n.* view that takes into account the rights and value of all creatures, not just humans. *cf.* anthropocentrism.

biochemical oxygen demand (BOD) measurement of the amount of organic pollution in water, measured as the amount of oxygen taken up from a sample containing a known amount of oxygen kept at 20 °C for five days. A low BOD indicates little pollution, a high BOD indicates increased activity of heterotrophic microorganisms and thus heavy pollution. *alt.* biological oxygen demand.

biochemistry *n.* the chemistry of living organisms and its study.

biochore *n.* (1) boundary of a floral or faunal region; (2) climatic boundary of a floral region.

biocide *n.* any agent that kills living organisms.

bioclimatology *n.* study of the relationship between living organisms and climate.

biocoen *n.* (1) the living parts of an environment; (2) the biosphere *q.v.*

biocoenosis *n.* community of organisms inhabiting a particular biotope.

bioconcentration *n.* build-up of a pollutant in an aquatic organism by direct uptake from the surrounding water. Bioaccumulation (*q.v.*) is often used synonymously with bioconcentration but technically the former includes uptake from food as well as from the ambient water. *cf.* biomagnification.

bioconcentration factor (BCF) concentration of a pollutant within an organism on a wet weight basis, divided by the concentration in the ambient water, a quantitative measure of the degree of bioconcentration *q.v.*

biocontrol biological control *q.v.*

bioconversion *n.* conversion of organic wastes, e.g. animal dung and crop residues, into energy resources, e.g. biogas (*q.v.*), by microbial activity.

biocycle *n.* one of the three main divisions of the biosphere: marine, freshwater or terrestrial habitat.

biodegradable *a. appl.* materials that can be broken down into small molecules, such as carbon dioxide and water, by microorganisms. *cf.* non-biodegradable.

biodegradation *n.* the breakdown of materials by living organisms, mainly microorganisms.

biodemography *n.* the science dealing with the integration of ecology and population genetics.

biodiversity biological diversity *q.v.*

Biodiversity Convention United Nations Convention on Biological Diversity *q.v.*

bioelements *n.plu.* chemical elements that are necessary for the normal growth and development of an organism. *alt.* essential elements.

bioenergetics *n.* (1) the energy flow in an ecosystem; (2) study of energy transformation in living organisms.

bioengineering *n.* (1) use of artifical replacements for body organs; (2) use of technology in the biosynthesis of economically important compounds. *see also* biotechnology, genetic engineering.

biofertilization *n.* increase in soil fertility through biological means, e.g. through addition of nitrogen-fixing microorganisms.

biofertilizers organic manures *q.v.*

biofilm *n.* a thin layer of microorganisms that forms, e.g., on the surface of soil particles.

bioflavonoids *n.plu.* group of flavonoids present in citrus and other fruits, such as paprika, that have biological activity in animals due to their reducing and chelating properties, e.g. citrin.

biofuel *n.* gas, such as methane, or liquid fuel, such as ethanol (ethyl alcohol), made from organic waste material, usually by microbial action.

biogas *n.* gas with a high methane content produced by the microbial anaerobic digestion of organic wastes, such as crop residues, animal dung or sewage sludge. Biogas can be used for a variety of purposes, including heating, lighting and electricity generation.

biogas digester sealed container used for the production of biogas by the anaerobic digestion of organic residues.

biogenic, biogenetic *a.* originating from living organisms, *appl.* deposits such as coal, oil and chalk.

biogenically reworked zone sediment that has been disturbed by the actions of living organisms.

biogeochemical cycle closed circuit described by an essential element [e.g. carbon (C), sulphur (S) or oxygen (O)] as it passes from within organisms (biotic phase) into the physical, i.e. geochemical, environment (abiotic phase), and back again. *alt.* nutrient cycle.

biogeochemistry *n.* the study of the distribution and movement of chemical elements present in living organisms in relation to their geographical environment, and the movement of elements between living organisms and their non-living environment.

biogeocoenosis *n.* a community of organisms in relation to its special habitat.

biogeographical province an area of the Earth's surface defined by the endemic species it contains.

biogeographical realms, biogeographical kingdoms the major geographical divisions of the terrestrial environment characterized by their overall flora and fauna, *see* floral realm, zoogeographical kingdom.

biogeography *n.* that part of biology dealing with the geographical distribution of plants (phytogeography) and animals (zoogeography).

bioleaching *n.* extraction of metals (mainly copper and uranium) from their ores using microorganisms. This technique can be used to recover metals from mined deposits, old mine dumps and directly from orebodies deep underground.

biological *a. pert.* to living organisms or to their study.

biological amplification, bioamplification biomagnification *q.v.*

biological clocks hypothetical mechanisms underlying the regular metabolic and behavioural rhythms seen in many cells and organisms, the biochemical basis for which is as yet unknown.

biological containment in genetic engineering, the use of non-infectious, enfeebled and exceptionally nutritionally fastidious strains of microorganism, which cannot survive outside the laboratory, as vehicles in which to clone recombinant DNA in order to minimize risk in case of accident.

biological control control of pests and weeds by other living organisms, usually other insects, bacteria or viruses, or by biological products, such as hormones or toxins.

biological diversity as defined by the United Nations Convention on Biological Diversity: 'the variability among living organisms from all sources, including, *inter alia*, terrestrial, marine, and other aquatic ecosystems and the ecological complexes of which they are part. This includes diversity within species, between species and of ecosystems.' The number of different species is estimated at between 40 and 80 million,

most of them still undiscovered and uncharacterized, each species containing yet further genetic diversity. The rapid loss of biological diversity that is occurring as a result of species loss due to habitat destruction and fragmentation is considered one of the most pressing environmental problems. Biological diversity is colloquially known as biodiversity. *see also* alpha diversity, beta diversity, genetic diversity, species diversity, ecological diversity.

biological filter trickling filter *q.v.*

biological indicator *see* indicator species.

biological monitoring any one of several techniques used to measure the effect of environmental change on the biological species present, in terms of their numbers, diversity, etc.

Biological Monitoring Working Party (BMWP) score biotic index widely used in the UK to indicate the water quality of its rivers. Aquatic macroinvertebrates are sampled in a standardized manner and then identified to family level. Each family is given a score on a scale of 1–10, with those perceived to be most tolerant to pollution, e.g. oligochaetes (worms) and chironomid larvae, scoring lowest. The total of the individual scores is the BMWP score, which may be divided by the number of taxa to give the Average Score Per Taxon (ASPT). In both scores, a high number indicates low levels of pollution.

biological oxygen demand biochemical oxygen demand *q.v.*

biological pest control *see* biological control.

biological races genetically distinct strains of a species that are alike morphologically but differ in some physiological way, such as a parasite or saprophyte with particular host requirement, or a free-living organism with a food or habitat preference.

biological resources genetic resources, organisms or parts of organisms, populations or any other biotic components of ecosystems with actual or potential use or value to humans.

biological rhythms inbuilt periodic, daily or seasonal behaviours or metabolic changes in living organisms, which will continue, at least for some time, even when the environmental rhythm (e.g. cycle of light

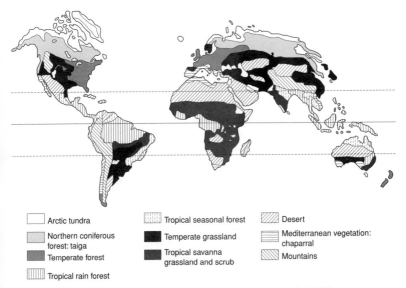

Fig. 3 The terrestrial biomes of the world (after Cox *et al.*, 1976).

Legend:
- Arctic tundra
- Northern coniferous forest: taiga
- Temperate forest
- Tropical rain forest
- Tropical seasonal forest
- Temperate grassland
- Tropical savanna grassland and scrub
- Desert
- Mediterranean vegetation: chaparral
- Mountains

and dark) to which they are entrained is absent. *see* circadian rhythm, circannual rhythm, diurnal. *alt.* biorhythms.

biological shield the structure designed to prevent the escape of gamma and neutron radiation from the core of a nuclear reactor. It is generally fabricated from concrete, several metres thick. In most designs, there is a pressure vessel (*q.v.*) between the biological shield and the core. There are reactors, however, in which the pressure vessel and biological shield are one and the same steel-lined prestressed concrete container.

biological species population of individuals that can interbreed, i.e. a true species.

biological weathering organic weathering *q.v.*

biology *n.* the science dealing with living organisms, a term coined by J. B. de Lamarck in 1802.

bioluminescence *n.* the production of light by living organisms, which is the result of an enzyme-catalysed biochemical reaction in which an inactive precursor is converted into a light-emitting chemical.

biomagnification *n.* increase in concentration of pollutants, e.g. fat-soluble pesticides such as DDT, in the bodies of living organisms at successively higher levels in the food chain. *cf.* bioconcentration. *alt.* bioaccumulation, biological amplification.

biomanipulation *n.* the deliberate manipulation of the species composition of an ecosystem, e.g. to try and regenerate a hypertrophic lake, after the organic pollution itself has been ameliorated.

biomass *n.* (1) total weight, volume or energy equivalent of organisms in a given area, usually expressed as dry weight per given area; (2) plant materials and animal wastes used as a source of fuel or other industrial products; (3) in biotechnology, the microbial matter in the system.

biomass energy useful energy derived from various types of biomass, e.g. wood and wood-processing residues, animal wastes, crop residues, municipal wastes and energy crops *q.v.*

biome *n.* a continental-scale regional ecosystem type characterized by its distinctive vegetation and climate, e.g. boreal forest, tropical rain forest, tundra, taiga, grassland, desert. *see* Fig. 3.

biometeorology *n.* the study of the effects of the weather on plants and animals.

biometrics, biometry *n.* statistical study of living organisms and their variations.

biomineralization *n.* the production of partly or wholly mineralized internal or external structures by living organisms.

biophage *n.* an organism feeding upon other living organisms.

biophyte *n.* a parasitic plant.

bioplastic *see* biopolymer.

biopolymer *n.* biodegradable polymer produced by living organisms, e.g. polysaccharide gums (xanthans) produced by the bacterium *Xanthomonas* and other spp., and bioplastics, such as poly-β-hydroxybutyric acid and other poly-β-hydroxyalkanoates, produced by the bacterium *Alcaligenes eutrophus*.

bioregion *n.* a unique area with distinctive soils, landforms, climate and indigenous plants and animals.

bioremediation *n.* any technique used to enhance the recovery of a contaminated site by the use of living organisms, e.g. the application of limiting nutrients to oiled beaches in order to encourage the growth of microorganisms capable of using the oil as an energy source, thereby removing contamination.

biorhythm biological rhythm *q.v.*

bios *n.* living organisms.

biosensor *n.* any organism, microorganism, enzyme system or other biological structure used as an assay or indicator.

bioseries *n.* a succession of changes of any single heritable character.

bioseston *n.* plankton or particulates of organic matter suspended in water.

biospecies biological species *q.v.*

biospeleology *n.* the biology of cave-dwelling organisms and its study.

biosphere *n.* the part of the planet containing living organisms, the living world. It consists of the hydrosphere (water), the surface of the lithosphere and the lower atmosphere.

biosphere reserve area protected under the Biosphere Reserve Programme (*q.v.*). Such reserves are put to multiple use and zoned into different areas, namely a fully protected central core area, a buffer zone in which limited human activity, e.g. traditional resource use, is allowed, and a transition zone, which protects the whole reserve from encroachment. *see* Fig. 4.

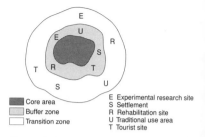

Core area | E Experimental research site
Buffer zone | S Settlement
Transition zone | R Rehabilitation site
| U Traditional use area
| T Tourist site

Fig. 4 Schematic plan of a biosphere reserve.

Biosphere Reserve Programme conservation programme introduced in the early 1970s by UNESCO under its Man and the Biosphere Programme. Under this, an international network of protected areas, known as biosphere reserves (*q.v.*), was established.

biostasis *n.* the ability of living organisms to withstand environmental changes without being changed themselves.

biostatics *n.* study of the structure of living organisms in relation to function.

biostratigraphic unit rock stratum or collection of strata that is distinguishable by the fossils it contains, without recourse to geological or physical features, and which is established by world-wide correlations. *alt.* biostratigraphic zone.

biostratigraphy *n.* the study of the relative chronology of successive rock formations and their index fossils.

biostratinomy *n.* the study of the whole range of events that must have occurred from the death of an organism to its inclusion in a sediment where it became fossilized.

biosynthesis *n.* formation of organic compounds by living organisms.

biosystem ecosystem *q.v.*

biosystematics *see* systematics.

biota *n.* (1) the total fauna and flora of a region; (2) the population of living organisms in general.

biotechnology *n.* the use of living cells or microorganisms (e.g. bacteria) in industry and technology to manufacture drugs and chemicals, break down waste, etc. In recent years esp. refers to the use of genetically modified cells and microorganisms in such processes.

biotic *a. pert.* life and living organisms.

biotic climax a plant community that is maintained in the climax state by some biotic factor such as grazing. *alt.* plagioclimax.

biotic community the whole community of plants and animals that share a particular habitat or region.

biotic environment the part of an organism's environment produced by its interaction with other organisms.

biotic factors the influence of organisms and their activities on other organisms and the environment.

biotic index a measure of the ecological quality of the environment, generally with regard to organic pollution, using assessments of the number and abundance of key indicator species present. For example, for rivers, biotic indices based on the type of invertebrate community present are used. A high biotic index, representing high diversity and the presence of pollution-sensitive species, indicates a clean river; a low biotic index, reflecting the presence of only a few pollution-tolerant species, indicates heavy organic pollution. *cf.* diversity index.

biotic phase *see* biogeochemical cycle.

biotic potential highest possible rate of population increase (r_{max}), resulting from maximum rate of reproduction and minimum mortality.

biotic province a major ecological region of a continent.

biotic pyramid ecological pyramid *q.v.*

biotic succession the part of a succession that is controlled by the activities and interactions of the species present rather than by the physical environment.

biotite *n.* $K(Mg,Fe)_3(AlSi_3O_{10})(OH,F)_2$, one of the mica minerals *q.v.*

biotite granite *see* granite.

biotite–muscovite granite *see* granite.

biotope *n.* an area or habitat of a particular type, defined by the organisms (plants, animals, microorganisms) that typically inhabit it, e.g. grassland, woodland, etc., or, on a smaller scale, a microhabitat.

biotroph *n.* any organism that feeds on other living organisms, e.g. parasitic or symbiotic bacteria and fungi, carnivores, herbivores, etc. *a.* biotrophic.

biotype *n.* group of individuals with similar genetic constitution.

biparous *a.* bearing two young at a time.

biped *n.* a two-footed animal. *a.* bipedal.

bipolar *a.* having, located at or *pert.* two ends or poles.

biradial *a.* symmetrical both radially and bilaterally, as some coelenterates.

bird cherry aphid *see* cereal aphids.

bird lice common name for an order of insects, the Mallophaga or biting lice, which are ectoparasites of birds.

birds *see* Aves (*q.v.*) and Appendix 7.

bird's foot delta *see* delta.

bird's nest fungi common name for the Nidulariales, an order of gasteromycete fungi that have fruiting bodies resembling minute bird's nests (the peridium) full of eggs (the peridioles containing the spores).

birth control the deliberate attempt to prevent conception.

birth rate number of live births within a population over a set period, usually a year. The crude birth rate is calculated for human populations as the annual number of live births per 1000 population in a given geographical area, with the population number usually taken at the midpoint of the year in question. A more accurate measure is the annual number of live births per 1000 females of reproductive age (15–44 years), known as the standardized birth rate or fertility rate.

bisect (1) *n.* a stratum transect chart with root system as well as shoot included. (2) *v.* to divide into two equal halves.

bisexual *a.* having both male and female organs.

bismuth (Bi) *n.* shiny metallic element (at. no: 83, r.a.m. 208.98) that is found naturally in the ore bismuthinite. It is a good absorber of gamma rays and is used as a gamma-ray filter.

bismuth glance bismuthinite *q.v.*

bismuthinite *n.* Bi_2S_3, a bismuth ore mineral. *alt.* bismuth glance.

biting lice *see* bird lice.

bituminous coal medium-rank, soft coal with a fixed carbon content of 46–86% and a calorific value of 19–36 MJ kg^{-1}, used for domestic heating and electricity production.

bivalent *a.* (1) (*chem.*) having a valency (*q.v.*) of two. *alt.* divalent; (2) (*biol.*) *appl.* antibody with two antigen-binding sites.

bivalent *n.* (*biol.*) (1) chromosome that has duplicated to form two sister chromatids still held together at the centromere; (2) a pair of duplicated homologous chromosomes held together by chiasmata at meiosis. *alt.* tetrad.

Bivalvia, bivalves *n.*, *n.plu.* class of bilaterally symmetrical molluscs that are laterally flattened and have a shell made of two hinged valves, e.g. clams, mussels, scallops, cockles.

bivoltine *a.* having two broods in a year.

Bk symbol for the chemical element berkelium *q.v.*

black bean aphid blackfly *q.v.*

black bent *Agrostis gigantea*, a serious weed pest of arable land.

black body (*phys.*) hypothetical body that acts both as a perfect absorber and as a perfect radiator of electromagnetic radiation. *cf.* grey body.

black coal collective term for bituminous coal and anthracite. *alt.* hard coal.

black corals common name for an order of stony corals that are colonial and have black or brown skeletons.

Black Death plague *q.v.*

black earth chernozem *q.v.*

blackfly *n. Aphis fabae*, an insect pest of bean crops that also attacks mangolds and sugar beet. *alt.* black bean aphid.

black grass *Alopecurus myosuroides*, a serious weed pest of temperate cereal crops.

black ice *see* glazed frost.

Black Jack sphalerite *q.v.*

Black List list produced by the European Community of the most dangerous toxic chemicals, e.g. organophosphorus compounds, and mercury and its compounds. Substances are included in this list on the basis of their toxicity, persistence in the environment and capacity to bioaccumulate. *cf.* Grey List. *alt.* List 1.

black rust stem rust *q.v.*

black vomit yellow fever *q.v.*

bladderworm *n.* larval stage of some Cestoda (tapeworms) in the intermediate host, which is in the form of a bladder containing an inverted scolex, *see* cysticercoid.

blade *n.* thin flattened part of leaf, bone, etc.

blank *n.* (*chem.*) in instrumental analysis, this is a mixture of all of the reagents and solvents with which the sample and standards are treated, but which contains no sample or added analyte.

blanket *n.* a layer of fertile isotopes placed around the core of a fast breeder reactor (*q.v.*) in order to capture neutrons as they leave, thereby maximizing the yield of bred fissile isotopes.

blanket bog, blanket mire, blanket peat acid peat bog covering large stretches of country that develops in cold, very wet climates (where precipitation is very high and evaporation is low) in upland regions where drainage is impeded and the soil is acid. It is composed of a layer of dead but undecomposed organic matter (peat) overlying waterlogged, acidic, nutrient-poor ground.

blende sphalerite *q.v.*

blight *n.* insect or fungus disease of plants and also the agent causing it.

blizzard *n.* severe snow storm.

bloat *n.* a ruminant disorder involving an abnormal build-up of gases within the rumen, associated with a lack of fibre in the diet.

blockfield felsenmeer *q.v.*

blocking anticyclone cell of high pressure that impedes the flow of eastward-moving depressions, resulting in abnormal weather conditions in the mid-latitudes. Such cells originate in the polar or subtropical high-pressure zones.

blocky *a.* (*soil sci.*) appl. soil peds (*q.v.*) that are roughly cubic in shape. *cf.* subangular blocky.

blocky lava a viscous, slow-moving lava flow that has a surface layer composed of relatively smooth-surfaced blocks of lava. *cf.* aa.

blood stone heliotrope *q.v.*

bloodworm *n.* (1) any reddish aquatic larva of the midge family Chironomidae, e.g. *Chironomus riparius*, which is tolerant of heavy organic pollution; (2) reddish oligochaete worm living in river mud; (3) a red bristle worm found on muddy shores.

bloom *n.* (1) waxy layer on surface of certain fruits, such as grapes, some berries, etc.; (2) blossom or flower; (3) seasonal dense growth of algae or phytoplankton.

blowdown *n.* extensive toppling of trees by wind in a given area.

blow outs deflation hollows *q.v.*

blubber *n.* insulating layer of fat in whales, seals, etc., lying between skin and muscle layer.

blue asbestos crocidolite *q.v.*

blue-baby syndrome the term used to denote the symptoms of the severe blood disorder methaemoglobinaemia (*q.v.*) in infants, which include a characteristic bluish tinge to the skin and lips.

blue coral common name for corals of the genus *Heliopora*, an individual having a solid calcareous skeleton with vertical tubular cavities containing polyps.

blue-green algae common name for the cyanobacteria *q.v.*

blue light receptor receptor in plants for light in the blue region of the spectrum (wavelength 400–500 nm) that is involved in phototropism, inhibition of stem growth, promotion of leaf expansion and induction of gene expression. The nature of the receptor(s) is still unknown but it is thought to be a flavoprotein. *alt.* cryptochrome.

bluff line a pronounced slope that marks the edge of a floodplain.

BMWP score Biological Monitoring Working Party score *q.v.*

BNFL British Nuclear Fuels plc, a company wholly owned by the British government.

BOD biochemical oxygen demand *q.v.*

body burden amount of radioactive or toxic material present in the body at any given time.

body lice common name for members of the insect order Anoplura that are ectoparasitic on mammals (e.g. body louse, bed bug). Some are carriers of diseases such as typhus.

body waves (*geog./geol.*) seismic waves that travel within the Earth. Two types are recognized: S waves (also known as shake or shear waves), which displace objects at 90° to their direction of movement and P waves (also known as push waves), which move objects parallel to their direction of movement. *cf.* surface waves.

boehmite *n.* AlO(OH), an aluminium ore mineral, one of the constituents of bauxite *q.v.*

bog *n.* (1) characteristic plant community developing on wet, very acid peat, containing e.g., sundews (*Drosera* spp.), sphag-

num moss and bog myrtle (*Myrica gale*), *see* blanket bog, raised bog; (2) sometimes also refers to alkaline bogs developing in valleys. *cf.* fen.

bog burst sudden rupture of a bog and the consequent release of water and organic matter.

boiling water reactor (BWR) a type of fission nuclear reactor that utilizes Zircaloy-clad fuel pellets of enriched uranium oxide. It is cooled and moderated by light water *q.v.*

bole *n.* main trunk of a tree.

boletes *n.plu.* basidomycete fungi of the genus *Boletus* and related genera, similar in general form to agarics, but in which the hymenium lines pores and not gills on the underside of the cap.

boll *n.* a capsule or spherical fruit containing a cottony mass of seeds, as in cotton plant.

boll weevil *Anthonomus grandis*, an insect pest that feeds on the bolls of cotton plants, thus destroying the valuable fibre.

bolt *v.* (*agric.*) to run to seed prematurely.

bone-beds *n.plu.* deposits formed largely by remains of bones of fishes and reptiles, such as Liassic bone-beds.

bonemeal *n.* ground, dried animal bones used as both a fertilizer and an animal feedstuff component.

bonitation *n.* evaluation of the numerical distribution of a species in a particular locality or season, esp. in relation to agricultural, veterinary, or medical implications.

Bonn Convention Convention on the Conservation of Migratory Species of Wild Animals, an international agreement made in 1979 under the auspices of UNEP.

bony fishes common name for the Osteichthyes, a class of fishes with bony skeletons, usually possessing a swimbladder or lung, and gills covered by an operculum, and that includes teleosts, lungfishes and crossopterygians.

book lice common name for some members of the order Psocoptera, small wingless insects with a globular abdomen and incomplete metamorphosis, often living on paper.

Bora *n.* cold, gusty north-easterly wind located on the Adriatic coast.

boralf *n.* a suborder of the order alfisol of the USDA Soil Taxonomy system, consisting of alfisols of cool or cold climatic regions.

borax *n*. $Na_2B_4O_5(OH)_4.8H_2O$, an evaporite mineral used as a flux, as an antiseptic and in the manufacture of glass and ceramics.

Bordeaux mixture fungicide consisting of a mixture of copper(II) sulphate, lime (calcium oxide) and water, used as a spray to control, e.g., tomato blight and potato blight.

boreal *a. pert.* (1) or *appl.* climate with snow in winter and short dry summers; (2) or *appl.* northern biogeographical region; (3) post-glacial age from about 7500 to 5500 BC with a continental type of climate.

boreal forest the northern coniferous forest (taiga) growing across northern North America and northern Asia.

Boreal kingdom phytogeographical kingdom comprising all of the northern hemisphere north of latitude 30° N.

bornhardts *see* inselberg.

boroll *n*. a suborder of the order mollisol of the USDA Soil Taxonomy system consisting of mollisols of cool or cold climatic regions.

boron (B) *n*. metalloid element (at. no. 5, r.a.m. 10.81), forming a yellow-brown powder or crystals. It is extracted chiefly from borax (sodium borate). Boron isotopes are used in control rods and shields in nuclear reactors as boron is a good absorber of neutrons. It is an essential micronutrient for plants.

bosset *n*. the beginning of antler formation in deer in the first year.

botany *n*. the branch of biology dealing with plants.

Botrytis *n*. genus of fungi, members of which are responsible for grey mould (botrytis) and other diseases of crop and ornamental plants, e.g. neck rot in onions and chocolate spot in beans.

bottle banks receptacles used for the deposition of waste domestic glass (bottles, jars, etc.) for recycling purposes.

bottleneck *n*. (1) (*biol.*) a sudden decrease in population size with a resulting decrease in genetic variability within the succeeding population; (2) (*geog.*) any part of a road system that is frequently the site of congestion.

bottom load bedload *q.v.*

botulinum toxin protein produced by the bacterium *Clostridium botulinum* under anaerobic conditions, e.g. in inadequately sterilized canned or bottled food, a powerful poison affecting the nervous system and causing botulism.

boulder clay till *q.v.*

boundary *n*. in thermodynamic analysis, the boundary is the borderline that separates the system from its surroundings.

bournes *n.plu.* seasonal streams that flow in those valleys, in chalk landscapes, that are normally dry. *alt.* bourns.

bournonite *n*. $PbCuSbS_3$, an ore mineral of copper, lead and antimony. *alt.* wheel-ore.

Bovidae, bovids *n*., *n.plu.* mammalian family of the Artiodactyla comprising cattle, sheep, goats and bison, which have four-chambered stomachs and horns that are not shed.

bovine spongiform encephalopathy (BSE) a neurological disease of cattle, of recent origin. The development of BSE has been linked to the presence in compound cattle feed of protein from sheep infected with scrapie (*q.v.*). *alt.* mad cow disease.

box-jellies Cubozoa *q.v.*

box model a method of portraying the biogeochemical cycle of an element or chemical species. In this system, each individual reservoir (*q.v.*) is represented by a box, while the direction of fluxes between reservoirs is indicated by arrows connecting the boxes.

bp base pair, used as a unit of length in DNA. *see* base pairing.

b.p. boiling point.

BP, b.p. before present.

BPEO best practicable environmental option, *q.v.*

BPM best practicable means *q.v.*

Bq becquerel *q.v.*

brachial *a. pert.* arms, arm-like.

brachiate *v*. to move along by swinging the arms from one hold to another, as in the gibbon. *n*. brachiation.

Brachiopoda, brachiopods *n*., *n.plu.* a small phylum of coelomate animals, the lamp shells, superficially resembling the bivalve molluscs but different in symmetry of the shell and internal structure. A characteristic structure is the lophophore, consisting of coiled tentacles (brachia) surrounding the mouth.

brachy- prefix from Gk *brachys*, short.

Brachycera *n.* the short-horned flies, a sub-order of Diptera with short stout antenna, which include the blood-sucking horse-flies and gadflies, the metallic-coloured soldierflies, the slender, long-legged snipe-flies, the bee-flies, which have a strong superficial resemblance to bees and hover and feed in flowers, the robberflies, and other families.

brackish water water with a salinity of between 0.5 and 30 parts per thousand total dissolved solids. *cf.* seawater.

bract *n.* (1) modified leaf in whose axil an inflorescence or flower arises; (2) a floral leaf; (3) a leaf-like structure.

bradytelic *a.* evolving at a rate slower than the standard rate.

braided stream a stream that has been divided into numerous intertwining smaller streams by the presence of midstream islands and bars.

branch *n.* (*biol.*) a taxonomic group used in different ways by different specialists but usually referring to a level between subphylum and class.

branchicolous *a.* parasitic on fish gills.

Branchiopoda, branchiopods *n.*, *n.plu.* the water fleas, brine shrimps and their allies, a subclass of mainly freshwater crustaceans whose carapace, if present, forms a dorsal shield or bivalve shell and that have broad lobed trunk appendages fringed with hairs.

brand *n.* (*biol.*) a burnt appearance of leaves, caused by rust and smut fungi.

brand fungi a common name for the rust and smut fungi.

brandling worms *Eisenia foetida*, *see* wormery.

brassicas *n.plu.* plants of the genus *Brassica*, which includes several cultivated crops, e.g. cabbage (*Brassica oleracea*), oilseed rape (*Brassica napus*) and turnip (*Brassica campestris*).

bread wheat the hexaploid wheat, *Triticum aestivum. see* wheat.

breakbone fever dengue *q.v.*

break crop a crop, e.g. oilseed rape, whose cultivation temporarily interrupts the otherwise continuous cultivation of a main crop, such as wheat or barley.

breakwater *n.* man-made barrier, e.g. of concrete or boulders, that stretches into the sea to protect coasts and harbours from wave action.

breccia *n.* a sedimentary rock composed of angular or jagged rock fragments set in a finer-grained matrix.

breeder reactor a nuclear reactor designed to convert fertile isotopes into fissile ones (a process called breeding). *see* fast breeder reactors.

breeding *n.* (1) (*biol.*) human manipulation of heritable characteristics in domesticated plants or animals by artificial selection (*q.v.*); (2) (*phys.*) the conversion of non-fissile but fertile (*q.v.*) nuclides into fissile material. This occurs within fast breeder reactors *q.v.*

breeding season a period each year, for animals that do not breed all the year round, in which animals court, mate and rear their young.

breeding size the number of individuals in a population, per generation, actually involved in reproduction.

breeding system the extent and mode of interbreeding within a species or group of closely related species.

brevi- prefix from L. *brevis*, short.

brigalow forest acacia forest covering large areas in Australia, in which the dominant species is the brigalow (*Acacia harpophylla*).

bright-field microscopy technique of optical microscopy in which a living cell is viewed by direct transmission of light through it.

brine shrimp small marine crustacean of the subclass Branchiopoda, usually refers to *Artemia*.

bristletails *n.plu.* common name for certain insects of the orders Thysanura and Diplura, small wingless insects characterized by a single or two-pronged bristle at the tail end.

bristle worms common name for the Polychaeta, a class of annelid worms.

British thermal unit (Btu) a unit of heat quantity equal to 1055.06 joules. Formerly defined as the quantity of heat needed to warm 1 lb of water through 1 °F.

brittle stars common name for the Ophiuroidea, a class of echinoderms with each member having five slender arms clearly marked off from the central disc.

broad-leaf, broad-leaved *a. appl.* (1) trees, name commonly given to the angiosperm trees of temperate climates, characterized by thin flat leaves (as opposed to the needle-bearing conifers); (2) dicot weeds with flat broad leaves, as opposed to the narrow-leaved grasses.

broiler *n.* chicken reared for meat, generally killed when it has attained a live weight of 1.8–2.0 kg.

bromacil *n.* contact and soil-acting herbicide used, e.g., to control weeds on non-cropped areas.

Bromeliales, bromeliads *n., n.plu.* order of terrestrial and epiphytic monocots with reduced stem and rosette of fleshy water-storing leaves and comprising the family Bromeliaceae (pineapple).

brominated hydrocarbons hydrocarbons (*q.v.*) with one or more of the hydrogen atoms replaced by atoms of bromine. Gaseous brominated hydrocarbons entering the atmosphere may reach the stratosphere, where they undergo photolysis, liberating bromine atoms. The bromine atoms released enter into catalytic cycles, destroying some of the ozone that shields the biosphere from harmful ultraviolet light. As well as acting as ozone-destroying catalysts, Br synergistically enhances the rate of chlorine-induced ozone depletion. The main atmospheric brominated hydrocarbon, bromomethane (CH_3Br), is mostly of natural origin, although it is also used as a fumigant. Its atmospheric concentration is low (about 0.013 p.p.b.) but is rising fast (by 12–15% per year) (1992 data).

bromine (Br) *n.* halogen element (at. no. 35, r.a.m. 79.90), a red-brown volatile corrosive liquid in elemental form (Br_2). Bromine occurs naturally, as bromides, in seawater and salt deposits. It has many industrial uses. Bromomethane is used as a fumigant against insects, worms and rodents.

bromomethane *n.* soil sterilant mainly used under glass to control nematodes and soil fungi, also used as a fumigant.

It is a potential ozone-destroying chemical if it reaches the stratosphere. *alt.* methyl bromide. *see also* brominated hydrocarbons.

bromophos *n.* contact and stomach-acting organophosphate insecticide, used to control soil pests, such as wireworms and cutworms. Also used as a sheep dip in the UK.

bromovirus group plant virus group named after the type member, brome mosaic virus, which is a small isometric single-stranded RNA virus, causing mosaic symptoms (mottling of leaves). Bromoviruses are multicomponent viruses in which four genomic RNAs are encapsidated in three different virus particles.

bromoxynil *n.* contact herbicide used to control broad-leaved weeds mainly in cereal crops.

bronchitis *n.* inflammation of the membranes of the bronchial tubes, causing persistent coughing and over-production of mucus.

Brønsted–Lowry acid a donor of H^+ ions, i.e. protons. For example, in the reaction represented by the following equation, ethanoic acid (CH_3COOH) acts as a Brønsted–Lowry acid because it donates a proton to the water (H_2O), forming the ethanoate anion (CH_3COO^-) and the hydroxonium ion (H_3O^+):

$$CH_3COOH_{(aq)} + H_2O_{(l)} \rightleftharpoons CH_3COO^-_{(aq)} + H_3O^+_{(aq)}$$

alt. Brønsted acid.

Brønsted–Lowry base an acceptor of H^+ ions, i.e. protons. For example, in the reaction represented by the following equation, ammonia (NH_3) acts as a Brønsted–Lowry base because it accepts a proton from the water (H_2O), forming the ammonium cation (NH_4^+) and the hydroxide anion (OH^-):

$$NH_3 + H_2O \rightleftharpoons NH_4^+ + OH^-$$

alt. Brønsted base.

brood *n.* (1) the offspring of a single birth event or clutch of eggs; (2) any young animals being cared for by adults.

brood parasite an animal that lays its eggs in the nest of another member of the same species (intraspecific brood parasite) or of a different species (interspecific brood parasite), who then rears them.

brown algae common name for the Phaeophyta, mainly marine algae (seaweeds) containing the brown pigment fucoxanthin.

brown asbestos amosite (*q.v.*). *see also* asbestos.

brown coal sub-bituminous coal *q.v.*

brown earths, brown forest soils dark brown friable soils associated with areas of the Earth's land surface originally covered with deciduous forest.

brown fat highly vascularized adipose tissue rich in mitochondria, the cytochromes of which help to give it a brown colour, and which is involved in thermoregulation in hibernators and in young mammals generally. It typically occurs around the shoulder blades, neck, heart, large blood vessels and lungs. It is specialized for heat generation as a result of uncoupling of fatty acid oxidation and electron transfer in mitochondria from ATP synthesis. *alt.* brown adipose tissue (BAT).

brownfield site *n.* site, within a city, cleared for redevelopment. *cf.* greenfield site.

Brownian movement movement of small particles, such as pollen grains, bacteria, etc., when suspended in a colloidal solution, due to their bombardment by molecules of the solution.

brown planthopper *Nilaparvata lugens*, a serious pest of rice.

brown podzolic soil acid forest soil with a layer of litter over a greyish-brown organic and mineral layer and a pale leached layer below.

brown rust a disease affecting barley and wheat, caused by the fungi *Puccinia hordei* and *Puccinia recondita*, respectively.

brown soils soils similar to chernozems (*q.v.*) but found in warmer and drier areas and supporting short grassland.

browse *n.* tender parts of woody plants, such as young shoots and leaves, eaten by animals. *v.* to eat such material. *cf.* graze.

brucite *n.* a mineral with the formula $Mg(OH)_2$.

Brundtland Report report of the World Commission on Environment and Development (*q.v.*) published in 1987, which led to the convening of the 1992 United Nations Conference on Environment and Development *q.v.*

bryocole *n.* animal living among moss.

bryology *n.* the branch of botany that deals with mosses and liverworts.

Bryophyta, bryophytes *n.*, *n.plu.* a division of the plant kingdom containing the mosses (Musci), liverworts (Hepaticae) and hornworts (Anthocerotae). They are small non-vascular plants either thalloid in form (hornworts and some liverworts) or differentiated into stems and leaves (mosses and some liverworts) and attached to the substrate by rhizoids. They have a well-marked alternation of generations, the independent plant being the gametophyte, which produces motile male gametes that fertilize single egg cells contained in flask-shaped archegonia. The sporophyte (the capsule) grows out from the fertilized egg and produces spores from which new plants develop. *alt.* Bryopsida.

Bryopsida Bryophyta *q.v.*

Bryozoa, bryozoans moss animals, *see* Ectoprocta.

BSE bovine spongiform encephalopathy *q.v.*

Btu British thermal unit *q.v.*

bubonic plage *see* plague.

bud *n.* (1) (*bot.*) structure from which shoot, leaf or flower develops; (*zool.*) (2) outgrowth on an adult parent organism that develops into a new individual, as in hydra; (3) incipient outgrowth, as limb buds in embryo from which limbs develop.

budding *n.* (1) the production of buds; (2) (*zool.*) method of asexual reproduction common in sponges, coelenterates and some other invertebrates, in which new individuals develop as outgrowths of the parent organism, and may eventually be set free; (3) (*hort.*) artificial vegetative propagation by insertion of a bud within the bark of another plant; (4) (*mycol.*) cell division by the outgrowth of a new cell from the parent cell; (5) (*virol.*) release of certain animal viruses from the cell by their envelopment in a piece of plasma membrane that subsequently pinches off from the cell.

budget *n.* *see* energy budget, time and energy budget.

buffer *n.* (1) salt solution that minimizes changes in pH when an acid or alkali is added; (2) any factor that reduces the impact of external changes on a system.

buffer species species that is usually of only secondary importance as a food source, but which becomes the primary food source in adverse conditions.

buffer zone an area of lesser ecological value surrounding a core area of particular interest that protects it from external influences. *see* Fig. 4, p. 50.

bugs *n.plu.* (1) common name for insects of the order Hemiptera (*q.v.*); (2) colloquial term for disease-causing bacteria and viruses.

builder *n.* basic component of synthetic detergents whose presence is required to provide a source of hydroxide ions (OH⁻) necessary for effective detergent action.

built environment that part of the environment comprising buildings, roads, etc.

bulb *n.* (1) (*bot.*) specialized underground reproductive organ of a plant consisting of a short stem bearing a number of swollen fleshy leaf bases or scale leaves, the whole enclosing next year's bud; (2) any part or structure of a plant or animal resembling a bulb, a bulb-like swelling.

bulbil *n.* (1) a fleshy axillary bud that may fall and produce a new plant, as in some lilies; (2) any small bulb-shaped structure or swelling.

bundle sheath cells in some tropical plants, the cells in photosynthetic tissues in which carbon dioxide incorporated into aspartate and malate in the C4 pathway is released and enters the Calvin cycle.

bunds *n.plu.* embankments built between low-lying fields of paddy rice to help retain the seasonal flood water needed for wetland cultivation.

bunt fungi smut fungi *q.v.*

Bunyaviridae *n.* enveloped, spherical single-stranded, segmented RNA viruses, including the Bunyamwera virus and Rift Valley fever viruses.

bupirimate *n.* systemic fungicide used to control powdery mildew on, e.g., blackcurrants and apples.

burden *n.* of parasites, the total number or mass of parasites infecting an individual.

burner reactors thermal reactors *q.v.*

burning *see* combustion.

burnup *n.* the energy liberated within a nuclear fission reactor per unit mass of fuel per unit time.

bush *n.* (1) small shrub; (2) vegetation cover composed of grassland and shrubs.

bush fallow a system of cultivation practised in the humid and semi-humid tropics, esp. in South-East Asia and West Africa, in which periods of cultivation are alternated with periods of fallow, of varying length. This differs from shifting cultivation (*q.v.*) in a number of ways, e.g. tillage is by hoe rather than digging stick, the periods of fallow are shorter and the communities that farm in this way are sedentary rather than itinerant. *alt.* rotational bush fallow.

bush layer the horizontal ecological stratum of a plant community comprised of shrubs, which is higher than the field or herb layer and lower than the tree layer. *alt.* shrub layer.

but- (*chem.*) the prefix used to indicate the presence of four carbon atoms in a chain in an organic compound (e.g. butane $CH_3CH_2CH_2CH_3$) or radical.

butte *n.* a small mesa *q.v.*

butterflies *n.plu.* common name for members of the order Lepidoptera (*q.v.*) that have clubbed antennae.

buttress roots branch roots given off above ground, arching away from stem before entering the soil, forming additional support for trunk.

BWR boiling water reactor *q.v.*

by abbreviation for a billion (10^9) years.

byproduct *n.* a useful product that is made as a consequence of a manufacturing or agricultural process whose primary purpose is to produce another product. For example, the pulverized fuel ash formed during the generation of electricity in coal-fired power stations may be used as additives to concrete.

byssinosis *n.* a disabling lung disease caused by excessive inhalation of cotton dust.

bytownite *n.* one of the plagioclase (*q.v.*) minerals.

C

C (1) symbol for the chemical element carbon (*q.v.*); (2) Calorie (equals 1000 calories); (3) coulomb (*q.v.*); (4) cysteine (*q.v.*); (5) cytosine (*q.v.*); Simpson dominance index *q.v.*

C$_g$ Morisita's similarity index *q.v.*

C3 pathway carbon dioxide fixation in plants via the Calvin cycle (*q.v.*) to produce a three-carbon sugar. *cf.* C4 pathway.

C3 plant any plant in which carbon dioxide fixation is solely via the Calvin cycle (*q.v.*), as in most temperate plants. *cf.* C4 plant.

C4 pathway alternative pathway of carbon dioxide fixation present in many tropical plants in which CO_2 is incorporated into four-carbon compounds, first oxaloacetate, then malate and aspartate, which are transported to chloroplasts of bundle sheath cells where the CO_2 is released and enters the Calvin cycle, ensuring an adequate and continuous supply of CO_2 for photosynthesis under tropical conditions. *alt.* Hatch–Slack pathway.

C4 plant any of a diverse group of chiefly tropical plants adapted to high temperatures and low humidity that possess the alternative C4 pathway (*q.v.*) of carbon dioxide fixation in which various metabolic pathways concerned with photosynthesis are compartmented between mesophyll cells and bundle sheath cells in the leaf. *cf.* C3 plant.

^{12}C–^{13}C ratio *see* carbon isotope ratio.

Ca symbol for the chemical element calcium *q.v.*

caatinga *n.* thorn woodland, characterized by shrubs and low trees with small evergreen leaves, located in north-east Brazil.

cabbage root fly *Erioiscia brassicae*, an insect pest of cabbages. The larvae feed on roots and tunnel into stems; young plants are particularly at risk.

cacao *n.* plant (*Cacao theobroma*) whose beans provide the raw material for cocoa and chocolate manufacture.

Cactales, cacti *n.*, *n.plu.* an order of succulent dicots, found in arid and semi-arid regions, mainly in tropical America, and adapted to hot, dry conditions. The leaves are absent or much reduced and the fleshy stems often bear clusters of spines. Contains one family, the Cactaceae.

Cactoblastis cactorum a moth used in the control of prickly pear cactus *q.v.*

cadavericole *n.* animal that feeds on carrion. *alt.* carrion feeder.

caddis flies common name for the Trichoptera, an order of insects somewhat resembling moths, with weak flight and mouth parts adapted for licking, with aquatic larvae (caddis worms) that construct protective cases.

cadmium (Cd) *n.* metallic element (at. no. 48, r.a.m. 112.41), forming a silvery-bluish metal. It is found in nature together with other elements, especially zinc. It is a good absorber of neutrons and is used in the control rods of some nuclear reactors. It has numerous other industrial uses. Cadmium and many of its compounds are poisonous.

caecilians *see* Apoda.

Caenozoic Cenozoic *q.v.*

caesium (Cs) *n.* metallic element (at. no. 55, r.a.m. 132.91), forming a soft silvery metal that reacts vigorously with many non-metals and also with water. It is used in solar photoelectric cells and in the caesium atomic clocks that are used as international time standards. The radioactive isotopes ^{134}Cs and ^{137}Cs (half-life 33 years), are produced in nuclear explosions and are likely to be present in accidental radioactive discharges. The common non-radioactive isotope of caesium is ^{133}Cs. *alt.* cesium.

caespitose, cespitose *a.* (1) having low, closely matted stems; (2) growing densely in tufts.

caffeine *n.* 1,3,7-trimethylxanthine, a purine with a bitter taste, found in coffee, tea, maté and kola nuts, which is a stimulant of the central nervous system and a diuretic.

Cainozoic Cenozoic *q.v.*

cal calorie *q.v.*

calamine (1) smithsonite (*q.v.*); (2) hemimorphite *q.v.*

calcareous *a.* (1) composed chiefly of calcium carbonate (lime); (2) growing on limestone or chalky soil; (3) *pert.* limestone.

calcareous mudstone a homogeneous fine-grained sedimentary rock that is essentially calcite.

calcareous sponges sponges of the class Calcarea, with skeletons of one-, three- or four-rayed spicules composed chiefly of calcite (calcium carbonate).

calcicole *n.* plant that thrives in soil rich in lime or other calcium salts. *a.* calcicolous, *appl.* grassland.

calciferous *a.* containing or producing calcium salts.

calcification *n.* (1) deposition of calcium salts in tissue; (2) (*soil sci.*) the build-up of calcium carbonate ($CaCO_3$) in the B-horizon of soils, typically those of continental interiors with low rainfalls, e.g. Asian steppes.

calcifuge *n.* plant that thrives only in soils poor in lime and usually acid. *a.* calcifugous, *appl.* grassland.

calcimorphic soil any type of soil developed from highly calcareous parent material, e.g. rendzina.

calciphile calcicole *q.v.*

calcite *n.* $CaCO_3$, a crystalline form of calcium carbonate, the mineral that is the main constituent of limestone (including chalk) and marble. It is also found in the calcareous igneous rocks called carbonatites. It is used, after being quarried as limestone, as a building material, a raw material in cement manufacture, a flux in smelting and as a fertilizer. It is also used in some flue gas desulphurization (*q.v.*) processes. It is one of the constituents of mollusc shells

and the skeletons of calcareous sponges. *alt.* calcspar. *cf.* aragonite.

calcium (Ca) *n.* metallic element (at. no. 20, r.a.m. 40.08), forming a soft silvery-grey metal that tarnishes in air. It occurs in many rocks (e.g. limestone, chalk) as calcium carbonate and in seawater. It is an essential macronutrient for living organisms, where it is required for many cellular activities and also for the formation of shells, bones and teeth. *see also* calcium cycle.

calcium cycle the movement of calcium from inorganic sources in the soil and water first into plants and microorganisms, and then through the food chain, and its return to the inorganic environment.

calcium nitrate $Ca(NO_3)_2$, (*agric.*) an inorganic salt used as a nitrogenous fertilizer, especially in continental Europe.

calcrete caliche *q.v.*

calc-silicate rock a rock formed by the metamorphism of limestone that contains shaly or sandy impurities. It consists of calcite together with a selection of other minerals including calcium silicate minerals (e.g. diopside, wollastonite, grossular garnet).

calcspar calcite *q.v.*

caldera *n.* large circular depression (with a diameter greater than 2 km) formed when the depletion of a subterranean magma chamber (either through eruption or drainage of the magma) leads to the subsidence and collapse of the walls of a volcano. A caldera can also be formed when a particularly violent eruption removes the entire summit of a volcano.

caliche *n.* duricrust (*q.v.*) formed by the deposition of calcium carbonate ($CaCO_3$). *alt.* calcrete.

Caliciviridae, caliciviruses *n., n.plu.* family of icosahedral, single-stranded RNA viruses including vesicular exanthema of swine.

californium (Cf) *n.* a transuranic element [at. no. 98, r.a.m. (of its most stable known isotope) 251].

callose *a.* having hardened thickened areas on skin or bark.

callose *n.* amorphous polysaccharide of glucose, usually found on sieve plates in phloem but also in parenchyma cells after injury.

callunetum *n.* plant community dominated by the heather *Calluna vulgaris*.

callus *n.* (1) small hard outgrowth or swelling; (2) mass of hard tissue that forms over cut or damaged plant surface; (3) mass of undifferentiated cells that initially arises from plant cell or tissue in artificial culture.

calobiosis *n.* in social insects, when one species lives in the nest of another and at its expense.

calomel *n.* common name for mercury(I) chloride (Hg_2Cl_2), a highly poisonous fungicide used to control onion white rot and club root in brassicas. *alt.* mercurous chloride.

calorie *n.* a unit of heat quantity equal to 4.1855 J. It is the quantity of heat needed to warm 1 g of water through 1 °C (from 14.5 to 15.5 °C for the 15 °calorie). Symbol cal. Note, 1000 calories = 1 Calorie.

Calorie *n.* the kilocalorie, kilogram-calorie, kcalorie or large calorie equals 1000 calories. Symbol kcal or C.

calorific *a.* heat-producing.

calorific value (of a fuel) specific enthalpy of combustion *q.v.*

calorific value, gross *see* gross calorific value.

calorific value, net *see* net calorific value.

calorigenic *a.* promoting oxygen consumption and heat production.

calorimetry *n.* the measurement of heat, in animal physiology used to measure heat production and thus metabolic rate.

Calvin cycle the cycle of reactions in the stroma of chloroplasts in which ATP and NADPH produced during the light reaction of photosynthesis provide energy and reducing power for the incorporation of carbon dioxide into carbohydrate. The first reaction is that of ribulose-1,5-bisphosphate with carbon dioxide to form 3-phosphoglycerate. This is converted in several stages to reform ribulose-1,5-bisphosphate, producing in the process the three-carbon sugar glyceraldehyde-3-phosphate, which is the precursor of starch, amino acids, fatty acids and sucrose.

calving *n.* (*geol.*) the break-up of the outer edges of ice shelves and the detachment of huge blocks of ice (known as calves) into the sea.

Calycerales *n.* an order of herbaceous dicots comprising the family Calyceraceae.

calyx *n.* (1) the sepals collectively, forming the outer whorl of the flower, *see* Fig. 12, p. 157; (2) various structures resembling the calyx of a flower, as the cup-like body of crinoids.

CAM crassulacean acid metabolism *q.v.*

cambium *n.* layer of meristematic tissue forming a sheath around the main axis of roots and stems in plants and from which radial growth occurs, the cambium producing xylem from one face and phloem from the other. The bark of trees is also produced from a cambial layer.

Cambrian *a. pert.* or *appl.* geological period lasting from *ca.* 590 to 505 million years ago and during which many phyla of multicellular animals first arose.

Camelford incident, the event in which 20 tonnes of aluminium sulphate was accidently placed into the water supply mains of Camelford, Cornwall, UK, on 6 July 1988.

Camellia genus of shrubs and small trees that includes the tea plant, *C. sinensis*.

Campaign for Nuclear Disarmament (CND) a UK pressure group, working for the elimination of nuclear weapons.

Campanulales, Campanulatae *n.* an order of mainly herbaceous dicots, often with latex vessels, and including the families Campanulaceae (bellflower), Goodeniaceae and Lobeliaceae (lobelia).

can *n.* in a fission nuclear reactor, sealed container made of metal (stainless steel, Magnox or Zircaloy) for holding the fuel.

canalizing selection selection for phenotypic characters that is largely unaffected by environmental fluctuations and genetic variability.

can banks receptacles used for the deposition of empty food and beverage cans for recycling.

cancer *n.* malignant, ill-regulated proliferation of cells, causing either a solid tumour or other abnormal conditions and usually fatal if untreated. Cancer cells are abnormal in many ways esp. in their ability to multiply indefinitely, to invade underlying tissues and to migrate to other sites in the body and multiply there (metastasis). Most cancers arise as a result of an accumulation of mutations in the cell. *see also* carcinogen.

Cancer, Tropic of *see* Tropic of Cancer.

cancrinite *n.* $(Na,Ca)_7Al_6Si_6O_{24}(CO_3,SO_4, Cl)_{1.5-2}.1-5H_2O$, one of the feldspathoid minerals *q.v.*

candela (cd) *n.* the SI unit of luminous intensity. It is defined as the luminous intensity in the perpendicular direction of the black-body radiation from an area of 1/600 000 s m^2 at the temperature of freezing platinum under a pressure of 101 325 Pa.

candu reactor a type of fission nuclear reactor that utilizes Zircaloy-clad fuel pellets of natural uranium oxide. It uses heavy water as the coolant and moderator.

canids *n.plu.* members of the mammalian family Canidae, which includes the dogs, wolves, foxes, jackals and coyotes.

canine *a. pert.* to a dog, or to the genus *Canis.*

canine *n.* the tooth next to the incisors.

canker *n.* general term for any disease in a plant that causes decay of the bark and wood.

Cannabis sativa hemp, varieties of which are cultivated for fibre and for the drug cannabis (marijuana).

cannibalistic *a.* eating the flesh of one's own species.

canoids *n.plu.* mammals of the dog, hyaena, bear, panda and related families.

canopy *n.* the cover formed by the branches and leaves of trees in a wood or forest.

canopy closure stage in growth of a wood or forest in which the leaves of neighbouring trees or shrubs start to overlap to form a continuous canopy.

canopy cover the vertical projection of the crowns of trees and shrubs onto the ground, the amount of ground shaded by the leaves of trees and shrubs in a given habitat.

canyon *n.* deep, narrow river valley with steep sides. *alt.* gorge.

capillovirus group plant virus group composed of filamentous single-stranded RNA viruses, type member apple stem grooving virus.

capital *see* Earth capital.

capitulum *n.* flowerhead like that of a

dandelion or daisy, composed of numerous stalkless florets crowded together.

Capparales *n.* order of dicot herbs, shrubs, small trees and lianas, including the families Brassicaceae (Cruciferae) (mustard, etc.), Capparaceae (caper) and Resedaceae (mignonette).

Capricorn, Tropic of *see* Tropic of Capricorn.

caprification *n.* the pollination of flowers of fig trees by chalcid wasps.

Caprimulgiformes *n.* an order of birds including the nightjars.

Capsicum genus of the Solanaceae that includes the sweet peppers and chillies.

capsid *n.* (1) the external protein coat of a virus particle; (2) (*zool.*) common name for a bug of the family Capsidae.

capsule *n.* (1) a sac-like membrane enclosing an organ; (*bot.*) (2) any closed box-like vessel containing spores, seeds or fruits; (3) a one- or more celled, many-seeded dehiscent fruit; (4) (*bact.*) thick slime layer surrounding certain bacteria, composed of polysaccharides or, more rarely, polypeptides.

captafol *n.* fungicide used in the control of potato blight.

captan *n.* fungicide used, e.g., in the control of black spot on roses and scab on pears and apples.

captive breeding conservation practice in which threatened animal species are bred in captivity to increase their population sizes, with the ultimate aim of releasing stock into the wild.

Captorhinida, captorhinids *n., n.plu.* the earliest and most primitive order of extinct reptiles known, which evolved from the amphibians in the Late Carboniferous and is found until the end of the Triassic. *alt.* cotylosaurs.

capture–recapture method method of estimating population size by marking and releasing a sample of individuals, allowing them to mingle with the population, then taking another sample, and using the ratio of marked to unmarked animals in this sample to estimate total population size. This ratio is known as the Lincoln index.

carapace *n.* (1) bony plates beneath the horny shell of tortoises and other chelonians; (2)

chitinous covering in crustaceans starting behind the head and covering the whole or part of the trunk.

carbaryl *n.* non-persistent contact carbamate insecticide used, e.g., to kill pea moth on peas and earthworms in turf.

carbendazim *n.* systemic fungicide used in the control of diseases such as powdery mildew on fruit and eyespot on barley and wheat.

carbetamide *n.* foliar and soil-acting carbamate herbicide used to control weeds, e.g. wild oats, in field crops such as brassicas and legumes.

carbofuran *n.* systemic carbamate insecticide and nematicide used in the control of many soil pests.

carbohydrates *n.plu.* compounds of carbon, oxygen and hydrogen, of general formula $C_x(H_2O)_y$, including sugars (monosaccharides and disaccharides) and their derivatives, and polysaccharides such as starch and cellulose.

carbolic acid phenol *q.v.*

carbon (C) *n.* non-metallic element (at. no. 6, r.a.m. 12.01), found naturally as diamond and graphite. It occurs as various isotopes, of which ^{12}C is by far the most common. Carbon atoms (valency 4) can bond to each other and to other elements to form rings and chains, producing a vast variety of large carbon-based molecules. All life on Earth is based on carbon compounds (organic compounds), in which carbon is combined mainly with hydrogen, oxygen, nitrogen and phosphorus. *see also* radiocarbon.

carbon-14 *see* radiocarbon.

carbonaceous *a.* consisting of or containing carbon.

carbonaceous coal anthracite *q.v.*

carbonation *n.* type of chemical weathering involving the action of carbonic acid on rocks and soils, esp. important in shaping limestone landscapes.

carbon budget the amounts of carbon (as carbon compounds) in various reservoirs (e.g. ocean surface, ocean sediments, soil, atmosphere, land biota) within the global ecosystem or a specified part of it, and the flows between such reservoirs.

carbon cycle the various biological processes by which carbon from atmospheric carbon dioxide, CO_2, enters the biosphere by photosynthetic fixation of carbon into organic compounds, circulates within the biosphere as organic carbon, and is eventually returned to the atmosphere as CO_2, chiefly by respiration of living organisms, but also by burning of wood and fossil fuels.

carbon dating *see* radiocarbon.

carbon dioxide CO_2, a gas present in the troposphere at a concentration of about 362 p.p.m. (1993 data). Currently, its concentration is increasing at a rate of *ca.* 0.5% per year. This is a cause for concern as CO_2 is a major greenhouse gas (*q.v.*). The main sources of atmospheric CO_2 are biological respiration and comparable non-biological reactions, combustion and outgassing from the oceans. Its sinks are photosynthesis and dissolution in seawater. Human activity has enhanced the atmospheric concentration of CO_2 by the burning of fossil fuels and biomass. *see also* greenhouse effect.

carbon dioxide compensation concentration, carbon dioxide compensation point the ambient concentration of carbon dioxide at which, when light is not limiting, photosynthesis just compensates for respiration. At 25 °C and 21% O_2 this value is about 45 p.p.m. for C3 plants. *see also* compensation point.

carbon fixation (1) the pathway incorporating carbon dioxide into carbohydrates that occurs in the stroma of chloroplasts; (2) any reaction in which carbon dioxide is incorporated into organic compounds.

carbon, fixed (1) that carbon that has been the subject of photosynthesis; (2) in proximate analysis of fuels (*q.v.*), it is the difference between the total fuel and that attributable to ash, volatile matter and moisture (expressed as a percentage).

carbonic acid H_2CO_3, a very weak acid formed when carbon dioxide dissolves in water, e.g. atmospheric CO_2 in rain water, an agent of chemical weathering.

carbonicolous *a.* living on burnt soils or burnt wood.

carboniferous *a.* coal-bearing, *appl.* rocks.

Carboniferous *a. pert.* period of late Palaeozoic era, lasting from *ca.* 360 to 286 million years ago, and during which the Coal Measures were formed.

carbon isotope ratio $^{12}C/^{13}C$ ratio, a ratio of the proportion of ^{12}C to ^{13}C in a geological deposit, etc., that determines whether or not it has originated from living matter. The difference arises because enzymes involved in photosynthesis discriminate between the two isotopes, preferentially incorporating ^{13}C into living matter from carbon dioxide. The two isotopic forms of carbon exist in atmospheric carbon dioxide and other inorganic carbon compounds in a stable ratio.

carbonization n. (1) form of fossilization in which the organism, esp. with respect to plant material, is turned into coal or into a thin film of carbon within a rock; (2) coalification (q.v.); (3) the heating of solid fuel (esp. coal), in essentially anoxic conditions, generally with the aim of producing coke and coal-gas, or smokeless fuel. Coal tar is a byproduct of this process.

carbon monoxide CO, a colourless, odourless and toxic gas (at room temperature and pressure). It forms a stable compound with haemoglobin and is lethal in large amounts. Produced from natural sources and by the atmospheric oxidation of hydrocarbons and the incomplete combustion of fossil fuels (e.g. vehicle exhausts) and biomass. As well as its immediate toxic effect as a pollutant, elevated emissions of CO are likely to reduce tropospheric concentrations of the hydroxyl (q.v.) free radical (OH). This, in turn, will reduce the removal rates of many other pollutants, as reaction with the OH radical is a major sink for numerous contaminants. CO may also contribute to global warming as it and the CO_2 to which it is oxidized are greenhouse gases q.v.

carbon sink any part of the biosphere in which carbon is absorbed and immobilized faster than it is released, e.g. ocean sediments and tropical rain forests.

carbon source (1) any carbon-containing compound that can be utilized as a source of carbon by an organism; (2) in the carbon cycle, a source of carbon dioxide, such as respiration.

carbon tax a tax levied with the aim of decreasing the amount of carbon dioxide released into the atmosphere. Under such a regime, on the basis of useful energy available, gas would be taxed less than oil, and oil less than coal.

carbon tetrachloride (CTC) CCl_4, an industrial solvent, which has also been used in dry-cleaning. Prolonged exposure may lead to liver and kidney damage, and it is also a carcinogen, and is now replaced by other chemicals. It is a potential ozone-destroying chemical covered by the Montreal Protocol q.v.

carbonyl n. any metal complex of carbon monoxide.

carbonyl group the C=O moiety, found in aldehydes and ketones.

carbophenothion n. organophosphate pesticide used as an acaricide on citrus crops, an insecticide against wheat bulb fly (as a seed treatment) and as sheep dip.

carbosulfan n. systemic carbamate insecticide used in the control of pests of both sugar-beet and brassicas.

carboxin n. systemic fungicide principally used as a seed treatment, usually in combination with organomercurial compounds.

carboxyl group -COOH, a univalent radical that consists of a carbon atom to which is bonded both a hydroxyl group and, via a double bond, an oxygen atom.

carboxylic acid organic molecule that contains at least one carboxyl group q.v.

carcinogen n. any agent capable of causing cancer in humans or animals. a. carcinogenic.

carcinology n. the study of crustaceans.

Caribbean Floral Region Central American Floral Region q.v.

Caribbean karst type of limestone landscape that is flat and low-lying, characterized by numerous round lakes formed in sinkholes. This landscape is restricted to a few localities, e.g. the Yucatan Peninsula in southern Mexico.

carlavirus group plant virus group composed of rod-shaped single-stranded RNA viruses, type member, carnation latent virus.

carmovirus group plant virus group composed of isometric single-stranded RNA viruses, type member, carnation mottle virus.

carnallite n. $KMgCl_3.6H_2O$, an evaporite mineral that is exploited as a source of potassium compounds.

carnelian n. SiO_2, red to reddish-brown varieties of chalcedony (q.v.). cf. sard.

Carnivora *n.* order of flesh-eating mammals containing the suborders Fissipedia comprising the terrestrial carnivores and Pinnipedia, the aquatic carnivores (seals, walruses and sea-lions).

carnivore *n.* animal that feeds on other animals, esp. the flesh-eating mammals (Carnivora) such as dogs, cats, bears and seals.

carnivorous *a.* (1) flesh-eating; (2) (*bot.*) *appl.* certain plants that trap and digest insects and other small animals.

Carnot efficiency the maximum theoretically obtainable ratio (often expressed as a percentage) of the work output (W) of a heat engine (*q.v.*) to the heat absorbed from its heat source (Q). This may be calculated from a knowledge of the absolute temperature (*q.v.*) of both the heat source (T_{source}) and the heat sink (T_{sink}) (to which the waste heat is rejected) using the Heat Engine Rule. This states that the theoretical maximum value of $W/Q = (T_{source} - T_{sink})/T_{source}$. In practice, any heat engine is less efficient than the limit expressed by its Carnot efficiency.

carnotite *n.* $K_2(UO_2)_2(VO_4)_2.3H_2O$, a uranium ore mineral.

carotenoid *n.* any of a group of widely distributed orange, yellow, red or brown fat-soluble pigments, synthesized in plants, involved in photosynthesis as accessory pigments and also found in flowers and fruits, consisting of two groups: carotenes (e.g. beta-carotene) and xanthophylls (e.g. lutein, violaxanthin, fucoxanthin).

carpal *n.* a wrist bone.

carpel *n.* female reproductive structure in angiosperm flowers, consisting of an ovary containing one or more ovules, style and stigma, the carpels together making up the pistil or gynoecium. *see* Fig. 12, p. 157.

carr *n.* fen woodland, usually dominated by alder or willow.

carrier *n.* (1) individual infected with a transmissible pathogen, who does not suffer from the disease but can transmit the pathogen to others; (2) (*genet.*) individual heterozygous for a recessive allele, esp. allele responsible for a genetic disease, who shows no symptoms of disease but can pass the disease allele on to their offspring.

carrot fly *Psilia rosae*, an insect pest that mainly attacks carrots. In affected plants, the foliage turns red and dies and the roots are riddled with larvae.

carrying capacity (K) (1) the maximum number of individuals of a particular species that can be supported indefinitely by a given part of the environment; (2) the number of grazing animals a piece of land can support without deterioration; (3) the level of use an environment or resource can sustain without being destroyed or suffering an unacceptable deterioration.

cartilaginous fishes common name for fishes of the class Selachii, an individual having a cartilaginous skeleton, a spiral valve in the gut and no lungs or air bladder, which includes the sharks and their allies.

Caryoblastea *n.* phylum of protists containing one species, the giant multinucleate amoeba *Pelomyxa palustris*, which lacks mitochondria and other organelles characteristic of eukaryotic cells and whose cells divide without mitosis.

Caryophyllales *n.* an order of herbaceous dicots, rarely shrubs or trees, containing betalain pigments and including the families Amaranthaceae (amaranth), Caryophyllaceae (pink), Chenopodiaceae (goosefoot), Phytolaccaceae (pokeweed) and others.

caryopsis *n.* achene with pericarp and seed coat inseparably fused, as in grasses.

cash crops crops, e.g. coffee and cocoa, that are cultivated primarily for export to earn hard currency, often at the expense of growing subsistence food crops for local consumption.

cassava *n. Maniot esculenta*, tropical plant of the family Euphorbiaceae, whose starchy roots are used as food (tapioca) in parts of the world and that have to be processed before being turned into flour to remove toxic compounds.

cassiterite *n.* SnO_2, the principal tin ore mineral. *alt.* tinstone.

caste *n.* one of the distinct forms found among social insects, e.g. worker, drone, queen.

caste polyethism the division of labour between different castes in social insects.

casual *n.* plant that has been introduced but has not yet become established as a wild plant, although occurring uncultivated.

casual society a temporary and highly unstable group formed by individuals within a society, such as for play, feeding, etc.

Casuariformes *n.* an order of flightless birds including the cassowaries and emus.

Casuarinales *n.* an order of dicot trees or shrubs with whorls of branches and that comprises the family Casuarinaceae (she-oak).

catabolism *n.* the breaking down of complex molecules by living organisms with release of energy. *a.* catabolic.

catabolite *n.* any substance that is the product of catabolism.

catadromous *a.* (1) tending downward; (2) *appl.* fishes that migrate from fresh to salt water for spawning.

catalepsis *n.* a shamming dead reflex, as in spiders.

catalysis *n.* the acceleration (or more rarely the slowing down) of a reaction due to the presence of a substance (catalyst) that is itself not changed during the reaction.

catalyst *n.* a substance that alters the rate of a chemical reaction and that is still present, unchanged, after the reaction. The term catalyst is most commonly applied to substances that speed up reactions.

catalytic converter device fitted to the exhaust system of an internal combustion engine in order to diminish the concentration of specific pollutants within the waste gas stream. It operates by providing catalysts to facilitate reactions between the chemical species present. Most catalytic converters are designed for use in conjunction with petrol (i.e. gasoline) engines running on unleaded fuel. There are also those that diminish the levels of unburnt hydrocarbons and soot emitted by diesel engines. *see also* oxidation catalyst-based catalytic converter, reduction catalyst-based catalytic converter, three-way catalytic converter. *alt.* catalytic reactor.

catalytic reactor catalytic converter *q.v.*

cataplexis *n.* (1) condition of an animal feigning death; (2) maintenance of a postural reflex induced by restraint or shock.

catarobic *a. appl.* water containing organic matter that is decomposing slowly enough to avoid depletion of dissolved oxygen.

catarrhines *n.plu.* the Old World monkeys, the apes and humans.

catastrophe *n.* any sudden violent event, such as a volcanic eruption, flood or forest fire, that causes drastic changes to an ecosystem.

catastrophic speciation rapid evolution of new species resulting from sudden radical changes to the environment.

catch crop a quick-growing crop cultivated in the interim between two main crops.

catchment drainage basin *q.v.*

catena *n.* sequence of soil types, of similar age and from similar parent material, whose characteristics differ as a result of variations in drainage and relief, often observed on slopes. *alt.* hydrologic sequence, toposequence.

catenate *v.* (*chem.*) to form covalent bonds between atoms of the same element, thereby forming chains or rings. Carbon has a much greater propensity to catenate than any other element.

caterpillar *n.* fleshy thin-skinned larva, esp. of Lepidoptera, having segmented body, true legs and also prolegs on abdomen and no cerci.

catfishes *n.plu.* a group of mainly tropical, mainly freshwater bony fish (the order Siluriformes) often with long whisker-like barbels from which they take their name. (The marine fish *Anaricas*, commonly called the catfish in the UK, is not a member of this group.)

cation *n.* positively charged ion that moves towards the cathode or negative pole, e.g. K^+, Na^+, Ca^{2+}. *cf.* anion.

cation adsorption capacity cation exchange capacity *q.v.*

cation exchange the replacement of cations held on the surface of a solid which has a negative charge with cations in a solution in contact with that solid. This is an important process in soils, where the solids concerned are clays and humus and the solution is the soil solution. It also forms the basis of certain water purification treatments and water softening treatments. *see also* deionized water.

cation exchange capacity (CEC) quantification of the ability of a negatively charged solid to exchange cations held on its surface with cations in the surrounding solution. It is commonly expressed in milliequivalents per 100 g of solid (meq. per 100 g or meq. $100\ g^{-1}$) (where, in this context, an equivalent is a mole of charge).

More recently, it is expressed in centimoles of charge per kilogram of solid (cmol kg^{-1}). When *appl.* a soil, it indicates the extent to which the soil can absorb and hold exchangeable cations (e.g. Ca^{2+}, Mg^{2+}) on the surfaces of soil particles. The value gives an indication of whether the soil can hold enough nutrients long enough for them to be useful to plants. *alt.* base exchange capacity, cation adsorption capacity, total exchange capacity.

catkin *n.* inflorescence consisting of a hanging spike of small unisexual flowers interspersed with bracts, as in willows, poplars and the male flowers of hazel.

caudal *a.* (1) of or *pert.* a tail, e.g. caudal fin of fishes; (2) towards the tail end of the body.

Caudata Urodela *q.v.*

cauliflory *n.* condition of having flowers arising from axillary buds on the main stem or older branches, and thus appearing as if growing directly out of the branch. *a.* cauliflorous.

caulimovirus group plant virus group containing isometric double-stranded DNA viruses, type member cauliflower mosaic virus.

cavernicolous *a.* cave-dwelling.

cavitation *n.* bubble formation and implosion caused by pressure differences in swiftly flowing liquids. This form of hydraulic action may be involved in fluvial erosion, e.g. at the foot of waterfalls and in rapids.

cavity collapse *see* subsidence.

CBA cost–benefit analysis *q.v.*

CCAMLR Convention on the Conservation of Antarctic Marine Living Resources, an international agreement made in 1980.

CCW Countryside Council for Wales *q.v.*

cd candela *q.v.*

Cd symbol for the chemical element cadmium *q.v.*

cDNA complementary DNA *q.v.*

cDNA clone a DNA clone derived from a complementary DNA (cDNA) transcript of an mRNA.

Ce symbol for the chemical element cerium *q.v.*

CEC cation exchange capacity *q.v.*

Celastrales *n.* order of dicot trees, shrubs or vines, rarely herbs and including the families Aquifoliaceae (holly), Celastraceae (staff tree) and others.

celestine, celestite *n.* SrSO$_4$, a mineral used as a source of strontium.

cell *n.* (1) the basic structural building block of living organisms, consisting of protoplasm delimited by a cell membrane, and in plants, bacteria and fungi also surrounded by a non-living rigid cell wall. Some organisms consist of a single cell (bacteria, protozoans and some algae), others of cells of a few different types, and the more complex animals and plants of billions of cells of many different types. Bacterial (prokaryotic) cells have a relatively simple internal structure in which the DNA is not enclosed in a discrete nucleus and the cytoplasm is not differentiated into specialized organelles. Cells from all other living organisms (eukaryotic cells) typically are comprised of a nucleus enclosing the DNA, which is organized into chromosomes, and a cytoplasm containing a cytoskeleton of fine protein tubules and filaments and specialized membrane-bounded organelles such as mitochondria and (in photosynthetic plants and algae) chloroplasts. *see* Fig. 5; (2) a small cavity or hollow; (3) space between veins of insect wing.

cell culture cells growing outside the organism in nutrient medium in the laboratory.

cell division splitting of a cell into two complete new cells: by binary fission in bacteria and other prokaryotes, and by division of both nucleus and cytoplasm in eukaryotic cells. *see also* meiosis, mitosis.

cell membrane plasma membrane *q.v.*

cell sap the fluid in vacuoles in plant cells, being a solution of small organic molecules and ions in water.

cellular *a. pert.* or consisting of cells.

cellular slime moulds a group of simple eukaryotic soil microorganisms (Acrasiomycetes) typified by *Dictyostelium discoideum*. Free-living unicellular amoebae (myxamoebae) aggregate to form a multicellular fruiting body differentiated into stalk and sporehead (a compound sporangium). *cf.* acellular slime moulds, plasmodial slime moulds.

cellulolytic *a.* able to degrade cellulose.

cellulose *n.* a linear polysaccharide made up of glucose residues joined by beta-1,4 linkages, the most abundant organic

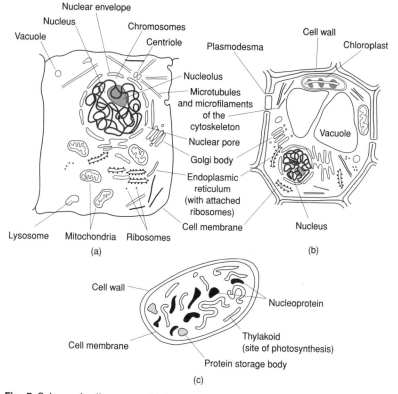

Fig. 5 Schematic diagrams of (a) a generalized animal cell, (b) a generalized plant cell and (c) a prokaryotic cell (the example shown is a cyanobacterium).

compound in the biosphere, comprising the bulk of plant and algal cell walls, where it occurs as cellulose microfibrils, and also found in certain tunicates.

cell wall non-living semi-rigid structure surrounding the plasma membrane of algal, plant, fungal and bacterial cells. In green plants and green algae, it is predominantly composed of cellulose fibrils, and also contains many other polysaccharides. Many fungal cell walls are composed largely of chitin. Those of bacteria are composed of peptidoglycans.

Celsius scale metric scale for measuring temperature, in which the freezing point and boiling point of water are set at 0 and

100 °C, respectively. *alt.* centigrade scale. *cf.* Fahrenheit scale.

cementation *n.* (*geol.*) one of the processes involved in the formation of sedimentary rocks. In cementation, minerals (e.g. calcite) dissolved in the pore water are deposited on the surfaces of sediment grains, causing them to stick together.

Cenozoic *a. pert.* geological era following the Mesozoic, commencing *ca.* 65 million years ago. In some usages it is followed by the Quaternary era whereas in others it is considered to last until the present and is subdivided into the Tertiary and Quaternary epochs (periods). *alt.* Cainozoic, Coenozoic.

censer mechanism method of seed dispersal by which seeds are shaken out of fruit by wind action.

census *n.* a complete counting of a whole population with respect to the variable under study. *cf.* sampling.

centi- (c) SI prefix indicating that the unit it applies to is multiplied by 10^{-2}, e.g. 1 centilitre = 1×10^{-2} l (one-hundredth of a litre).

centigrade scale Celsius scale *q.v.*

centimetre–gram–second *see* c.g.s. units.

centipedes *n.plu.* common name for the Chilopoda, a group of arthropods having numerous and similar body segments each with one pair of walking legs, except the first segment, which bears a pair of poison claws.

Central American Floral Region part of the Neotropical Realm comprising Central America, the southern tips of California and Florida, the islands of the Caribbean and the northern part of South America. *alt.* Caribbean Floral Region.

central dogma the principle that the transfer of genetic information from DNA to RNA by transcription and from RNA to protein by translation is irreversible, now modified to take into account the transfer of information from RNA to DNA by reverse transcription carried out by some viruses.

central eruption the term for any volcanic eruption in which magma is released through a single pipe-like vent. *cf.* fissure eruption.

central receiver system (CRS) a type of high-temperature solar thermal technology where many solar collectors (known as heliostats *q.v.*) reflect and concentrate the Sun's rays onto an elevated central receiver. The temperature of the fluid in the receiver (often water) becomes very high (~1700 °C). The heated fluid may then be used to drive a turbine for the production of electricity.

centres of diversity regions identified originally by the Russian botanist N. Vavilov in which large numbers of different strains or races of a particular cultivated plant occur. In some cases, this diversity may indicate the geographical place of origin of the crop plant.

centripetal drainage *see* drainage pattern.

centrosome *n.* organelle in plant and animal cells from which the spindle develops in mitosis and meiosis. The centrosome is situated near the nucleus and in animal cells contains the centrioles.

Centrospermae Caryophyllales *q.v.*

-cephalic, cephalo- word elements derived from Gk *kephalē*, head.

cephalic index one hundred times maximum breadth of skull divided by maximum length.

Cephalochordata, cephalochordates *n., n.plu.* subphylum of small cigar-shaped aquatic chordates commonly called lancelets and including amphioxus (*Branchiostoma*). They have a persistent notochord in the adult and a large sac-like pharynx with gill slits for food collection and respiration.

Cephalopoda, cephalopods *n., n.plu.* class of marine molluscs including octopus, squid and nautiloids, an individual having a well-developed head, large brain and eyes superficially resembling those of vertebrates in structure. The head is surrounded by prehensile tentacles and the animals can move very rapidly by jet propulsion, squirting water out of the large mantle cavity, which communicates with the exterior by a siphon or funnel.

cephalosporins *n.plu.* beta-lactam antibiotics produced by streptomycetes and strains of *Cephalosporium*.

cephalothorax *n.* (1) body region formed by fusion of head and thorax in crustaceans; (2) in arachnids, the prosoma.

cerargyrite chlorargyrite *q.v.*

ceratomorphs *n.plu.* suborder of mammals that contains the rhinoceroses and tapirs.

cercaria *n.* a heart-shaped, tailed, larval stage of a trematode (fluke) produced in the snail host, which is released from the snail, sometimes then encysting, and subsequently infects a vertebrate host. *plu.* cercariae.

Cercidiphyllales *n.* order of dicot trees comprising the family Cercidiphyllaceae, with a single genus *Cercidiphyllum*.

cercopithecoid *a. appl.* monkeys of the superfamily Cercopithecoidea, the Old World monkeys (e.g. baboons), which together with apes and man are the only primates with a fully opposable thumb.

cereal *n.* any plant of the family Gramineae (the grasses) whose seeds are used as food.

cereal aphids various species of greenfly that attack cereal crops. Examples are bird cherry aphid, grass aphid and grain aphid. Cereal aphids are instrumental in the spread of viral diseases such as barley yellow dwarf virus.

cereal cyst nematode, cereal root eelworm *Heterodera avenae*, a parasitic roundworm that feeds on the roots of barley, wheat and esp. oats.

cerium (Ce) *n.* a lanthanide element (at. no. 58, r.a.m. 140.12), a soft steel-grey metal when in elemental form. It occurs in several rare minerals, e.g. monazite (*q.v.*), and is used, in the form of cerium dioxide (CeO_2), in glass polishing.

cerussite *n.* $PbCO_3$, a lead ore mineral.

cervids *n.plu.* members of the mammalian family Cervidae, the deer.

cesium *alt.* spelling of caesium *q.v.*

Cestoda, cestodes *n.*, *n.plu.* tapeworms, a class of platyhelminths (flatworms) that are internal parasites of humans and animals. An individual has a long flattened ribbon-like body lacking a gut or mouth, usually divided into many identical segments (proglottids), and attaches itself to the wall of the gut through an attachment organ at the anterior end. Reproduction is via mature proglottids that detach to form a new reproductive unit. The complex life cycle involves two or more hosts.

Cetacea, cetaceans *n.*, *n.plu.* order of wholly aquatic placental mammals, including the whales and dolphins, having bodies highly adapted for swimming, with the fore limbs modified as flippers and the hind limbs often hardly developed and invisible externally.

cetology *n.* study of whales and dolphins.

Cf symbol for the chemical element californium *q.v.*

CFCs chlorofluorocarbons *q.v.*

c.g.s. units units that conform to the metric centimetre–gram–second system. Replaced in technical and scientific usage by SI units *q.v.*

chabazite *n.* $(Ca,Na_2)Al_2Si_4O_{12}.6H_2O$, one of the zeolite minerals *q.v.*

Chaetognatha, chaetognaths *n.*, *n.plu.* arrow worms, a phylum of small predatory marine coelomate animals found in swarms in plankton, having an elongated transparent body with head, trunk and tail.

chaetopods *n.plu.* the annelid worms that bear chaetae (bristles): the Polychaeta and Oligochaeta.

chaff *n.* (*agric.*) the husks of cereal grains separated by winnowing or threshing.

chain behaviour a series of actions, each being induced by the antecedent action and being an integral part of a unified performance.

chain reaction any nuclear or chemical reaction that is self-sustaining because its products are involved in the continuance of the reaction.

chalcedony *n.* SiO_2, (1) a group of silica minerals (*q.v.*) characterized by a compact matrix of minute crystals of quartz with incorporated pores of submicroscopic dimensions; (2) uniformly coloured varieties of chalcedony, *cf.* agate.

chalcocite *n.* a widespread and valuable copper ore mineral of composition Cu_2S. *alt.* chalcosine, copper glance.

chalcogens *n.plu.* the elements oxygen (O), sulphur (S), selenium (Se), tellurium (Te) and polonium (Po), i.e. Group 16 of the periodic table. *see* Appendix 2.

chalcopyrite *n.* the most common copper mineral. It is an important copper ore mineral and has the composition $CuFeS_2$. *alt.* copper pyrites.

chalcosine chalcocite *q.v.*

chalicotheres *n.plu.* extinct family of ungulates that had clawed feet.

chalk *n.* a pure limestone formed by the build-up of the skeletal parts of pelagic (floating) organisms. It is essentially calcite and commonly contains flint and marcasite.

chalybite siderite *q.v.*

chamaephyte, chamaeophyte *n.* perennial woody plant having overwintering buds at or just above ground level.

chaparral *n.* type of vegetation found in areas with a mediterranean climate, dominated by evergreen shrubs with broad, hard leaves.

Chapman mechanism a sequence of naturally occurring reactions that cause ozone (O_3) to be created and destroyed in the stratosphere throughout the hours of daylight. *see* Fig. 6.

$$O_2 + h\nu \ (\lambda < 240 \, nm) \rightarrow 2O \ \} \text{ Ozone}$$
$$O + O_2 + M \rightarrow O_3 + M \ \ \ \ \text{production}$$

$$O_3 + h\nu \ (\lambda < 325 \, nm) \rightarrow O_2 + O \ \} \text{ Ozone}$$
$$O + O_3 \rightarrow 2O_2 \ \ \ \ \ \ \ \ \ \ \ \ \text{destruction}$$

where $h\nu$ represents a photon and M is a third body, which may be a molecule or a solid surface. It is needed to absorb energy given out during new bond formation, so stabilizing the product.

Fig. 6 The Chapman mechanism.

characins *n.plu.* group of tropical freshwater bony fish (the Characinoidei) that includes the tetras and piranhas. Characterized by complex teeth bearing five to seven cusps, strong jaws and a scaly body.

character *n.* (*genet.*) any genetically determined feature or trait.

character convergence the condition in which two species interact in such a way that one or both converges to the other. *alt.* social mimicry.

character displacement, character divergence the condition in which two newly evolved species with overlapping ranges interact in such a way that both diverge physically further from each other to minimize competition.

characteristic species plant species that are almost always found within a particular association.

character state (*genet.*) any one of the range of values or expressions of a particular character.

Charadriiformes *n.* a large diverse group of shore and wading birds including the gulls and terns, auks, oystercatchers, plovers, curlews, snipes and waders.

charcoal *n.* black, carbonaceous residue produced by heating organic materials, esp. wood, in essentially anoxic conditions.

charcoal, activated *see* activated carbon.

Charophyceae *n.* a class of algae of the Chlorophyta that are encrusted with calcium carbonate so are commonly called stoneworts. They have a filamentous or thalloid body bearing lateral branches in whorls. Sometimes considered as a division, the Charophyta.

chasmogamous *a. appl.* flowers that open when mature to ensure fertilization. *cf.* cleistogamous.

chasmophyte *n.* a plant that grows in rock crevices.

chats *n.plu.* small potato tubers.

cheating *n.* in animal behaviour, any behaviour (e.g. exaggeration of body size) intended to mislead a rival or potential mate into an incorrect estimate of the animal's strength or genetic fitness.

chela *n.* the large claw borne on certain limbs of arthropods, as the pincers of a crab. *plu.* chelae.

chelate *a.* (*biol.*) claw-like or pincerlike.

chelate *n.* (*chem.*) a complex in which the Lewis acid (*q.v.*) (usually a metal ion) forms more than one dative bond (*q.v.*) with a molecular Lewis base (*q.v.*) via more than one donor atom (*q.v.*) (often N, O or S). In doing this, the Lewis acid forms part of one or more ring structures, each of which usually contains five or six atoms/ions. *v.* to form a chelate.

chelating agent compound that can react with a metal ion and form a stable compound, called a chelate, with it.

Chelicerata, chelicerates *n., n.plu.* a class or subphylum of arthropods (*q.v.*) an individual having a body generally in two parts, a prosoma bearing the paired chelicerae (poison jaws) and sensory pedipalps, and a posterior opisthosoma bearing usually four pairs of walking legs. The Chelicerata include the arachnids (e.g. spiders, ticks, mites, scorpions), pycnogonids (sea spiders), horseshoe crabs and extinct eurypterids. *see also* Arachnida, Merostomata, Pycnogonidae.

Chelonia, chelonians *n., n.plu.* turtles and tortoises, an order of reptiles where the individual has a short broad trunk protected by a dorsal shield (carapace) and ventral shield (plastron) composed of bony plates overlain by epidermal plates of tortoiseshell.

cheluviation *n.* the combined processes of chelation and eluviation, i.e. the formation of metal chelates (involving organic ligands) and their subsequent leaching down through the soil profile.

chemical bond a strong force of attraction that holds atoms together in a molecule or crystal. *see also* covalent bonds, ionic bonds, non-covalent bonds.

chemical defences unpalatable or toxic chemicals, such as astringent tannins and toxic alkaloids, produced by plants in their tissues to deter herbivores.

chemical ecology the study of the secondary chemical compounds produced by plants and animals (e.g. antibiotics, alkaloids, unpalatable and/or toxic compounds, etc.) and their effect on the interaction of the organism with other animals and plants in the ecosystem, especially in respect of plants' defences against herbivores and animals' defences against predators.

chemical element *see* element.

chemical equations representations of chemical reactions that take the form

$$a\text{A} + b\text{B} + \ldots \to w\text{W} + x\text{X} + \ldots$$

or

$$a\text{A} + b\text{B} + \ldots \rightleftharpoons w\text{W} + x\text{X} + \ldots$$

where A and B are the chemical formulae of the reactants, W and X are the chemical formulae of the products and a, b, w and x are the numbers of molecules, atoms or ions of each required to balance the equation (coefficients of stoichiometry). The arrow ($\to$) is used to show a reaction that essentially goes to completion (i.e. the reaction proceeds until one or all of the reactants are consumed), whereas the symbol $\rightleftharpoons$ denotes that the reaction reaches a dynamic equilibrium before any one of the reactants is fully depleted. For example, the equation representing methane (CH_4) burning in oxygen (O_2) to form carbon dioxide (CO_2) and water (H_2O) would be:

$$1CH_4 + 2O_2 \to 1CO_2 + 2H_2O$$

or, more usually

$$CH_4 + 2O_2 \to CO_2 + 2H_2O$$

as coefficients of stoichiometry equal to one are not normally shown. The dissociation of ethanoic acid (CH_3COOH) in water would be represented thus:

$$CH_3COOH + H_2O \rightleftharpoons CH_3COO^- + H_3O^+$$

chemical fertilizers inorganic fertilizers *q.v.*

chemical formula a representation of a chemical species (*q.v.*). It takes the form of the elemental symbols of the elements present, each followed by a subscripted number to represent the stoichiometry of the species (if no number is written, then the number one is assumed). The overall charge on the species is superscripted after the last elemental symbol. For example, water, made up of hydrogen (H) and oxygen (O) in a 2 : 1 stoichiometry with no overall charge, is written as H_2O. The sulphate ion, made up of sulphur and oxygen in a 1 : 4 ratio with an overall charge of -2, is given the formula SO_4^{2-}. State symbols are sometimes placed in parentheses (usually subscripted) after a chemical formula to show the physical state that the species is in. The state symbols in common use are: s for solid, l for liquid, g for gas, aq for aqueous solution and, occasionally, solv for dissolved in a solvent, org for dissolved in an organic solvent or c for condensed phase (i.e. liquid or solid). Hence liquid water would be represented as $H_2O_{(l)}$. *plu.* chemical formulae.

chemical fossils supposed chemical traces of life, such as alkanes and porphyrins, found in rocks older than the earliest true fossil-bearing rocks.

chemical kinetics the study of the rate at which chemical reactions proceed.

chemical oxygen demand (COD) a chemical test for the degree of organic pollution of water that measures the amount of oxygen taken up from a sample of water by the organic matter in the sample, expressed as parts per million of oxygen taken up from a solution of boiling potassium dichromate in 2 h.

chemical sedimentary rocks non-clastic rocks formed by the precipitation of dissolved minerals from seawater or fresh water, generally by evaporation. e.g. rock salt.

chemical speciation the collection of chemical species (*q.v.*) in which a given element is found within a given environment. *alt.* speciation.

chemical species any atom, ion or molecule with identifiable chemical properties. *alt.* species.

chemical weathering disintegration of rocks *in situ* by chemical processes, e.g. hydrolysis, oxidation, hydration and carbonation. This type of weathering involves the formation of new minerals.

chemiluminescence *n.* light production during a chemical reaction at ordinary temperature, such as bioluminescence *q.v.*

chemisorption *see* adsorption.

chemoattractant *n.* any chemical that attracts cells or organisms to move towards it. *cf.* chemorepellant.

chemoautotroph *n.* any organism using inorganic sources of carbon, nitrogen, etc., as starting materials for biosynthesis, and an inorganic chemical energy source. *a.* chemoautotrophic.

chemoheterotroph *see* chemotroph, heterotroph.

chemokinesis *n.* a movement in response to the intensity of a chemical stimulus, including that of scent.

chemonasty *n.* a nastic movement in response to diffuse or indirect chemical stimuli.

chemo-organotroph chemoheterotroph *q.v.*

chemorepellant *n.* any chemical that causes cells or organisms to move away from it. *cf.* chemoattractant.

chemosensory *a.* (1) sensitive to chemical stimuli; (2) *pert.* sensing of chemical stimuli.

chemosynthesis *n.* the oxidation of inorganic compounds as the energy source for biosynthesis, used in some bacteria. *a.* chemosynthetic.

chemotaxis *n.* reaction of motile cells or microorganisms to chemical stimuli by moving towards or away from the source of the chemical. *a.* chemotactic.

chemotroph *n.* any organism obtaining energy by taking in and oxidizing chemical compounds, either inorganic chemical sources of energy as chemoautotrophs, or by the breakdown of complex organic compounds (food) as chemoheterotrophs. *a.* chemotrophic. *cf.* phototroph.

chemotropism *n.* curvature of plant or plant organ in response to chemical stimuli.

chenier plain a coastal plain consisting of a series of linear sand ridges (cheniers) separated by muddy marsh areas, formed approximately parallel to the coast. Such plains are found, e.g., in Australia and New Zealand.

cheniers *see* chenier plain.

Chenopodiales Caryophylales *q.v.*

Chernobyl nuclear accident the worst accident in the history of nuclear power. It occurred at the V. I. Lenin nuclear power plant near Chernobyl in the Ukraine on 26 April 1986. The failure to adhere to reactor safety procedures during an unauthorized experiment at low power, combined with serious design faults in the reactor itself, led to a catastrophic explosion and near-meltdown of the reactor core. The release of radioactivity into the atmosphere caused widespread contamination in the then Soviet Union and over northern Europe, affecting thousands of square kilometres.

chernozem *n.* black soil, formed under continental climatic conditions and characteristic of subhumid to temperate grasslands. *alt.* black earth.

chersophilous *a.* thriving on dry waste land.

chert *n.* SiO_2, an opaque chalcedony (*q.v.*) that is bedded and massive, fractures in a conchoidal fashion and is generally dull grey-black in colour.

chessylite azurite *q.v.*

chestnut soils dark-brown soils of semi-arid steppe lands, fertile under adequate rainfall or when irrigated.

chickpea *n. Cicer arietinum*, a seed legume (family Leguminosae) grown for food and fodder.

Chile saltpetre nitratine *q.v.*

Chilopoda, chilopods *n., n.plu.* in some classifications, a class of arthropods comprising the centipedes, which have numerous and similar body segments each with one pair of walking legs, except the first segment, which bears a pair of poison claws. Considered as a subclass or order of class Myriapoda in some classifications.

china clay kaolin *q.v.*

Chinook *see* föhn.

chionophyte *n.* a snow-loving plant.

chironomid *a. appl.* midges of the family Chironomidae. Their aquatic larvae are used as indicator species in assessing biotic indices for freshwater habitats. Abundance of the larva of *ironomus riparius* (a bloodworm) only indicates heavy organic pollution.

Chiroptera, chiropterans *n.*, *n.plu.* order of placental mammals including the small mainly insectivorous bats (microchiroptera) and the fruit-eating flying foxes (megachiroptera), having their fore limbs modified for flight supporting a wing membrane stretched between limbs, and their bodies and hind limbs reduced.

chiropterophilous *a.* pollinated by the agency of bats.

chisel teeth chisel-shaped incisors of rodents.

chitin *n.* long-chain polymer of *N*-acetyl-glucosamine units, the chief polysaccharide in fungal cell walls, and also found in the exoskeletons of arthropods.

chitons *n.plu.* common name for some members of the Polyplacophora, a class of marine molluscs with elongated bodies bearing a shell of calcareous plates and a muscular foot on which they crawl about.

chlamydiae *n.plu.* a group of prokaryotic microorganisms, obligate intracellular parasites, responsible for a variety of human and animal diseases.

chloanthite *n.* a mineral that is essentially nickel arsenide, but may contain cobalt. Its composition may be expressed as $(Ni,Co)As_{3-x}$. It is an ore mineral for both nickel and cobalt and is closely related to skutterudite (*q.v.*) and smaltite *q.v.*

chloracne *n.* a persistent acne-like skin rash caused in humans by exposure to certain chlorinated organic compounds, including polychlorinated biphenyls *q.v.*

chlor-alkali process electrolysis of brine to produce chlorine (Cl_2) and sodium hydroxide (NaOH). The use of mercury electrodes in this industrial process has been responsible for the contamination of the environment by this metal. Since 1970, better housekeeping and the introduction of mercury-free techniques based on diaphragm cells has significantly decreased this problem.

chloramphenicol *n.* antibiotic produced by the actinomycete *Streptomyces venezuelae*, which blocks translation in bacteria and mitochondria.

chlorapatite *see* apatite.

chlorargyrite *n.* AgCl, a secondary silver mineral. *alt.* cerargyrite, horn silver.

chlorbromuron *n.* contact and soil-acting herbicide used to control annual weeds

in vegetable crops, e.g. potatoes and carrots.

chlordane *n.* a persistent chlorinated hydrocarbon pesticide that has been used to kill earthworms in turf. Its use is now banned or restricted in many countries.

chlorenchyma *n.* plant tissue containing chlorophyll.

chlorfenprop-methyl *n.* translocated herbicide used to control wild oats in cereal crops.

chlorfenvinphos *n.* contact organophosphate insecticide used, e.g., in the control of frit fly, wheat bulb fly and carrot fly.

chloridazon *n.* contact and soil-acting herbicide used to control weeds in field crops such as sugar-beet. *alt.* pyrazon.

chlorinated hydrocarbons hydrocarbons (*q.v.*) in which one or more of the hydrogen atoms have been replaced by atoms of chlorine. They include many pesticides and industrial pollutants such as polychlorinated biphenyls (PCBs). They are persistent and toxic, entering food chains and often accumulating in organisms higher up the chain. Examples are DDT, PCBs, dieldrin, lindane and heptachlor. The use of some of these compounds is no longer permitted in many countries. Gaseous chlorinated hydrocarbons entering the atmosphere may reach the stratosphere, where they undergo photolysis, liberating chlorine atoms. The chlorine atoms enter into catalytic cycles, destroying some of the ozone that shields the biosphere from harmful ultraviolet light. For this reason, there is concern about anthropogenic emissions of compounds such as chlorofluorocarbons (CFCs), carbon tetrachloride (CCl_4) and 1,1,1-trichloroethane (CCl_3CH_3, methyl chloroform). *see also* Montreal Protocol, stratospheric ozone.

chlorine (Cl) *n.* halogen element (at. no. 17, r.a.m. 35.45), a highly toxic, irritating, yellow-green gas (Cl_2) in the free state. It occurs naturally as sodium chloride (NaCl) in seawater and in rock salt and as chlorides of other metals. Widely used as a disinfectant for water supplies and swimming pools, and has many industrial uses. *see also* chlorinated hydrocarbons, chlorofluorocarbons.

chlorite *n.* $(Mg,Fe,Al)_6(Si,Al)_4O_{10}(OH)_8$. A hydrous phyllosilicate (*q.v.*) mineral with a composition reminiscent of the mica minerals (*q.v.*). It is a clay mineral *q.v.*

chlormequat *see* plant growth regulators.

chloroethene vinyl chloride *q.v.*

chlorofluorocarbon (CFC) *n.* any of a group of compounds of carbon, chlorine and fluorine. They may also contain hydrogen, in which case they may be considered as HCFCs (hydrochlorofluorocarbons). CFCs are chemically inert and are used as the working fluid in refrigerators, as solvents, as blowing agents in the manufacture of foamed plastics and as propellants in aerosol sprays. They are entirely man-made chemicals that, since their introduction in the 1930s, have been present in the atmosphere in trace amounts. The two most abundant CFCs are CCl_3F (CFC-11) and CCl_2F_2 (CFC-12). Atmospheric concentrations of these gases are currently 0.280 and 0.484 p.p.b., respectively, each rising at *ca.* 4% per year (1992 data). CFCs have long atmospheric lifetimes of tens to hundreds of years and thus pass across the tropopause into the stratosphere. Within the stratosphere, CFCs are slowly degraded by photolysis, yielding chlorine atoms. These atoms then enter into catalytic cycles, destroying some of the ozone that shields the biosphere from harmful ultraviolet light. Because of their deleterious effect on stratospheric ozone, the use of CFCs is now restricted in many countries. *see also* chlorofluorocarbon nomenclature, halons, hydrochlorofluorocarbons, Montreal Protocol, stratospheric ozone.

chlorofluorocarbon nomenclature a system by which chlorofluorocarbons (CFCs) are given code names, each CFC being represented by a symbol, e.g. CFC-11. To find the composition of a given CFC, add 90 to the numerical part of its symbol. The three digits of the resulting number stand for the numbers of carbon, hydrogen and fluorine atoms in the molecule, respectively. The remaining atoms are chlorine. For example, for CFC-11: 11 + 90 = 101, i.e. one carbon, zero hydrogens and one fluorine atom. This carbon atom has only one of its four valencies satisfied (by the fluorine) and so CFC-11 must also contain three chlorine atoms. CFC-11 therefore has the formula CCl_3F. Some CFCs,

such as CFC-142, have several isomers (*q.v.*). These are differentiated by the letters a, b, c, etc. (e.g. CFC-142b represents the isomer CH_3CF_2Cl).

Chlorophyceae Chlorophyta *q.v.*

chlorophyll *n.* principal photosynthetic pigment of green plants and algae, consisting of a porphyrin (tetrapyrrole) ring with magnesium at the centre and esterified to a long-chain aliphatic alcohol (phytol), different chlorophylls having different side chains. It is located in the thylakoid membranes of chloroplasts, where it traps light energy, absorbing mainly in red and violet-blue regions of the spectrum, chemically distinct forms having different absorption maxima. Chlorophylls *a* and *b* are found in higher plants, chlorophylls *c* and *d* in algae. *see also* bacteriochlorophyll.

Chlorophyta *n.* the green algae, in some modern classifications a division of the kingdom Protista, in more traditional classifications regarded as a division of the kingdom Plantae. Mostly freshwater, but some marine, they include unicellular and multicellular groups, have chlorophylls and carotenoids similar to those of vascular plants and appear green, store food as starch and have cellulose cell walls.

chloroplast *n.* semi-autonomous organelle found in the cytoplasm of cells of all green plants and photosynthetic algae, and in which the reactions of photosynthesis take place. It contains the green pigment chlorophyll and other pigments involved with photosynthesis and also contains DNA that specifies rRNAs, tRNAs and some chloroplast proteins. *alt.* plastid. *see also* photosynthesis.

chloroplast pigments chlorophylls, carotene and xanthophyll.

chlorosis *n.* abnormal condition in plants characterized by lack of green pigment owing to lack of light, or to magnesium or iron deficiency, or to genetic deficiencies in chlorophyll synthesis. *a.* chlorotic.

chloroxuron *n.* soil-acting herbicide used to control germinating weeds in chrysanthemums and strawberries.

chlorpyrifos *n.* contact organophosphate insecticide used for a broad spectrum of fruit, cereal and vegetable root pests, and also in sheep dip.

chlortoluron *n.* soil-acting, translocated herbicide used to control weeds, e.g. black-grass, in cereal crops such as barley and winter wheat.

Choanichthyes Sarcopterygii *q.v.*

chocolate spot a disease affecting bean crops, caused by the fungi *Botrytis cinerea* and *B. fabae*.

cholera *n.* acute intestinal disease caused by the bacterium *Vibrio cholerae*, spread by contaminated drinking water or food.

Chondrichthyes *n.* class of fishes known from the Devonian to the present day, commonly known as the cartilaginous fishes, having a cartilaginous skeleton, spiral valve in the gut and no lungs or air bladder, and including the sharks, dogfishes, skates and rays. *cf.* Osteichthyes.

chondrodite *n.* $Mg(OH,F)_2.2Mg_2SiO_4$, a mineral of the humite series *q.v.*

chondrophores *n.plu.* colonial hydrozoans of the order Chondrophora, showing a degree of division of labour and cooperation between zooids approaching that of siphonophores.

Chondrostei, chondrosteans *n.*, *n.plu.* group of primitive actinopterygian bony fishes including the extant bichirs of the Nile, which have lungs, and paddlefishes and sturgeons, as well as many extinct groups. The bony skeleton has largely been substituted by cartilage. An individual usually has a spiral valve in the gut and retains the spiracle: paddlefishes and sturgeons have heterocercal tails. *see also* palaeoniscids.

Chordata, chordates *n.*, *n.plu.* a phylum of coelomate animals having a notochord and gill clefts in the pharynx at some point in their life history, and a hollow nerve cord running dorsally with the anterior end usually dilated to form a brain. The chordates include the vertebrates, in which the notochord does not persist beyond the embryonic stage, the cephalochordates (e.g. amphioxus) and the urochordates (e.g. sea squirts).

chore *n.* an area showing a unity of geographical or environmental conditions.

C-horizon inorganic layer of unconsolidated, chemically and physically weathered parent material that may be present beneath the B-horizon in soils. *see* Fig. 28, p. 385. *see also* regolith, soil horizons.

chorology *n.* study of the geographical distribution of plants and animals.

choronomic *a.* external, *appl.* influences of geographical or regional environment.

CHP plants combined heat and power plants *q.v.*

chromatid *n.* one of the two copies of a replicated chromosome, while still associated with each other, as in the early stages of mitosis or meiosis.

chromatography *n.* separation of compounds from a mixture on the basis of their individual affinities for, and rates of migration with, a solvent running through a solid support such as paper or a column packed with adsorbing material. In paper chromatography, for example, the individual components of the mixture separate out to form discrete spots or bands at particular positions on the paper. *see also* affinity chromatography, gel-filtration chromatography, partition chromatography.

chromatophore *n.* (1) cell organelle containing pigment; (2) a pigment cell or group of cells in the skin of an animal, which, under the control of hormones or the nervous system, can be altered in shape and thus cause an alteration in the colour of the animal.

chromite *n.* $FeCr_2O_4$, the major chromium ore mineral. Chromite is used to make refractory bricks (bricks resistant to very high temperatures).

chromium (Cr) *n.* metallic element (at. no. 24, r.a.m. 52), forming a hard white metal. Its principal ore is chromite ($FeCr_2O_4$). Chromium has numerous uses both as a non-corroding surface coating, in the manufacture of stainless steel, and, as its compounds, in pigments.

chromosomal incompatibility failure to interbreed due to differences in chromosome composition.

chromosome *n.* one of the small, rod-shaped, deeply staining bodies that become visible under the light microscope in the eukaryotic cell at mitosis and meiosis and that carry the genes. In the interphase stage of the cell cycle, before DNA replication has taken place, each chromosome consists of a single very long molecule of DNA associated with histones and other proteins to form chromatin. Each species has a characteristic set of chromosomes, which can often be distinguished under the microscope from those of other species in shape, size and number. The single DNA molecule in prokaryotic cells is also often called a chromosome. *see also* DNA, gene.

chromosome complement karyotype *q.v.*

chromosome number the constant number of chromosomes per cell that is characteristic of a particular species: usually specified as haploid chromosome number (*n* or *x*) or diploid chromosome number (2*n*).

chromosome pairs the pairs of homologous chromosomes in a diploid cell, one derived from the maternal, one from the paternal parent, which become associated at meiosis.

chromosome races races of a species that differ in number of chromosomes, or in number of sets of chromosomes.

chronic *a. appl.* medical conditions or diseases, of slow progress and long duration. *cf.* acute.

chronological *a.* (1) arranged in order of time; (2) *pert.* chronology.

chronology *n.* (1) a time-related list of dates, events and/or objects in order of occurrence; (2) the establishment of lists of this type.

chronosequence *n.* sequence of related soils that differ primarily as a result of age.

chrysalis *n.* the pupa of insects with complete metamorphosis, enclosed in a protective case, which is sometimes itself called the chrysalis.

Chrysophyta, **chrysophytes** *n.*, *n.plu.* in some classifications (1) a phylum of protists containing the golden-yellow algae (class Chrysophyceae or division Chrysophyta of the plant kingdom in other classifications). Mainly freshwater, the marine groups are, however, an important constituent of the marine nanoplankton. They are unicellular or colonial. At some stage of the life cycle the cells bear two unequal undulipodia. Chrysophytes contain large amounts of carotenoid pigments, reserves of oil and the polysaccharide chrysolaminarin. Some Chrysophyta possess siliceous or organic skeletons or scales; (2) a division of the plant kingdom containing the golden-brown algae (class Chrysophyceae), the diatoms (class Bacillariophyceae), and the yellow-green algae (class Xanthophyceae). *see* Bacillariophyta, Xanthophyta.

chrysoprase *n.* SiO_2, apple-green variety of chalcedony *q.v.*

chrysotile *n.* a variety of the mineral serpentine (*q.v.*). It is a form of asbestos (*q.v.*) when highly fibrous, known as white asbestos.

chylophyllous *a.* with fleshy leaves, *appl.* succulent plants adapted to dry conditions, such as stonecrops and sedums.

Chytridiomycota, Chytridiomycetes, chytrids *n.*, *n.*, *n.plu.* phylum of saprobic or parasitic protists living in soil or fresh water and formerly classified as fungi. They are unicellular or coenocytic, with cell walls containing chitin. Sexual reproduction is oogamous and they also reproduce asexually by motile zoospores. Sperm and zoospores all bear a single posterior whiplash flagellum, distinguishing them from the hyphochytrids.

Ci curie *q.v.*

Ciconiiformes *n.* an order of wading birds with long necks, legs and bill, including herons, bitterns, storks and flamingoes.

Ciliata, Ciliatea, Ciliophora, ciliates *n.*, *n.*, *n.*, *n.plu.* class of free-living and sessile protozoans of complex cellular structure (e.g. *Paramecium*, *Stentor* and *Tetrahymena*), bearing cilia often in rows on the surface or grouped into compound structures. Common in marine and fresh water, and in rumen of cattle and other ruminants, where they digest cellulose. Almost all ciliates possess two nuclei, a macronucleus that directs vegetative growth, and a diploid micronucleus involved in sexual reproduction.

cinder cone a small volcanic landform composed entirely of tephra *q.v.*

cinnabar *n.* HgS, the most important mercury ore mineral.

circadian rhythm metabolic or behavioural rhythm with a period of about 24 h.

circannual rhythm a rhythm or cycle of behaviour with a period of approximately one year.

circaseptan rhythm biological rhythm with a period of around 7–8 days.

circum- prefix derived from L. *circum*, around, surrounding.

circumboreal *a.* surrounding the northern (boreal) regions.

circumpolar *a. appl.* (1) flora and fauna of polar regions; (2) distributions of plants and animals in northerly parts of the northern hemisphere that extend through Asia, Europe and North America.

cirque *n.* an armchair-shaped hollow, with a steep backwall (known as the headwall) and a bowl-shaped rock basin, usually situated at the head of a deep valley. These landforms are produced by a combination of glacial erosion and periglacial action. *alt.* corrie (Scotland, UK), cwm (Wales, UK), coombe, coomb, combe or comb (esp. northern England, UK).

cirque glaciers glaciers, larger than niche glaciers (*q.v.*), that occur in armchair-shaped hollows in mountains. *alt.* corrie glaciers, cwm glaciers.

Cirripedia, cirriped(e)s *n.*, *n.plu.* a subclass of aquatic crustaceans, commonly called barnacles, which as adults are stalked or sessile sedentary animals with the head and abdomen reduced and the body enclosed in a shell of calcareous plates.

cirrocumulus *n.* high-altitude cloud composed of an interconnected series of white, fleecy, rounded masses, earning it the description 'mackerel sky'.

cirrostratus *n.* high-altitude cloud forming a whitish, veiled cover.

cirrus *n.* white, high-altitude cloud, wispy in appearance. Feathery filaments (known as 'mares' tails') may be present, indicating strong winds in the upper atmosphere.

Cistales Violales *q.v.*

cistern epiphyte an epiphyte lacking roots and gathering water between leaf bases.

CITES Convention on International Trade in Endangered Species *q.v.*

citric acid cycle tricarboxylic acid cycle *q.v.*

citrine *n.* SiO_2, yellow transparent quartz.

Citrus genus of small trees and shrubs that includes the oranges, lemons, limes, etc.

city climate urban climate *q.v.*

CJD Creutzfeldt–Jakob disease *q.v.*

Cl symbol for the chemical element chlorine *q.v.*

clade *n.* a branch of a phylogenetic tree containing the set of all organisms descended from a particular common ancestor that is not an ancestor of any non-member of the group.

cladistic *a. pert.* similarities between species or other taxa due to recent origin from a common ancestor.

cladistics *n.* method of classification of living organisms that makes use of lines of descent only, rather than phenotypic similarities, to deduce evolutionary relationships and that groups organisms strictly on the relative recency of common ancestry. Cladistic methods of classification only permit taxa in which all the members share a common ancestor which is also a member of the taxon and that include all the descendants of that common ancestor.

cladodont *a.* having or *appl.* teeth with prominent central and small lateral cusps.

cladogenesis *n.* (1) branching of evolutionary lineages so as to produce new types; (2) evolutionary change as a result of multiplication of species at any one time and their subsequent evolution along different lines. *cf.* anagenesis.

cladogram *n.* tree-like diagram showing the evolutionary descent of any group of organisms. Cladograms are constructed strictly according to the relative times of divergence between different groups, as deduced, for example, from molecular data. *cf.* phenogram. *see* Fig. 7.

clan phratry *q.v.*

clarifier *n.* mechanical device employed to extract solid material from water, used, e.g., in the treatment of effluent from paper mills.

class *n.* taxonomic group into which a phylum or a division is divided, and that is itself divided into orders. *see* Appendices 5–9.

classical smog London-type smog *q.v.*

classification *n.* (*biol.*) the arrangement of living organisms into groups on the basis of observed similarities and differences, modern classifications of plants and animals attempting wherever possible to reflect degrees of evolutionary relatedness. The smallest group in classification is usually the species, although subspecies, races and varieties (in cultivated plants) below

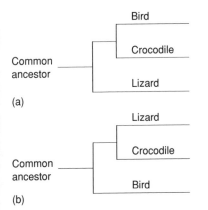

(a)

(b)

Fig. 7 Cladogram and phenogram: alternative ways of grouping crocodiles, lizards and birds. The cladogram (a) groups these animals on the basis of their most recent common ancestor, the phenogram (b) groups them on the basis of phenotypic resemblances.

the level of the species are often recognized. Species are grouped into genera, genera into families, families into orders, orders into classes, classes into phyla (for animals) or divisions (for plants), and phyla and divisions into kingdoms. There are also intermediate categories such as superfamilies and infraclasses. *see* Appendices 5–9.

clastic rocks sedimentary rocks formed from fragments derived, as a consequence of weathering and erosion, from pre-existing rocks. The vast majority of sedimentary rocks are of this type. They may be classified on the basis of sediment size into rudaceous rocks (coarse sediments, e.g. breccias), arenaceous rocks (medium-sized sediments, e.g. sandstones and arenites) and argillaceous rocks (fine sediments, e.g. shales and siltstones). *alt.* mechanically formed sedimentary rocks.

clay *n.* (*soil sci.*) (1) that fraction of the mineral component of soil that is made up of particles with diameters no bigger than 2×10^{-6} m (i.e. 0.002 mm). Mineralogically, this portion of the soil is generally dominated by weathering products including the clay minerals (*q.v.*) and/or hydrated

oxides of iron(III), aluminium, titanium and/or manganese. However, it may contain grains of quartz and sometimes mica (typically these grains are larger than 5×10^{-7} m, i.e. 0.0005 mm in diameter); (2) a soil textural class (*see* texture) into which fall soils with a fine earth (*q.v.*) that contain a large proportion of clay-sized particles. Typically, soils of this class have high moisture- and nutrient-holding capacities but tend to be poorly drained and slow to warm in the spring; (3) (*geol.*) a soft and uncompacted sedimentary rock that is made up of grains of less than 0.004 mm in diameter. *cf.* mudstone, shale.

clay–humus complexes combinations of clay particles and humus found in soil, notable for their adsorption and retention of plant nutrients. *alt.* adsorption complexes.

clay minerals a group of hydrous aluminosilicate minerals that are produced either by weathering or by hydrothermal alteration. With the exception of the relatively rare sepiolite–attapulgite series (*q.v.*) and related minerals, they have a layer-lattice structure. They are composed of sheets of oxide ions, together with sheets of hydroxide ions and/or a mixture of oxide and hydroxide ions, bound together by cations that occupy either octahedral or tetrahedral sites between these sheets. Octahedral sites are typically occupied by Al^{3+} or Mg^{2+}, while tetrahedral sites are occupied by Si^{4+}, sometimes replaced by Al^{3+}. The resulting octahedral (alumina) and tetrahedral (silica) layers are then coupled by shared oxygens (part of an oxide/hydroxide layer) to produce the laminae that make up the overall clay structure. Clay minerals may be classified as 1 : 1 clay minerals (*q.v.*) or 2 : 1 clay minerals (*q.v.*) depending on the composition of the laminae, and also include minerals containing intercalated units from different named clay minerals, which are called mixed layer or interstratified clay minerals. Such interstratification may be either regular (as in chlorite *q.v.*) or irregular (as in illite–montmorillonite). *alt.* silicate clays.

1 : 1 clay minerals, 1 : 1 group of clay minerals, 1 : 1-type clay minerals a group of clay minerals (*q.v.*), typified by kaolinite (*q.v.*), in which each lamina has one tetrahedral layer coupled to one octahedral layer, producing laminae with a sheet of oxygen atoms on the side occupied by the tetrahedral layer and a sheet of hydroxyl groups on the other side. The laminae are hydrogen bonded to one another through these groups.

2 : 1 clay minerals, 2 : 1 group of clay minerals, 2 : 1-type clay minerals a group of clay minerals (*q.v.*) in which each lamina consists of two tetrahedral layers bound together by one octahedral layer (hence 2 : 1), producing laminae with a sheet of oxygen atoms on each side, which therefore cannot directly hydrogen bond to one another. In 2 : 1-type expanding clay minerals [exemplified by montmorillonite (*q.v.* and vermiculite (*q.v.*)], the laminae are held together by hydrogen-bonded water molecules and exchangeable cations. In contrast, the laminae in 2 : 1-type non-expanding clay minerals (e.g. illite) are more tightly bound to each other by potassium ions.

claypan *n.* hardpan formed by the build-up of clay particles in the B-horizon.

claystone mudstone *q.v.*

clear-cutting, clear-felling *n.* the forestry practice of cutting down all the trees in a given area at the same time. *cf.* selective cutting.

cleavage *n.* (*geol.*) (1) a rock structure generated by deformation during dynamic metamorphism that causes the rock to exhibit anisotropic (*q.v.*) tendencies to split; (2) attribute of mineral crystals that break along one or more planes that are related to the crystal structure and are parallel to potential faces of the crystal; (3) (*biol.*) phase of early embryonic development in which the fertilized egg is rapidly split into many small cells.

cleistogamic, cleistogamous *a. appl.* flowers that never open up when mature, and are self-pollinated. Cleistogamous flowers are often small and inconspicuous. *cf.* chasmogamous.

cleptobiosis *see* kleptobiosis.

cleptoparasitism *see* kleptoparasitism.

climacteric *n.* (1) a critical phase, or period of changes, in living organisms; *appl.* (2) change associated with menopause or with recession of male function; (3) (*bot.*) phase of increased respiratory activity at ripening of fruit.

climate *n.* long-term view of the weather patterns of a particular location, generally defined in terms of either its average weather conditions and/or the frequency of occurrence of a particular atmospheric feature, such as thunderstorms.

climate change, climatic change a distinct and long-term change from one type of climate to another (i.e. excluding short-term fluctuations in weather conditions), such as the changes from non-glacial to ice age conditions that have occurred in the past. In contemporary environmental discussion the term often refers specifically to the changes in climate that are foreseen as a result of global warming (*q.v.*). *see also* greenhouse effect, Intergovernmental Panel on Climate Change, stratospheric ozone.

climatic *a.* relating to climate.

climatic climax regional plant climax (*q.v.*) community developed and maintained by climatic factors.

climatology *n.* scientific study of climates.

climatype *n.* a biotype resulting from selection in particular climatic conditions.

climax *n.* the mature or stabilized stage in a successional series of plant communities, in which the dominant species are completely adapted to the prevailing environmental conditions. In the absence of human interference or climatic change, a climax community will tend to remain stable over long periods of time. *a.* climactic.

climbing dune an irregular dune that forms on the windward side of a large topographic feature such as a cliff or an escarpment.

climosequence *n.* series of related soils that differ mainly because of climatic variations during their formation.

cline *n.* a graded series of different forms of the same species, usually distributed along a spatial dimension.

clinohumite *n.* $Mg(OH,F)_2.4Mg_2SiO_4$, a mineral of the humite series *q.v.*

clinopyroxenes *n.plu.* pyroxene minerals (*q.v.*) that crystallize in the monoclinic system. *cf.* orthopyroxenes.

clinozoisite *n.* $Ca_2Al_3Si_3O_{12}(OH)$, a mineral widespread in metamorphic rocks of medium to low grade.

clints *see* limestone pavement.

clisere *n.* succession of plant communities that results from a changing climate.

clod *n.* compact lump of soil formed by tillage.

clonal *a. pert.* a clone.

clonal dispersal the vegetative spread of an organism by outgrowth of new individuals without their actual detachment from the parent.

clone *n.* a group of genetically identical individuals or cells derived from a single cell or single individual by repeated asexual divisions, *see also* DNA clone.

clone *v.* to produce a set of identical organisms, individual cells or DNA molecules from a single starting cell or molecule.

cloning *see* clone (*q.v.*), DNA cloning *q.v.*

clonotype *n.* a specimen of an asexually propagated part of a type specimen (i.e. holotype).

Clophen a trade name under which polychlorinated biphenyls (PCBs) were sold.

closed canopy forest where the crowns of the trees shade over 20% of the ground area. *cf.* open canopy.

closed community ecological community into which further colonization is prevented as all niches are occupied.

closed forest forest where the crowns of the trees touch and produce a closed canopy for all or part of the year.

closed system a term used in thermodynamic analysis to denote a system (*q.v.*) that can exchange energy but not matter with its surroundings.

closterovirus group plant virus group containing very long thread-like single-stranded RNA viruses, with two subgroups: (1) aphid-transmitted viruses, type member beet yellows virus; (2) no known vectors, type member apple chlorotic leafspot virus.

clostridia *n.plu.* bacteria of the genus *Clostridium*, strictly anaerobic, spore-forming rods widely distributed in soil, some species of which produce powerful toxins and that include *Cl. tetani*, the causal agent of tetanus, *Cl. botulinum*, whose toxin formed in contaminated food causes botulism, and *Cl. histolyticum*, the cause of gas gangrene.

cloud *n.* visible mass of minute ice crystals or water droplets suspended in the atmosphere. Clouds are categorized according to their altitude and their general shape, structure and vertical extent. For low-altitude clouds, *see* cumulonimbus, cumulus, nimbostratus, stratocumulus, stratus. For medium-altitude clouds, *see* altocumulus, altostratus. For high-altitude clouds, *see* cirrocumulus, cirrostratus, cirrus.

cloud seeding method whereby substances, e.g. silver iodide or solid carbon dioxide, are deliberately introduced into the atmosphere to provide artificial hygroscopic nuclei that act as foci for water condensation and so induce precipitation.

club moss common name for a member of the genus *Lycopodium* of the division Lycophyta (*q.v.*), a group of seedless vascular plants. Their leaves are arranged in tight whorls on aerial stems. Sporangia arise either from the leaf axils or on separate club-like structures called strobili.

Club of Rome, the multinational study group set up in 1968 in order to examine the human predicament within the context of finite resources and to evaluate the impact of current policies on a wide range of issues (e.g. unemployment, pollution and urban planning) and to suggest alternative policies.

club root a disease that affects brassicas (*q.v.*), caused by the soil-borne fungus *Plasmodiophora brassicae*. Infected plants have stunted growth and swollen roots. *alt.* finger and toe.

clumped distribution the spatial arrangement of organisms in a community when they are closer together than they would be if arranged randomly or equally dispersed.

clutch *n.* number of eggs laid by a female at one time.

Cm symbol for the chemical element curium *q.v.*

CND Campaign for Nuclear Disarmament *q.v.*

Cnidaria *n.* phylum of aquatic, mostly marine, invertebrate animals containing corals, sea fans and sea anemones (class Anthozoa), the hydroids and milleporine corals (Hydrozoa) and the jellyfishes (Scyphozoa). They include both colonial and solitary forms. Individuals are generally radially symmetrical, with only one opening (mouth) to the gut and a simple two-layered body with a primitive nerve net between the two layers. Cnidaria have hydroid (polyp) and/or medusa forms, and bear stinging cells (cnidoblasts) on the tentacles fringing the mouth. With the phylum Ctenophora, the Cnidaria form the large grouping known as the coelenterates.

Cnidospora *n.* in some classifications a subphylum of parasitic protozoans, e.g. *Myxobolus*, which have a multinucleate sporing stage and that cause disease in fish and some other animals.

Co symbol for the chemical element cobalt *q.v.*

coacervate *n.* inorganic colloidal particle, e.g. clay, on which organic molecules have been adsorbed. Such particles may have been important in prebiotic evolution to bring together organic molecules in sufficient concentration for life to evolve.

coaction *n.* the reciprocal activity of organisms within a community.

coadaptation *n.* the correlated variation and adaptation displayed in two mutually dependent organs, organisms, etc.

coadapted *a. appl.* a set of genes all involved in effecting a complex process.

coagulation *n.* (1) curdling or clotting, *appl.* milk and blood; (2) the change from a liquid to a viscous or solid state by chemical reaction.

coal ball a more-or-less spherical aggregate of petrified plant structures found in certain coal measures.

coalescence process (*clim.*) process during which larger water droplets grow at the expense of smaller ones, in clouds with above freezing temperatures, until droplets are large enough to fall as rain. This process is responsible for the formation of most rain in tropical latitudes. *cf.* ice-crystal process.

coal-gas *n.* a gas, formerly widely used as a fuel, formed when coal is heated in essentially anoxic conditions, generally at *ca.* 1300 °C. On a volume-by-volume basis, it is typically made up of 50% hydrogen, 30% methane, 8% carbon monoxide, 4% hydrocarbons and 8% nitrogen, oxygen and carbon dioxide. *alt.* town gas.

coal gasification process whereby coal is converted into a gaseous fuel, through its reaction with steam and air or oxygen. *see also* syngas.

coalification *n.* progressive conversion of organic compounds within ancient deposits of partially decomposed vegetable matter (e.g. peat) to carbon during the formation of coal. *alt.* carbonization.

coal liquefaction process whereby coal is converted, physically or chemically, into a liquid fuel.

Coal Measures series of rock strata that contain coal seams deemed economically viable to exploit.

coal series successive stages of lignite, then sub-bituminous coal, then bituminous coal and finally anthracite formed during the progressive burial of ancient deposits of partially decomposed vegetable matter (e.g. peat) under accumulating layers of sediments. With each successive stage, the carbon content, density and hardness of the material increase, while the moisture and volatile substance contents decrease.

coal tar dark-coloured tar produced by the distillation of bituminous coal, used, e.g., in the production of pesticides and drugs.

coancestry coefficient of kinship *q.v.*

coarse-grained *see* grain size.

coarse-textured soil a soil with a high proportion of coarse particles. Such soils are generally free-draining and quick to warm but often have low moisture- and nutrient-holding capacities.

coastal wetlands land that is flooded with salt water for at least part of the year, such as salt marshes and mangrove swamps.

coastal zone the zone that extends from the high-tide mark on land to the edge of the continental shelf, comprising the flora and fauna of the beach and rocks and of the relatively warm, nutrient-rich shallow waters over the continental shelf.

cobalt (Co) *n.* magnetic metallic element (at. no. 27, r.a.m. 58.93), forming a hard silvery-grey metal. It occurs in combination with sulphur and arsenic. Commonly used in an alloy with other metals, as in steel, and in pigments. The isotope ^{60}Co is radioactive and is used in cancer therapy. An essential micronutrient for living organisms, it is a prosthetic group for some enzymes and a constituent of the vitamin cobalamin (vitamin B_{12}).

cobaltite *n.* (Co,Fe)AsS, a cobalt ore mineral of relatively minor abundance.

coca *n.* the dried leaves of the coca plant *Erythroxylon coca*, which contains cocaine.

cocaine *n.* an alkaloid obtained from coca leaves, which has been used as a local anaesthetic, is taken as a stimulant and can result in addiction.

cocarcinogen *n.* an agent that assists a carcinogen in the development and spread of cancer but is not itself carcinogenic. *alt.* promoter.

coccids *n.plu.* the scale insects, minute bugs with winged males, scale-like females that are attached to the infested plant, and young that suck the sap.

coccolith *n.* thin calcareous plate formed on an organic base, which forms a cellular covering in golden algae of the Haptophyta *q.v.*

coccolithophorids *n.plu.* resting form of some species of motile golden algae, having calcareous plates (coccoliths) covering the cells, and possessing a haptoneme, which may be used to anchor the cell to a surface. Such algae, which have a chrysophyte-like planktonic life phase, are sometimes classified as a separate phylum of the Protista, the Haptophyta or Prymnesiophyta.

cockpit karst a type of karst (*q.v.*) terrain found in the humid tropics. It consists of irregular or star-shaped closed depressions (known in Jamaica as cockpits), often separated by residual rounded hills (known as cones or kegels). *alt.* cone karst, kegelkarst.

cockroaches *n.plu.* common name for many members of the Dictyoptera *q.v.*

cocoon *n.* (1) the protective case of many insect larval forms before they become pupae; (2) silky or other covering formed by many animals for their eggs.

Cocos nucifera the coconut palm, a tropical tree of considerable economic importance.

COD chemical oxygen demand *q.v.*

codeine *n.* an alkaloid found with morphine in opium and having effects similar to but weaker than morphine.

codominance *n.* property shown by alleles which, when present in a heterozygous state, produce a different phenotype from that produced by either allele in the homozygous state.

codominant *a. appl.* (1) (*genet.*) alleles showing codominance (*q.v.*); (2) (*bot.*) two

species being equally dominant in climax vegetation.

coefficient of community measurement of the degree of resemblance of two plant communities, based upon their species compositions. It may be calculated in various ways using the ratios of number of common species to the total number of species in each. *alt.* similarity index. *see* Gleason's index, Jaccard index, Kulezinski index, Morisita's similarity index, Simpson's index of floristic resemblance, Sørensen similarity index.

coefficient of consanguinity coefficient of kinship *q.v.*

coefficient of genetic relatedness coefficient of relationship *q.v.*

coefficient of inbreeding *see* inbreeding coefficient.

coefficient of kinship (*f***)** of two individuals, the probability that two gametes taken at random, one from each individual, carry alleles at a given locus that are identical by virtue of common descent.

coefficient of relationship (*r***)** probability that a gene in one individual will be identical by virtue of common descent to a gene in a particular relative, e.g. for monozygotic twins $r = 1$, for parents and offspring $r = 0.5$, for full siblings $r = 0.5$, for grandparents and grandchildren $r = 0.25$, etc. r also gives the fraction of genes identical by common descent between two individuals.

coefficient of selection a measure of the strength of natural selection, calculated as the proportional reduction in contribution of one genotype to the gametes compared with a standard genotype, and that may have any value from zero to one.

coefficients of stoichiometry *see* stoichiometry, coefficients of.

coelacanth *see* Crossopterygii.

coelenterates *n.plu.* the animal phyla Cnidaria (corals, sea anemones, jellyfish and hydroids) and Ctenophora (sea combs or sea gooseberries) collectively. They have radial symmetry, a single body cavity (coelenteron) opening in a mouth, and a simple body wall of endoderm (gastrodermis) and ectoderm (epidermis) separated by usually non-cellular gelatinous mesogloea, and include both solitary and colonial (e.g. corals) forms. Contractile musculo-epithelial cells and a simple nerve net are present.

coelomates *n.plu.* animals possessing a true coelom, comprising the phyla Mollusca, Annelida, Arthropoda, Phoronida, Bryozoa, Brachiopoda, Echinodermata, Chaetognatha, Hemichordata and Chordata.

coenobium *n.* colony of unicellular organisms having a definite form and organization, which behaves as an individual and reproduces to give daughter coenobia. *alt.* cenobium.

coenocyte *n.* fungal or algal tissue in which constituent protoplasts are not separated by cell walls. *a.* coenocytic, *alt.* aseptate.

coenogamodeme *n.* a unit made up of all interbreeding subunits that can, under specified conditions, exchange genes.

coenopopulation *n.* an aggregate of individuals in an assemblage of plants living in a particular locality.

coenosis *n.* random assemblage of organisms with similar ecological preferences. *cf.* community.

coenosite *n.* an organism habitually sharing food with another.

coenospecies *n.* a group of taxonomic units, such as species, ecospecies or varieties, which can intercross to form hybrids that are sometimes fertile, the group being equivalent to a subgenus or superspecies.

coevolution *n.* (1) evolution of two species in relation to each other, such as a predator and its prey; (2) the evolution of two identical genes together so that they are both maintained in the original functional form and do not markedly diverge.

cofactor *n.* any non-protein substance required by a protein for its biological activity, such as prosthetic groups and, esp. for enzymes, other compounds such as NAD, NADP, flavin nucleotides.

Coffea genus of small shrubs that includes *C. robusta* and *C. arabica*, whose seeds (coffee beans) are used to make coffee.

co-generation plants combined heat and power plants *q.v.*

cognition *n.* those higher mental processes in humans and animals, such as the formation of associations, concept formation and insight, whose existence can only be inferred and not directly observed.

cohort *n.* a group of individuals of the same age in a population.

cohort life table life table giving data for just one particular cohort within the population.

coinage metals a collective term for the chemical elements copper (Cu), silver (Ag) and gold (Au), i.e. Group 11 of the periodic table. *see* Appendix 2.

coir compost processed fibrous material from the husks of coconuts, used as a substitute for horticultural peat.

coition, coitus *n.* sexual intercourse, copulation.

coke *n.* a solid formed when coal is heated in essentially anoxic conditions, generally at *ca.* 1300 °C. It is used as a fuel, and in iron smelting within blast furnaces.

col *n.* gap that bisects a range of mountains or hills, formed, e.g., by glacial erosion or river capture. *alt.* saddle.

cold-based glacier a glacier whose basal ice layer is below its pressure melting point (*q.v.*). This type of glacier occurs in both temperate and polar regions.

cold-blooded *a.* a colloquial term *appl.* animals whose source of body heat is primarily external. The terms ectothermic (*q.v.*) or poikilothermic (*q.v.*) should be used to describe such animals.

cold deserts deserts characterized by cold winters, such as the Gobi Desert in Asia and the Californian Desert.

cold front borderline between an advancing cold air mass undercutting a warm air mass, found to the rear of the warm sector in a depression.

cold occlusion type of occluded front (*q.v.*) where the overtaking cold front is colder than the air beyond the warm front.

colemanite *n.* a mineral of formula CaB_3O_4-$(OH)_3.H_2O$.

Coleoptera *n.* a very large order of insects, commonly called beetles, having complete metamorphosis, with the fore wings modified as hard wing-cases (elytra) and covering membranous hind wings that may be reduced or absent. They include some serious plant pests, attacking both standing crops and stored grain.

coliform *n.* any of a group of colon bacteria typified by *Escherichia coli*. Their presence is used as a standard indicator of faecal contamination of water.

Coliiformes *n.* an order of birds including the colies or mousebirds.

collagen *n.* one or all of a family of fibrous proteins found in all animals in extracellular matrices, forming insoluble fibres of high tensile strength, and that contains the unusual amino acids hydroxyproline and hydroxylysine. It is rich in glycine but lacks cysteine and tryptophan, and has an unusually regular amino acid sequence.

collapse dolines *see* doline.

collectivization *n.* process that took place in the early part of this century in the then USSR whereby peasant farmers were forced to join their small farms into larger-scale collective farms.

Collembola *n.* order of insects containing the springtails, small wingless insects with only six segments in the abdomen and with two long projections from the abdomen that spring out from a folded position to make the animal jump. Common in soil and leaf litter.

colloid *n.* (1) a substance composed of two homogeneous parts or phases, one of which is dispersed in the other; (2) substance of high molecular weight that does not readily diffuse through a semipermeable membrane.

collophane *see* apatite.

colonial *a. appl.* organisms that live together in large numbers, esp. where the individual organisms form part of a larger structure, as in some algae and in soft corals, reef-building corals, gorgonians and other anthozoans, and in the siphonophores.

colonization *n.* (1) the invasion of a new habitat by a species; (2) the establishment of a bacterial flora in intestine, etc.

colonizer *n.* organism that is among the first to establish itself or regrow in a new habitat, esp. an organism that is adapted to this role. Colonizers are usually robust, undemanding species able to disperse over a wide area. *alt.* early-successional species, pioneer species.

colony *n.* (1) a group of individuals of the same species living together in close proximity. The colony may be an aggregate structure with organic connections between the individual members (e.g. in sponges, corals) or, as in the social insects (e.g. bees,

ants, termites), consist of large numbers of free-living individuals that habitually live together in a strictly organized social group; (2) a group of animals or plants living together and somewhat isolated, or recently established in a new area; (3) a coenobium (*q.v.*); (4) (*microbiol.*) a discrete aggregate formed by certain microorganisms (esp. bacteria and fungi) when growing on solid media. Bacterial colonies comprise the progeny of a single cell, those of fungi and actinomycetes the mycelium arising from a single spore or fragment of hypha. Colony colour, shape and texture on various media are important features in identification.

colony fission the production of new colonies by departure of some members while leaving the parent colony intact.

colostrum *n.* clear fluid secreted from mammary glands at the end of pregnancy and differing in composition from the milk secreted later.

colour index (*geol.*) the volume percentage of mafic minerals (*q.v.*) in an igneous rock *q.v.*

colour phase an unusual but regularly occurring colour variety of a plant or animal, e.g. the white colour forms of many plants.

Columbiformes *n.* an order of birds including the doves and pigeons.

columbite *n.* a mineral with a composition $(Fe,Mn)(Nb,Ta)_2O_6$ in which niobium (Nb) predominates over tantalum (Ta). A niobium and tantalum ore mineral. *see also* tantalite.

columnar *a. pert.* or like a column. (*biol.*) *appl.* (1) cells longer than broad; (2) epithelium composed of such cells; (3) (*geol.*) *appl.* jointing that breaks some thick lava flows into polygonal columns. The long axes of these are parallel to one another and usually vertical. Particularly common in basalts; (4) (*soil sci.*) *appl.* soil peds (*q.v.*) that are vertical pillars with rounded tops. *cf.* prismatic.

Columniferae Malvales *q.v.*

columns *see* speleothems.

combination colours colours produced by structural features of a surface in conjunction with pigment.

combined cycle plants electricity generating installations that consist of a gas turbine system powered by either natural gas or oil. The waste heat is used to drive a steam turbine system. Both turbine systems are used to generate electricity. The total thermal efficiency of such plants is *ca*. 50%, compared with a maximum of *ca*. 40% achievable from coal-fired steam plant.

combined heat and power plants (CHP plants) installations that burn fuel to produce power (usually in the form of electricity) and heat (generally conveyed by steam or hot water). Such installations do not maximize the amount of power generated. However, the total fraction of the fuel energy that is converted into useful heat and power is typically about 80%. This is greater than would be produced by plants producing heat and power separately. *alt*. co-generation plants.

comb jellies common name for the Ctenophora, a phylum of marine coelenterates, also called sea gooseberries, which are free-swimming and biradially symmetrical with eight meridional rows of ciliated ribs (swimming plates) by which the organism propels itself.

combustible *a*. capable of combustion, i.e. burning.

combustion *n*. burning, the chemical combination of a substance with oxygen, accompanied by the production of light, heat and flame, e.g. the burning of fossil fuels.

Commelinales *n*. order of herbaceous monocots including the family Commelinaceae (tradescantia) and others.

commensal *n*. a partner, usually the one that benefits, in a commensalism. *a*. *pert*. commensalism.

commensalism *n*. the association between two organisms of different species that live together and share food resources, one species benefiting from the association and the other not being harmed.

commercial extinction the depletion of the population of a wild species used as a natural resource to such a level that it is no longer profitable to harvest it.

commercial farming type of farming geared towards making maximum profits from the sale of crops, livestock and/or livestock products (milk, wool, etc.). *cf*. subsistence farming.

common-property resource *see* commons.

commons *n*. (1) originally and specifically, land or some other resource that is not privately owned and is available to a particular community for common use, usually according to mutually agreed unwritten rules; (2) in present-day environmental usage, environmental resources such as the atmosphere, water, the deep oceans, etc., that are not owned by anyone, are in principle available to everyone, which are difficult to exclude people from using or to exert any control over their overuse or abuse. *alt*. common-property resource, public good. *cf*. private good, private-property resource, public-property resource. *see also* tragedy of the commons.

common salt sodium chloride ($NaCl$).

common wild oat *Avena fatua*, a major weed pest of temperate cereal crops.

community *n*. (1) (*biol*.) a well-defined interdependent assemblage of plants and/or animals covering a given area, clearly distinguishable from other such assemblages; (2) (*geog*.) an assemblage of people resident within a particular geographical area.

community biomass total weight per unit area of the organisms in a community.

community ecology the study of the ecology of a community as a whole.

community production primary production, the quantity of dry matter formed by the vegetation covering a given area.

commuter *n*. person who lives in one community and routinely travels considerable distances to and from work in another.

commuter village village in which a high percentage of the workforce is not employed locally but commutes considerable distances to work in other location(s), usually large towns or cities. *alt*. dormitory village, suburbanized village.

comovirus group plant virus group containing small isometric single-stranded RNA viruses, type member cowpea mosaic virus. They are multicomponent viruses in which two genomic RNAs are encapsidated in three different virus particles, one of which lacks nucleic acid.

compaction (1) (*geol*.) one of the processes involved in the formation of sedimentary rocks, in which pressure exerted by accumulating sediment forces the grains in the deeper sediment layers to move more closely together; (2) (*soil sci.*) consolidation of soil by the action of vehicles, farm machinery and/or animals.

compactor *n*. mechanical device used to reduce the volume of waste materials by compression.

compatible *a*. (1) able to cross-fertilize; (2) having the capacity for self-fertilization; (3) able to coexist; (4) in plant pathology, *appl.* an interaction between host and pathogen that results in disease.

compensation point (1) the point at which respiration and photosynthesis are balanced at a given temperature, as determined by the intensity of light or the concentration of carbon dioxide, *see* carbon dioxide compensation point, light compensation point; (2) limit of lake or sea depth beyond which green plants and algae lose more by respiration than they gain by photosynthesis. *alt.* compensation depth or level.

compensatory population growth increased population growth that results when a population is reduced below its carrying capacity.

competition *n*. (1) active demand by two or more organisms for a material or condition, e.g. plants competing for light, water, etc., *cf.* amensalism; (2) active demand by two or more substances for the same binding site on enzymes, receptor molecules, cells, etc.

competition coefficient the inhibitory effect of one species on the population growth rate of another species with which it is in competition.

competitive exclusion principle the principle that two different species living in the same area cannot indefinitely occupy the same ecological niche, one eventually being eliminated. *alt.* Gause's principle.

competitive release the increase in size of a population and its spread into a broader range of niches when the competitive effect of other species is removed. *alt.* ecological release.

complemental air volume of air that can be taken in in addition to that drawn in during normal breathing.

complemental male pygmy male *q.v.*

complementary DNA (cDNA) single-stranded DNA synthesized *in vitro* on an RNA template by reverse transcriptase, used widely in genetic engineering to make DNA for cloning using mRNAs isolated from cells as the template RNAs.

complete metamorphosis insect metamorphosis in which the young are usually different from the adult and are called larvae, and go through a resting stage, the pupa, before reaching the adult stage (the imago), the wings being developed inside the body during the pupal stage. *cf.* incomplete metamorphosis.

complex *n*. (1) (*mol. biol.*) generally refers to two or more molecules held together by non-covalent bonding as opposed to covalent bonding; (2) (*chem.*) a compound or ion that contains at least one dative bond (i.e. a covalent bond formed by the sharing of a pair of electrons that originated from one of the two atoms bonded together); (3) (*ecol.*) the meeting of several distinct communities related to each other by certain shared species.

complex, activated *see* activated complex.

complexation reactions (*chem.*) reactions that result in the formation of dative bonds *q.v.*

complex dunes dunes composed of two or more different basic dune types.

Compositae, composites *n*., *n.plu.* a very large family of dicotyledonous flowering plants, typified by dandelions, daisies and thistles, in which the inflorescence is a capitulum made up of many tiny florets.

composite volcanoes volcanoes composed of different layers of lava and compacted pyroclasts from successive eruptions. Such volcanoes are typically steep-sided and highly symmetrical in form. *alt.* stratovolcanoes.

compost *n*. nutrient-rich material produced by the biodegradation of organic matter under aerobic conditions and used to enhance the condition and fertility of soil.

compound *a*. (1) made up of several elements; (*biol.*) *appl.* (2) e.g. flowerheads made up of many or several individual flowers, eyes (of insects) made up of many identical structural elements; (3) leaves made up of several leaflets.

compound *n*. (*chem.*) a substance made up of more than one chemical element. In the vast majority of compounds, the different elements are present in definite proportions by mass. The atoms of the compound are held together by ionic and/or covalent bonds. Any given compound has its own characteristic set of properties (e.g. melting point, density) that are generally very different from those of the elements that it contains. *cf.* mixture.

compound dunes dunes composed of at least two dunes of the same basic type, either merged or superimposed.

compound eye eye characteristic of insects and most crustaceans, made up of many identical photoreceptive units.

compound fertilizers manufactured fertilizers that contain a mixture of nitrogen (N), phosphorus (P) and potassium (K), generally in the form of inorganic compounds. A code is used to represent their composition, thus a compound fertilizer coded 10 : 25 : 25 would contain the equivalent of 10% N, 25% P_2O_5 and 25% K_2O. *cf.* straight fertilizers. *alt.* mixed fertilizers.

compound nest nest containing colonies of two or more species of social insects, up to the point where the galleries of the nest run together. Adults sometimes intermingle but the broods are kept separate.

Comprehensive Soil Classification System (CSCS) *see* Soil Taxonomy.

compression ratio *appl.* internal combustion engines. It is the ratio of the volume of the cylinder when the piston is at the bottom of its stroke to the volume when it is at the top of its stroke. The fuel efficiency of an internal combustion engine can be enhanced by an increase in its compression ratio.

compression wood in conifers, reaction wood formed on the lower sides of crooked stems or branches, and having a dense structure and much lignification.

concentrate and contain waste management technique in which wastes are stored in a concentrated form, rather than being allowed to dissipate within the environment. *cf.* dilute and disperse.

concentration *n.* the amount of a given substance in a given volume or amount of another substance.

concentration factor the factor by which the level of a toxic pollutant, e.g. a heavy metal or pesticide, accumulates in the tissue of a living organism compared with its concentration in the environment, or with its concentration in the previous organism in the food chain.

concerted evolution coevolution *q.v.*

conchoidal a. (*geol.*) *appl.* (1) rock or mineral fracture that produces fragments with curved, shell-shaped surfaces; (2) rocks and minerals that exhibit this type of fracture.

conchology *n.* that branch of zoology dealing with molluscs and their shells.

concordant intrusion *see* intrusion.

Concorde effect in animal behaviour, the case where future behaviour is determined by past investment rather than future prospects.

concretion *n.* (*geol.*) (1) nodule (*q.v.*) with a concentric structure; (2) discrete portion of a sediment, the grains of which are more firmly cemented together than in the surrounding sediment; (3) (*soil sci.*) group of soil particles permanently cemented together by a localized accumulation of an insoluble material.

condensation *n.* (1) process whereby atmospheric water vapour is transformed into a liquid (at temperatures >0 ˚C) or a solid (at temperatures <0 ˚C), generally as a result of cooling to its dewpoint (saturation level); (2) (*chem.*) the formation of a larger molecule from the polymerization of smaller units accompanied by the elimination of the elements of a single small molecule (often water) during the formation of each bond. It occurs, e.g., during the formation of proteins from amino acids and nucleic acids from nucleotides; (3) (*biol.*) of chromosomes, the compaction in the chromatin that takes place as chromosomes prepare to enter mitosis or meiosis.

condensation nuclei hygroscopic nuclei *q.v.*

conducting *a.* conveying, *appl.* tissues or structural elements that convey material or a signal from one place to another.

conduction *n.* (1) (*bot.*) the transfer of soluble material from one part of a plant to another; (2) (*zool.*) the conveyance of an electrical impulse along a nerve fibre; (3) (*phys.*) thermal conduction, the transfer of heat, either within a substance or between two substances that are in direct physical contact, via the transfer of kinetic energy between the atoms, ions, free electrons and/ or molecules that make up the substance(s) concerned. During this process, heat flows from a region of higher temperature (i.e. one where the constituent particles have more kinetic energy) to one of lower temperature (i.e. one where the constituent particles have less kinetic energy).

cone *n.* (1) the reproductive structure in certain groups of plants, e.g. conifers, cycads, *see* strobilus; (2) type of cell in vertebrate retina responsible for colour vision and vision in good light, individual cone cells being sensitive to red, green and blue wavelengths.

cone karst cockpit karst *q.v.*

confidence interval in statistics, an interval that has a specific probability of containing the real value of an unknown parameter. For example, a 95% confidence interval has a 95% probability of containing the parameter being estimated.

confined aquifer *see* aquifer.

confinement *n.* the separation of the thermonuclear plasma from the interior surface of the reactor during a controlled nuclear fusion reaction. *alt.* containment.

confinement time the approximate time spent by ions within the field that contains the thermonuclear plasma of a controlled nuclear fusion reaction. *alt.* containment time.

conflict *n.* situation in which two motivations compete for dominance in the control of behaviour, as when an animal is deciding which of two objects to approach (or avoid) or whether to approach or run away from an object.

conformer *n.* an animal whose internal environment is influenced by external factors. *cf.* regulator.

congeneric *a.* belonging to the same genus.

congenital *a.* present at birth, *appl.* physiological or morphological defects, not necessarily inherited.

conglomerate *n.* a clastic sedimentary rock composed of rounded gravels, pebbles or boulders set in a finer-grained matrix.

conifer *n.* (1) a cone-bearing tree, sometimes used to include all temperate gymnosperm trees (as opposed to broad-leaved angiosperm trees) even when (as yews, junipers, etc.) they do not bear cones; (2)

sometimes refers just to the needle-leaved, cone-bearing gymnosperms, such as pines, cypresses, spruces, larches, etc., with reproductive organs as separate male and female cones.

Coniferales, Coniferae *n.* an order of gymnosperm trees, with reproductive organs as separate male and female cones, and usually needle-shaped leaves (e.g. pines, larches, cypresses).

Coniferophyta *n.* one of the five main divisions of extant seed-bearing plants, commonly called conifers. They have simple, often needle-like leaves, and bear their megasporangia usually in compound strobili (cones). In some classifications called Coniferopsida.

coniferous *a.* (1) cone-bearing; (2) *pert.* conifers or other cone-bearing plants.

conjugate acid (*chem.*) the chemical species formed from a Brønsted–Lowry base on the acceptance of a proton from a Brønsted–Lowry acid. For example, in the following equation, BH^+ represents the conjugate acid of the base B:

$$HA + B \rightleftharpoons A^- + BH^+$$

where HA is a Brønsted–Lowry acid and A^- is its conjugate base.

conjugate base (*chem.*) the chemical species formed from a Brønsted–Lowry acid on the donation of a proton to a Brønsted–Lowry base. For example, in the following equation, A^- represents the conjugate base of the acid HA:

$$HA + B \rightleftharpoons A^- + BH^+$$

where B is a Brønsted–Lowry base and BH^+ is its conjugate acid.

Connarales *n.* order of dicots comprising the family Connaraceae.

connate water fossil water *q.v.*

conodonts *n.plu.* abundant tooth-like fossils from the Palaeozoic and early Triassic, thought to be the jaw elements of invertebrates of uncertain affinities and generally placed in a separate phylum, Conodonta.

consanguineous *a.* related, in human genetics *appl.* matings between relatives.

consciousness *n.* (1) an awareness of one's actions or intentions, and having a purpose and intention in one's actions;

(2) sometimes defined as the presence of mental images and their use by an animal to regulate its behaviour, although the question of whether animals possess consciousness is controversial.

conservation *n.* management of the environment and its natural resources with the aim of protecting it from the damaging effects of human activity. *see also* energy conservation, nature conservation, preservation, recycling, soil conservation.

conservation area an area of particular ecological, landscape, historical or architectural interest, which is set aside or subject to strict planning regulations for the purposes of preserving its character.

conservationist *n.* someone who believes resources should be managed and protected so that they will not be degraded and wasted and will be available to future generations, esp. someone who is active in promoting these views.

conservation-tillage farming *see* reduced cultivation.

conservative *a.* (*biol.*) *appl.* (1) characters that change little during evolution; (2) taxa retaining many ancestral characters.

conservative boundary transform boundary *q.v.*

conservative properties (*oceanogr.*) *appl.* water masses (*q.v.*), properties of seawater, such as salinity, that are affected only by physical processes. Salinity, e.g., is changed solely by mixing with water of different salinities. *cf.* non-conservative properties.

conserved *a.* (1) conservative (*q.v.*); (2) (*mol. biol.*) *appl.* biochemical pathways, proteins, genes, DNA sequences, etc., that are identical or very similar in different organisms. Also *appl.* stretches of DNA sequences that are very similar in different genes.

consociation *n.* (1) a unit of plant association; (2) a climax community characterized by a single dominant species.

consocies *n.* a consociation representing a stage in the process of succession.

consortium *n.* (*biol.*) a kind of symbiosis in which both gain benefit, as of the alga and fungus in a lichen.

conspecific *a.* belonging to the same species.

constancy *n.* (1) the ability of a living system such as a population to maintain a certain size; (2) (*ecol.*) the frequency with which a particular species occurs in different samples of the same association.

constant *n.* (*ecol.*) a species that occurs in at least 95% of samples taken at random from within a community.

constructive boundary divergent boundary *q.v.*

constructive waves (*oceanogr.*) waves responsible for the deposition of marine material on the beach. These occur when the swash is stronger than the backwash. *cf.* destructive waves.

consumed water water that is lost from the water supply system after use. *cf.* withdrawn water.

consumer *n.* (*biol.*) heterotrophic organism, i.e. one that must consume resources provided by autotrophic organisms. *see also* primary consumer, secondary consumer.

consummatory act action at the end of an instinctive behaviour such as eating. *alt.* consummatory behaviour.

consumption *n.* (1) the use of a resource, e.g. energy, food, etc.; (2) tuberculosis *q.v.*

consumption efficiency the ratio of energy ingested by one trophic level in an ecosystem to the energy produced as productivity by the previous trophic level.

contact cooling radiation cooling *q.v.*

contact herbicide any herbicide, such as glyphosate, that kills a plant on immediate contact with leaves, etc. *cf.* systemic herbicide.

contact metamorphic aureole the area of altered rocks that surrounds an igneous intrusion.

contact metamorphism the formation of metamorphic rocks in the vicinity of igneous intrusions. The zone of altered rocks is known as a contact metamorphic aureole and its extent is dependent on the size of the intrusion and temperature of the magma (*q.v.*). Contact metamorphism is both thermal and metasomatic in nature.

contact metasomatism the movement of hot liquids and gases through country rock during an intrusion leading to alteration in its bulk elemental composition. Occasionally this will lead to the formation of orebodies within the contact aureole. *alt.* pyrometasomatism.

contact pesticide pesticide that kills by direct contact with the pest. This may occur, e.g., when the pest passes through treated soil or when it absorbs some of the pesticide vapour.

contagious *a. appl.* disease that can be transmitted by direct or indirect contact.

containment (1) confinement (*q.v.*); (2) biological containment (*q.v.*); (3) physical containment *q.v.*

containment time confinement time *q.v.*

contamination *n.* the presence of raised concentrations of substances in water, sediments or organisms, not necessarily resulting in a deleterious effect. *cf.* pollution.

contest competition type of competition in which the successful competitor gains sufficient resources for survival and reproduction whereas the unsuccessful competitors gain nothing or insufficient for survival.

continental climate a climate characterized by some or all of the features of weather associated with continental interiors, namely hot summers and cold winters with a wide temperature range between extremes, short spring and autumn seasons, marked rainy and dry seasons.

continental drift the hypothesis that the present-day continents were once part of a huge supercontinent named Pangaea that started to break up and drift apart about 135 million years ago. According to this theory, first proposed by Alfred Wegener in 1915, Pangaea initially split into Laurasia in the north and Gondwanaland in the south. About 65 million years ago, further splitting occurred, Laurasia forming Eurasia and North America and Gondwanaland fragmenting into South America, Africa, Australia, Antarctica and India. *see also* plate tectonics.

continental effect the lack of a maritime effect (*q.v.*) on the climate, as experienced in continental interiors. *alt.* continentality.

continentality continental effect *q.v.*

continental margin submerged region consisting of the continental shelf and the continental slope.

continental rise gently sloping sea-floor zone formed by the deposition of continental sediments and located at the foot of a continental slope.

continental shelf gently sloping submerged plain bordering a continental landmass. The world's continental shelves have an average width of *ca.* 65 km.

continental shelf break boundary between the continental shelf and continental slope, usually found at depths *ca.* 150 m.

continental shields large relatively flat, expanses of ancient rocks, e.g. the Laurentian Shield, northern North America.

continental slope steep slope with an angle of slope up to 45° located at the periphery of the continental shelf and separating it from the abyssal (deep-sea) plain.

continuous culture methods of growing bacterial cultures so that nutrients and space do not become exhausted and the culture is always in the rapidly multiplying phase of growth.

continuous permafrost a type of permafrost (*q.v.*) characterized by being unbroken in the horizontal plane. It can extend to depths in excess of 1000 m and occurs mainly within the Arctic Circle.

continuous stocking set stocking *q.v.*

continuous variation variation between individuals of a population in which differences are slight and grade into each other, continuously variable phenotypic characters being those that are determined by a large number of genes and/or considerable environmental influence.

continuum *n.* (*bot.*) a form of vegetation in which one type passes almost imperceptibly into another and no two types are repeated exactly.

contour ploughing ploughing along the contours of a hill and not up and down its slope, an agricultural practice designed to promote soil conservation.

contour ridging practice, widespread in tropical agriculture, of building ridges, usually by hand, along the contours of a hill in order to prevent soil erosion.

contour strip cropping *see* strip cropping.

contour strip mining mining by cutting a series of terraces along the side of a hill to remove a near-surface mineral such as coal.

contour tillage the preparation of land, for the cultivation of crops, in lines perpendicular to the incline of a slope. This is done in order to decrease sheet erosion (*q.v.*) and rill erosion *q.v.*

contraception *n*. the deliberate prevention of conception and pregnancy, generally by chemical (e.g. birth control pills) or physical means (e.g. condoms, intrauterine devices).

contraceptive *n*. any physical, chemical or biological means used to avoid pregnancy.

contract farming arrangement between a farmer and a buyer in which the farmer agrees to supply produce, e.g. cereals, livestock, fruit or vegetables, of suitable quality, in the amount, and at the time, stipulated by the buyer.

control *n*. an experiment or test carried out to provide a standard against which experimental results can be evaluated.

controlled tipping the practice of depositing domestic refuse and/or hazardous wastes in suitable landfill sites in a regulated manner. *alt*. sanitary landfill, secured landfill.

control rods rods that contain neutron-absorbing elements (such as cadmium, boron or hafnium) that can be moved into the core of a nuclear reactor, so causing it to become subcritical and shut down. *alt*. safety rods.

conurbation *n*. extensive built-up area, often formed by the merger of a city with its neighbouring towns.

convection *n*. transfer of heat through fluids, such as air or water, bought about by the movement of the fluid in question. For example, in the atmosphere, when air is warmed by proximity to a heated ground surface, it rises as a consequence of its reduced density. This rising air is supplanted by colder, denser sinking air, which is then warmed in turn, thus establishing streams of moving air termed convection currents. The rising column of air so produced is termed a convection cell.

convectional rain rain formed when moist air, warmed by its proximity to a heated ground surface, is forced to rise. The subsequent condensation of water vapour in the ascending air mass produces heavy precipitation. This type of rain is associated with conditions of atmospheric instability and the development of thunderstorms.

convection cell *see* convection.

convection currents *see* convection.

Convention *see* United Nations Conventions, and Conventions listed below.

conventional behaviour any behaviour by which members of a population reveal their presence and allow another organism to assess their numbers.

conventional cultivation the traditional method of cultivation used in Western agriculture in which the top 20–25 cm of soil is thoroughly disturbed during the preparation of the seedbed. *cf*. reduced cultivation. *alt*. conventional tillage.

conventional tillage conventional cultivation *q.v.*

Convention Concerning the Protection of the World Cultural and Natural Heritage full title of the World Heritage Convention *q.v.*

Convention on International Trade in Endangered Species (CITES) convention adopted in 1975, developed by the World Conservation Union (IUCN) and administered by the UN Environment Programme, which seeks to provide protection for certain endangered species. It lists species that cannot be commercially traded as live specimens or wildlife products because they are endangered or threatened.

Convention on Wetlands of International Importance Especially as Waterfowl Habitat full title of the Ramsar Convention *q.v.*

convergence *n*. (1) similarity between two organs or organisms (or protein or DNA molecules) due to independent evolution along similar lines rather than to descent from a common ancestor, *alt*. convergent evolution; (2) coordinated movement of eyes when focusing a near point.

convergence zone *see* intertropical convergence zone.

convergent boundary (*geol*.) type of plate boundary that occurs when two plates move towards each other and collide. Convergent plate boundaries are associated with intense crustal activity such as mountain building. *alt*. destructive boundary.

convergent evolution *see* convergence.

convivium *n*. a geographically isolated and distinct population within a species, usually a subspecies or ecotype.

coolant *n*. any fluid used in a system to remove heat, found, e.g., in refrigerators, nuclear reactors.

cool-season plant plant that makes most of its growth during the cooler seasons of the year.

coombe, coomb, combe, comb *n.* (1) steep-sided, narrow stream valley, dry for much of the year, characteristic of chalk landscapes; (2) cirque *q.v.*

coombe ridges arêtes *q.v.*

cooperation *n.* in animal behaviour, the sharing of a task between different animals, as in hunting by a wild dog pack, the sharing of child care by relatives of the mother in chimpanzees, feeding of nestlings by other members of the community in some communal birds. *see also* symbiosis.

cooperative breeding a breeding system in which parents are assisted in the care of their offspring by other adults.

coordinate bond dative bond *q.v.*

coordination compound (*chem.*) a compound that is, or contains, a complex *q.v.*

copal *n.* resin exuding from various tropical trees and hardening to a colourless, yellow, red or brown mass. Used in varnishes.

Copepoda, copepods *n., n.plu.* subclass of free-living or parasitic small crustaceans that form a large part of the marine zooplankton and are also found in fresh water. They have no carapace and have one median eye in the adult.

Cope's rule rule stating that body size tends to increase during the evolution of a species.

copiotrophic *a.* able to grow only in conditions of good nutrient supply.

copper (Cu) *n.* metallic element (at. no. 29, r.a.m. 63.54), forming a malleable reddish metal that is an excellent conductor of electricity. It is found as the native element (*q.v.*) and as compounds, notably the oxide Cu_2O (cuprite) and the sulphides Cu_2S (chalcocite) and $CuFeS_2$ (chalcopyrite). In its elemental form, it is used as an electrical conductor and for the manufacture of steam boilers and water pipes. It has many alloys of commercial importance and, as are its compounds, is used as a fungicide and pigment. Roasting of copper sulphide ores, during metal winning, liberates sulphur dioxide (SO_2), thus causing atmospheric pollution. Copper is an essential micronutrient for living organisms because it is a prosthetic group for some

essential enzymes and other proteins. Some copper compounds are toxic.

copper glance chalcocite *q.v.*

copper pyrites chalcopyrite *q.v.*

coppicing *n.* woodland management practice in which trees such as willow, hazel and sweet chestnut are regularly cut back almost to the ground every few years so that they develop a low 'stool' from which many long straight shoots arise, which can be harvested for fuelwood and other uses. *cf.* pollarding.

coprolite, coprolith *n.* fossilized faeces.

coprophage *n.* organism that feeds on dung. *a.* coprophagous.

coprophagy *n.* (1) habitual feeding on dung; (2) the deliberate reingestion of own faeces by animals, such as rabbits, *alt.* refection.

coprophil, -ic, -ous *a.* growing in or on dung, *appl.* certain fungi, bacteria and other microorganisms.

coprophyte *n.* plant that grows on dung.

coprozoic *a.* living in faeces, as some protozoans.

coprozoite *n.* any animal that lives in or feeds on dung.

copulation *n.* sexual union.

Coraciformes *n.* an order of birds including the kingfishers, rollers and hornbills.

coral *n.* member of a group of colonial coelenterates of the class Anthozoa composed of individual polyps connected by living tissue. Some forms secrete a stony matrix binding the colony together and some build extensive reefs. *see* soft corals, stony corals.

coralliferous *a.* (1) coral-forming; (2) containing coral.

coralligenous *a.* coral-forming.

coralline *a.* (1) resembling a coral, *appl.* to some lime-encrusted red algae; (2) composed of or containing coral; (3) *appl.* zone of coastal waters at about 30–100 m.

coral reef a submarine ridge formed by the progressive deposition of calcium carbonate by colonial corals and calcareous algae. Coral reefs are vibrant and highly productive ecosystems that are widely distributed in clear, shallow tropical and subtropical waters. There are three main types of coral reef, *see* atoll, barrier reef, fringing reef.

Cordaitales *n.* order of fossil conifers, being mostly tall trees with slender trunks and a crown of branches, with spirally arranged simple grass-like or paddle-like leaves and with mega- and microsporangia in compound strobili.

cordierite *n.* $(Mg,Fe)_2Al_4Si_5O_{18}.nH_2O$, a mineral found in some aluminous metamorphic rocks and igneous rocks which incorporate aluminous material of sedimentary origin.

core *n.* (1) (*geol.*) the innermost region of the Earth. The core is subdivided into a solid inner core (radius 1220 km) and a liquid outer core with a thickness of 2250 km; (2) that part of a fission nuclear reactor where the nuclear reaction occurs. It contains the fuel in a matrix that allows access by the coolant and control rods (*q.v.*). In thermal reactors (but not in fast breeder reactors) it also contains the moderator *q.v.*

core area (1) *see* biosphere reserve; (2) that part of the home range of an animal in which it spends most of its time.

core electrons those electrons of an atom that are not in its valence shell. These electrons are not directly involved in the formation of chemical bonds.

coregonid *n.* freshwater fish of the genus *Coregonus*, found only in unpolluted waters, e.g. houting, powan and vendace.

core temperature (*biol.*) temperature of an animal's body measured at or near the centre.

Coriolis force an apparent force that applies to objects moving across the surface of a plane (e.g. a disc) that is rotating about an axis that is perpendicular to its surface (*see* Fig. 8). This force, which is apparent from any fixed point on the surface of the rotating plane, operates at right angles to the direction of motion of the object, to the right over an anticlockwise rotating surface and to the left over one that is rotating clockwise. The northern hemisphere of the Earth, when viewed from space along its axis, approximates to an anticlockwise rotating disc. When viewed from the Earth, air or sea currents moving over its surface therefore experience a movement to the right. Similarly, air or sea currents moving across the southern hemisphere will be

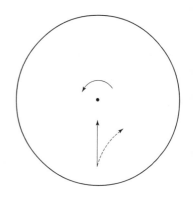

•   Axis of rotation

↙   Direction of rotation

↑   Initial direction of movement of a projectile

↗   Path of projectile across the surface of the disc, showing the deflection to the *right* produced over an *anti-clockwise* rotating surface. Deflection to the *left* is produced over a *clockwise* rotating surface (not shown)

Fig. 8 The Coriolis force.

deflected to the left (*see* Fig. 9). The spinning disc approximation is best at the poles. It becomes progressively less accurate as the globe is traversed in an equatorial direction. The Coriolis force is therefore maximal at the poles and zero at the Equator.

cork *n.* external layer of plant tissue composed of dead cells filled with suberin, forming a seal impermeable to water. Present on woody stems and derived from the cork cambium. *alt.* phellem.

corm *n.* enlarged, solid underground stem, rounded in shape, composed of two or more internodes, and covered externally with a few thin scales or leaves.

corn *n.* (1) maize (*q.v.*), US usage; (2) wheat (*q.v.*), UK usage.

Cornales *n.* order of dicot trees, shrubs and herbs, with leaves often much divided, and including the families Umbelliferaceae (carrot, etc.), Araliaceae (ginseng), Cornaceae (dogwood), Davidiaceae (dove tree) and others.

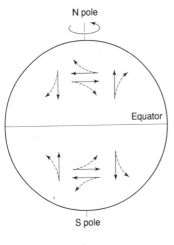

N pole

Equator

S pole

→ Initial path
----→ Path under influence of the
Coriolis force

Fig. 9 The direction of airflow or sea-current deflection as a consequence of the Coriolis force.

corolla *n.* the petals of a flower collectively, esp. when joined into a tube.

corona *n.* (1) (*bot.*) frill at mouth of corolla tube formed by union of scales on petals, as in the trumpet of a daffodil; (*zool.*) (2) the theca and arms of a crinoid; (3) ciliated disc or circular band of certain animals such as rotifers; (4) the head or upper portion of any structure.

Coronaviridae, coronaviruses *n.*, *n.plu.* family of medium-sized, enveloped, single-stranded RNA viruses covered with protein petal-like projections, which include human coronavirus and avian infectious bronchitis virus.

corrasion abrasion *q.v.*

correlation coefficient (r) a statistical measure of the likelihood that the value of one variable is dependent on the value of another.

corridor *n.* strip of suitable habitat that connects otherwise isolated habitats, allowing wildlife to move between them. *alt.* green corridor, habitat corridor, migratory pathway, wildlife corridor.

corrie cirque *q.v.*

corrie glaciers cirque glaciers *q.v.*

corrosion *n.* (1) (*chem.*) the destructive attack on an elemental metal or alloy by its environment; (2) (*geog.*) *see* fluvial erosion.

cortex *n.* (1) the outer layer of a structure or organ, as brain; (2) in vascular plants, the tissue in stem and root surrounding but not part of the vascular bundles; (3) in cells, the outermost layer of cytoplasm, just under the cell membrane; (4) in bacterial spores, the envelope between spore wall and spore coat. *plu.* cortices.

corundum *n.* Al$_2$O$_3$, a mineral used industrially as a refractory and an abrasive and, in some forms, decoratively as a gemstone. *see also* emery, ruby, sapphire.

corvine *a. appl.* birds of the Corvidae (crow family), e.g. ravens, jackdaws, crows, magpies. *n.* corvid.

corymb *n.* a flowerhead in the form of a raceme with lower flower stalks elongated so that the top of the flowerhead is nearly flat.

corynebacteria *n.plu.* bacteria of the family Corynebacteriaceae, which are characterized by irregularly shaped cells dividing by 'snapping fission' and that include the human pathogen *Corynebacterium diphtheriae*, the cause of diphtheria. *sing.* corynebacterium.

cosere *n.* a series of plant successions on the same site.

cosmopolitan *a.* world-wide in distribution.

cosmopolite *n.* a species with a world-wide distribution.

cost *n.* the decrement in an animal's inclusive fitness that results from a particular behaviour. *cf.* benefit.

cost–benefit analysis (CBA) economic evaluation technique in which the benefits and costs (in environmental, commercial and social terms) of a particular course of action are assessed. The term is also used in analyses of animal behaviour and ecological relationships. *alt.* benefit–cost analysis.

cost function the combination of the various costs of an animal's behaviour, which is used to evaluate all aspects of the animal's state and behaviour.

coterie *n.* a social group that defends a common territory against other coteries.

coterminous *a.* (1) of similar distribution; (2) bordering on; (3) having a common boundary.

cotton *n.* (1) *Gossypium* spp., a dual-purpose crop grown mainly for its fibre, used as a textile, but also for its oil-rich seed, which is processed into edible oil and animal cake; (2) the fibre obtained from the seeds of spp. of *Gossypium*, whose seed-coats are covered with long cellulose fibres.

cottony-cushion scale *Icerya purchasi*, an insect pest, native of Australia, which attacks citrus trees. Its accidental introduction into California in the late nineteenth century threatened the entire Californian citrus industry. A natural predator, an Australian ladybird beetle called the vedalia (*Rodolia cardinalis*), was successfully used to control the outbreak.

cotyledon *n.* (1) (*bot.*) the first leaf or leaves of a seed plant, found in the embryo, which may form the first photosynthetic leaves or may remain below ground; (2) (*zool.*) a patch of villi on mammalian placenta. *a.* cotyledonary.

cotylosaurs Captorhinida *q.v.*

cotype syntype *q.v.*

couch grass *Agropyron repens*, a serious weed pest of temperate cereal crops. *alt.* twitch grass.

coulomb (C) *n.* the derived SI unit of electric charge, defined as the quantity of electricity transported by one ampere in one second.

coumarin *n.* substance found in many plants, esp. clover, having a scent of new-mown hay. It is used in perfumery and to make the anticoagulant dicoumarin (dicumarol).

counteracting selection the operation of selection pressures on two or more levels of organization, e.g. individual, family or population, in such a way that certain genes are favoured at one level but disfavoured at another. *cf.* reinforcing selection.

counteradaptation *n.* a character trait or behaviour that evolves in response to adaptations of newly encountered organisms.

counterevolution *n.* the evolution of traits in one population in response to adverse interactions with another population, as between prey and predator.

counterion *n.* ion of opposite charge to that of a given chemical species. The presence of the counterion is required to maintain electrical neutrality.

countershading *n.* condition of an animal being dark dorsally and pale ventrally, so that when lighting is from above the ventral shadow is obscured and the animal appears evenly coloured and inconspicuous.

country park an area of countryside, usually in close proximity to a town or city, set aside to provide leisure and recreational opportunities for urban dwellers.

country rock existing rock into which there has been an intrusion of igneous rock or mineral veins. *alt.* host rock.

Countryside Commission government agency responsible for nature conservation and countryside policy in England, UK, including the designation of National Parks and Areas of Outstanding Natural Beauty (AONB).

Countryside Council for Wales (CCW) government agency responsible for nature conservation and countryside policy in Wales, UK, including the designation of National Parks and Areas of Outstanding Natural Beauty (AONB).

court *n.* a small area of a lek (*q.v.*) used by an individual male for display.

courtship *n.* behaviour pattern preceding mating in animals, often elaborate and ritualized.

covalent bond a bond that holds together the adjacent atoms of a molecule via the sharing of electrons between the atoms concerned.

covalent compound a compound that is composed of molecules, i.e. one in which the atoms are held together by covalent bonds not ionic bonds.

cover *n.* the percentage of ground surface covered with plants.

cover crops crops such as alfalfa (*Medicago* spp.) or clover (*Trifolium* spp.) that are planted immediately after harvesting to protect the soil from erosion.

Coxsackie virus a picornavirus (*q.v.*) that multiplies chiefly in the intestinal tract, but which can cause aseptic meningitis and other conditions.

cracking *n.* chemical decomposition by the action of heat (pyrolysis), esp. of heavy oils, where the larger molecules are 'cracked' into the smaller ones of lighter oils.

crag-and-tail *n*. a large mass of resistant rock (crag) that protects the softer rocks on its leeward side (tail) from glacial erosion.

cranial *a*. (1) *pert*. skull, or that part that encloses the brain; (2) *appl*. nerves arising from the brain.

Craniata, craniates *n*., *n.plu*. in some classifications, the Vertebrata *q.v.*

cranium *n*. the skull, more particularly that part enclosing the brain.

crassulacean acid metabolism (CAM) metabolic pathway in some succulent plants (Crassulaceae and Cactaceae) in which carbon dioxide taken in during the night is first fixed non-photosynthetically into carboxylic acids (chiefly malate), which are then mobilized in the light and used as CO_2 sources for the Calvin cycle. This enables desert plants to keep their stomata closed during the day to reduce evaporation.

crater *n*. depression (1) found in the centre of a volcanic cone; (2) created on the surface of a planet, esp. Earth, or a moon by the impact of a meteorite.

creationism *n*. the doctrine that the different types of living organisms have all arisen independently by divine creation. It has now been superseded as a serious doctrine in biology by the overwhelming evidence for evolution, but is still held by many non-scientists on religious grounds, and by a tiny minority of practising biologists.

creep *n*. an opening in a fence that allows only young livestock access to new areas. This forms the basis of a grazing system known as creep-grazing.

creep-grazing *see* creep.

creep movement the slowest type of mass movement (*q.v.*), occurring in both soil and rock. Although imperceptible, the presence of this type of movement may be revealed by the down-slope tilt of trees and fence posts in affected areas.

creodonts *n.plu*. a group of extinct placental mammals of the Cretaceous–Pliocene, which were probably archaic carnivores.

crepitation *n*. in insects, a defence mechanism in which fluid is discharged with an explosive sound.

crepuscular *a*. *pert*. dusk, *appl*. animals flying before sunrise or at twilight. *alt*. vespertine.

Cretaceous *n*. last period of the Mesozoic era, from *ca*. 144 to 65 million years ago, during which chalk was being laid down, occurring after the Jurassic and before the Tertiary.

Creutzfeldt–Jakob disease (CJD) a rare spongiform encephalopathy (a type of neurological disease) that affects man, caused by the same class of disease agent as bovine spongiform encephalopathy (BSE) in cattle and scrapie in sheep.

crevasse *n*. deep crack formed by ice fracture in the surface of a glacier or ice sheet.

crevasse-splay deposit a fan-shaped deposit of relatively coarse sediment found in the backswamp zone (*q.v.*) of a floodplain. This deposit is formed when a levee barrier is breached during a period of high river discharge. The water escaping through the crevasse rapidly deposits its load in a characteristic splayed formation.

crickets *n.plu*. common name for many members of the Orthoptera *q.v.*

Crinoidea, crinoids *n*., *n.plu*. a class of echinoderms, commonly called sea lilies and feather stars, present since the Cambrian, having a cup-shaped body with feathery arms, attached to the substratum by a stalk in the case of sea lilies.

critical *a*. *appl*. nuclear fission chain reactions in which the number of neutrons absorbed by the fissile material in each generation is just sufficient to maintain the reaction at a steady rate. Under these conditions, the number of neutrons produced in one generation will be the same as in the previous generation. This is the normal running condition of a nuclear reactor. *cf*. subcritical, supercritical.

critical depth *see* compensation point.

critical factor (*biol*.) that factor that is nearest to a tolerance limit for a particular species at a given time, *see* Shelford's law of tolerance.

critical group a taxonomic group containing organisms that cannot be divided into smaller groups, as in apomictic species.

critical links organisms in a food chain that are responsible for primary energy capture and nutrient assimilation and that are critical in the transformation of nutrients into forms that can be used by organisms at higher trophic levels in the chain.

critical load the threshold value for, e.g., use of a resource or release of a pollutant that can be sustained by the environment without any deleterious effects.

critical mass amount of fissile material needed to sustain a nuclear fission chain reaction.

critical minimum area, critical minimum size minimum continuous area of land required to sustain a viable population of a given species.

critical pollutants the seven substances that cause most air pollution, being carbon monoxide, sulphur dioxide, particulates, hydrocarbons, nitrogen oxides, ozone and lead.

crocidolite *n.* a fibrous variety of the amphibole mineral riebeckite (*q.v.*). The form of asbestos (*q.v.*) that is most hazardous to health. *alt.* blue asbestos.

Crocodylia, crocodilians *n., n.plu.* order of reptiles found first in the Triassic, typified by the present-day crocodiles and alligators, which are armoured, have front limbs shorter than hind and bodies elongated for swimming.

crofting *n.* type of small-scale subsistence farming practised in the Highlands and Islands of Scotland, UK, usually in combination with other remunerative activities such as fishing or weaving.

Cro-Magnon man early type of modern man, *Homo sapiens*, whose fossils were first found at Cro-Magnon in the Dordogne in France.

crop plant a plant that is cultivated for food or other use.

crop rotation the cultivation of a repeated sequence of different crops on the same land. Typically, a rotation incorporates a cereal crop, a break crop (such as potatoes or field beans) and a grass/legume sward for the grazing of livestock. This practice forms the basis of mixed farming. *see also* Norfolk four-course rotation.

cross *n.* an organism produced by mating parents of different strains or breeds. *v.* to hybridize.

cross-breeding *n.* the mating together of plants or animals from different strains or varieties.

cross-fertilization *n.* the fusion of male and female gametes from different individuals, especially of different genotypes. *alt.* allogamy, allomixis.

cross-media pollution pollution in one medium caused by pollutants transferred to that medium from another. For example, water bodies may become contaminated by pollutants present in the atmosphere.

Crossopterygii, crossopterygians *n., n.plu.* a group of mainly extinct bony fishes, of which the coelacanth (*Latimeria*) is the only known living member, also called tassel-finned or lobe-finned fishes.

cross-pollination *n.* transfer of pollen from the anther of one flower to the stigma of another, especially on a different plant.

cross-section *n.* a section cut across a specimen. For an organism, a section cut at right angles to the anterior–posterior or apical–basal axis.

crotovina *n.* obsolete animal burrow in one soil horizon that has become infilled with material from another soil layer. *alt.* krotovina.

crown *n.* (1) crest; (2) the top of head; (3) the exposed part of a tooth, especially the grinding surface; (4) head, cup and arms of a crinoid; (*bot.*) (5) leaves and branches of the upper part of a tree, above the main trunk; (6) short rootstock with leaves.

crown fire highly destructive form of forest fire so intense it spreads from the crown of one tree to that of the next. *cf.* ground fire.

crown gall a plant tumour caused by the bacterium *Agrobacterium tumefaciens*.

crown of thorns starfish a starfish that eats hard corals and is associated with tropical coral reefs. An increase in numbers since the 1960s has led to serious damage on some reefs.

CRS central receiver system *q.v.*

crucifer *n.* any member of a large family of dicotyledons, the Cruciferae, whose flowers have four petals in the form of a cross, and which includes cabbage, turnip, wallflower, mustard, cress, etc. *a.* cruciferous.

crude birth rate annual number of live births per 1000 persons within a specified population and geographical area, with the population number usually taken at the midpoint of the year in question.

crude death rate annual number of deaths per 1000 persons within a specified population and geographical area, with the population number usually taken at the midpoint of the year in question.

crude density the total population count or total biomass divided by the total area of the ecosystem under study.

crude fibre the cellulose–lignin component of animal feedstuffs.

crude oil naturally occurring, brownish-black, viscous liquid composed of many different organic compounds, predominantly hydrocarbons, found in association with a number of geological structures, *see* oil traps. It is refined by the process of fractional distillation to yield a number of commercially important products, such as petrol, diesel oil and tar. *alt.* mineral oil, petroleum.

crude protein the protein component of animal feedstuffs.

crumb *a.* (*soil sci.*) *appl.* soil peds (*q.v.*) that are spheroidal in shape and very porous. *cf.* granular.

crust *n.* the thin, outermost layer of the Earth. Composed of rocks, the crust represents less than 1% of the Earth's total mass and only about 0.5% of its radius. It is not of uniform thickness, varying from an average of about 35 km in the continental regions to about 5–10 km under the oceans.

Crustacea, crustaceans *n.*, *n.plu.* subphylum of arthropods (*q.v.*), considered as a class in older classifications. They are mainy aquatic, gill-breathing animals, such as crabs, lobsters and shrimps, each having a body divided into a head bearing five pairs of appendages (two pairs of preoral sensory feelers and three pairs of postoral feeding appendages) and a trunk and abdomen bearing a variable number of often biramous appendages that serve as walking legs and gills. Crustacea often have a hard carapace or shell.

cryand *see* andisol.

cryogenic scrap recovery a highly efficient scrap steel recovery method that involves the cooling of mixed waste (e.g. scrapped cars) with liquid nitrogen (b.p. = $-195.7\ °C$). The waste is then hammered, which, at these low temperatures, shatters the mild steel component, which can then be isolated using magnets.

cryolite *n.* Na_3AlF_6, a fluoride mineral. Used industrially as a flux, especially in the electrolytic production of aluminium metal.

cryopediments *n.plu.* level or nearly level platforms developed on the lower slopes of hills or at the base of valley sides as a result of periglacial processes. *cf.* cryoplanation terraces.

cryophil(ic) *a.* thriving at low temperature.

cryophyte *n.* plant, alga, bacterium or fungus that lives in snow and ice.

cryoplanation terraces level or nearly level platforms cut into bedrock at the top of hills or on their upper slopes in periglacial areas. *alt.* altiplanation terraces. *cf.* cryopediments.

cryoplankton *n.* plankton found around glaciers and in the polar regions.

cryosphere *n.* the Earth's ice system.

crypsis *n.* the phenomenon of being camouflaged to resemble part of the environment.

cryptic *a.* *appl.* (1) colouring, protective colouring making concealment easier; (2) variation, genetic variation due to the presence of recessive genes.

cryptic growth survival of soil microorganisms in nutrient-poor conditions by subsistence on the lysis products of dead cells.

cryptic species (1) species extremely similar in external appearance but that do not normally interbreed; (2) plants and animals that either live hidden in holes and crevices or are small and inconspicuous.

cryptofauna *n.* animals living in concealment in protected situations, such as crevices in coral reefs.

cryptogams *n.plu.* archaic term for plants reproducing by spores, such as the mosses and ferns, often also used for plants without flowers, or without true stems, roots or leaves.

cryptogram *n.* a method of expressing in standard form a collection of data used in classification.

Cryptophyta, cryptomonads *n.*, *n.plu.* (1) protists with an ovoid flattened cell and a gullet from which arise two unequal flagella; (2) a group of protists including the zooxanthellae.

cryptophyte *n.* a perennial plant persisting over winter by means of rhizomes, corms, or bulbs underground, or by underwater buds.

cryptosphere *n*. the leaf litter and twig layer on the soil surface, when considered as a habitat.

cryptovirus group plant virus group containing isometric double-stranded RNA viruses that produce no symptoms on infected plants. There are two subgroups: A, with smooth isometric particles, type member white clover cryptic virus 1; B, with isometric particles with prominent subunits, type member white clover cryptic virus 2.

cryptozoa *n*. (1) (*ecol*.) term for the small terrestrial animals that live on the ground but above the soil in leaf litter and twigs, among and under pieces of bark, stones, etc., this habitat being called the cryptosphere; (2) animals that live in crevices. *a*. cryptozoic.

crystalloid *n*. (1) substance that in solution readily diffuses through a semipermeable membrane, *cf*. colloid; (2) protein crystals found in some plant cells.

Cs symbol for the chemical element caesium *q.v.*

CS conditioned stimulus *q.v.*

CSCS Comprehensive Soil Classification System. *see* Soil Taxonomy.

Ctenophora, ctenophores *n*., *n.plu*. phylum of coelenterates containing the sea gooseberries or comb jellies. They are free-living and biradially symmetrical with eight meridional rows of ciliated ribs (ctenes) by which they propel themselves. They have no nematoblasts (stinging cells).

C-terminus the carboxy-terminus of a polypeptide chain, the end that carries the free COO⁻ group.

CTP cytidine triphosphate *q.v.*

Cu symbol for the chemical element copper *q.v.*

Cubozoa *n*. class of Cnidaria comprising the box-jellies *q.v.*

Cuculiformes *n*. an order of birds including the cuckoos, roadrunners and touracos.

cucumovirus group plant virus group containing small isometric single-stranded RNA viruses, type member cucumber mosaic virus. They are multicomponent viruses in which four genomic RNAs are encapsidated in three different virus particles.

cucurbit *n*. any member of the Cucurbitaceae, a family of dicotyledonous plants including the marrows, cucumbers and gourds.

Cucurbitales *n*. order of dicots, herbs and small trees, often climbing by tendrils and comprising the family Cucurbitaceae (gourds, melon, cucumber, etc.).

cuesta *n*. asymmetric ridge, formed by differential erosion, with a very gentle slope (dip slope) and a steep slope (scarp slope). The latter of these cuts across the slightly inclined bedding plane of the rock that underlies this feature. *see also* escarpment.

cullet *n*. broken or refuse glass (colour-graded) suitable for remelting.

culling *n*. (1) a conservation technique in which individual members of a given species are selectively killed in order to stabilize its own population, or that of its prey; (2) (*agric*.) the selective removal of old or inferior animals from a breeding population.

culm *n*. the flowering stem of grasses and sedges.

culmicole, culmicolous *a*. living on grass stems.

cultivar *n*. plant variety found only in cultivation, conventionally denoted by the species name followed by the abbreviation cv. followed by the cultivar name, e.g. *Rosa foetida* cv. Persian Yellow, or by placing the name in single quotes, e.g. *Rosa foetida* 'Persian Yellow'.

cultural eutrophication overenrichment of waterbodies as a result of human activity such as agriculture, urbanization and sewage discharge. *alt*. accelerated eutrophication. *see* eutrophication.

cultural environments artificial environments, such as arable land, in which the vegetation cover is deliberately determined by humans. *cf*. natural environments, semi-natural environments.

cultural inheritance the transmission of particular traits and behaviours from generation to generation by learning rather than genetic inheritance.

culture *n*. (*biol*.) microorganisms, cells or tissues growing in a nutrient medium in the laboratory. *v*. to isolate and grow microorganisms, cells or tissues as above.

culture collection a reference collection of different species and strains of microorganisms or cultured cells.

culvert *n*. pipe conveying the water of a ditch or stream beneath, e.g., a road, canal or railway embankment.

cummingtonite *n.* an amphibole mineral with the formula $(Mg,Fe)_7Si_8O_{22}(OH)_2$, the magnesium-rich member of the cummingtonite–grunerite series (*q.v.*) of minerals. The highly fibrous variety of this mineral is known as amosite or brown asbestos.

cummingtonite–grunerite series a series of amphibole minerals (*q.v.*) with the formula $(Mg,Fe)_7Si_8O_{22}(OH)_2$. *see also* cummingtonite, grunerite.

cumulonimbus *n.* a low-altitude cloud with massive vertical development, often spreading laterally at the top in the semblance of an anvil at the end of their development. These enormous, dense, grey clouds are associated with thunderstorms. *alt.* anvil cloud, thunderhead.

cumulus *n.* low-altitude cloud with a characteristic domed shape and flat base. White cumulus clouds are portents of fair weather. *alt.* fair weather cloud.

cup fungi common name for the Discomycetes *q.v.*

cuprite *n.* Cu_2O, a copper ore mineral. *alt.* red copper ore.

curie (Ci) *n.* unit of radiation corresponding to an amount of radioactive material producing 3.7×10^{10} disintegrations per second, which is the activity of radium. It is being replaced by the derived SI unit, the becquerel (Bq). 1 Ci = 3.7×10^{10} Bq.

curium (Cm) *n.* a transuranic element [at. no. 96, r.a.m. (of its most stable known isotope) 247], a silvery-white metal when in elemental form.

currant gall type of small spherical gall found on oak leaves or catkins caused by larvae of the gall wasp *Neuoterus quercus baccarum*.

current *see* ocean current.

current-bedding (*geol.*) bedding that occurs by deposition on an inclined surface (e.g. a sand dune's lee slope or a deltaic frontal slope). It is generally characterized by inclined lenticular or gently curving laminae. Typically, these laminae asymptotically approach a lower surface (that is nearer the horizontal than the incline upon which the deposition occurred) and are truncated by a very roughly parallel upper surface. This upper surface represents the lower surface of the next group of lam-

inae, etc. Current-bedding patterns can be used to establish which way up a disturbed bedding sequence lies.

cursorial *a.* having limbs adapted for running, *appl.* birds, bipedal dinosaurs, etc.

cusp *n.* (1) a prominence, as on molar teeth; (2) a sharp point. *alt.* tubercle.

cuspate delta *see* delta.

cuticle *n.* (*biol.*) (1) an outer skin, sometimes referring to the epidermis as a whole, esp. when impermeable to water; (2) (*zool.*) an outer protective layer of material, of various composition, produced by the epidermal cells, that covers the body of many invertebrates; (3) (*bot.*) layer of waxy material, cutin, on the outer wall of epidermal cells in many plants, making them fairly impermeable to water. *a.* cuticular.

cutoff oxbow lake *q.v.*

cuttlefish *n.* common name for a group of cephalopod molluscs, e.g. *Sepia*, characterized by a shell of unusual structure (cuttlebone), and having eight short arms around the mouth and long tentacles.

C value the total amount of DNA in the haploid genome of a species, either measured directly in picograms or expressed in base pairs or daltons.

cwm cirque *q.v.*

cwm glaciers cirque glaciers *q.v.*

cyanazine *n.* foliar and soil-acting herbicide used to control annual meadow grass and broad-leaved weeds in maize and peas.

cyanidin *n.* violet flavonoid pigment present in many flowers.

Cyanobacteria, cyanobacteria *n.*, *n.plu.* a group of prokaryotic, photosynthetic, non-flagellate, unicellular, filamentous or colonial microorganisms, found in aquatic and terrestrial environments either free-living or in symbiotic association with fungi as lichens. They have an oxygen-evolving type of photosynthesis resembling that of algae and green plants. Some species (e.g. *Anabaena, Nostoc*) can fix atmospheric nitrogen. Commonly (but misleadingly) known as the blue-green algae and formerly classified in the plant kingdom as the Cyanophyta or Cyanophyceae. They contain the pigments chlorophyll (contained in infolded membranes), alpha- and beta-carotenes, phycoerythrin and phycocyanin and often appear blue. Some cyanobacteria produce toxins that can become a health hazard in conditions where cyano-bacterial 'algal blooms' appear. *see* Fig. 5, p. 68.

cyanogenesis *n.* the elaboration of hydro-cyanic acid by some plants. *a.* cyanogenic.

cyanogenic glycosides plant glycosides that, when fully hydrolysed, liberate hydro-gen cyanide, which is toxic to most cells. Found in species of *Sorghum*, *Prunus* and *Linum*.

Cyanophyceae,   Cyanophyta *see* cyanobacteria.

cyanosis *n.* bluish tinge to the skin caused by lack of oxygen.

Cycadeoidophyta, cycadeoids *n.*, *n.plu.* division of fossil cycad-like plants that had massive stems, pinnate leaves and were usually monoecious, with sporophylls ar-ranged in a flower-like structure. They have been considered as possible ancestors of angiosperms.

Cycadophyta, cycads *n.*, *n.plu.* one of the five main divisions of extant seed-bearing plants, commonly called cycads or sago palms. They are palm-like in appearance with massive stems that may be short or tree-like. Microsporangia and megasporan-gia are borne on sporophylls arranged in cones, male and female cones being borne on separate plants. In some classifications called the Cycadopsida or Cycadales.

Cyclanthales *n.* order of often palm-like monocots, also climbers and large herbs, comprising the family Cyclantheraceae.

cycloate *n.* soil-acting carbamate herbicide used to control annual weeds in crops such as beet.

cyclodiene insecticides a group of organo-chlorine insecticides, including dieldrin, aldrin and chlordane.

cyclogenesis *n.* (*clim.*) development of cyclones (low-pressure cells). *cf.* anti-cyclogenesis.

cycloheximide *n.* antibiotic produced by the actinomycete *Streptomyces griseus*, which inhibits protein synthesis in eukaryotic cells only, by inhibiting translation.

cyclone dust separator device used to remove particulate matter from waste gas streams, prior to their discharge into the atmosphere. During operation, dust-laden gases enter the top of this essentially cylin-drical device, at a tangent. These move downwards in a helical fashion, generat-ing centripetal forces that drive the particu-lates to the walls. These then fall to exit the cyclone at the bottom. The cleaned gases then leave the top of the cyclone via a pipe at its centre. *see also* electrostatic precipitator, fabric filter.

cyclones depressions *q.v.*

cyclonic rain rain formed when two air masses of different temperatures meet and the warmer, moister, less dense air is forced to override the colder, denser air, thus leading to cloud formation as it ascends. This type of rain is associated with tropical cyclones and, in temperate latitudes, with the passage of depressions. *alt.* frontal rain.

cyclostomes *n.plu.* lampreys and hagfish, primitive jawless fishes.

cyhexatin *n.* organotin compound used as a contact acaricide to control, e.g., red spider mite on fruit crops.

cyme *n.* any repeatedly branching determin-ate inflorescence in which each growing point ends in a flower, with the oldest flowers at the end of the branch.

Cynodontia, cynodonts *n.*, *n.plu.* group of extinct mammal-like reptiles, found from the Late Permian to the middle of the Jurassic, possessing a secondary bony palate, complex crowns on the cheek teeth, and other mammalian features and that are believed to be the direct ancestors of mammals.

Cyperales *n.* order of herbaceous monocots with rhizomes and solid stems triangular in cross-section and comprising the fam-ily Cyperaceae (sedge).

cypermethrin *n.* synthetic pyrethroid used as a contact insecticide to control many pests of fruit and vegetables.

cyprinids, cyprinoids *n.plu.* group of fresh-water fish (the Cyprinoidei) widespread in Europe, Asia, Africa and North America, and including the carps and minnows.

cyprinodonts *n.plu.* an order of small, mainly tropical fishes, the Cyprinidonti-for mes, including the toothed carps.

cypsela *n.* an inferior achene composed of two carpels, as in Compositae.

Cys cysteine *q.v.*

cyst *n.* (1) protective coat surrounding rest-ing cells of, e.g., soil amoebae and some bacteria, that is formed in dry conditions; (2) a bladder or air vesicle in certain seaweeds; (3) an abnormal fluid-filled sac developing in tissues; (4) any bladder- or sac-like structure.

cysteine (Cys, C) *n.* a sulphur-containing amino acid, constituent of many proteins, where it forms disulphide bonds cross-linking the protein chain.

cystine (Cys–Cys) *n.* amino acid residue formed in proteins by oxidation of the sulphydryl groups of two cystine residues to form a disulphide bridge cross-linking the protein chain.

cyt cytochrome *q.v.*

cytidine *n.* a nucleoside made up of the pyrimidine base cytosine linked to ribose.

cytidine triphosphate (CTP) cytosine nucleotide containing a triphosphate group, one of the ribonucleotides required for RNA synthesis; also acts as an energy donor in metabolic reactions in a manner analogous to ATP.

cytocidal *a.* cell-destroying.

cytodeme *n.* a local interbreeding unit in a taxon differing cytologically, usually in chromosome number, from the rest of the taxon.

cytogenesis *n.* development or formation of cells.

cytokinins *n.plu.* plant growth hormones, derivatives of adenine, and including naturally occurring zeatin and isopentenyladenine, and kinetin, which probably does not occur in nature. They act in concert with IAA (an auxin) to promote rapid cell division, and are widely used to induce the formation of plantlets from callus tissue in culture. *alt.* kinin, phytokinin.

cytomegaloviruses *n.plu.* a group of DNA viruses of the herpesvirus family, infecting vertebrates, the infection characterized by the formation of large inclusion bodies in infected glandular cells (often of salivary gland in humans), which can be fatal in very young children.

cytoplasm *n.* all the living part of a cell inside the cell membrane, excluding the nucleus. *a.* cytoplasmic, *pert.* or found in the cytoplasm.

cytosine (C) *n.* a pyrimidine base, constituent of DNA and RNA, which is the base in the nucleoside cytidine. In DNA and RNA it pairs with guanine.

cytoskeleton *n.* internal proteinaceous framework of a eukaryotic cell. It is composed of actin microfilaments, intermediate filaments and microtubules and gives shape to a cell, provides support for cell extensions, such as villi and axons of nerve cells, is involved in cell movement, in interactions with the substratum on which the cell is lying, and in intracellular transport.

cytosol *n.* the ground substance of the cytoplasm other than the various membrane-bounded organelles.

cytostatic *a. appl.* any substance suppressing cell growth and multiplication.

cytostome *n.* the specialized region acting as a 'mouth' of a unicellular organism, as in some protozoans.

cytotaxis *n.* movement of cells to or away from a stimulus.

cytotoxic *a.* attacking or destroying cells.

D

δ, Δ delta, for headwords with prefix δ- or Δ- refer to delta-.

d (1) prefix attached to abbreviations for nucleotides (e.g. dATP, dCTP, dGTP) to indicate they contain the sugar deoxyribose rather than ribose; (2) symbol for the prefix deci-*q.v.*

D (1) aspartic acid (*q.v.*); (2) dalton (*q.v.*); (3) deuterium (*q.v.*); (4) Simpson diversity index *q.v.*

2,4-D 2,4-dichlorophenoxyacetic acid, a synthetic auxin widely used as a translocated herbicide for controlling broad-leaved weeds in grassland and cereals.

D-, L- prefixes denoting particular molecular configurations, defined according to convention, of certain optically active compounds, esp. monosaccharides and amino acids, the L configuration being a mirror image of the D.

Da, dal dalton *q.v.*

dalapon *n.* translocated herbicide used to control annual and perennial grasses in, e.g., fruit crops that grow on canes, bushes or trees.

dalmatian coastline a submerged coastline characterized by inlets that run parallel to the coast, and chains of islands.

DALR dry adiabatic lapse rate *q.v.*

dalton (**D, Da, dal**) *n.* a mass unit equal (by definition) to one-twelfth of the mass of a single atom of carbon-12, *ca.* 1.66×10^{-27} kg. *alt.* atomic mass unit.

daminozide *n.* plant growth regulator used, e.g., in fruit trees to retard vegetative growth.

damselflies *see* Odonata.

dance, of bees a series of movements performed by honeybees on their return to the hive after finding a food source, which informs other bees in the hive of the location of the food. *see also* waggle dance.

daphnid *n.* any of various small water fleas, esp. of the genus *Daphnia*.

dark-field microscopy type of light or optical microscopy used for studying living cells, which produces an illuminated object on a dark background.

dark reactions in photosynthesis, reactions occurring in the stroma of chloroplasts, for which light is not required, in which carbon dioxide is reduced to carbohydrate. *alt.* Calvin cycle, carbon dioxide fixation, photosynthetic carbon reduction cycle (PCR cycle).

dark ruby silver pyrargyrite *q.v.*

Darwinian evolution *see* Darwinism.

Darwinian fitness the fitness of a genotype measured by its proportional contribution to the gene pool of the next generation.

Darwinism *n.* the theory of evolution by means of natural selection, put forward by Charles Darwin in his book *Origin of Species* published in 1859 but formulated some years earlier. The theory was based on his observation of the genetic variability that exists within a species and the fact that organisms produce more offspring than can survive. Under any particular set of environmental pressures, those heritable characteristics favouring survival and successful reproduction would therefore be preferentially passed on to the next generation (natural selection). Selection for particular aspects of life-style or in different environmental conditions could therefore eventually lead to two populations differing in many ways from the original and to the development of complex adaptations to a particular mode of life or environment. A very similar theory was proposed independently by Alfred Russel Wallace. In the 1930s and 1940s the theory of Mendelian genetics was incorporated into Darwin's original theory to produce a modern version, the neo-Darwinian synthesis. *see also* natural selection, neo-Darwinism.

dating technique any technique used to date an object or event. Absolute dating techniques are used to give absolute ages [i.e. age with respect to the present, e.g. dendrochronology (*q.v.*), potassium–argon dating (*q.v.*), radiocarbon dating (*q.v.*)] while relative dating techniques, e.g. pollen analysis (*q.v.*), are used to locate samples within a chronological sequence of events.

Datiscales Begoniales *q.v.*

dative bond (*chem.*) a covalent bond formed by the sharing of a lone pair of electrons that originated from one of the two atoms bonded together. *alt.* co-ordinate bond.

daughter *n.* (*biol.*) progeny of cell, nucleus, etc., arising from division, with no reference to sex, as daughter cell, daughter nucleus.

daughter nucleus (1) (*phys.*) the atomic nucleus formed when an unstable parent nucleus undergoes radioactive decay. Many daughter nuclei are themselves unstable and therefore radioactive. *alt.* daughter product; (2) (*biol.*) *see* daughter.

daylighting *n.* the use of natural light to illuminate the interior of commercial buildings such as offices and shopping centres.

day-neutral *a. appl.* plants in which flowering can be induced by either a long or a short photoperiod or by neither.

dazomet *n.* soil sterilant used as a fumigant to control nematodes and soil fungi under glass, and to check weed growth.

dB decibel *q.v.*

DBH diameter at breast height (1.4 m), the standard measurement of a trunk of a tree.

DDE dichlorodichlorophenylethylene, a compound formed by the dehydrochlorination of the organochlorine insecticide DDT during its metabolism in animals. Also produced in the environment by the slow degradation of DDT.

DDT dichlorodiphenyltrichloroethane, a persistent organochlorine insecticide, first manufactured in 1940 for Allied use in the Second World War, where it was used to control tropical malaria (carried by mosquitoes) and typhus (transmitted by body lice). Post-war, its use was extended to the control of insect pests of crops. By the 1970s, concern over its persistence in the environment, and its biomagnification up the food chain, led to its use (and that of other related organochlorine pesticides, such as lindane and dieldrin) being banned in many of the more-developed countries.

de- prefix from L. *de*, away from, denoting removal of.

death point temperature, or other environmental variable, above or below which organisms cannot exist.

death rate number of deaths within a population over a set period, usually a year. The crude death rate is calculated for human populations as the annual number of deaths per 1000 population in a given geographical area, with the population number usually taken at the midpoint of the year in question. *alt.* mortality rate.

debris flow *see* flow movement.

debt-for-nature swap practice whereby conservation organizations buy debt obligations of developing countries from lenders, such as banks, usually at a considerable discount. An offer to cancel the debt is then made to the debtor country, on condition that they agree to take certain conservation measures, such as the establishment of nature reserves.

deca- (1) prefix derived from Gk *deka*, ten, denoting having ten of, divided into ten, etc.; (2) SI prefix denoting that the unit it applies to is multiplied by ten, abbreviation da.

decalcify *v.* to treat with acid to remove calcareous parts.

Decapoda, decapods *n., n.plu.* (1) order of freshwater, marine and terrestrial crustaceans having five pairs of legs on the thorax and a carapace completely covering the throat, and including the prawns, shrimps, crabs and lobsters; (2) an order of cephalopods having two retractile arms as well as eight normal arms, including the squids and cuttlefish.

decay *n*. (1) the decomposition of dead organic matter; (2) radioactive decay *q.v.*

deception *n*. any behaviour or feature that deceives a predator or other animal and that ranges from physical mimicry to apparently cognitive deceptions. *cf*. honest behaviour.

deci- (d) SI prefix used to indicate that the unit it applies to is multiplied by 10^{-1}, e.g. 1 decilitre = 1×10^{-1} l (one-tenth of a litre).

decibel (dB) *n*. unit used to measure the intensity of sound on a logarithmic scale. It is based on measurements of sound intensity in watts per square metre.

deciduous *a*. (1) falling at the end of growth period or at maturity; *appl*. (2) teeth, milk teeth; (3) trees, those having leaves that all fall at a certain time of the year.

decline phase of population growth *see* S-shaped growth curve.

decommissioning process process (as yet untried) whereby a nuclear power plant is finally closed down. A three-stage process is proposed in the event of decommissioning, i.e. mothballing (an initial period to allow for the natural subsidence of radiation), entombment (the permanent encasement of radioactive parts with suitably lined reinforced concrete) and dismantlement (the total removal of radioactive material).

decomposer *n*. organism that feeds on dead plant and animal matter, breaking it down physically and chemically and recycling elements and organic and inorganic compounds to the environment. Decomposers are chiefly microorganisms and small invertebrates.

decomposer food chain pathway within a food web that is based on the consumption of dead organic matter (in the form of plant and animal remains and faecal waste) by decomposers. *cf*. grazer food chain.

deep ecology belief that Earth has finite resources and that continuing population growth, production and consumption will put severe stress on the natural processes that maintain the natural resource base of air, water and soil. People should thus work with nature by controlling population growth, reducing unnecessary use and waste of matter and energy resources and not cause the premature extinction of other species. The belief that technology will be able to solve the problems of environmental overload is considered to be misplaced.

deep litter system an intensive housing system used for rearing poultry, usually for breeding or meat production. The litter layer is 15–30 cm deep and composed of chopped straw or wood shavings.

deep mining subsurface extraction of mineral materials (e.g. coal, diamonds and gold) from deposits lying deep underground. Workings typically consist of a vertical shaft from which a number of horizontal tunnels radiate at different levels. *cf*. surface mining.

deep-sea plain abyssal plain *q.v.*

deep-sea trenches steep-sided troughs, V-shaped in cross-section, cut into abyssal plains. These trenches are associated with intense volcanic activity.

deepshaft effluent treatment system sewage effluent treatment system that utilizes a sealed shaft *ca*. 135 m deep. The sewage is made to flow down a central cavity within the shaft, returning to the surface between the wall of the cavity and that of the shaft. During its passage through the system, the sewage is aerated with compressed air, facilitating the breakdown of its organic content by aerobic microorganisms. This process compares favourably with conventional systems in terms of cost and efficiency and, esp. the area of land occupied.

defaunation *n*. removal of animal life from an area.

deficiency diseases pathological conditions in plants and animals due to lack of some vitamin, trace element or other minor nutrient, e.g. crown rot in sugar-beet due to boron deficiency, vitamin deficiency diseases in mammals such as scurvy due to lack of vitamin C, beriberi due to lack of vitamin B_1, rickets due to lack of vitamin D.

definitive host host organism in which a parasite lives during its adult phase, and in which it may reproduce.

deflation *n*. process whereby fine particles are carried away by wind sweeping along a ground surface. This process is particularly active in arid and semi-arid regions.

deflation hollows shallow basins found in desert and semidesert regions, formed by deflation (*q.v.*). *alt.* blow outs.

deflocculation *n.* separation of clumps of particles (flocs) into their individual component particles.

defoliants *n.plu.* herbicides that cause plants to lose their leaves prematurely. *see* Agent Orange.

defoliation *n.* the removal of all leaves from a plant or tree.

deforestation *n.* complete and permanent removal of forest or woodland and its associated undergrowth.

deglaciation *n.* melting and retreat of a glacier or ice sheet.

degradation *n.* (*geog.*) process whereby a land surface is lowered and rendered smooth by the action of weathering and erosion. *cf.* aggradation.

degree of relatedness coefficient of relationship *q.v.*

dehiscent *a. appl.* a fruit or spore capsule that opens spontaneously along predetermined lines or in a particular direction.

dehydration *n.* the removal of water from a substance.

deimatic display frightening behaviour consisting of adoption of a posture by one animal to intimidate another.

de-inking *n.* industrial process used to remove ink from printed paper, prior to paper recycling.

deionized water water that has been purified by passage through a cation exchanger in H^+ form and an anion exchanger in OH^- form. The former replaces cations present (e.g. , $Ca^{2+}_{(aq)}$, $Na^+_{(aq)}$) with $H^+_{(aq)}$, while the latter replaces anions present (e.g. $Cl^-_{(aq)}$, $SO_4^{2-}_{(aq)}$) with $OH^-_{(aq)}$. The $H^+_{(aq)}$ and $OH^-_{(aq)}$ released then combine to form water.

delivered energy that energy (in the form of gas, electricity, etc.) that is delivered to the consumer's premises. It differs from primary energy in that account has been taken of losses in extraction, processing, conversion and transmission.

delphinology *n.* the study of dolphins.

delta *n.* accumulation of fine sediment deposited by a river where it enters a lake or sea. Three main types of delta are recognized on the basis of shape: bird's

foot delta (lobe-shaped) e.g. Mississippi; arcuate delta (fan-shaped) e.g. Nile; and cuspate delta (tooth-shaped) e.g. Tiber.

delta front a delta's shoreline plus the submerged zone of the delta found offshore.

delta G (ΔG) Gibbs' free energy change *see* free energy.

deltamethrin *n.* synthetic pyrethroid used as a contact insecticide to control numerous types of pests of vegetables and fruit.

delta plain the landward portion of the delta behind the delta front. This plain is usually traversed by many small distributary channels.

deme *n.* (1) (*bot.*) assemblage of individuals of a given taxon, usually qualified by a prefix as ecodeme (*q.v.*), gamodeme (*q.v.*), topodeme (*q.v.*); (2) (*zool.*) a gamodeme, a local population unit of a species within which breeding is completely random.

demephion *n.* systemic organophosphate acaricide and insecticide used to control red spider mites and aphids on a variety of crops.

demersal *a.* living on or near the bottom of sea or lake.

demersed *a.* growing under water, *appl.* parts of plants.

demeton-S-methyl *n.* systemic organophosphate acaricide and insecticide used to control red spider mites and aphids on most crops.

demographic society a society that is relatively stable throughout time, being relatively closed to newcomers and whose composition is therefore the result largely of the demographic processes of birth and death.

demographic transition the transition from high birth and death rates to low birth and death rates, usually associated with the general development and industrialization of a society.

demography *n.* the study of numbers of organisms in a population and their variation over time.

Demospongia *n.* a class of sponges, Porifera (*q.v.*), which may have silica spicules in the form of simple needles or a four-armed spicule whose points describe a tetrahedron, or may have no spicules and that often have the body wall strengthened by a tangled mass of spongin fibres (e.g.

in the bath sponge *Spongia*). They are found on shores, and down to depths of more than 5000 m.

denatant *a*. swimming, drifting or migrating with the current.

dendritic drainage *see* drainage pattern.

dendrochronology *n*. the study of the age of trees and timber artefacts, generally by counting the annual growth rings, and the study and analysis of tree-rings in relation to changes in climate over time.

dendroclimatology *n*. the study of the annual growth rings of trees in relation to past climate.

Dendrogaea *n*. a biogeographical region including all the neotropical region except temperate South America.

dendrogram *n*. any branching tree-like diagram illustrating the relationship between organisms or objects.

dendrology *n*. the study of trees.

dendrometer *n*. a device for measuring small changes in the diameter of a tree trunk, such as the minute amounts of shrinkage and swelling that accompany the daily fluctuations in transpiration.

dengue *n*. an acute viral disease spread by mosquitoes. Symptoms of this (rarely fatal) disease include fever and severe joint pain. *alt*. breakbone fever.

denitrification *n*. (1) the conversion of nitrate to nitrite and nitrite to molecular nitrogen leading to the loss of nitrogen from the biosphere, carried out by a few genera of anaerobic bacteria; (2) the reduction of nitrates to nitrites and ammonia, as in plant tissues. *see* nitrogen cycle. *see* Fig. 18, p. 277.

denitrifier, denitrifying bacteria *n*. any of a group of diverse anaerobic bacteria capable of converting nitrate to nitrite and nitrite to molecular nitrogen (denitrification), e.g. *Pseudomonas*, *Achromobacter*, *Thiobacillus* and *Micrococcus* spp.

density *n*. (1) mass of a substance per unit volume, expressed, in SI units, in kg m^{-3}; (2) number of individuals inhabiting a given area.

density currents *see* ocean currents.

density-dependent *a*. *appl*. (1) factors limiting the growth of a population that are dependent on the existing population density and that are generally effects of other species in the form of competition,

predation or parasitism; (2) selection that either favours or disfavours the rarer forms of individual within a population.

density-independent *a*. *appl*. factors limiting the growth of a population that are independent of the existing density of the population, which are generally abiotic factors such as temperature, light intensity, wind.

density overcompensation the situation when an increase in population is matched by an even greater decrease in the birth rate or increase in the death rate than that necessary to compensate, leading to an eventual decrease in population size.

dental formula method of representing the number of each type of tooth in a mammal, consisting of a series of fractions, the numerators representing the number of each type of tooth in one-half of the upper jaw, and the denominators the number in the corresponding lower jaw.

dentition *n*. the type, number and arrangment of teeth.

denudation *n*. (*geog*.) process by which rock is exposed as a result of the removal of overlying material, by a combination of weathering and erosion.

deoxyribonuclease (DNase) *n*. any of various enzymes that cleave DNA into shorter oligonucleotides or degrade it completely into its constituent deoxyribonucleotides. *alt*. nuclease.

deoxyribonucleic acid *see* DNA and associated entries.

deoxyribonucleotide *n*. a nucleotide containing the sugar deoxyribose. The four deoxyribonucleotides required for DNA synthesis are dATP, dCTP, dGTP and dTTP.

deoxyribose *n*. a pentose sugar similar to ribose but lacking an oxygen atom, present in DNA.

deplasmolysis *n*. re-entry of water into a plant cell after plasmolysis and reversal of shrinkage of protoplasm.

depleted *a*. *appl*. (1) a natural resource that is being used up faster than it is being replaced, or to a non-renewable resource that is being used up; (2) material with a smaller than normal level of a particular isotope; (3) uranium from which the fissile isotope uranium-235 has been removed, leaving essentially pure uranium-238; (4) spent fuel from nuclear reactors.

depletion time the time it takes to use a given fraction, usually taken as 80%, of a known or estimated supply of a non-renewable resource at an assumed rate of use.

deposit feeder aquatic organism that swallows sediments in order to feed on the associated microorganisms.

depressions *n.plu.* (*clim.*) essentially circular areas of low pressure, with diameters of up to 3000 km, which usually develop at fronts and typically last for 4–7 days. They are common atmospheric features of the mid-latitudes and are usually associated with significant precipitation. *cf.* nonfrontal depressions. *alt.* cyclones, extratropical cyclones, frontal depressions, lows.

Derbyshire neck goitre *q.v.*

derived *a. appl.* character or character state not present in the ancestral stock.

Dermaptera *n.* an order of insects, commonly called earwigs, having cerci modified as forceps, small leathery fore wings and membranous hind wings. Undergo a slight metamorphosis.

dermatomycosis *n.* a fungal infection of human or animal skin.

dermatophyte *n.* any fungal parasite of skin.

derris *n.* broad-spectrum, non-persistent insecticide and acaricide, extracted from the roots of certain legumes. Used mainly in gardens to control, e.g., caterpillars and aphids. It is toxic to fish, but not to mammals or birds. *alt.* rotenone.

DERV (1) diesel-engined road vehicles; (2) the (low-sulphur) grade of fuel used by such vehicles.

DES diethylstilboestrol *q.v.*

desalination *n.* the removal of salt from saline water by freezing, distillation, or electrolysis followed by reverse osmosis. *alt.* desalinization.

desalinization *n.* (1) the removal of salt from water, *alt.* desalination; (2) the removal of salt from soil.

desert *n.* a biome where the average amount of precipitation is erratic and less than 25 cm per annum, and evaporation exceeds precipitation. Such areas have sparse highly adapted vegetation, e.g. cacti, succulents and spiny shrubs. Hot deserts such as the Sahara have very high daytime temperatures, cold deserts such as the Gobi Desert and the Northern Californian Desert have very low winter temperatures.

deserticolous *a.* living in the desert.

desertification *n.* the conversion of pastureland and crop land into desert, or the gradual enlargement and encroachment of deserts into formerly marginal arid lands, caused by climatic factors such as prolonged drought and by overgrazing and overcultivation.

desert pavement a type of ground surface found in arid regions where a coarse layer of gravel or larger-sized particles (known as lag deposits) covers finer sediments. *alt.* stone pavement.

desert varnish a coating layer (10–30 μm thick) found on rock surfaces in arid regions. This varnish, composed of iron and manganese oxides and clay minerals, takes thousands of years to develop, apparently as a result of microbial activity. *alt.* rock varnish.

desiccant *n.* a substance used as a drying agent.

desirability quotient in risk–benefit analysis, a number produced by dividing the estimate of the benefits to society of using a particular product or process by the estimated risk.

desmetryne *n.* translocated contact and soil-acting herbicide used to control annual broad-leaved weeds in brassicas.

desmids *n.plu.* a group of unicellular or colonial green freshwater algae whose cells are typically almost divided in two by a narrow constriction of the cell wall.

despotism *n.* social system in animals in which one individual dominates the rest of the flock, which are all equally subservient to him and of equal rank with each other. *see also* dominance systems.

desquammation exfoliation *q.v.*

destructive boundary convergent boundary *q.v.*

destructive waves *n.plu.* (*oceanogr.*) waves responsible for beach erosion. These occur when the backwash is stronger than the swash. *cf.* constructive waves.

detergents *n.plu.* manufactured cleaning agents that primarily act by facilitating the dispersal of dirt and grease in water. The formulation of a detergent typically includes a surfactant and a builder, together with such substances as optical brighteners, perfumes, bleaches, enzymes and foam regulators. Detergents high in phosphates (used as builders) have led to problems of eutrophication (*q.v.*), primarily in lakes.

detergent swans mounds of foam generated at waterfalls and weirs downstream from sewage outlets, caused by the presence of non-biodegradable branch-chained organic detergents (hard detergents). The introduction in the 1960s of biodegradable linear organic detergents (soft detergents) unable to survive secondary sewage treatment has alleviated this problem of foaming.

determinate error systematic error *q.v.*

detoxification *n.* rendering a poison or toxin harmless by chemical or heat denaturation or by removing its toxic constituents.

detritivore *n.* organism that feeds on detritus. *alt.* detritiphage, detritus feeder.

detritus *n.* (1) small pieces of dead, cast-off and decomposing plants and animals and their wastes; (2) detached and broken down fragments of a structure. *a.* detrital.

detritus feeders soil animals that extract nutrients from detritus (*q.v.*), e.g. earthworms. *see also* decomposers.

deuterium *n.* the isotope of hydrogen that has a mass number of two, it is symbolized ^2_1H, ^2H or D.

deuterium oxide *see* heavy water.

Deuteromycota, Deuteromycotina, deuteromycetes *n.*, *n.*, *n.plu.* large group of fungi, known only in the asexual, conidia-bearing form, but which display strong affinities to the ascomycetes. They may be ascomycetes that have lost their ascus stage or in which the sexual phase has not yet been discovered. *alt.* Fungi Imperfecti.

deuterostomes *n.plu.* collectively, all animals with a true coelom, radial cleavage of the egg, and in which the blastopore becomes the anus: pogonophorans, hemichordates, echinoderms, urochordates and chordates. *cf.* protostomes.

deuterotoky *n.* parthenogenesis where both sexes are produced.

deuterotype *n.* the specimen chosen to replace the original type specimen for designation of a species.

development *n.* in biology, the changes that occur as a multicellular organism develops from a single-celled zygote, from the first cleavage of the fertilized ovum until maturity. *alt.* ontogeny.

developmental *a. pert.* or involved in development, *appl.* genes, hormones, etc., specifically active during development.

Devonian *a. pert.* or *appl.* geological period lasting from *ca.* 400 to 360 million years ago.

dew *n.* fine droplets of water formed on cool surfaces (vegetation, ground, etc.) as a result of the condensation of atmospheric water vapour. Overnight cooling of air near to the ground is the usual mechanism by which dew is formed.

dew-point temperature at which the partial pressure of water vapour present in moisture-laden air equals the saturated vapour pressure of water. Air cooled to this temperature will become saturated and condensation of water will begin if there are suitable surfaces (e.g. those of dust particles) on which this can occur. *alt.* saturation level.

dexiotropic *a.* turning from right to left, as whorls, *appl.* shells.

dextral *a.* on or *pert.* the right.

dextran *n.* any of a variety of storage polysaccharides (usually branched) made of glucose residues joined by alpha-1,6 linkages, found in yeast and bacteria.

D-horizon the unweathered parent rock, or bedrock, at the base of a soil. *see* Fig. 28, p. 385. *alt.* R-horizon.

diabase *n.* (1) in USA, *alt.* dolerite; (2) in UK, dolerite that has been altered.

diachronous *a.* dating from different periods, *appl.* fossils occurring in the same geological formation but of different ages.

diadromous *a.* migrating between fresh and sea water.

diagenesis lithification *q.v.*

diageotropism *n.* a growth movement in a plant organ so that it assumes a position at right angles to the direction of gravity.

diagnostic *a.* (1) distinguishing; (2) differentiating the species, genus, etc., from others similar.

diaheliotropism *see* diaphototropism.

dialect *n*. local variant of bird songs, mating calls, bee waggle dances, etc.

di-allate *n*. soil-acting herbicide used to control wild oats and blackgrass in beet and brassicas.

dialysis *n*. separation of colloids (such as proteins) from small molecules and ions by the inability of the larger molecules to pass through a semipermeable membrane.

diameter at breast height *see* DBH.

dianthovirus group plant virus group containing isometric single-stranded RNA viruses, type member carnation ringspot virus.

diapause *n*. a spontaneous state of dormancy occurring in the lives of many insects, esp. larval stages.

Diapensales *n*. order of dicot trees and herbs including the families Ebenaceae (ebony), Sapotaceae (sapote), Styracaceae (styrax) and others.

diaphototropism *n*. (1) a growth movement in plant organs to assume a position at right angles to rays of light; (2) when the light is sunlight known as diaheliotropism.

diaphragm cells mercury-free electrochemical cells used as an alternative to those employing mercury cathodes in the chloralkali process. *alt*. membrane cells.

diapsid *a*. having a skull with both dorsal and ventral temporal openings on each side.

diapsids *n.plu*. reptiles of the subclass Diapsida, with diapsid skulls, known from the Late Carboniferous, to which most modern reptiles belong, and also including extinct forms such as the dinosaurs.

diaspore *n*. (1) (*geol*.) AlO(OH), an aluminium ore mineral, one of the constituents of bauxite (*q.v.*); (2) (*biol*.) any spore, seed, fruit, or other part of plant when being dispersed and able to produce a new plant.

diastrophism *n*. extensive tectonic deformation of the Earth's crust, e.g. by folding and faulting processes, orogenesis (mountain building) and epeirogenesis (*q.v.*). *alt*. tectonism.

diatom *n*. common name for a member of the Bacillariophyta, a group of algae characterized by delicately marked thin double shells of silica.

diatomaceous *a*. containing the shells of diatoms, *appl*. earth.

diatropism *n*. tendency of organs or organisms to place themselves at right angles to line of action of stimulus.

diazinon *n*. contact organophosphate insecticide used in the control of carrot fly and cabbage root fly.

diazotroph *n*. organism able to fix elemental nitrogen to ammonia.

dicalcium phosphate ($CaHPO_4$) (*agric*.) inorganic phosphorus fertilizer supplied in granular form.

dicamba *n*. translocated herbicide used to control broad-leaved weeds in grassland and cereal crops.

dichlobenil *n*. soil-acting herbicide used to control weeds, e.g., in forests and orchards.

dichlofluanid *n*. fungicide used in the control of diseases such as botrytis on, e.g., soft fruits.

dichlorodichlorophenylethylene DDE *q.v.*

dichlorofluoromethane *see* chlorofluorocarbons.

2,4-dichlorophenoxyacetic acid 2,4-D *q.v.*

dichlorodiphenyltrichloroethane DDT *q.v.*

dichloropropene *n*. organochlorine nematicide used as a fumigant to sterilize nematode-infested soil, either in the field or under glass.

dichlorprop *n*. translocated herbicide used in the control of many broad-leaved weeds.

dichlorvos *n*. organophosphate pesticide used as a fumigant in glasshouses, in the form of impregnated resin strips, to control pests such as aphids, whiteflies and red spider mites.

dichotomy *n*. (1) branching that results from division of growing point into two equal parts; (2) repeated forking. *a*. dichotomous.

dichromatism *n*. condition in which members of a species show one of only two distinct colour patterns. *a*. dichromatic.

dickite *n*. $Al_2Si_2O_5(OH)_4$, one of the kandite group (*q.v.*) of clay minerals (*q.v.*). A polymorph of kaolinite.

diclofop-methyl *n*. foliar-acting herbicide used to control wild oats in a number of crops.

dicofol *n.* contact organochlorine acaricide used to control red spider mites on fruit and glasshouse crops.

dicots dicotyledons *q.v.*

Dicotyledones, dicotyledons *n., n.plu.* a class of flowering plants having an embryo with two cotyledons (seed leaves), parts of the flower usually in twos or fives or their multiples, leaves with net veins, and vascular bundles in the stem in a ring surrounding a central pith. *alt.* Magnoliopsida.

dicotyledonous *a. pert.* (1) dicotyledons (*q.v.*); (2) an embryo with two cotyledons (seed leaves).

Dictyoptera *n.* order of insects including the cockroaches and praying mantises, winged but often non-flying, with long antennae, biting mouthparts, tough narrow forewings and broad membranous hindwings.

Dicyemida(e) *n.* class of Mesozoa having a body that is not annulated.

Didymelales *n.* order of dicot trees comprising the family Didymelaceae with a single genus *Didymeles*.

dieback *n.* (1) population crash (*q.v.*); (2) death of stems of woody plants from the tip backwards; (3) gradual death of trees over a wide area, *see* recent forest decline.

diecious *alt.* spelling of dioecious *q.v.*

diel *a.* (1) during or *pert.* 24 h; (2) occurring at 24-h intervals.

dieldrin *n.* contact and stomach-acting organochlorine insecticide used in the control of soil pests, and also as a sheep dip. Its use is now restricted due to its environmental persistence.

diesel engine a compression ignition internal combustion engine.

diesel oil gas oil *q.v.*

diethylstilboestrol (DES) *n.* synthetic growth-promoting hormone used for cattle in the USA, implicated as a carcinogen.

difenzoquat *n.* translocated herbicide used to control wild oats in cereals.

diffraction colours colours produced not by pigment but by unevenness on the surface of an organism resulting in the diffraction of light reflected from it.

diffuse *a.* (1) widely spread; (2) not localized; (3) not sharply defined at the margin.

diffuse-porous *appl.* wood in which vessels of approximately the same diameter tend to be evenly distributed in a growth ring. *cf.* ring-porous.

diffusion *n.* the free passage of molecules, ions, etc., from a region of high concentration to a region of low concentration.

diffusion pressure deficit suction pressure *q.v.*

diflubenzuron *n.* insecticide used to control caterpillars, e.g., on fruit crops.

digenean *a. appl.* parasitic flatworms of the order Digenea, which include liver, blood and gut flukes such as *Schistosoma*, the cause of schistosomiasis in humans. As adults they are endoparasites of many vertebrates. They have complex life cycles with larval stages in molluscs and sometimes also in several other different hosts.

digenesis *n.* alternation of sexual and asexual generations.

digenetic *a.* (1) *pert.* digenesis; (2) requiring an alternation of hosts, *appl.* parasites.

digestibility *n.* of feed, the proportion of a feed that is digested by an animal. This is often expressed in the form of a coefficient where a value approaching 1.0 represents high digestibility.

digestion *n.* (*biol.*) the process whereby nutrients are rendered soluble and capable of being absorbed by the organism or by a cell, by the action of various hydrolytic enzymes that break down proteins, large carbohydrates, fats, etc.

digger wasps solitary insects of the superfamilies Pompiloidea and Sphecoidea of the order Hymenoptera, somewhat resembling the true wasps in appearance, and including species that nest in burrows they dig in the ground.

digit *n.* terminal division of limb, such as finger or toe, in vertebrates other than fishes.

digitalin, digitonin, digitoxin *n.* glycosides from leaves of foxglove *Digitalis purpurea*.

digitigrade *a.* walking with only digits touching the ground.

digoneutic *a.* breeding twice a year.

dihybrid *n.* (1) the progeny of a cross in which the parents differ in two distinct characters; (2) an organism heterozygous at two distinct loci.

dikaryon *n.* (1) a pair of nuclei situated close to one another and dividing at the same time, as in some fungal hyphae; (2) the hypha containing such a pair of nuclei. *a.* dikaryotic.

dike *see* dyke.

Dilleniales *n.* order of woody, often climbing dicots, comprising the families Crossosomataceae and Dilleniaceae.

dilute and disperse (1) any waste disposal method that relies on the dissipation of the waste stream within the wider environment, e.g. the use of tall chimneys to ameliorate the effects of stack gases; (2) the waste disposal method in which liquid (industrial) waste is allowed to percolate down through domestic refuse and the rock strata beneath and, in the process, is assumed to become relatively environmentally benign. *cf.* concentrate and contain.

diluvial *a.* produced by a flood, *appl.* soil deposits.

dimeric *a.* (1) having two parts; (2) bilaterally symmetrical.

dimethoate *n.* systemic organophosphate pesticide used as an insecticide against aphids and as an acaricide against red spider mites on a wide range of crops.

dimixis *n.* fusion of two kinds of nuclei in heterothallic organisms.

dimorphism *n.* (1) having two different forms, as when a substance crystallizes in two different forms; (2) condition of having two distinct forms within a species, of having two distinct sizes of stamens, two different kinds of leaves, etc. *a.* dimorphic.

dinitrogen *n.* molecular nitrogen, N_2, as found in the atmosphere.

dinitrogen oxide (N_2O) nitrous oxide *q.v.*

dinocap *n.* pesticide used as a fungicide to control powdery mildews on, e.g., hops and apples, and as an acaricide to control red spider mites.

Dinoflagellata, dinoflagellates *n.*, *n.plu.* phylum of unicellular protists (class Phytomastigophorea in animal classification or Pyrrophyta (Dinophyta) in plant classification), having two flagella, one pointing forwards, the other forming a girdle around the body. A major component of marine and freshwater plankton. Some are autotrophic and photosynthetic, some are heterotrophic.

Dinophyta dinoflagellates *q.v.*

Dinornithiformes *n.* an order of flightless birds from New Zealand, including the moas.

dinosaurs *n.plu.* members of either of two orders of reptiles that flourished during the Mesozoic, the Saurischia, the lizard-hipped dinosaurs, or the Ornithischia, the bird-hipped dinosaurs. The Saurischia included both bipedal carnivores and very large quadrupedal herbivores. The Ornithischia were mostly quadrupedal and all herbivorous.

dinoseb *n.* contact and soil-acting herbicide used, e.g., to control broad-leaved annual weeds in leguminous crops.

dioecious *a.* (1) having the sexes separate; (2) having male and female flowers on different individuals. *n.* dioecism.

dioestrus *n.* the quiescent period between periods of heat in animals with more than one period of fertility each year.

diopside *n.* $Ca(Mg,Fe)Si_2O_6$ where $45\% < Ca/(Ca + Mg + Fe) < 50\%$ and $Mg/(Mg + Fe) < 50\%$, one of the pyroxene minerals (*q.v.*) of the clinopyroxene subgroup. *cf.* hedenbergite.

diorite *n.* an intrusive igneous rock with a colour index of 40 to 90 that is medium- to coarse-grained or pegmatitic. Its mineralogy is essentially plagioclase feldspar (andesine or oligoclase) and amphibole (hornblende). Pyroxene and/or mica (biotite) may also be present, as may quartz ($<20\%$) and alkali feldspar ($<10\%$).

dioxins *n.plu.* a colloquial term for polychlorinated di-benzo-p-dioxins (*q.v.*). *see also* TCDD.

dip *n.* (*geol.*) the angle measurable between a given geological plane (e.g. bedding layer, fault, foliation) and the horizontal through a vertical plane that is at right angles to the strike (*q.v.*) of the feature concerned. *see also* dip, true; dip, apparent.

dip, apparent (*geol.*) any one angle, other than the maximum angle, measurable between a given geological plane (e.g. bedding layer, fault, foliation) and the horizontal through a vertical plane that is at right angles to the strike (*q.v.*) of the feature concerned. *cf.* dip, true.

diphasic *a.* having two distinct states, *appl.* life cycle, etc.

diphyletic *a. pert.* or having origin in two separate lines of descent.

diplobiont *n.* organism characterized by at least two kinds of individual in its life cycle, such as sexual and asexual. *cf.* haplobiont.

diploblastic *a.* having only two germ layers, e.g. endoderm and ectoderm, as coelenterates. *cf.* triploblastic.

diploid *a. appl.* organisms whose cells (apart from the gametes) have two sets of chromosomes, and therefore two copies of the basic genetic complement of the species, designated $2n$. *n.* a diploid organism or cell.

diploidy *n.* the diploid state.

diplophase *n.* stage in life history of an organism in which nuclei are diploid, *alt.* sporophyte phase.

Diplopoda, diplopods *n., n.plu.* in some classifications, a class of arthropods commonly called millipedes, having numerous similar apparent segments each in fact made up of two segments and therefore bearing two pairs of legs. In some classifications considered a subclass or order of the class Myriapoda.

Diplura *n.* order of wingless insects with a pair of cerci and two 'tails' on last segment, sometimes called two-pronged bristletails. Minute white insects with no eyes, found in soil and under stones.

Dipnoi, dipnoans *n., n.plu.* group of bony fishes, commonly called lungfish, known from the Devonian, possessing lungs and crushing broad toothplates. The three genera of modern lungfish (found in Australia, South America and Africa) are air-breathing and live in tropical areas with a dry season, and have a reduced skeleton.

dipole *n.* (1) a point positive charge a small distance from a point negative charge of the same magnitude (electric dipole); (2) a north point magnetic pole separated by a short distance from an equal but opposite (i.e. south) point magnetic pole (magnetic dipole).

dipole–dipole interaction (*chem.*) a weak intermolecular attraction between two polar molecules arranged such that the positive pole of one is near the negative pole of the other.

Dipsacales *n.* order of dicot herbs and shrubs, rarely small trees, comprising the families Adoxaceae (moschatel), Caprifoliaceae (honeysuckle), Dipsacaceae (teasel) and Valerianaceae (valerian).

dip slope *see* cuesta.

Diptera, dipterans *n., n.plu.* the true flies, a large order of two-winged insects including, e.g., the housefly and other similar flies, hoverflies, craneflies, mosquitoes, midges, soldier-flies, snipe-flies, bee-flies, horseflies, robber-flies and fruit flies, all having one pair of wings only, the second pair being reduced to small halteres. The eruciform larvae (grubs) undergo a complete metamorphosis.

dip, true (*geol.*) the maximum angle measurable between a given geological plane (e.g. bedding layer, fault, foliation) and the horizontal through a vertical plane that is at right angles to the strike (*q.v.*) of the feature concerned. *cf.* dip, apparent.

diquat *n.* contact herbicide used, e.g., as an aquatic herbicide.

direct drilling the sowing of crop seeds directly into untilled soil. This practice helps to minimize the use of agricultural machinery and the associated problems of soil compaction. *alt.* no-till agriculture, zero tillage (USA).

directional selection selection that acts on one extreme of the range of variation in a particular character, and therefore tends to shift the entire population to one end of this range.

direct precipitation precipitation that results from the interception of low-lying stratus clouds by ground surfaces.

direct recycling type of recycling in which waste materials produced *in situ* re-enter the manufacturing process without first being used, e.g. broken or substandard bottles in the glass manufacturing industry. *cf.* indirect recycling.

disaccharide *n.* a carbohydrate composed of two sugars, e.g. sucrose, which is composed of glucose and fructose.

disassortative mating mating between organisms of unlike phenotype.

discharge *n.* of a river, the unit volume of water flowing in a river per unit time, usually expressed as cubic metres per second $(m^3 s^{-1})$.

disclimax *n.* a subclimax stage in plant succession replacing or modifying true climax, usually due to animal and or human agency, e.g. cultivated crops.

Discomycetes *n.* a group of ascomycete fungi, commonly called cup fungi, and including also the earth tongues, truffles and morels, in which the fruiting body is in the form of an apothecium, usually black or brightly coloured, which may be open and cup-shaped or disc-like, or closed and subterranean (truffles).

disconformity *n.* a name sometimes given to an unconformity (*q.v.*) in which the new and old sets of rocks have parallel bedding patterns.

discontinuity *n.* (1) occurrence (e.g. of a species) in two or more separate areas of a geographical region; (2) thermocline *q.v.*

discontinuous distribution pattern of geographical distribution where the same or similar species are found in widely separated parts of the world. Such distribution may be due to a species formerly distributed over the whole area having become extinct in the intervening regions.

discontinuous permafrost a type of permafrost (*q.v.*) that characteristically occurs in patches, in the horizontal plane, separated by unfrozen ground. This occurs to the south of the continuous permafrost found in the Arctic.

discontinuous variation variation between individuals of a population in which differences are marked and do not grade into each other, brought about by the effects of different alleles at a few major genes. *alt.* qualitative inheritance. *cf.* continuous variation.

discordant intrusion *see* intrusion.

discrete generations succeeding generations where one finishes before the other begins, a situation rare in nature.

discrete source point source *q.v.*

discount rate the economic value a resource will have in the future compared with its present value.

disease *n.* any pathological condition in which normal function is impaired, and which may be caused by physiological malfunction, infection, poor nourishment or pollution.

disjunct *a. appl.* a distribution of a species in which potentially interbreeding populations are geographically separated by sufficient distance to preclude gene flow.

disjunctive symbiosis a mutually helpful condition of symbiosis although there is no direct connection between the partners.

dislocation metamorphism the formation of metamorphic rocks in the vicinity of major earth movements. In this type of metamorphism, directed pressure is the main cause of the alteration, hence dynamic metamorphism predominates.

dismantlement of nuclear reactors *see* decommissioning process.

disoperation *n.* (1) co-actions resulting in disadvantage to individual or group; (2) indirectly harmful influence of organisms upon each other.

dispersal *n.* (*biol.*) the active or passive spread of organisms, seeds or spores away from each other, esp. away from parents.

dispersant *n.* compound (1) that may be added to a solution of an organic compound to aid its uniform delivery as fine droplets; (2) used to break up oil after an oil spill.

dispersion *n.* (*biol.*) pattern of spatial distribution of organisms within a population.

dispersion force London force *q.v.*

displacement activity the performance of a piece of behaviour, usually in moments of frustration or indecision, that is not directly relevant to the situation at hand.

display *n.* series of stereotyped movements, sounds, etc., that cause a specific response in another animal, usually of the same species, often used in courtship or territorial defence.

disruptive coloration colour patterns that obscure the outline of an animal and so act as camouflage and protection against predators.

disruptive selection selection that operates against the middle range of variation in a particular character, tending to split populations into two showing the extreme at either end of the range.

dissimilation *n.* the breakdown of nutrients to provide energy and simple compounds for intermediary metabolism. *a.* dissimilatory.

dissolved load material in solution carried by a stream or river.

dissolved organic carbon (DOC) the fraction of carbon bound in organic compounds in water that is made up of particles smaller than 0.45 µm, which is separated out from total organic carbon (*q.v.*) by filtration.

dissolved oxygen level (DO) amount of oxygen gas (O_2) dissolved in a given volume of water, and usually expressed as the concentration of oxygen in parts per million at a given temperature and pressure. Water with less than 5 p.p.m dissolved oxygen does not support fish and similar organisms. *see also* biological oxygen demand.

dissolved solids total dissolved solids *q.v.*

distal *a*. (1) far apart, distant, *appl.* bristles, etc.; (2) *pert*. the end of any structure furthest away from the middle line of an organism or point of attachment. *cf.* proximal.

disthene kyanite *q.v.*

distillation *n*. process whereby a liquid is converted into a vapour, which is subsequently condensed and collected. This technique may be used, e.g., in desalination (*q.v.*) or in the separation of a mixture of liquids with different boiling points (known as fractional distillation).

distraction display behaviour in female birds that distracts an enemy from the eggs or chicks, and often takes the form of feigning injury to entice the predator away.

distributaries *n.plu.* small streams formed where the accumulation of sediment at the mouth of a river (known as a delta) causes the major river channel to divide into several smaller channels.

distribution *n*. geographical range of a species or group of species.

district heating scheme whereby domestic heating for a district is supplied from a single location, e.g. from a purpose-built boiler plant, or (in the form of waste heat) from a power station.

disturbance *n*. (*ecol*.) any perturbation (either natural or caused by human activities) experienced by an ecosystem.

disturbance climax disclimax *q.v.*

disulfoton *n*. systemic organophosphate insecticide used to control aphids on beans, potatoes, sugar-beet and brassicas, and also to control carrot fly.

ditalimfos *n*. organophosphate fungicide used to control powdery mildews on both apples and spring barley.

dithianon *n*. fungicide used to control scab in pears and apples.

ditocous, ditokous *a*. producing two eggs or two young at one time.

diurnal *a*. (1) occurring every day; (2) active in the daytime; (3) opening in the daytime.

diurnal rhythm metabolic or behavioural rhythm with a cycle of about 24 h.

diurnal tides a tidal regime with one high tide and one low tide approximately every 24 h, common in South-East Asia.

diuron *n*. soil-acting herbicide used, e.g., in orchards.

divalent *a*. (*chem*.) having a valency (*q.v.*) of two. *alt*. bivalent.

divergence *n*. of protein or nucleotide sequences, the percentages of amino acid or nucleotide residues that are different in corresponding proteins or genes from different species or related proteins or genes from the same species, etc.

divergent *a*. (1) separated from another, having tips further apart than the bases; (2) *appl*. evolutionary change tending to produce differences between two organisms, genes, etc.

divergent boundary (*geol*.) type of plate boundary that occurs when two lithospheric plates move apart. Oceanic divergent boundaries are associated with the formation of mid-ocean ridges while continental divergent boundaries are associated with rift valleys. *alt*. constructive boundary.

diversification *n*. increase in variety.

diversion behaviour (1) distraction display (*q.v.*); (2) behaviour likely to confuse an enemy, e.g. squids ejecting a cloud of black 'ink'.

diversity *see* alpha diversity, beta diversity, biological diversity (biodiversity), ecological diversity, gamma diversity, genetic diversity, habitat diversity, species composition, species diversity, species richness.

diversity gradient a geographical gradient (e.g. in altitude, latitude) along which a change in species diversity is found.

diversity index a measure of the biological diversity (generally the species diversity) within an environment. There are various types of diversity index, which are calculated in various ways from the number of species present and their relative abundance (*see*, e.g., Simpson diversity index). Such indices can be used to detect ecological changes due, e.g., to stress on an environment. *cf.* biotic index.

divide watershed *q.v.*

division *n.* a major taxonomic grouping in plants, corresponding to a phylum in animals. Examples: Bryophyta (mosses and liverworts), Pterophyta (ferns, etc.), Spermatophyta (seed-bearing plants, the gymnosperms and angiosperms).

dizygotic *a.* originating from two fertilized ova, as non-identical or fraternal twins (DZ twins).

DM dry matter *q.v.*

DNA deoxyribonucleic acid, a very large linear molecule that acts as the store of genetic information in all cells. A DNA molecule is composed of two chains of covalently linked deoxyribonucleotide subunits, and a single molecule may be millions of nucleotides in length. The four types of deoxyribonucleotide subunits in DNA contain the bases adenine (A), thymine (T), cytosine (C) and guanine (G), respectively. Genetic information is encoded in the order of the bases A, T, C and G in the chains. The two chains are exactly complementary to each other in base sequence and are held together by specific hydrogen bonding between A on one chain and T on the other and between C and G. The two chains are wound round each other to form the Watson–Crick right-handed 'double helix'. *see also* other DNA entries, chromosome, complementary DNA, gene, genetic code, nucleic acid, recombinant DNA.

DNA clone a piece of DNA incorporated into a bacterial plasmid or phage such that many identical copies can be made by replication in an appropriate host cell.

DNA cloning the isolation and multiplication of a piece of DNA by incorporating it into a specially modified phage or plasmid and introducing it into a bacterial cell where the DNA of interest is replicated along with the phage or plasmid DNA and can subsequently be recovered from the bacterial culture in large amounts.

DNA fingerprinting, DNA profiling, DNA typing method of ascertaining individual identity, family relationships,

etc., by means of DNA analysis. The DNA fingerprint consists of a pattern of DNA fragments obtained on analysis of certain highly variable repeated DNA sequences within the genome, whose number and arrangement are virtually unique to each person or animal. DNA fingerprints can be obtained from a tiny quantity of blood, tissue or semen, and are widely used in forensic work, and also in ecological studies.

DNA hybridization (1) technique for determining the similarity of two DNAs (or DNA and RNA) by reassociating single strands from each molecule and determining the extent of double-helix formation (indicating similar base sequences); (2) a general method involving reassociation of complementary DNA or RNA strands used widely to identify and isolate particular DNA or RNA molecules from a mixture. *see also in situ* hybridization.

DNA library a collection of cloned DNA fragments, usually representing an entire genome (genomic library) or copies of the mRNAs present in a particular tissue (cDNA library).

DNA profiling DNA fingerprinting *q.v.*

DNA repair various biochemical processes by which mistakes that occur during DNA replication or damage caused by the action of chemicals or irradiation can be repaired. Altered and incorrectly matched bases are recognized by enzymes that excise them and new DNA is then synthesized by reference to the undamaged strand. Breaks in DNA are repaired by DNA ligases. The repair of certain types of damage may result in alteration of the base sequence of the DNA and thus in mutation.

DNA replication the process by which a new copy of a DNA molecule is made. The two strands of the double helix are separated and each acts as a template for the enzymatic synthesis of a new complementary strand, resulting in two new identical double-stranded DNA molecules. This type of replication is termed semiconservative. It is catalysed by the enzyme DNA polymerase.

DNase, DNAse deoxyribonuclease *q.v.*

DNA sequence order of nucleotides or base pairs in a DNA molecule, and which in protein-coding DNA determines the order of amino acids in the proteins specified. *alt.* base sequence, nucleotide sequence.

DNA transcription *see* transcription.

DNA tumour viruses a group of unrelated DNA viruses that can cause cancers by various means. They include adenovirus, SV40 (simian virus 40) and polyoma, which are only tumorigenic in certain highly susceptible newborn animals, but which are widely studied in cultured cells as models of tumorigenesis, and also viruses such as Epstein–Barr virus, hepatitis B and certain human papilloma viruses, which are implicated in the development of some types of cancers in humans.

DNA typing DNA fingerprinting *q.v.*

DNA viruses viruses containing DNA as the genetic material and including the Adenoviridae, Herpesviridae, Poxviridae, Papovaviridae, Parvoviridae and Iridoviridae amongst vertebrate viruses, and the cauliviruses and geminiviruses amongst the plant viruses. *see* Appendix 10.

DO dissolved oxygen level *q.v.*

Dobson unit a measure of the total ozone (O_3) content of the air column. This is assessed in terms of the thickness of a notional layer of pure ozone at sea-level. A Dobson unit is equivalent to a thickness of 1×10^{-4} m at standard temperature and pressure. Typically, the air column contains about 360 Dobson units, although under the ozone hole (*see* stratospheric ozone) readings of 170 Dobson units have been made.

DOC dissolved organic carbon *q.v.*

docking disorder a complex disorder found in sugar beet grown on sandy soils, mainly caused by ectoparasitic nematodes (*Longidorus* spp., *Trihodorus* spp.). Affected plants show abnormal root development and stunted growth.

dodine *n.* fungicide used to control scab in pears and apples.

doldrums *n.* (1) near calm conditions, with only light, variable winds, experienced in the intertropical convergence zone (*q.v.*); (2) the geographical location associated with these conditions.

dolerite *n.* a medium-grained basic igneous rock with a mineralogy similar to that of gabbro (*q.v.*). It is typically found in minor intrusions (dykes or sills).

doline *n.* small, closed depression, typically oval or round in plan, characteristic of karst landscapes. These may be formed through the collapse of a cave roof (collapse dolines) or through the dissolution of limestone at or near the surface (solution dolines). *alt.* sinkhole (USA), swallowhole (UK).

Dollo's law, Dollo's rule that evolution is irreversible and that structures and functions once lost are not regained.

dolomite *n.* (1) $CaMg(CO_3)_2$, an abundant rock-forming mineral; (2) a rock that has a high proportion of the mineral dolomite. Such rocks are exploited industrially as building materials, in the manufacture of refractory bricks and as a source of carbon dioxide.

dolomitization *n.* the process by which the mineral dolomite (*q.v.*) is formed. In most cases, this is generally believed to occur by the *in situ* action of magnesium-rich solutions on the calcium carbonate (calcite) of limestone.

dolphin *n.* a member of the marine family Delphinidae or of the Platanistidae (river dolphins) of the suborder Odontoceti (toothed whales) of the Cetacea (*q.v.*). They are slim and fast-moving with a prominent elongated snout or 'beak'.

DOM dissolved organic material, in waters.

domatium *n.* a crevice or hollow in some plants, serving as a lodging for insects or mites.

dome dune a circular dune of significant size that has no external slip-face *q.v.*

domestication *n.* deliberate selection of and breeding from wild plants and animals by humans, a process instrumental in the development of early agriculture. *a.* domesticated, *appl.* animals.

domestic refuse solid waste, e.g. paper, plastics, ashes and vegetable matter, generated by households (UK term). *alt.* garbage (USA).

Domin scale scale (1–10) used to indicate the approximate percentage cover of individual plant species in a given area, with 1 corresponding to insignificant cover, 3 corresponding to 1–5% cover, 8 to 50–75% cover, and 10 equal to 100% cover. + indicates organisms occurring singly.

dominance *n.* (1) (*ecol.*) the extent to which a particular species predominates in a community and affects other species; (2) (*behav.*) *see* dominance systems; (3) (*genet.*) property possessed by some alleles of solely determining the phenotype for any particular gene when present as one member of a heterozygous pair, when they mask the effects of the other allele (the recessive allele) to give a phenotype identical to that when the dominant allele is homozygous. This phenomenon is known as complete dominance. Incomplete dominance is exhibited when the effects of the other allele are not completely masked, *see also* codominance.

dominance–diversity curve graph plotting the importance values, density, abundance or other measure of relative importance of the species in a community along the *y*-axis so that the species are arranged in descending order along the *x*-axis.

dominance frequency proportion of samples in which a particular species is predominant.

dominance hierarchy *see* dominance systems.

dominance systems social systems in which certain individuals aggressively dominate others. In the case where one individual dominates all the others with no intermediate ranks it is known as a despotism. In the more common dominance hierachies or social hierarchies there are distinct ranks, with individuals of any rank dominating those below them and submitting to those above them.

dominant *a. appl.* (1) plants that by their numbers and extent determine the biotic conditions in an area; (2) species most prevalent in a particular community, or at a given period; (3) an individual that is high ranking in the social hierarchy or peck order; (4) (*genet.*) a phenotypic character state or an allele that masks an alternative character state or allele when both are present in a hybrid, *see also* heterozygote. *cf.* recessive.

donor atom (*chem.*) atom with a lone pair of electrons that is capable of forming a dative bond. Ligands with one such atom are said to be unidentate, those with two, bidentate and those with many, multidentate.

dormancy *n.* (*biol.*) resting or quiescent condition with reduced metabolism, as in seeds, spores, buds, etc., under unfavourable conditions for germination or growth. *a.* dormant.

dormant volcano a volcano that although currently inactive is thought to have the potential to erupt.

dormitory *a. appl.* settlement esp. town or village in which a high percentage of the workforce is not employed locally but commutes considerable distances to work in other location(s), usually large towns or cities.

dorsal *a.* (1) *pert.* or nearer the back (not hind end) of an animal, which is usually the upper surface; (2) *appl.* upper surface of leaf, wing, etc. *cf.* ventral.

dorso-ventral *a. appl.* axis, from back to belly of the animal body, from upper to lower surface of a limb, leaf, etc.

dosage compensation regulation of the dosage of genes carried on the sex chromosomes in the sex carrying two (or more) copies of the same chromosome. In mammals, e.g., one copy of the X chromosome is permanently inactivated in the somatic cells of the female.

dose *n.* (1) the amount of radiation received, or the amount of a drug or toxic compound taken in, over a given time period, *alt.* dose rate; (2) the amount of radiation received at a single exposure or the amount of a drug administered at one time.

dose equivalent a standardized measure of the effect of ionizing radiation on tissue, which is the measured absorbed dose multiplied by weighting factors for particular tissues and types of radiation. It represents the risk to health from that amount of radiation if it had been absorbed uniformly throughout the body. It is expressed in sieverts (SI unit).

dose rate *see* dose.

dose–response curve for any assay, the relation between the concentration of active agent (virus, hormone, enzyme, etc.) in the sample, and the quantitative response in that particular assay.

dose–response method the use, e.g. in environmental impact assessment, of techniques relating differing levels of pollution (dose) to differing levels of damage to the environment (response).

dosimeter *n.* instrument for measuring the amount of radiation received.

double bond a covalent bond (*q.v.*) that consists of two pairs of electrons shared between the bonded atoms. Diagrammatically, this is represented as a double line between the bonded atoms, e.g. the double bond in O_2 may be shown thus O=O.

double recessive cell or organism homozygous for a recessive allele and thus showing the recessive phenotype.

doubling time the time it takes for the quantity of something growing exponentially, e.g. a population of living organisms, to double.

down, downland *n.* vegetation typical of the chalk grasslands of southern England, which is generated and maintained by continuous grazing (by sheep and rabbits), and which is short turf rich in small flowering herbs.

downdraught *n.* (*clim.*) downward air movement found, e.g., in association with thunderstorms and in the lee of obstacles.

downthrown *a. appl.* a block of rock that is bounded by a fault, and that moves downwards as a result of tectonic activity.

down-valley sweep the progressive, downstream migratory movement of meanders *q.v.*

downy mildew plant disease characterized by felty fungal growth on the plant surface, caused by certain species of oomycetes.

DPD diffusion pressure deficit. *see* suction pressure.

draa *n.* large-scale dune, up to 400 m high and 4 km in length. *alt.* megadune.

dracunculiasis *n.* tropical disease caused by the parasitic nematode *Dracunculus medinensis* (guinea-worm). Freshwater copepod crustaceans act as intermediate hosts. *alt.* guinea-worm disease.

dragonflies *see* Odonata.

drainage basin an area of land drained by a main (trunk) river and its tributaries. *alt.* catchment.

drainage density the average length of stream channel per unit area of drainage basin *q.v.*

drainage divide watershed *q.v.*

drainage pattern the arrangement of a river and its tributaries within a drainage basin. The underlying geological structure is the most important factor in determining the drainage pattern present. Several different drainage patterns are recognized, namely parallel, dendritic, radial, centripetal, annular, trellis and rectangular drainage, *see* Fig. 10. Parallel drainage is the simplest pattern and is indicative of the presence of closely spaced faults or isoclinal folds. Dendritic drainage develops in basins composed of a single rock type, e.g. extensive batholiths of uniform hardness. In radial drainage (characteristic of volcanic domes) streams flow outwards from the centre while in centripetal drainage (characteristic of craters and calderas), streams flow inwards towards the centre. Annular drainage has a circular arrangement of the main rivers and develops on eroded domes with alternating weak and resistant sediments. Trellis and rectangular drainage both have a regular and orderly arrangement, although this is less developed in the latter. The trellis pattern often develops where folded or tilted sedimentary rocks of different degrees of hardness are present, while the rectangular pattern indicates the presence of jointing or faulting.

draught animals animals, such as horses, oxen and mules, that are used to pull implements, e.g. ploughs.

dravite *n.* a variety of the mineral tourmaline *q.v.*

dredge *v.* to remove sediment from natural or man-made waterbodies, e.g. rivers, canals, harbours, esp. in order to make them navigable.

dredge corn a mixture of cereal crops grown together, e.g. barley and oats.

dredging *n.* (1) the removal of sediment from, e.g., rivers, harbours and canals; (2) type of surface mining in which chains of buckets drag up, e.g., sand and gravel or other surface deposits that are covered with water.

dreikanters *see* ventifacts.

dressing-out percentage the carcass weight (kg) of a livestock animal divided by its live weight at slaughter (kg) multiplied by 100. *alt.* killing-out percentage.

drift *see* genetic drift, continental drift.

drift mining type of mining in which a coal seam, for example, exposed on the side of a hill is excavated horizontally.

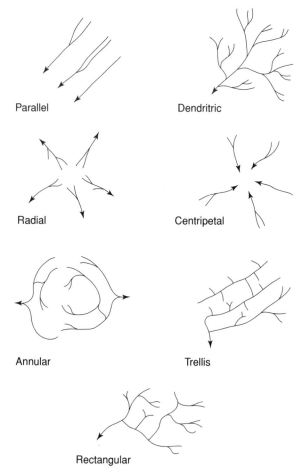

Parallel

Dendritric

Radial

Centripetal

Annular

Trellis

Rectangular

Fig. 10 Drainage patterns.

drift-netting type of marine fishing using huge nets, up to 65 km long, that drift vertically in the water. Fish become caught in the nets by their gills. Drift net fishing is considered environmentally undesirable as it rapidly depletes fisheries, is highly indiscriminate, and kills marine mammals and other animals, such as dolphins, turtles, seals and sharks, that become entangled in the nets.

drip irrigation irrigation method in which plastic pipes punctured at intervals with small holes are positioned on the ground close to crop plants to ensure effective water delivery. *alt.* trickle irrigation.

drive *n.* the motivation of an animal resulting in its achieving a goal or satisfying a need.

drizzle *n.* rain consisting of water droplets <0.5 mm in diameter.

drone *n.* male social insect, esp. honey bee.

drought *n.* situation in which a region does not receive its normal amount of water because of decreased rainfall, increased evaporation due to higher than normal temperatures or a combination of both. In temperate countries that usually receive sufficient rainfall, a drought is technically declared after a certain number of consecutive days without rain.

drug-resistance factors or plasmids (R factors) plasmids in enterobacteria and other medically important bacteria that carry genes for resistance to various commonly used antibiotics.

drumlin *n*. smooth elliptical hummock composed of glacial till (*q.v.*), whose long axis lies parallel to the direction of ice flow. Drumlins usually occur in clusters (drumlin fields), creating a characteristic 'basket-of-eggs' landscape.

drumlin fields *see* drumlin.

drupe *n*. a more-or-less fleshy fruit with one compartment and one or more seeds. The pericarp is differentiated into a thin skin (epicarp), a fleshy mesocarp and a hard stony endocarp. Plums, cherries and other stone fruits are drupes.

drupel *n*. one of a collection of small drupes forming an aggregate fruit in, e.g., raspberry, blackberry. *alt.* drupelet.

dry *a. appl.* sheep, cows, etc., not lactating (i.e. not producing milk).

dry adiabatic lapse rate (DALR) rate of temperature decrease of an ascending air mass, cooling adiabatically, that is unsaturated with water vapour. This is calculated at *ca.* 1 °C decline per 100 m ascent.

dry bulb temperature *see* psychrometer.

dry bulb thermometer *see* psychrometer.

dry deposition the direct transfer of material from the atmosphere onto wet or dry solid surfaces. *cf.* wet deposition.

dry farming cultivation of grain crops in semi-arid regions without the use of irrigation.

dryland rice *see* rice.

dry mass *see* dry matter.

dry matter (DM) the material left after removal of water from organic matter, such as plant biomass or soil, obtained by heating to constant weight in an oven at 90–95 °C.

dryopithecids, dryopithecines *n.plu.* a group of Miocene ape-like fossils from India and Africa, including the genera *Dryopithecus* and *Proconsul*, dating from around 20 to 8 million years ago, and which are thought to include the ancestors of modern apes.

dry weight the weight or mass of organic matter or soil after removal of water by heating to constant weight. *alt.* dry mass. *see* dry matter.

dry weight rank method technique for estimating the contribution each plant species makes to the total yield of a community, such as a pasture.

duetting *n*. rapid antiphonal calling back and forth between mated pairs in some birds, presumed to be a recognition and bonding device.

dulosis *n*. slavery among ants, in which those of one species are captured by another species and work for them, an extreme example of social parasitism.

dune *n*. an accumulation of sand-sized particles deposited by the wind. A number of different types of dune are recognized, taking into account such factors as their size and shape, their formation relative to the prevailing wind direction, and whether they form freely or in response to an impeding obstacle. *see* aklé, barchan, climbing dune, complex dune, compound dune, dome dune, draa, falling dune, fixed dune, linear dune, lunette, nebkha, parabolic dune, relict dune, reversing dune, seif, star dune, transverse dune.

dung fuel fuel composed of dried animal excrement, used extensively in developing countries (mainly for cooking).

dunite *n*. an ultramafic igneous rock that is medium- to coarse-grained and that is >90 wt % olivine in composition.

duricrust *n*. hard, cemented layer developed in the upper horizons of some soils, most frequently those of semi-arid regions. Duricrusts form as a result of the deposition of minerals precipitated from the soil solution. *see* caliche, gypcrete, salcrete, laterite.

durum wheat *Triticum durum*, a species of wheat whose seeds have a hard, vitreous endosperm. This fragments during milling to produce a coarse material (semolina) that is used for the production of pasta. Durum wheat is also used in breakfast cereals.

dust *n*. air-suspended solid particles, each with a diameter between 2 and 500 μm.

dust bowl any semi-arid region where wind deflation is responsible for the removal of soil. *see also* Dust Bowl.

Dust Bowl term used to describe conditions in the Great Plains, USA, during the 1930s, when severe winds stripped the topsoil from vast areas (mainly of agricultural land). This soil erosion created huge dust storms that traversed the length of the country.

dust dome a dome of heated air that surrounds a town or city and traps and keeps pollutants, esp. particulates, in suspension.

dust lice *see* Psocoptera.

duststorm *n.* cloud of dust-sized particles (2–500 μm in diameter) blown into the air by strong winds. Duststorms are common in overgrazed and desert areas.

Dutch elm disease a disease of elm trees caused by the fungus *Ceratocystis ulmi*, which is spread by species of elm bark beetle.

D-value (*agric.*) the proportion of digestible organic matter in a feedstuff, expressed as a percentage of its dry matter content. A measure of herbage digestibility.

DW dry weight *q.v.*

dwarf male a small, usually simply formed male individual in many classes of animal, either free-living or carried by the female. *alt.* pygmy male.

Dy symbol for the chemical element dysprosium *q.v.*

dyke *n.* (1) minor intrusion, vertical or steeply inclined, which cuts across existing rock strata; (2) a ditch; (3) an embankment built to protect susceptible land from flooding. *alt.* dike.

dynamic equilibrium *see* equilibrium, dynamic.

dynamic metamorphism the formation of metamorphic rocks by the action of directed pressure or stress. In isolation, dynamic metamorphism alters the structure of the rock but its mineralogical composition remains essentially unchanged.

dynamic rejuvenation *see* rejuvenation.

dynamic steady state as *appl.* biogeochemical cycles, the condition of a reservoir that is of constant size because the total flux into it equals the total flux out of it.

dysgenesis *n.* infertility of hybrids in matings between themselves, although fertile with individuals of either parental stock.

dysgenic *a. appl.* traits inimical to the propagation of the organism, such as sterility, chromosomal aberrations, mutations and abnormal segregation of chromosomes at meiosis.

dysphotic *a.* (1) dim; (2) *appl.* zone, waters at depths between 80 and 600 m, between euphotic and aphotic zones.

dysprosium (Dy) *n.* a lanthanide element (at. no. 66, r.a.m. 162.5), a soft silver-coloured metal when in elemental form. It is used, e.g., to absorb neutrons in nuclear reactors.

dystrophic *a.* (1) wrongly or inadequately nourished; (2) inhibiting adequate nutrition; (3) *pert.* faulty nutrition; (4) *appl.* lakes rich in undecomposed organic matter so that nutrients are scarce.

DZ twins dizygotic twins *q.v.*

E

e- prefix derived from L. *ex*, out of, or *ex* without, often denoting a lack of, e.g., ebracteolate, lacking bracts, ecaudate, without a tail.

e⁻ *see* electron.

E glutamic acid *q.v.*

E (1) energy *q.v.*; (2) evenness *q.v.*

E = *mc*² Einstein's mass–energy relation *q.v.*

EA (1) Environment Agency (*q.v.*); environmental assessment *q.v.*

ear *n.* (1) sense organ in vertebrates concerned with hearing and gravity detection; (2) (*bot.*) flower- or seed-containing head of grasses, usually *appl.* cereals (barley, wheat, rye, etc.).

earth ball common name for basidiomycete fungi of the genus *Scleroderma* and their relatives, with hard tuberous unstalked fruiting bodies that crack open to release a mass of spores.

Earth capital Earth's natural resources and processes that sustain humans and all other life.

earthflow *see* flow movement.

earth history geological timescale *q.v.*

earthquake *n.* sudden release of pressure that has built up over time within the rocks of the Earth's crust. Shock waves (known as seismic waves) radiate outwards into the surrounding rock and, if these are sufficiently large, may be felt as vibrations at the Earth's surface.

Earth Resources Technology Satellite (ERTS) any satellite launched by NASA as part of a programme (the ERTS programme) to monitor the use of environmental resources such as land, water, minerals and forests. Renamed Landsat in 1975.

Earthscan an agency supplying news and information on environmental matters.

Earth Summit United Nations Conference on Environment and Development *q.v.*

earthworms *n.plu.* common name for a number of terrestrial oligochaete worms (*Lumbricus* spp. and others) inhabiting the soil, which contribute to soil aeration through their tunnels and to soil fertility by bringing humus-containing soil to the surface.

earwigs *n.plu.* common name for insects of the order Dermaptera *q.v.*

East African Floral Region part of the Palaeotropical Realm comprising Africa from north of Lake Victoria south to southern Mozambique and westward to include southern Angola.

easterlies polar easterlies *q.v.*

easterly equatorial jet stream jet stream (*q.v.*) that develops during the summer months (extending from the South China Sea across South-East Asia and central India to Saharan Africa) and influences the summer monsoon of the Indian subcontinent.

eavesdropping *n.* behavioural strategy in which rival males are attracted to a female by another male's courtship display.

EC (1) effective concentration (*q.v.*); (2) European Commission.

ecad *n.* (1) a plant or animal form modified by the environment; (2) habitat form. *alt.* ecophenotype. *cf.* ecotype.

eccritic *a.* (1) causing or *pert.* excretion; (2) preferred, *appl.* temperature or other environmental state.

ecdemic *a.* not native.

ecdysis *n.* moulting, the periodic shedding of cuticular exoskeleton in insects and some other arthropods, to allow for growth. *a.* ecdysial.

ecesis *n.* the invasion of organisms into a new habitat, or adjustment to new conditions.

Echinodermata, echinoderms *n.*, *n.plu.* a phylum of marine coelomate animals that are bilaterally symmetrical as larvae but show five-rayed symmetry as adults and have a calcareous endoskeleton and a water vascular system. It includes the classes Crinoidea (sea lilies and feather stars), Asteroidea (starfish), Ophiuroidea (brittle stars), Echinoidea (sea urchins) and Holothuroidea (sea cucumbers).

Echinoidea *n.* class of echinoderms, commonly called sea urchins, having a typically globular body with skeletal plates fitting together to form a rigid test.

Echiura, echiurans *n.*, *n.plu.* phylum of unsegmented coelomate marine worms, with soft plump bodies, which live in U-shaped tubes or in rock crevices down to abyssal depths. They have an extensible proboscis with a ciliated groove for collecting food. *alt.* spoon worms.

echolocation *n.* locating objects by sensing the echoes returned by very high frequency sounds emitted by the animal, as used, e.g., by bats.

echoviruses, ECHO viruses group of picornaviruses (*q.v.*) infecting the intestinal tract and that may also cause respiratory illnesses and meningitis.

eclipse *n.* (1) plumage assumed after spring moult, as in drake; (2) the period of multiplication of a virus when it is not easily detectable in the host cell.

eclogite *n.* an extremely dense silicate rock with a characteristic garnet–pyroxene mineralogy. It is found as blocks, lenses or xenoliths within metamorphic or igneous rocks and is believed to have been transported to these locations from either the mantle or the base of the crust.

eclosion *n.* hatching from egg or pupa case.

eco-audit *see* environmental audit.

ecobiotic *a. appl.* adaptation to a particular mode of life within a habitat.

ecoclimatic *a. appl.* adaptation to the physical and climatic conditions in a particular region.

ecocline *n.* a continuous gradient of variation of ecotypes in relation to variation in conditions.

ecodeme *n.* a deme occupying a particular ecological habitat.

ecological *a. pert.* or concerned with ecology.

ecological balance the balance of nature, the state in which the relative population numbers of different species remain fairly constant due to the interactions between species.

ecological buffer zone *see* buffer zone.

ecological community group of species inhabiting a particular area.

ecological diversity (1) the number of different species, and their relative abundance and distribution, in a given area or ecosystem, *see* alpha diversity, beta diversity, gamma diversity; (2) the diversity of ecosystems, e.g. forest, desert, grassland, oceans, etc., within a given region.

ecological efficiency (1) ratio of the energy ingested by one trophic level to the energy ingested by the previous trophic level; (2) the percentage of energy transferred from one trophic level to the next; (3) the efficiency of use of energy by an ecosystem, which is described in terms of several energy coefficients: the energy coefficient of the first order (production, P, divided by consumption, C), the energy coefficient of the second order (P divided by assimilated energy, A), and the assimilation efficiency (A/C).

ecological energetics the study of the flow of energy between the different living components of an ecosystem.

ecological equivalents different species, which do not share a common ancestor, but which fulfil essentially the same ecological role in different communities, e.g. the African potto and the South American three-toed sloth. Ecological equivalents are often similar in form and function through adaptation to the same type of ecological conditions (an example of convergent evolution *q.v.*).

ecological evaluation (1) use of animal and/or plant indicator species (*q.v.*) to assess the quality of a particular environment, often used as part of an environmental impact assessment; (2) methods used to determine the conservation needs of a species, habitat, community, etc., by means of ecological criteria.

ecological gradient *see* environmental gradient.

ecological growth efficiency the ratio of the energy contained in the biomass pro-

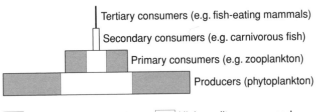

Tertiary consumers (e.g. fish-eating mammals)

Secondary consumers (e.g. carnivorous fish)

Primary consumers (e.g. zooplankton)

Producers (phytoplankton)

▨ Energy lost in respiration   ☐ High-quality energy used

Fig. 11 Typical pyramid of energy flow through a river ecosystem.

duced to the energy ingested or absorbed (by plants). *alt.* gross growth efficiency.

ecological impact the changes to the ecology of an area as a result of human activity.

ecological niche the role of an organism in a community in terms of the habitat it occupies, its interactions with other organisms and its effect on the environment. A given niche, e.g. the small herbivore niche, may be occupied by different species in different ecosystems and different parts of the world. *alt.* niche. *see also* realized niche.

ecological pyramid diagram showing the biomass, numbers or energy levels of individuals of each trophic level in an ecosystem, starting with the primary producers (photosynthetic or chemosynthetic organisms) at the base. *see* Fig. 11.

ecological release *see* competitive release.

ecological succession the natural process whereby communities of plant and animal species are replaced by others, usually more complex, over time as a mature ecosystem develops, *see* climax, primary succession, secondary succession, succession.

ecological time a time period corresponding to about ten generations of an organism's lifespan, usually of the order of years to hundreds of years. *cf.* evolutionary time.

ecology *n.* (1) the interrelationships between organisms and their environment and each other; (2) the study of these interrelationships.

ecomorphotype *n.* morphological form characteristic of a particular habitat. *a.* ecomorphotypic.

economic density of a population, the number of individuals per unit of inhabited area.

economic depletion generally refers to exhaustion of 80% of a non-renewable resource, because finding and extracting the remaining 20% is usually not economically worthwhile. It may also refer to depletion of certain renewable resources, such as fish stocks or timber.

economic good any material item or service that satisfies people. *alt.* good. *see also* private good, public good.

economic resources natural resources, the capital goods manufactured from them, i.e. tools, industrial machinery, transportation, etc., and the labour that is used in an economy to produce material goods and services.

ecoparasite *a.* parasite restricted to a specific host or small group of host species.

ecophenotype ecad *q.v.*

ecophysiology *n.* the study of animal physiology in relation to life-style and adaptation to environment.

ecospecies *n.* a group of individuals associated with a particular ecological niche and behaving like a species, but capable of interbreeding with neighbouring populations of the same species.

ecosphere *n.* the planetary ecosystem, consisting of the living organisms of the world and the components of the environment with which they interact.

ecosystem *n.* a community of interdependent species together with their non-living environment, which is relatively self-contained in terms of energy flow, and is distinct from neighbouring communities. Different types of ecosystems are defined by the collection of organisms found within them, e.g. forest, soil, grassland. The term ecosystem strictly describes a concept rather than an actual area or any one particular community of organisms. Continuous ecosystems covering very large areas, such as the northern coniferous forest or the steppe grassland, are known as biomes *q.v.*

ecosystem services the natural processes carried out by ecosystems and that support human and other life on Earth, such as atmospheric gas regulation, climate regulation, water supply, erosion control, soil formation, nutrient cycling, pollination, food production, etc., esp. those that are difficult to value or not commonly valued by conventional commercial methods. *see also* commons, tragedy of the commons.

ecotaxation *n.* taxation policies deliberately designed to benefit the environment by making an environmentally damaging activity more expensive or by encouraging environmentally friendly activities by tax advantages. *see* carbon tax, energy tax.

ecotone *n.* zone where two ecosystems overlap, which supports species from both ecosystems as well as species found only in this zone.

ecotope *n.* (1) a particular kind of habitat within a region; (2) the total relationship of an organism with its environment, being the complex interaction of niche, habitat and population factors.

eco-tourism *n.* tourism with the particular purpose of visiting areas of unspoilt natural beauty and rich in wildlife.

ecotoxicology *n.* scientific study of the environmental fate of pollutants and their potential harmful effects.

ecotype *n.* a genetically distinct subspecific form within a true species resulting from selection within a particular habitat and therefore adapted genetically to that habitat, but which can interbreed with other members of the species. *a.* ecotypic. *alt.* ecodeme, ecological race, physiological race. *cf.* ecad.

ect-, ecto- word elements derived from Gk *ektos*, outside.

ectocommensal *n.* commensal living on the surface of another organism.

ectocrine *a. appl.* chemicals released by organisms, esp. decomposing organisms, that affect the growth of neighbouring organisms.

ectogenic *a.* of external origin, not produced by organisms themselves.

ectogenous *a.* (1) able to live an independent life; (2) originating outside the organism.

ectomycorrhiza *n.* type of mycorrhiza in which the fungal hyphae form a superficial covering and do not extensively penetrate the root, found on both coniferous and broad-leaved forest trees, the infecting fungi being chiefly higher basidiomycetes.

ectoparasite *n.* parasite that lives on the surface of an organism.

ectophagous *a.* feeding on the outside of a food source.

ectophyte *n.* any external plant parasite of plants and animals. *a.* ectophytic.

ectopic *a.* not in normal position, *appl.* organs, etc.

Ectoprocta, ectoprocts *n., n.plu.* phylum of small marine and freshwater colonial animals, which superficially resemble mosses, hence the common name of moss animals, composed of zooids each bearing a crown of ciliated tentacles (a lophophore), and which live in horny, calcareous or gelatinous cases. *alt.* (formerly) Bryozoa, bryozoans, Polyzoa.

ectothermic *a. appl.* regulation of body temperature by animals using heat from the surrounding environment, e.g. as in fish, lizards, insects. Such animals are known as ectotherms. *cf.* endothermic. *n.* ectothermy. *see also* poikilothermic.

ectotrophic *a.* (1) finding nourishment from outside; (2) *appl.* mycorrhizas in which the fungal hyphae form a superficial covering on the host roots and do not extensively penetrate the root itself.

ectotropic *a.* tending to curve or curving outwards.

ecumene *n.* (1) the portion of the Earth's surface permanently inhabited by humans, as distinguished from the uninhabited or temporarily inhabited portions; (2) the centre of maximum activity of a state, having the densest population and closest network of transportation routes.

eczema *n.* allergic skin condition, characterized by a very itchy, scaly, red rash, caused by exposure to one or more allergens, e.g. wheat, dairy products, washing powders or animal fur.

edaphic *a. pert.* or influenced by conditions of soil or substratum.

Edentata, edentates *n.*, *n.plu.* an order of placental mammals known from the Paleocene, extant members of which include the armadillo, anteater and two-toed sloth, having reduced teeth and often an armoured body.

edentate *a.* without teeth or tooth-like projections.

edge effect tendency to have greater variety and density of organisms in the boundary zone between communities.

edge species species living primarily or most frequently or numerously at junctions of communities.

Ediacaran fauna a fauna of soft-bodied animals of problematic affinities present in late Precambrian strata, and bearing little resemblance to any later organisms.

EDTA ethylenediamine tetraacetic acid, a synthetic chelating agent that binds strongly to a number of metal ions, including calcium and magnesium. It is widely used in titrimetric analysis for these and other ions. It is added to detergents to improve their efficiency. As a consequence, it now occurs in rivers and lakes, where it is the cause of concern because it forms soluble chelates with toxic metals, such as mercury and lead, often increasing their bioavailability.

EEC European Economic Community.

eelworms *n.plu.* group of soil nematode worms, some of which cause serious damage to crop plants.

EFA essential fatty acids *q.v.*

effective concentration (EC) the concentration of a toxic substance that is sufficient to cause adverse symptoms (in cases where effects other than death are being studied) within a given period, and which is expressed as e.g. 48-h EC_{50}, the concentration required to cause symptoms in 50% of the animals tested within 48 h.

effective population number the number of individuals in an ideal, randomly breeding population with a 1 : 1 sex ratio that would have the same rate of heterozygosity decrease as the actual population under consideration.

effector *n.* any organ, cell, etc., that reacts to a stimulus by producing something, carrying out a specific set of functions, or doing mechanical work, such as muscle,

electric and luminous organs, chromatophores, glands, plasma cells and mature T cells of the immune system, etc.

efferent *a.* conveying from, *appl.* motor nerves carrying impulses outwards from central nervous system, *appl.* vessels conveying blood or lymph outwards from an organ or lymph node, respectively. *cf.* afferent.

efficiency *n.* the measure of how much output, e.g. of product or of energy, is produced for a certain input.

efflorescence *n.* (1) flowering; (2) time of flowering; (3) bloom as on surface of grapes and other fruits.

effluent *n.* domestic, industrial or agricultural waste discharged into the environment, usually in liquid form.

effodient *a.* having the habit of digging.

eft *n.* juvenile phase in life cycle of a newt.

egesta *n.plu.* (1) the sum total of material and fluid discharged from the body; (2) material passed out of the body in egestion.

egestion *n.* (1) the process of ridding the body of any waste material as by defaecation and excretion; (2) specifically, the excretion of material that has never been taken out of the gut, as defaecation.

egg *n.* (1) ovum (*q.v.*); (2) in certain animals, e.g. reptiles, birds, amphibians and insects, the fertilized ovum together with its nutritive and protective tissues, and from which a young individual emerges.

egg pulling an approach to the preservation of critically endangered bird species that involves collecting eggs laid in the wild, by the remaining breeding pairs, and hatching them in captivity.

egocentric *a. appl.* behaviour that benefits the survival of the individual that exhibits it.

E-horizon *see* A-horizon, podzol and Fig. 28, p. 385.

EIA environmental impact assessment *q.v.*

Eichler's rule groups of hosts with more variation are parasitized by more species than taxonomically uniform groups.

einkorn *n.* a primitive cultivated diploid wheat (*Triticum monococcum*) first cultivated in the Near East and South-West Asia around 11 000 years ago, which is derived from the wild *T. boeoticum*.

Einstein's mass–energy relation the relationship that quantifies the transformation of mass into energy, or *vice versa*. It takes the form $E = mc^2$, where E = energy (released when mass is lost), m = the change in mass and c numerically equals the speed of light (3.00×10^8 m s^{-1}). If m is in kilograms and c in metres per second, then E is in joules.

einsteinium (Es) *n.* a transuranic element [at. no. 99, r.a.m. (of its most stable known isotope) 254].

EIS environmental impact statement *q.v.*

Ekman spiral the change in wind direction (clockwise in the northern hemisphere, anticlockwise in the southern hemisphere) with increasing altitude that occurs within the friction layer (*q.v.*) of the atmosphere as a consequence of the decrease in the influence of the frictional force with altitude.

elaborate *v.* to form complex organic substances from simple materials.

Elaeagnales *n.* an order of shrubs, often with leathery leaves and thorns, comprising the family Elaeagnaceae (oleaster).

elaioplankton *n.* planktonic organisms rendered buoyant by oil globules in their cells.

elaioplast *n.* colourless organelle in plant cells in which oils or fats are formed and stored.

elaiosome *n.* fleshy oil-containing appendages present on seeds that are to be dispersed by ants, such as those of castor oil plant.

elasmobranchs *n.plu.* a group of cartilaginous fishes including the sharks, dogfishes, skates and rays, which have an outer covering of minute bony tooth-like scales, a spiracle and no covering over the gill openings.

elbow of capture *see* river capture.

electric dipole *see* dipole.

electrochemical gradient the gradient across a membrane with respect to an ion or other solute, which comprises both the concentration gradient and the gradient of electrical charge across the membrane.

electrochemical reduction any process in which the oxidation state of an element is decreased by the application of electrical energy. May be used to win certain metals, e.g. aluminium, from their ores.

electrolyte *n.* a compound that will, when molten or in solution, provide mobile ions and thereby conduct electricity. The passage of an electric current through a molten or dissolved electrolyte will cause it to be decomposed, i.e. electrolysed.

electromagnetic radiation radiation generated by the acceleration of an electric charge. It is associated with two fields, one electric, the other magnetic. These are mutually at right angles and are both at right angles to the direction of propagation. These fields need no medium to support them, hence this radiation can move through a vacuum, which it does at a speed of 2.9979×10^8 m s^{-1} (this constant is given the symbol c). This radiation has both wave-like and particle-like properties and can be pictured as consisting of discrete 'packets', called photons, the energy of each (E) being related to the frequency (ν) of the radiation by the equation $E = h\nu$, where h is the Planck constant, $6.626\ 196 \times 10^{-34}$ J s. The relationship between the frequency of a wave of electromagnetic radiation travelling in a vacuum and its wavelength (λ) is $\nu = c/\lambda$. *see also* electromagnetic spectrum.

electromagnetic senses senses that detect electric fields and that are used by some animals, e.g. some fish, to detect distortions in the Earth's electric field caused by objects in the locality. *see also* magnetotaxis.

electromagnetic spectrum the complete range of electromagnetic radiation, from gamma rays at wavelengths of *ca.* 10^{-13} m to radio waves at wavelengths of many metres. Visible light falls in the region from 360 (violet) to 760 nm (red).

electron *n.* (1) a subatomic particle with a mass of 5.5×10^{-4} a.m.u. and an electric charge designated -1 (equal to a negative charge of 1.602×10^{-19} C). Symbol e$^-$. All atoms contain at least one electron, found in a diffuse, yet ordered cloud outside the nucleus; (2) less commonly, an umbrella term for both the positive electron (positron) and the negative electron (negatron).

electron acceptor substance that receives an electron transferred from another substance in a chemical reaction, and thus becomes reduced.

electron capture a spontaneous nuclear reaction during which an atomic nucleus captures an orbital electron from the atom's innermost shell, so converting a proton to a neutron.

electron donor substance that loses an electron, which is transferred to another substance (an electron acceptor), in a chemical reaction and thus becomes oxidized.

electronegative *a.* (*chem.*) *appl.* an element the atoms of which, when in molecules, strongly attract the electrons of a covalent bond, also *appl.* the atoms themselves. Electronegativity, the ability to attract the electrons of a bond, is given the symbol χ. The most electronegative elements are, in order of decreasing electronegativity, F, O, N, Cl and Br. *cf.* electropositive.

electron microscopy type of microscopy in which a beam of electrons focused by magnets (lenses) is used to create the image. It has a resolution much greater than optical microscopy but living material cannot be viewed. *see* scanning electron microscope, transmission electron microscope.

electron-volt a unit of energy, symbol eV. 1 eV = $1.602\ 10 \times 10^{-19}$ J.

electrophoresis *n.* technique for separating molecules in an electric field on the basis of their net charge by their differential migration through paper, or through a polyacrylamide or agarose gel (gel electrophoresis). Gel electrophoresis is the main technique used for separating a mixture of proteins or nucleic acid fragments. As in the conditions of gel electrophoresis, such molecules all carry a similar net electrical charge, they migrate differentially according to their mass and thus are separated on the basis of size.

electropositive *a.* (*chem.*) *appl.* an element, the atoms of which, when in molecules, do not strongly attract the electrons of a covalent bond, also *appl.* the atoms themselves. The metallic elements are all, to varying degrees, electropositive, i.e. they exhibit electropositivity. *cf.* electronegative.

electrosensory *a. appl.* sensory systems in animals, such as the ampullae Lorenzini in dogfish (*Scyliorhinus*), that detect changes in electric fields.

electrostatic bond bond formed by electrostatic interaction between oppositely charged groups, important in biological

molecules. *alt.* ionic bond, ion pair, salt bond, salt linkage.

electrostatic precipitator (ESP) device used to remove particulate matter from waste gas streams, prior to their discharge into the atmosphere. The passage of the effluent stream through an intense electric field charges the particles. These are then attracted to collector plates, of opposite polarity to themselves, and subsequently removed. *see also* cyclone dust separator, fabric filter.

electrotaxis *n.* orientation of movement of a cell or organism within an electric field.

electrotropism *n.* plant curvature in an electric field.

element, chemical any substance that cannot be broken down into a simpler substance by chemical means. An element is composed of atoms, all of which have the same number of protons and electrons, i.e. they all have the same atomic number. Each chemical element has a name and is represented by a symbol, *see* Appendix 1. *alt.* element.

elephantiasis *n.* disease caused by the parasitic nematode *Wuchereria bancrofti*, transmitted to man by infected mosquitoes. In humans, severe infestation can cause blockage of the lymph vessels resulting in abnormal growth and swelling of the host tissue, especially in the legs. *alt.* lymphatic filariasis.

elephants *see* Proboscidea.

elephant-tooth shells common name for the Scaphopoda *q.v.*

elfinwood *n.* scrub composed of dwarfed trees found in extreme conditions as at edge of treeline in mountains.

elicitor *n.* in plant pathology, a compound that induces a plant defence response, and which may either be derived from a plant pathogenic microorganism (a biotic elicitor) or may be an inorganic material such as mercury and other heavy metals.

elittoral *a. appl.* zone out from the coast where light ceases to penetrate to the sea bottom.

El Niño warm current that periodically flows along the western coast of South America, usually forming around December–January.

El Niño–Southern Oscillation (ENSO) temporary reversal of airflows and surface ocean currents in the equatorial Pacific Ocean. This results in abnormal warming of the surface waters off the coast of Peru and in the disturbance of global weather patterns. This abnormal situation recurs every 2–10 years, and may last for over a year.

ELR environmental lapse rate *q.v.*

Eltonian pyramid pyramid of numbers *q.v.*

elutriation *n.* (1) the washing away of fine particles within a soil; (2) the sorting of soil particles by size using a current of air or water.

eluvial *a. appl.* (1) layer in soils that is leached, above the illuvial layer, *alt.* A-horizon; (2) gravels formed by breakdown of rocks *in situ.*

eluviation *n.* the process in which water removes soil materials, either in suspension or solution, usually from an upper soil horizon to a lower one. *cf.* illuviation. The removal of dissolved soil material by water is often referred to as leaching.

EM electron microscope, electron microscopy. *see* scanning electron microscope, transmission electron microscope.

embiids *n.plu.* common name for the Embioptera *q.v.*

Embioptera *n.* order of insects with soft flattened bodies, and incomplete metamorphosis, commonly called embiids or foot spinners, which live in groups in silken tunnels and have wingless females and winged males.

embryo *n.* animal or plant in the earliest stages of development, when it is entirely dependent directly or indirectly on resources provided by the parent (e.g. in seed or egg). The embryo stage lasts from the fertilized egg (zygote) up to the stage of a free-living miniature adult or larva in animals, or a photosynthesizing seedling in plants. In human development, the conceptus is technically known as an embryo until the main parts of the body and main internal organs have started to take shape at around the seventh week of gestation. After this it is called a foetus.

embryogenesis *n.* the development of an embryo from a fertilized ovum.

embryogeny *n.* formation of the embryo, used esp. in botany.

embryology *n.* the study of the formation and development of embryos.

embryonated *a. appl.* egg in which an embryo has developed.

embryonic *a. pert.* an embryo.

emergence *n.* (1) hatching, of embryo or larva; (2) an outgrowth from subepidermal tissue; (3) an epidermal appendage.

emergent *a. appl.* plants whose roots are in shallow water but whose stems and leaves rise above the surface of the water.

emersed *a.* rising above surface of water, *appl.* leaves.

emery *n.* a mixture of corundum (*q.v.*), magnetite (*q.v.*) and spinel (*q.v.*), which is used as an abrasive.

emigration *n.* (*ecol.*) the movement of a population of a species, or of individuals, out of a particular region or habitat.

emission *n.* discharge of a pollutant, esp. into the atmosphere, from, e.g., a furnace or a factory (known as the emission source).

emission inventory area map showing the distribution of emissions, together with the location of major emission sources.

emission source *see* emission.

emission standard maximum amount of a particular pollutant that may be legally discharged from a specific pollution source.

emmenophyte, emmophyte *n.* a water plant without any floating parts.

emmer *n.* a primitive cultivated tetraploid wheat (*Triticum dicoccum*) first domesticated in the Near East and South-West Asia around 11 000 years ago and that is thought to be derived from wild emmer (*T. dicoccoides*), which is a hybrid between *T. monococcum* and goat-grass (*Aegilops*).

emphysema *n.* chronic disease of the lungs causing shortness of breath, which is exacerbated by cigarette smoke and air pollution.

empirical *a. pert.* knowledge based on observation or experiment, not theory.

en echelon faults a series of parallel or near-parallel faults.

enantiomer *n.* one of two isomers of a molecule that are mirror images of each other.

enantiomorphic *a.* (1) similar but contraposed, as mirror image; (2) deviating from normal symmetry.

enargite *n.* Cu_3AsS_4, a copper ore mineral.

Encarsia formosa a minute parasitic wasp used for biological control of whitefly (*Trialeurodes vaporariorum*) in glasshouses.

enchytracheids *n.plu.* group of small oligochaete worms, living in soil.

enclosure *n.* (1) an area of land surrounded by a physical barrier; (2) the conversion of areas of common land into privately owned land, *see* enclosure movement.

enclosure movement process during which areas of open land, e.g. heathland and forest, intended for agricultural use, were permanently surrounded by walls or hedges. This took place in England and Wales, UK, between the sixteenth and seventeenth centuries, esp. after 1750, when it was supported by a private Act of Parliament.

encyst *v.* of a cell or small organism, to surround itself with an outer tough coat or capsule. *n.* encystment.

endangered *a.* IUCN definition *appl.* species or larger taxa whose numbers have become so low, or whose habitats have been so drastically reduced, that they are thought to be in immediate danger of extinction in the wild in the foreseeable future if there is no change in circumstances, *see also* rare, rarity, vulnerable.

end moraine *see* moraine.

endemic *a.* (1) indigenous to, and restricted to, a certain region or part of region; (2) *appl.* disease, present at relatively low levels in the population all the time. *n.* endemism.

Endemic Bird Areas priority areas for bird species conservation identified by the International Council for Bird Preservation.

endergonic *a.* absorbing or requiring energy, *appl.* metabolic reactions.

endo- prefix derived from Gk *endon*, within, signifying within, inside, acting inside, opening to the inside, etc.

endobiont *n.* organism living within the cells of another organism, apparently without harming it.

endobiotic *a.* living within a substratum or within another living organism.

endodeme *n.* a gamodeme composed of predominantly inbreeding dioecious plants or bisexual animals.

endogamy *n.* (1) zygote formation within a cyst by the fusion of two of the products of preceding division; (2) self-pollination; (3) inbreeding.

endogenous *a.* (1) originating within the organism, cell, or system being studied;

(2) originating from a deep-seated layer, *appl.* lateral roots; (3) *appl.* metabolism, biosynthetic and degradative processes in tissues. *cf.* exogenous.

endogenous rhythm a metabolic or behavioural rhythm that originates within the organism and persists although external conditions are kept artificially constant. There may be some slight change in the periodicity under constant conditions, when the rhythm is said to be free-running, such as a circadian rhythm changing to 25 or 28 h. Endogenous circadian rhythms are maintained at the regular 24-h cycle by an external stimulus (the zeitgeber), such as light or temperature.

endogenous virus viruses (e.g. C-type viruses) carried permanently in an inactive proviral state in the genome and inherited from generation to generation, which may become activated on infection with another virus. *alt.* provirus.

endolithic *a.* burrowing or existing in a stony substratum, as algal filaments.

endomixis *n.* self-fertilization in which a male and female nucleus from the same individual fuse.

endomycorrhiza *see* mycorrhiza.

endoparasite *n.* parasite that lives inside another cell or organism.

endophagous *a.* feeding inside a food source.

endophyllous *a.* (1) sheathed by a leaf; (2) living within a leaf.

endophyte *n.* bacterium, fungus, alga or other plant living inside the body or cells of another organism. *alt.* endosymbiont.

endoplasmic reticulum (ER) extensive, convoluted internal membrane in eukaryotic cells, continuous with the outer nuclear membrane and enclosing a continuous internal space (lumen). Involved in the synthesis and transport of membrane proteins and lipids and of proteins destined for secretion from the cell. The cytoplasmic face of the membrane may be studded with protein-synthesizing ribosomes, when it is known as rough endoplasmic reticulum (RER), or lack ribosomes, when it is called smooth endoplasmic reticulum (SER).

endopolyploidy *n.* polyploidy resulting from repeated doubling of chromosome number without normal mitosis.

Endoprocta, endoprocts Entoprocta *q.v.*

endopsammon *n.* microscopic fauna living in sand or mud.

Endopterygota, **endopterygotes** *n.,* *n.plu.* a division of insects having complete metamorphosis, with wings developing internally and larvae different from adults. Includes the Coleoptera, Neuroptera, Mecoptera, Trichoptera, Lepidoptera, Diptera, Siphonaptera, Hymenoptera and other orders. *cf.* Exopterygota.

endorheic drainage term applied to rivers that drain into inland lakes.

endorhizosphere *n.* the epidermis and cortex of normal healthy roots that are invaded by soil microorganisms.

endoskeleton *n.* any internal skeleton or supporting structure.

endosperm *n.* the nutritive tissue surrounding the embryo in most seeds. *alt.* albumen.

endospore *n.* (1) any spore formed within a sex organ or sporangium; (2) (*microbiol.*) thick-walled heat- and drought-resistant asexual spore formed within the cells of certain bacteria e.g. *Bacillus* and *Clostridium* spp., in unfavourable growing conditions, and which can remain dormant for many years.

endosulfan *n.* contact and stomach-acting organochlorine insecticide and acaricide, persistent in the environment and therefore of limited use.

endosymbiont *see* endosymbiosis.

endosymbiosis *n.* symbiosis in which one partner (the endosymbiont) lives inside the cells of the other to their mutual benefit, e.g. photosynthetic cyanobacteria living in the cells of non-photosynthetic dinoflagellates. *a.* endosymbiotic.

endothelium *n.* a squamous epithelium that lines internal body surfaces such as in heart and blood and lymphatic vessels and other fluid-filled cavities and glands.

endothermic *a. appl.* (1) chemical or physical changes for which $\Delta H > 0$, *see* enthalpy; (2) the regulation of body temperature by an animal by producing heat within its own tissues, as do birds and mammals. Such animals are known as endotherms. *cf.* ectothermic. *n.* endothermy. *see also* homoiothermic.

endotoxin *n.* toxin produced by bacteria that is an integral component of the bacterial cell and remains within the cell or attached to the cell, such as bacterial lipopolysaccharide. *cf.* exotoxin.

endotrophic *a. appl.* mycorrhizas in which the fungal hyphae extensively penetrate the cells of the host's roots.

endozoic *a.* (1) living within an animal; (2) involving passage through an animal, as in the distribution of some seeds.

endozoochory *n.* dispersal of seeds by ingestion by animals and subsequent passage through their digestive tracts.

endrin *n.* organochlorine insecticide that is extremely toxic and persistent in the environment. Its use is banned or restricted in many countries.

energy *n.* (1) the capacity to do work or supply heat. Symbol E. Its SI unit is the joule (J); (2) the energy obtained for industrial and domestic use by the burning of fossil fuels and biomass or from nuclear energy. *see also* delivered energy, primary energy, useful energy.

energy budget (1) analysis of the quantity of energy used by different living components or levels of a biosystem, or of the different components of the flow of energy through an individual organism; (2) the balance of energy input and use in a biological system, expressed as consumption (C) = production (P) + respiration (R) + rejecta (faeces and urine) (F + U). *see also* assimilated energy.

energy capital fossil and nuclear sources of energy.

energy conservation the reduction in the absolute amount of energy consumed by using it more efficiently or by cutting down on its use. Conservation measures include insulation of houses to reduce heat loss, better design of industrial processes so that they use less energy, etc.

energy consumption (1) a term that is in common usage although it is a misnomer as, according to the first law of thermodynamics, the energy content of a system of constant mass can be neither increased nor decreased. However, it can be taken to mean the transformation of energy in a potentially useful form, e.g. chemical energy, to a relatively useless one (ultimately, random thermal motion at ambient temperatures); (2) in non-technical usage it means the use of energy in the form of fossil fuels and electricity by humans. *cf.* primary energy consumption.

energy crisis any situation in which there is a significant shortage of energy that causes hardship.

energy crops crops that are grown specifically for use as a biomass energy resource. These may be used directly as fuels for cooking, heating or electricity generation or they may be converted into other fuels. In the USA for example, grain (especially maize) is grown for fermentation into ethanol. This is used as a supplement in petrol (gasoline) and the resultant mixture is known as gasohol. *alt.* fuel crops.

energy efficiency in any biological or industrial process, the percentage of the total energy put into the system that can be converted into useful work (or production) and not lost as heat. *see* assimilation efficiency, ecological efficiency, ecological growth efficiency.

energy flow (*biol.*) the transfer of energy from one organism to another through the trophic levels in an ecosystem. About 90% of the chemical energy is lost at each transfer.

energy intensity a term used in environmental economics to express the effectiveness of energy utilization. For a country, it may be defined as the amount of energy consumed (*see* energy consumption) per unit gross domestic product (GDP). Typical units used are megajoules per GDP in dollars (MJ/$GDP).

energy production a term that is in common usage although it is a misnomer as, according to the first law of thermodynamics, the energy content of a system of constant mass can be neither increased nor decreased. However, it can be taken to mean the conversion, by human endeavour, of energy from a source found in nature to one that is useful to man.

energy pyramid diagram representing the loss of energy at each trophic level in a biological community. *see* Fig. 11, p. 127.

energy quality ability of a form of energy to do useful work. The concentrated chemical energy in fossil fuels is considered as high-quality energy, while low-temperature heat is considered as low-quality energy.

energy services those functions, e.g. heating a house or powering a car, that useful energy (*q.v.*) performs.

energy tax proposed tax that would be levied on non-renewable fuels, as part of energy conservation strategies. *see also* carbon tax.

engine *n.* any device that is used to convert energy from one form (often chemical) to another (usually mechanical).

englacial debris glacial debris transported within a glacier.

English Nature official UK agency that establishes and maintains National Nature Reserves, advises Government on the designation of Sites of Special Scientific Interest (SSSIs) and provides advice to the Secretary of State for the Environment on matters of ecology and nature conservation in England. There are separate agencies for Wales and Scotland. *see* Countryside Council for Wales, Scottish Natural Heritage.

enhalid *a.* (1) containing salt water, *appl.* soils; (2) growing in saltings or on loose soil in salt water, *appl.* plants.

enhanced greenhouse effect the increased warming of the troposphere caused by the anthropogenic emission of greenhouse gases (*q.v.*). *see also* greenhouse effect.

enhanced recovery *see* oil recovery.

Enopla *n.* class of proboscis worms having the mouth anterior to the brain, and usually an armed proboscis.

enphytotic *a.* afflicting plants, *appl.* diseases restricted to a locality.

enriched fuel nuclear fuel that contains uranium-235 at higher than natural abundance, i.e. >0.715% of total uranium present is uranium-235.

enrichment culture a technique for isolating a particular microorganism or microorganisms from a natural mixed population by culture in conditions particularly favourable for their growth, so that they become the predominant form in the resulting culture.

ensilage *v.* to make silage *q.v.*

ENSO El Niño–Southern Oscillation *q.v.*

enstatite *n.* $Mg_2Si_2O_6$, one of the pyroxene minerals (*q.v.*) of the orthopyroxene subgroup.

enteric bacteria *see* enterobacteria.

enteric fever typhoid fever *q.v.*

enterobacteria *n.plu.* bacteria belonging to the family Enterobacteriaceae, Gramnegative facultatively anaerobic rods, which include some species living in soil, some normal inhabitants of the mammalian intestine, e.g. *Escherichia coli*, and some which cause disease, e.g. *Salmonella* spp.

Enteropneusta *n.* class of solitary, worm-like burrowing hemichordates, having many gill slits and no lophophore, commonly called acorn worms.

enterotoxin *n.* toxin produced by certain bacteria infecting the intestine.

enteroviruses *n.plu.* picornaviruses (*q.v.*) that typically occur in the intestine and stomach, and which include, e.g., Coxsackie viruses, echoviruses, poliovirus.

enterozoon *n.* any animal parasite inhabiting the intestines.

enthalpy *n.* a thermodynamic term (symbol *H*). It is a property of a system that equates to $U + PV$, where U is the internal energy of the system, P is its pressure and V is its volume. If only pressure–volume work (*q.v.*) occurs, changes in enthalpy (ΔH) at constant pressure equate to the heat flow in or out of the system. Chemical or physical changes that cause a system to absorb heat from its surroundings under these conditions have positive values for ΔH and are said to be endothermic, while those that release heat are exothermic (ΔH is negative).

enthalpy density the heat interaction that occurs when a given volume of a given fuel is burned at a constant pressure (usually 1 atm., i.e. 101 325 pascals). Enthalpy densities are quoted in kJ dm^{-3}, at a specific temperature (usually 298 K), and, unlike most other enthalpies, are defined as positive for exothermic processes.

enthetic *a.* (1) introduced; (2) implanted.

entisol *n.* an order of the USDA Soil Taxonomy system comprising recently formed soils with little or no development of distinctive soil horizons, found in a wide variety of different environments. *see* aquent, arent, fluvent, orthent, psamment.

entobranchiate *a.* having internal gills.

entombment of nuclear reactors *see* decommissioning process.

entomochory *n.* dispersal of seeds or spores through the agency of insects. *a.* entomochoric.

entomofauna *n.* the insects of a particular environment or region.

entomogenous *a.* growing in or on insects, as certain fungi.

entomology *n.* the study of insects.

entomophagous insectivorous *q.v.*

entomophilous *a.* pollinated by insects. *n.* entomophily.

entomophyte *n.* any fungus growing on or in an insect.

entomostracans *n.plu.* large group of small crustaceans, including branchiopods, *Branchiura* (fish lice), copepods, barnacles and ostracods.

entopic *a.* in the normal position. *cf.* ectopic.

Entoprocta Ectoprocta *q.v.*

entozoic *a.* living inside the body of another animal or plant.

entrainment *n.* (1) (*biol.*) process by which a free-running endogenous circadian rhythm is synchronized to an exact 24-h cycle; (2) (*phys.*) the act of trapping and transporting material within a moving fluid, e.g. glacial debris within a glacier or fly ash within a flue gas.

entropy *n.* a thermodynamic term describing the disorder or randomness of a system.

E number classification for food additives used by the European Community. Food additives are classified as colouring agents, preservatives, antioxidants, emulsifiers and stabilizers, acids and bases, anti-caking agents, and flavour enhancers and sweeteners.

environment *n.* the sum total of external influences acting on an organism.

Environment Agency (EA) a statutory body within the UK set up, in 1996, to undertake those functions formerly carried out by Her Majesty's Inspectorate of Pollution (HMIP), the National Rivers Authority (NRA) and waste regulation authorities.

environmental assessment (EA) an evaluation of the likely environmental consequences of a proposed development project or piece of legislation and of the measures to be taken to minimize adverse effects, which is now a legal requirement in some countries. *alt.* environmental impact assessment. *see also* environmental impact statement.

environmental audit a survey of all aspects of an environment, including, e.g., the geology, hydrology, habitats and species present, and human influences.

environmental degradation the destruction or impoverishment of an environment by using its resources at a faster rate than they can be replenished.

environmental gradient a gradual change in environmental conditions between two extremes, e.g. from a hot to a cold climate.

environmental impact the changes in the total environment, both in terms of the ecology and the social impact, caused by human activities.

environmental impact assessment (EIA) *see* environmental assessment.

environmental impact statement (EIS) in the USA and in some other countries, a comprehensive environmental assessment (*q.v.*) that must be filed by all federal agencies for any proposed major development or piece of legislation. It must include the purpose and need for the proposed action, the probable environmental impacts of the development and of its alternatives, any adverse environmental effects that could not be avoided should the project be implemented, the relationships between the probable short-term and long-term impacts on environmental quality, the irreversible and irretrievable commitments of resources involved, objections raised by reviewers on the preliminary draft of the statement, the names and qualifications of the people primarily responsible for preparing the environmental impact statement, references necessary to back up all statements, and a conclusion. *see also* environmental audit.

environmentalism *n.* active concern with protecting and conserving the integrity of the environment, both living and non-living, and in promoting an ideal of human society that works with Earth's natural processes rather than continually trying to override them and change them. Such a course of action is considered, by its proponents, not only morally right but essential for the continued ability of the Earth to support human societies. *see* sustainability.

environmentalist *n.* (1) someone who espouses the idea of environmentalism (*q.v.*); (2) anyone who is concerned with protecting the environment, esp. with regard to the prevention of pollution and the conservation of biological diversity and the quality of soil, air and water. *see also* conservationist.

environmental lapse rate (ELR) rate at which atmospheric temperature changes with altitude at a given time and place, averaging an 0.6 °C decline per 100 m ascent.

Environmental Protection Agency (EPA) US federal government agency, established in 1970 to provide a national framework for pollution control. It is concerned with such issues as clean air, clean water and waste disposal.

environmental quality standard (EQS) a measure used in setting regulatory limits for potentially polluting emissions, which is the level of the substance in the environment that is consistent with the maintenance of that environment in a healthy, unaffected state.

environmental resistance (1) the factors limiting the population growth of a species in a given environment; (2) the product of the equations for simple population growth and environmental resistance, expressed as $(K - N)/(K)$, where K is the carrying capacity of the environment for the species and N the number of individuals present.

environmental science the study of how humans and other species interact with their non-living and living environments.

environmental services *see* ecosystem services.

environmental stress any physical factor, e.g. temperature or pollution, that has an adverse effect on a community or ecosystem.

environmental taxation *see* carbon tax, ecotaxation, energy tax.

enzootic *a.* (1) afflicting animals; (2) *appl.* disease in animals restricted to a locality.

enzyme *n.* any of a large and diverse group of (mainly) proteins that function as biological catalysts in virtually all biochemical reactions, essential in all cells, different enzymes being highly specific for a particular chemical reaction and reactants. Enzymes are classified according to the type of chemical reaction catalysed and substrate acted on, e.g. hydrolases, carboxylases, oxidoreductases, nucleases, proteinases. Certain RNAs can also act as enzymes, *see* ribozyme.

enzyme polymorphism the occurrence in a species, or in a population of a species, of two or more forms of an enzyme, each specified by different alleles of the gene.

Eocene *n.* early epoch of the Tertiary period, between the Paleocene and Oligocene, lasting from around 55 million years ago to 40 million years ago. *see* Appendix 4.

Eogaea *n.* a zoogeographical division including Africa, South America and Australasia.

eolian aeolian *q.v.*

eon *n.* in geological time, the largest unit of timescale. *see* Appendix 4.

EPA Environmental Protection Agency *q.v.*

epanthous *a.* living on flowers, *appl.* certain fungi.

epeirogenesis, epeirogeny *n.* the uplift of extensive areas of the Earth's surface with no significant faulting or folding of the crustal rocks.

ephedrine *n.* a sympathomimetic alkaloid obtained from *Ephedra* spp., having the same bronchial smooth muscle relaxant effects as adrenaline and used as a nasal decongestant and to treat hayfever and asthma.

ephemeral *a.* (1) short-lived; (2) taking place once only, *appl.* plant movements, such as expanding buds; (3) completing life cycle within a brief period. *n.* a short-lived plant or animal species.

ephemeral stream a stream whose flow is intermittent, a common feature of desert regions.

Ephemeroptera, ephemerids *n., n.plu.* order of insects including the mayfly, whose adult life is very short, sometimes less than a day. They have an aquatic nymph (larval) stage. They have three appendages, 'tails', projecting from the end of the abdomen, and the hindwing is smaller than the forewing.

epi- prefix derived from *Gk epi*, upon, signifying above or upon.

epibenthos *n.* fauna and flora of the sea bottom between low-water mark and the 200 m line.

epibiotic *a.* (1) *appl.* endemic species that are relics of a former flora or fauna; (2) growing on the exterior of living organisms; (3) living on a surface, as of sea bottom.

epicentre *n.* point on the Earth's surface located immediately above the focus (place of origin) of an earthquake. At this point, the most severe of the shock waves are usually experienced.

epicormic *a.* growing from a dormant bud.

epideictic display a suggested territorial display by which members of a population reveal their presence and allow others to assess the density of the population.

epidemic *a.* affecting a large number of individuals at the same time, *appl.* disease. *n.* an outbreak of epidemic disease.

epidemic spawning the phenomenon in which a large number of individuals all release their gametes at the same time.

epidemiology *n.* (1) the study of the occurrence of infectious diseases, their origins and pattern of spread through the population; (2) the cause and pattern of spread of a disease; (3) the study of the incidence of non-infectious disease (e.g. cancer) with a view to finding causes, e.g. the causal link between smoking and lung cancer was found by epidemiological studies.

epidermis *n.* (1) outer layer or layers of the skin, derived from embryonic ectoderm. In vertebrates, a non-vascular stratified tissue, often keratinized; (2) outer epithelial covering of roots, stems and leaves in plants. *a.* epidermal.

epidote *n.* a group of minerals with the formula $A_2B_3Si_3O_{12}(OH)$, where, usually, A is Ca and B is Al or Fe.

epifauna *n.* (1) animals living on the surface of the ocean floor; (2) any encrusting fauna.

epigamic *a.* (1) *appl.* any trait related to courtship and sex other than the essential sexual organs and copulatory behaviour; (2) tending to attract the opposite sex, e.g. *appl.* colour displayed in courtship.

epigeal, epigean, epigeous *a.* (1) borne above ground; *appl.* (2) type of germination in which cotyledons are carried above ground as the shoot grows; (3) insects, living near the ground.

epigenetic *a. appl.* the chain of processes linking genotype and phenotype, other than the initial gene action.

epigenous *a.* developing or growing on a surface.

epilimnion *n.* upper water layer, above thermocline in lakes, rich in oxygen.

epilithic *a.* attached to rocks, *appl.* algae, lichens.

epimeletic *a. appl.* animal behaviour relating to the care of others.

epimorphic *a.* maintaining the same form in successive stages of growth.

epimorphosis *n.* type of regeneration in animals in which proliferation of new material precedes development of new part.

epinasty *n.* the more rapid growth of upper surface of an organ, e.g. a leaf, thus causing unrolling or downward curvature.

epinekton *n.* small animals that cannot swim but are attached to actively swimming organisms.

epineuston *n.* those animals living at the surface of water, in the air.

epiorganism *n.* a group of organisms that functions as a colony.

epiparasite ectoparasite *q.v.*

epipedon *n.* diagnostic surface horizon of a soil.

epipelagic *a. pert.* (1) deep-sea water between surface and bathypelagic zone; (2) marine organisms that live in oceanic waters not exceeding 200 m in depth (the epipelagic zone).

epiphenomenon *n.* something produced as a side-effect of a process.

epiphyllous *a.* growing upon leaves. *n.* epiphyll.

epiphyte *n.* a plant that lives on the surface of other plants but does not derive water or nourishment from them.

epiphyton *n.* community of plants living attached to other plants, as many algae in aquatic environments.

epiphytotic *a. pert.* disease epidemic in plants.

epiplankton *n.* that portion of plankton from surface to about 200 m.

epirhizous *a.* growing upon a root.

episematic *a.* aiding in recognition, *appl.* coloration, markings. *cf.* aposematic, parasematic, sematic.

episode *n.* a single event or a single instance of a phenomenon, such as a pest infestation.

epistasis, epistasy *n.* the suppression or masking of the effect of a gene by another different gene. *a.* epistatic. *cf.* dominance (3).

epithelium *n.* sheet of cells tightly bound together, lining any external or internal surface in multicellular organisms, e.g. the epidermis, surfaces of mucous membranes, the lining of the gut, and the linings of ducts and glands. Epithelia variously serve protective, secretory or absorptive functions. *plu.* epithelia. *a.* epithelial.

epitreptic *a. appl.* animal behaviour causing another animal of the same species to approach.

epizoic *a.* (1) living on or attached to the body of an animal; (2) having seeds or fruits dispersed by being attached to the surface of an animal.

epizoite *n.* animal that lives on the body or shell of another but is not parasitic upon it.

epizoochory *n.* dispersal of seeds by being carried on the body of an animal.

epizoon *n.* animal that lives on the body of another animal.

epizootic *n.* epidemic disease among animals.

epoch *n.* in geological time, the subdivision of a period.

epontic *a. appl.* aquatic microorganisms that live attached to the surface of objects.

eponym *n.* name of a person used in the designation of, e.g., a species, organ, law, disease.

epsom salts epsomite *q.v.*

epsomite *n.* a mineral of formula $MgSO_4.7H_2O$. *alt.* epsom salts.

EQS environmental quality standard *q.v.*

equations, chemical *see* chemical equations.

equatorial low intertropical convergence zone *q.v.*

equilateral *a.* (1) having the sides equal; (2) *appl.* shells symmetrical about a transverse line drawn through the umbo.

equilibrium constant (K) a parameter that expresses the extent to which a chemical reaction has approached completion at the point at which it has reached dynamic equilibrium.

For a reaction $aA + bB \rightleftharpoons cC + dD$, where A and B are the reactants; C and D are the products; and a, b, c and d are their coefficients of stoichiometry:

$$K = \frac{\{C\}^c \times \{D\}^d}{\{A\}^a \times \{B\}^b} \qquad (1)$$

where the curly brackets represent the activity of the reactant or product that they enclose. K remains constant for a given reaction, at a given temperature, irrespective of the starting concentration of the reactants or products. At low concentrations:

$$K \simeq \frac{[C]^c \times [D]^d}{[A]^a \times [B]^b} \qquad (2)$$

where the square brackets represent the concentration in moles per cubic decimetre (mol dm^{-3}). The partial pressure, P, of an ideal gas is proportional to its concentration in mol dm^{-3}. Therefore, partial pressures can be used instead of molar concentrations in Equation 2 and activities based on partial pressures can be used instead of those based on molar concentrations in Equation 1. However, care must be exercised as, in general, the value of K obtained using partial pressures will be different numerically from one based on concentrations.

equilibrium, dynamic a state that represents a system at a thermodynamic minimum, i.e. $\Delta G = 0$, in which no net change is observable and that is maintained by a balance between equal and opposite processes.

equimolecular *See* isotonic.

equinoctial *a. appl.* flowers that open and close at definite times.

equinox *n.* one of the two days during each year when, at noon, the Sun's rays hit the Earth vertically at the Equator and day and night are of equal length throughout the world. In the northern hemisphere, the spring (vernal) equinox falls on 21 March and the autumnal (fall) equinox falls on 23 September.

Equisetales horsetails. *see* Sphenophyta.

equitability (H′) *n.* the degree of similarity in the population numbers of different species in a community, the greatest equitability being achieved when all species are represented by the same number of individuals.

Er symbol for the chemical element erbium *q.v.*

ER endoplasmic reticulum *q.v.*

era *n.* a main division of geological time, such as Palaeozoic, Mesozoic, Cenozoic, and divided into periods. *see* Appendix 4.

eradication *n.* the extinction of a species in a particular area.

erbium (Er) *n.* a lanthanide element (at. no. 68, r.a.m. 167.26), a silver-coloured metal when in elemental form. It is used, e.g., to absorb neutrons in nuclear reactors.

erect *a.* (1) directed towards summit of ovary, *appl.* ovule; (2) growing straight up, *appl.* plants.

Eremian *a. appl.* or *pert.* part of the Palaearctic region including the deserts of North Africa and Asia.

eremic *a. pert.*, or living in, deserts.

eremobic *a.* (1) growing or living in isolation; (2) having a solitary existence.

eremophyte *n.* a desert plant.

erg *n.* (1) (*phys.*) the c.g.s. unit of energy or work (1 erg = 10^{-7} J); (2) (*geog.*) a large expanse of desert covered with sand. In total, ergs occupy about 12% of the world's arid lands, *alt.* sand sea.

ergonomics *n.* the anatomical, physiological and psychological study of humans in their working environment.

ergonomy *n.* (1) the differentiation of functions; (2) physiological differentiations associated with morphological specialization.

ergot *n.* the hardened mycelial mass (sclerotium) of the fungus *Claviceps purpurea* that replaces the grain of infected rye and some other grasses and that contains poisonous alkaloids that cause ergotism (abortion, hallucinations and sometimes death) in animals and humans who eat the infected grain.

ericaceous *a.* of or *pert.* the Ericaceae, the heather family, which includes the heathers and rhododendrons.

Ericales *n.* order of dicot shrubs, rarely trees or herbs, and including the families Actinidiaceae, Empetraceae (crowberry), Ericaceae (heathers and rhododendrons), Monotropaceae (Indian pipe) and Pyrolaceae (wintergreen).

ericilignosa *n.* vegetation dominated by heathers (members of the genus *Erica*).

ericoid *a. appl.* endomycorrhizas formed on members of the Ericaceae, lacking a fungal sheath and with extensive penetration of the cell of the root cortex, with formation of intracellular hyphal coils.

Eriocaules *n.* order of monocot herbs comprising the family Eriocaulaceae (pipewort).

eriophyid *n.* any of a large family (Eriophyidae) of minute plant-eating mites that have two pairs of legs at front and no respiratory system.

erosion *n.* the transportation of particles produced by weathering by a variety of agents including moving ice, flowing water, wind and mass movement. *see also* soil erosion.

erosion surface planation surface *q.v.*

erratics *n.plu.* rock fragments or boulders that have been transported by glaciers and deposited several kilometres from their source area. Where these are precariously positioned, they are termed perched blocks or rocking stones.

error, absolute *see* absolute error.

error, relative *see* relative error.

ERTS Earth Resources Technology Satellite *q.v.*

eruciform *a.* (1) *appl.* insect larvae with a more-or-less cylindrical body and stumpy legs on the abdomen as well as the true thoracic legs, such as caterpillars and grubs; (2) having the shape of a caterpillar or grub.

erythrism *n.* abnormal presence or excessive amount of red colouring matter, as in petals, feathers, hair, eggs.

erythroaphins *n.plu.* red pigments formed by the post-mortem enzymatic transformation of yellow plant pigments in aphids.

erythromycin *n.* an antibiotic synthesized by the actinomycete *Streptomyces erythreus*, which inhibits bacterial protein synthesis.

erythrophyll *n.* a red anthocyanin, such as found in some leaves and red algae.

Es symbol for the chemical element einsteinium *q.v.*

escape *a. appl.* behaviour in which an animal moves away from an unpleasant stimulus.

escape *n.* plant or animal originally domesticated and now established in the wild.

escarpment *n.* (1) cuesta (*q.v.*); (2) the scarp slope of a cuesta; (3) a steep slope created by erosion or faulting.

eserine *n.* plant alkaloid, specific inhibitor of cholinesterase. *alt.* physostigmine.

esker *n.* a narrow sinuous ridge composed of stratified coarse sands and gravel, which may be over 100 km long. This type of fluvioglacial deposit may be laid down by, e.g., water flowing in a long ice tunnel at the base of a melting glacier.

esoteric *a.* arising within the organism.

ESP electrostatic precipitator *q.v.*

ESS evolutionarily stable strategy *q.v.*

essential amino acids amino acids that cannot be synthesized by the body, or only in insufficient amounts, and must be supplied in the diet: for humans these are Arginine, histidine, isoleucine, leucine, lysine, methionine, phenylalanine, threonine, tryptophan, valine.

essential elements chemical elements, such as carbon (C), nitrogen (N), sulphur (S) and oxygen (O), supplied by nutrients, necessary for the normal growth of organisms. *alt.* bioelements.

essential fatty acids (EFA) those fatty acids that cannot be synthesized by mammals and must be present in the diet, e.g. linoleic acid and linolenic acid.

essential mineral (*geol.*) any one of a relatively small number of minerals that make up the bulk of most igneous rocks. These minerals belong to the following mineral groups: amphiboles, feldspars, feldspathoids, micas, olivines, pyroxenes and silica (in the form of quartz). The classification of igneous rocks is, to a large extent, based on the presence and relative proportions of these essential minerals.

essential oils mixtures of various volatile oils derived from benzenes and terpenes found in plants and producing characteristic odours, and having various functions such as attracting insects or warding off fungal attacks.

establishment *n.* (*biol.*) the stage following immigration or invasion of a region or habitat by a new species or individual, when they become permanent members of the community.

estancia *n.* a beef-cattle ranch in the pampas region of South America (Uruguay and northern Argentina).

ester *n.* the organic product of a condensation reaction between an alcohol and a carboxylic acid.

estival aestival *q.v.*

estivation aestivation *q.v.*

estrogen oestrogen *q.v.*

estrous oestrus *q.v.*

estuarine *a.* living in the lower part of a river or estuary where fresh water and seawater meet.

estuary *n.* (1) semi-enclosed body of water, containing a mixture of fresh water and seawater, formed where a river enters a sea; (2) less commonly, a sea inlet.

eth- (*chem.*) the prefix used to indicate the presence of two carbon atoms in a chain in an organic compound (e.g. ethane CH_3CH_3) or radical.

ethanol *n.* CH_3CH_2OH, colourless, flammable alcohol that is used, e.g., as a solvent, a fuel and the intoxicating component of alcoholic beverages. For consumption, it is made by the microbial fermentation of starch or sugars by yeasts and other organisms in the brewing and wine-making industries. Fuel alcohol is also made from organic wastes by fermentation. For industrial use, ethanol can be made by the treatment of ethene with sulphuric acid, yielding ethyl hydrogen sulphate, which is then hydrolysed to liberate the ethanol. *alt.* ethyl alcohol, grain alcohol.

ethene *n.* the standard name for ethylene *q.v.*

ether *n.* an organic compound that contains an oxygen atom bonded to two hydrocarbon radicals. Although this is a general term, the compound diethyl ether (ethoxyethane), $(C_2H_5)_2O$, is often referred to simply as ether.

ethereal *a. appl.* (1) a class of odours including those of ethers and fruits; (2) fragrant oils in many seed plants.

Ethiopian *a. pert.* (1) a zoogeographical region including Africa south of the Sahara and southern Arabia and divisible into African and Malagasy subregions; (2) a floral region that is part of the Palaeotropical Floral Realm, comprising Ethiopia and the south-west tip of the Arabian peninsula.

ethirimol *n.* systemic fungicide used to control mildews in crops such as barley.

ethnobotany *n.* the study of the use of plants by humans.

ethnography *n.* the description and study of human races.

ethnology *n.* science dealing with the different human races, their distribution, relationship and activities.

ethnozoology *n.* the study of the use of animals by humans.

ethofumesate *n.* soil-acting and translocated herbicide used, e.g., to control broad-leaved weeds and weed grasses in grass crops.

ethogram *n.* a catalogue of the natural behaviours of an animal and the contexts in which they occur.

ethological isolation the prevention of interbreeding between species as the result of behavioural differences.

ethology *n.* the study of the behaviour of animals in their natural habitats. *a.* ethological.

ethyl alcohol ethanol *q.v.*

ethylene *n.* C_2H_4, a gas produced by plants in minute amounts that has various developmental effects as a hormone, including regulation of fruit ripening. *alt.* ethene.

ethylenediamine tetraacetic acid EDTA *q.v.*

etiolation *n.* the appearance of plants grown in the dark, having no chlorophyll, chloroplasts not developing, internodes being greatly elongated so the plants are tall and spindly, and having small, rudimentary leaves. *a.* etiolated.

etiology aetiology *q.v.*

etioplast *n.* (1) chloroplast formed in the absence of light, found in etiolated leaves, lacking thylakoid membranes and chlorophyll, that will develop into a functional chloroplast on illumination; (2) chloroplast precursor.

-etum (*ecol.*) a suffix used to indicate a plant community dominated by a particular species, e.g. a callunetum, a community dominated by heather or ling (*Calluna vulgaris*).

Eu symbol for the chemical element europium *q.v.*

EU European Union *q.v.*

eubacteria *n.plu.* bacteria other than the archaebacteria. They are generally unicellular, prokaryotic microorganisms possessing cell walls, with cells in the form of rods, cocci or spirilla, many species being motile with cells bearing one or more flagella. Now sometimes classified as a superkingdom, the Bacteria. *see* Appendix 9.

Eucalyptus genus of trees, native to Australia, that are now cultivated widely in warm climates because of their fast growth and their valuable timber, gums and oils.

eucaryote, eucaryotic *alt.* spelling of eukaryote, eukaryotic *q.v.*

euchroic *a.* having normal pigmentation.

euclase *n.* BeAl(SiO$_4$)(OH), a rare cyclosilicate mineral that is occasionally used as a gemstone.

Eucommiales *n.* an order of dicot trees comprising a single family Eucommiaceae with one genus *Eucommia*.

eudominant *n.* a dominant species that is more or less restricted to a particular climax vegetation.

eugenics *n.* a discredited pseudoscientific philosophy at its height in Europe and the USA in the early to mid-twentieth century that aimed to 'improve' the genetic quality of the human population and that eventually led to abuses such as compulsory sterilization of those deemed 'unfit' and persecution of racial groups.

Euglenophyta, euglenoids *n.*, *n.plu.* division of unicellular flagellate protists, typified by *Euglena*, which have no rigid cell wall, have chlorophyll and carotenoid pigments (although pigment may be absent) and store food as fat or as the polysaccharide paramylon. In older botanical classifications they are treated as a division of the algae. In zoological classifications they are included in the Mastigophora.

euhalabous *a.* living in salt waters (salinity of 35–40 parts per thousand), *appl.* plankton.

euhaline *a.* (1) *appl.* seawater, or water of comparable salinity, i.e. *ca.* 35 parts per thousand of sodium chloride (salt); (2) living only in saline waters.

euhyponeuston *n.* organisms living in the top 5 cm of the water column for the whole of their lives.

Eukarya *n.* one of three proposed 'super-kingdoms', the others being the Archaea and the Bacteria, into which all living organisms can be divided, the Eukarya to include all organisms with eukaryotic cells.

Eukaryota Eukarya *q.v.*

eukaryotes *n.plu.* organisms with cells possessing a membrane-bounded nucleus in which the DNA is complexed with histones and organized into chromosomes. A eukaryotic cell also has an extensive cytoskeleton of protein filaments and tubules, and many cellular functions are sequestered in membrane-bounded organelles such as mitochondria, chloroplasts, endoplasmic reticulum and Golgi apparatus. The eukaryotes comprise protozoans, algae, fungi, slime moulds, plants and animals. *a.* eukaryotic. *cf.* prokaryotes.

eumetazoa *n.* the multicellular animals excluding the sponges.

Eumycota *n.* the 'true' fungi, comprising the classes Ascomycotina, Basidiomycotina, Deuteromycotina, Mastigomycotina and Zygomycotina, and excluding the slime moulds (Myxomycota). *see* Appendix 6.

euphausiid *n.* a member of the order Euphausiacea. They are small, usually luminescent, shrimp-like crustaceans (including krill), forming an important part of the marine plankton.

Euphorbiales *n.* order of dicot trees, shrubs and occasionally herbs, including the families Buxaceae (boxwood), Euphorbiaceae (spurge), Simmondsiaceae (jojoba) and others.

euphotic *a.* (1) well illuminated, *appl.* zone of surface waters to depth of around 80–100 m; (2) upper layer of photic zone.

euphotometric *a. appl.* leaves oriented to receive maximum diffuse light.

euplankton *n.* the plankton of open water, organisms that spend almost the whole of their life cycle as plankton.

euploid *a.* (1) having an exact multiple of the haploid number of chromosomes, e.g. being diploid, triploid, tetraploid, etc.; (2) *appl.* chromosomal abnormalities that do not disrupt relative gene dosage, *alt.* balanced. *n.* euploidy.

eupotamic *a.* thriving both in streams and in their backwaters, *appl.* plankton.

Euptales *n.* order of dicot trees and shrubs comprising the family Eupteleaceae with a single genus *Euptelea*.

Euro-Siberian Floral Region part of the Holarctic Realm comprising the whole of

Europe from southern Scandinavia to northern Spain, and Asia north of the Caspian Sea to Northern Japan.

European Economic Community (EEC) a regional institution made up of signatory European countries. It was established in 1958 in order to promote economic stability and economic expansion within the Community, while facilitating closer sociopolitical relationships between member nations. The 1991 Treaty on European Union decreed that the EEC would hitherto be known as the European Union *q.v.*

European Union (EU) a regional body (consisting, as of 1996, of Austria, Belgium, Britain, Denmark, Finland, France, Germany, Greece, Irish Republic, Italy, Luxembourg, Netherlands, Portugal, Spain and Sweden) that exists in order to promote, for its member countries, economic growth and trade, together with intergovernmental cooperation in a number of areas including security policy and justice and home affairs.

europium (Eu) *n.* a lanthanide element (at. no. 63, r.a.m. 151.96), a silver-coloured metal when in elemental form.

eurybaric *a. appl.* animals adaptable to great differences in altitude or pressure.

eurybathic *a.* having a wide range of vertical distribution.

eurybenthic *a. pert.* organisms that live on the ocean floor within a wide range of depths in the sea.

eurychoric *a.* widely distributed.

euryhaline *a. appl.* marine organisms adaptable to a wide range of salinity.

euryhygric *a. appl.* organisms adaptable to a wide range of atmospheric humidity.

euryoecious *a.* having a wide range of habitat selection.

euryphotic *a.* adaptable to a wide range of illumination.

Eurypterida, eurypterids *n., n.plu.* subclass (or order) of giant (2 m long) fossil predatory aquatic arthropods of the class Merostomata, present in the Ordovician, having a short non-segmented prosoma and a long segmented opisthosoma, and resembling scorpions.

eurythermal, eurythermic *a. appl.* organisms adaptable to a wide range of temperature. *alt.* eurythermous.

eurytopic *a.* having a wide range of geographical distribution.

euryxerophilous *a. appl.* plants adaptable to a wide range of dry conditions within a temperate climate.

eusocial *a. appl.* social insects that display cooperative care of the young, reproductive division of labour with more-or-less sterile individuals working on behalf of those engaged in reproduction, and an overlap of at least two generations able to contribute to colony labour. They include all ants and termites and some bees and wasps.

eustatic *a. appl.* global sea-level change.

Eustigmatophyta *n.* division of mainly unicellular algae, found in fresh water and soil, forming uniflagellate zoospores, individuals having a large orange carotenoid eyespot at the anterior end.

Eutheria, eutherians *n., n.plu.* an infraclass of mammals, including all mammals except the monotremes and marsupials, which are viviparous with an allantoic placenta, and have a long period of gestation, after which the young are born as immature adults. *alt.* placental mammals.

eutrophic *a.* of water bodies, rich in plant nutrients and therefore usually highly productive, with very large numbers of plankton, often dominated by cyanobacteria, and often with turbid water in summer. Eutrophic waters suffer frequent algal blooms. Coarse fish (e.g. perch, roach and carp) are dominant. Larger aquatic plants may be absent as the water can become depleted of dissolved oxygen through the decay of large amounts of organic matter. *n.* eutrophy. *see* eutrophication.

eutrophication *n.* the enrichment of bodies of fresh water by inorganic plant nutrients (e.g. nitrate, phosphate). It may occur naturally but can also be the result of human activity (cultural eutrophication from fertilizer runoff and sewage discharge) and is particularly evident in slow-moving rivers and shallow lakes. The biomass of plankton and herbivorous zooplankton increases and species diversity decreases. The water becomes turbid in summer, the growth of larger aquatic plants may eventually be suppressed and algal blooms are frequent. The water may become anoxic through the decay of large amounts of organic matter. Increased sediment deposition can eventually raise the level of the lake or river bed, allowing land plants to colonize the edges, and eventually converting the area to dry land.

eutropous *a. appl.* (1) insects adapted to visiting special kinds of flowers; (2) flowers whose nectar is available to only a restricted group of insects.

euxerophyte *n.* plant that shows adaptations to and thrives in very dry conditions.

evanescent *a.* (1) disappearing early; (2) *appl.* flowers that fade quickly.

evaporation *n.* process by which a liquid, e.g. water, changes into a gas, e.g. water vapour. *alt.* vaporization.

evaporative water loss (EWL) (1) (*bot.*) *see* evapotranspiration; (2) (*zool.*) in mammals and birds, the loss of heat from the body through the evaporation of water, which may occur from the body surface (sweating) in some mammals and/or from the respiratory tract (thermal panting) in some mammals and birds. Although EWL is actively employed by some birds and animals for cooling, in small mammals it could lead to dehydration and is minimized.

evaporite *a. pert.* material precipitated from natural waters held in lakes, lagoons or seas as a direct consequence of the evaporation of that water, *appl.* mineral, rock.

evaporites *n.plu.* sedimentary rocks (nonclastic) formed by the precipitation of minerals dissolved in seawater or fresh water, as a result of evaporation.

evapotranspiration *n.* loss of water from the soil by evaporation from the surface and by transpiration from the plants growing thereon.

even-aged *a. appl.* community composed of individuals of about the same age, which most usually occurs for very new communities.

even-aged stand forest in which the trees are all about the same age, as occurs in a plantation or in forest regrown after clear-felling.

evenness (E) *n.* the extent to which different species are equally abundant within a community. *see also* equitability. *cf.* dominance.

even-toed ungulates artiodactyls *q.v.*

evergreen *a. appl.* vascular plants that do not shed all their leaves at the same time and therefore appear green all the year round. *n.* such a plant.

evolution *n.* the development of new types of living organisms from pre-existing types by the accumulation of genetic differences over long periods of time. It is studied by reference to the fossil record and to the anatomical, physiological and genetical differences between extant organisms. Present-day views on the process of evolution are based largely on the theory of evolution by natural selection formulated by Charles Darwin and Alfred Russel Wallace in the nineteenth century. Darwin's theory has undergone certain modifications to incorporate the principles of Mendelian genetics, unknown in his day, and the more recent discoveries of molecular biology, but still remains a basic framework of modern biology. *see also* creationism, Darwinism, natural selection, neo-Darwinism, macroevolution, microevolution, molecular evolution.

evolutionarily stable strategy (ESS) in evolutionary theory, a behaviour pattern or strategy that, if most of the population adopt it, cannot be bettered by any other strategy and will therefore tend to become established by natural selection. Using games theory the results of various different strategies (e.g. in contests between males) can be worked out and a theoretical ESS determined and compared with actual behaviour.

evolutionary clock *see* molecular clock.

evolutionary grade the level of development of a structure, physiological process or behaviour occupied by a species or group of species, not necessarily related.

evolutionary taxonomy taxonomic philosophy and method that utilizes both phenotypic characters and lines of descent in the classification of organisms. One difference from the cladistic method of classification is the acceptance of taxa that do not contain all the descendants of the common ancestor, e.g. the class Reptilia, which does not contain the birds, which are descendants of a reptile ancestor. *cf.* cladistics.

evolutionary time period of time comprising hundreds of successive generations of a population, generally some hundreds to millions of years, a time period sufficient for random mutation to generate evolutionary change.

evolutionary transformation series pair of homologous characters one of which is derived from the other.

evolutionary tree branching tree-like diagram used to represent the evolutionary relationships between groups of organisms.

evolve *v.* to undergo evolution.

ewe *n.* a female sheep.

EWL evaporative water loss *q.v.*

ex- prefix derived from Gk *ex*, without.

exact compensation the situation when population increases result in a matching decrease in birth rate or increase in death rate, so that the population size is kept constant.

excess density compensation density overcompensation *q.v.*

exchangeable cations positively charged ions in soil (e.g. Ca^{2+} and Mg^{2+}) that are adsorbed onto the surfaces of soil particles and that may replace one another as soil conditions change.

excitability *n.* the capability of a living cell or tissue to respond to an environmental stimulus.

excitation *n.* (1) act of producing or increasing stimulation; (2) immediate response of a cell, tissue or organism to a stimulus.

exclusive species a species that is confined to one community.

excreta *n.plu.* waste material eliminated from body or any tissue thereof.

excretion *n.* the elimination of waste material from the body of a plant or animal, specifically the elimination of waste materials produced by metabolism.

excretory *a. pert.* or functioning in excretion, *appl.* organs, ducts, etc.

exergonic *a.* releasing energy, *appl.* metabolic reactions.

exfoliation *n.* (*biol.*) (1) the shedding of leaves or scales from a bud; (2) shedding in flakes, as of bark; (3) (*geol.*) weathering process in which the surface layers of bedrock, boulders or rocks are peeled away, *alt.* desquamation, onion-skin weathering, spheroidal weathering.

exhalant, exhalent *a.* carrying from the interior outwards.

exhaustible resources non-renewable resources *q.v.*

exine *n.* tough and durable outer layer of wall of pollen grain, often intricately sculptured, composed mainly of sporopollenin.

exo- prefix derived from Gk *exo*, without, signifying outside, acting outside, opening to the outside, etc.

exobiology *n.* the study of possible life originating outside the Earth.

exobiotic *a.* living on the exterior of a substrate or the outside of an organism.

exogamy *n.* (1) outbreeding; (2) cross-pollination; (3) disassortative mating.

exogenous *a.* (1) originating outside the organism, cell or system being studied; (2) originating in a different area from that in which found, *appl.* plankton brought into an area by ocean currents; (3) developed from superficial tissue, such as the superficial meristem; (4) *appl.* metabolism concerned with motor and sensory activities, hormone production and action, temperature control, etc. *cf.* endogenous.

exogenous rhythm a metabolic or behavioural rhythm that is synchronized by some external factor and that ceases to occur when this factor is absent.

exomixis *n.* union of gametes derived from different sources.

exophytic *a.* on, or *pert.* exterior of plants.

Exopterygota *n.* major division of the insects including those with only slight metamorphosis and no pupal stage. Includes the Anoplura (*q.v.*), Dermaptera (*q.v.*), Dictyoptera (*q.v.*), Embioptera (*q.v.*), Ephemeroptera (*q.v.*), Hemiptera (*q.v.*), Isoptera (*q.v.*), Mallophaga (*q.v.*), Phasmida (*q.v.*), Plectoptera (*q.v.*), Psocoptera (*q.v.*), Odonata (*q.v.*), Orthoptera (*q.v.*), Thysanoptera *q.v.*

exoskeleton *n.* hard supporting structure secreted by and external to the epidermis, as the calcareous exoskeletons of some sponges and the chitinous exoskeleton of arthropods.

exosphere *n.* the outermost layer of the atmosphere, found above the thermosphere, which extends from *ca.* 500 to 2000 km. Air atoms may escape from this layer into space as the likelihood of downward deflection as a result of collision decreases with increasing altitude.

exoteric *a.* produced or developed outside the organism.

exothermic *a.* (1) *appl.* chemical or physical changes for which $\Delta H < 0$, *see* enthalpy; (2) (*biol.*) ectothermic *q.v.*

exotic *n.* (1) a species of plant or animal not native to the region; (2) plant species that has not fully acclimatized or naturalized.

expiration *n.* (1) the act of emitting air or water from the respiratory organs; (2) emission of carbon dioxide by plants and animals.

exploitation *n.* (1) the consumption or use of a resource (e.g. food, water) by an organism; (2) the consumption or use (or overuse) of natural resources by humans.

explosive *a. appl.* (1) flowers in which pollen is suddenly discharged on decompression of stamens by alighting insects, as of broom and gorse; (2) fruits with sudden dehiscence, seeds being discharged to some distance; (3) evolution, rapid formation of numerous new types; (4) speciation, rapid formation of species from a single species in one locality.

exponential growth type of growth in which numbers increase by a fixed percentage of the total population in a given time period and that gives a J-shaped curve when numbers are plotted over time. *alt.* logarithmic growth. *cf.* linear growth.

exponential growth phase phase of maximum population growth. *see* exponential growth, J-shaped growth curve, S-shaped growth curve.

exposure limits, occupational *see* occupational control limits.

expressivity *n.* the degree to which a gene produces a phenotypic effect.

exsiccation *n.* drying of soil as a result of groundwater removal, as when wetlands are drained or when deforestation leads to excess evaporation.

extant *a.* still in existence.

extended phenotype the concept of the phenotype of an organism extended to include its behaviour, and its relations with its family group, which share some of its genes, and other members of its own species.

extensive agriculture agriculture where the active management role played by humans is minimal and agricultural pro-

ductivity is low. The two main types are pastoral farming (*q.v.*) and shifting cultivation *q.v.*

extensive property (*phys.*) a physical property that varies with the amount, i.e. extent, of material present, e.g. mass. *cf.* intensive property.

exterior *a.* situated on side away from axis or definitive plane.

external *a.* (1) outside or near the outside; (2) away from the medial plane.

external benefit a beneficial social effect of producing and using an economic good that is not included in the market price of that good.

external cost a harmful social effect of producing and using an economic good that is not included in the market price of that good and remains uncompensated for.

external fertilization fertilization that takes place outside the body.

externalities *n.plu.* social benefits and costs that are not included in the market price of an economic good.

external respiration respiration considered in terms of the gaseous exchange between organism and environment and the transport of gases to and from cells. *cf.* internal respiration.

extinct *a.* (1) no longer in existence, *appl.* biological species; (2) *appl.* a volcano that has not erupted in historic time.

extinction *n.* (1) the complete disappearance of a species from the Earth; (2) (*phys.*) absorbance *q.v.*

extinction point the minimum level of illumination below which a plant is unable to survive in natural conditions.

extirpate *v.* (1) to render extinct; (2) to destroy totally.

extra- prefix derived from L. *extra*, outside, signifying located outside, etc.

extracellular *a.* (1) occurring outside the cell; (2) *appl.* material secreted by or diffused out of the cell.

extracellular matrix macromolecular ground substance of connective tissue, secreted by fibroblasts and other connective tissue cells, and which generally consists of proteins, polysaccharides and proteoglycans.

extractive metallurgy processing of metal ores to recover the desired metal. *see*

147

pyrometallurgy, hydrometallurgy, electro-chemical reduction.

extratropical cyclones depressions *q.v.*

extrinsic *a.* (1) acting from the outside; *appl.* (2) muscles not entirely within the part or organ on which they act; (3) membrane proteins that are embedded in the outer layer of a biological membrane.

extrinsic isolating mechanism an environmental barrier that isolates potentially interbreeding populations.

extrusive rocks *n.plu.* fine-grained or glassy igneous rocks formed on the Earth's surface when volcanic lava cools rapidly and solidifies. *alt.* volcanic rocks.

exudate *n.* any substance released from a cell, organ or organism by exudation, e.g. sweat, gums, resins.

exudation *n.* the discharge of material from a cell, organ or organism through a membrane, incision, pore or gland.

exuviae *n.plu.* the cast-off skins, shells, etc., of animals.

Exxon Valdez incident a major oil spill caused when the supertanker *Exxon Valdez* ran aground on reefs in Prince William Sound, Alaska, on 24 March 1989. A total of 30 000 tonnes of crude oil were lost, causing the death of numerous seabirds (esp. guillemots), sea mammals (esp. sea otters) and fish. At the height of the disaster, the oil slick covered an area of 2500 km^2 and a total of 1700 km of shoreline was contaminated with oil.

eye *n.* (1) light-sensitive organ, the organ of sight or vision, taking various forms in different groups of animals. Insects and most crustaceans have compound eyes as well as simple single eyes in some cases, made up of many separate units. The vertebrate eye consists of a jelly-filled ball, the back of which is lined with a photosensitive layer, the retina. Light is focused onto the retina through a single transparent lens. The amount of light entering the eye is regulated by varying the size of the pupil. Cephalopods also have a very similar type of eye; (2) (*bot.*) the bud of a tuber; (3) of a hurricane, *see* hurricane.

eyespot *n.* (1) small cup-shaped pigmented spot of sensory tisssue in invertebrates, and also in some vertebrates, which has a light-detecting or visual function; (2) orange carotenoid-containing structure in some flagellates; (3) eye-like marking on wings of some butterflies and moths, or on the bodies of other animals, which are exposed to distract predators when the animal is disturbed, *alt.* stigma; (4) soil-borne disease affecting wheat, barley and oats, caused by the fungus *Cercosporella herpotrichoides*. Eyespot damages the tillers and weakens the stems of infected cereals, and young crops are particularly at risk.

eye wall boundary of the open vertical tube (eye) of a hurricane and the site of the heaviest precipitation and strongest winds.

F

f coefficient of kinship *q.v.*

F (1) symbol for the chemical element fluorine (*q.v.*); (2) farad (*q.v.*); (3) phenylalanine (*q.v.*); (4) inbreeding coefficient *q.v.*

F_1, F1 denotes first filial generation, hybrids arising from a first cross, successive generations arising from this one being denoted by F_2, F_3, etc. P_1 denotes the parents of the F_1 generation, P_2 the grandparents, etc.

Fabales *n.* an order of dicots, also known as legumes, whose fruit is a pod, and whose roots contain nitrogen-fixing bacteria of the genus *Rhizobium*. They include the families Mimosaceae, Cesalpinaceae and Papilionaceae (Fabaceae), and are also known as the Leguminosae.

fabavirus group group of isometric single-stranded RNA plant viruses similar in structure to comoviruses, but producing different symptoms. The type member is broad bean wilt virus. They are multi-component viruses in which two genomic RNAs are encapsidated in three different virus particles, one of which lacks nucleic acid.

fabric *n.* (*geol.*) collectively, those characteristics of rock that are caused by deformation and that pervade the rock body. The individual features (e.g. cleavage, grain shape) that together make up the fabric are called fabric elements.

fabric filter a device constructed from closely woven material for the removal of particulates from a gaseous waste stream. The filter may be of many designs, although tubular constructions are common. During use, the dust burden is periodically removed by mechanical shaking and/or the reversal of the direction of gas flow. The terms fabric filter and bag filter are generally used interchangeably.

faciation *n.* (1) formation or character of a facies; (2) grouping of dominant species within an association; (3) geographical differences in abundance or proportion of dominant species in a community.

facies *n.* (1) a surface, in anatomy; (2) aspect, as superior or inferior; (3) (*bot.*) a grouping of dominant plants in the course of a successional series; (4) (*geol.*) one of different types of deposit in a geological series or system, and the palaeontological and lithological character of a deposit.

facies fossils fossils characterizing the environment prevailing in a particular area of sedimentation. *cf.* index fossils.

facilitation *n.* (1) any alteration to the environment caused by the early species in a succession that means that later species establish themselves more easily; (2) in animal behaviour, an improvement in a pre-existing capability in response to a particular stimulus. *see also* social facilitation.

facilitation theory the classical explanation for plant succession. This hypothesizes that species replacement occurs because existing species modify their habitat making it less suitable for themselves but more suitable for other species, which then outcompete and replace them. *see* inhibition theory, tolerance theory.

factor *n.* (1) any agent (biological, climatic, nutritional, etc.) contributing to a result or effect; (2) in physiology and cell biology, may be used for any ill-defined endogenous substance that appears to have a physiological effect.

factory farming a term sometimes applied to intensive methods of livestock production, esp. of poultry. *see* battery production.

factory fishing highly organized method of fishing in which several well-equipped vessels locate and catch fish, which are then transferred to a factory ship (*alt.* mother ship) for immediate processing and freezing. Such highly intensive harvesting can lead to the exhaustion of fish stocks.

factory ship *see* factory fishing.

facultative *a.* having the capacity to live under different conditions, e.g. *appl.* aerobes, anaerobes, symbionts, parasites, etc. Such organisms can live in this way but are not obliged to and may under certain conditions adopt another mode of life. *cf.* obligate.

FAD flavin adenine dinucleotide, a derivative of riboflavin, which acts as a carrier of hydrogen and high-energy electrons in biochemical reactions. The reduced form is FADH$_2$.

faeces *n.* excrement from alimentary canal.

Fagales *n.* an order of dicots including many deciduous forest trees such as beech, birch, oak, sweet chestnut, hazel and hornbeam. *see also* Betulales.

Fahrenheit scale temperature scale (used in the USA) in which the freezing point and boiling point of water are set at 32 and 212 °F, respectively. *cf.* Celsius scale.

fair weather cloud cumulus *q.v.*

Falconiformes *n.* the birds of prey, an order of birds with strong clawed feet and hooked beaks, with a well-developed ability to soar. It includes eagles, hawks, buzzards, kites, kestrels, falcons and vultures.

fall autumn *q.v.*

fall equinox *see* equinox.

falling dune an irregular dune that develops on the leeward side of a topographic obstruction such as a cliff or hill.

fall movement (*geol.*) the fastest type of mass movement (*q.v.*) involving the falling of slope material through the air under the influence of gravity.

fallout *n.* (1) deposition on the Earth's surface of atmospheric particles such as ash, soot or radioactive material; (2) also *appl.* to the particles themselves.

fallow *see* fallowing.

fallowing *n.* practice of leaving agricultural land uncultivated in order to conserve nutrients and water, and so restore productivity. Land in this resting state is known as fallow.

false scorpions pseudoscorpions *q.v.*

false yeasts yeast-like fungi that have no known ascus stage and are therefore classified in the Deuteromycetes. They include *Torulopsis*, which has been utilized as a source of protein for food, and several human pathogens.

familial *a.* (1) *pert.* family; *appl.* (2) traits that tend to occur in several members and subsequent generations of a family. They may not necessarily be genetically based but may be due to the family environment; (3) disease, usually signifies a genetically based heritable condition, as opposed to the spontaneous and sporadic occurrence of the same condition, *see* genetic disease.

family *n.* (1) parents and their offspring, together with other relatives, depending on context; (2) taxonomic group of related genera, related families being grouped into orders. Familial names usually end in -aceae in plants and -idae in animals.

family planning the provision of information and services to help people control the number of children they wish to have and when to have them.

family, radioactive radioactive series *q.v.*

famine *n.* acute food shortage, resulting in widespread starvation and loss of life.

FAO Food and Agriculture Organisation *q.v.*

farad (F) *n.* the derived SI unit of capacitance (the ability to store an electrical charge), being the capacitance of a capacitor between whose plates there appears a potential difference of 1 V when it is charged with 1 C of electricity.

farm forestry agroforestry *q.v.*

farmyard manure (FYM) the traditional solid form of manure consisting of livestock dung, urine and bedding material (usually straw).

far red light light of wavelength between 700 and 800 nm, which reverses the effect of red light on phytochrome.

farrow *v.* to give birth, *appl.* pigs.

fasciated *a.* (1) banded; (2) arranged in bundles or tufts; (3) *appl.* stems or branches malformed and flattened.

fasciation *n.* (1) the formation of bundles; (2) the coalescent development of branches of a shoot system, as in cauliflower; (3) abnormal development of flattened, malformed fused stems or branches.

fascicle *n*. (1) a bundle, as of pine needles; (2) a small bundle, as of muscle fibres or nerve fibres; (3) tuft, as of leaves.

fast breeder reactors nuclear fission reactors that do not contain a moderator and therefore use fast neutrons (*q.v.*) to maintain the fission chain reaction. Fast neutrons are not very efficient inducers of fission in fissile material. Consequently, the fuel must contain a high proportion of fissile material. Some of the neutrons produced during the chain reaction are available for generating fissile isotopes from non-fissile uranium-238 or thorium-232, a process called breeding. Fast breeder reactors are designed to breed more fissile material than is consumed during fission.

fast fission nuclear fission induced by fast neutrons *q.v.*

fast neutrons the fast-moving (i.e. high kinetic energy) neutrons liberated by nuclear fission reactions. Such neutrons are slowed down by collisions. However, they are usually still described as fast if their individual kinetic energies are >0.1 MeV, or, in the context of fast fission (*q.v.*), >1.5 MeV. *cf.* slow neutrons.

fast reactor a nuclear fission reactor that contains no moderator to slow the neutrons and therefore uses fast neutrons (*q.v.*) to maintain the chain reaction. *see* fast breeder reactors.

fat index ratio of dry weight of total body fat to that of non-fat.

fat lamb *n*. lamb deemed ready to go for slaughter.

fats *n.plu.* compounds of fatty acids and glycerol (triacylglycerols) having a large proportion of saturated fatty acids and being solid at 20 °C. They contain carbon, oxygen and hydrogen, but no nitrogen. Fats are stored in animal and plant cells where they provide a concentrated source of energy. Adipose tissue in animals consists of cells filled with globules of fats. When energy is required, fats are hydrolysed by lipases to fatty acids and glycerol and the fatty acids metabolized in the mitochondria. *cf.* oils.

fat-soluble *a. appl.* hydrophobic organic compounds, such as hydrocarbons, that are soluble in lipids. Such compounds can be taken up into fatty tissue and remain there, which is one of the ways in which pesticides accumulate to high levels in the food chain.

fattening *n*. (*agric.*) the process of feeding store animals so that they quickly gain muscle and, more particularly, fat, to reach a condition satisfactory for slaughter. *alt.* finishing.

fatty acid long-chain organic acid of the general formula $CH_3(C_nH_x)COOH$, where the hydrocarbon chain is either saturated ($x = 2n$) (e.g. palmitatic acid, $C_{15}H_{31}COOH$) or unsaturated (e.g. oleic acid, $C_{17}H_{33}COOH$). Fatty acids are constituents of some lipids, including the phospholipids of cell membranes, and are also used as a source of energy for respiration.

fault *n*. a fracture in the Earth's crustal rocks along which significant movement has occurred. This involves the displacement of the rock to one side of the fault plane with respect to that on the other. A number of different types of faults are recognized. *see* normal fault, reverse fault, thrust fault, transcurrent fault. *cf.* jointing.

fault line the surface trace of a fault.

fault-line scarp a scarp, along a fault line, that has been exposed by erosion and not directly produced by tectonic activity.

fault scarp a steep slope produced when vertical movement takes place during faulting. A fault scarp may be formed by either a reverse fault or a normal fault.

fauna *n*. the animals peculiar to a country, area, specified environment or period, microscopic animals usually being called the microfauna.

faunal collapse local extinction of an animal species or a number of animal species.

faunal region (1) area characterized by a special group or groups of animals; (2) zoogeographical region *q.v.*

fayalite *n*. an olivine (*q.v.*) mineral with the formula Fe_2SiO_4. This is one of the end members in the continuous solid-solution series of formula $(Mg,Fe)_2SiO_4$, the other being forsterite *q.v.*

FBA furnace bottom ash *q.v.*

FBC fluidized bed combustion *q.v.*

FDNB fluorodinitrobenzene.

Fe symbol for the chemical element iron *q.v.*

feces faeces *q.v.*

fecundity *n*. (1) the capacity of an individual or a species to multiply rapidly; (2) in a stricter sense, the number of eggs produced by an individual; (3) fertility *q.v.*

fecundity schedule fertility schedule *q.v.*

feed additives (*agric*.) substances, such as antibiotics, flavouring agents and growth promoters, that are added to the feedstuffs of more intensively reared farm livestock. *see also* food additive.

feedback mechanism general mechanism operative in many biological and biochemical processes, in which once a product or result of the process reaches a certain level it inhibits or promotes further reaction.

feed–conversion ratio (*agric*.) in domestic livestock, the amount of feed consumed relative to the amount of weight gained.

feedlot system a system of livestock production, widely used in the USA, whereby high densities of cattle or pigs are kept in open pens and fattened before going to market.

feldspar minerals the most abundant group of minerals found in the Earth's crust. They are framework aluminosilicate minerals that share the formula $X(Al,Si)_4O_8$, where, in common feldspars, X is K, Na or Ca. The minerals orthoclase ($KAlSi_3O_8$), albite ($NaAlSi_3O_8$) and anorthite ($CaAl_2Si_2O_8$) represent the end-members of a solid-solution series that contains the common feldspars. However, the composition of most feldspars approximates to some position on either the solid solution between albite and orthoclase (the alkali feldspars) or that between albite and anorthite (the plagioclase feldspars). Twinning (*q.v.*) is common in feldspars and there are a number of polymorphs *q.v.*

feldspathoid minerals a group of minerals that have no clear definition other than that they are reminiscent of the feldspar minerals (*q.v.*) and that they are characteristic of alkaline igneous rocks that are undersaturated with silica. They are like feldspars in that they are aluminosilicates (generally of potassium or sodium, sometimes with calcium) but contain less silica. The following are generally accepted as members of the feldspathoid group: leucite ($KAlSi_2O_6$, may contain some Na), members of the nepheline ($NaAlSiO_4$)–kalsilite ($KAlSiO_4$) solid-solution series, cancrinite ($(Na,Ca)_7Al_6Si_6O_{24}(CO_3,SO_4,Cl)_{1.5-2}\cdot1-5H_2O$) and members of the sodalite group *q.v.*

felid *n*. member of the mammalian family Felidae, the cats.

fell *n*. uncultivated open hillside or mountainside.

feloids *n.plu.* mammals of the cat (Felidae) or mongoose (Viverridae) families.

felsenmeer *n*. a surface layer of jagged rock fragments, often found in alpine and arctic regions, attributed to the physical process of frost weathering (*q.v.*) *alt*. rock sea, blockfield.

felsic minerals a collective term for quartz, the feldspars and feldspathoids.

felsic rock igneous rock that has one or more of the felsic minerals (*q.v.*) as a major part of its volume.

female *n*. individual whose reproductive organs produce only female gametes, ♀.

feminization *n*. the production (e.g. by hormones) of secondary female sexual characteristics in genetic males.

femto- (f) SI prefix indicating that the unit it is associated with is multiplied by 10^{-15}.

fen *n*. plant community on marshy low-lying ground, on alkaline, neutral or slightly acid peat or mineral soil, covered with shallow, usually stagnant, water originating from groundwater. Characterized by tall herbaceous plants, e.g. reeds, reed canary grass. *cf*. bog.

fenitrothion *n*. contact organophosphate insecticide used, e.g., in the control of rice borers and of a wide range of pests of other crops, especially fruit.

fenoprop *n*. translocated herbicide used for weed control in forests and grassland.

fen peat type of peat (alkaline, neutral or slightly acidic) formed where depressions allow the accumulation of drainage water and the rate of growth of grasses, sedges and trees exceeds that of plant decomposition, which is usually substantial. *alt*. basin peat.

fenpropimorph *n*. systemic fungicide used on cereal crops to control, e.g., rusts and mildews.

fentin *n*. contact organotin fungicide used, e.g. in a mixture with maneb (*q.v.*), to control potato blight.

feral *a.* wild, or escaped from domestication and reverted to wild state.

fermentation *n.* anaerobic breakdown of carbohydrates by living cells, esp. microorganisms, often with the production of heat and waste gases (as in alcoholic fermentation in yeasts) and a wide variety of end-products (e.g. ethanol, lactic acid).

fermium (Fm) *n.* a transuranic element [at. no. 100, r.a.m. (of its most stable known isotope) 257].

fern *n.* common name for a member of the Pterophyta *q.v.*

ferrallitic soils deep red soils, acid in reaction, found on freely drained sites in humid tropical regions.

ferrallitization *n.* soil-forming process, operative in humid tropical regions, which results in a build-up of sesquioxides of aluminium (Al^{3+}) and iron (Fe^{3+}).

Ferrel cell hypothetical cell of the primary atmospheric circulation system located in each hemisphere between the subtropical high-pressure belt (30° latitude) and the upper-mid-latitude low-pressure belt (60° latitude) (*see* Fig. 32, p. 419). *alt.* mid-latitude cell.

ferrod *n.* suborder of the order spodosol of the USDA Soil Taxonomy system, consisting of spodosols with an illuvial horizon rich in iron oxides.

fertile *a.* (1) producing viable gametes; (2) capable of producing living offspring; *appl.* (3) eggs or seeds capable of developing; (4) a soil containing the necessary nutrients for plant growth; (5) (*phys.*) non-fissile nuclides that can be converted to fissile nuclides by bombardment with fast neutrons, a process that occurs in fast breeder reactors *q.v.*

fertility *n.* (1) the ability to reproduce; (2) reproductive performance of an individual or population, measured as the number of viable offspring produced per unit time, *alt.* fecundity.

fertility rate *see* birth rate.

fertility schedule demographic data giving the average number of female offspring that will be produced by a female at each particular age. *alt.* fecundity schedule.

fertilization *n.* the union of male and female gametes, e.g. sperm and egg, to form a zygote.

fertilize *v.* (1) to bring about the fusion of a male and female gamete, either naturally or by artificial means; (2) to add nutrients to the soil.

fertilizers *n.plu.* (*agric.*) substances applied before or during the cultivation process to provide essential nutrients for the crops. Two classes of fertilizer are generally recognized: inorganic fertilizers (*q.v.*) and organic manures *q.v.*

fetal, fetus *alt.* spelling of foetal, foetus *q.v.*

fetch *n.* (*oceanogr.*) distance of open water over which a wind blows. The height and period of the waves generated are determined in part by the extent of the fetch.

FFA free fatty acid *q.v.*

FGD flue gas desulphurization *q.v.*

F₁ hybrid in horticulture and experimental genetics, the first cross between two pure-breeding lines.

Fibonacci series the unending sequence 1, 1, 2, 3, 5, 8, 13, 21, 34, . . . , where each term is defined as the sum of its two preceding terms.

fibre *n.* (1) an elongated cell or aggregation of cells forming a strand of, e.g. muscle, nerve or connective tissue; (2) protein filament as of keratin in wool and hair; (*bot.*) (3) a delicate root; (4) a tapering elongated sclerenchyma cell providing mechanical strength in stem.

fibric *a. appl.* organic soil material in the early stages of decomposition.

fibril *n.* (1) small thread-like structure or fibre; (2) component part of a fibre; (3) root hair.

fibrist *see* histosol.

fibrolite sillimanite *q.v.*

fibrosis *n.* scarring or thickening of connective tissue, e.g. of lungs.

fibrous *a.* (1) composed of fibres, *appl.* tissue, mycelium, etc.; (2) forming fibres, *appl.* proteins, such as collagen, elastin, keratin, fibrin, fibroin, etc.

fibrous root system a root system in which the roots form a branched mass of roots all much the same size, without a main tap root.

fidelity *n.* (1) (*ecol.*) the degree of limitation of a species to a particular habitat; (2) (*biochem.*) of DNA replication, transcription and translation, the probability of an error being made during the copying of

DNA into DNA or RNA, or the translation of RNA into protein.

field bean *Vicia faba*, a seed legume (family Leguminosae) grown for food and fodder. It is also used as a break crop and/or green manure.

field capacity the maximum amount of water that a particular soil can hold on to against the forces of gravity. This point is reached when a soil, saturated by rain, drains under the influence of gravity until only the capillary water, held by surface tension, remains. *alt.* field moisture capacity.

field layer herb layer *q.v.*

field metabolic rate the metabolic rate as measured in a freely ranging animal, most commonly by the doubly labelled water method, in which water labelled with either deuterium (^{2}H) or tritium (^{3}H) and the oxygen isotope ^{18}O is injected at the start of the experiment and the decline of ^{18}O and labelled hydrogen in the blood after a period of days or weeks is measured and the concomitant rate of CO_2 production calculated.

field moisture capacity field capacity *q.v.*

field system the partition of land into fields for the grazing of livestock and the cultivation of crops.

Fijian Floral Region part of the Palaeotropical Realm comprising the islands of Fiji.

fijivirus group group of isometric double-stranded RNA plant viruses of the family Reoviridae, type member Fiji disease virus of sugar-cane, which are transmitted by plant-hoppers. The RNA genome is composed of ten different RNAs.

filament *n.* (1) any slender thread-like structure, such as a fungal hypha, a chain of algal cells, etc.; (2) the stalk of an anther, *see* Fig. 12, p. 157.

filial generation *see* F$_1$.

filial imprinting imprinting (*q.v.*) resulting in attachment of an offspring to parents or foster parents.

Filicales *n.* group of ferns (Pterophyta), including most extant species, mainly terrestrial, with typically large compound leaves and rhizomatous roots, which produce spores on the undersides of the leaves.

filter feeders organisms that feed on small organisms in water or air, straining them out of the surrounding medium by various means.

filtrate *n.* clear liquid obtained by filtration.

filtration *n.* separation process in which a liquid is passed through a porous material in order to separate out any suspended particles or solids present.

fin *n.* (1) fold of skin supported by bony or cartilaginous rays in fishes and used for locomotion, balancing, steering, display, etc. Most fishes have an upright dorsal fin on the back, a caudal fin at the end of the tail, an anal fin on the underside just anterior to the anus, a pair of pelvic fins on the underside and a pair of pectoral fins just behind the gills. The pectoral and pelvic fins represent the fore and hind limbs of other vertebrates; (2) any similarly shaped structure in other aquatic animals.

fine earth (*soil sci.*) that part of air-dry soil that, once it has been gently ground, will pass through a sieve with holes of 2 mm diameter. *cf.* stones, gravel.

fineness count a measure used to appraise the fineness of wool fibre. This scale uses, as its basis, the number of yarn lengths (511.84 m) spun from 0.454 kg weight of wool made ready for spinning.

finger and toe club root *q.v.*

finger lakes long, narrow lakes that occupy the over-deepened portions of glacial troughs, e.g. Windermere, Lake District, UK. *alt.* ribbon lakes.

fingerprinting *see* DNA fingerprinting.

fine-grained *see* grain size.

fine-textured soil a soil with a high proportion of fine particles. Such soils, in temperate regions, have high moisture- and nutrient-holding capacities but a tendency to be poorly drained and slow to warm in spring.

finishing fattening *q.v.*

fiord fjord *q.v.*

fire climax plant community maintained as climax vegetation by natural or man-made fires that destroy the plants that would otherwise become dominant.

Firmibacteria *n.* a class of prokaryotes including most Gram-positive bacteria, e.g. staphylococci, bacilli, streptococci.

Firmicutes *n.* major division of prokaryotes including the Gram-positive bacteria and the Actinomycetes and their relatives.

firn *n.* compacted snow, an intermediate stage during the conversion of freshly fallen snow into solid ice. *alt.* névé.

firn line snow line *q.v.*

first filial generation F_1 *q.v.*

first law of thermodynamics *see* thermodynamics, the laws of.

First World, the term used for industrialized countries with market economies, namely Australia, New Zealand, Japan and those of Western Europe and North America.

Fisher's sex-ratio theory theory that states that when daughters and sons are of equal reproductive value, parents should adjust the ratio of sons to daughters produced so that the average fitness costs of sons and daughters are equal.

fishery *n.* (1) stock of fish or other aquatic or marine organisms harvested for food; (2) an area where fish are caught.

fishes *n.plu.* a group of aquatic limbless vertebrates, the Pisces, breathing mainly by means of gills, with streamlined bodies and fins and with the body covered in scales (in bony fishes), and comprising the Chondrichthyes (*q.v.*) (cartilaginous fishes) and the Osteichthyes (*q.v.*) (bony fishes). The Agnatha (*q.v.*) (jawless vertebrates) are also sometimes called fishes. *see* Appendix 7.

fish farming type of aquaculture in which fish are commercially bred and reared in freshwater and/or marine environments.

fishing *n.* harvesting of freshwater or marine animals, esp. fish, for commercial or recreational purposes. *see* drift netting, line fishing, seine netting, trawling.

fish ranching type of aquaculture in which fish such as salmon are reared in captivity for the first years of their lives and then released into the wild and caught when they return to their birthplace to spawn.

fissile *a. appl.* materials that undergo nuclear fission upon bombardment with slow neutrons. The fuels used in nuclear power generation, $^{235}_{92}U$, $^{233}_{92}U$ and $^{239}_{94}Pu$, are all fissile.

fissionable *a. appl.* materials that undergo induced nuclear fission *q.v.*

fission bomb *see* nuclear bomb.

fission–fusion bomb *see* nuclear bomb.

fission, nuclear *see* nuclear fission.

fissiped *a.* with digits of feet separated, as the Fissipedia, the name for the carnivores, such as cats, dogs, bears, etc., in some zoological classifications.

Fissipedia *n.* in some classifications the name given to the order of terrestrial carnivorous mammals containing the cats, dogs, bears, hyaenas, etc. (the marine carnivores, the seals, etc., being placed in the order Pinnipedia).

fissure eruption any volcanic eruption in which magma is released through long cracks in the Earth's lithosphere. This type of volcanic eruption results in the deposition of extensive sheets of lava. *cf.* central eruption.

fissures *n.plu.* (*soil sci.*) voids present in the soil that are roughly planar in shape. *cf.* pores.

fitness *n.* Darwinian fitness, the fitness of an individual defined as the contribution of its genotype to the next generation relative to the contributions of other genotypes. Its fitness is determined by the number of offspring it manages to produce and rear successfully. *see also* inclusive fitness.

fixation *n.* (1) of carbon dioxide and nitrogen *see* carbon dioxide fixation, nitrogen fixation; (2) (*genet.*) of an allele, its spread throughout a population until it is the only allele found at that locus.

fixation index in population genetics, a measure of genetic differentiation between subpopulations, being the proportionate reduction in average heterozygosity compared with the theoretical heterozygosity if the different subpopulations were a single randomly mating population.

fixed *a. appl.* (1) carbon that was once carbon dioxide (CO_2) but that has been incorporated into a living, or once living, organism via the process of photosynthesis; (2) nitrogen that was once atmospheric N_2 but that has been reduced to ammonia, NH_3 (or ammonium, NH_4^+), *see* nitrogen fixation; (3) phosphate ions or potassium ions that, although within the soil, are not available for plant growth because of the insoluble nature of the chemical species within which the ions concerned reside; (4) (*genet.*) allele that has become the only allele found at a locus.

fixed dune a dune, often found in coastal regions, that has become stabilized by the growth of vegetation upon it.

fjord *n.* long, steep-sided coastal inlet formed when a narrow glacial trough became submerged by a relative rise in sea-level. *alt.* fiord. *cf.* ria.

flagellates *n.plu.* a highly diverse group of unicellular eukaryotic microorganisms, including photosynthetic species and non-photosynthetic, heterotrophic species, and classified in various schemes as protozoans, protists or algae. They are motile in the adult stage, swimming by means of flagella, and include both free-living marine and freshwater species and some important commensals, such as those living in the guts of ruminants, and human parasites, such as trypanosomes. In zoological classifications they are often placed in a superclass, Mastigophora, and are divided into the non-photosynthetic Zoomastigophora and the photosynthetic Phytomastigophora. In botanical classifications the photosynthetic flagellates are variously classed within the divisions Chrysophyta, Euglenophyta, Eustigmatophyta, Pyrrophyta (Pyrrophyceae), Prymnesiophyta and Xanthophyta. *see* Appendices 5, 7 and 8.

flammable *a.* easily ignited, i.e. set on fire. *alt.* inflammable.

flamprop-isopropyl *n.* translocated herbicide used to control wild oats in barley and wheat.

flamprop-methyl *n.* translocated herbicide used to control wild oats in wheat.

flare *n.* flame produced when waste gases, e.g. from a landfill site or oil refinery, are burned in the open air.

flash colours the sudden flash of colour displayed by some species during an attempt to escape from a predator, which may startle and distract the predator or deceive it into thinking the prey has gone.

flash point minimum temperature at which the vapour given off by a flammable liquid, e.g. fuel oil, will ignite, in the presence of a small flame.

flask fungi a common name for the Pyrenomycetes *q.v.*

flatworms Platyhelminthes *q.v.*

flavone *n.* any of a group of pale yellow flavonoid plant pigments, with the C_3 part of the molecule forming an oxygen-containing ring.

flavonoid *n.* any of various compounds containing a $C_6C_3C_6$ skeleton, the C_6 parts being benzene rings and the C_3 part varying in different compounds, often forming an oxygen-containing ring, which include many water-soluble plant pigments.

flavonol *n.* any of a group of pale yellow flavonoid plant pigments.

flax *n.* (1) any of several long-strawed varieties of the plant *Linum usitatissimum*, a member of the family Linaceae, cultivated for the fibres present in their stems; (2) the fibre extracted from *L. usitatissimum*, which is used to make linen. *cf.* linseed.

F-layer *see* mor humus. *see also* Fig. 28, p. 385.

fleas *n.plu.* common name for the Siphonaptera (Aphaniptera) *q.v.*

fledgling *n.* a young bird that is acquiring the feathers necessary for flight, up to the time it begins to fly.

flies *n.plu.* common name for members of the insect order Diptera *q.v.*

flint *n.* SiO_2, the dull grey-black opaque variety of chalcedony (*q.v.*) that is found as nodules in many chalk deposits.

Flixborough accident a serious industrial accident that occurred at a large chemical works at Flixborough, on the River Humber, UK. A massive explosion fuelled by cyclohexane, followed by more explosions and a fire, claimed the lives of 29 people on 1 June 1974.

floc *n.* solid lump of material formed in a liquid suspension as in a culture of microorganisms, sewage, etc.

flocculation *n.* coagulation of finely divided particles into clumps (known as flocs), e.g. of clay particles in alkaline soils, or of fine particulate matter during the sewage treatment process.

flocculence *n.* adhesion in small flakes, as of a precipitate.

flock *n.* group of birds or herbivorous animals kept together by social interactions.

flood *n.* the over-running of land with water, e.g. from a river channel due to a sudden increase in the river's discharge *q.v.*

flood alleviation schemes engineering of river channel and banks in order to prevent or reduce flooding.

flood banks artificial levees *q.v.*

flood irrigation surface method of irrigation, often used in developing countries, in which water diverted from rivers is used, in a controlled manner, to flood relatively large areas of land.

flood meadow *see* water meadow.

floodplain *n.* a low-lying and extensive landform bordering a river, formed by the progressive deposition of river-borne sediments (alluvium) during episodes of flooding.

flor *n.* covering of yeasts and bacteria and other microorganisms that forms on the surfaces of some wines during fermentation.

flora *n.* (1) the plants peculiar to a country, area, specified environment or period; (2) a book giving descriptions of these plants; (3) the microorganisms that naturally live in and on animals and plants, e.g. gut flora, skin flora.

floral *a. pert.* (1) flora of a country or area; (2) flowers.

floral diagram a conventional way of representing a flower, indicating the position of the parts relative to each other.

floral formula an expression summarizing the number and position of parts of each whorl of a flower.

floral kingdom *see* phytogeographical kingdoms.

floral realm (1) a large geographical area of the world distinguished by a particular flora; (2) in one system of phytogeographical classification, the highest level recognized in the geographical grouping of plants, a realm being divided into regions, which are further divided into provinces or domains. Four floral realms are generally recognized: the Holarctic, the Neotropical, the Palaeotropical and the Austral Realms (*see* individual entries). *see also* phytogeographical kingdoms.

floret *n.* (1) one of the small individual flowers of a crowded flowerhead, such as a capitulum of Compositae; (2) individual flower of grasses.

floristic *a.* (1) *pert.* the species composition of a plant community; (2) *appl.* flora.

floristics *n.* the study of an area of vegetation in terms of the species of plants in it.

flow *a.* (*geol.*) *appl.* a rock texture (*q.v.*) in which elongate or tabular crystalline mineral grains are aligned so that their long axes are parallel to the direction of flow of the magma that formed the rock. *alt.* fluidal.

flower *n.* the reproductive structure of Anthophyta (*q.v.*), being derived evolutionarily from a leafy shoot in which leaves have become modified into petals, sepals and calyx, and into the carpels and stamens in which the gametes are formed. Although flowers can take many different forms, they can all be represented by concentric whorls of different parts inserted on a base (the receptacle). The outermost whorl of sepals (often green) forms the calyx, inside that is a whorl of often brightly coloured petals, next is a ring of stamens (the male reproductive organs) and in the centre are the carpels (the female reproductive organs). *see* Fig. 12.

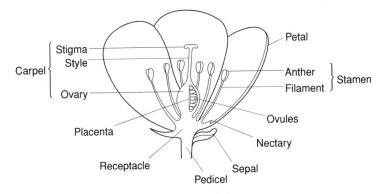

Fig. 12 Cross-section through a generalized flower.

flowering plant *see* Anthophyta.

flow movement a type of mass movement (*q.v.*) that is faster than creep movement (*q.v.*) and usually involves abundant water. Four categories of flow are recognized depending on their main components: mudflow (clay particles), earthflow (sand-sized particles), debris flow (rock fragments) and avalanche (mainly snow and ice). Solifluction is the term used for the slowest type of flow, that of saturated soil. This can occur where the angle of slope is less than 1°. In periglacial environments, the abundant water provided by the thawing of the surface above the permafrost (*q.v.*) frequently gives rise to solifluction, known in this context as gelifluction.

flow resources renewable resources *q.v.*

flue *n.* chimney, or, more usually, a duct within or leading to a chimney.

flue gas, flue gases the gaseous waste carried by a flue.

flue gas desulphurization (FGD) process in which sulphur dioxide (SO_2) is removed from flue gases prior to their release into the atmosphere, using alkaline absorbents such as lime or limestone.

fluidal flow *q.v.*

fluidized bed bed of inert material, e.g. sand, kept in a state of agitation, i.e. fluidized, by a strong updraught of air. This technology can be used in waste segregation, e.g. to separate metals from less dense materials such as plastics, and in combustion systems. *see* fluidized bed combustion.

fluidized bed combustion (FBC) a highly efficient combustion system, suitable for burning certain wastes and/or low-grade fuel. These are added to a bed of inert material, such as sand or coal ash. This is kept fluidized by a strong updraught of air, which acts as the oxidant.

flukes *n.plu.* a group of parasitic flatworms (platyhelminths), including the Monogenea, which are ectoparasitic on the skin and gills of fishes, and the Trematoda, which include endoparasitic blood, liver and gut flukes, such as *Fasciola*, the common liver fluke of sheep, and *Schistosoma*, the cause of schistosomiasis in humans. *see also* digenean, Platyhelminthes, Trematoda.

Fluon trade name for polytetrafluoroethene (PTFE).

fluorapatite *see* apatite.

fluorescence microscopy technique for locating particular molecules in cells by labelling with specific antibodies tagged with fluorescent dyes to make them visible under the microscope.

fluorescyanine *n.* a mixture of pterins with a yellow or blue fluorescence, found in the eyes, eggs and luminous organs of some insects.

fluoridation *n.* practice of adding trace quantities of fluoride (F^-), in the form of a suitable salt, to drinking water in order to reduce dental decay among the general population.

fluorine (F) *n.* highly reactive halogen element (at. no. 9, r.a.m. 19.00), a toxic, corrosive and explosive yellow-green gas (F_2) in the free state. It occurs in the minerals fluorite (fluorspar, CaF_2) and cryolite. *see also* chlorofluorocarbons.

fluorite *n.* CaF_2, a mineral used as a flux in smelting and as a source of fluorine and its compounds. It is also used, to a lesser extent, decoratively. *see also* fluorspar.

fluorosis *n.* chronic fluoride poisoning.

fluorspar *n.* (1) fluorite (*q.v.*); (2) a rock that contains enough fluorite for it to be commercially valuable.

flush *n.* (1) a patch of ground where water lies but does not run into a channel; (2) a period of growth, esp. in a woody plant; (3) a rapid increase in the size of a population.

flushing *n.* the washing of dissolved substances upwards in the soil so that they are deposited near the surface. *cf.* leaching.

fluvent *n.* a suborder of the order entisol of the USDA Soil Taxonomy system consisting of entisols found, e.g., in floodplains where there has been recent deposition of alluvial sediments.

fluvial *a.* pert. running water.

fluvial erosion *n.* erosion of the bed and banks of a stream or river through the action of flowing water and the material it transports. This can involve hydraulic action (the sheer force of the water alone), abrasion (the wearing down of the bedrock by material dragged along by the streamflow) and/or corrosion (the removal of certain minerals and rocks dissolved by the stream water). *alt.* river erosion, stream erosion.

fluvial transportation carrying of material by a stream or river. *see* bedload, dissolved load, suspended load.

fluviatile *a.* (1) growing in or near streams; (2) inhabiting and developing in streams, *appl.* certain insect larvae; (3) (*geol.*) caused by rivers, *appl.* deposits.

fluvioglacial *a. pert.* meltwater from glaciers, *appl.* deposits and landforms left by such water.

fluvioglacial deposit stratified drift *q.v.*

fluvioterrestrial *a.* found in streams and in the land beside them.

flux *n.* (1) flow; (2) as *appl.* biogeochemical cycles, the rate of movement of material from one reservoir to another, generally expressed in kilograms per annum (kg per year) or moles per annum (mol per year).

fly *n.* common name for a member of the Diptera *q.v.*

fly ash fine particles of fossil fuel ash (esp. coal ash) found in flue gases.

flyggbergs *see* roche moutonée.

fly tipping illegal and indiscriminate dumping of waste.

flyway *n.* a fixed route along which waterfowl migrate.

Fm symbol for the chemical element fermium *q.v.*

FMN flavin mononucleotide, a derivative of riboflavin, which acts as a carrier of hydrogen and high-energy electrons in biochemical reactions. The reduced form is $FMNH_2$.

focus *n.* (*geol.*) place of origin of an earthquake.

fodder crops crops, such as roots, cereals and grasses, that are grown and harvested for use as animal feedstuffs. *cf.* forage crops.

foetal *a. pert.* a foetus.

foetus *n.* (1) a mammalian embryo after the stage at which it becomes recognizable; (2) technically, the human embryo from seven weeks after fertilization.

fog *n.* suspension of water droplets in the lower atmosphere that reduces visibility to less than 1 km (greater than 1 km, this is known as mist). *see* advection fog, radiation fog, steam fog.

fog-basking *n.* behaviour of some small insects in desert environments in which nocturnal fogs regularly occur, by which they can collect drinking water from the condensation of the fog on the body surface.

föhn *n.* warm, dry, blustery adiabatic wind that blows down the lee slopes of mountains. Originally applied to winds in the Alps (where it may be written Föhn), the term föhn is used for any adiabatic wind, e.g. the Zonda of the Andes, the Chinook of the Rockies, the Samun in Iran and the Nor'wester in New Zealand.

folding *n.* (*geol.*) process by which rocks (usually layered sedimentary rocks) bend and buckle in response to compressional stresses.

fold mountains mountains generated by the tectonic compression of geosynclines, e.g. the Himalayas.

foliage *n.* the leaves of a plant or plant community collectively.

foliage height density formula for representing the density of foliage at various heights above the ground, and used to represent the degree of vertical stratification or layering in a plant community.

foliation *n.* (*bot.*) (1) the process of leaf formation; (2) the state of having leaves; (3) (*geol.*) any feature of a rock mass that is planar, i.e. leaf-like, in nature and penetrative or repeated. An example is the parallel structure given to some high-grade metamorphic rocks, such as gneiss, by the banded arrangement of minerals, caused by the realignment during metamorphism of minerals present in the original rock.

folicolous *a.* growing on leaves, *appl.* certain fungi.

folist *see* histosol.

folivorous *a.* leaf-eating.

following response the innate response shown by the young of many species (e.g. chicks, ducklings) that will indiscriminately follow moving objects.

fonofos *n.* contact organophosphate insecticide used in the control of soil insect pests, especially cabbage root fly.

food additive natural and synthetic substances that are added to food as colouring agents, preservatives, antioxidants, emulsifiers and stabilizers, acids and bases, anticaking agents, and flavour enhancers and sweeteners.

Food and Agriculture Organisation (FAO) a United Nations specialized

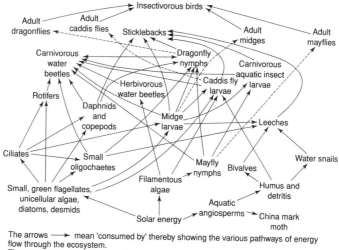

The arrows ——→ mean 'consumed by' thereby showing the various pathways of energy flow through the ecosystem.
The arrows ---→ mean 'become'.

Fig. 13 A food web from a freshwater habitat (after Popham, 1955).

agency, established in 1945. The aim of this organization is to foster action that will enhance the production and distribution of all agricultural products, raise living standards and levels of nutrition, and improve the quality of life for the peoples of rural areas.

food bodies Beltian bodies *q.v.*

food chain a sequence of organisms within an ecosystem in which each is the food of the next member in the chain. A chain starts with the primary producers, which are photosynthetic organisms (e.g. algae, plants, some bacteria) or chemoautotrophic bacteria. These are eaten by herbivores (primary consumers), which are in turn eaten by carnivores (secondary consumers). Small carnivores may be eaten by larger carnivores. *see also* food web.

food chain efficiency ecological efficiency *q.v.*

food pollen pollen present in flowers to provide food for visiting insects, instead of or as well as nectar, and that may be sterile and produced in special anthers.

food web the interconnected food chains in an ecosystem. *see* Fig. 13.

forage *n.* the vegetation eaten by grazing and browsing animals. *v.* to search for food.

forage crops crops, such as grasses, legumes and some roots, that are cultivated for grazing by domestic livestock. *cf.* fodder crops.

foraging *n.* the collection of food by animals.

foraging strategy method by which animals seek out food. *see also* optimal foraging strategy.

Foraminifera, foraminiferans *n., n.plu.* order of mainly marine protozoans (of class Sarcodina), having a calcareous, siliceous or composite shell through which project fine pseudopodia, and a highly vacuolated outer layer of cytoplasm. Chalk is largely composed of foraminiferan shells and they are also major components of many deepsea oozes. *see* globigerina ooze.

foraminiferous *a.* containing shells of foraminiferans.

forb *n.* herbaceous plant, esp. a pasture plant other than grasses.

foreshore *n.* the zone of the beach between the high- and low-water marks. *see also* seashore.

forest *n.* the biome consisting of continuous or semicontinuous tree cover, which may develop in areas where the average annual precipitation is sufficient (>75 cm) to support the growth of trees and shrubs. Broad-leaved forest, coniferous forest, pine forest, etc., describe forest in which the named types of tree comprise at least 80% of the canopy. *see also* ancient forest, mixed forest, monsoon forest, old-growth forest, rain forest, secondary forest, taiga, temperate rain forest, tropical rain forest. *cf*. desert, grassland.

forest dieback increased death of trees not attributable solely to disease or insect pests and that is thought to be due to air pollution.

forest floor (1) ground level in a forest; (2) the surface of the soil, including its litter covering, in a forest ecosysem.

forestry *n*. the management of forests, either natural or planted, generally for timber.

Forestry Commission a UK organization that replants and manages forests, established in 1919.

form *n*. (1) a taxonomic unit consisting of individuals that differ from those of a larger unit by a single character, therefore being the smallest category of classification, *alt*. forma; (2) one of the kinds of a polymorphic species; (3) the concealed resting place of a hare.

formaldehyde methanal *q.v.*

formalin *n*. aqueous solution of methanal (formaldehyde), used as a preservative for biological specimens.

formation *n*. (1) the vegetation proper to a definite type of habitat over a large area, such as tundra, coniferous forest, prairie, tropical rain forest, etc.; (2) (*geol.*) a sequence of beds that are similar to one another in composition and that may be used as a delimitable unit in geological mapping.

formation constant stability constant *q.v.*

form genus a genus whose species may not be related by common ancestry.

formicarian *a*. (1) *pert*. ants; (2) *appl*. plants that attract ants by means of sweet secretions.

formicarium *n*. ants' nest, particularly an artificial arrangement for purposes of study.

formothion *n*. contact organophosphate acaricide and insecticide used mainly on horticultural crops to control mites and insects, such as aphids.

form species the members of a form genus *q.v.*

formula, chemical *see* chemical formula.

formula weight (fw) the weight (in grams) of one mole of the substance concerned. Units g mol^{-1} (frequently not stated).

forsterite *n*. an olivine (*q.v.*) mineral with the formula Mg_2SiO_4. This is one of the end members in the continuous solid-solution series of formula $(Mg,Fe)_2SiO_4$, the other being fayalite *q.v.*

fossil *n*. the remains or trace of an organism that lived in the past. Fossils are found embedded in rock either as petrified hard parts of the organism or as moulds, casts or tracks.

fossil assemblage a group of fossils found together, representing organisms that lived together in a particular environment and time.

fossil diagenesis the changes (due, e.g., to pressure and temperature) that occur in the course of the fossilization of an organism.

fossil fuels the ancient, altered remains of organisms that are extracted from the surface/near-surface parts of the Earth's crust and used as fuel. Such fuels are predominantly coal, natural gas and oil, and collectively account for *ca*. 87% of the world's primary energy consumption.

fossiliferous *a*. containing fossils.

fossilization *n*. the process of fossil formation, e.g., by carbonization, permineralization or recrystallization of the remains of organisms.

fossil water water deposited in sedimentary rocks at the time of their formation, an important groundwater source. *alt*. connate water.

fossorial *a*. adapted for digging.

founder effect genetic differences between an original population and an isolated offshoot due to alleles in the founder members of the new population being unrepresentative of the alleles in the original population as a whole.

Fourth World, the term used for the least-developed countries, most of which are African, and also for very poor communities within wealthier nations.

Fr symbol for the chemical element francium *q.v.*

fractional distillation *see* distillation.

fragile ecosystem any ecosystem, e.g. tropical rain forest, that is particularly vulnerable to disturbance, whether natural or of human origin.

fragipan *n.* (*geol.*) compact, brittle, cemented horizon located between the soil zone and the underlying bedrock, often formed as a result of periglacial processes.

fragmentation *n.* (1) type of asexual reproduction in which the organism breaks into smaller pieces, each of which can develop into a new individual, as in some algae; (2) (*ecol.*) *see* habitat fragmentation, species–area effect.

frame silk one type of silk produced by spiders, which forms the supporting frame and radii of a typical orb web. *see also* viscid silk.

francium (Fr) *n.* radioactive metallic element (at. no. 87, r.a.m. 223). The most stable isotope is ^{223}Fr, which has a half-life of 21 min.

frankincense *n.* a balsam obtained from plants of the genus *Boswellia*.

franklinite *n.* (Zn, Mn^{2+}, Fe^{2+})(Fe^{3+}, Mn^{3+})$_2O_4$, a rare zinc and manganese ore mineral. Found in the zinc deposits of Franklin, NJ, USA.

fraternal *a. appl.* twins produced from two separately fertilized eggs.

free atmosphere the atmosphere above the friction layer *q.v.*

free energy (G) Gibbs free energy, a thermodynamic function used to describe chemical reactions: the change in free energy, ΔG, being given by the equation $\Delta G = \Delta H - T\Delta S$ for a system undergoing change at constant temperature (T) and constant pressure, where ΔH is the change in enthalpy of the system and ΔS is the change in entropy of the system. A reaction can only occur spontaneously if ΔG is negative.

free-living *a. appl.* an organism that is able to live independently of another living organism. *cf.* colonial, parasitic.

freemartin *n.* a sterile female or intersex twin born with a male, the abnormality being due to sharing of blood circulation in the uterus and consequent masculinization of the female by male hormones.

free radical a molecule, or esp. a fragment of a molecule, that contains one or more unpaired electrons. *alt.* radical.

free-range system method of poultry keeping in which hens, kept for egg-laying, are given access to the outside, and allowed to roam over a relatively large area. *cf.* battery production.

free-running *a. appl.* an endogenous rhythm unaffected by any external influence.

freeway motorway *q.v.*

freeze–thaw *see* frost weathering.

freezing nuclei (*clim.*) tiny atmospheric particles, such as dust and salt, around which ice crystals form by the sublimation of water vapour or the freezing of liquid water. *cf.* hygroscopic nuclei.

freezing point temperature at which a substance changes state from a liquid to a solid at a given pressure (usually standard atmospheric pressure).

freezing rain rain that freezes on impact with the ground.

Freon trade name for some chlorofluorocarbons *q.v.*

frequency-dependent selection selection occurring when the fitness of particular genotypes is related to their frequency in the population. When rare, the particular genotype is at an advantage compared with the other possible genotypes, but when it is common it is at a disadvantage.

freshwater *a. pert.* or living in water containing less than 0.5 parts per thousand dissolved salt (sodium chloride), such as that in rivers, ponds and lakes. *cf.* brackish water.

frictional force (*clim.*) a force that operates on flowing air in direct opposition to the direction of movement and that is exerted by the surface over which the air is moving.

friction layer (*clim.*) that part of the atmosphere within which friction with the ground influences the flow of air. Above this layer is the free atmosphere.

Friends of the Earth international pressure group which, through its many activities, seeks to raise the profile of environmental issues and problems.

fringing reef a coral reef that is connected to land.

frit fly *Oscinella frit*, an insect larval pest of spring oats, maize, barley, wheat and pasture grass.

frog *n.* common name for a member of the amphibian order Anura. *see* Amphibia.

frond *n.* (1) a leaf, esp. of fern or palm; (2) flattened thallus of certain seaweeds or liverworts; (3) a leaf-like outgrowth from thallus, as of a lichen; (4) any leaf-like structure.

front *n.* (*clim.*) zone of rapid temperature change across the horizontal plane, formed when two air masses with different temperature characteristics meet. *see* cold front, occluded front, polar front, stationary front, warm front.

frontal depressions (*clim.*) depressions *q.v.*

frontal rain cyclonic rain *q.v.*

frost *n.* ice particles deposited on surfaces, usually during the night. *see* glazed frost, hoar frost, rime.

frostbite *n.* condition in which the flow of blood to the skin becomes frozen as a consequence of exposure to extreme cold. In severe cases, this can lead to gangrene of the affected part.

frost creep the down-slope movement of rock fragments in the active layer (*q.v.*) as a result of gravity.

frost-free days *see* growing season.

frost heaving the vertical, upward movement of rock fragments within the active layer (*q.v.*) due to the formation of ice.

frost hollow topographic depression, e.g. the bottom of a valley, in which cold air accumulates.

frost shattering frost weathering *q.v.*

frost thrusting the horizontal movement of rock fragments within the active layer (*q.v.*) due to the formation of ice.

frost weathering, frost wedging type of physical weathering mediated by water that has entered cracks and voids in the rock, brought about by freezing and thawing (a process known as freeze–thaw). *alt.* frost shattering.

froth flotation technique used to separate a mixture of finely divided minerals, involving agitation with water and a frothing agent.

fructose *n.* hexose sugar found in many plants, which is sweeter than sucrose. *alt.* fruit sugar.

frugivore *n.* a fruit eater. *a.* frugivorous.

fruit *n.* the developed ovary of a flower containing the ripe seeds, and any associated structures. In mosses, the spore-containing capsule is often called the fruit.

frustule *n.* the siliceous two-part wall of diatoms.

frutescent *a.* becoming shrub-like.

fruticose *a.* shrubby, *appl.* lichens that grow in the form of tiny shrubs.

fry *n.* newly hatched fish.

fucoid *a. pert.* or resembling a seaweed.

fucoxanthin *n.* brown xanthophyll carotenoid pigment found in brown algae, diatoms and Chrysophyta.

fuel *n.* a material used to produce heat by the transformation of either chemical energy (during combustion) or nuclear energy. The combustion process requires the presence of an oxidizing agent [usually molecular oxygen (O_2) from the air] and a reducing agent. It is the reducing agent that is referred to as the fuel. *see also* nuclear fuel.

fuel bundle group of nuclear fuel rods, typically 63 in number, as used within the core of a nuclear reactor.

fuel cell an electrochemical device that converts fuel to electricity directly. It contains an anode and a cathode. These are connected by an electrolytically conducting substance within the cell and an electrically conducting circuit outside the cell. During operation, the fuel is supplied to the anode and the oxidant to the cathode. The usual fuel is hydrogen and/or carbon monoxide, while the oxidant is air. The reactions within the cell produce both electricity and heat. The heat can be used either directly or to generate electricity via a conventional gas or steam power cycle.

fuel crops energy crops *q.v.*

fuel rods *n.plu.* rods containing nuclear fuel, used in nuclear reactors. In thermal reactors, these usually contain uranium (U) or uranium oxide (UO_2) incorporating a small percentage of fissile ^{235}U.

fuelwood *n.* timber felled to provide fuel for cooking and domestic heating. Demand for fuelwood has outstripped supply in much of the Third World, resulting in a 'fuelwood crisis'.

fugacious *a.* withering or falling off very rapidly.

fugitive species a species of newly disturbed habitats, which has a high ability to disperse, and which is usually eliminated from established habitats by interspecific competition.

Fuller's earth a natural clay used industrially as an absorbent. Mineralogically, its principal constituents are smectite group (*q.v.*) minerals although, in the USA, the term Fuller's earth is applied to minerals of the sepiolite–attapulgite series *q.v.*

fulvic acid fraction remaining in a solution of humus in weak alkali after removal of humin and humic acid.

fulvous *a.* deep yellow, tawny.

fumarole *n.* volcanic vent from which volcanic gases are emitted, independent of any volcanic eruption.

fume *n.* (1) airborne particulate matter of minute size, i.e. diameter <1 mm; (2) also *appl.* to vapours of volatile liquids and solids.

fume fever fever induced by exposure to the fumes produced by heating certain metals or PTFE (polytetrafluoroethene). *see also* metal fever.

functional groups covalently bonded groups of atoms with recognizable structural characteristics that exhibit certain chemical properties almost irrespective of the molecule within which they are found.

function of state any property of a substance that does not depend on the method used to prepare that substance. *alt.* state function, state property.

fundamental niche the largest niche an organism could occupy in the absence of competition or other interacting species.

fund resources non-renewable resources *q.v.*

fungal *a.* of or *pert.* fungi.

Fungi, fungi *n.*, *n.plu.* a kingdom of heterotrophic, non-motile, non-photosynthetic and chiefly multicellular organisms that absorb nutrients from dead or living organisms. Although traditionally grouped with plants they are now considered an independent evolutionary line. Multicellular terrestrial fungi comprise three main groups, the Zygomycotina (zygomycetes) (e.g. bread moulds), the Ascomycotina (ascomycetes) (e.g. yeasts and sac fungi) and the Basidiomycotina (basidiomycetes) (e.g. mushroooms and toadstools). The lower fungi include the slime moulds and water moulds, which are sometimes alternatively classified as protists. Many fungi are serious plant pathogens. Multicellular fungi grow vegetatively as a mycelium, a mat of thread-like hyphae from which characteristic fruiting bodies arise (e.g. the blue or black spore-heads of many common moulds, the mushrooms and toadstools of agarics, and the brackets of bracket fungi). In some fungi the hyphae are divided into uninucleate or binucleate cells by transverse partitions (septa). Hyphae possess rigid cell walls that differ in composition from those of plants, chitin rather than cellulose being a main constituent in most fungi. *see* Appendix 6.

fungicide *n.* any chemical or other agent that kills fungi.

fungicolous *a.* living in or on fungi.

Fungi Imperfecti Deuteromycotina *q.v.*

fungistatic *a.* inhibiting the growth of, but not killing, fungi.

fungivore *n.* an organism that feeds on fungi.

fungivorous *a.* feeding on fungi.

fungoid, fungous *a.* (1) *pert.* fungi; (2) with the character or consistency of a fungus.

fungus *sing.* of fungi.

fur *n.* (*chem.*) the solid deposit, which is essentially calcium and magnesium carbonates ($CaCO_3$ and $MgCO_3$), precipitated when water containing temporary hardness is boiled. *see* hard water. *alt.* scale.

furans a near-colloquial term for polychlorinated dibenzofurans (PCDFs) *q.v.*

Furious Fifties *see* Roaring Forties.

furnace bottom ash (FBA) ash collected from the bottom of the furnace in power stations fuelled by finely ground coal. *cf.* pulverized fuel ash.

furovirus group group of fungus-transmitted rod-shaped single-stranded RNA plant viruses, type member soil-borne wheat mosaic virus. They are multicomponent viruses in which two genomic RNAs are encapsidated in two different-sized virus particles.

furrow *n.* a narrow groove made in the soil with a plough, usually in preparation for seed sowing.

furrow irrigation surface method of irrigation, used in temperate agriculture, in which water is allowed to flow down parallel furrows, thus reaching the root or vegetable crops planted in the intervening ridges, through seepage.

fuscous *a.* of a dark brown, almost black, colour.

fusion bomb *see* nuclear bomb.

fusion, nuclear *see* nuclear fusion.

fw formula weight *q.v.*

FYM farmyard manure *q.v.*

fynbos *n.* South African term for temperate evergreen scrubland, a type of chaparral *q.v.*

G

γ gamma, for headwords with prefix γ- refer to gamma-.

g abbreviation for gram.

G (1) Gibbs' free energy, *see* free energy; (2) glycine (*q.v.*); (3) guanine *q.v.*

G geometric mean *q.v.*

Ga symbol for the chemical element gallium *q.v.*

GA (1) gibberellic acid or gibberellin (*q.v.*); (2) Golgi apparatus *q.v.*

gabbion *n.* in soil and water conservation technology, a stone and rock-filled bolster used to protect a surface vulnerable to erosion.

gabbro *n.* a basic igneous rock with a colour index of 30–90 that is coarse-grained or pegmatitic. Its mineralogy is essentially plagioclase feldspar (bytownite or labradorite) and pyroxene.

GAC granular activated carbon, *see* activated carbon.

gadfly *see* Brachycera.

Gadidae, gadids *n., n.plu.* large and economically important family of marine bony fishes including the cod, whiting and haddock.

gadolinium (Gd) *n.* a lanthanide element (at. no. 64, r.a.m. 157.25), a soft silver-coloured metal when in elemental form. It is used, e.g., to absorb neutrons in nuclear reactors.

GAG glycosaminoglycan *q.v.*

Gaia hypothesis a hypothesis, formulated by James Lovelock in 1979, and named for the Greek goddess of the Earth, Gaia, which proposes that the biosphere operates, in some respects, as a single organism, modifying the non-living environment in order to optimize the conditions necessary for its continued existence.

gain-of-function *a. appl.* mutant alleles directing overexpression of normal gene product.

galactan *n.* polygalactose, galactose residues linked together in a chain, a hydrolysis product of, e.g., pectins and plant gums.

galactose *n.* six-carbon aldose sugar, a constituent with glucose of the disaccharide lactose. Also found in various complex carbohydrates such as the pectins of plant cell walls and in some glycolipids and glycoproteins.

galena *n.* PbS, the most important lead ore mineral.

gall *n.* (1) an abnormal outgrowth from plant stem or leaf caused by the presence of young insects (e.g. gall wasps or gall mites) in the tissues (as e.g. oak-apple) or by infection by certain fungi or bacteria (e.g. crown gall caused by *Agrobacterium*); (2) bile *q.v.*

gall flower in fig trees, an infertile female flower in which the fig wasp lays its eggs.

gallicolous *a.* living in plant galls.

Galliformes *n.* an order of heavy-bodied, chicken-like land birds, including grouse, partridges, pheasants, quails and domestic fowl.

gallinaceous *a.* resembling the domestic fowl, *appl.* birds of the same family.

gallium (Ga) *n.* metallic element (at. no. 31, r.a.m. 69.74), which as gallium arsenide is used in semiconductors.

gall wasps minute hymenopteran insects belonging to the superfamily Cynipoidea that lay their eggs in the leaf and stem tissue of oak and a few other plants (e.g. roses), inducing the formation of a gall, inside which the grub develops, feeding on the gall tissue.

game *n.* birds, animals and fish traditionally hunted for sport and for food, such as pheasants, partridges, grouse, hares, wild boars, deer, sharks.

gametangium *n*. any structure producing gametes. *plu*. gametangia.

gamete *n*. haploid reproductive cell produced by sexually reproducing organisms that fuses with another gamete of opposite sex or mating type to produce a zygote, the male gamete being variously called a spermatozoon, spermatozoid or antherozoid and the female gamete an ovum or egg. *alt*. germ cell, sex cell, sexual cell.

game theory mathematical theory concerned with determining the optimal strategy in situations of competition or conflict. This theory can be applied to the relationships within a community, and the computer simulation of such relationships to determine winning strategies can help to throw light on ecological and social relationships and their evolution.

gametophyte *n*. the haploid gamete-forming phase in the alternation of generations in plants. *cf*. sporophyte.

gamma decay a type of radioactive decay in which excited nuclei, i.e. those with high potential energies, relax, i.e. lose energy, lowering their potential energies, via the liberation of gamma rays. Gamma decay frequently accompanies both beta and alpha decay processes.

gamma diversity the diversity of species within a given geographical area.

gamma-HCH lindane *q.v.*

gamma radiation, gamma rays electromagnetic radiation of wavelength 1×10^{-10} to 1×10^{-13} m emitted during gamma decay (*q.v.*). *alt*. γ-rays.

gamodeme *n*. a deme forming a relatively isolated interbreeding community.

Gamophyta, gamophytes *n*., *n.plu*. in some classifications, a division of conjugating green algae, consisting of filamentous or unicellular freshwater green algae that reproduce by conjugation and the fusion of amoeboid gametes. They include filamentous algae, such as *Spirogyra* and *Mougeotia*, and the unicellular desmids.

gangrene *n*. death and decay of part of the body resulting from defective blood supply. This may be caused, e.g., by frostbite.

gangue *n*. unwanted minerals and rock material in a mined ore deposit that is physically separated from the ore mineral prior to processing.

garbage (1) domestic refuse (*q.v.*); (2) wastes (animal and vegetable) generated during food preparation.

garden city town or city planned to combine low-density housing with plentiful open space, in order to provide healthy living conditions, e.g. Welwyn Garden City, Hertfordshire, UK (established in 1920).

garnet minerals a group of minerals with the formula $A_3B_2Si_3O_{12}$, where A is a divalent cation (commonly Mg^{2+}, Fe^{2+}, Mn^{2+} or Ca^{2+}) and B is a trivalent cation (Al^{3+}, Fe^{3+} or, less commonly, Cr^{3+}). There are named end-members of solid-solution series within the garnet group. Pyrope ($A = Mg^{2+}$, $B = Al^{3+}$), almandine ($A = Fe^{2+}$, $B = Al^{3+}$) and spessartine ($A = Mn^{2+}$, $B = Al^{3+}$) form one continuous solid-solution series; while grossular ($A = Ca^{2+}$, $B = Al^{3+}$), uvarovite ($A = Ca^{2+}$, $B = Cr^{3+}$) and andradite ($A = Ca^{2+}$, $B = Fe^{3+}$) form another. Some garnets are used as gemstones.

garrigue *n*. scrub vegetation common in mediterranean regions, characterized by open areas of tufted grass interspersed with areas of low-growing, hard-leaved shrubs.

Garryales *n*. in some classifications, an order of woody dicots with opposite evergreen leaves and flowers in hanging catkin-like panicles.

gas *n*. matter in a form that fully occupies any container into which it is placed and that can be readily compressed. *cf*. liquid, solid.

gas bladder swimbladder *q.v.*

gaseous exchange the exchange of gases between an organism and its surroundings, including uptake of oxygen and release of carbon dioxide in respiration in animals and plants, and the uptake of carbon dioxide and release of oxygen in photosynthesis in plants.

gasification of coal *see* coal gasification.

gasohol *n*. a mixture of gasoline (petrol) and ethanol.

gas oil liquid containing C13–C25 hydrocarbons (b.p. 220–350 ˚C) produced by the fractional distillation of crude oil and used as a fuel for diesel engines. *alt*. diesel oil.

gasoline petrol *q.v.*

Gasteromycetes *n*. a group of basidiomycete fungi in which the hymenium is completely enclosed in a basidiocarp and never exposed. It comprises the puffballs, earth stars, stinkhorns and bird's nest fungi.

Gasteropoda, gasteropods Gastropoda, gastropods *q.v.*

gastroenteritis *n.* inflammation of the stomach and intestines, accompanied by vomiting, diarrhoea and abdominal pain. This can be caused, e.g., by consuming food contaminated with bacteria of the genus *Salmonella* or with enterotoxigenic strains of *Escherichia coli.*

Gastropoda, gastropods *n.*, *n.plu.* class of molluscs including the winkles and whelks, sea slugs, water snails, and land snails and slugs. They are characterized by a large flat muscular foot on which they crawl about, and, where a shell is present, it is a rounded or conical spirally coiled shell in one piece.

Gastrotricha *n.* a phylum of marine and freshwater microscopic pseudocoelomate animals that have an elongated body and move by ventral cilia.

gas turbine a motor that allows the gaseous products of the combustion of a fuel to expand through a turbine (*q.v.*), thereby converting the chemical energy liberated during combustion into mechanical energy.

GATT General Agreement on Tariffs and Trade *q.v.*

Gause's principle an ecological principle stating that usually only one species may occupy a particular niche in a habitat. *alt.* competitive exclusion principle.

Gaussian curve the symmetrical curve representing the frequency distribution of a normally distributed population.

Gaussian distribution a symmetrical distribution about the mean. *alt.* normal distribution.

Gaviiformes *n.* an order of birds, including divers and loons.

Gd symbol for the chemical element gadolinium *q.v.*

GDP gross domestic product *q.v.*

Ge symbol for the chemical element germanium *q.v.*

gel electrophoresis *see* electrophoresis, polyacrylamide gel electrophoresis.

gel-filtration chromatography technique for separating molecules on the basis of size by passage through a column of beads of an insoluble, highly hydrated polymer (e.g. Sephadex), larger molecules being unable to enter the spaces within the beads and thus being more rapidly eluted from the column.

gelifluction *see* flow movement.

GEM genetically engineered microorganism.

geminivirus group the only group of plant viruses with single-stranded circular DNA genomes and that have 'twinned' particles composed of two isometric virus particles attached to each other. There are three subgroups: A, leafhopper-transmitted viruses with genomes composed of a single type of DNA, infecting monocotyledonous plants, type member maize streak virus; B, whitefly-transmitted viruses with two types of DNA, infecting dicotyledonous plants, type member African cassava mosaic virus; C, leafhopper-transmitted viruses with genomes composed of a single type of DNA, infecting dicots, type member beet curly top virus.

gemma *n.* a bud or outgrowth from a plant or animal that develops into a new organism. *plu.* gemmae.

GEMS Global Environmental Monitoring System *q.v.*

gene *n.* the basic unit of inheritance, by which hereditary characteristics are transmitted from parent to offspring. At the molecular level a single gene consists of a length of DNA (or in some viruses, RNA) that exerts its influence on the organism's form and function by encoding and directing the synthesis of a protein, or in some cases, a tRNA, rRNA or other structural RNA. Each living cell carries a full complement of the genes typical of the species, carried in linear order on the chromosomes.

genealogy *n.* (1) study of the evolutionary descent of organisms; (2) study of family lineage in humans.

gene bank gene library, *see* DNA library. *see also* germplasm, seed bank.

gene centres geographical regions in which certain species of cultivated plant are represented in their greatest number of different varieties or forms and that may in some cases correspond to their centre of origin and initial domestication.

gene cloning *see* DNA cloning.

genecology *n.* ecology studied in relation to the population genetics of the organisms concerned.

gene expression the realization of genetic information encoded in the genes to produce a functional protein or RNA. In its broadest sense it encompasses both transcription and translation, but often refers to transcription only.

gene family set of genes with similarities in their nucleotide sequences and that is thought to be descended by duplication and subsequent variation from the same ancestral gene.

gene flow the spread of particular alleles in a population and between populations resulting from outbreeding and subsequent intercrossing.

gene-for-gene resistance in plant pathology, a type of resistance in which a so-called avirulence gene in the pathogen is matched by a resistance gene in the plant. The outcome of an infection depends on which alleles of each gene are present.

gene frequency the frequency of a particular variant (allele) of a gene in a population.

gene locus the position of a gene on a chromosome, a position that may be occupied by any of the alleles of that gene. *alt.* genetic locus.

gene manipulation genetic engineering *q.v.*

gene pool all the genes, and their various alleles, present in an interbreeding population.

genera *plu.* of genus *q.v.*

General Agreement on Tariffs and Trade (GATT) an international, multilateral treaty signed in 1947 with the express aim of enhancing international trade, esp. via the removal or reduction of trade barriers, both tariff and non-tariff.

general fertility rate demographic measurement of the age structure of a population and the resulting fertility. For human populations it is calculated from the crude birth rate multiplied by the percentage of women in the age group 15–45 years.

generalist, generalist species organism or species with a very broad ecological niche, such as humans, rats and house-sparrows, which can tolerate a wide range of environmental conditions and eat a variety of foods. *cf.* specialist species.

generalized *a.* combining characteristics of two groups, as in many fossils, not specialized.

generation *n.* (1) in a line of descent, individuals that share a common ancestor and are all the same number of broods away from that ancestor; (2) in human populations often refers to all individuals in the same age group.

generation time the average span of time between the birth of parents and the birth of their offspring.

generic *a.* (1) common to all species of a genus; (2) *pert.* a genus.

gene superfamily a set of genes that are thought originally to have derived from a single ancestral gene, but which have diverged to such an extent that they now encode proteins with many different functions and roles, as the immunoglobulin superfamily and the serine proteinase superfamily.

genet *n.* (1) a unit or group of individuals deriving by asexual reproduction from a single zygote; (2) civet cat.

genetic *a.* (1) *appl.* anything involving, caused by or *pert.* genes; *pert.* (2) genetics; (3) genesis. *see also* gene.

genetic adaptation adaptation to a new habitat or changed environmental conditions as a result of genetic change.

genetically engineered microorganisms (GEMs) *see* genetic engineering. *alt.* genetically modified microorganisms (GMMs).

genetic code rules by which the amino acid sequence of a polypeptide chain is specified by the order of bases in DNA, in which consecutive groups of three bases (triplets or codons) each specify one of the 20 amino acids of which proteins are composed. The code is redundant in that each amino acid can be specified by more than one triplet.

genetic conservation preservation of genetic resources (*q.v.*) through the use of DNA libraries, seed banks, captive breeding, and the protection of geographical areas with a high species diversity.

genetic control the control of plant pests and diseases by breeding genetically determined resistance into the plants.

genetic damage damage to the chromosomes and DNA causing mutations with deleterious effects. Genetic damage can be caused by exposure to radiation or mutagenic chemicals or by spontaneous mutation.

genetic diversity variability within a species or a population of that species due to genetic differences between individuals.

genetic drift (1) random changes in gene frequency in small isolated populations owing to factors other than natural selection, such as sampling of gametes in each generation, *alt.* Sewall Wright effect; (2) random nucleotide changes in a gene not subject to natural selection.

genetic engineering any change in the genetic constitution of a living organism that has been brought about by artificial means, i.e. not by conventional breeding, and which usually would not occur in nature, such as the introduction of a gene from another species. This was first accomplished in bacteria and other microorganisms, and strains of bacteria and yeast altered in this way are used to produce large quantities of valuable proteins, such as hormones, viral proteins for vaccines, etc., that are difficult to obtain by other means. Genes can also be introduced into mammalian and other animal cells in culture and into plant cells. In mammals a permanently altered strain of animal can be produced by introducing the required gene into the fertilized egg or very early embryo *in vitro*, and then replacing it in the uterus. The genetic constitution of some other animals can also be altered in various ways. Genetically engineered plants are produced by introducing new genes into a single protoplast or piece of cultured plant tissue, from which a complete new plant can be regenerated. *see also* recombinant DNA. *see* Fig. 14.

genetic equilibrium condition in a population where gene frequencies stay constant from generation to generation.

genetic fingerprinting *see* DNA fingerprinting.

genetic fitness *see* fitness.

genetic isolation lack of interbreeding between groups of a population or between different populations due to geographic isolation or, in humans, to cultural preferences also.

genetic load (1) the proportion of mutant alleles with deleterious effects in a population; (2) the average number of mutant alleles that reduce fitness per individual in a population, considered as an accumulated decrease in fitness from a theoretical optimum.

genetic locus gene locus *q.v.*

genetic manipulation genetic engineering *q.v.*

genetic material any material of plant, animal or microbial origin that contains functional genes.

genetic polymorphism the stable, long-term existence of multiple alleles at a gene locus. Technically, a locus is said to be polymorphic if the most common homozygote occurs at a frequency of less than 90% in the population.

genetic recombination the process in sexually reproducing organisms by which DNA is exchanged between homologous chromosomes by chromosome pairing and crossing-over during meiosis. Thus alleles from both the maternal and paternal chromosomes of a pair may become recombined on the same chromosome, and be passed on to a gamete in this new combination. Genetic recombination produces most of the variation between individuals in sexually reproducing populations.

genetic resistance resistance to a pest, disease, or stressful environmental conditions that is genetically determined.

genetic resources genetic material of actual or potential value to humans, e.g. wild species of plants that may carry genes for resistance to pests and diseases that could be bred or genetically engineered into crop plants. *see also* genetic conservation, germplasm, seed bank.

genetics *n.* (1) that part of biology dealing with heredity and variation and their physical basis in DNA, the genetic material; (2) of an organism, the physical basis of its inherited characteristics, i.e. the sequence and arrangement of its genes.

genetic variation heritable variation in a population as a result of the presence of different variants (alleles) of any gene and, in eukaryotes, their shuffling into new combinations by sexual reproduction and recombination.

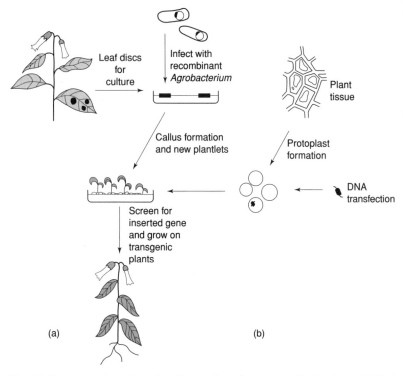

Fig. 14 Plant genetic engineering. Generation of a transgenic plant from (a) leaf discs infected *in vitro* with recombinant *Agrobacterium*, or (b) protoplasts transfected with DNA.

gene transfer the introduction of genes from one species into another, using recombinant DNA techniques. *see* genetic engineering, recombinant DNA, transgenic, transgene.

genital *a. pert.* region of the reproductive organs.

genocline *n.* a gradual reduction in the frequency of various genotypes within a population in a particular spatial direction.

genodeme *n.* a deme differing from others genotypically, but not necessarily phenotypically.

genoholotype *n.* a species defined as typical of its genus.

genome *n.* (1) the genetic complement of a living organism or a single cell, more specifically the total haploid genetic complement of a diploid organism or the total number of genes carried by a prokaryotic microorganism or a virus; (2) sometimes used for the total DNA content of any nucleus. *a.* genomic.

genome library DNA library (*q.v.*) consisting of fragments of chromosomal DNA (as opposed to fragments of cDNA).

genospecies *n.* a group of bacterial strains capable of gene exchange.

genosyntype *n.* a series of species together defined as typical of their genus.

genotype *n.* the genetic constitution of an organism, which acting together with environmental factors determines phenotype. *a.* genotypic. *cf.* phenotype.

genotypic variance a statistical measure used in determining the heritability of a particular trait, being that part of the phenotypic variance that can be attributed to differences in genotype between individuals.

Gentianales *n.* an order of dicot trees, shrubs or herbs and including the families Asclepiadaceae (milkweed), Gentianaceae (gentian), Rubiaceae (madder) and others.

genus *n.* taxonomic group of closely related species, similar and related genera being grouped into families. Generic names are italicized in the scientific literature, e.g. *Homo* (man), *Quercus* (oaks), *Canis* (wolves and dogs), *Salmo* (salmon and relatives). *plu.* genera. *a.* generic.

geo *n.* narrow inlet cut into a cliff along a line of weakness, such as a joint or a fault.

geobiotic terrestrial *q.v.*

geobotany phytogeography *q.v.*

geochronology *n.* science dealing with the measurement of time in relation to the Earth's evolution.

geocline *n.* a gradual and continuous change in a phenotypic character over a considerable area as a result of adaptation to changing geographical conditions.

geocoles *n.plu.* organisms that spend part of their life in the soil and affect it by aeration, drainage, etc.

geocryptophyte *n.* a plant with dormant parts hidden underground.

geographical information systems (GIS) computer software that enables geographically referenced data (e.g. on populations, species records) to be linked graphically to map features.

geographical isolation the separation of two populations originally of the same species from each other by a physical barrier such as mountains, oceans, rivers, etc. Eventually this may lead to such differences evolving in one or both of the isolated populations that they are unable to interbreed and a new species is formed.

geographical race a population separated from other populations of the same species by geographical barriers such as mountain ranges or oceans, and showing little or no differences from the rest of the population, and which can usually interbreed to produce fertile offspring if brought into contact with the rest of the species.

geographical speciation the evolution of a new species from a geographically isolated population.

geological column geological timescale *q.v.*

geological timescale chronology of the Earth's geological history, constructed with reference to various crustal rock strata and the fossils they contain. This timescale is subdivided first into eras, then periods and then epochs. *see* Appendix 4. *alt.* earth history, geological column, stratigraphical column.

geology *n.* science dealing with the physical structure, activity and history of the Earth.

geometric growth, geometric rate of increase growth (or rate of increase) in which there is a constant ratio between successive quantities. *alt.* exponential growth.

geometric mean (G) a measure of the 'average' or central location of a series of numbers that increase by multiplication rather than addition. In general, the geometric mean of *n* numbers is the *n*th root of their product. *see also* mean.

geomorphology *n.* scientific study of land surface features and the processes that created them.

geonasty *n.* a curvature, usually a growth curvature, in response to gravity. *a.* geonastic.

geophilous *a.* (1) living in or on the soil; (2) having leaves borne at soil level on short stout stems.

geophyte *n.* (1) a land plant; (2) a perennial herbaceous plant with dormant parts (tubers, bulbs, rhizomes) underground.

geosere *n.* (1) series of climax formations developed through geological time; (2) the total plant succession of the geological past.

geosmins *n.plu.* volatile terpene derivatives that contribute to the 'earthy' smell of soil, produced by actinomycetes.

geostationary orbit orbit described by a geostationary satellite (*q.v.*). *alt.* geosynchronous orbit, stationary orbit, synchronous orbit.

geostationary satellite satellite placed in a high-altitude orbit (*ca.* 35 000 km) above the Equator, which revolves at the same angular velocity as the Earth and thus is stationary above a given location on the Earth's surface.

geostrophic wind wind that flows parallel to the isobars, produced when the Coriolis force (*q.v.*) and the pressure gradient force (*q.v.*) balance each other. This type of windflow is common in the upper atmosphere, where friction with the Earth's surface does not intervene.

geosynchronous orbit geostationary orbit *q.v.*

geosyncline *n.* elongated trough in which great thicknesses of sediments have accumulated over time. The subsequent folding of geosynclines has created fold mountains such as the Himalayas.

geotaxis *n.* movement in response to gravity. *a.* geotactic.

geothermal energy the thermal energy present in rocks and fluids deep within the Earth's crust.

geothermal gradient the natural increase in temperature of the Earth's crust with increasing depth. In areas of geological stability, the geothermal gradient is about 30 °C km^{-1}.

geotropism *n.* movement or growth in relation to gravity, either in the direction of gravity (positive geotropism, as in plant roots) or away from the ground (negative geotropism, as in plant shoots).

Geraniales *n.* order of dicot trees, shrubs or herbs including the families Balsaminaceae (balsam), Erythrhoxylaceae (coca), Geraniaceae (geranium), Linaceae (flax), Tropaeolaceae (nasturtium) and others.

germ *n.* (1) vernacular for a microorganism, esp. one that causes disease; (2) the embryo of a seed.

germanium (Ge) *n.* brittle white metalloid element (at. no. 32, r.a.m. 72.59) found in zinc ores. It is used as a semiconductor.

germ cell (1) a reproductive cell; (2) cell giving rise to gamete, often set aside early in embryonic life.

germicide *n.* chemical that kills microorganisms.

germination *n.* (1) resumption of growth of plant embryo in favourable conditions after maturation and dispersal of seed, and emergence of young shoot and root from the seed. Germination is taken to be complete when photosynthesis commences and the plant is no longer dependent on the food stored in the seed. Germination requires a suitable temperature and the

presence of sufficient water and oxygen; (2) emergence of thallus, hypha, etc., from spore.

germ line line of cells from which gametes, e.g. ova and sperm, are produced and through which hereditary characteristics (as determined by the genes) are passed on from generation to generation in animals.

germplasm *n.* originally, a term coined by the nineteenth-century biologist, A. Weismann, to denote the idea of protoplasm that was transmitted unchanged from generation to generation in the germ cells (as opposed to the inheritance of acquired characteristics). Nowadays usually denotes cells from which a new plant or animal can be regenerated, as in collections of plant seeds in seed banks. *alt.* germ plasm.

gerontology *n.* the study of senescence and ageing.

gestalt *n.* (1) an organized or fixed response to an arrangement of stimuli; (2) coordinated movements or configuration of motor reactions; (3) a mental process considered as an organized pattern, involving explanation of parts in terms of the whole; (4) a pattern considered in relation to background or environment, *appl.* morphology irrespective of taxonomic or phylogenetic relationships.

gestation *n.* the period between conception and birth in animals that give birth to live young.

geyser *n.* volcanic vent from which superheated water and steam periodically and violently erupt.

giant-grass community grassland with grasses up to 4 m high.

gibberellins (GA) *n.plu.* a group of substances affecting plant growth that was first discovered in the fungus *Gibberella fujikuroi*, and later in many flowering plants, where the substances are believed to act as natural growth hormones. When applied to plants they increase both cell division and cell elongation in stems and leaves, producing long spindly plants, or in dwarf mutant plants increasing stem growth so that it resembles a normal plant. Gibberellins are able to break dormancy in some seeds and are used to induce uniform germination in malt barley. They also induce stem elongation and flowering in some plants. In germinating seeds, gibberellin stimulates growth by inducing the synthesis of enzymes that break down the starch stored in the endosperm into sugars.

Gibbs' free energy *see* free energy.

gibbsite *n.* $Al(OH)_3$ an aluminium ore mineral, one of the constituents of bauxite (*q.v.*). *alt.* hydrargillite.

giga- (G) SI prefix indicating that the unit it is associated with is multiplied by 10^9.

gigantism *n.* growth of an organ or organism to an abnormally large size.

gilgai *n.* type of microrelief consisting of shallow basins separated by groups of hummocks/ridges, found in areas with clay-rich soils.

gill *n.* (1) respiratory organ of many aquatic animals (e.g. crustaceans, fishes, amphibians), a plate-like or filamentous outgrowth well supplied with blood vessels at which gas exchange between water and blood occurs; (2) (*mycol.*) one of the radial spore-bearing lamellae on the underside of caps of toadstools and mushrooms.

gilt *n.* a young female pig that has not yet produced a litter.

gimmer *n.* a young female sheep, 12–18 months old.

Ginkgophyta *n.* one of the five main divisions of extant seed-bearing plants, containing only one living genus, *Ginkgo*, characterized by its distinctive fan-shaped leaves. Ovules are borne in pairs on long stalks and microsporophylls in cones on separate plants.

girdling *n.* the removal of a complete ring of bark from around the trunk of a tree, thus preventing the flow of solutes from leaves to roots through the phloem, which is removed with the bark. It eventually kills the tree.

GIS geographical information systems *q.v.*

glabrous *a.* (1) with a smooth even surface; (2) without hairs.

glacial *a.* (1) of a glacier; (2) *pert.* or caused by the movement of a glacier or advancing ice sheet; (3) *pert.* or *appl.* any period in the geological past during which a drop in the mean global air temperature to below *ca.* 13 °C caused large parts of the land surface to be covered with ice (glaciation); (4) *pert.* or *appl.* the Pleistocene epoch of the Quaternary period, characterized by periodic glaciation.

glacial *n.* any period in the geological past during which the mean global air temperature fell below *ca.* 13 °C, with the advance of glaciers and polar ice caps to cover large parts of the land surface.

glacial abrasion type of erosion caused by the advance of a glacier or ice sheet in which hard rock fragments embedded in the ice scour the valley floor and sides.

glacial budget the difference between the total accumulation and total ablation (melting) experienced by a glacier calculated over the balance year *q.v.*

glacial creep type of ice movement in a glacier in which the ice crystals, which align themselves in the direction of the glacier's movement, slip over one another.

glacial drift any material (clay, sand, gravel or boulders) laid down under glacial conditions. These deposits may be subdivided into till (*q.v.*) and stratified drift *q.v.*

glacial erosion continual removal of bedrock and soil by glaciers. *see* glacial abrasion, plucking.

glacial sliding basal sliding *q.v.*

glacial striations deep scratches in a bedrock surface caused by abrasion by a moving glacier or ice sheet. The scratches lie roughly parallel to the direction of ice movement.

glacial surge a sudden increase in the rate of flow of a glacier, which is usually sustained over a period of several months.

glacial till till *q.v.*

glacial trough U-shaped valley with steep sides and a flat floor, produced by the erosion of a pre-existing V-shaped river valley by a glacier.

glaciation *n.* the state of being covered with ice.

glacier *n.* a body of ice and snow, developed on land, which will deform and flow under its own weight. A number of different types of glacier are recognized on the basis of size and shape. *see* cirque glaciers, ice caps, ice sheets, niche glaciers, piedmont glaciers, valley glaciers.

glasshouse cultivation greenhouse cultivation *q.v.*

glass sponges hexactinellid sponges *q.v.*

glauconite *n.* $(K,Na,Ca)_{0.5-1}(Fe,Al,Mg)_2$-$(Si,Al)_4O_{10}(OH)_2.nH_2O$, one of the mica minerals (*q.v.*). Found in some sedimentary rocks as a consequence of authigenesis (*q.v.*) and as a clay mineral *q.v.*

glaucophane *n.* $Na_2Mg_3Al_2Si_8O_{22}(OH)_2$, one of the amphibole minerals *q.v.*

Glaucophyta *n.* division of algae composed of unicellular eukaryotes lacking chlorophyll but carrying modified cyanobacteria (blue-green algae) as endosymbionts.

glazed frost *n.* type of frost formed when rain or dew turns to ice on contact with a freezing surface, e.g. on roads, where it is known as 'black ice'.

glean *v.* to gather food or fodder (esp. ears of corn) left behind after the crop has been harvested.

Gleason's index an index of the similarity between two communities, *see* coefficient of community (*q.v.*), that is a modified version of the Jaccard index (*q.v.*). It takes account of the dominance or importance of species common to both communities and of unique species, rather than simply measuring the number of common and unique species. *cf.* Jaccard index, Kulezinski index, Morisita's similarity index, Simpson's index of floristic resemblance, Sørensen similarity index.

gleying *n.* reduction of red-coloured iron(III) compounds to blue-green/grey-coloured iron(II) compounds in a soil where anaerobic conditions prevail. This occurs either as a result of a high water table (groundwater gleying) or impeded drainage (surface water gleying or stagnogleying).

gley soil a soil formed under conditions of poor drainage and waterlogged all or part of the time. Its characteristic blue-green/grey colour is due to the presence of iron(II) compounds formed by the process of gleying.

Gln glutamine *q.v.*

global distillation the redistribution polewards of moderately volatile pollutants (e.g. PCBs, gamma-HCH) as a consequence of volatilization followed by transportation by wind to higher, colder latitudes, and subsequent condensation.

Global Environmental Monitoring System (GEMS) organization created by UNEP to collect global data on climate change and changes in renewable resources and oceans, and to monitor long-distance movements of pollutants.

global positioning system (GPS) a network of satellites developed by the US Department of Defense and now available for public use as surveying and navigational aids. By use of this system a site on the Earth's surface can be pinpointed to within several metres.

global warming the increase in the temperature of the lower atmosphere that has occurred over the past 100 years and that is considered to result principally from the increased emission of CO_2 into the atmosphere as a result of fossil fuel burning and deforestation. *see also* greenhouse effect, greenhouse gases.

global warming potential (GWP) an assessment of the relative warming effect on the atmosphere of the different greenhouse gases. A concept introduced by the Intergovernmental Panel on Climate Change in its 1990 report.

globigerina ooze mud formed largely from the shells of foraminiferans, esp. the calcareous shells of *Globigerina*.

globulin *n.* any of a large group of compact proteins that are water-insoluble but soluble in dilute salt solution, from which they can be salted out. They comprise a major class of proteins in blood serum and other secretions, and are found as storage proteins in plant seeds.

Gloger's rule general principle that as the climate gets cooler from the Equator to the North and South Poles, animals become less pigmented, i.e. paler in colour.

Glu (1) glucose (*q.v.*); (2) glutamic acid *q.v.*

glucose *n.* a hexose sugar found in all living cells, especially plant sap and in the blood and tissue fluids of animals, being the chief end-product of carbohydrate digestion, and as glucose 6-phosphate being the chief substrate for cellular respiration and taking part in many other metabolic pathways. *alt.* dextrose, grape sugar, starch sugar, blood sugar.

glume *n.* a dry chaffy bract, present in a pair at the base of spikelets in grasses.

glume blotch a disease affecting wheat and barley, caused by the fungus *Septoria nodorum*. Irregular blotches appear on the glumes, followed by blackened ears (as a result of secondary infection).

Glumiflorae Poales (*q.v.*), the grasses.

glutamic acid (Glu, E) *n.* amino acid, negatively charged form of alpha-aminoglutaric acid, constituent of protein and also acts as a neurotransmitter in the central nervous system.

glutamine (Gln, Q) *n.* amino acid, uncharged monoamide derivative of glutamic acid, constituent of protein, key component in control of nitrogen metabolism.

gluten *n.* the protein present in wheat grain that makes flour milled from it suitable for bread-making. The gluten gives elasticity to the bread dough, thus allowing it to rise. Varieties of wheat with high gluten values are known as hard wheats.

Glx glutamic acid or glutamine *q.v.*

Gly glycine *q.v.*

glycan *n.* (1) any carbohydrate polymer, *see* glycosaminoglycan, polysaccharide; (2) the oligosaccharide or polysaccharide portions of any macromolecule containing a considerable amount of carbohydrate, as proteoglycans and glycoproteins.

glycine (Gly, G) *n.* the simplest amino acid, having a hydrogen atom as its side chain. It is a constituent of protein and also acts as an inhibitory neurotransmitter in the central nervous system.

Glycine max soya bean *q.v.*

glycogen *n.* a branched-chain polysaccharide made up of glucose units that acts as a storage substance in vertebrate liver and muscle and that is also found in invertebrates. It is broken down in cells to form glucose 1-phosphate, and can also be chemically hydrolysed via dextrins and maltose to glucose.

glycolysis *n.* anaerobic breakdown of glucose to pyruvate in living cells, with the production of ATP, the first step in the oxidative breakdown of glucose in cellular respiration in almost all cells.

glycophyte *n.* a plant unable to tolerate saline conditions.

glycoprotein *n.* any of a large class of proteins that contain carbohydrate in the form of chains of monosaccharide units attached to specific amino acid residues. Ubiquitous components of cell membranes and cell secretions.

glycosaminoglycan (GAG) *n.* any of a group of polysaccharides made up of repeating disaccharide units of amino sugar derivatives, and including hyaluronate, chondroitin sulphate, keratan sulphate and heparin, which are found in the proteoglycans of connective tissue. *alt.* (formerly) mucopolysaccharide.

glycoside *n.* any of a class of compounds that on hydrolysis give a sugar and a non-sugar (aglutone) residue. Widely distributed in plants and including anthocyanins and anthoxanthins.

glyphosate *n.* translocated herbicide used, e.g., in the control of weeds such as couch grass, and as an aquatic herbicide.

GMMs genetically modified microorganisms *see* genetic engineering.

Gnathostomata, gnathostomes *n.*, *n.plu.* (1) jawed vertebrates; (2) group of irregularly shaped sea urchins.

Gnathostomulida *n.* phylum of tiny acoelomate marine worms living in sediments and on plant and algal surfaces in shallow waters.

gneiss *n.* a coarse-grained metamorphic rock formed by high-grade regional metamorphism from a number of different parental rock types, such as impure limestone and shale. It consists mainly of feldspar and quartz, and has a distinctive layered appearance. *see* foliation.

Gnetophyta *n.* one of the five divisions of extant seed-bearing plants, comprising only three genera, *Gnetum*, *Ephedra* and *Welwitschia*, that differ considerably in form and reproduction. They are considered as gymnosperms although having some angiosperm features.

gnotobiosis *n.* the rearing of laboratory animals in a germ-free state or containing only a known prespecified flora of microorganisms.

GNP gross national product *q.v.*

GOES geostationary orbiting environmental satellite *q.v.*

goethite *n.* FeO(OH), a naturally occurring hydrated oxide of iron. A secondary iron mineral and a common constituent of soils.

goitre *n.* a deficiency disease caused by lack of iodine, which is characterized by the slow enlargement of the thyroid gland in the neck. *alt.* Derbyshire neck.

gold (Au) *n.* highly unreactive soft yellow metal (at. no. 79, r.a.m. 197.00). It is found as a native element (*q.v.*) and is valued for its rarity, appearance, resistance to chemical attack, ductility and malleability. It is mined in several parts of the world, notably South Africa (which dominates world production), the former USSR, Canada, USA and Brazil. Environmental concerns over the exploitation of gold reserves often centre on the use of toxic materials (especially cyanide and mercury) in ore processing.

golden-brown algae common name for the Chrysophyta *q.v.*

Golgi apparatus (GA) stacks of flattened membrane sacs present in eukaryotic cells, and which are involved in directing membrane lipids and proteins and secretory proteins to their correct destination in the cell, and also in trimming and adding to the sugar side chains of glycoproteins. *alt.* Golgi bodies, dictyosomes.

Golgi body one of the stacks of the Golgi apparatus. *alt.* (plants) dictyosome.

gonad *n.* organ in which reproductive cells are produced in animals, i.e. the ovary and testis, or in lower animals an ovotestis.

Gondwanaland *n.* a southern land mass composed of South America, Africa, India, Australia and Antarctica, before continental drift moved these continents to their present positions.

good *n.* any material item or service that satisfies people. *alt.* economic good. *see also* private good, public good.

gordian worms Nematomorpha *q.v.*

gorge canyon *q.v.*

gorgonians *n.plu.* the sea fans, horny corals with large, delicate, fan-shaped, branching colonies anchored by a 'stalk' to the substrate.

Gossypium *see* cotton.

gout fly *n. Chlorops pumilionis*, an insect larval pest, esp. of barley. Symptoms of attack include distorted, swollen leaf sheaths around the ear and poorly developed grain.

GPP gross primary production *q.v.*

GPS global positioning system *q.v.*

graben *n.* a depressed fault block sunk between a pair of approximately parallel faults.

gracile *a.* lightly built, small and slender, *appl.* australopithecines: *Australopithecus africanus.*

Gracilicutes *n.* division of the kingdom Monera or Prokaryotae or of the superkingdom Bacteria, including all Gram-negative eubacteria (e.g. enterobacteria, pseudomonads, Gram-negative cocci, spirilla, Gram-negative rods, etc.) and photosynthetic bacteria (i.e. green and purple photosynthetic sulphur bacteria and the cyanobacteria).

grade *n.* (*biol.*) a taxonomic category representing a level of morphological organization in which a group of organisms share a number of characteristics in common but may not owe them to a common ancestor, e.g. the protozoa.

graded bedding (*geol.*) bedding that exhibits an even diminution in the grain size of the rock as a given bed is traversed from the bottom to the top. Graded bedding patterns can be used to establish which way up a disturbed bedding sequence lies.

graded river river in which a profile of equilibrium has been reached, i.e. the river's load may be transported without degradation or aggradation at any point along the river profile *q.v.*

grade of metamorphism (*geol.*) metamorphism (*q.v.*), and the rocks thereby generated, may be categorized as being either high-, medium- or low-grade, depending on the intensity of the metamorphism.

graft *n.* cells, tissue, organ or part of an organism inserted into and uniting with a larger part of the same or another organism.

graft *v.* to insert scion into stock in plants, or to transfer animal cells or tissue from donor to recipient in a transplant.

grain *n.* (*bot.*) naked seed of grasses, usually *appl.* cereals (barley, wheat, rye, etc.).

grain alcohol ethanol *q.v.*

grain aphid *see* cereal aphids.

grain legumes seed legumes *q.v.*

grain size (*geol.*) the average diameter of the mineral grains of a rock. On this basis, those rocks that are not glassy may be described as fine-grained, medium-grained, coarse-grained or, for the most coarse-grained igneous rocks, pegmatitic. More than one scale is used for this categorization. The grain size of rocks with a porphyritic (*q.v.*) texture is based on their groundmass only. *see also* texture.

gramicidin S cyclic peptide antibiotic produced by a *Bacillus* sp., and which is not synthesized by the mRNA–tRNA–ribosome system.

graminaceous *a*. (1) *pert*. grasses, *appl.* members of the grass family (Gramineae), such as cereals; (2) grass-coloured, *appl.* insects.

Gramineae *n*. the grasses, a large and ubiquitous family of monocotyledonous plants, mostly herbaceous but with some woody species (e.g. bamboos). The linear leaves arise from the nodes of jointed stems. Flowering stems bear clusters of wind-pollinated or cleistogamous flowers. The Gramineae contain many important crop plants including wheat (*Triticum* spp.), barley (*Hordeum vulgare*), rye (*Secale cereale*), oats (*Avena sativa*), maize (*Zea mays*), rice (*Oryza sativa*) and sugar-cane (*Saccharum officinale*). Valuable temperate pasture grasses include cocksfoot (*Dactylis*), fescues (*Festuca* spp.) and rye-grasses (*Lolium* spp.). In many grasses, the growing shoot remains near or under the ground with the leaves arising mainly from ground level. Leaves can therefore be continually cropped to near the ground by grazing without damaging the plant.

graminicolous *a*. living on grasses.

graminivorous *a*. grass-eating.

graminology *n*. the study of grasses.

Gram stain a staining procedure for bacteria devised by the Danish physician H.C.J. Gram, which correlates well with certain morphological features of bacteria and is widely used in identification, consisting essentially of staining with crystal violet and subsequent treatment with alcohol, those bacteria that are decolorized by alcohol being termed Gram-negative and those retaining the stain Gram-positive, a property related to the structure of the cell wall.

granite *n*. a coarse-grained intrusive igneous rock with a colour index of 0–30. It contains quartz ($\geq$20% v/v) and alkali feldspar. It may contain sodium-rich plagioclase (generally oligoclase) but this is never the dominant feldspar (*cf.* granodiorite). The micas biotite and muscovite are common constituents. They may occur singly, generating biotite granite and muscovite granite, or together, in biotite–muscovite granite. Hornblende may also be present

(typically *ca*. 15–18% v/v). A wide range of accessory minerals are known within granites. *see also* migmatite, porphyritic granites.

granite mosses a small group of mosses, class Andreaeidae, sometimes called rock mosses, small blackish-green or olive-brown tufted plants, in which the gametophyte arises from a plate-like protonema and spores are shed from slits in the capsule wall.

granite pegmatites very coarse- or giant-grained igneous rocks that are formed during the latter stages of the solidification of acidic intrusions. While pegmatites have essentially the same mineralogy as the rest of the intrusion, they may contain significant concentrations of minerals rich in less abundant elements (e.g. Be, Li, Sn, W, Rb, Cs, Nb, Ta, the rare earths, U). Pegmatite deposits are worked both for their metallic content and the valuable industrial minerals that they may contain.

granivorous *a*. feeding on grain.

granoblastic *a*. *appl*. the texture of metamorphic rocks constituted from essentially equisized mineral grains.

granodiorite *n*. a coarse-grained igneous rock with essentially the same mineralogy as granite except that >67% of its feldspar component is plagioclase. *cf*. granite.

granular *a*. *appl*. (1) (*soil sci.*) soil peds (*q.v.*) that are spheroidal in shape and porous, *cf*. crumb; (2) (*geol.*) a rock texture (*q.v.*) in which all of the grains are essentially of equant shape and are roughly equal in size.

granular activated carbon *see* activated carbon.

granulite *n*. a rock believed to have been formed at great depths within the continental crust by regional metamorphism at very high temperatures and pressures.

graphic granite granite with a texture formed by the intergrowth of quartz and alkali feldspar that is reminiscent of ancient cuneiform writing.

Graptolita, graptolites *n*., *n.plu*. a group of fossil invertebrates from the Palaeozoic, of doubtful affinity but thought to be allied with the hemichordates.

grass aphid *see* cereal aphids.

grasses *n.plu*. common name for members of the large monocot family Gramineae *q.v.*

grasshoppers *n.plu.* common name for many members of the Orthoptera *q.v.*

grassland *n.* a biome found in regions where the average annual precipitation (around 25–76 cm) is sufficient to support the growth of grasses and other herbaceous plants but generally insufficient to support continuous tree cover, which may also be suppressed as a result of grazing by herbivores. *see* alpine grassland, prairie, savanna, short-grass community, steppe, tall-grass community, veldt.

grass staggers staggers *q.v.*

grats arêtes *q.v.*

gravel *n.* (*soil sci.*) according to the International or Atterberg, and USDA systems, that part of air-dry soil that, once it has been gently ground, will not pass through a sieve with holes of 2 mm diameter. Synonymous with the word stones used by the Soil Survey of England and Wales, British Standards and the Massachusetts Institute of Technology. *cf.* fine earth.

gravel pits pits resulting from extraction of sand or gravel, which when disused and filled with water are often valuable wildlife habitats.

gravid *a. appl.* (1) female with eggs; (2) pregnant uterus.

gravimeter *n.* instrument for measuring the Earth's gravitational field.

gravimetric *a. appl.* chemical analysis methods that are founded on the measurement of mass.

graviperception *n.* the sensation or perception of gravity.

gravitational *a.* (*soil sci.*) *appl.* water in excess of soil capacity, which sinks under the action of gravity and drains away.

gravitaxis *n.* geotaxis *q.v.*

gravitropism geotropism *q.v.*

gray (Gy) *n.* the derived SI unit of the dose of ionizing radiation absorbed by living tissue, being equal to 1 J of energy imparted to 1 kg of mass, which replaces the non-SI unit, the rad. 1 Gy = 100 rad.

graze *v.* to eat non-woody plants. *cf.* browse.

grazer food chain, grazing food chain pathway within a food web that is based on the consumption of green plants or algae by herbivorous animals or microorganisms. *cf.* decomposer food chain.

grazing *n.* the consumption of green plant material or algae by animals and microorganisms such as some protozoa.

green *a.* adjective that has become applied to anything that is connected or supposedly connected with a concern for the environment, such as green taxes, green politics.

green algae common name for the Chlorophyta *q.v.*

green belt designated zone around a city within which special planning regulations exist in order to control urban sprawl and conserve countryside areas for farming and recreation purposes.

green corridor *see* corridor.

green deserts areas of little native wildlife and ecological interest despite their 'green' appearance, e.g. extensive areas of crop monocultures, heavily fertilized pastureland replanted with alien grass species, and large areas of closely mown grass in urban parks.

greenfield site agricultural land, usually on the periphery of an urban area, earmarked as a site for possible future residential or industrial development. Also *appl.* such a site during and immediately after development. *alt.* virgin site. *cf.* brownfield site.

greenfly *n.* a colloquial term for any green-coloured aphid *q.v.*

greenhouse cultivation type of horticulture in which vegetables, fruit and flowers are intensively cultivated, at temperatures above ambient, under glass. This facilitates protection from frost and enables tender crops to be produced early or grown out of season. *alt.* glasshouse cultivation.

greenhouse effect warming of the troposphere (esp. near to ground level) caused by the absorption of terrestrial radiation by greenhouse gases (*q.v.*). Without this effect, the mean temperature of the atmosphere at the surface of the Earth would fall from its current value of 15 °C to *ca.* −17 °C or lower, probably making the world uninhabitable. Outputs of greenhouse gases (esp. carbon dioxide, CO_2) due to human activity have caused an enhanced greenhouse effect, which will probably lead to significant global warming and concomitant climate change.

greenhouse gases those gases capable of absorbing terrestrial radiation (i.e. radiation emitted by the Earth). With respect to their influence on climate, the main greenhouse gases are water vapour (H_2O), carbon dioxide (CO_2), methane (CH_4), nitrous oxide (N_2O) and the wholly anthropogenic chlorofluorocarbons (CFCs).

green manures the name given to green crops, such as the field bean (*Vicia faba*), that are grown as part of a crop rotation and deliberately incorporated into the soil while still green. The use of green manures has a number of benefits including the addition of organic matter and if the plant is a legume, the addition of nitrogen.

Greenpeace an international non-governmental organization (NGO), founded in 1971, committed to challenging those activities, such as whaling and the dumping of nuclear wastes at sea, that threaten the global environment.

Green Revolution the introduction of scientifically bred high-yielding varieties of staple crop plants, such as wheat, rice and maize, from the 1960s onwards, into areas of the world where lower-yielding traditional varieties adapted to local conditions were formerly grown, and which has considerably increased food production, particularly in the countries of South-East Asia and India.

Greens general name for political parties that are primarily concerned with the environment and the overexploitation of the natural world, and who have as their central ethos the concern that people must live in harmony with nature. *see also* environmentalism.

green soiling zero grazing *q.v.*

green sulphur bacteria photoautotrophic bacteria, mainly aquatic, which oxidize sulphide to sulphur.

green taxes colloquial name for taxes designed to benefit the environment. *see* carbon tax, ecotaxation, energy tax.

gregarines *n.plu.* a group of parasitic protozoans of invertebrates, in which only the adults live outside the host cell, and in which the male and female gametes are smaller than the normal cell.

gregarious *a.* (1) tending to herd together; (2) growing in clusters; (3) colonial.

grey body *n.* (*phys.*) body that is neither a perfect absorber nor a perfect radiator of electromagnetic radiation, e.g. the Earth. *cf.* black body.

Grey List list produced by the European Community of toxic chemicals considered to be less dangerous that those on the Black List (*q.v.*). This list includes cyanides, fluorides and metals such as lead, silver and nickel, and their compounds. *alt.* List 2.

grey mould *n.* fungus disease that affects flowers and soft fruit, especially raspberries and strawberries. *alt.* botrytis.

greywacke *n.* a dark-coloured sedimentary rock consisting of poorly sorted rock fragments and mineral grains. Many greywackes exhibit graded bedding, with each bedding unit having predominantly coarse-grained (e.g. sand) material at the bottom grading upwards to finer-grained particles (e.g. clay or silt). Many such units may be stacked one upon another, producing extremely thick sediments.

grikes *see* limestone pavement.

grinding teeth molars *q.v.*

griseofulvin *n.* antibiotic produced by some species of Penicillium, esp. P. griseofulvum, that is toxic to some fungi by inhibiting mitotic metaphase.

grit *n.* (1) (*geol.*) a stone similar to sandstone (*q.v.*) but with sharply angular grains (sandstone grains are rounded to subangular); (2) (*clim.*) air-suspended solid matter with a diameter >500 μm.

groins groynes *q.v.*

grooming *n.* the cleaning of fur or feathers by an animal (generally called preening in birds), performed either by itself (self grooming) or by another member of the same species (social grooming), functioning not only to keep the animal clean, but as a displacement activity or to improve the social cohesion of the group.

gross calorific value the specific enthalpy of combustion (*q.v.*) of a fuel measured when the fuel is burned, under standard conditions, in water vapour saturated oxygen.

gross domestic product (GDP) total monetary value of all the goods and services produced in a national economy, usually measured over a period of one year.

gross efficiency a measure of the efficiency of an animal in converting food consumed into body substance.

gross error an error that occurs for an unexpected reason. For example, an abrupt change in electrical supply voltage or an operator mistake, such as misreading a scale. Gross errors may affect a whole set of replicate data but, usually, produce a small number of data (called outliers) that vary greatly from the remainder of the data set.

gross growth efficiency ecological energy efficiency *q.v.*

gross national product (GNP) gross domestic product (GDP) (*q.v.*) adjusted to take into account trade figures for 'invisibles', i.e. primarily financial deals and services. GNP is used as a measure of economic success.

gross primary production (GPP) the total assimilation of inorganic nutrients in a plant community. *cf.* net primary production.

grossular *n.* one of the garnet minerals *q.v.*

ground cover any grass or other herbaceous plant that is planted to prevent erosion.

ground fire fire in which organic matter beneath the surface of the soil is burning.

ground layer moss layer *q.v.*

groundmass *n.* (*geol.*) the relatively fine-grained or glassy matrix within which large crystals are imbedded in rocks with porphyritic (*q.v.*) textures.

ground moraine *see* moraine.

groundnut *n. Arachis hypogea*, an important seed legume (family Leguminosae) grown for fodder, food and as a source of edible oil. *alt.* peanut.

ground tissue of plants, the parenchyma, collenchyma and sclerenchyma, i.e. all tissues other than the epidermis, reproductive tissues and vascular tissues. It arises from the ground meristem, a primary meristem derived from the apical meristem in the developing embryo.

groundwater (1) water that sinks down through soil and rock and collects in underground aquifers; (2) underground water in the zone of saturation, below the water table, *cf.* soil water.

groundwater gleying *see* gleying.

groups *n.plu.* (*chem.*) *appl.* vertical columns of elements within the periodic table (*q.v.*). *see* Appendix 2.

group selection selection that operates on two or more members of a lineage group as a unit, such that characters that benefit the group rather than the individual may be selected for.

growing point a part of the plant body at which cell division is localized, generally terminal and composed of meristematic cells.

growing season *n.* (1) that portion of the year during which climatic conditions, e.g. temperature and moisture, are favourable for plant growth and development; (2) in temperate agriculture, number of frost-free days, i.e. number of days between the last severe frost of spring and the first of autumn.

growth *n.* increase in mass and size by cell division and/or cell enlargement.

growth curve (*biol.*) generally a plot of log numbers of a population of living organisms against time. *see also* J-shaped growth curve, S-shaped growth curve.

growth efficiency *see* gross efficiency, gross growth efficiency.

growth factors (1) organic compounds other than those required as carbon and energy sources that are needed by many organisms for proper growth and development and that may include vitamins, amino acids, purines, pyrimidines, etc.; (2) (*cell biol.*) general term for specific peptides or proteins required by particular cells for division, differentiation or survival, e.g. nerve growth factor, epidermal growth factor.

growth promoters substances administered to domestic livestock, either in feedstuffs or, as is often the case with hormones, via ear implants, to promote growth.

growth rate (r) increase in the size of a population per unit of time.

growth rings (1) (*bot.*) the rings seen on the transverse cut trunk of some trees, each representing the growth of wood in one year. The width of each ring reflects climatic conditions (e.g. temperature and rainfall) in that year, wider rings being produced in more favourable years, *see also* dendrochronology; (*zool.*) (2) the layer of shell laid down in each growth period in various animals such as bivalve molluscs; (3) a layer of a scale in fishes.

growth substance generally refers to any compound, natural or artificial, that when present in small amounts has a marked effect on the growth and/or development of a plant. *alt.* plant growth substance.

growth yield the growth achieved from the amount of substrate used over a given time interval, measured as cell growth (in grams) per amount of substrate used (in grams).

groynes *n.plu.* artificial barriers built at intervals along a beach, usually at right angles to the shore, to help prevent sand loss through the process of longshore drift (*q.v.*). *alt.* groins.

grub *n.* legless larva of insects of the orders Diptera, Coleoptera and Hymenoptera.

Gruiformes *n.* an order of birds including cranes, bustards, rails, gallinules and coots and the hemipodes.

Gruinales Geraniales *q.v.*

grunerite *n.* the iron-rich members of the cummingtonite–grunerite series (*q.v.*) of minerals.

GTP guanosine triphosphate *q.v.*

guanine (G) *n.* a purine base, constituent of DNA and RNA, which is the base in the nucleoside guanosine and is also found as iridescent granules or crystals in certain chromatophores, *see* guanophore.

guano *n.* an earthy, friable phosphate rock that is formed from accumulated seabird droppings. It is rich in phosphates and has been used as fertilizer. It was formerly collected from islands off the west coast of South America.

guanosine *n.* a nucleoside made up of the purine base guanine linked to ribose.

guanosine triphosphate (GTP) guanine nucleotide containing three phosphate groups linked in series, one of the four ribonucleotides required for RNA synthesis, also acts as a source of phosphate groups and energy for metabolic reactions in a manner analogous to ATP.

guard cells two specialized epidermal cells that contain chloroplasts and surround the central pore of a stoma. Changes in the turgor of these cells open and close the stomatal opening.

guest *n.* an animal living and breeding in the nest of another, esp. an insect.

guild *n.* a group of species having similar requirements and foraging habits and so having similar roles in the community.

guinea-worm disease dracunculiasis *q.v.*

Gulf Stream the huge ocean current of warm water that originates in the Gulf of Mexico, runs parallel to the east coast of North America and then, at Newfoundland, flows northeastwards towards Europe, where it has a warming effect on the climate. *see also* North Atlantic Drift.

gully *n.* a deep channel cut into bedrock (e.g. clay or shale) or weakly consolidated sediments (e.g. loess or volcanic ash) by ephemeral streams formed during heavy rainstorms.

gully erosion type of soil erosion in which ephemeral streams, formed during heavy rainstorms, cut into bedrock or loosely consolidated sediments to form deep channels (called gullies). This process is often facilitated by the removal of vegetation cover. *alt.* gullying.

gullying gully erosion *q.v.*

gums *n.plu.* (1) various materials, composed largely of polysaccharides, resulting from breakdown of plant cell walls and exuding from plant wounds; (2) trees of the genus *Eucalyptus*; (3) (*zool.*) tissue investing the jaws.

gusher *n.* oil well in which the crude oil spurts to the surface upon drilling.

gut *n.* the alimentary canal or part of it.

Gutenberg discontinuity seismic-velocity boundary between the mantle and the core of the Earth, found at a depth of *ca.* 2900 km.

guttation *n.* (1) formation of drops of water on leaves, forced out of leaves through special pores (hydathodes) by root pressure, seen, e.g., on tips of leaves of many grasses in early morning; (2) formation of drops of water on surface of plant from moisture in air; (3) exudation of aqueous solutions, as by sporangiophores, nectaries, etc.

guyots *see* seamounts.

Gy the SI unit of ionizing radiation absorbed by tissue, the gray *q.v.*

gymno- prefix derived from Gk *gymnos*, naked.

Gymnolaemata *n.* a class of marine Ectoprocta (Bryozoa) in which individual zooids have a circular crown of tentacles (lophophore) and are enclosed in a box-like calcareous exoskeleton, found encrusting seaweeds, stones and shells.

Gymnomycota *n.* the cellular and acellular slime moulds.

Gymnophiona Apoda *q.v.*

gymnosperms *n.plu.* large group of seed-bearing woody plants, having seeds not enclosed in an ovary but borne on the surface of the sporophylls, either treated as a division (Gymnophyta or Pinophyta) of the seed plants, or as a grouping of four divisions, the cycads (Cycadophyta), *Ginkgo* (Ginkgophyta), the conifers (Coniferophyta) and the gnetophytes (Gnetophyta), taxa that have been given various ranks in other classifications. Extinct taxa of gymnosperms include the seed ferns, cycadeoids (Bennetiales) and cordaites (Cordaitales). *alt.* (formerly) Gymnospermae.

gymnotoids *n.plu.* small group of slender freshwater bony fish from Central and South America (the Gymnotodei) with very long anal fins, and an electric organ that sets up an electric field around the fish, commonly called American knife-fishes, and including the electric eel *Electrophorus.*

-gyn- word element derived from Gk *gynē*, woman, indicating female.

gynaecium gynoecium *q.v.*

gynandromorph *n.* an individual exhibiting a spatial mosaic of male and female characters. *alt.* gynander, sex mosaic.

gynecium gynoecium *q.v.*

gynodioecious *a.* having female and hermaphrodite flowers on different plants.

gynoecious *a.* having female flowers only.

gynoecium *n.* the female reproductive organ of a flower, consisting of one or more carpels forming one or more ovaries, together with their stigmas and styles. *alt.* pistil.

gynomonoecious *a.* having female and hermaphrodite flowers on the same plant.

gypcrete *n.* duricrust (*q.v.*) formed by the deposition of calcium sulphate.

gypsophil(ous) *a.* thriving in soils containing chalk or gypsum.

gypsum *n.* $CaSO_4.2H_2O$, a mineral of industrial importance. Used in making plasterboards and cements. Heating gypsum to 120–130 °C yields plaster of Paris ($CaSO_4.O.5H_2O$). There are several named varieties of this mineral, notably alabaster (*q.v.*), satin spar (*q.v.*) and selenite *q.v.*

Gyr gigayear, 10^9 years.

gyration *n.* (1) rotation, as of a cell; (2) a whorl of a spiral shell.

gyre *n.* (*oceanogr.*) any one of the huge open-ocean circulation cells, composed of several major currents, caused by prevailing winds in conjunction with the Coriolis force (*q.v.*) and, in some cases, influenced by the shape of neighbouring land masses. Some gyres run clockwise, while others operate in an anticlockwise direction.

H

H symbol for the chemical element hydrogen *q.v.*

H (1) symbol for enthalpy (*q.v.*); (2) Henry's law constant *q.v.*

H′ symbol for equitability *q.v.*

ha hectare.

Ha symbol for the chemical element hahnium *q.v.*

Haber process a process for the artificial fixation of atmospheric nitrogen via the reaction represented by the equation

$$N_{2(g)} + 3H_{2(g)} \rightleftharpoons 2NH_{3(g)}$$

This is carried out under pressure (between 150 and 600 atm.), at relatively high temperatures (400–600 °C), in the presence of a catalyst. The ammonia generated is used for the production of fertilizers, plastics and explosives.

habilines *n.plu.* fossil hominids from East Africa, dated at around two million years ago, assigned by some palaeoanthropologists to the species *Homo habilis*, thus making them possibly the earliest human fossils yet known.

habit *n.* (1) (*bot.*) the external appearance or way of growth of a plant, e.g. climbing, bushy, erect, shrubby, etc.; (2) (*zool.*) the normal or regular behaviour of an animal; (3) (*geol.*) the characteristic form of the crystals that make up a particular mineral.

habitat *n.* the place where an organism or species normally lives, characterized by the physical characteristics of the environment and/or the dominant vegetation or other stable biotic characteristics. Examples of habitats can be as general as lakes, woodland, soil, etc., or more specific, such as mudflats, the bark of an oak tree, chalk downland, etc. *cf.* niche. *see also* Shelford's law of tolerance.

habitat breadth the range of different types of habitats an organism can inhabit.

habitat corridor strip of land, e.g. motorway embankment, canal towpath, whose presence allows the free movement of animals between otherwise isolated areas of suitable habitat. *alt.* wildlife corridor.

habitat diversification beta diversity *q.v.*

habitat form the way a plant grows resulting from conditions in that particular habitat.

habitat fragmentation the breaking up of a continuous area of habitat into ever smaller areas physically isolated from each other by, e.g., urban development, intervening areas of intensive agriculture or by a different type of habitat. This leads to species loss over the original habitat as a whole (*see* species–area effect), reduced population sizes that are more vulnerable to extinction, and a loss of genetic diversity within these smaller populations. Habitat fragmentation is currently one of the greatest threats to nature and biological diversity.

habitat 'island' isolated area or fragment of a particular habitat completely surrounded by a different habitat.

habitat loss the destruction of suitable living space for plants and animals as the result of human activity.

habitat space the habitable part of a space or area available for establishing a population.

habitat type a group of plant communities that produce similar habitats.

habituation *n.* (1) adjustment, by a cell or an organism, by which subsequent contacts with the same stimulus produce diminishing effects; (2) a form of learning in which reflex behaviour is extinguished when the animal finds it has no adaptive value.

hadal *a. appl.* or *pert.* abyssal deeps below 6000 m.

Hadley cell cell of primary atmospheric circulation system that operates in each hemisphere between the equatorial low-pressure belt (ITCZ) and the subtropical high-pressure belt (30° latitude). *see* Fig. 32, p. 419.

haematite hematite *q.v.*

haematophagous *a.* feeding on or obtaining nourishment from blood.

hafnium (Hf) *n.* metallic element (at. no. 72, r.a.m. 178.49) closely resembling zirconium and found in zirconium minerals. It is a strong neutron absorber and is used for control rods in nuclear reactors.

hagfishes *n.plu.* common name for the Myxiniformes, a small order of bottom-living jawless marine fish, eel-like in shape, lacking pectoral or pelvic fins and with no scales.

hahnium (Ha) *n.* a transuranic element (at. no. 105).

hail *n.* frozen raindrops, usually 5–50 mm in diameter. These may be formed within cumulonimbus clouds or in association with cold fronts.

hair *n.* (*bot.*) (1) trichome (*q.v.*); (2) plant root hair (*q.v.*); (3) (*zool.*) in mammals, a thread-like epidermal structure consisting of cornified epithelial cells that grows by cell division from a hair follicle at its base.

halarch succession halosere *q.v.*

Haldane's rule a rule stating that when offspring of one sex produced from a cross are unviable or infertile, it is always the heterogametic sex.

half-hardy *a. appl.* ornamental plants that can withstand some cold but not severe winters.

half-life *n.* (1) (*phys.*) of radioactive elements, time required for the radioactive decay of half the original amount of material; (2) (*physiol.*) time required for the disappearance of half the original quantity of a given substance, e.g. from the circulation, assuming that the substance disappears at a regular rate.

half-sibs *n.plu.* individuals having only one parent in common.

half value depth in soil, depth above which more than half of a given pollutant or other chemical applied to the soil remains after a given time.

halite *n.* natural sodium chloride (NaCl). Occurs as an evaporite mineral. Exploited as a raw material for both the chemical and food industries. May be referred to as rock salt *q.v.*

halite crust salcrete *q.v.*

halleflinta *n.* a metamorphic rock formed from tuffs imbued with secondary silica.

halloysite *n.* $Al_2Si_2O_5(OH)_4.4H_2O$, one of the kandite group (*q.v.*) of clay minerals (*q.v.*). Its structure is that of kaolinite plus intercalated water.

halobacteria halophilic bacteria *q.v.*

halobenthos *n.* marine benthos.

halobios, halobiota *n.* (1) sum total of organisms living in sea or any salt water; (2) animals living in sea or any salt water. *a.* halobiotic.

halocline *n.* the sizable step in the salinity profile of a typical body of seawater.

halogenated hydrocarbon hydrocarbon also containing halogen atoms (fluorine, bromine, chlorine or iodine), e.g. chlorinated hydrocarbon insecticides and CFCs.

halogenation *n.* process whereby one or more halogen atoms are incorporated into a chemical compound, by substitution or addition.

halogens *n.plu.* elements of Group 17 of the periodic table (*see* Appendix 2), i.e. fluorine (F), chlorine (Cl), bromine (Br), iodine (I) and astatine (At).

halomorphic *a. appl.* soils containing an excess of salt or an alkali.

halons *n.plu.* brominated analogues of chlorofluorocarbons, which (*q.v.*) like CFCs, are entirely man-made. Halons are frequently given a four-digit code, prefixed by H for halon, e.g. H1211. The digits stand, respectively, for the number of carbon, fluorine, chlorine and bromine atoms in the molecule (CF_2ClBr in this case), any other atoms present are hydrogen. Halons are used as fire extinguishers and therefore enter the atmosphere. These compounds release bromine atoms when photolysed in the stratosphere. The bromine atoms then enter into catalytic cycles, destroying some of the ozone that shields the biosphere from harmful ultraviolet light. As well as their action as ozone-destroying catalysts, the Br atoms synergistically enhance the rate of chlorine-induced ozone depletion. Atmospheric concentrations of halons are low but are rising fast. Concentrations of $CBrClF_2$ and $CBrF_3$ are 1.7×10^{-3} and 2.0×10^{-3} p.p.b., respectively, each rising by 12–15% per year (1992 data).

halophilic *a.* (1) salt-loving; (2) thriving in the presence of salt.

halophilic bacteria group of archaebacteria (e.g. *Halobacterium*, *Halococcus*) adapted to life in saturated salt solutions, e.g. salt lakes and brine pools. *alt.* halobacteria, halobacters.

halophobe *n.* a plant intolerant of salt.

halophyte *n.* (1) a seashore plant; (2) a plant that can thrive on soils impregnated with salt. *a.* halophytic.

haloplankton *n.* the organisms drifting in the sea.

Haloragales Hippuridales *q.v.*

halosere *n.* a plant succession originating in a saline area, as in salt marshes.

Hamamelidales *n.* order of woody dicots including the families Hamamelidaceae (witch hazel), Myrothamnaceae and Platanaceae (plane).

hamlet *n.* tiny settlement in the country made up of a few houses and without a parish church.

hammada *n.* a bare rock surface, sometimes covered in angular rock debris produced by weathering, common in many arid regions.

handling time the time it takes a predator to catch and eat its prey.

hanger *n.* a wood situated on a hillside.

hanging valleys tributary glacial valleys that are left high above the level of the main glacial valley when the glacier melts. These features arise due to the slower rate of erosion of the tributary glaciers compared with that of the main valley glacier into which they feed.

haplobiont *n.* organism characterized by only one type of individual in its life-cycle. *cf.* diplobiont.

haplodiploid *a. appl.* species in which sex is determined by the male being haploid, the female diploid, as in bees and wasps. *n.* haplodiploidy.

haplodiplont *n.* (1) an organism exhibiting the haplodiploid condition; (2) a plant with haploid and diploid vegetative phases.

haploid *a. appl.* cells having one set of chromosomes representing the basic genetic

complement of the organism, usually designated *n.* *n.* a haploid organism or cell.

haploidy *n.* the state of being haploid.

haplont *n.* any organism having haploid somatic nuclei or cells.

haplophase *n.* stage in life history of an organism when nuclei are haploid.

haplophyte *n.* a haploid plant or gametophyte.

haplotype *n.* the set of alleles borne on one of a pair of homologous chromosomes, esp. in relation to complex loci such as the major histocompatibility complex (MHC).

haptonasty *n.* a plant movement elicited by touching, as the drooping of the leaves of mimosa.

Haptophyta, haptomonads *n., n.plu.* group of small motile golden algae, mainly marine, bearing a coiled thread-like haptoneme between the two flagella, which acts as a holdfast, and a cell surface covered with thin scales. Many species have a resting stage when the surface becomes covered with elaborate calcareous scales (coccoliths), when the organisms are known as coccolithophorids. *alt.* Prymnesiophyta.

haptotropism *n.* response by curvature to a contact stimulus, as in tendrils or stems that twine around a support. *alt.* thigmotropism.

hard coal black coal *q.v.*

hard detergents *see* detergent swans.

hardening *n.* (1) the process of acclimatizing plants to cold; (2) gradually exposing seedlings raised inside to outdoor temperatures.

hardiness *n.* (1) the ability of a plant or animal to withstand unfavourable conditions, esp. cold; (2) strictly, the range of minimum winter temperatures a plant will survive.

hardpan *n.* hard layer developed in the B-horizon of a soil, which restricts drainage and root growth. Hardpans form as a result of the illuviation of chemical compounds from the A-horizon. *see* claypan, ironpan, moorpan.

hard shoulder *see* motorway.

hard water water that does not readily form a lather with soap due to the presence of dissolved calcium and magnesium salts esp. hydrogen carbonates, chlorides and sulphates. The hydrogen carbonates are decomposed on boiling, thus:

$$M(HCO_3)_{2(aq)} \rightarrow MCO_{3(s)} + CO_{2(g)} + H_2O_{(l)}$$

where M = Ca or Mg. The hardness that can be removed from solution by this process is called temporary hardness, while that left in solution (essentially magnesium and calcium sulphates and/or chlorides) is called permanent hardness.

hard wheats *see* gluten.

hardwood *n.* wood of broad-leaved trees, although not all hardwoods are in fact physically harder than all softwoods (wood from coniferous trees).

Hardy–Weinberg law in a large randomly mating population, in the absence of migration, mutation and selection, allele frequencies stay the same from generation to generation according to the following rule (the Hardy–Weinberg rule): this states that if the allele frequency of one of the alleles at a locus is p and that of the other is q then the frequencies of the two homozygotes and the heterozygote are given by p^2, q^2 and $2pq$, respectively. The ratio $p^2 : 2pq : q^2$ is the Hardy–Weinberg ratio.

harem *n.* a group of breeding females that a dominant male mates with and guards from rival males.

harmotome *n.* $BaAl_2Si_6O_{16}.6H_2O$, one of the zeolite minerals *q.v.*

Hartig net network of fungal hyphae within the epidermis and outer cortex of roots of plants with mycorrhizas, the hyphae not penetrating into the endodermis, and only rarely entering the root cells.

harvest index in a cereal crop, the ratio between the weight of the grain and the weight of the total plant.

harvestmen *n.plu.* common name for the Opiliones, an order of arachnids having very long legs and with the prosoma and opisthosoma forming a single structure.

harzburgite *see* peridotite.

Hatch–Slack pathway C4 pathway *q.v.*

haulm *n.* stem or stalk of many plants including potatoes, beans, peas, hops and grasses.

haustorium *n.* an outgrowth of stem, root or hyphae of certain parasitic plants or fungi, through which they obtain food from the host plant. *a.* haustorial. *plu.* haustoria.

haüyne *n.* $(Na,Ca)_{4-8}(Al_6Si_6O_{24})(SO_4)_{1-2}$, one of the sodalite group (*q.v.*) of feldspathoid minerals *q.v.*

Hawaiian Floral Region part of the Palaeotropical Realm comprising the Hawaiian Islands.

hay *n.* fodder material consisting of legumes, such as lucerne, and grasses, which have been dried for storage.

hayfever *n.* common name for an allergic condition that affects the mucous membranes of the eyelids and nose, causing a runny nose and swollen and watering eyes. It is caused by exposure to airborne allergens such as pollen, mould spores and particles from animal skin. *alt.* allergic rhinitis.

hazardous substances chemicals that can cause harm to humans and the environment because they are explosive, inflammable, corrosive or irritant.

hazardous waste waste that contains toxic chemicals or radioactive material, or which is highly inflammable, corrosive, unstable or explosive.

haze *n.* a fine aerosol of water vapour and particulate matter in the atmosphere, often indicative of air pollution.

H-bomb *see* nuclear bomb.

HCFCs hydrochlorofluorocarbons *q.v.*

He symbol for the chemical element helium *q.v.*

heading date (*agric.*) in grasses and legumes, the mean date of ear emergence after 1 April. This is used in classification to distinguish between varieties of a particular species.

headland *n.* (1) (*agric.*) a piece of land left undisturbed during ploughing at the end of the furrows, to allow the plough to be turned more easily; (2) (*geog.*) a point of land that projects into an expanse of water, usually the sea, *alt.* promontory.

headward erosion the process by which a river valley extends up-slope towards its watershed and beyond. This may result in river capture *q.v.*

headwaters *n.plu.* source streams of a river system.

heap-leach extraction extraction technique used to separate gold from very low-grade ore. Huge heaps of crushed ore are placed on an impervious pad and sprayed with cyanide solution, which dissolves the gold as it trickles through. The resultant effluent is then removed and processed to recover the gold.

heart *n*. hollow muscular organ that by rhythmic contractions pumps blood around the body. The mammalian heart consists of four chambers, an upper thin-walled atrium and lower thick-walled ventricle on each side. Venous blood from the body enters the right atrium and leaves for the lungs from the right ventricle, oxygenated blood re-enters the heart through the left atrium and leaves the heart from the left ventricle. The muscular contraction of the heart is self-sustaining and is synchronized by pacemaker cells located in the sinuatrial node.

heart rot fungal decay of the wood in the centre of a tree.

heartwood *n*. the darker, harder, central wood of trees, containing no living cells. *cf*. sapwood.

heat *n*. energy that is transferred by virtue of a difference in temperature. The heat interaction (*q.v.*) is what occurs when an object comes into contact with another that is either hotter or cooler than itself. Units: joules (J) in the SI system, calories (cal) in the c.g.s. system and British thermal units (Btu) in the Imperial system.

heat engine any device that converts heat into work, e.g. an internal combustion engine.

heat engine rule *see* Carnot efficiency. *alt*. HER.

heath *n*. vegetation developing on poor, usually acid, sandy or gravelly soils in the lowlands and dominated by gorse (*Ulex*) and heathers (*Calluna* and *Erica*) or other narrow-leaved plants.

heathland *n*. ecosystem with a predominant cover of heath *q.v.*

heating value of a fuel, specific enthalpy of combustion *q.v.*

heat island, urban *see* urban heat island.

heat pump device that, with the imput of energy, will transfer heat from a place at one temperature to another at a higher temperature.

heat sink a term used in the thermodynamic analysis of heat-to-work conversions to denote a low-temperature reservoir of heat. *cf*. heat source.

heat source a term used in the thermodynamic analysis of heat-to-work conversions to denote a high-temperature reservoir of heat. *cf*. heat sink.

heavy metal a rather vague term for any metal (in whatever chemical form) with a fairly high relative atomic mass, esp. those that are significantly toxic (e.g. lead, cadmium, mercury). They persist in the environment and can accumulate in plant and animal tissues. Mining and industrial wastes and sewage sludge are potential sources of heavy metal pollution.

heavy oil high-sulphur tar-like oil found in deposits of crude oil, tar sands and oil shale.

heavy spar baryte *q.v.*

heavy water water that contains deuterium, symbol D, (i.e. the $^2_1$H isotope) in place of hydrogen. It has the formula DHO or D_2O, the latter being also known as deuterium oxide.

hecto- (h) prefix attached to an SI unit to indicate that the associated unit is multiplied by 10^2.

hedenbergite *n*. [$Ca(Mg,Fe)Si_2O_6$ where $45\% < Ca/(Ca + Mg + Fe) < 50\%$ and $Mg/(Mg + Fe) > 50\%$] one of the pyroxene minerals (*q.v.*) of the clinopyroxene subgroup. *cf*. diopside.

hedgerow *n*. (1) row of trees and shrubs forming a hedge around a field; (2) line of trees separating a field from a road.

Heidelberg man type of primitive man known from fossils found near Heidelberg in Germany, which is now considered as a subspecies of *Homo erectus*.

heifer *n*. a young cow that has not yet calved.

heliophil, heliophilic, heliophilous *a*. adapted to a relatively high intensity of light.

heliophobe *n*. plant that thrives in shade.

heliophyll *n*. plant with leaves of similar structure on both sides and arranged vertically.

heliophyte *n*. a plant requiring full sunlight to thrive.

heliosis *n*. production of discoloured spots on leaves through sunlight.

heliostats *n.plu*. solar collectors, with mirror-like surfaces, that are programmed to track the path of the Sun across the sky. They are used in central receiver systems *q.v.*

heliotaxis *n*. movement in response to the stimulus of sunlight.

heliothermism *n*. method of regulating body heat adopted by some ectothermic animals, which adjust their body orientation throughout the day to change the amount of solar radiation received.

heliotrope *n.* SiO_2, a variety of chalcedony (*q.v.*) that is green with red spots. *alt.* blood stone.

heliotropism *n.* a plant growth movement in response to the stimulus of sunlight.

helioxerophil *n.* plant that thrives in full sunlight and in arid conditions.

heliozoan *n.* member of the Heliozoa, an order of mostly freshwater protozoans of the Sarcodina, having a radially symmetrical body, stiff slender pseudopodia and often a skeleton of spicules.

helium (He) *n.* an inert element (at. no. 2, r.a.m. 4.003), a colourless odourless gas, lighter than air. It occurs in some natural gas deposits, and in bubbles within some radioactive ores, and also in the atmosphere in one part per 200 000. It is used in meteorological balloons, and has many other commercial uses.

helminth *n.* parasitic flatworm (fluke and tapeworm) or roundworm.

helminthology *n.* (1) the study of worms; (2) the study of parasitic flatworms and roundworms.

helotism *n.* symbiosis in which one organism enslaves another and forces it to labour on its behalf, e.g. in some species of ants.

hematite *n.* Fe_2O_3, a major iron ore mineral. A constituent of tropical soils. *alt.* haematite.

hemi- prefix derived from Gk *hēmi*, half.

Hemiascomycetae, Hemiascomycetidae *n.* subclass of unicellular and simple mycelial Ascomycetes including the budding yeasts and leaf-curl fungi, which bear naked asci not enclosed in an ascocarp.

hemibiotroph *n.* fungus that is only partly dependent on a living host. *cf.* biotroph.

hemic *a. appl.* organic soil material in a stage of decomposition intermediate between fibric and sapric.

hemicellulose *n.* any of a diverse group of polysaccharides found in plant cell walls and as storage carbohydrates in some seeds and that contain a mixture of sugar residues including xylose, arabinose, mannose, glucose, glucuronic acid and galactose, and which include xylans, glucomannans, arabinoxylans, xyloglucans.

Hemichordata, hemichordates *n.*, *n.plu.* a phylum of marine, worm-like, coelomate invertebrate animals, which have pharyngeal gill slits and a body divided into three regions.

hemicryptophyte *n.* herbaceous perennial plant in which the perennating parts are at soil level, often protected by the dead leaves.

hemiepiphyte *n.* a plant that does not spend its whole life cycle as a complete epiphyte: either a plant whose seeds germinate on another plant, but which later sends roots to the ground, or a plant that begins life rooted but later becomes an epiphyte.

hemimetabolous *a. appl.* the orders of insects having an incomplete metamorphosis, with no pupal stage in the life history, comprising the orders Orthoptera (crickets, locusts), Dictyoptera (cockroaches), Plecoptera (stoneflies), Dermaptera (earwigs), Ephemeroptera (mayflies), Odonata (dragonflies), Embioptera (foot spinners), Isoptera (termites), Psocoptera (book lice), Anoplura (biting and sucking lice), Thysanoptera (thrips) and Hemiptera (bugs).

hemimorphite *n.* [$Zn_4Si_2O_7(OH)_2.H_2O$] a secondary zinc mineral that, in several deposits, can be a minor ore mineral of this metal. *alt.* calamine (a name also applied to smithsonite *q.v.*).

hemiparasite *n.* (1) an individual that is partly parasitic but that can survive in the absence of its host; (2) a parasitic plant that develops from seeds germinating in the soil rather than in the host body; (3) a parasite that can exist as a saprophyte.

hemipodes *n.plu.* the Gruiformes, an order of small quail-like birds, related to cranes and rails.

Hemiptera, hemipterans *n.*, *n.plu.* order of sucking insects commonly known as bugs, and including water boatmen and pond skaters as well as blood-sucking bugs parasitic on mammals and the sap-sucking aphids, scale insects, leaf-hoppers, mealy bugs and cicadas. Some blood-sucking bugs transmit disease, and the aphids are important vectors of plant viral diseases.

hemisaprophyte *n.* (1) a plant living partly by photosynthesis, partly by obtaining food from humus; (2) a saprophyte that can also survive as a parasite.

hemist *see* histosol.

hemizygous *a. appl.* genetic locus present in only one copy, either as a gene in a haploid organism, or a sex-linked gene in the heterogametic sex (e.g. X-linked genes in human males), or a gene in a segment of chromosome whose partner has been deleted.

hemp *n.* (1) *Cannabis sativa*, an annual herbaceous plant grown for its fibre, also termed hemp, which is used for making cords and strong fabrics; (2) narcotic drugs made from certain varieties of *Cannabis sativa*; (3) kenaf or ambari hemp, the plant *Hibiscus cannabinus* (unrelated to *Cannabis*), which is cultivated for its fibres; (4) Manila hemp, a tough fibre obtained from the leaf stalks of *Musa textilis*.

Henry's law constant (K_H or k_H or H) an equilibrium constant that relates to the dissolution of gases in solvents, a process that may be represented by the general equation $X_{(g)} \rightleftharpoons X_{(solv)}$. Henry's law constant is expressed as $K_H = [X_{(solv)}]/P_{X(g)}$, where P stands for partial pressure measured in atmospheres and the square brackets, [], represent molar concentrations.

HEP hydroelectric power *q.v.*

Hepadnaviridae *n.* family of single-stranded DNA viruses, which includes the hepatitis B virus, with an unusual mode of replication, in which viral DNA is synthesized from an RNA template. Chronic infection with hepatitis B virus has been implicated in the increased incidence of liver cancer in areas where the virus is widespread.

hepatic *a. pert.* (1) like, or associated with the liver; (2) (*bot.*) liverworts. *n.* a liverwort.

Hepaticae, Hepaticopsida Hepatophyta *q.v.*

hepatitis *n.* inflammation of the liver.

Hepatophyta *n.* a division of non-vascular, spore-bearing green plants commonly called liverworts, which with the hornworts and mosses are known as the bryophytes (*q.v.*). Liverworts are small, generally inconspicuous, plants, growing in low clumps on the ground or on rocks, tree bark, etc. The photosynthetic gametophyte is thalloid or leafy, most liverworts having stems with three rows of leaves. The sporophyte is a stalked capsule growing from the gametophyte. The plant is anchored to the ground by fine unicellular rhizoids. Liverworts lack specialized conducting tissue (with a few possible exceptions), a cuticle and stomata, and are the simplest of multicellular plants. *alt.* hepatics.

hepta- prefix derived from Gk *hepta*, seven, and denoting having seven of, or arranged in sevens, etc., as in heptaploid, having seven times the haploid number of chromosomes, heptamerous, having seven of each part, of flowers.

heptachlor *n.* a chlorinated hydrocarbon insecticide related to aldrin and dieldrin. Persistent and toxic, it is now banned in some countries.

heptose *n.* any sugar having the formula $(CH_2O)_7$, e.g. sedoheptulose.

HER heat engine rule. *see* Carnot efficiency.

herb *n.* any seed plant with non-woody green stems.

herbaceous *a.* (1) *appl.* seed plants with non-woody green stems; (2) soft, green, with little woody tissue, *appl.* plant organs.

herbarium *n.* a collection of dried or preserved plants, or of their parts, and the place where they are kept.

herbicide *n.* a chemical that kills plants.

herbivore *n.* an animal that feeds exclusively on plants. *a.* herbivorous.

herbivory *n.* feeding on plants.

herb layer a horizontal ecological stratum of a plant community comprising the herbaceous plants. *alt.* field layer.

herbosa *n.* vegetation composed of herbaceous plants.

herd *n.* group of animals, esp. large herbivores, that feed and travel together.

hereditary *a. appl.* characteristics that can be transmitted from parent to offspring, i.e. characters that are genetically determined.

heredity *n.* (1) the genetic constitution of an individual; (2) the transmission of genetically based characteristics from parents to offspring.

heritability *n.* (1) capacity for being transmitted from one generation to the next; (2) in quantitative genetics can be used in two senses: (i) broad-sense heritability, which is the proportion of phenotypic variation in a particular trait that is attributable to differences in genotype between individuals, (ii) narrow-sense heritability, which is used, e.g., to determine the amount of improvement possible in a particular trait by selective breeding.

heritable *a*. able to be inherited, *appl*. character, disease, trait.

Her Majesty's Inspectorate of Pollution (HMIP) a statutory pollution regulation agency in the UK, covering environmentally damaging industrial processes, whose functions, as of 1996, have been subsumed within the Environment Agency (*q.v.*) and the Scottish Environment Protection Agency *q.v.*

hermaphrodite *n*. (1) an organism with both male and female reproductive organs, as occurs normally in some groups of animals and many plants, *alt*. bisexual; (2) in mammals and some other groups of animals, an individual with a mixture of male and female organs arising as a result of a developmental abnormality, more properly called a pseudohermaphrodite. *a*. hermaphrodite, *alt*. androgynous, bisexual.

heroin *n*. an addictive alkaloid obtained from morphine by acetylation, and which acts as a narcotic.

Herpesviridae, herpesviruses *n.*, *n.plu.* family of DNA viruses including the various herpesviruses, which cause cold sores, genital herpes, etc., and the Epstein–Barr virus, which causes glandular fever, and is also involved in Burkitt's lymphoma in children in Africa and nasopharyngeal carcinoma in China and South-East Asia.

herpetology *n*. that part of zoology dealing with the study of reptiles.

hertz (Hz) *n*. the SI unit for frequency. 1 Hz = 1 cycle per second.

hesperidium *n*. type of indehiscent fruit exemplified by oranges and lemons, formed from a superior, multilocular ovary, having epicarp and mesocarp joined together, with endocarp projecting into the interior as membranous partitions that divide the pulp into segments.

heterecious, heterecism *alt*. spelling of heteroecious, heteroecism *q.v.*

hetero- prefix from Gk *heteros*, other, indicating difference in structure, from different sources, of different origins, containing different components, etc.

Heterobasidiomycetidae, heterobasidiomycetes *n.*, *n.plu.* a group of basidiomycete fungi including the rusts, smuts and jelly fungi.

heterodont *a*. having teeth differentiated for various purposes.

Heterodontiformes *n*. order of selachians, including the hornsharks, having the notochord only partially replaced in extant species.

heteroecious *a*. (1) passing different stages of life history in different hosts; (2) requiring two hosts to complete its life cycle, *appl*. some rust fungi.

heterogametic sex the sex possessing a pair of non-homologous sex chromosomes (e.g. the male, which is XY, in mammals, and the female, which is WZ, in birds), and therefore producing some gametes possessing one type of sex chromosome and some possessing the other.

heterogeneous *a*. consisting of dissimilar parts or composed of a mixture of different components. *cf*. homogeneous.

heterogeneous summation, law of rule that the different independent features of an environmental stimulus (e.g. the shape, size and coloration of eggs) are additive in their effect on an animal's behaviour.

heterogenetic *a*. descended from different ancestral stock.

heterogenous *a*. (1) having a different origin, not originating in the body; (2) heterogeneous *q.v.*

heterogynous *a*. with two types of females.

heterokaryon *n*. a cell containing two (or more) genetically different nuclei, formed naturally in various organisms, such as many fungi, or artificially in culture by the fusion of two animal cells or plant protoplasts. *a*. heterokaryote, heterokaryotic.

heterologous *a*. (1) of different origin; (2) derived from a different species; (3) differing morphologically, *appl*. alternation of generations.

heterophagous *a*. having very immature young.

Heteroptera *n*. in some classifications, an order of insects including the water boatmen, capsids and bed bugs.

heterosis *n*. (1) cross-fertilization; (2) hybrid vigour *q.v.*

heterostyly *n*. condition in which individual plants within a species differ in the length of their styles, thus preventing self-fertilization.

heterotherm ectotherm *q.v.*

heterotroph *n.* any organism only able to utilize organic sources of carbon, nitrogen, etc., as starting materials for biosynthesis, using either sunlight as a primary energy source (photoheterotrophs: a few bacteria and algal flagellates) or energy from chemical processes (chemoheterotrophs: all animals, fungi, most bacteria, some parasitic plants). *a.* heterotrophic.

heterozygosity *n.* proportion of heterozygotes for a given locus in a population.

heterozygote advantage the case where the heterozygote for a given pair of alleles is of superior fitness than either of the two homozygotes.

heterozygote *n.* a heterozygous organism or cell. *alt.* hybrid.

heterozygous *a. appl.* diploid organism that has inherited different alleles (of a given gene) from each parent, i.e. carries different alleles at the corresponding sites on homologous chromosomes, *appl.* also to cells or nuclei from such an organism. *alt.* hybrid.

heulandite *n.* $(Na,Ca)_{4-6}Al_6(Al,Si)_4Si_{26}O_{72}.24H_2O$, one of the zeolite minerals *q.v.*

Hevea brasiliensis the rubber-tree.

hex- (*chem.*) the prefix used to indicate the presence of six carbon atoms in a chain in an organic compound (e.g. hexane $CH_3CH_2CH_2CH_2CH_2CH_3$) or radical.

hexa- prefix derived from Gk *hex*, six, signifying having six of, arranged in sixes, etc.

Hexactinellida *n.* a class of Porifera, the glass sponges or hexactinellid sponges, typically radially symmetrical with a skeleton of large six-rayed spicules of silica, often fused to form a three-dimensional network.

hexaploid *a.* having six sets of chromosomes. *n.* an organism having six times the haploid chromosome number.

hexapod *a.* having six legs. *n.* an insect.

Hexapoda older name for Insecta *q.v.*

hexose *n.* a monosaccharide containing six carbon atoms (formula $C_6H_{12}O_6$), e.g. glucose, fructose, galactose, mannose.

Hf symbol for the chemical element hafnium *q.v.*

HFAs hydrogen fluoroalkanes *q.v.*

Hg symbol for the chemical element mercury *q.v.*

hibernaculum *n.* (1) structure that protects the embryo or growing shoot tip in plants, such as a bud; (2) shelter for an overwintering animal.

hibernal *a.* of the winter.

hibernating glands former term for brown adipose tissue *q.v.*

hibernation *n.* the condition of passing the winter in a resting state of deep sleep, when metabolic rate and body temperature drop considerably. Only a few small mammals, e.g. some rodents, hedgehogs, bats and other small insectivores, undergo a 'true' hibernation. Obligate hibernators enter hibernation spontaneously as the result of a circannual behavioural rhythm. Facultative hibernators enter hibernation when food becomes scarce and temperatures drop below a certain level. Related conditions include winter torpor in reptiles and winter lethargy in larger mammals, e.g. bears, badgers, skunks and racoons. *see also* aestivation.

hiddenite *n.* a variety of spodumene (*q.v.*) used as a gemstone.

hidrosis *n.* sweating, perspiration.

hiemal *a. pert.* winter, *appl.* aspect of a community.

hiemilignosa *n.* monsoon forest composed of small-leaved trees and shrubs that shed their leaves in the dry season.

hierarchy *n.* (1) *see* dominance systems; (2) a natural classification system in which organisms are grouped according to the number of characteristics they have in common and ranked one above another.

high-energy bond a somewhat misleading term denoting any chemical linkage whose breakage releases a large amount of free energy, such as the bond between the two most terminal phosphate groups in ATP.

high grading type of selective cutting in which the most valuable trees in an area are removed without any regard for the consequent destruction of the surrounding forest and without making any provision for regeneration. A practice most prevalent in the harvesting of valuable tropical hardwoods.

high-level radioactive waste radioactive waste produced by the nuclear industry, which may be spent fuel or wastes derived from fuel reprocessing. It is produced in small amounts, is hot and intensely radio-active, contains radioactive elements with very long half-lives and must be isolated from the environment for up to 1000 years. At present it is stored in tanks, but there are techniques under development to incorporate it into a solid glass-like material. *see also* intermediate-level radioactive waste, low-level radioactive waste.

high-quality energy refers to forms of energy and sources of energy that have a high capacity to do useful work, such as high-temperature heat, electricity, coal, oil, petrol, sunlight and nuclear energy. *cf.* low-quality energy.

high-rank coal coal that has been greatly affected by the action of burial or tectonic activity, e.g anthracite. *cf.* low-rank coal.

highs anticyclones *q.v.*

high-temperature reactor (HTR) a type of fission nuclear reactor that utilizes carbon–silicon carbide clad fuel granules of enriched uranium carbide or oxide. It has helium as the coolant and a graphite moderator.

high-tide platform a type of shore platform (*q.v.*) that develops near the high-tide mark, mainly as a result of water-layer weathering *q.v.*

highway motorway *q.v.*

high-yielding varieties (HYVs) *see* Green Revolution.

hippomorphs *n.plu.* a group of the Perissodactyla including the horses (family Equidae) and the extinct brontotheres.

Hippuridales *n.* order of dicots, land, marsh or water plants, comprising the families Gunneraceae (gunnera), Haloragaceae and Hippuridaceae (mare's tail).

Hirudinea, Hirudinoidea *n.* class of carnivorous or ectoparasitic annelids, commonly called leeches, which have 33 segments, circumoral and posterior suckers and usually no chaetae.

His histidine *q.v.*

histidine (His, H) *n.* amino acid, constituent of protein, probably non-essential in the human diet, precursor of histamine.

histogen *n.* zone of tissue in apical meristems in plants from which new tissue develops.

histogram *n.* type of graphical representation in which data are grouped in some

way (e.g. per month, for rainfall data) and represented as columns, the height of each column being the amount or frequency of the data item (e.g. millimetres or centimetres of rainfall) in the group.

histology *n.* the study of the detailed structure of living tissue, using microscopy.

histosol *n.* an order of the USDA Soil Taxonomy system consisting of organic soils that develop in water-saturated environments. Four suborders are recognized based on the state of decomposition of the organic material present: fibrist (slight decomposition), folist (mass of leaves in early stage of decomposition), hemist (intermediate stage of decomposition) and saprist (most advanced stage of decomposition).

H-layer *see* mor humus.

HMIP Her Majesty's Inspectorate of Pollution *q.v.*

Ho symbol for the chemical element holmium *q.v.*

hoar frost term used for ice crystals formed on surfaces when nocturnal cooling of the Earth (brought about by terrestrial radiation) causes the temperature of the air in contact with the surface to drop below its dew-point. Note that hoar frost only occurs if the dew-point is <0 °C (if the dew-point is >0 °C, dew will result).

hogback *n.* a long, steep-sided, roughly symmetrical ridge, formed by the differential erosion of steeply inclined (*ca.* 45°) rock strata.

Holarctic *a. appl.* or *pert.* (1) (*zool.*) Holarctica (*q.v.*); (2) (*bot.*) the floral realm consisting of the northern hemisphere south to the Tropic of Cancer and consisting of eight floral regions: Arctic, Atlantic North American, Euro-Siberian, Hudsonian, Irano-Turanian, Mediterranean, Pacific North American and Sino-Japanese (*see* individual entries).

Holarctica *n.* zoogeographical region comprising the Nearctic and Palaearctic regions.

holdfast *n.* adhesive region by which an organism can attach itself to a surface, used esp. for the adhesive disc by which members of the brown seaweeds attach themselves to rocks.

holistic *a. appl.* explanations that attempt to explain complex phenomena in terms of the properties of the system as a whole. *n.* holism. *cf.* reductionist.

holmium (Ho) *n.* a lanthanide element (at. no. 67, r.a.m. 164.93), a soft silver-coloured metal when in elemental form.

holobenthic *a.* living on sea bottom or in depths of sea throughout life.

Holocene *n.* recent geological epoch following Pleistocene, began *ca.* 10 000 years ago. *alt.* Recent.

holocephalian *a. pert.* cartilaginous fishes of the subclass Holocephali, the rabbit fishes, with crushing teeth, a whip-like tail and an operculum covering the gills. *alt.* chimaeras, rat-fish.

holometabolous *a. appl.* the orders of insects that undergo full metamorphosis, with a four-stage life history (egg, larva, pupa, adult), and comprising the orders Neuroptera (alderflies, lacewings), Mecoptera (scorpion flies), Trichoptera (caddis flies), Lepidoptera (butterflies and moths), Coleoptera (beetles), Strepsiptera, Hymenoptera (ants, bees and wasps), Diptera (two-winged flies) and Siphonaptera (fleas).

holoplankton *n.* organisms that complete their life cycle in the plankton.

Holostei, holosteans *n., n.plu.* group of bony fishes present from the Mesozoic but now represented only by the garpike and bowfin.

Holothuroidea, holothurians *n., n.plu.* class of sausage-shaped echinoderms commonly called sea cucumbers, having minute skeletal plates embedded in the fleshy body wall.

Holotrichia, holotrichans *n., n.plu.* a group of ciliate protozoans having no obvious zone of composite cilia around the mouth, and swimming by cilia distributed all over the body.

holotype type specimen *q.v.*

holozoic *a.* obtaining food in the manner of animals, by ingesting solid food material and then digesting it.

home range territory *q.v.*

homeostasis *n.* (1) maintenance of the constancy of internal environment of the body or part of body; (2) maintenance of equilibrium between organism and environment; (3) the balance of nature, *see* ecological balance. *a.* homeostatic. *alt.* homoeostasis.

homeotely *n.* evolution from homologous parts, but with less close resemblance.

homeothermic homoiothermic *q.v.*

homing *n.* ability of animals, esp. birds, to navigate their return to their place of origin over great distances.

hominid *n.* a member of a human (*Homo* spp.) or human-like (*Australopithecus*) species characterized by upright posture and other features distinguishing it from the ape lineage (the pongids).

Hominidae, hominids *n., n.plu.* the family of Primates that comprises true humans (*Homo* spp.) and human-like hominids (*Australopithecus* spp.).

hominoid *a.* having similarities to humans, *appl.* African apes and various ape-like fossils, as well as early hominids.

Hominoidea, hominoids *n., n.plu.* the superfamily of primates that includes the families Hominidae (humans and human-like hominids), Pongidae (great apes) and Hylobatidae (gibbons).

homiothermic homoiothermic *q.v.*

homo- prefix from Gk *homo*, the same: indicating similarity of structure, from the same source, of similar origins, containing similar components, etc.

Homo the genus of true humans, including several extinct forms (*H. habilis, H. erectus, H. neanderthalensis*) and modern man, *H. sapiens*, who are or were primates characterized by completely erect stature, bipedal locomotion, reduced dentition and above all by an enlarged brain.

Homobasidiomycetae, homobasidiomycetes *n., n.plu.* the mushrooms and toadstools, bracket fungi, coral fungi, puffballs, earthstars, stinkhorns and bird's nest fungi: i.e. fungi producing their basidiospores on typically club-shaped, non-septate basidia (holobasidia). *alt.* Holobasidiomycetae.

homodont *a.* having teeth all alike, not differentiated.

homoeologous *a.* partly homologous, *appl.* sets of chromosomes in a polyploid species that are not identical to and not of the same origin as the other chromosome sets in the genome.

homoeostasis homeostasis *q.v.*

homogametic sex the sex possessing a pair of homologous sex chromosomes and therefore producing gametes all possessing the same sex chromosome. In mammals it is the female, which is XX. In birds, reptiles and lepidopterans the homogametic sex (ZZ) is the male.

homogenetic *a*. having the same origin.

homogeneous, homogenous *a*. composed of identical or similar components.

homoiosmotic *a. appl.* organisms with constant internal osmotic pressure.

homoiothermic *a. appl.* animals that maintain a more-or-less constant body temperature regardless of external temperature variations, such as birds and mammals. Although virtually all homoiotherms are also endothermic (*q.v.*), the two terms are not synonyms and describe different aspects of thermoregulation. *n*. homoiotherm. *cf.* poikilothermic.

homokaryon *n*. a hypha or mycelium having more than one haploid nucleus of identical genetic constitution per cell. *a*. homokaryotic.

homologous *a*. (1) resembling in structure and origin; *appl.* (2) chromosomes in a diploid cell that contain the same sequence of genes but are derived from different parents; (3) genes determining the same character; (4) DNAs of identical or very similar base sequence; (5) structures having the same phylogenetic origin but not necessarily the same final structure or function, e.g. wings and legs in insects, which are examples of the phenomenon of homology.

homologous series (*chem.*) a group of compounds that all have stoichiometries that can be described by a general formula. The series can be generated, in principle, by the successive addition of a given group of atoms to the simplest member of the series. Many of the properties of the members of any one homologous series will vary incrementally each time an addition is made of the extra group of atoms. For example, alkanes have the general formula C_nH_{2n+2}. They can be generated from the simplest member, methane (CH_4), by the successive addition of CH_2 groups.

homologue *n*. any structure of similar evolutionary and developmental origin to another structure, but serving different functions. *alt.* homolog.

homology *n*. resemblance by virtue of common descent, as structures, DNA sequences, even though they may now have different functions.

homonym *n*. a name that has been given to two different species. When a case is discovered the second named species must be renamed.

homoplasy *n*. resemblance in form or structure between different organs or organisms due to evolution along similar lines rather than common descent. *alt.* homoplasty, convergent evolution.

Homoptera *n*. group of insects, including the plant bugs, aphids, cicadas and scale insects.

homotaxis, homotaxy *n*. similar assemblage or succession of species or types in different regions or strata, not necessarily contemporaneous. *a*. homotaxial.

homozygote *n*. a homozygous organism or cell.

homozygous *a. appl.* diploid organism that has inherited the same allele (of any particular gene) from both parents, i.e. carries identical alleles at the corresponding sites on homologous chromosomes, also *appl.* cells or nuclei from such an organism. *n*. homozygosity. *cf.* heterozygous.

honest behaviour behaviour that conveys the individual's real intentions to another individual.

honey bee generally refers to *Apis mellifera*, the hive bee, *see also* bees.

honeydew *n*. (1) sugary exudate on leaves of many plants; (2) sweet liquid secreted by aphids.

honey guides nectar guides *q.v.*

hoof and horn meal byproduct of the meat-processing industry (containing 7–16% nitrogen) used as a slow-release nitrogen fertilizer.

hookworms *n.plu.* parasitic nematode worms that cause severe disease in humans, and including *Ancylostoma duodenale* and *Necator americanus*. Common and widespread in tropical areas, the larvae enter the body through the skin, and the adult worm lives in the intestine, abrading the intestinal walls and eventually causing severe anaemia and general debilitation.

hordein *n*. a simple storage protein present in barley grains.

hordeivirus group plant virus group containing rigid rod-shaped single-stranded RNA viruses, type member barley stripe mosaic virus. They are multicomponent viruses in which three genomic RNAs are encapsidated in different virus particles.

horizon *n.* (1) (*soil sci.*) soil layer of more-or-less well-defined character; (2) (*geol.*) a layer of deposit characterized by definite fossil species and formed at a definite time.

hormone weedkillers weedkillers containing high concentrations of synthetic auxins (a group of plant growth hormones) e.g. 2,4-D and 2,4,5-T.

horn *n.* (1) (*biol.*) a hollow projection on the head of many ruminants, consisting of layers of keratinized epidermis (outer layer of skin) laid down on a bony base; (2) (*geog.*) a pyramidal mountain peak formed by the intersection of three or more developing cirques *q.v.*

hornblende *n.* $(Ca,Na)_{2-3}(Mg,Al,Fe)_5$-$Si_6(Si,Al)_2O_{22}(OH)_2$, a widespread rock-forming mineral with a wide range of composition. One of the amphibole minerals *q.v.*

hornblende lamprophyre a fine- to medium-grained, generally porphyritic (sometimes granular), igneous rock. Phenocrysts are of hornblende. These are in a ground-mass of hornblende and feldspars (sodic plagioclase or orthoclase).

hornfels *n.* a generally fine- to medium-grained rock that may be porphyroblastic (*q.v.*) or poikiloblastic (*q.v.*), generated within a contact aureole by thermal metamorphism. *plu.* hornfelses.

horn silver chlorargyrite *q.v.*

hornworts *n.plu.* common name for members of the plant division Anthocerophyta, a group of small spore-bearing non-vascular green plants with a thalloid gametophyte and rosette-like habit of growth, whose cells typically contain a single large chloroplast with a pyrenoid, and which often carry symbiotic photosynthetic cyanobacteria in their intercellular spaces. The sporophyte is typically an upright elongated sporangium on a stalk (the foot) growing from the gametophyte. *alt.* Anthoceratopsida, Anthocerotae, Anthocerotales.

horny corals another name for the gorgonians *q.v.*

horsehair worms Nematomorpha *q.v.*

horse latitudes belts of high pressure located in the subtropics (latitudes 30° N

and 30° S), characterized by calm weather conditions.

horseshoe crabs common name for the Xiphosura, also called king crabs, a group of aquatic arthropods with affinities to the arachnids rather than the crustaceans, and often placed in the separate class Merostomata. They have a heavily chitinized body with the cephalothorax covered by a horseshoe-shaped carapace.

horst *n.* a raised fault block elevated between a pair of approximately parallel faults.

horsetails *n.plu.* common name for the Sphenophyta *q.v.*

horticulture *n.* intensive cultivation of fruit, vegetables and flowers, i.e. horticultural crops, esp. for commercial production.

host *n.* (1) any organism in which another spends part or all of its life, and from which it derives nourishment or gets protection; (2) the recipient of grafted or transplanted tissue.

host rock country rock *q.v.*

hot dry rock technique technique (still at the experimental stage) used to harness geothermal energy. In this, water is injected into previously fractured rocks deep within the Earth's crust and the resultant superheated steam extracted for use as an energy source.

hot spots (*geol.*) areas of anomalous volcanic activity not associated with subduction zones (*q.v.*) or mid-oceanic ridges (*q.v.*). Hot spots occur where very hot mantle is present at the base of the lithosphere.

hot springs springs with a water temperature greater than 20 °C, associated with areas of geothermal activity. *alt.* thermal springs.

HPLC high-pressure liquid chromatography.

HTR high-temperature reactor *q.v.*

Hudsonian Floral Region part of the Holarctic Realm comprising North America from southern Alaska and the northern shores of the Great Lakes north to northern Alaska and Labrador.

humic acid fraction that precipitates from a solution of humus in weak alkali on addition of acid.

humicolous *a.* living in the soil. *n.* humicole.

humidity *n.* measure of the amount of water vapour present in a specified portion of the atmosphere.

humification *n.* the production of humus in the soil by the action of microorganisms on plant and animal residues.

humin *n.* black insoluble residue left when humus is dissolved in dilute alkali.

humite *n.* $Mg(OH,F)_2.3Mg_2SiO_4$, a mineral of the humite series *q.v.*

humite series a group of nesosilicate minerals of the formula $Mg(OH,F)_2.1-4Mg_2SiO_4$. The minerals in this group are clinohumite [$Mg(OH,F)_2.4Mg_2SiO_4$], humite [$Mg(OH,F)_2.3Mg_2SiO_4$], chondrodite [$Mg(OH,F)_2.2Mg_2SiO_4$] and norbergite [$Mg(OH,F)_2.Mg_2SiO_4$].

humod *n.* a suborder of the order spodosol of the USDA Soil Taxonomy system, consisting of spodosols with an illuvial horizon rich in humus.

humult *n.* a suborder of the order ultisol of the USDA Soil Taxonomy system, consisting of ultisols with high or very high organic matter contents.

Humulus lupulus the hop plant, whose flowerheads are used as a flavouring for beer.

humus *n.* stable pool of organic matter in soil that results from activities of decomposer organisms on plant and animal residues, esp. on complex materials such as lignin, but is itself degraded only very slowly. It is a black organic material of complex composition consisting of humic acids, fulvic acids and humin.

hunter–gatherer individual who subsists by hunting wild animals, catching fish and collecting berries, insects, wild plants, etc., e.g. Pygmies of central Africa and the Australian Aborigines.

hurricane *n.* violent tropical storm consisting of a whirling vortex of enormous clouds, heavy precipitation and very strong winds encircling a central low-pressure area, calm and cloudless in character (known as the eye). *alt.* tropical cyclone, typhoon, willy-willy (Australian term).

husk *n.* (1) covering found on certain seeds and fruits; (2) parasitic bronchitis, a disorder of cattle (mainly) caused by lungworms and characterized by severe coughing.

hybrid *n.* (1) any animal or plant resulting from the crossing of two pure-breeding lines; (2) heterozygote (*q.v.*); (3) any macromolecule (esp. DNA) composed of two or more portions of different origins. *v.* hybridize. *a.* hybrid.

hybrid cline the serial arrangement of characters or forms produced by crossing species.

hybridization *n.* (1) formation of a hybrid (*q.v.*); (2) state of being hybridized; (3) cross-fertilization. *see also* DNA hybridization.

hybrid sterility sterility in an individual arising from the fact that it is a hybrid.

hybrid swarms populations consisting of descendants of species hybrids, as at borders between geographical areas populated by these species.

hybrid vigour the phenomenon often seen in crosses between two pure-breeding lines of plants or animals, that the hybrid is more vigorous than either of its parents, presumably owing to increased heterozygosity. *alt.* heterosis. *see also* overdominance.

hybrid zone (*biol.*) a geographic area in which two populations, once separated by a geographical barrier, hybridize after the barrier has broken down, in the absence of reproductive isolation.

hydathode *n.* an epidermal structure in plants specialized for secretion, or for exudation, of water.

hydrargillite gibbsite *q.v.*

hydrate *n.* compound incorporating combined water, often *appl.* salts having water of crystallization. *a.* hydrated.

hydration *n.* (1) formation of a hydrate (*q.v.*); (2) one of the processes of chemical weathering.

hydraulic action *see* fluvial erosion.

hydric *a.* having an abundant supply of moisture.

hydrobiology *n.* the study of aquatic plants and animals and their environment.

hydrobiont *n.* an organism living mainly in water.

hydrocarbon *n.* any organic compound consisting solely of hydrogen and carbon atoms that are covalently bound together.

Hydrocharitales *n.* order of aquatic herbaceous monocots comprising the family Hydrocharitaceae (frog's-bit).

hydrochlorofluorocarbons (HCFCs) *n.plu.* covalent compounds that contain carbon, hydrogen, fluorine and chlorine, e.g. HCFC-21 ($CHFCl_2$). The potential of HCFCs for destroying stratospheric ozone is approximately one-fifth that of those CFCs that do not contain hydrogen, as the bulk of the chlorine contained in HCFCs is liberated into the troposphere, not the stratosphere. HCFCs are therefore viewed as less damaging alternatives. Confusingly, some HCFCs have been labelled as HFAs, *see* hydrogen fluoroalkanes.

hydrochoric *a.* (1) dispersed by water; (2) dependent on water for dissemination. *n.* hydrochory.

hydrocoles *n.plu.* animals living in water or a wet environment.

hydroelectric power (HEP) electricity generated from flowing water via turbines. Suitable rivers may be used directly as the energy source but in most cases, HEP production involves the damming of a river and the flooding of a valley to provide a reservoir. The flow of water from the reservoir through a pipe in the dam is used to drive the turbine and can be closely regulated.

hydrofuge *a.* water-repelling.

hydrogen (H) *n.* the lightest element (at. no. 1, r.a.m. 1.008). A colourless odourless flammable gas in the free state (H_2). A constituent of all organic molecules, hydrogen is one of the essential elements for living organisms. It has many industrial uses, and has been proposed as a liquid fuel. *see also* deuterium, pH, tritium.

hydrogenation *n.* process whereby hydrogen is made to chemically combine with another substance.

hydrogenation of coal process used in the manufacture of artificial mineral oil from coal in which hydrogen is made to combine chemically with the carbon present in the coal to produce hydrocarbons.

hydrogen bomb *see* nuclear bomb.

hydrogen bond the attraction between an electronegative atom with a lone pair of electrons and a hydrogen atom that is covalently bonded to another electronegative atom, e.g. –O . . . H–N–. It is the strongest of the weak intra- and intermo-lecular attractions and is of great importance in biology as it is one of the main forces governing, e.g., the folding of a protein chain into its final functional three-dimensional structure, and the interactions of proteins with each other and with small molecules such as enzyme substrates. Hydrogen bonds also hold together the two DNA strands of a DNA molecule. Hydrogen bonding between water molecules is responsible for the high melting and boiling points of water (compared with those of, e.g., methane) and its high surface tension.

hydrogen fluoroalkanes (HFAs) covalent compounds that contain carbon, hydrogen and fluorine, e.g. HFA-134 (CHF_2CHF_2). Unlike CFCs, these compounds do not contain chlorine and do not pose a threat to stratospheric ozone. They are therefore viewed as potential alternatives to CFCs. Confusingly, some of the compounds labelled as HFAs are in fact HCFCs, *see* hydrochlorofluorocarbons.

hydrogen ion concentration *see* pH.

hydrogen sulphide (H_2S) a highly toxic, colourless gas with a smell of rotten eggs, produced naturally by, e.g., the anaerobic decomposition of the organic sulphides present in organic matter.

hydrograph *n.* a graph showing the discharge of a river over a period of time.

hydroid *n.* one of the forms of individuals in the Hydrozoa, a class of solitary and colonial coelenterates, having a hollow cylindrical body closed at one end and with a mouth at the other surrounded by tentacles, *alt.* polyp.

hydrological cycle water cycle *q.v.*

hydrologic sequence catena *q.v.*

hydrology *n.* the study of water, its properties, occurrence, circulation and distribution.

hydrolysis *n.* the addition of the hydrogen and hydroxyl ions of water to a molecule, with its consequent splitting into two or more simpler molecules, with the creation of a new element–oxygen bond.

hydrometallurgy *n.* type of chemical processing involving the use of water in the extraction of metals from their ores.

hydrometeors precipitation *q.v.*

hydromorphic *a. appl.* soils containing excess water.

hydronasty *n.* plant movement induced by changes in atmospheric humidity.

hydronium ion hydroxonium ion *q.v.*

hydrophilic *a.* water-attracting or attracted to water, as polar groups on compounds. Hydrophilic substances tend to dissolve in water.

hydrophilous *a.* pollinated by the agency of water. *n.* hydrophily.

hydrophobia rabies *q.v.*

hydrophobic *a.* water-repelling or repelled by water, as non-polar groups on compounds, which tend to aggregate, excluding water from between them.

hydrophyte *n.* (1) an aquatic plant living on or in the water; (2) aquatic perennial herbaceous plant in which the perennating parts lie in water.

hydroponics *n.* cultivation of plants without soil in nutrient-rich water, which is usually irrigated over some inert medium such as sand.

hydropower *n.* the harnessing of flowing waters to provide energy in a useful form. Originally, hydropower was used in water mills to grind corn, etc. Today, it is used almost exclusively for the production of electricity, *see* hydroelectric power.

hydrosere *n.* a plant succession originating in a wet environment.

hydrosphere *n.* the portion of the planet that is water, i.e. the oceans, rivers, lakes, streams, etc., and including soil water.

hydrostatic equilibrium (*clim.*) the balance between the vertical component of the pressure gradient force (*q.v.*) and the force of gravity.

hydrotaxis *n.* a movement or locomotion in response to the stimulus of water.

hydrothermal *a.* (*geol.*) *pert.* conditions that involve hot water as an active agent in the formation of new rocks and/or minerals by dissolution and precipitation, *appl.* rocks and minerals so formed, the solution from which these materials precipitate and the processes involved in their formation.

hydrothermal deposition (1) the formation of any mineral by precipitation from a hydrothermal solution; (2) the enrichment process responsible for orebody formation in which the movement of hot water through rocks results in the concentrated deposition of valuable minerals.

hydrothermal vent community community of organisms living around volcanic vents in the sea-floor at great depths (hydrothermal vents), in which the primary producers include chemoautotrophic sulphide-oxidizing bacteria that use the energy of sulphide oxidation to fix CO_2, and which include free-living species (e.g. *Beggiatoa*) and intracellular symbionts living in giant vestimentiferan tubeworms, such as *Riftia pachyptila*.

hydrotropic *a. appl.* the curvature of a plant organ towards a greater degree of moisture.

hydroxide *n.* a compound that contains either the hydroxide ion (OH^-) or the hydroxyl (-OH) group.

hydroxide ion the OH^- ion.

hydroxonium ion (*chem.*) the H_3O^+ ion that is formed when water acts as a Brønsted–Lowry base, accepting a proton from a Brønsted–Lowry acid. *alt.* hydronium ion.

hydroxyl *n.* (*chem.*) (1) the -OH group, found in all alcohols (*q.v.*), which consists of an oxygen atom bonded to a hydrogen atom; (2) the OH free radical. This is a highly reactive oxidizing species that is involved in many of the tropospheric transformations of reduced atmospheric species (including pollutants), particularly during daylight.

hydroxylapatite *see* apatite.

hydroxyl radical (*chem.*) the OH group. The use of this term usually implies that this group is neither bonded to anything nor charged, in which case the phrase 'hydroxyl free radical' is more precise but a little old-fashioned.

Hydrozoa, hydrozoans *n., n.plu.* class of coelenterates often having two body forms, hydroid (polyp) and medusa, generally occurring as different stages of the life cycle. They include solitary forms such as *Hydra*, branching colonial forms, and the siphonophores such as the Portuguese man o' war, which are colonies of several different types of modified polyps and medusae.

hygric *a.* (1) humid; (2) tolerating, or adapted to humid conditions.

hygrokinesis *n.* movement induced by a change in humidity.

hygrometer *n.* instrument used for measuring atmospheric humidity.

hygrophilic, hygrophilous *a.* inhabiting moist or marshy places.

hygrophyte *n.* plant that thrives in plentiful moisture, but is not aquatic.

hygroscopic *a.* (1) sensitive to moisture; (2) absorbing water.

hygroscopic nuclei (*clim.*) tiny atmospheric particles, such as dust and salt, upon which water vapour condenses to produce clouds, fog and precipitation. *cf.* freezing nuclei. *alt.* condensation nuclei.

hygrotaxis *n.* movement in response to moisture or humidity.

hygrotropism *n.* a plant growth movement in response to moisture or humidity.

Hymenomycetes *n.* a group of basidiomycete fungi bearing their basidia in a well-defined layer (hymenium) that becomes exposed while the basidia are still immature. It comprises, in modern classifications, those fungi commonly called mushrooms and toadstools and bracket fungi.

Hymenoptera, hymenopterans *n., n.plu.* order of insects, including solitary and social species, comprising the ants, bees and wasps. They have two pairs of wings, and many have a pronounced waist between the second and third abdominal segments. Males are haploid and females diploid, males developing from unfertilized ova. In colonial forms, a colony usually contains one reproductive female (the queen), sterile female workers, a few reproductive males and (in ants) sterile soldiers.

Hyp hydroxyproline *q.v.*

hypabyssal rocks intrusive igneous rocks formed at relatively shallow depths. These largely medium-grained rocks are found in minor intrusions such as dykes and sills.

hyperactivity *n.* abnormal activity in a child that is characterized by aggressive behaviour, restlessness and insomnia and may be caused by lead poisoning or food allergies.

hyperosmotic *a. appl.* a solution of higher osmotic concentration than a given reference solution.

hyperparasite *n.* an organism that is a parasite of, or in, another parasite.

hyperphagia *n.* eating disorder in which an animal consumes excessive amounts of food and also ingests other non-food substances.

hypersthene *n.* (Mg,Fe)SiO$_3$, one of the pyroxene minerals (*q.v.*) of the orthopyroxene subgroup.

hypertely, hypertelia *n.* (1) excessive imitation of colour or pattern, being of problematical utility; (2) overdevelopment of canines of babirusa, an East Indian pig, the male of which has four large tusks.

hyperthermia *n.* a rise in body temperature above normal, which is used adaptively by some animals living in hot climates as a water-conserving mechanism.

hypertonic *a.* having a higher osmotic pressure than that of another fluid, so that if the two solutions were separated by a semipermeable membrane, water would flow into the hypertonic solution from the other.

hypertrophic *a. appl.* (1) waters grossly enriched with plant nutrients; (2) a structure that arises from excessive growth.

hypha *n.* a tubular filament that is the basic structural unit of a mycelium, the vegetative growth phase of fungi that is produced by hyphal extension and branching. Hyphae may comprise continuous tubes of multinucleate protoplasm or may be partially or completely subdivided along their length by transverse partitions (septa) into uninucleate or binucleate compartments. Fungal hyphae are covered by a rigid cell wall, which in most cases contains chitin as well as or instead of cellulose. The acellular filaments produced by the prokaryotic actinomycetes are also known as hyphae, as sometimes are similar filamentous vegetative structures in algae. *plu.* hyphae. *a.* hyphal.

Hyphochytridiomycota, Hyphochytridiomycetes, hyphochytrids *n., n., n.plu.* phylum of freshwater protists commonly known as water moulds and formerly classified as fungi. They are parasitic on algae or fungi or saprobic on plant and insect debris, growing as fine threads and producing motile zoospores, each with one anterior tinsel flagellum.

hypo- prefix from the Gk *hypo*, under. In anatomical terms, often denoting situated under, in physiological and biochemical terms denoting a decrease in.

hypobenthos *n.* the fauna of the sea bottom below 1000 m.

hypocalcaemia *see* milk fever.

hypogeal, hypogean *a.* (1) living or growing underground; (2) *appl.* germination when cotyledons remain underground.

hypogenous *a.* growing on the undersurface of anything.

hypolimnion *n.* the water between the thermocline and bottom of lakes.

hypomagnesaemia *see* staggers.

hyponasty *n.* the state of growth in a flattened structure when the underside grows more vigorously than the upper.

hyponeuston *n.* organisms swimming or floating immediately under the water surface.

hyponym *n.* (1) generic name not founded on a type species; (2) a provisional name for a specimen.

hypo-osmotic *a. appl.* a solution of lower osmotic concentration than a given reference solution.

hypoplankton *n.* plankton found near the bottom of a body of water.

hyporheic zone zone around a river, esp. those with gravel beds, in which river water and its microflora and fauna extends as groundwater throughout the surrounding land.

hypothermia *n.* a drop in body temperature below normal limits, which leads to death if an external heat source is not applied and which is usually distinguished from the decreased body temperature that occurs in hibernating animals.

hypothesis *n.* a proposition constructed to explain observed phenomena and that can be tested by experiment.

hypotonic *a.* having a lower osmotic pressure than that of another fluid, so that if the two solutions were separated by a semipermeable membrane, water would flow from the hypotonic solution to the other.

hypoxia *n.* low levels of oxygen. *a.* hypoxic.

hyracoids *n.plu.* a group of placental mammals, including the hyrax, which have a rodent-like body and skull, but digits over a pad and bearing nails like elephants.

hysteresis *n.* (1) a lag in one of two associated processes or phenomena; (2) lag in adjustment of external form to internal stresses.

hythergraph *n.* plot of mean monthly temperature against mean monthly precipitation (or humidity).

HYVs high-yielding varieties. *see* Green Revolution.

Hz hertz *q.v.*

I

I (1) symbol for the chemical element iodine (*q.v.*); (2) inosine (*q.v.*); (3) isoleucine (*q.v.*); (4) Morisita's index of dispersion *q.v.*

IAA indole-3-acetic acid, a natural plant hormone. *see* auxin.

IAEA International Atomic Energy Agency *q.v.*

IAMS International Association of Microbiological Societies.

IAN indole acetonitrile, a naturally occurring auxin.

IAPT The International Association of Plant Taxonomy.

IBP International Biological Programme *q.v.*

IBPGR International Board for Plant Genetic Resources *q.v.*

IBRD International Bank for Reconstruction and Development *q.v.*

ICBN International Code of Botanical Nomenclature.

ICBP International Council for Bird Preservation.

ice age *n.* any period in the geological past during which ice covered considerably more of the surface of the globe than it currently does.

Ice Age Pleistocene *q.v.*

iceberg *n.* large portion of a glacier or ice sheet that has become detached and carried out to sea.

ice caps ice sheets smaller than 50 000 km^2.

ice-crystal process (*clim.*) process during which ice crystals grow at the expense of water droplets in a cloud containing both (mixed cloud). This process is responsible for the formation of most rain and snow in the mid-latitudes. *alt.* Bergeron–Findeisen process. *cf.* coalescence process.

ice fog *see* steam fog.

ice sheets the largest of the glaciers, e.g. the Antarctic ice sheet. These extend outwards from central domes, covering areas in excess of 50 000 km^2. When they occur floating offshore but still attached to land, they are known as ice shelves.

ice shelves *see* ice sheets.

ichneumon flies insects of the Ichneumonidae, a family of hymenopterans that are parasitoids, laying their eggs in the larvae of other insects, especially butterflies and moths.

ichthyic, ichthyoid *a. pert.*, characteristic of, or resembling fishes.

ichthyofauna *n.* the fishes of a particular region, area or habitat.

ichthyology *n.* the study of fishes.

Ichthyosauria, ichthyosaurs *n., n.plu.* group of Mesozoic aquatic reptiles having spindle-shaped bodies with fins and fin-like limbs.

ichthyostegalians, ichthyostegids *n.plu.* group of primitive extinct amphibians from the Devonian–Carboniferous, having many fish-like characteristics and sometimes considered to be primitive labyrinthodonts.

ICNB International Committee on Nomenclature of Bacteria.

ICNV International Committee on Nomenclature of Viruses.

iconotype *n.* a representation, drawing or photograph of a type specimen.

ICRP International Commission on Radiation Protection *q.v.*

ICSB International Committee on Systematic Bacteriology.

ICSU International Council of Scientific Unions *q.v.*

icterus jaundice *q.v.*

ICTV International Committee on Taxonomy of Viruses.

ICZN International Commission on Zoological Nomenclature.

IDA International Development Association *q.v.*

IGBP International Geosphere–Biosphere Programme *q.v.*

igneous rocks rocks formed from the cooling and crystallization of magma (*q.v.*). Igneous rocks formed within the crust are known as intrusive rocks while those formed on the Earth's surface are termed extrusive rocks. *see also* metamorphic rocks, sedimentary rocks.

ignimbrite *n.* a pyroclastic rock consisting of small (generally <1 cm across), frequently flattened, bits of pumice in a matrix of smaller-sized glass fragments.

ignition temperature the temperature required to start a nuclear fusion reaction.

ilarvirus group plant virus group containing quasi-isometric single-stranded RNA viruses, type member tobacco ringspot virus. They are multicomponent viruses in which four genomic RNAs are encapsidated in three different virus particles.

Ile isoleucine *q.v.*

Illicidales *n.* an order of woody dicots comprising shrubs, climbers and small trees, and including the two families Iliaceae (star anise) and Schisandraceae.

illite *n.* $K_3Al_4(Al_3Si_5O_{20})(OH)_4$, one of the illite group (*q.v.*) of clay minerals *q.v.*

illite group *n.* a group of clay minerals (*q.v.*) that includes illite, glauconite (*q.v.*), hydromicas and phengite.

illite–montmorillonite *n.* a mixed-layer clay mineral. *see* clay minerals.

illuvial *a. appl.* layer of deposition and accumulation below the eluvial layer in soils. *alt.* B-horizon.

illuviation *n.* process by which suspended and/or dissolved soil material, from an upper soil horizon, is deposited in a lower soil horizon by the action of water. *cf.* eluviation.

ilmenite *n.* $FeTiO_3$, a titanium ore mineral.

imago *n.* the last or adult stage of insect metamorphosis, the perfect insect.

imbibition *n.* the passive uptake of water, esp. by substances such as cellulose and starch, as in uptake of water by seeds before germination.

imbricate *a.* overlapping, of scales, etc.

IMCO Intergovernmental Maritime Consultative Organization.

imitative *a. appl.* form, habit, colouring, etc., assumed by an organism for protection or aggression when it imitates that of another organism.

immature *a.* (1) still not fully developed; (2) not yet adult. *cf.* mature.

immature sediments sediments constituted of grains that differ greatly from one another in terms of size, *cf.* mature sediments.

immature soil soil in the early stages of formation, which has undeveloped or poorly developed soil horizons. *cf.* mature soil.

immigrant species species that migrate into an ecosystem or are introduced accidentally or deliberately by humans.

immigration *n.* movement of individuals into an established population.

immiscible *a.* incapable of being mixed, *appl.* liquids, e.g. oil and water.

immobilization *n.* the locking up of elements essential for plant nutrition in organic matter in soil by soil microorganisms so that they are not available for plant growth.

imperfect fungi Deuteromycetes *q.v.*

imperfect stage in ascomycete fungi, the asexual reproductive phase in their life history in which they bear conidia.

impermeable rock *see* permeability (of rock).

importance value a measure of the role of a species within a community, obtained by adding together its relative density, its relative dominance and the relative frequency at which it is found.

impoverishment *n.* the depletion of resources of an environment or the reduction in the diversity of an ecosystem.

imprinting *n.* a process usually occurring shortly after birth in animals in which a particular stimulus normally provided by a parent becomes permanently associated with a particular response, e.g. when young birds follow any large moving object, usually the parent bird, or when young mammals follow their mother in response to the smell of her milk.

In symbol for the chemical element indium *q.v.*

Inarticulata *n.* class of brachiopods in which the shells are joined by muscles only and not by a hinged joint. *cf.* Articulata.

inbreeding *n.* (1) matings between related individuals, *alt.* consanguineous matings; (2) successive crossing between very closely related individuals, in laboratory animals, plants, etc., leading to the establishment of pure-breeding strains or varieties in which individuals are homozygous at a large proportion of their loci, or at selected loci.

inbreeding coefficient (F) (1) of an individual, the probability that a pair of alleles carried by the male and female gametes that produced it are identical by descent from a common ancestor, as a result of inbreeding; (2) a measure of the reduction of heterozygosity as a result of inbreeding, given by $F_S = (H_I - H_S)/H_S$, where H_I and H_S are heterozygosity among an inbred and outbred group of individuals of the same population, respectively.

inbreeding depression loss of vigour following inbreeding, due to the expression of numbers of deleterious genes in the homozygous state.

inceptisol *n.* an order of the USDA Soil Taxonomy system comprising young soils whose horizons have formed rather quickly, largely through alteration of parent material rather than as a result of extreme weathering. Horizons containing accumulations of iron and aluminium oxides and clay are not present in this soil order. *see* aquept, ochrept, plaggept, tropept, umbrept.

incidence *n.* (1) occurrence of an event; (2) rate at which something occurs.

incineration *n.* a waste disposal method based on combustion under controlled conditions with the aim of destroying pathogens and/or toxic chemical species, while reducing the mass and/or volume of material left to be disposed of.

incipient lethal level concentration of a toxin at which 50% of the population of test organisms can live for an indefinite time.

incipient species populations that are diverging towards the point of becoming separate species, but which can still interbreed although they are prevented from doing so by some geographical barrier.

incised meander meander that has been cut down into the underlying bedrock either as a result of tectonic uplift or a drop in the river's base level (i.e. sea-level). Two types are recognized, *see* ingrown meander, intrenched meander.

included niche the case where the niche of a species occurs completely within the niche space of another species.

inclusive fitness the sum of an individual's own fitness plus all its influence on fitness in its relatives other than direct descendants.

incompatibility *n.* genetically determined inability to mate successfully.

incompatible *a.* (1) *see* incompatibility; (2) in plant pathology, *appl.* an interaction between host and pathogen that does not result in disease.

incomplete dominance codominance *q.v.*

incomplete penetrance the lack of expression of a genetic trait in some individuals possessing the genotype associated with the character.

incubation period the time between exposure to a disease organism and the time at which symptoms of the disease first occur.

indefinite *a.* not limited in size, number, etc.

indehiscent *a. appl.* fruits, capsules, sporangia, etc., which do not open to release the seeds or spores, the whole structure being shed from the organism. *cf.* dehiscent.

independent assortment the second of Mendel's laws, which describes the fact that each allele will be inherited independently of another, i.e. the gametes may contain any combination of the parental alleles. This law has had to be modified by the subsequent discovery of linkage, i.e. that alleles carried on the same chromosome tend to be inherited together.

indeterminate *a.* (1) indefinite (*q.v.*); (2) undefined; (3) not classified.

indeterminate error random error *q.v.*

indeterminate growth (1) (*bot.*) growth from an apical meristem that forms an unrestricted number of lateral organs, such as stems, branches or shoots, indefinitely, not stopped by the development of a terminal bud; (2) indefinite prolongation and subdivision of an axis.

index fossils fossils representing organisms present only within a geologically brief, defined period, but which are geographically widely distributed, enabling rock strata at different sites to be correlated chronologically.

index species an organism that lives only within a narrow range of environmental conditions and whose presence therefore indicates places where those conditions exist. *alt.* indicator species.

Indian Floral Region part of the Palaeotropical Realm comprising most of the Indian subcontinent except for the extreme north-west.

indicator species (1) species characteristic of climate, soil and other conditions in a particular region or habitat; (2) the dominant species in a biotype; (3) species whose disappearance or disturbance gives early warning of the degradation of an ecosystem.

indifference curves iso-utility curves *q.v.*

indifferent species species that can live in many different environments and is not found in any particular community.

indigenous *a.* (1) belonging to the locality; (2) not imported; (3) native.

indirect competition limitation of population size, fertility, etc., of two or more organisms because of competition for the same limited resource.

indirect recycling type of recycling in which used and subsequently discarded waste material is reclaimed. For various reasons, such reclaimed material is not always put to its original use but, often after extensive reprocessing, is converted into a lower-grade product, e.g. plastic containers ground up for use in fence posts or pallets. *cf.* direct recycling.

indium (In) *n.* soft silvery metallic element (at. no. 49, r.a.m. 114.82), used in the manufacture of semiconductors and to monitor neutron fluxes near nuclear reactors.

individual distance the distance around a bird in a flock, which it defends while feeding, etc.

individualism *n.* symbiosis in which the two parties together form what appears to be a single organism.

individuation *n.* (1) the formation of interdependent functional units, as in a colonial organism; (2) process of developing into an individual.

indole-2-acetic acid an auxin *q.v.*

Indo-Malaysian subkingdom a subdivision of the Palaeotropical kingdom comprising the Indian subcontinent and the Himalayas, China, Japan, South-East Asia, the Malaysian archipelago, Indonesia and Papua-New Guinea.

induced mutations mutations occurring as a result of deliberate treatment with a mutagen.

induced nuclear fission nuclear fission (*q.v.*) that occurs as a consequence of bombardment with neutrons or, less commonly, rapidly moving charged particles or gamma-ray photons (photofission). *cf.* spontaneous nuclear fission.

industrialized agriculture agriculture that uses large inputs of energy, water, fertilizers and pesticides to produce large quantities of crops and livestock. *alt.* intensive agriculture.

industrial melanics dark-coloured forms of otherwise light-coloured moths and other insects that have increased in industrial areas since the Industrial Revolution, selected by the need for camouflage against soot-blackened walls, trees, etc.

Industrial Revolution the use of new sources of fossil fuels and new technologies to manufacture goods, and the introduction of scientifically based methods for increasing food production. In Europe and the USA this transformation of the means of production and the consequent transformation of society occurred mostly during the nineteenth century.

inert *a.* (1) slow to react; (2) unreactive; (3) slowly reaches dynamic equilibrium; (4) difficult to alter by chemical reaction.

inert gases noble gases *q.v.*

inertia *n.* (*biol.*) the ability of a living system to resist being disturbed.

inertial confinement fusion an approach to controlled nuclear fusion in which a laser or high-energy particle beam is directed at a small, spherical capsule of fusible material. The sudden increase in temperature induced in the capsule is intended to cause it to explode. If the conditions are right, the ion density in the centre of the site of explosion will increase to reach the Lawson Criterion (*q.v.*) at temperatures at which fusion is possible. *alt.* pellet fusion.

INES International Nuclear Events Scale *q.v.*

infant mortality rate in humans, the death rate of babies of less than one year old, calculated as the number of deaths in this age group per 1000 live births.

infauna *n.* animals that live in the bottom sediments of the ocean floor.

infection *n.* invasion by endoparasites, i.e. bacteria, viruses, fungi, protozoans, etc.

infection thread structure formed by invagination of root hair cell via which the nitrogen-fixing rhizobia enter host tissue.

infectious *a.* (1) capable of causing an invasion; (2) capable of being transmitted from one organism to another.

infectivity *n.* a quantitative measure of the ability of a virus, bacterium or other parasite to produce an infection.

infertile *a.* (1) not fertile; (2) unable to produce viable young; (3) non-reproductive.

infestation *n.* invasion by ectoparasites.

infield *see* infield–outfield system.

infield–outfield system farming system, found, e.g., in Ireland, in which land use is differentiated on the basis of its proximity to the farm buildings. Land close by (the infield) is used for continuous arable cultivation while outlying land (the outfield) is less intensively farmed, e.g. rough grazing of livestock.

infiltration *n.* the transfer of water from the surface of the ground into the soil beneath. The rate of this process is dependent on a number of factors such as the slope of the surface, the nature and extent of the vegetation present and the moisture content of the soil.

inflammable flammable *q.v.*

inflorescence *n.* (1) flowerhead in flowering plants; (2) in mosses and liverworts the area bearing the antheridia and archegonia.

influents *n.plu.* the animals present in a plant community, or those primarily dependent and acting upon the dominant plant species.

infra- (1) prefix derived from Gk *infra*, below; (2) in biological classification, denotes a group just below the status of a subgroup of the taxon following it, as in infraclass, the group below the subclass.

infraclass *n.* taxonomic grouping between subclass and order, e.g. Eutheria (mammals lacking pouches) and Metatheria (marsupials that bring forth live young) are infraclasses in the subclass Theria.

infradian *a. appl.* a biological rhythm with a period of less than 24 h.

infralittoral *a. appl.* (1) depth zone of lake that is permanently covered with rooted or floating macroscopic vegetation; (2) upper subdivision of the marine sublittoral zone.

infraneuston *n.* the animals living on the underside of the surface film of water, e.g. some mosquito larvae.

infrared radiation (IR) electromagnetic radiation with wavelengths from *ca.* 0.8 μm to *ca.* 1 mm.

infraspecific *a.* (1) occurring within a species, *appl.* variation, competition, etc.; (2) *pert.* a subdivision of a species, as subspecies and varieties.

ingesta *n.* sum total of material taken in by ingestion.

ingestion *n.* the swallowing or taking in of food by an animal into the gut or food cavity. *v.* ingest.

ingrown meander type of incised meander, asymmetrical in cross-section, which forms when lateral meander migration is coupled with relatively slow incision by the river. *cf.* intrenched meander.

inheritance *see* heredity, Mendelian inheritance, non-Mendelian, polygenic inheritance, simple Mendelian traits.

inhibition *n.* (1) prevention or checking of an action, process or biochemical reaction; (2) (*ecol.*) tendency of early successional species to resist the establishment of later ones. *v.* inhibit.

inhibition theory a theory of plant succession. This hypothesizes that the success of any species in establishing itself depends entirely on it arriving first, no one species is competitively superior to another and short-lived species are eventually replaced by longer-lived species. *see* facilitation theory, tolerance theory.

inhibitor *n.* (1) any agent that checks or prevents an action or process; (2) a substance that reversibly or irreversibly prevents the normal action of an enzyme without destroying the enzyme. Competitive inhibitors act by binding to the active site and preventing binding of substrate, non-competitive inhibitors act by binding to other parts of the enzyme.

inhibitory *a.* tending to prevent an action, event or response occurring, *appl.* stimuli, cells, chemical compounds, etc. *cf.* excitatory.

initial *n.* self-renewing cell in a plant meristem, which remains undifferentiated and in the meristem and continues to divide, producing both further initials, and sister cells that go on to leave the meristem and differentiate.

inland wetland land that is covered all or part of the year with fresh water, as swamps, marshes, fens, bogs.

inlier *n.* (*geol.*) area of older rocks completely surrounded by younger rocks. *cf.* outlier.

innate *a.* (1) inborn, *appl.* behaviour that does not need to be learned; (2) *appl.* immunity, non-specific defence mechanisms against disease-causing microorganisms.

innate releasing mechanism (IRM) an internal and instinctive mechanism in an animal that is activated to produce a response by some external stimulus.

innominate *a.* nameless.

inoculation *n.* (1) the administration of a vaccine in order to induce protective immunity; (2) the introduction of bacteria or other microorganisms, or plant and animal cells, into nutrient medium to start a new culture; (3) introduction of a pathogen into a host.

inorganic *a. appl.* material or molecules that do not contain carbon.

inorganic chemistry the chemistry of substances other than those containing carbon.

inorganic fertilizers materials used to supply essential plant nutrients that are either produced industrially (as end-products or byproducts of chemical processes) or mined from natural deposits. *alt.* artificial fertilizers, chemical fertilizers, mineral fertilizers.

input *n.* material or energy that enters a biological system.

inquiline *n.* animal living in the home of another and getting a share of its food, without having any obvious detrimental effect on that other species.

inquilinism *n.* type of symbiosis in which an animal lives in the nest of another, is tolerated by the host and shares its food.

Insecta, insects *n.*, *n.plu.* very large class of arthropods (*q.v.*), found as fossils from the Devonian onwards, containing some three-quarters of all known extant species of animal. The insects include flies, bees and wasps, ants, butterflies and moths, beetles, dragonflies, grasshoppers and crickets and many other orders. The segmented body is divided into distinct head, thorax and abdomen. The head bears one pair of antennae and paired mouthparts, and the thorax bears three pairs of walking legs and usually one or two pairs of wings. Other types of appendage may be present on the abdomen. The life history usually includes metamorphosis. *alt.* Hexapoda. *see* Appendix 7 for orders.

insecticide *n.* chemical that kills insects.

Insectivora, insectivores *n.*, *n.plu.* a large order of primitive insect-eating and omnivorous placental mammals known from Cretaceous times to the present, including hedgehogs, moles, shrews.

insectivorous *a.* insect-eating, *appl.* certain animals and carnivorous plants. *alt.* entomophagous. *n.* insectivore.

inselberg *n.* massive outcrop of solid rock rising steeply above the surrounding plain, commonly found in semi-arid and arid regions of the tropics and subtropics. Large, domed inselbergs are known as bornhardts while smaller, degraded outcrops are known as kopjes or tors.

insessorial *a.* adapted for perching.

insets phenocrysts *q.v.*

insight learning learning involving reasoning, as in humans and to a lesser extent in some animals.

in situ in the original place.

in situ conservation the conservation of ecosystems and natural habitats and the maintenance of viable populations of species in their natural surroundings, and in the case of domesticated or cultivated species, in the surroundings in which they have developed their distinctive characteristics.

in situ hybridization technique for locating the site of a specific DNA sequence on a chromosome, by treating mitotic cells with a radioactive nucleic acid probe exactly complementary for that sequence and that binds to it. Its position can then be detected by autoradiography. The technique is also applied to detect and locate the synthesis of specific mRNAs in tissues.

insolation *n.* (1) exposure to the Sun's rays; (2) amount of solar radiation that falls on the Earth's surface per unit horizontal area, measured in watts per square metre (W m^{-2}).

insolation weathering a type of physical weathering that involves thermal contraction and expansion of the rock. This process is thought to occur in deserts, where diurnal temperature fluctuations are characteristically high.

inspiration *n.* the act of drawing air or water into the respiratory organs.

instability *n.* (*clim.*) characteristic of air that when forced to rise (or fall) tends to continue to move away from its original position once it is set in motion. *cf.* stability.

instar *n.* insect or other arthropod at a particular stage between moults.

instinct *n.* behaviour that occurs as an inevitable stereotyped response to an appropriate stimulus, sometimes equivalent to species-specific behaviour. *a.* instinctive.

integrated pest management (IPM) combined use of biological, chemical and cultivation methods to keep pests at an acceptable level.

integrated pollution control (IPC) a concept that recognizes the interrelationships between water, air and land, and takes this cross-media element into account in the control of particular pollution sources.

intellectual property rights rights over knowledge that is capable of being sold for profit. Legally recognized and enforceable intellectual property rights already exist in the form of copyrights, breeders' rights and patents, but the term is also used to describe the moral right of, e.g., the native inhabitants of an area to profit from the exploitation by others of their knowledge of the medicinal or agricultural value of their native plants.

intensive agriculture agriculture characterized by high levels of management and high productivity. Such agriculture is highly mechanized and places a heavy reliance on the use of pesticides and inorganic fertilizers. *alt.* industrialized agriculture.

intensive property (*phys.*) a physical property that does not vary with the amount of material present, e.g. density, temperature. *cf.* extensive property.

intentional behaviour behaviour that involves a mental representation of a goal that guides the behaviour.

intention movement preparatory motions an animal goes through before a complete behavioural response, e.g. the snarl before the bite.

inter- prefix derived from L. *inter*, between.

interaction *n.* in thermodynamic analysis, an interaction is an observable change in one part of the universe that is correlated with a corresponding change in another that has been brought about by work done or energy supplied to one of these parts.

interbreed *v.* to cross different varieties, species, or genera of plants and animals.

interbreeding *a. appl.* a population whose members can breed successfully with each other.

intercropping mixed cropping *q.v.*

intercross *n.* the crossing of heterozygotes of the F$_1$ generation among themselves.

interculture *n.* type of mixed cropping in which arable crops are cultivated underneath perennial crops, e.g. cereal crops under permanently planted fruit trees, as practised in the Mediterranean basin.

interdemic selection selection of entire breeding groups (demes) as the basic unit.

interference colours colours produced by optical interference between reflections from different layers of the surface.

interference competition competition between two organisms in which one physically prevents the other from occupying a particular habitat and thus gaining access to resources.

interference, matrix matrix effect *q.v.*

interfertile *a.* able to interbreed.

interfluve *n.* a hill that separates two river valleys.

intergeneric *a.* between genera, *appl.* hybridization.

interglacial *n.* period of time between times during which ice covered large parts of the Earth's land surface (glacials), particularly those occurring during the Pleistocene. *a. appl.* or *pert.* such periods.

Intergovernmental Panel on Climate Change (IPCC) interagency programme established in 1988 (jointly sponsored by WMO and UNEP) to prepare a wide-ranging, authoritative report on climate change. This scientific assessment included aspects such as global warming, the greenhouse effect, greenhouse gases and changes in sea-level. Two scientific assessments have been published, one in 1990 and the other in 1995.

interlittoral *n.* shallow marine zone to a depth of around 20 m.

interlocking spurs (*geog.*) alternating projections of the sides of a river valley, characteristic of its upper course.

intermediary metabolism metabolic pathways by which the basic molecular building blocks in a cell (e.g. monosaccharides, amino acids, nucleotides, etc.) are interconverted and incorporated into larger molecules.

intermediate *a.* (*geol.*) *appl.* igneous rocks that contain 52–66% silica (SiO_2).

intermediate host a host in which a parasite lives for part of its life cycle but in which it does not become sexually mature.

intermediate-level radioactive waste radioactive wastes produced by the nuclear industry and consisting of materials used to clean gases and liquids before discharge from nuclear plants, sludges from cooling ponds where spent fuel is stored before reprocessing, and materials contaminated with plutonium. They are bulky and contain radioactive elements with long half-lives but are not extremely hot like high-level wastes. They are disposed of after incorporating in concrete or some other solid and at present are stored on land, although in the past they have been dumped at sea.

intermediate technology the use of relatively unsophisticated, reliable and robust technology in machinery that can be produced, maintained and repaired in, e.g., rural areas in less developed countries.

intermolecular *a.* between molecules, *appl.* hydrogen bonds, distances, etc.

internal combustion engine a device in which a fuel is burned within a chamber that is closed by a piston, so that some of the energy liberated during burning is converted into work in the movement of the piston, e.g. the Otto engine, the diesel engine.

internal cost the direct cost of an economic good paid by the producer and purchaser, which is reflected in its market price.

internal energy the internal energy (symbol U) of a system is the total energy of all of its constituent particles.

internal respiration the biochemical, intracellular reactions of respiration. *cf.* external respiration.

International Atomic Energy Agency (IAEA) an independent organization (established in 1957) that works, under the aegis of the UN, to promote the peaceful use of nuclear energy world-wide.

International Bank for Reconstruction and Development (IBRD) specialized agency (part of the World Bank group) set up in 1945 to give financial assistance to nations destroyed by the Second World War. Now involved in encouraging economic growth in less developed member countries. Sometimes referred to as the World Bank.

International Biological Programme (IBP) an international research programme (1964–74), mounted by the International Council of Scientific Unions (ICSU), during which biological productivity data were gathered in different biomes, and systems models of ecological processes designed.

International Board for Plant Genetic Resources international organization, a member of the Consultative Group for International Agricultural Research, which recommends and advises on the establishment of seed banks and the conservation of plant genetic resources under threat, esp. of crop plants and their relatives.

International Commission on Radiation Protection (ICRP) a non-governmental scientific organization responsible for setting the maximum permissible dose and dose limits of ionizing radiation for radiation workers and members of the general public.

International Council of Scientific Unions (ICSU) world-wide institution, influential in environmental matters, with permanent unions in major disciplines, responsible for mounting major programmes such as the International Biological Programme (IBP) and the International Geosphere–Biosphere Programme (IGBP).

international date line imagined line on the Earth's surface that follows the 180° meridian through the Pacific Ocean, except where it is deflected to east or west to avoid areas of land. If the line is crossed towards the east, a day is repeated, whereas if it is crossed towards the west, a day is jumped.

International Development Association (IDA) specialized agency (part of the World Bank group) established in 1960 to offer long-term loans on concessionary terms, i.e. with little or no interest, to less developed nations.

International Geosphere–Biosphere Programme (IGBP) a major programme originated by the International Council of Scientific Unions in the early 1980s, focused on the Earth's system, including such aspects as the regulatory role of biological, chemical and physical processes, and the influence of human activities.

International Nuclear Events Scale (INES) international scheme, set up in 1990, to assess the seriousness of nuclear incidents and accidents.

International Union for Conservation of Nature and Natural Resources (IUCN) *see* World Conservation Union.

International Unit unit used to measure vitamin content of foods.

International Whaling Commission (IWC) international association established to provide for the conservation of whale populations through the regulation of commercial whaling. Established under the 1946 International Convention for the Regulation of Whaling, it has set annual quotas for the different species since 1949, but has no authority to enforce its recommendations.

intersex *n*. (1) an organism with characteristics intermediate between a typical male and a typical female of its species; (2) an organism first developing as a male or female, then as an individual of the opposite sex.

interspecific *a*. between distinct species, *appl*. crosses, as mule, hinny, cattalo, tigron.

interspecific competition competition between the members of different species for the same resource. *cf*. intraspecific competition.

intersterility *n*. incapacity to interbreed.

interstitial *a*. occurring in interstices or spaces, *appl*. (1) (*biol*.) flora and fauna living between sand grains or soil particles; (2) (*geol*.) texture of igneous rock that is characterized by one mineral type filling the spaces between other mineral grains.

intertidal *a*. *appl*. shore organisms living between high- and low-water marks.

intertidal platform a type of shore platform (*q.v.*) that lies between the high- and low-water marks and is formed mainly by wave abrasion and quarrying.

intertropical convergence zone (ITCZ) zone near to the Equator where the northeast and south-east trade winds converge. This low-pressure belt of rising air produces a wide cloud and precipitation band that can be seen from space. *alt*. equatorial low.

intervarietal *a*. *appl*. crosses between two distinct varieties of a species.

intolerant *a*. incapable of living in a particular set of conditions.

intra- prefix derived from L. *intra*, within, and signifying within, inside.

intrademic selection selection within a local breeding group.

intrageneric *a*. among members of the same genus.

intragenic *a*. within a gene, *appl*. recombination, mutation, etc.

intramolecular *a*. within a molecule, *appl*. bonds, hydrogen bonds, etc.

intramolecular respiration formation of carbon dioxide and organic acid by normally aerobic organisms if deprived of oxygen.

intrasexual *a*. *appl*. selection between competing individuals of the same sex.

intraspecific *a*. (1) within a species; (2) *appl*. variation, *appl*. competition between the members of the same species for the same resource, e.g. mates, food, territory, etc.

intrazonal *a*. within a zone, *appl*. locally limited soils, differing from prevalent or normal soils of the region or zone.

intrenched meander *n.* type of incised meander, symmetrical in cross-section, which develops when rapid downcutting by the river is coupled with minimal lateral migration. *cf.* ingrown meander.

intrinsic *a.* (1) inherent; (2) inward; *appl.* (3) inner muscles of a part or organ; (4) rate of natural increase in a population having a balanced age distribution; (5) membrane proteins that span the whole membrane. *cf.* extrinsic.

intrinsic isolating mechanism any genetic mechanism preventing interbreeding.

intrinsic rate of increase (*r***)** the fraction by which a population is growing at each instant of time, symbolized by *r*.

introduced *a. appl.* plants and animals not native to the country and thought to have been brought in by man.

introgression *n.* the gradual diffusion of genes from the gene pool of one species into another when there is some hybridization between them as a result of incomplete genetic isolation.

intrusion *n.* a mass of intrusive rock. Intrusions that cut across existing rock strata (e.g. dykes) are termed discordant while those that do not disrupt existing rock strata (e.g. sills) are known as concordant intrusions.

intrusive rocks medium- or coarse-grained igneous rocks formed when magma cools slowly within the crust. *see* hypabyssal rocks, plutonic rocks.

intussusception *n.* (*biol.*) growth in surface extent or volume by intercalation of new material among that already present. *cf.* accretion, apposition.

inversion *see* temperature inversion.

invertebrates *n.plu.* a general term for all animals without a backbone, i.e. all animal groups except the vertebrates.

investment *n.* the outer covering of a part, organ, animal or plant.

in vitro *appl.* biological processes and reactions occurring in (1) cells or tissues grown in culture; (2) cell extracts or synthetic mixtures of cell components.

in vitro fertilization (IVF) the fertilization of an ovum outside the mother's body, generally followed by replacement of the fertilized egg into the mother or into a pseudopregnant foster mother, where it develops normally.

in vivo *appl.* biological processes occurring in a living organism.

involucre *n.* a circle of bracts at the base of a compact flowerhead.

involuntary *a.* not under the control of the will, *appl.* movements, etc.

iodine (I) *n.* halogen element (at. no. 53, r.a.m. 126.90) that forms a black shiny volatile solid in the elemental state (I_2). Present in small amounts in seawater and concentrated in seaweed. Essential micronutrient for humans and other vertebrates as it is a constituent of some thyroid hormones. Radioactive isotopes of iodine are produced in nuclear reactors and may be released in nuclear accidents. The radioisotope ^{131}I (half-life 8.6 days) is used as a tracer to diagnose and treat thyroid gland disorders.

ion *n.* any chemical species that has either more electrons than protons, and therefore one or more negative charges overall (an anion), or fewer electrons than protons, and therefore one or more positive charges overall (a cation).

ion–dipole interaction (*chem.*) a weak attraction between an ion and that end of a polar molecule that carries a partial charge of the opposite sign.

ion exchange the replacement of ions held, by electrostatic attraction, on the surface of a solid with ions in a solution that is in contact with that solid. Two types of ion exchange are recognized, anion exchange (*q.v.*) and cation exchange (*q.v.*). Ion exchange has many uses, e.g. in water softening and solution mining.

ion-exchange chromatography separation of molecules, such as proteins, on the basis of their net charge by differential binding to a column of carboxylated polymer, positively charged molecules binding to the column.

ionic *a. pert.* ions, *appl.* compounds, bonds, etc.

ionic bond the electrostatic interaction that holds ions of opposite charge together.

ionic product of water (K_w) an equilibrium constant that relates to the autoionization of water, a process that may be represented by the equation

$$2H_2O \rightleftharpoons H_3O^+ + OH^-$$

This equilibrium constant is expressed as $K_w = \{H_3O^+\}\{OH^-\} = 1 \times 10^{-14}$ at 25 °C, where the curly brackets, $\{\ \}$, represent activity based on molar concentration. In dilute solutions, the activity of a solute approximates to its molar concentration.

ionization *n.* formation of ions.

ionizing radiation short-wavelength high-energy radiation, such as gamma-rays, and fast-moving particles, such as alpha-particles and beta-particles, emitted by radioisotopes, which cause the formation of ions in tissues, thus contributing to DNA and tissue damage.

ionophoresis *n.* movement of ions under the influence of an electric current. *a.* ionophoretic.

ion pair electrostatic bond *q.v.*

ioxynil *n.* contact herbicide used to control weeds in, e.g., newly sown turf.

IPC integrated pollution control *q.v.*

IPCC Intergovernmental Panel on Climate Change *q.v.*

IPM integrated pest management *q.v.*

ipsilateral *a. pert.* or situated on the same side. *cf.* contralateral.

IR infrared radiation *q.v.*

Irano–Turanian Floral Region part of the Holarctic Realm comprising Central Asia north of the Himalayas and from the western edge of the Black Sea to central China.

Iridales *n.* an order of herbaceous monocots and including the families Corsiaceae, Iridaceae (iris) and others.

iridium (Ir) *n.* metallic element (at. no. 77, r.a.m. 192.22), forming an unreactive silvery metal. Relatively rare on Earth, it is more common in meteorites, and its localized presence at levels greater than normal (the iridium layer) has been taken as evidence of the impact of a large meteorite.

Iridoviridae *n.* family of enveloped, double-stranded DNA viruses including that causing African swine fever. Most are insect viruses.

Irish potato famine *see* potato blight.

IRM innate releasing mechanism *q.v.*

iron (Fe) *n.* magnetic metallic element (at. no. 26, r.a.m. 55.85), forming a strong, hard, but malleable greyish-white metal. The most important ores are hematite (Fe_2O_3), magnetite (Fe_3O_4), siderite ($FeCO_3$) and limonite (hydrated iron oxide), and

iron is also found in many other minerals, such as pyrite (fool's gold, FeS_2). An essential micronutrient for living organisms, as Fe atoms are prosthetic groups in many enzymes and other proteins.

ironpan *n.* hardpan in which iron oxide is the main cementing agent.

iron pyrites *n.* FeS_2, a sulphide mineral of very widespread distribution. *alt.* pyrite.

ironstone rock that is at least 15% iron.

irradiation *n.* (1) exposure to radiation; (2) any technique in which material is deliberately exposed to radiation in order to confer upon it desired characteristics, used, e.g., in food preservation, and in the sterilization of male insects for pest control purposes.

irrigation *n.* conveyance of water from a source of supply to an area of land where it is needed for the cultivation of crops. *see* drip irrigation, flood irrigation, furrow irrigation, sprinkler irrigation.

irritability *n.* capacity to receive external stimuli and respond to them.

irruptive growth Malthusian growth *q.v.*

island arc curved line of islands associated with lithospheric plate boundaries and characterized by seismic and volcanic activity, e.g. West Indies.

island biogeography the study of the biogeography of the flora and fauna of islands with a view to understanding the nature and evolution of biological diversity in an isolated environment.

isoalleles *n.plu.* alleles that are identical in their gross phenotypic effects but can be distinguished by biochemical means at protein or DNA level.

isobar *n.* line used on a weather map to connect points of equal atmospheric pressure.

isochronous *a.* (1) having equal duration; (2) occurring at the same rate.

isodemic *a.* (1) with, or *pert.*, populations composed of an equal number of individuals; (2) *appl.* lines on a map that pass through points representing equal population density.

isodynamic *a.* (1) of equal strength; (2) providing the same amount of energy, *appl.* foods.

isoelectric point (IEP) the pH at which an amphoteric molecule, such as a protein, carries no net charge, being a definite value for each protein.

isoenzyme *n.* any one of several different allelic forms in which some enzymes may be found, each having a similar enzyme specificity but differing in amino acid sequence and thus in properties such as optimum pH or isoelectric point. *alt.* isozyme.

Isoetales *n.* order of Lycopsida having linear leaves, and a 'corm' with complex secondary thickening, and including the quillworts.

isogenes *n.plu.* lines on a map that connect points where the same gene frequency is found.

isogenetic *a.* (1) arising from the same or a similar origin; (2) of the same genotype.

isogenous *a.* of the same origin.

isohaline *a.* having equal levels of salinity.

isolate *n.* (1) a breeding group limited by isolation; (2) the first pure culture of a microorganism derived from soil, tissues, etc.

isolated system a term used in thermodynamic analysis to denote a system (*q.v.*) that cannot exchange either matter or energy with its surroundings.

isolateral *a.* having equal sides, *appl.* leaves with palisade tissue on both sides.

isolating mechanisms mechanisms that prevent breeding between two populations and eventually lead to speciation, chiefly including geographical isolation, but also the development of genetic, anatomical and behavioural barriers to successful interbreeding.

isolation *n.* prevention of mating between breeding groups owing to spatial, topographical, ecological, morphological, physiological, genetic, behavioural, or other factors.

isoleucine (Ile, I) *n.* an amino acid, stereoisomer of leucine, constituent of proteins, and essential in diet of man and other animals.

isomer *n.* one of two or more chemical compounds each having the same kind and number of atoms but differing in arrangement of the atoms and in physical and (sometimes) chemical properties. *see also* enantiomer, optical isomer, stereoisomer, tautomerism.

isomorphic, isomorphous *a.* (1) superficially alike; (2) *appl.* alternation of haploid and diploid phases in morphologically similar generations.

isonym *n.* a new name, of species, etc., based upon oldest name or basinym.

iso-octane *n.* 2,2,4-trimethylpentane [$CH_3C(CH_3)_2CH_2CH(CH_3)CH_3$], an isomer of octane.

iso-osmotic *a. appl.* two solutions of the same osmotic concentration. *see* isotonic.

isophane *n.* a line connecting all places within a region at which a biological phenomenon, e.g. flowering of a plant, occurs at the same time.

isophene *n.* a contour line delimiting an area corresponding to a given frequency of a variant form.

isoplankt *n.* a line representing on a map the distribution of equal amounts of plankton, or of particular species.

Isopoda, isopods *n., n.plu.* group of marine, freshwater and terrestrial malacostracan crustaceans, including the woodlice and water slaters, having a dorso-ventrally flattened body and no carapace.

isoprenoid *n.* any of a large and varied group of organic compounds built up of five-carbon isoprene units and including carotenoids, terpenes, natural rubber and the side chains of, e.g., chlorophyll and vitamin K.

Isoptera *n.* order of social insects comprising the termites and their relatives, which live in large organized colonies containing reproductive forms (the queen and the king) and non-reproductive wingless soldiers and workers, all offspring of the king and queen. Termites have gut flora that enable them to digest wood, and they can be serious pests, devouring wooden buildings, trees and paper.

isostructural *a.* having the same structure.

isotherm *n.* line used on a weather map to connect points of equal temperature.

isotonic *a.* (1) of equal tension; (2) of equal osmotic pressure, *appl.* solutions.

isotopes *n.plu.* atoms that have the same atomic number as each other but different mass numbers. The full symbol for an isotope takes the form $^A_Z E$ (where E is the elemental symbol), frequently given in an abbreviated form, $^A E$, as Z is implied by E.

For example, the element oxygen (atomic number 8) has three stable isotopes, oxygen-16, oxygen-17 and oxygen-18. These are given the symbols $^{16}_{8}O$, $^{17}_{8}O$, $^{18}_{8}O$ or ^{16}O, ^{17}O, ^{18}O, respectively. *a.* isotopic.

isotropic, isotropous *a.* (1) symmetrical around longitudinal axis; (2) not influenced in any one direction more than another, *appl.* growth rate; (3) without predetermined axes, as some ova. *n.* isotropy.

iso-utility curves in the study of animal behaviour and ecology, curves joining all points of equal utility or benefit, and which show how different behaviours, e.g. feeding behaviours, can result in the same quantitative benefit. *alt.* indifference curves.

isoxanthopterin *n.* a colourless pterin in the wings of cabbage butterflies and in eyes and bodies of other insects. *alt.* leucopterin B.

isozoic *a.* inhabited by similar animals.

isozyme isoenzyme *q.v.*

***itai-itai* disease** a degenerative bone disease, characterized by severe joint pain, caused by the accumulation in the body of the highly toxic metal, cadmium.

ITCZ intertropical convergence zone *q.v.*

iteration *n.* repetition, as of similar trends in successive branches of a taxonomic group.

iteroparity *n.* production of offspring by an organism in successive groups. *a.* iteroparous. *cf.* semelparity.

IU International Unit *q.v.*

IUCN International Union for Conservation of Nature and Natural Resources, *see* World Conservation Union.

IVF *in vitro* fertilization *q.v.*

ivory *n.* dentine of teeth, usually that of tusks, as of elephant and narwhal.

IWC International Whaling Commission *q.v.*

J

J joule *q.v.*

Jaccard index, Jaccard coefficient of similarity a measure of the similarity (*see* coefficient of community) between two plant communities based on their species composition. It can be calculated from the formula $J = c/(a + b - c)$ where a = no. of species at site A, b = no. of species at site B, and c = no. of species common to each site. J will lie between zero and one. The greater the value of J, the more similar the sites are in their species composition. *cf.* Gleason's index, Kulezinski index, Morisita's similarity index, Simpson's index of floristic resemblance, Sørensen similarity index.

jade *n.* a semiprecious green ornamental stone that mineralogically is either jadeite (*q.v.*) or nephrite (*q.v.*). Other minerals, similar in appearance but less hard, e.g. serpentine (*q.v.*), are frequently traded as jade.

jadeite *n.* $NaAlSi_2O_6$, one of the pyroxene minerals (*q.v.*) of the clinopyroxene subgroup. Prized as an ornamental semiprecious stone. *see also* jade.

jamesonite *n.* $Pb_4FeSb_6S_{14}$, a lead ore mineral of relatively minor importance.

jasper *n.* SiO_2, opaque, coloured varieties of chalcedony (*q.v.*), esp. red ones.

jaundice *n.* condition in which the skin and the whites of the eyes take on a yellow coloration due to the presence of the bile pigment (bilirubin) in the blood, e.g. as a result of hepatitis. *alt.* icterus.

Java man fossil hominid found in Java and originally called *Pithecanthropus erectus*, now called *Homo erectus*, and dating from mid-Pleistocene.

jellyfish *n.* common name for the Scyphozoa *q.v.*

jelly fungi common name for a group of basidiomycete fungi whose fruiting bodies are typically of a jelly-like consistency, e.g. the funnel-shaped *Tremella* and the ear-shaped *Auricularia*.

jet streams narrow bands of extremely fast-moving air, which may reach speeds in excess of 230 km h^{-1}, located high in the troposphere. Five different jet streams are recognized, two of which are of particular importance (*see* polar front jet stream and subtropical jet stream) and one is of seasonal significance (*see* easterly equatorial jet stream).

joint (*geol.*) *see* jointing.

jointing *n.* (*geol.*) the development of small fractures (joints) within rocks. This process is not accompanied by any significant movement and is found in all three rock types, i.e. igneous, sedimentary and metamorphic. *cf.* fault.

Jordan's laws (1) species most closely related to each other will be located nearest to each other, but separated by a geographical barrier; (2) in a particular species of fish, individuals living in cold environments tend to develop more vertebrae than those living in warm environments.

joule (J) *n.* the derived unit of energy or work in the SI system, which is used instead of the calorie. One joule of work is done whenever a force of 1 N (equivalent to 1 kg m s^{-2}) is sustained over 1 m in that force's direction. Therefore, 1 J is equivalent to 1 kg m^2 s^{-2}. 1 J = 0.238 92 calories.

J-shaped growth curve type of growth curve (*q.v.*) exhibited by populations that grow exponentially and then crash to very low numbers, typical of organisms with many generations per year, e.g. planktonic algae. *see* Fig. 15. *see also* exponential growth. *cf.* S-shaped growth curve.

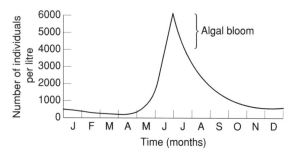

Fig. 15 J-shaped growth curve of planktonic algae.

Juglandales *n.* order of dicot trees, often aromatic, with pinnate leaves and comprising two families, Juglandaceae (walnut) and Rhoiptelaceae.

Juncaceae *n.* the rushes, a family of monocotyledons that live mainly in wet or cold habitats, with small flowers having a perianth of six brown or green segments.

Juncales *n.* order of herbaceous monocots with long narrow channelled or grass-like leaves and comprising the families Juncaceae (rushes) and Thurniaceae.

jungle *n.* dense tropical forest with abundant undergrowth, used to describe the first phase of regeneration of tropical forest after disturbance.

junk DNA a vernacular term for the large amounts of DNA in the genomes of plants and animals that is composed of simple repetitive non-coding sequences and appears to have no effect on phenotype.

Jurassic *a. pert.* or *appl.* geological period lasting from *ca.* 213 to 144 million years ago, after the Triassic and before the Cretaceous.

jute *n.* (1) either of the plants *Corchorus olitorius* and *C. capsularis*, members of the family Tiliaceae, grown for the fibre content of their bark; (2) the fibre extracted from *C. olitorius* or *C. capsularis*, used for making mats, sacks, etc.

juvenal, juvenile *a.* youthful, *appl.* plumage replacing nestling down of first plumage.

juvenile *n.* a young bird or other animal, before it has acquired full adult plumage or form.

juvenile hormone insect hormone that maintains larval state and prevents premature metamorphosis into an adult. Some are used in pest control.

juvenile water magmatic water *q.v.*

K

K (1) symbol for the chemical element potassium (*q.v.*); (2) lysine (*q.v.*); (3) kelvin *q.v.*

K (1) symbol for the carrying capacity (*q.v.*) of the environment; (2) equilibrium constant (*q.v.*); (3) partition coefficient *q.v.*

K_a acid ionization constant *q.v.*

K_b base ionization constant *q.v.*

K_H, k_H Henry's law constant *q.v.*

K_m Michaelis constant, *see* Michaelis–Menten kinetics.

K_{ow} octanol–water partition coefficient *q.v.*

K_{sp} solubility product *q.v.*

K_{stab} stability constant *q.v.*

K_w ionic product of water *q.v.*

kainit *n.* inorganic potassium fertilizer of variable chemical composition with a potassium content equivalent to 12–16% K_2O, supplied as a powder.

kalsilite *n.* $KAlSiO_4$, one of the feldspathoid minerals (*q.v.*). An end-member of a solid-solution series, the other end-member being nepheline ($NaAlSiO_4$).

kames *n.plu.* undulating mounds of sorted glacial drift (sands and gravels) that are deposited by meltwater in an irregular pattern along the front of a stationary or slowly retreating ice sheet.

kame terraces narrow ridges of sorted glacial drift (sands and gravels) that occur at the sides of valleys. These fluvioglacial landforms are deposited by meltwater streams that flow in the trough between the glacier and the valley side.

kanamycin *n.* antibiotic produced by *Streptomyces kanamyceticus* that interferes with bacterial protein synthesis. A gene for kanamycin resistance is widely used as a selectable marker gene in the construction of recombinant DNA.

kandite group *n.* a group of clay minerals (*q.v.*) that includes kaolinite (*q.v.*), dickite

and nacrite (both less common polymorphs of kaolinite) and halloysite (*q.v.*). *alt.* kaolinite group, kaolinites.

Kaneclor a trade name under which polychlorinated biphenyls (PCBs) were sold.

kaolin *n.* a mineral product of industrial importance, constituted mainly of the clay mineral kaolinite (*q.v.*). *alt.* china clay.

kaolinite *n.* $Al_2Si_2O_5(OH)_4$, the most prominent of the 1 : 1 clay minerals (*q.v.*) found in soils. An alteration product of aluminous silicates. The main constituent of kaolin *q.v.*

kaolinite group, kaolinites kandite group *q.v.*

karren *n.* a German word used to describe minor solutional features formed on limestone. *alt.* lapiés (French). There is no equivalent English word.

karst *n.* distinctive limestone landscape found in many parts of the world, characterized by surface depressions, underground river systems with caves and a lack of surface water. Three main types are recognized: Caribbean karst (*q.v.*), temperate karst (*q.v.*), tropical karst (*q.v.*). *see also* cockpit karst, thermokarst, tower karst.

karstification *n.* the process of development of a karst (*q.v.*) landscape.

karyomixis *n.* mingling or union of nuclear material of gametes.

karyotype *n.* representation of the chromosome complement of a cell, with individual mitotic chromosomes arranged in pairs in order of size.

kasugamycin *n.* antibiotic that blocks initiation of translation in bacteria.

katabatic wind wind that flows down the slopes of mountain valleys during the night. This localized airflow develops during calm, clear weather conditions when air above the valley is cooled, becomes more dense and consequently sinks down the valley sides. *alt.* mountain wind. *cf.* anabatic wind.

katharobic *a.* living in clean waters, *appl.* protists.

kb, kbp kilobase, kilobase-pair. Equal to 1000 bases or base-pairs of DNA.

kDa, kdal kilodalton *q.v.*

kegelkarst cockpit karst *q.v.*

kelp *n.* common name for seaweeds of the Laminariales, marine multicellular brown algae with a large broad-bladed thallus attached to the substratum by a tough stalk and holdfast.

kelvin (K) *n.* the SI unit of temperature. *see* absolute temperature.

kemp *n.* very coarse, non-wool fibre present in the fleece of sheep.

kenaf *see* hemp.

keratinization *n.* intracellular deposition of the protein keratin to form an inert horny material, e.g. nails, claws, horns, outer layers of skin. *alt.* cornification.

kerogen *n.* a form of organic matter found in certain rocks, e.g. oil shale, which can be broken down by heating to produce material that is similar to crude oil.

kerosene paraffin oil *q.v.*

ketone *n.* (*chem.*) an organic compound that contains a carbonyl (keto) group (C=O), the carbon atom of which is bonded to two hydrocarbon radicals. The simplest ketone is propanone (also called acetone) $CH_3C(O)CH_3$.

ketose *n.* any monosaccharide containing a keto (C=O) group. *cf.* aldose.

kettle lakes *see* kettles.

kettles *n.plu.* enclosed steep-sided depressions formed when large blocks of ice, buried in the outwash plain of glaciers, melt. Kettles often contain water, forming kettle lakes (*alt.* kettle hole lakes). *alt.* kettle holes.

key *n.* (1) a means of identifying objects or organisms by a series of questions with alternative possible answers, each answer leading either to another question or to a positive identification; (2) (*bot.*) winged nutlet hanging in clusters, as in ash, *alt.* samara.

keystone species species that have a key role in an ecosystem, affecting many other species, and whose removal leads to a series of extinctions within the system.

kg kilogram *q.v.*

Khamsin *n.* hot, dry southerly or southeasterly wind found in North Africa and Arabia, which blows for a period of 50 days during March–May.

kibbutz *n.* Israeli rural settlement that operates through the cooperation of its members, who voluntarily pool labour, resources and income.

killing-out percentage dressing-out percentage *q.v.*

killing power (k-value) the effects of intraspecific competition on mortality, fecundity and growth of a population obtained by comparing the initial density to the final density of the population.

kilo (k) SI prefix indicating that the unit it applies to is multiplied by 10^3, e.g. $1 \text{ kW} = 1 \times 10^3 \text{ W}$.

kilobase (kb), kilobase-pair (kbp) unit of length used for polynucleotides and nucleic acids, corresponding to 1000 bases or base-pairs.

kilodalton (kDa, kdal) unit of mass equal to 1000 Da, or 1000 a.m.u., also sometimes abbreviated to K (e.g. a 30 K protein), used chiefly for proteins.

kilogram (kg) *n.* the basic SI unit of mass, defined as being equal to the mass of the international prototype kept by the Bureau International des Poids et Mesures, near Paris. 1 kg = 1000 g. *alt.* kilogramme.

kimberlite *n.* a serpentinized fine- to coarse-grained intrusive, ultrabasic igneous rock that is generally porphyritic in texture and that incorporates xenoliths. It is commonly found as volcanic pipes and, occasionally, as dykes. It contains olivine, pyrope, phlogopite, chrome diopside and orthopyroxene. Kimberlites are the main commercial source of diamond, which is present as an accessory mineral.

kindred *n.* in human genetics, a group of people related by marriage or ancestry.

kinesis *n.* (1) random movement; (2) an orientation movement in which the organism swims at random until it reaches a better environment, the movement depending on the intensity, not the direction, of the stimulus.

kinetic *a.* (1) *pert.* movement; (2) active; (3) *appl.* the energy possessed by a body by virtue of its motion.

kinetics, chemical *see* chemical kinetics.

kinetin *n.* synthetic plant growth substance that causes cell division.

king crabs horseshoe crabs *q.v.*

kingdom *n.* in taxonomy, a primary division of the living world, five kingdoms being generally recognized: Monera or Prokaryotae (bacteria and other prokaryotes), Protista (simple eukaryotic organisms such as the protozoans and algae), Fungi, Plantae (multicellular green plants other than algae) and Animalia (multicellular animals). Older classifications placed the fungi, algae and bacteria in the Plantae and protozoans in the Animalia. Kingdoms are divided into phyla, for animals, and divisions for plants and other organisms. *see also* superkingdom. *see* Appendices 5–9.

Kinorhyncha *n.* a phylum of microscopic marine pseudocoelomate animals having a body of jointed spiny segments and a spiny head.

kin selection the selection of genes due to individuals favouring or disfavouring the survival of relatives, other than offspring, who possess the same genes by common descent.

kinship *n.* possession of a common ancestor in the not too distant past.

kinship, coefficient of *see* coefficient of kinship.

kirromycin *n.* antibiotic that inhibits elongation of peptide chain during translation in bacteria by inhibiting elongation factor EF–Tu.

Kjeldahl analysis a technique widely used to determine the total nitrogen (N) content of a tissue.

Klamath weed *Hypericum perforatum*, a noxious weed found in overgrazed pastureland in temperate regions throughout the world. Leaf-eating beetles, esp. *Chrysolina quadrigemina*, have been successful in the biological control of this weed. *alt.* St John's wort.

kleptobiosis *n.* the robbing of food stores or scavenging in the refuse piles of one species by another that does not live in close association with it.

kleptoparasitism *n.* type of parasitism in which the female searches out the prey or stored food of another female, usually of a different species, and takes it for her own offspring. *alt.* cleptoparasitism.

klinokinesis *n.* (1) movement in which an organism continues to move in a straight line until it meets an unfavourable environment, when it turns, resulting in its remaining in a favourable environment, the frequency of turning depending on the intensity of the environmental stimulus; (2) change in rate of change of direction or angular velocity due to intensity of stimulation.

klinotaxis *n.* a taxis in which an organism orients itself in relation to a stimulus by moving its head or whole body from side to side symmetrically in moving towards the stimulus, and so compares the intensity of the stimulus on either side.

knickpoint *n.* (1) an oversteepened reach; (2) a discontinuity in a river's longitudinal profile caused by a fall in the river's base level, i.e. sea-level.

knocking *n.* the pre-ignition of the fuel–air mixture in Otto engines that leads to excessive noise, loss of efficiency and accelerated engine wear. *alt.* pinging, pinking.

kolkhoz *n.* collective farm in the former USSR.

kopje *see* inselberg.

Kornberg cycle glyoxylate cycle *q.v.*

Kr symbol for the chemical element krypton *q.v.*

Krakatoa (natural disaster) a major volcanic eruption that occurred on the uninhabited Indonesian island of Krakatoa in 1883. Although no-one was killed in the explosion itself, the generation of seismic sea waves (tsunami) swept over other islands and the neighbouring coasts of Java and Sumatra with an estimated loss of 40 000 lives. *alt.* Krakatau.

krasnozems *n.plu.* deep friable red loamy soils found in the subtropics and developed from base-rich parent materials.

Krebs cycle tricarboxylic acid cycle *q.v.*

krill *n.* planktonic crustaceans, which are abundant in the oceans and form the principal food of the whalebone (filter-feeding) whales.

krotovina crotovina *q.v.*

krummholtz *n.* stunted trees found growing above the timberline in mountains. From the German for crooked wood.

krypton (Kr) *n.* inert colourless gas (at. no. 36, r.a.m. 83.80), one of the noble gases. It is found in minute quantities in the atmosphere.

***K*-selected species, *K*-strategist** species selected for its superiority in a stable environment, typically with slow development, relatively large size and producing only a small number of offspring at a time.

***K* selection** selection favouring superiority in stable, predictable environments in which rapid population growth is unimportant. *cf. r* selection.

Ku symbol for the chemical element kurchatovium *q.v.*

Kulezinski index index of similarity (*see* coefficient of community) between two plant communities based upon their species composition. It is calculated by dividing the number of species in common by the sum of the total number of species in each community. *cf.* Gleason's index, Jaccard index, Morisita's similarity index, Simpson's index of floristic resemblance, Sørensen similarity index.

kunzite *n.* a variety of spodumene (*q.v.*) used as a gemstone.

Kupfernickel *n.* natural nickel arsenide (NiAs), an important nickel ore mineral. *alt.* Kupfer-nickel, niccolite, nickeline.

kurchatovium (Ku) *n.* transuranic element (at. no. 104).

k-value *see* killing power.

kwashiorkor *n.* deficiency disease caused by an insufficiency of protein.

kyanite *n.* Al_2SiO_5, a nesosilicate mineral of metamorphic rocks, typically gneisses and schists, generated by regional metamorphism. Used as a refractory. *alt.* disthene.

L

λ lambda, used as the symbol for the wavelength of light.

L leucine *q.v.*

L Avogadro's number *q.v.*

L-, D- prefixes denoting particular molecular configurations, defined according to convention, of certain optically active compounds esp. monosaccharides and amino acids, the L configuration being a mirror image of the D. In living cells such molecules usually occur in one or other of these configurations but not both (e.g. glucose as D-glucose, amino acids always in the L form in proteins).

La symbol for the chemical element lanthanum *q.v.*

labelled *a. appl.* molecule made detectable and traceable by incorporation of a radioactive element, or by linkage to some other detectable chemical tag.

labile *a.* (1) readily undergoing change; (2) unstable; (3) *appl.* genes that have a tendency to mutate; (*chem.*) (4) quick to react; (5) reactive; (6) rapidly reaching dynamic equilibrium.

lability *n.* in evolutionary theory, the ease and speed with which particular categories of traits evolve.

Laboulbeniomycetes *n.* a group of highly specialized ascomycete fungi, exoparasitic on insects and arachnids, which have an ascogonium with a trichogyne and fertilization by spermatia.

labradorite *n.* one of the plagioclase (*q.v.*) minerals.

Labyrinthodontia, labyrinthodonts *n., n.plu.* subclass of early amphibians, of the late Palaeozoic and Mesozoic, with labyrinthodont teeth (teeth with a complicated arrangement of dentine), now extinct. They included temnospondyls, anthracosaurs and ichthyostegalians.

Labyrinthulomycota, labyrinthulids *n., n.plu.* phylum of colonial protists that make slimy transparent colonies composed of lines of spindle-shaped cells moving in tunnels of secreted slime. They include the eel-grass parasite *Labyrinthula. alt.* slime nets.

lac *n.* a resinous secretion of the lac glands of certain insects, some types used to make shellac.

laccolith *n.* a minor intrusion in rock, similar to a sill (*q.v.*), which becomes thickened to produce a dome. The expanding dome pushes up the overlying rock strata without destroying it.

Lacertilia, lacertids *n., n.plu.* the suborder of reptiles containing the lizards (geckos, iguanas, agamas, skinks, bearded lizards, monitors, etc.). Most species are four-legged, some running on their hind legs, but some (e.g. slow-worms) are legless. They include insectivorous, herbivorous and carnivorous species, and are adapted to a wide range of habitats, including very dry regions.

lacewings *see* Neuroptera.

lachrymator *n.* tear-inducing substance. *alt.* lacrimator, lacrymator.

lactation *n.* (1) secretion of milk in mammary glands; (2) period during which milk is secreted.

lactic acid 3-carbon organic acid formed in animal cells, esp. muscle, when insufficient oxygen is supplied, and that is also produced by fermentation of sugars by certain bacteria, esp. lactobacilli.

lactic acid bacteria lactobacilli *q.v.*

lactiferous *a.* (1) forming or carrying milk; (2) (*bot.*) carrying latex.

lactobacillus *n.* any of a group of rod-shaped, Gram-negative bacteria (the genus *Lactobacillus*) characteristically producing lactic acid as an end-product of anaerobic respiration, responsible for the souring of milk. *plu.* lactobacilli.

lactose *n.* a disaccharide, esp. abundant in milk, composed of glucose and galactose residues.

lacustrine *a. pert.*, or living in or beside, lakes.

lag deposits *see* desert pavement.

Lagomorpha, lagomorphs *n., n.plu.* the rabbits, hares and pikas, an order of herbivorous mammals known from the Eocene, with skulls and dentition similar to rodents, but with a second pair of incisors, and with hind limbs modified for leaping.

lagoon *n.* a shallow body of water that, e.g., separates a barrier island (*q.v.*) from the mainland coast.

lag phase the first phase of growth of a bacterial culture (or of any other population) in which there is no appreciable increase in cell numbers. *see* S-shaped growth curve.

lahar *n.* (1) a mudflow formed from a mixture of water and volcanic debris; (2) also *appl.* the resultant rock. Lahars originate in a number of different ways, e.g. the deposition of lava or pyroclasts onto ice and snow. *see* Armero natural disaster.

LAI leaf area index *q.v.*

lake *n.* a large body of water, esp. fresh water, surrounded by land.

Lake Nyos (natural disaster) on 21 August 1986, an eruption of gas (probably carbon dioxide) occurred from the crater lake, Lake Nyos, in Cameroon, West Africa. The gas cloud affected an area of *ca.* 15 km^2 and claimed the lives of over 1500 people.

Lamarckism *n.* the theory of evolution chiefly formulated by the French scientist J. B. de Lamarck in the eighteenth century, which embodied the principle, now known to be mistaken, that characteristics acquired by an organism during its lifetime can be inherited.

lamellibranchs *n. plu.* the Lamellibranchia, a large subclass of bivalve molluscs, including clams, cockles, mussels, etc.

Lamiales *n.* an order of dicot herbs, shrubs and trees including Verbenaceae (verbena), Lamiaceae (mint, etc.) and others.

lamina *n.* (1) a thin layer, plate or scale; (*bot.*) (2) the blade of a leaf or petal; (3) the flattened part of a thallus; (4) (*geol.*) a layer of sedimentary rock <1 cm thick, *alt.* lamination. *plu.* laminae.

laminar *a.* (*soil sci.*) *appl.* soil peds (*q.v.*) that are plate-like in appearance, horizontally elongate with parallel sides. *cf.* lenticular.

laminar flow the smooth flow of a fluid (liquid or gas) without internal mixing, sometimes esp. along a streamlined surface. *cf.* turbulent flow.

laminarian *a. appl.* zone between low-tide line to about 30 m, i.e. the zone typically inhabited by *Laminaria* seaweeds.

lamination *see* lamina.

lampreys *n.plu.* common name for primitive fish-like freshwater and marine chordates of the order Petromyzoniformes, in the class Agnatha, which as adults have sucking and rasping mouthparts.

lamprophyre *n.* a type of igneous rock found in minor intrusions (dykes or sills). Several varieties are recognized, e.g. mica lamprophyre (*q.v.*) and hornblende lamprophyre *q.v.*

lamp shells common name for the Brachiopoda *q.v.*

lancelets *n.plu.* common name for the Cephalocordata *q.v.*

land breeze gentle wind that blows offshore during the night, i.e. from the land to the sea. This airflow is created by an increase in air pressure over the land due to its more rapid nocturnal cooling, compared with that of the adjacent sea. *cf.* sea breeze.

land devils *see* whirlwind.

landfill gas (LFG) mixture of gases, predominantly methane, that is produced in landfill sites by the anaerobic decomposition of refuse.

landfill site suitable land site, such as a disused quarry, into which household and/or industrial waste is deposited in a regulated manner.

landfill tax tax levied per unit of waste sent to landfill.

landform *n.* a single structure on the Earth's surface, e.g.volcanic dome, valley or mountain. Landforms are classified as either primary or secondary. Primary landforms are those produced directly by deformation of the Earth's crust, e.g. by faulting or folding, and include volcanic and seismic structures. Secondary landforms are those produced by the sculpting action of various types of erosion.

land inventory information, in the form of maps and/or tables, showing the land use of a particular area, e.g. of a farm. *alt.* land-use survey.

landnam *n.* primitive type of agriculture practised in northern Europe in which crops were grown in cleared patches of forest.

landrace *n.* variety of a crop plant that has arisen by selection over time by indigenous farmers, and is thus adapted to local conditions.

land reclamation any process whereby hitherto unproductive land is brought back into use. *see* polder.

Landsat *see* Earth Resources Technology Satellite.

landscape *n.* (*geog.*) a distinct region, e.g. coastal or volcanic, composed of individual landforms *q.v.*

landslide *see* slide movement.

land systems approach method, based on aerial photographs, satellite images and field surveys, by which an area of land is environmentally assessed (in terms of its geology, geomorphology, climate, vegetation, soils and topography) for reasons of planning, engineering or agriculture.

land-use classification (LUC) assignation of land into different categories depending on use, e.g. industrial, urban or agricultural.

land-use survey land inventory *q.v.*

lanthanides *n.plu.* a collective term for the elements with atomic numbers 58 (cerium) to 71 (lutetium) inclusive, although lanthanum (atomic number 57) is considered by some to be a lanthanide. *alt.* lanthanide series, lanthanoids, lanthanons, rare earth elements, rare earths (*see* Appendix 2).

lanthanum (La) *n.* a lanthanide element (at. no. 57, r.a.m. 138.91), a silver-coloured metal when in elemental form. It is used, e.g., in the petrochemical industry as a catalyst in the cracking of oil.

lapiés karren *q.v.*

lapilli *see* pyroclasts.

lapis-lazuli *n.* a blue rock with a high content of lazurite (*q.v.*), which is prized as a decorative stone.

lapse rate rate at which atmospheric temperature decreases with altitude, i.e. a positive lapse rate equates to a negative vertical temperature gradient (*q.v.*) of the same magnitude. *see* dry adiabatic lapse rate, environmental lapse rate, saturated adiabatic lapse rate.

LAR leaf area ratio *q.v.*

larva *n.* independently living, post-embryonic stage of an animal that is markedly different in form from the adult and that undergoes metamorphosis into the adult form, e.g. caterpillar, grub, tadpole. *plu.* larvae. *a.* larval.

Larvacea *n.* class of tunicates (urochordates) that retain the larval 'tadpole' form throughout their lives.

larvivorous *a.* larva-eating.

latent *a.* (1) lying dormant but capable of development under certain circumstances, *appl.* buds, resting stages; (2) *appl.* characteristics that will become apparent under certain conditions.

lateral bud bud arising in the axil of a leaf at a node of a stem and that will develop into a side shoot.

laterite *a. appl.* tropical red soils containing alumina and iron oxides and little silica owing to leaching under hot moist conditions.

laterite *n.* duricrust (*q.v.*) formed by the deposition of iron and aluminium oxides.

late wood wood formed in the later part of an annual ring, having denser and smaller cells than early wood. *alt.* summer wood.

latex *n.* thick milky or clear juice or emulsion of diverse composition present in plants such as rubber trees and spurges, and in certain agaric fungi.

lathyrism *n.* disease of animals, characterized by fragile collagen, caused by eating seeds of certain *Lathyrus* species that contain beta-aminopropionitrile, which inhibits essential post-translational modification of collagen.

latifundia *n.plu.* large estates in Central and South America.

latitude *n.* angular distance (in degrees north or south) of any point on the Earth's surface along a parallel from the Equator (the latter having a latitude of 0° by definition). Circles drawn parallel to the Equator join up points of equal latitude and are known as lines of latitude.

latosol *n.* a leached red or yellow tropical soil.

laumontite *n.* $CaAl_2Si_4O_{12}.4H_2O$, one of the zeolite minerals *q.v.*

Laurales *n.* an order of dicot trees, shrubs and climbers, with ethereal oils in cells, and including the families Calycanthaceae (calycanthus), Lauraceae (laurels) and others.

Laurasia *n.* the former northern land mass comprising present-day North America, Europe and northern Asia, before they were separated by continental drift. *cf.* Gondwanaland.

laurilignosa *n.* a type of subtropical forest and bush composed of laurel.

laurinoxylon *n.* fossil wood.

lava *n.* the term given to magma (*q.v.*) when it reaches the Earth's surface as a liquid. The composition and temperature of lava determines its viscosity and therefore the way it behaves. *see* acid lava, basic lava.

lava flow lava moving on the surface of the Earth as a stream.

law of faunal succession principle used in dating fossil-bearing rocks that states that fossil-bearing rocks at widely separated individual sites may be correlated when each of the rocks contains particular faunal assemblages that are known to be part of a chronostratigraphic sequence.

law of independent assortment *see* independent assortment.

law of limiting factors principle that each physical variable in an ecosystem has maximum and minimum levels outside of which a particular species cannot survive. Within these tolerance limits and under steady-state conditions, the variable in shortest supply will effectively limit the growth of the population. *see also* law of the minimum.

law of segregation *see* segregation of alleles.

law of the minimum the principle that the factor for which an organism or species has the narrowest range of tolerance or adaptability limits its existence. *see also* Liebig's law of the minimum.

Law of the Sea Convention international convention resulting from the 1982 UN Conference on the Law of the Sea, which established the International Seabed Authority, which regulates exploration and exploitation of mineral resources of the oceans. *see also* United Nations Conference on the Law of the Sea.

law of tolerance *see* Shelford's law of tolerance.

lawrencium (Lr) *n.* a transuranic element [at. no. 103, r.a.m. (of its only known isotope) 257].

Lawson criterion in order to extract useful fusion energy from a thermonuclear plasma, it is generally accepted that the product of the ion density (in ions per metre cubed) and the confinement time (*q.v.*) (in seconds) must exceed 10^{20} ions s m^{-3}. It is this condition that is called the Lawson Criterion.

layer *n.* (*bot.*) horizontal stratum in a plant community, i.e. the tree layer, comprising the canopy, the shrub layer comprising the shrubby understorey, the herb layer comprising grasses and herbaceous plants, and the ground (moss) layer comprising the ground surface and lichens and mosses.

layered *a.* (*geol.*) *appl.* structure of igneous rocks that exhibit bands that differ from each other in terms of both mineral composition and texture (*q.v.*) and/or colour, *alt.* banded.

lazurite *n.* a rare, azure-blue tectosilicate mineral, with a formula variously cited as either $(Na,Ca)_8(Al,Si)_{12}O_{24}(S,SO_4)$ or $(Na,Ca)_8(AlSiO_4)_6(SO_4,S,Cl)_2$. An important constituent of the rock lapis-lazuli, which is prized as a decorative stone.

lazy-bed cultivation type of mound cultivation practised in the Highlands and Islands of Scotland, UK, in which potatoes (mainly) are grown on ridges (dug by spade) separated by furrows.

LC lethal concentration *q.v.*

LC$_{50}$ lethal concentration 50, concentration of a toxic chemical that kills 50% of the organisms in a test population per unit time.

LD$_{50}$ lethal dose 50, a measure of infectivity for viruses, toxicity of chemicals, etc., the dose at which 50% of test animals die.

LDCs less developed countries *q.v.*

leachate *n.* any liquid that, having percolated through a particulate medium, contains material dissolved from that medium, e.g. water after it has seeped through a soil, spoil heap, landfill site or mine. Leachates contaminated with toxic chemical species may constitute a source of water pollution.

leaching *n.* the process by which chemicals in the upper layers of the soil are dissolved and carried down into lower layers.

lead (Pb) *n.* metallic element (at. no. 82, r.a.m. 207.20), forming an unreactive soft dense grey metal in elemental form. Extracted chiefly from the mineral galena (PbS). Lead compounds are used in many industries and are toxic. Metallic lead is used as a shield for radioactive materials. Combustion of petrol containing tetraalkyl lead compounds (antiknock additives) has introduced significant amounts of lead into the atmosphere. Environmental lead contamination can also arise through its use in paints, lead shot and water pipes. *see also* lead poisoning.

leaded petrol petrol to which tetraalkyl lead antiknock agents have been added.

leader–follower system (*agric.*) a grazing system used in beef production whereby young cattle are allowed to graze fresh areas of pasture first, followed in turn by older stock.

lead poisoning chronic poisoning resulting in muscle cramps, abdominal pain and nervous system damage, esp. in developing children, caused by inhalation, ingestion or absorption through the skin of lead compounds. Environmental sources of lead are vehicle exhausts (from antiknock compounds containing lead, such as tetraethyl lead), lead pigments in paint and lead in drinking water (from lead pipes and plumbing solder).

leaf *n.* an expanded outgrowth from a plant stem, usually green and the main photosynthetic organ of most plants.

leaf area index (LAI) of a given area of vegetation, the total area of photosynthetic leaf surface divided by the area of ground surface covered. It gives a measure of the photosynthetic potential of the area.

leaf area ratio (LAR) the ratio of the photosynthetic surface area of a leaf to its dry weight.

leaf blotch a disease affecting wheat and barley, caused by the fungus *Septoria tritici*.

leaf insects common name for some members of the Phasmida (*q.v.*) whose bodies mimic leaves in form.

leaf mosaic the arrangement of leaves on a plant that results in minimum overlap and maximum exposure to sunlight.

lean *a. appl.* mixtures of fuel and air in which the amount of air is in excess of that required for complete combustion of the fuel. *cf.* rich. Tuning an internal combustion engine from a rich to a lean mixture tends to decrease the amounts of unburnt hydrocarbons and carbon monoxide produced, while increasing the quantity of NO_x generated. *see also* lean burn technology.

lean burn technology an approach to the reduction in the pollution caused by internal combustion engines based on the use of lean (*q.v.*) fuel : air mixtures.

learning *n.* any process in an animal in which its behaviour becomes consistently modified as a result of experience.

leatherjackets *n.plu.* larvae of the cranefly (*Tipula* spp.). These grey-brown, tough-skinned larvae attack the roots of cereals and vegetable crops including potatoes and sugar beet.

Le Chatelier's principle a principle that states that if a system at dynamic equilibrium is perturbed by changes in concentration, pressure or temperature, it will alter its composition so as to minimize the change.

lectotype *n.* a specimen chosen from syntypes to designate type species.

leeches *n.plu.* common name for the Hirudinea *q.v.*

lee depressions *see* non-frontal depressions.

leeward side side of a mountain (or other area of high relief) facing away from the wind. *cf.* windward side.

Legionnaire's disease a bacterial disease caused by *Legionella pneumophila*, an opportunistic pathogen that thrives in water droplets in, e.g., air-conditioning cooling towers, but which can infect humans, causing pneumonia and sometimes death.

legume *n.* (1) type of fruit derived from a single carpel that splits down both sides at maturity, characteristic of the pea family; (2) a member of the Leguminosae, e.g. peas, beans, clovers, vetches, gorse, broom.

Leguminosae *n*. large family of dicotyledonous plants, commonly called legumes or leguminous plants, including trees, shrubs, herbs and climbers, with typical sweet-pea shaped flowers and fruit as pods, and including peas, beans, clovers, vetches, etc. *see also* Fabales.

leguminous *a*. *pert*. (1) Leguminosae; (2) legumes; (3) (*pert*.) or consisting of, peas, beans or other legumes.

Leishmania genus of parasitic protozoa, infecting humans and other mammals, with sandflies as the intermediate host and vector. *see* leishmaniasis.

leishmaniasis *n*. a disease, widespread in tropical countries, caused by parasitic protozoa of the genus *Leishmania*, transmitted to man, and other mammals, by infected sandflies. *L. donovani* causes the chronic and often fatal disease visceral leishmaniasis, while *L. tropica* causes cutaneous leishmaniasis or tropical sore.

Leitneriales *n*. an order of resinous dicot shrubs comprising the family Leitneriaceae with the single genus *Leitneria*.

lek *n*. a special arena removed from nesting and feeding grounds, used for communal courtship display (lekking) preceding mating in some birds, e.g. ruffs, many grouse species. The term is now also sometimes applied to similar areas used by other animals for communal displays.

lekking *n*. a highly ritualized sexual display by birds such as grouse, which takes place on a particular display ground, the lek, and that precedes mating.

lenacil *n*. soil-acting herbicide used to control annual weeds in crops such as beet.

lentic *a*. *appl*. (1) standing water; (2) organisms living in swamp, pond, lake or any other standing water.

lenticular *a*. (*soil sci.*) *appl*. soil peds (*q.v.*) that are plate-like in appearance, horizontally elongate with convex surfaces. *cf*. laminar.

lentiviruses *n.plu*. a subfamily of non-oncogenic retroviruses, which cause chronic infections that only become manifest years after infection, including human immunodeficiency virus (HIV).

lepidocrocite *n*. FeO(OH), a mineral similar to goethite (*q.v.*), to which it slowly converts. Formed in soils when iron(II) is oxidized in the presence of organic matter.

Lepidodendron a genus of fossil tree-ferns with small leaves producing scale-like leaf scars.

lepidolite *n*. $K(Li,Al)_{2-3}(Al,Si)_4O_{10}(O,OH,F)_2$, one of the mica minerals *q.v.*

lepidophyte *n*. fossil fern.

Lepidoptera *n*. order of insects commonly known as moths and butterflies. Their bodies and wings are covered by small scales, often brightly and variously coloured, forming characteristic patterns. They undergo complete metamorphosis, the larval (caterpillar) stage giving rise to a pupa in which metamorphosis occurs with development of adult structures such as the two pairs of membranous wings, the legs and compound eyes. Adult Lepidoptera feed largely on nectar, through a hollow proboscis. *a*. lepidopterous.

Lepidosauria, lepidosaurs *n*., *n.plu*. subclass of reptiles comprising the lizards, snakes and amphisbaenians, and the tuatara, with a diapsid skull, and with limbs and limb girdles unspecialized, reduced or absent.

leptospirosis *n*. acute infection of the central nervous system, liver and kidneys caused by bacteria of the genus *Leptospira*. These are carried in the kidneys of rodents (primary hosts) and are transmitted to humans when they come into contact with water contaminated with rodents' urine. A particularly serious form of leptospirosis is Weil's disease, caused by *Leptospira icterohaemorrhagiae*.

less developed countries (LDCs) non-industrialized nations, located mainly in the southern hemisphere, characterized by high population growth rates, high birth and death rates, and low per capita income. *cf*. more developed countries.

lessivage *n*. (*soil sci.*) the removal of clay and iron oxides from the A-horizon (eluvial layer) by essentially mechanical means and without chemical alteration.

lestobiosis *n*. the relation in which colonies of small species of insect nest in the walls of the nests of larger species and enter their chambers to prey on the brood or rob food stores.

lethal *a.* (1) causing death; *appl.* (2) a parasite, fatal or deadly in relation to a particular host; (3) (*genet.*) mutations or alleles that when present cause the death of the embryo at an early stage.

lethal concentration (LC) where death is the criterion of toxicity, the results of toxicity tests are expressed as a number (LC_{50}, LC_{70}) that indicates the percentage of test organisms killed at a particular concentration over a given exposure time, e.g. the 48-h LC_{70} is the concentration of a toxic material that kills 70% of the test organisms in 48 h.

lethal dose (LD) dose of a toxic chemical or of a pathogen that kills all the animals in a test sample within a certain time. *see also* LD_{50}. *cf.* median lethal dose.

lethality *n.* the ratio of fatal cases to the total number of cases affected by a disease or other harmful agent.

Leu leucine *q.v.*

leucine (Leu, L) *n.* alpha-amino isocaproic acid, an amino acid with a hydrocarbon side chain, constituent of protein, essential in human and animal diet.

leucite *n.* $KAlSi_2O_6$ (may contain some Na) one of the feldspathoid minerals *q.v.*

leuco- prefix derived from Gk *leukos*, white.

leucoanthocyanidins *n.plu.* a group of colourless flavonoids.

leucosin *n.* a storage polysaccharide forming whitish granules in some yellow-brown algae.

Levanter *n.* strong, easterly wind that affects western parts of the Mediterranean region such as Spain and the Straits of Gibraltar.

levees *see* artificial levees, natural levees.

Lewis acid (*chem.*) a chemical species that is capable of forming a dative bond with a Lewis base (*q.v.*). In order to be able to do this, it must possess a vacant orbital of an appropriate energy, so that it can accept a share of the lone pair of electrons donated to the bond by the Lewis base.

Lewis base (*chem.*) a chemical species that is capable of forming at least one dative bond with a Lewis acid (*q.v.*). In order to be able to do this, it must possess at least one lone pair of electrons, so that it can donate these to the bond.

ley *n.* temporary agricultural grassland, which is sown and used as a crop.

LFG landfill gas *q.v.*

LHC light-harvesting complex *q.v.*

lherzolite *see* peridotite.

L-horizon layer of undecomposed organic matter located on the soil surface.

Li symbol for the chemical element lithium *q.v.*

liana *n.* any woody climbing plant of tropical and semitropical forests.

Lias *n.* marine and estuarine deposits of the Jurassic period, containing remains of fossil cycads, insects, ammonites and saurians. *a.* Liassic.

lice *n.plu.* common name for various small insects of the orders Psocoptera (book lice) and Anoplura (the sucking and biting lice), and for crustaceans of the orders Isopoda (woodlice) and Branchiura.

lichen *n.* a composite organism formed from the symbiotic association of certain fungi and a green alga or cyanobacterium, forming a simple thallus, found encrusting rocks, tree trunks, etc., often in extreme environmental conditions.

lichen desert area devoid of lichens due to atmospheric pollution.

lichen heath type of ground cover composed entirely of lichens.

lichenicolous *a.* living or growing on lichens.

lichenization *n.* (1) production of a lichen by alga and fungus; (2) spreading or coating of lichens over a substrate; (3) effect of lichens on their substrates.

lichenology *n.* the study of lichens.

Liebig's law of the minimum (1) the food element least plentiful in proportion to the requirements of plants limits their growth; (2) law of the minimum *q.v.*

life *n.* living organisms are characterized by active metabolism, growth, reproduction and response to stimuli. In practical terms all known living organisms on this planet can be distingushed from other complex physico-chemical systems by their storage and transmission of molecular information in the form of nucleic acids, their possession of enzyme catalysts, their energy relations with the environment and their internal energy conversion processes (e.g. photosynthesis, respiration and other enzyme-catalysed metabolic activities), their ability to grow and reproduce, and their ability to respond to stimuli (irritability). Entities such as viruses, which satisfy only some of these criteria, are also sometimes considered as part of the living world.

life cycle the various phases an individual passes through from birth to maturity and reproduction.

life-cycle cost the initial cost of an economic good plus its lifetime operating costs.

life expectancy *see* average life expectancy.

life form the typical adult form of a species.

life span (1) the lifetime of an individual; (2) the longest time that an organism of a particular species can be expected to live; (3) of a species, the period of time from the first appearance of the species in the fossil record until its disappearance.

life tables demographic data required to calculate, e.g., the intrinsic rate of increase of a population. They comprise the survivorship schedule, which gives the number of individuals surviving to each particular age, and the fertility schedule, which gives the average number of female offspring that will be produced by a single female at each particular age. From these, the net reproductive rate, R_0, the average number of female offspring produced by each female during her lifetime, can be calculated. The intrinsic rate of increase, r, of the population can be computed from survivorship and fertility schedules using the Euler–Lotka equation.

lifetime *n.* (1) the period of time for which an individual organism lives or, e.g., a pollutant persists in the environment; (2) residence time *q.v.*

ligand *n.* (1) (*mol. biol.*) any molecule or ion that binds specifically to another molecule, usually used for signalling molecules binding to receptor proteins, regulatory molecules binding to enzymes, etc; (2) (*chem.*) chemical species that contain at least one donor atom (*q.v.*) and are therefore capable of forming dative bond(s) with a Lewis acid *q.v.*

light *n.* a term usually applied to visible light (*q.v.*), but may also be used to include

ultraviolet radiation (*q.v.*) and/or, less commonly, infrared radiation *q.v.*

light compensation point the level of light at which the rates of photosynthesis and respiration are equal.

light-field microscopy the simplest type of light or optical microscopy in which the specimen is illuminated directly by the light source and appears as a darker image against a light background.

light-harvesting complex (LHC) complex of chlorophyll, other pigments and proteins in plant chloroplasts that collects light energy and passes it on to the photosynthetic reaction centre.

light microscopy type of microscopy that uses visible light and optical lenses to create the magnified image. *alt.* optical microscopy. *cf.* electron microscopy.

lightning *n.* visible discharge of electricity between groups of clouds or between clouds and the ground, generated during thunderstorms.

light rapid transit system (LRTS) city railway system of light construction, usually powered by electricity, e.g. Tyneside Metro system, UK.

light reactions in photosynthesis, reactions occurring in the thylakoid membranes of chloroplasts, in which light energy drives the synthesis of NADPH and ATP, with the release of oxygen as a waste product. *cf.* dark reactions.

light ruby silver proustite *q.v.*

light saturation (1) the optimum intensity of light for maximum production in a particular ecosystem; (2) the amount of light that can be used by a species photosynthesizing at its maximum rate, *see* light compensation point.

light water normal water, i.e. water that contains the most common isotope of hydrogen, ^1_1H, the formula of this water being $^1_1H_2O$ or more simply H_2O. *cf.* heavy water.

lignicolous *a.* growing or living on or in wood.

lignification *n.* (1) wood formation; (2) the thickening of plant cell walls by deposition of lignin, which occurs in both primary and secondary walls.

lignin *n.* a hard material, highly resistant to degradation, found in the walls of cells of xylem and sclerenchyma fibres in plants, a very variable cross-linked polymer of phenylpropane units such as coniferyl alcohol (guiacyl), sinapyl alcohol (syringyl) or hydroxycinnamylalcohol, and which stiffens the cell wall.

lignite *n.* soft, brown-black, low-rank coal with a low carbon content and a high moisture and volatile content.

lignivorous *a.* eating wood, *appl.* various insects.

lignosa *n.* vegetation made up of woody plants.

Ligustrales Oleales *q.v.*

Liliales *n.* an order of monocot plants, growing from rhizomes or bulbs, mostly herbaceous, and including the families Liliaceae (lily), Agavaceae (agave), Alliaceae (onion), Amaryllidaceae (daffodil), Dioscoreaceae (yam) and others. *alt.* Liliiflorae.

Liliopsida *n.* in some plant classifications the name for the class containing the monocotyledons.

limb *n.* (1) branch; (2) arm; (3) leg; (4) wing; (5) (*bot.*) expanded part of calyx or corolla, the base of which is tubular.

lime *n.* (1) (*agric.*) a material containing the oxides, hydroxides and/or carbonates of calcium and/or magnesium, which is applied to acid soils for neutralization purposes; (2) (*chem.*) archaic collective term for calcium oxide, CaO (i.e. quicklime) and calcium hydroxide, $Ca(OH)_2$ (i.e. slaked lime).

limestone *n.* a non-clastic sedimentary rock composed mainly of calcite (calcium carbonate, $CaCO_3$). Some limestones are formed from the direct precipitation of calcium carbonate from water. However, the majority result from the accumulation of calcareous skeletons of aquatic organisms such as foraminifera, corals and molluscs.

limestone pavement a flat area of exposed limestone with a dissected surface as found, e.g., in the Craven district of Yorkshire, UK. The channels, roughly at right angles to each other, are produced by dissolution in the region of joint-fractures. These channels are known as grikes and the blocks that they separate are called clints.

limicolous *a.* living in mud.

limit cycle regular cyclic pattern oscillating between an upper and a lower limit,

such as the abundance of predators and prey in a predator–prey system.

limiting factor any single factor that limits, e.g., a biochemical process, the growth of an organism, or its abundance or distribution.

limivorous *a.* mud-eating, *appl.* certain aquatic animals.

limnetic *a.* (1) living in, or *pert.* marshes or lakes; (2) living in open water; (3) *appl.* zone of deep water between surface and compensation depth (depth at which photosynthesis cannot be supported owing to insufficient light).

limnium *n.* a lake community.

limnobiology *n.* the study of life in standing waters, i.e. ponds, marshes, lakes.

limnobios *n.* freshwater plants and animals collectively.

limnobiotic *a.* living in freshwater marshes.

limnology *n.* the study of the biological and other aspects of standing waters.

limnophyte *n.* a pond plant.

limnoplankton *n.* the floating microscopic life in freshwater lakes, ponds and marshes.

limonite *n.* natural hydrated iron(III) oxide ($Fe_2O_3.nH_2O$). An important iron ore mineral.

Limulus a genus of horseshoe crab *q.v.*

Lincoln index, Lincoln–Peterson index census formula used to estimate animal population size by the capture–recapture method *q.v.*

lindane *n.* the gamma isomer of hexachlorocyclohexane ($C_6H_6Cl_6$), an organochlorine insecticide used, e.g., as a seed dressing for cereals and as a fumigant to control insect pests of stored products. *alt.* gamma-HCH.

Lindemann efficiency ratio of energy assimilated at one trophic level to the energy assimilated by the previous trophic level.

lineage *n.* a line of common descent.

lineage group group of species allied by common descent.

linear dune an elongated dune with two approximately opposite slip-faces (*q.v.*). The origin of this type of dune is controversial, although it is usually believed to involve strong winds blowing (on a diurnal or seasonal basis) from two prevailing, but opposing directions.

linear growth type of growth in which the amount increases by the same amount over each set period (e.g. a year). *cf.* exponential growth.

line fishing type of marine fishing during which a single line, with numerous baited lines (called snoods) appended, is placed on the sea bed.

lines of latitude *see* latitude.

line transect the recording of types and numbers of plants or other organisms along a measured line. *see also* transect sampling.

linkage *n.* the tendency for some parental alleles to be inherited together, in opposition to Mendel's law of independent assortment, and which is due to their presence close together on the same chromosome.

linkage disequilibrium condition in which certain alleles at two linked loci are nonrandomly associated with each other, either because of very close physical proximity that virtually precludes recombination between them, or because the combination is under some form of selective pressure.

linkage group the genes carried on any one chromosome.

Linnean *a. pert.* or designating the system of binomial nomenclature and classification established by the eighteenth-century Swedish biologist Carl von Linné or Linnaeus. *see* binomial nomenclature.

linneon *n.* a taxonomic species distinguished on purely morphological grounds, esp. one of the large species described by Linnaeus or other early naturalists.

linseed *n.* short-strawed varieties of *Linum usitatissimum*, a member of the family Linaceae. These are cultivated for their seeds (also termed linseed), which have a high oil content. *cf.* flax.

Linum *see* (1) flax; (2) linseed.

linuron *n.* soil-acting and contact herbicide used for general weed control in crops such as carrots, potatoes and spring cereals.

lipid *n.* any of a diverse class of compounds found in all living cells, insoluble in water but soluble in organic solvents such as ether, acetone and chloroform, and which include fats and oils (triacylglycerols), fatty acids, glycolipids, phospholipids and steroids, some lipids being essential components of biological membranes, others

acting as energy stores and fuel molecules for cells.

lipid bilayer double layer of phospholipids, each molecule oriented with the hydrophilic group on the outside and the hydrophobic group to the interior of the layer, which is the basic structure of cell membranes.

lipid bodies lipid storage structures found in oil-rich plant seeds, composed of a large droplet of triacylglycerol surrounded by a single-layered membrane. *alt.* oil bodies.

lipogenesis *n.* synthesis of fatty acids and lipids.

lipogenous *a.* fat-producing.

lipolytic *a.* capable of digesting or dissolving fat, *appl.* enzymes.

lipophilic *a.* fat-soluble.

lipopolysaccharide (LPS) *n.* any molecule consisting of a lipid joined to a polysaccharide, one of the main constituents of the outer cell envelope of Gram-negative bacteria.

lipoprotein *n.* complex of lipid and protein.

liquefaction of coal *see* coal liquefaction.

liquefied natural gas (LNG) natural gas liquefied by a combination of lowered temperature and increased pressure for ease of transport or for use as an engine fuel.

liquefied petroleum gas (LPG) engine fuel composed of a mixture of petroleum gases, mainly butane and propane, liquefied by increased pressure.

liquid *n.* matter in a form that if placed into a container will fill it from the bottom up, moulding itself exactly to fit the container's inner walls and floor. *cf.* solid, gas.

liquid organic manure slurry *q.v.*

liquid paraffin a liquid distillate of crude oil suitable for human consumption, used, e.g., as a laxative, also known, esp. in North America, as mineral oil.

Lissamphibia *n.* in some classifications a subclass of amphibians containing all extant species and divided into three orders, i.e. Salientia (Anura), Urodela and Apoda.

List 1 Black List *q.v.*

List 2 Grey List *q.v.*

listeriosis *n.* an infection caused by the bacterium *Listeria monocytogenes*, transmitted to humans via contaminated foods esp. cheeses and patés.

List of Priority Pollutants list produced by the US Environmental Protection Agency (EPA) of the 129 substances considered to be most dangerously toxic.

lithification *n*. process whereby unconsolidated sediments are transformed, by compression, compaction and cementation, into sedimentary rocks. *alt*. diagenesis.

lithium (Li) *n*. metallic element (at. no. 3, r.a.m. 6.91), forming a light soft silvery-white metal.

lithocarp *n*. a fossil fruit.

lithodomous *a*. living in holes or clefts in rock.

lithogenous *a*. rock-forming or rock-building, as the reef-building corals.

lithophagous *a*. (1) stone-eating, as some birds; (2) rock-burrowing, as some molluscs and sea urchins.

lithophyll *n*. fossil leaf or leaf impression.

lithosequence *n*. series of related soils that differ from one another primarily as a consequence of differences between their parent materials.

lithosere *n*. a plant succession originating on rock surfaces.

lithosols *n.plu*. soils that develop at high altitudes on resistant parent materials that withstand weathering and result in a humus-rich, shallow, stony soil.

lithosphere *n*. (1) the non-living, non-organic part of the environment, such as rocks, the mineral fraction of soil, etc.; (2) the coherent solid layer formed from the Earth's crust and the upper part of the mantle.

lithospheric plates the essentially rigid plates of lithosphere that make up the Earth's surface. The plastic nature of the underlying asthenosphere allows these plates to move relative to one another. *alt*. tectonic plates.

lithotroph *n*. an autotroph, esp. a chemoautotroph (*q.v.*). *a*. lithotrophic.

litter *n*. (1) (*soil sci*.) material of which a little layer (*q.v.*) is composed; (2) (*zool*.) offspring produced at a single multiple birth.

litter layer undecomposed and partly decomposed plant residues (e.g. fallen leaves) on the surface of soil, mainly in woodlands, *see also* O-horizon and Fig. 28, p. 385. *alt*. L-layer.

litterfall *n*. amount of dead plant material added to the forest floor over a given period of time.

Little Ice Age period between 1430 and 1850 when global temperatures were significantly lower than present-day temperatures, resulting in glacier expansion in much of the world.

littoral *a*. growing or living near the seashore.

littoral zone (1) zone of seashore between high- and low-water marks (*see* Fig. 25, p. 369); (2) in lakes, *appl*. zone of shallow water and bottom above compensation depth (depth at which photosynthesis cannot be supported); (3) in lakes, *appl*. the zone of shore inhabited by rooted plants.

liverworts *n.plu*. common name for members of the plant division Hepatophyta *q.v.*

living fossil extant species of ancient lineage that has remained morphologically unchanged for a very long time, and whose only close relatives are fossils and that in some cases was itself thought to be extinct, such as the coelacanth, the ginkgo and the metasequoia.

lizardite *n*. a variety of the mineral serpentine *q.v.*

lizards *n.plu*. common name for members of the reptile suborder Lacertilia *q.v.*

llanos *n.plu*. vast treeless plains of South America, north of the Amazon.

L-layer litter layer *q.v.*

LNG liquefied natural gas *q.v.*

load *n*. (*geog*.) material carried by glaciers, wind or running water.

loam *n*. a rich friable soil consisting of a fairly equal mixture of sand and silt and a smaller proportion of clay.

lobe-finned fishes common name for lungfishes and crossopterygians.

local faciation *n*. local differences in abundance or proportion of dominant species.

local stability the tendency for a community to return to its original state after small disturbances.

loch *n*. Scottish (UK) word meaning lake. *see also* sea loch.

lociation local faciation *q.v.*

Loculoascomycetes *n*. a group of ascomycete fungi parasitic on plants and insects, and that bear their asci in cavities in a loose hyphal stroma. They include the sooty moulds and agents of various scab (e.g. *Venturia*, apple scab), leaf spot (e.g. *Mycosphaerella* on strawberries) and anthracnose diseases of plants.

locus *n.* physical position of a gene on a chromosome. For genes that occur in different variants (alleles) within the population, the locus is occupied by one of the alleles. *plu.* loci.

locusts *n.plu.* common name for many of the Orthoptera *q.v.*

lodes veins *q.v.*

lodged *a. appl.* cereal crops that are no longer standing upright.

loess *n.* (1) a sedimentary rock made up of fine particles deposited from suspension in the air; (2) fine-grained, silt-grade material transported and deposited by the wind, winnowed esp. from glacial terrains but also from deserts, river floodplains and deltas. Soils derived from loess deposits are typically very fertile when mixed with humus.

logarithmic phase, log phase the rapid stage in growth of a bacterial culture (or any other population) when increase follows a geometric progression.

logistic curve an S-shaped curve initially rising slowly, then steeply, and finally flattening out, and which is characteristic of the growth and stabilization of a population subject to some density-limiting factor such as food supply. *alt.* S-shaped growth curve.

log-normal distribution logarithmic-normal distribution, a statistical distribution in which the logarithm of a variable quantity plots as a bell-shaped (normal) curve.

London force the very weak attraction that occurs between any two atoms or molecules when in close proximity to one another. It originates from a sympathetic movement of electrons within each of the interacting particles, associated with an alignment in the orientations of the momentary and fluctuating electric dipoles present within them. The London force increases in magnitude with molecular weight and can be quite significant between large molecules. *alt.* dispersion force. *see also* van der Waals' forces.

London-type smog an intermittent, severe form of atmospheric pollution that consists of a combination of coal smoke and fog. Smogs of this type are initiated on cold winter nights under conditions that allow temperature inversions (*q.v.*) to occur. The main primary pollutants are sulphur dioxide (SO_2) and smoke particulates. The latter of these provide surfaces upon which the SO_2 may be oxidized to sulphuric acid. London-type smogs occur in many industrialized centres where large amounts of coal are burned and are now largely restricted to parts of the developing world. The last major smog in London, UK, occurred in December 1962. *alt.* classical smog, pea-souper (colloquial), sulphurous smog. *cf.* Los Angeles-type smog.

lone pair of electrons (*chem.*) two electrons in the valence shell of an atom or ion that are not part of a covalent bond. Lone pairs are capable of forming dative bonds (*q.v.*) or hydrogen bonds *q.v.*

long-day plants plants that will only flower if the daily period of light is longer than some critical length: they usually flower in summer. The critical factor is in fact the period of continuous darkness they are exposed to. *cf.* day-neutral plants, short-day plants.

longest day *see* solstice.

longevity *n.* (1) life span (*q.v.*); (2) length of life for an individual, a population or a species.

longitude *n.* angular distance (in degrees west or east) of any point on the Earth's surface along a meridian from the prime meridian (longitude 0˚).

longitudinal profile a study of something (e.g. a population) over time.

long-range forecast *see* weather forecasting.

longshore bar a submerged ridge of sand that forms parallel to the shoreline in the nearshore (*q.v.*) zone of the beach.

longshore current a water current that flows parallel to the shore.

longshore drift the common process by which coastal sediments, under the influence of longshore currents, move predominantly in one direction, parallel to the shore.

longshore trough a trough that occurs on the landward side of a longshore bar *q.v.*

long-wave radiation (*clim.*) in certain contexts used as a synonym for terrestrial radiation *q.v.*

lophodont *a.* having transverse ridges on the cheek–teeth grinding surface.

lophophore *n.* horseshoe-shaped crown of tentacles surrounding the mouth, characteristic of animals of the phyla Brachiopoda, Ectoprocta (Bryozoa) and Phoronida.

lopolith *n.* a basin-shaped, sheet-like igneous intrusion. Lopoliths frequently contain mafic rock.

Loricifera *n.* small phylum of tiny marine multicellular pseudocoelomate animals, with spiny heads, and abdomen covered with spiny plates called lorica, living in sediments.

Los Angeles-type smog a type of smog that occurs on warm, sunny days when traffic is busy. The main primary pollutants are NO_x (*q.v.*) (chiefly NO) and unburnt hydrocarbons emitted by motor vehicles. The concentrations of these pollutants are maximal during the morning and evening rush hours. On smoggy days, the primary pollutants present in the morning undergo photochemical reactions (*q.v.*), yielding the secondary pollutants that constitute the smog. The most significant secondary pollutants formed are nitrogen dioxide (NO_2) (*see* NO_x), oxidants (mainly ozone, O_3), partially oxidized organic species (including aldehydes and peroxyacyl nitrates *q.v.*) and an aerosol that contains both inorganic and organic matter. The generation of these noxious secondary pollutants occurs over several hours. Therefore, smogs of this type are at maximum intensity in the early afternoon. Cities currently worst afflicted by this phenomenon are Mexico City and Baghdad. *alt.* photochemical smog. *cf.* London-type smog.

loss-of-function *a. appl.* mutant alleles that produce no functional gene product.

lotic *a.* (1) *appl.* or *pert.* running water; (2) living in a brook or river.

Lotka–Volterra models two mathematical models put forward (independently) in the 1920s by Lotka and Volterra for (1) two-species competition and (2) predator–prey interactions.

Love Canal affair a case of environmental mismanagement concerning the Love Canal, Niagara Falls, New York state, USA. This canal, abandoned before completion, was used for the disposal of chemical wastes (including highly toxic pesticides) during the late 1940s and early 1950s. Once filled, with an estimated 20 000 tonnes of industrial waste, it was sealed with a clay cap. It was subsequently developed as a residential area. However, during the 1970s, toxic chemicals began to leach out of the dump into the basements of some nearby houses. This led to the evacuation and relocation of many residents, the sealing of the site and the initiation of a large and costly rehabilitation programme.

lower atmosphere that part of the Earth's atmosphere that is found within *ca.* 50 km of its surface, delimited by the stratopause, and encompassing the troposphere and stratosphere (*see* Fig. 16). The composition of the lower atmosphere is relatively constant, consisting of 78.08% nitrogen, 20.95% oxygen, 0.93% argon, 0.03% carbon dioxide and 0.01% trace gases (e.g. methane, hydrogen and helium) (all figures are volume percentages and refer to dry air) plus a variable amount of water vapour.

lower critical ambient temperature for an endothermic animal, the surrounding temperature below which the animal must generate heat in order to maintain its body temperature.

lower shore zone of seashore that extends from the lowest low-water level to the average low-water level (see Fig. 25, p. 369) and that is therefore only uncovered occasionally and for short periods.

lowland *n.* land up to about 700 m height above sea-level.

low-level radioactive waste radioactive wastes consisting of clothing, equipment and discarded materials from hospitals and laboratories where radioactive materials have been used, and slightly contaminated soil and rubble from buildings in which radioactive materials have been stored or used. It contains mainly radioactive elements with short half-lives and is disposed of by burial in the ground.

low-quality energy refers to energy that has little capacity to do useful work, e.g. low-temperature heat. *cf.* high-quality energy.

low-rank coal coal that is relatively unmodified by processes associated with

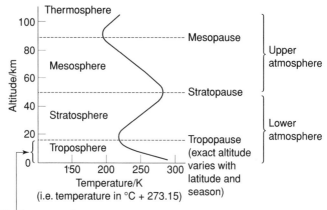

Weather occurs here

Fig. 16 The structure of the atmosphere (from Bunce, 1990, after Finlayson-Pitts and Pitts, 1986, *Atmospheric Chemistry*, reprinted by permission of John Wiley & Sons Inc.).

either burial or tectonic activity, e.g. lignite. *cf.* high-rank coal.

lows depressions *q.v.*

low-tide platform a type of shore platform (*q.v.*), which develops near the low-tide mark in calcareous rock strata, apparently through the processes of dissolution and biological erosion.

loxodont *a.* having molar teeth with shallow grooves between the ridges.

LPG liquefied petroleum gas *q.v.*

Lr symbol for the chemical element lawrencium *q.v.*

LRTS light rapid transit system *q.v.*

Lu symbol for the chemical element lutetium *q.v.*

lubricating oil liquid containing a mixture of hydrocarbons (C20 and above, b.p. >350 °C) produced by the fractional distillation of crude oil, and used for lubrication.

LUC land-use classification *q.v.*

lucerne *n.* a leguminous perennial (either *Medicago sativa* or *M. falcata*) that can be grazed but is usually grown for hay. *alt.* alfalfa.

luciferin *n.* light-emitting compound whose enzymatic oxidation by luciferase is responsible for bioluminescence in organisms such as many deep-sea fishes, coelenterates, fire-flies and glow-worms.

lumbricid *a. appl.* worms of the genus *Lumbricus* and close relatives, i.e. earthworms.

lumen *n.* internal space of any tubular or sac-like organ or subcellular organelle.

luminal *a.* within or *pert.* a lumen.

luminescent organs light-emitting organs, present in various animals.

lunar month the time interval from one new moon to the next (i.e. 29 days, 12 h and 44 min).

lunar rhythms physiological or behavioural patterns influenced by the lunar cycle, which occur in both marine and terrestrial organisms, *see also* tidal rhythms.

Lund tubes large tubes made of butyl rubber (diameter 45.5 m, depth 15 m) used in lakes to isolate columns of water plus bottom sediments, for the purpose of nutrient manipulation experiments.

lunettes *n.plu.* arc-shaped dunes composed of either clay aggregates and evaporite minerals, or sand particles. These form on the lee side of, e.g., pans (*q.v.*) or saline lakes.

lung *n.* organ specialized for the respiratory uptake of oxygen directly from air and release of carbon dioxide to the air. In vertebrates, lungs are present in air-breathing fishes (the lungfishes, Dipnoi) and tetrapods.

In mammals, the lungs are a paired mass of spongy tissue made up of finely divided airways lined with moist epithelium extending from the bronchi and ending in small sacs, the alveoli, providing a large surface area for gas exchange between air and bloodstream. The lung of terrestrial molluscs is a cavity under the mantle lined with vascular tissue.

lungfishes *n.plu.* lobe-finned, bony, airbreathing fish of the subclass Dipnoi, represented by only four extant genera.

Lusitanian *a. appl.* certain plants and animals that occur both in the Iberian Peninsula and in coastal regions of the far west of the British Isles, e.g. in western Ireland and Cornwall, and that are thought to be relicts of an interglacial period.

luteovirus group plant virus group containing isometric single-stranded RNA viruses that typically cause yellowing of the leaves, type member barley yellow dwarf virus.

lutetium (Lu) *n.* a lanthanide element (at. no. 71, r.a.m. 174.97), a silver-coloured metal when in elemental form. It is used as a catalyst, e.g., in polymerization and oil cracking.

L **waves** surface waves *q.v.*

Lycophyta, lycopods, lycophytes *n.*, *n.plu.*, *n.plu.* one of the four major divisions of extant seedless vascular plants, with 10 to 15 living genera comprising club mosses, *Selaginella* and the aquatic quillwort. They are characterized by a sporophyte with roots, stems and small leaves arranged spirally on the stem, with sporangia solitary and borne on or associated with a sporophyll. The gametophyte generation is reduced, and usually subterranean. Fossil lycophytes are found from the Devonian onwards and include extinct forms such as the woody tree-like lepidodendroids, which are the dominant plants of the Carboniferous Coal Measures. *alt.* Lycopodophyta.

Lycopodiales, Lycopodophyta, Lycopsida, lycopods *see* Lycophyta.

lyosphere *n.* a thin film of water surrounding a colloidal particle.

lymphatic filariasis elephantiasis *q.v.*

Lys lysine *q.v.*

Lysenkoism *n.* a doctrine promoted by the Soviet agriculturalist T. Lysenko in the 1930s and 1940s, which was based on the idea that variation was due to acquired characteristics caused by environmental influences that could be inherited. This mistaken theory became for some time the only officially permitted theory of genetics in the USSR and led to the suppression of the teaching of Mendelian genetics and work based on modern genetic concepts, and the persecution of geneticists who opposed Lysenkoism.

lysimeter *n.* instrument used for measuring water lost by evapotranspiration *q.v.*

lysine (Lys, K) *n.* diaminocaproic acid, a basic amino acid, constituent of protein, essential in the human diet.

lysosome *n.* an organelle in eukaryotic cells, esp. animal cells, that has a relatively acid interior and contains digestive enzymes that break down damaged cellular components and material taken in by phagocytosis. In plant cells, the function of the lysosome is performed by the vacuole. *a.* lysosomal.

M

m metre *q.v.*

M (1) methionine (*q.v.*); (2) molar, symbol for the units of concentration, mol l⁻¹ or mol dm⁻³. *see* molarity.

M_r relative molecular mass *q.v.*

MAB Man and the Biosphere programme *q.v.*

MAC maximum allowable concentration *q.v.*

Macaronesion Floral Region part of the Palaeotropical Realm comprising the islands off the west coast of Africa, e.g. Madeira, the Canary Islands and the Cape Verde Islands.

macaws *n.plu.* large, brilliantly coloured birds of the parrot family, found from Mexico south through tropical South America. Collecting has seriously depleted wild populations.

macchia maquis *q.v.*

machair *n.* herb-rich calcareous grassland on shell sand on the western Scottish (UK) coast.

machlovirus group plant virus group containing isometric single-stranded RNA viruses that typically cause chlorosis of the plant, type member maize chlorotic dwarf virus.

mackerel sky *see* cirrocumulus.

macro- prefix derived from Gk *makros*, large.

macroalga *n.* multicellular alga, e.g. a seaweed, large enough to be visible to the naked eye. *cf.* microalga.

macroarthropod *n.* medium- and large-sized arthropods whose size is measured in millimetres and centimetres rather than microscopic units.

macrobenthos *n.* the larger organisms that live on the sea-floor or lake bottom, in sediments or attached to objects.

macrobiota *n.* the population of organisms of a size larger than a few centimetres in any habitat or ecosystem, esp. applied to soil. *cf.* microbiota, mesobiota.

macroclimate *n.* (1) the climate over a relatively large area, generally synonymous with climate in the usual sense; (2) large-scale climatic region covering an extensive geographical area, e.g. mediterranean climate. *cf.* mesoclimate, microclimate.

macrocyclic *a. appl.* rust fungi that produce basidiospores, teleutospores and at least one other type of spore (aeciospores and/or uredospores) during their life cycle.

macroelement *n.* element required in large quantities in the nutrition of living organisms. *see* macronutrient.

macroevolution *n.* evolutionary processes extending through geological time, leading to the evolution of markedly new genera and higher taxa. *cf.* microevolution.

macrofauna *n.* animals whose size is measured in centimetres rather than microscopic units.

macrogamete *n.* the larger of the two gametes in organisms that have gametes of different sizes, usually considered equivalent to the ovum or egg.

macroinvertebrates *n.plu.* invertebrates and their larvae whose size is measured in millimetres or centimetres rather than microscopic units. Such species are one of the main groups of organisms sampled in surveys of water quality.

macromolecule *n.* very large organic molecule such as a protein, nucleic acid or polysaccharide.

macromutation *n.* (1) simultaneous mutation of several characters; (2) a hypothetical step involving a single large mutational change in an organism in some evolutionary theories.

macronutrient elements, macronutrients the nine elements required in relatively large amounts by living organisms for proper growth and development, being major constituents of living matter, i.e. oxygen (O), hydrogen (H), carbon (C), nitrogen (N), calcium (Ca), phosphorus (P), sulphur (S), potassium (K) and magnesium (Mg).

macrophagous *a.* feeding on relatively large masses of food. *cf.* microphagous.

macrophyll, macrophyllous megaphyll, megaphyllous *q.v.*

macrophyte *n.* large aquatic plant, e.g. water lily and water crowfoot, as opposed to the phytoplankton and small plants like duckweed.

macroplankton *n.* (1) the larger organisms drifting with the surrounding water, as jellyfish, sargassum weed, etc.; (2) macrozooplankton *q.v.*

macropores *see* pores.

macroscopic *a.* visible to the naked eye.

macrospecies *n.* a large polymorphic species, usually with several to many subdivisions.

macrosporophyll megasporophyll *q.v.*

macrosymbiont *n.* the larger of two symbiotic organisms.

macrotidal *see* tidal range.

macrozooplankton *n.* animal plankton generally larger than 1 mm in diameter, and large enough to be visible to the naked eye and to be caught in a plankton net.

macula *n.* (1) a spot or patch of colour; (2) a small pit or depression. *plu.* maculae.

Madagascan Floral Region part of the Palaeotropical Realm comprising the island of Madagascar and its neighbouring islands.

mad cow disease bovine spongiform encephalopathy *q.v.*

madrepore *n.* a branching, stony, reef-building coral of the order Scleractinia.

MAFF Ministry of Agriculture, Fisheries and Food, UK.

mafic minerals a collective term for pyroxenes, amphiboles, micas and olivines.

maggot *n.* common name for the worm-like larva of true flies (Diptera), having no appendages or distinct head.

magma *n.* molten or partially molten rock material formed from pre-existing solid rocks of the crust and/or the mantle.

magma chamber below-ground chamber in which magma gathers, a source of volcanic lava.

magmatic water water originating from deep sources of magma, which may issue at the surface in the form of hot springs. *alt.* juvenile water.

magnesite *n.* $MgCO_3$, a mineral used in the manufacture of refractories, cements and fertilizers.

magnesium (Mg) *n.* metallic element (at. no. 12, r.a.m. 24.31), forming a greyish metal of low density that tarnishes in air. Abundant in the environment in seawater (as $MgCl_2$) and in numerous minerals such as the magnesium ores dolomite and talc. It is an essential nutrient for living organisms.

magnetic confinement fusion an approach to controlled nuclear fusion in which a magnetic field is used to contain the plasma, while a large current is applied to raise its temperature to the ignition point.

magnetic dipole *see* dipole.

magnetite *n.* Fe_3O_4, an important iron ore mineral. Hydrated magnetite $(Fe_3O_4.nH_2O)$ is found in waterlogged soils.

magnetohydrodynamic power generation a method of electricity generation in which fuel is burned in a combustor to produce high-temperature exhaust gases. These gases are seeded with a material to induce further ionization (typically potassium carbonate is used for this purpose). The result is a plasma, at *ca.* 2500 °C. This is passed, at high speeds, through a stationary magnetic field. The electric current generated is extracted by electrodes in the channel through which the plasma is made to flow. *alt.* MHD power generation.

magnetotaxis *n.* directed movement within a magnetic field that is shown by some bacteria that contain small particles of magnetite within their cells. *a.* magnetotactic.

magnetotropism *n.* a tropism in response to lines of magnetic force.

Magnoliales *n.* an order of dicot trees and shrubs including the families Annonaceae (custard apple), Magnoliaceae (magnolia), Myristicaceae (nutmeg) and others.

Magnoliophyta *n.* in some classifications the name for the angiosperms.

Magnoliopsida *n.* in some classifications an alternative name for the dicotyledons *q.v.*

Magnox reactor the first commercial fission nuclear reactor used in the UK. It utilizes metallic natural uranium fuel (clad in a magnesium alloy called Magnox) cooled by carbon dioxide. The moderator is graphite.

maintenance behaviour animal behaviour involved in carrying out day-to-day activities such as the search for food, mating, reproduction or the avoidance of extreme environments.

maintenance energy energy required to support survival of a cell rather than its growth.

maintenance ration food required to maintain an animal when the production term (P) in the energy budget is zero.

maize *n. Zea mays*, a member of the Gramineae, originating and first domesticated in the Americas and now widely grown as a crop plant for human and animal feed. *alt.* corn (USA), sweetcorn.

major elements macronutrient elements *q.v.*

major gene a gene having a pronounced phenotypic effect, as distinguished from a modifying gene.

malachite *n.* $Cu_2CO_3(OH)_2$, a copper ore mineral that is also used as a decorative stone and to make green pigment.

malacology *n.* the study of molluscs.

malacophilous *a.* pollinated by the agency of gastropod molluscs, generally snails and slugs.

Malacostraca, malacostracans *n., n.plu.* a large subclass of aquatic (marine and freshwater) and terrestrial crustaceans containing the crabs, lobsters, crayfish, shrimps, woodlice and other groups. An individual generally has a head of five, a thorax of eight, and an abdomen of six segments, with well-developed, usually biramous appendages, and a carapace covering the thorax in most groups.

Malagasy *a. appl.* or *pert.* the zoogeographical subregion including Madagascar and adjacent islands.

malaria *n.* sometimes fatal disease characterized by recurrent fevers, caused by parasitic protozoa of the genus *Plasmodium*, of which *P. vivax* causes the most severe form, falciparum malaria. The intermediate hosts and vectors of *Plasmodium* spp. are anopheline mosquitoes.

malathion *n.* contact organophosphate pesticide used, e.g., to control aphids on crops and mites on livestock.

Malayan *a. appl.* or *pert.* the zoogeographical subregion including Malaysia, Indonesia west of Wallace's line and the Philippines.

Malaysian–Papuan Floral Region part of the Palaeotropical Realm comprising the southern part of the Malayan Peninsula, the islands of Indonesia and Papua-New Guinea.

male *n.* individual whose sex organs contain only male gametes, ♂.

maleic hydrazide translocated plant growth regulator used, e.g., to suppress the growth of grass.

male pronucleus the nucleus that the sperm contributes to the zygote.

male-sterile *a. appl.* mutants of normally hermaphrodite plants that do not produce viable pollen. Such mutants are of importance in plant breeding as they can be used to block self-fertilization and force cross-fertilization, thus avoiding the need for laborious emasculation of the plants used as the female parent. *see* cytoplasmic male sterility.

mallee scrub vegetation consisting of low bushes of *Eucalyptus* spp. typical of dry subtropical regions of south-east and south-west Australia.

Mallophaga *n.* in some classifications an order of insects known as bird or biting lice, which are ectoparasitic on birds, and have biting mouthparts, no wings and slight metamorphosis.

malnourishment malnutrition *q.v.*

malnutrition *n.* lack of sufficient nutrients, or lack of specific nutrients in sufficient quantity, caused by a lack of food or a poor diet, or by an inability to absorb and metabolize nutrients once ingested. *alt.* malnourishment.

Malthusian growth (*biol.*) rapid increase in population followed by a population crash due to exhaustion of food resources. *alt.* irruptive growth.

malting *n.* the process whereby barley grain is soaked, germinated and kiln-dried to produce malt (used in the beer-brewing and whisky-distilling industries).

maltose *n.* a disaccharide of glucose, produced by hydrolysis of starch with amylase. It does not occur widely in the free state, but is produced by germinating barley. Hydrolysed by maltase to glucose.

malt sugar maltose *q.v.*

Malvales *n.* order of dicot trees, shrubs and herbs, often mucilaginous, and including the families Malvaceae (mallow, cotton), Sterculiaceae (cocoa), Tiliaceae (linden), Bombacaceae (baobab, silk cotton) and others.

Mammalia, mammals *n.*, *n.plu.* a class of homoiothermic vertebrates, known from the late Triassic to the present, having the body covered with hair, a four-chambered heart, and mammary glands (mammae) producing milk with which the female suckles her young. Except in the egg-laying monotremes, the young develop inside the mother in the uterus and are born at a more or less mature stage.

mammal-like reptiles common name for reptiles of the extinct subclass Synapsida, living from the Carboniferous to the Triassic, with synapsid skulls (skulls with one ventral temporal fenestra on each side), including the orders Pelycosauria and Therapsida. The Therapsida were direct ancestors of the mammals.

mammalology *n.* the study of mammals.

mammary gland gland secreting milk in female mammals. *alt.* mamma.

Man and the Biosphere programme (MAB) a UNESCO programme, operative during the 1970s, which increased the scientific understanding of the relationships within and between world ecosystems, and was intended to facilitate their long-term conservation.

mancozeb *n.* zinc and maneb (*q.v.*) complex used as a fungicide, e.g. to control apple scab and potato blight.

mandible *n.* (1) in vertebrates, the lower jaw, either a single bone or comprised of several; (2) in arthropods, paired mouthpart, usually used for biting.

mandioca cassava *q.v.*

maneb *n.* fungicide used, e.g., to control potato blight and leaf mould on tomatoes.

manganese (Mn) *n.* metallic element (at. no. 25, r.a.m. 54.94), forming a hard

brittle reddish-white metal in elemental form, extracted mainly from the ore pyrolusite and also occurring in manganese-rich nodules on the seabed. It is an essential micronutrient for plants.

manganese nodules reddish-brown to black nodules, 5–200 mm in diameter, found over much of the deep-ocean floor. Rich in manganese minerals, these nodules also contain iron oxides and accessory minerals such as feldspars and clays. They may be exploited for their metal content as, in addition to manganese and iron, they also contain cobalt, copper, nickel, molybdenum, vanadium and zinc.

manganite *n.* MnO(OH), a manganese ore mineral.

mangolds *n.plu.* root plants related to beet, which are grown as a fodder crop for cattle. *alt.* mangels.

mangrove swamp ecosystem subtropical or tropical marine intertidal zone community dominated by mangroves, salt-tolerant evergreen shrubs and trees of the genera *Rhizophora* and *Avicennia*.

Manihot esculenta cassava *q.v.*

Manila hemp *see* hemp.

manioc cassava *q.v.*

manna *n.* the hardened exudate of bark of certain trees such as the European ash, *Fraxinus ornus*, and similar substances in other plants such as tamarisk, where its production is caused by infestation with scale insects.

mannan *n.* any of a group of polysaccharides composed of (predominantly) mannose residues linked together, found esp. in cell walls of conifers.

mannitol *n.* a sweet-tasting polyhydroxy-alcohol derivative of mannose or fructose found in many plants and some algae. Used commercially as a sweetening agent.

mannose *n.* a six-carbon aldose sugar found in many glycoproteins and polysaccharides, esp. those of plant cell walls.

manometer *n.* instrument used to measure gaseous pressure.

mantids *n.plu.* common name for many members of the Dictyoptera *q.v.*

mantle *n.* (1) (*bot.*) external sheath of fungal mycelium covering the plant rootlets in a mycorrhiza; (*zool.*) (2) fold of soft tissue underlying shell in molluscs, barnacles and brachiopods and that usually encloses a space, the mantle cavity, between it and the body proper; (3) body wall of ascidians; (4) feathers of bird between neck and back; (5) (*geol.*) the region of the Earth that lies between the crust and the core. The mantle is subdivided into the lower mantle (a solid region with a thickness of about 2200 km) and the upper mantle (a largely viscous layer with a thickness of about 670 km). The uppermost zone of the upper mantle is solid and together with the overlying crust forms the lithosphere.

manures *see* animal manures, green manures.

maquis *n.* dense, low-growing, hard-leaved (i.e. sclerophyllous) vegetation found in the Mediterranean area, composed of small trees (e.g. wild olive and pine), shrubs and aromatic herbaceous plants. *alt.* macchia, mattoral.

marafivirus group plant virus group containing isometric single-stranded linear RNA viruses, type member maize rayado fino virus.

marasmus *n.* a deficiency disease caused by general undernutrition, generally found in infants.

marble *n.* a rock formed from limestone by either contact or regional metamorphism. It is essentially calcite, although dolomite may also be present in appreciable amounts. It may also contain a variety of other minerals, which can give it attractive structures and colours. It is used as a decorative building material.

marble gall type of gall found on oak and caused by larvae of a species of gall wasp.

marcasite *n.* FeS_2, an iron sulphide mineral.

mares' tails *see* cirrus.

marialite *n.* $Na_4Al_3Si_9O_{24}Cl$, one of the scapolite group (*q.v.*) of minerals. An end-member of a solid-solution series, the other end-member being meionite, $Ca_4Al_6Si_6O_{24}(SO_4,CO_3)$.

mariculture *n.* farming of the marine environment to obtain animal (esp. fish) and/or plant products.

marine *a. pert.* or *appl.* seas and oceans and to the organisms living there.

marine abrasion platform shore platform *q.v.*

marine ecology the ecology of the seas and oceans.

marine snow clearly visible macroscopic aggregates (several millimetres to several centimetres across) of living organisms, detritus and inorganic particulates found within the marine environment.

maritime effect moderating influence of the ocean on the climate of coastal areas. These experience milder winters and cooler summers compared with inland areas at equivalent latitudes.

marker *n.* an identifying feature used to identify or track a particular DNA, cell or organism.

marram grass *Ammophila arenaria*, a grass that grows on sand dunes, helping to stabilize them, and that is often planted for this purpose.

marsh *n.* lowland plant community developing on wet but not peaty soil occasionally covered with water, dominated by reeds, rushes and sedges, and with no or few woody plants.

marsh gas *n.* a gas that is predominantly methane (CH_4), produced by the anaerobic decomposition of organic material.

marshland *n.* an extensive area of marsh.

Marsupialia, marsupials *n.*, *n.plu.* the only order in the Metatheria, a group of mammals found only in Australia and South America, in which the placenta does not develop or is not as efficient as that of therian mammals, so that the young are born in a very immature state. They then migrate to a pouch (marsupium), where they are suckled until relatively mature. Marsupials include kangaroos, wallabies and opossums and, in Australia, many other species adapted to fill ecological niches filled by therian mammals elsewhere. *alt.* metatherians.

masculinization *n.* the production, e.g. by hormones, of male secondary sexual characteristics in genetic females.

mashlum *n.* a mixture of cereals and beans or peas grown together, either for grain or silage.

mass burn incineration of unsorted solid refuse.

mass–energy relation, Einstein's *see* stein's mass–energy relation.

mass extinction any of the various episodes in evolutionary history in which numerous groups of organisms disappear from the fossil record over a relatively short time (as the extinction of the dinosaurs and other groups at the end of the Cretaceous). Mass extinctions are generally explained by sudden changes in climate due to various possible causes.

mass flow a theory of the way in which materials are translocated through phloem, which proposes that the cause of movement is the difference in the hydrostatic pressure at each end of a sieve tube, resulting in the flow of contents along the tube.

mass mortality death of most of a population, or of many species within a community.

mass movement movement of slope material *en masse* under the influence of gravity. Movements may be categorized according to their speed. From slowest to fastest they are: creep movement (*q.v.*), flow movement (*q.v.*), slide movement (*q.v.*), fall movement (*q.v.*). *alt.* mass wasting.

mass number (**A**) the total number of nucleons (i.e. protons plus neutrons) within an atom. *alt.* nucleon number.

mass provisioning the act of storing all of the food required for the development of a larva at the time the egg is laid.

mass wasting mass movement *q.v.*

mast *n.* the fruit of beech and some related trees.

Mastigomycota *n.* a major division of the Fungi including the classes Chytridiomycetes, Hyphochytridiomycetes, Plasmodiophoromycetes and Oomycetes, i.e. simple, generally aquatic, often microscopic fungi with motile flagellate zoospores and/or gametes.

Mastigophora *n.* a superclass of protozoans including the plant and animal flagellates.

mastitis *n.* (*agric.*) a disorder of dairy cows (and other mammals) in which the udder becomes inflamed, usually as a result of microbial infection.

material resource a resource whose quantity can be measured and whose supply is limited, e.g. coal, oil, iron.

materials balance a model of the economy as an open system pulling in useful materials and energy from the environment and eventually releasing the equivalent amount of waste (useless matter and low-quality energy) back into the environment.

maternal inheritance (1) the inheritance of genes carried by mitochondria, chloroplasts and any other cytoplasmic genes through the maternal line only, as only the egg contributes cytoplasm to the zygote; (2) preferential survival in a cross of genetic markers provided by one parent.

mating factors protein pheromones secreted by cells of different mating types in yeasts and other unicellular organisms, which attract cells of an appropriate mating type and induce conjugation.

mating system the pattern of mating within a population. *see* monogamy, polyandry, polygamy, polygyny.

mating type genetically determined property of bacteria, ciliates, fungi and algae, determining their ability to conjugate and undergo sexual reproduction with other individuals in the population. In some cases, such as fungi, individuals that can conjugate are said to belong to opposite mating types. In some other organisms, e.g. ciliates, individuals that can conjugate are said to belong to the same mating type.

matriclinal *a.* with inherited characteristics more maternal than paternal.

matric potential of soil, a measure of the availability of the water in the soil to plants, measured as the tension or suction pressure (in pascals) required to withdraw water from the soil.

matrifocal *a. pert.* a society in which most of the activities and personal relationships are centred on the mothers.

matrilineal *a.* passed from a mother to her offspring.

matrix *n.* (1) medium in which a substance is embedded; (2) (*chem.*) that part of a sample that is not the analyte (*q.v.*); (*biol.*) (3) ground substance of connective tissue; (4) part beneath body and root of nail; (5) body upon which a lichen or fungus grows; (6) substance within which a fossil is embedded; (7) in mitochondria, the inner region enclosed by the inner mitochondrial membrane.

matrix effect, matrix interference a source of error in instrumental analytical chemistry that occurs when a chemical species within the matrix, but not present in the standard, brings about an enhancement or attenuation of the instrument's response.

matrix matching a technique used in analytical chemistry in which the standards used to calibrate the instrument are made to be as similar as possible to the sample to be analysed. This is done to minimize matrix effects *q.v.*

matter *n.* anything that takes up space, exists through time and has mass. It is a form of energy that is related to other forms of energy (*E*, measured in joules) by Einstein's equation, $E = mc^2$, where *m* is the mass (in kilograms) of matter converted to *E*, and *c* is the speed of light in metres per second.

mattoral *n.* scrub woodland found in mediterranean-type semi-arid climates.

maturation *n.* (1) ripening; (2) the process of becoming mature, fully differentiated and fully functional, *appl.* divisions by which gametes are produced from primary gametocytes, during which meiosis occurs; (3) the automatic development of a behaviour pattern that becomes progressively more complex as the animal matures and that does not involve learning.

mature sediments sediments constituted of grains that differ little from one another in terms of size. *cf.* immature.

mature soil stable soil with a profile composed of well-developed horizons. *cf.* immature soil.

maximum allowable concentration (MAC) of pollutants, the concentration deemed in regulations to be safe to healthy adults in the work-place, assuming they are not exposed to the pollutant outside working hours.

maximum permissible body burden the concentration of a radioisotope that will not deliver more than the maximum permissible dose to any organ, if inhaled or ingested at a normal rate.

maximum permissible dose the dose of ionizing radiation, accumulated over a given time, that is considered not to result in any harmful effects to the individual over their lifetime or to cause genetic damage that might affect their descendants.

maximum sustainable yield the maximum crop or yield that can be harvested from a plant or animal population each year without harming it.

maxithermy *n.* the maintenance of body temperature at a maximum for as long as possible. *a.* maxithermic, *appl.* animals that can do this.

mayflies *n.plu.* common name for the Ephemeroptera *q.v.*

MCI multiple cropping index *q.v.*

MCPA translocated herbicide, widely used to control annual weeds.

MCPB translocated herbicide used to control broad-leaved annual and perennial weeds.

Md symbol for the chemical element mendelevium *q.v.*

MDCs more developed countries *q.v.*

M discontinuity Mohorovičić discontinuity *q.v.*

meadow *n.* permanent grassland, esp. one that is mown for hay and not grazed in summer. *cf.* ley, pasture, water meadow.

mealworm *n.* larva of the beetle *Tenebrio*, which lives in grain stores.

mealy bug scale insect *q.v.*

mean *n.* an average. Three types of mean are recognized, the arithmetic mean, the geometric mean and the harmonic mean. For a set of *n* data points a, b, c, . . . , the arithmetic mean equals (a + b + c + . . .)/*n*, the geometric mean (applicable only to positive numbers) equals (a × b × c × . . .)$^{1/n}$, and the harmonic mean equals $n/(1/a + 1/b + 1/c + . . .)$. For example, the arithmetic mean of 3, 4, 5 equals (3 + 4 + 5)/3 = 4, the geometric mean equals (3 × 4 × 5)$^{1/3}$ = 3.915, and the harmonic mean equals 3/(1/3 + 1/4 + 1/5) = 3.83. The terms average and mean are generally taken to be synonymous with the arithmetic mean.

mean date of ear emergence *see* heading date.

meandering stream a stream with a sinuosity (*q.v.*) greater than *ca.* 1.5. Such streams are found in both alluvial and tidal environments.

meanders *n.plu.* significant bends in a river channel. These become progressively more pronounced as material is eroded from the concave outside of the bend and deposited on the convex inside of the next bend. This increasing sinuosity may lead to the formation of oxbow lakes.

mean intensity of infection the average number of parasites per host in the population.

mechanically formed sedimentary rocks clastic rocks *q.v.*

mechanical weathering physical weathering *q.v.*

mechanical work mechanical work (*w*) (sometimes simply referred to as work) is done whenever an object is moved by a force (*F*) over a distance (*L*). The amount of work done = *FL* cos θ, where θ is the angle between the displacement and the force's line of action. Unit: joule (J) in the SI system.

mechanisms of succession means by which the process of succession (*q.v.*) occurs. Several explanatory theories have been proposed. *see* facilitation theory, inhibition theory, tolerance theory.

mecoprop *n.* translocated herbicide used in crops such as cereals to control broadleaved weeds.

Mecoptera *n.* an order of slender carnivorous insects with complete metamorphosis, commonly called scorpion flies, having biting mouthparts, long slender legs and membranous wings lying along the body in repose.

Medfly Mediterranean fruit fly *q.v.*

medial *a.* situated in the middle. *n.* the main middle vein of insect wing.

medial moraine *see* moraine.

median *a.* (1) lying or running in an axial plane; (2) intermediate; (3) middle.

median *n.* the middle variate when a set of variates are arranged in order of magnitude.

median lethal dose (LD$_{50}$) dose of a toxic chemical or of a pathogen that kills 50% of the animals in a test sample within a certain time.

Medio-Columbian Sonoran *q.v.*

mediolateral *a. appl.* axis, from the median plane outwards to both sides.

mediterranean *a. appl.* subtropical climate characterized by warm or hot dry summers and mild wet winters.

Mediterranean Floral Region that part of the Palaearctic Realm comprising southern Europe and North Africa, around the Mediterranean Sea.

Mediterranean fruit fly *Ceratitis capitata*, a serious pest of many different fruits and vegetables. This fruit fly originated in the Mediterranean region and is now a major problem in California, USA. *alt.* Medfly.

medium *n.* nutritive material in which microorganisms, cells and tissues are grown in the laboratory. *plu.* media.

medium-grained *a.* (*geol.*) *see* grain size.

medium-range forecast *see* weather forecasting.

medium-textured soil a soil with a fairly even mixture of particles of various sizes. Such soils are generally free-draining and quick to warm, with high moisture- and nutrient-holding capacities. They are therefore ideal for the growth of most plants.

medusa *n.* one of the forms of individuals of coelenterates of the classes Hydrozoa (hydroids) and Scyphozoa (jellyfish). It is bell-shaped, with a tube hanging down in the centre ending in a mouth, and tentacles around the edge of the bell. It forms the free–swimming sexual reproductive stage of most hydrozoans, and is large and conspicuous in jellyfish. *plu.* medusae.

mega- prefix derived from Gk *megas*, large.

megachiroptera *n.* fruit-eating bats (flying foxes), of the mammalian order Chiroptera. *cf.* microchiroptera.

megacity megalopolis *q.v.*

megadune draa *q.v.*

megafauna animals larger in size than a few centimetres.

megagametophyte *n.* in heterosporous plants, the female gametophyte, which develops from a megaspore.

megalopolis *n.* urban area with a population exceeding ten million, commonly formed by the merging of several large cities that existed in close proximity. *alt.* megacity, supercity.

megaphanerophyte *n.* a tree exceeding 30 m in height.

megaplankton macroplankton *q.v.*

megaspore *n.* (1) in heterosporous plants, the spore that is formed in a megasporangium and that gives rise to the female gametophyte; (2) in any organism that produces two types of spore, the larger spore.

megasporophyll *n.* leaf or leaf-like structure on which a megasporangium develops, in flowering plants known as a carpel.

megatherm *n.* plant thriving in moist tropical temperatures between 20 and 35 °C.

megawatt (MW) one million watts.

megistotherm *n.* a plant that thrives at a more or less uniformly high temperature.

meio- prefix derived from Gk *meion*, less.

meiobenthos *n.* organisms of medium size (0.1–0.5 mm) living on the sea-floor, or lake or river bottom.

meiofauna mesofauna *q.v.*

meionite *n.* $Ca_4Al_6Si_6O_{24}(SO_4,CO_3)$, one of the scapolite group (*q.v.*) of minerals. An end-member of a solid-solution series, the other end-member being marialite, $Na_4Al_3Si_9O_{24}Cl$.

meiosis *n.* a type of nuclear division that results in daughter nuclei each containing half the number of chromosomes of the parent, i.e. chromosome number is reduced from diploid to haploid. It comprises two distinct nuclear divisions, the first and second meiotic divisions, which may be separated by cell division, the actual reduction in chromosome number taking place during the first division. *cf.* mitosis. *see also* reduction division.

meiotherm *n.* a plant that thrives in a cool temperate environment.

meiotic *a. pert.* or produced by meiosis.

meiotic drive any mechanism that operates during meiosis in heterozygotes to produce a disproportionate representation of one member of a chromosome pair in the gametes. Usually due to certain mutations, e.g. the SD allele at the segregation distorter locus in *Drosophila*.

Melanconia *n.* a morphological class of deuteromycete fungi that reproduce by conidia borne in acervuli and that includes many plant pathogenic fungi causing anthracnose diseases (e.g. *Colletotrichum*, *Marssonina*). *alt.* Melanconiales.

melanin *n.* any of a range of black or brown pigments produced from tyrosine by the enzyme tyrosinase and giving colour to animal skin, etc., and also found in some plants.

melanism *n.* excessive development of black pigment. *see also* industrial melanism.

melanotic *a. appl.* animals and plants that are much darker than the usual colour, as a result of unusual or excessive production of melanin.

melittophile *n.* an organism that must spend at least part of its life cycle with bee colonies.

melliferous *a.* honey-producing.

mellisugent *a.* honey-sucking.

mellivorous *a.* feeding on honey.

meltdown *n.* the melting of the core of a nuclear reactor.

melting point (m.p.) temperature at which a given substance at a given pressure (usually standard atmospheric pressure) is present in both its liquid and solid forms, and these forms are in dynamic equilibrium.

membrane *n.* (1) a thin film, skin or layer of tissue covering a part of an animal or plant, or separating different layers of tissue; (2) of cells, an organized layer a few molecules thick forming the boundary of the cell (the cell membrane or plasma membrane) and of intracellular organelles, and composed of two oriented lipid layers in which proteins are embedded, and which acts as a selective permeability barrier.

membrane cells diaphragm cells *q.v.*

MeNA alpha-napththylacetic acid methyl ester, a volatile compound used as an artifical auxin.

menarche *n.* (1) first menstruation; (2) age at first menstruation.

mendelevium (Md) *n.* a transuranic element [at. no. 101, r.a.m. (of its most stable known isotope) 258].

Mendelian inheritance inheritance of genes or characters according to Mendel's laws. *cf.* non-Mendelian inheritance.

Mendel's laws laws of inheritance first proposed by Gregor Mendel in the nineteenth century from breeding experiments in plants, which describe some basic principles of inheritance in sexually reproducing organisms, the first law being that of the segregation of alleles, the second being the law of independent assortment of genes. *see* segregation of alleles, independent assortment.

Mendosicutes *n.* a division of the kingdom Monera or Prokaryotae, comprising a single class, the Archaebacteria. Now sometimes considered as one of three superkingdoms, the Archaea.

menotaxis *n.* (1) compensatory movements to maintain a given direction of body axis in relation to sensory stimuli, esp. light, but not necessarily moving towards or away from it; (2) maintenance of visual axis during locomotion.

menstrual *a.* (1) monthly; (2) lasting for a month, as of flower.

mental *a. pert.* (1) the mind; (2) or in the region of the chin, *appl.* nerve, spines, tubercle, muscle.

mercaptans thiols *q.v.*

merchantable timber trees of a suitable species and size for felling and sale for wood.

mercurous chloride calomel *q.v.*

mercury (Hg) *n.* metallic element (at. no. 80, r.a.m. 200.59), forming a dense silvery liquid in elemental form. The most important mercury ore is cinnabar (HgS). Mercury pollution from industrial uses is an important environmental contaminant as mercury, and its organic compounds, are persistent and highly toxic.

mercury barometer instrument used to measure air pressure. It consists of a mercury-filled tube, closed at the top but open at the bottom, inverted in a mercury-filled container. Changes in air pressure are shown by changes in the height of the mercury column in the tube. *alt.* Fortin barometer.

mercury(I) chloride calomel *q.v.*

mercury fungicides highly toxic organic compounds containing mercury that have been used as fungicides and soil disinfectants. *see also* calomel.

meridian (terrestrial) *n.* imagined circle drawn around the Earth that goes through both geographical poles.

meridional *a.* running from pole to pole of a structure, as along a meridian.

meridiungulates *n.plu.* extinct group of South American hoofed mammals, present during the Tertiary, which included the camel-like litopterns and the notoungulates.

meristem *n.* plant tissue capable of undergoing mitosis and so giving rise to new cells and tissues. It is located at the growing tips of shoots and roots (apical meristems), in the cambium and cork cambium encircling some plant stems (lateral meristems) and in leaves, fruits, etc. *a.* meristematic.

mermaid's purse the horny egg case of elasmobranch fishes.

meromorphosis *n.* regeneration of a part with the new part less than that lost.

meroplankton *n.* (1) temporary plankton, consisting of eggs and larvae; (2) seasonal plankton.

Merostomata *n.* a class of aquatic arthropods, the horseshoe crabs, which breathe by gills and have chelicerae (claws) and walking legs on the prosoma, and an opisthosoma with some segments lacking appendages.

mesa *n.* a flat-topped, steep-sided landform. This type of structure may develop when a sill (*q.v.*), more resistant to weathering that the sedimentary rock layers between which it was originally formed, forms the capping layer.

mesic *a. appl.* (1) environment with a moderate rainfall and temperature; (2) plants, those that require a reasonable amount of moisture to grow.

meso- prefix derived from Gk *mesos*, middle, signifying situated in the middle, intermediate, neither to one end nor the other of a range of conditions.

mesobenthos *n.* animal and plant life of the sea bottom at depths between about 200 and 1000 m.

mesobiota *n.* the population of organisms in an ecosystem, habitat, etc. (esp. soil) that range in size from approx. 200 μm to 1 cm, i.e. larger than bacteria and unicellular algae, and smaller than the large soil organisms, such as earthworms, and including, e.g., mites and nematode worms.

mesoclimate *n.* distinct climatic region covering several square kilometres, e.g. mountain climate. *cf.* macroclimate, microclimate.

mesofauna *n.* animals ranging in size from 200 μm to 1 cm. *alt.* meiofauna.

mesohalabous *a. appl.* plankton living in brackish water.

mesohaline *a. appl.* brackish water of salinity 5–15 parts per thousand.

mesohydrophytic *a.* growing in temperate regions but requiring much moisture.

mesomicrotherm *n.* plant living in a temperate climate that can resist low winter temperatures.

mesomorphic *a.* of normal or average structure, form or size, or intermediate between extremes. *n.* mesomorph (usually *appl.* animals).

mesopause *n.* the boundary between the mesosphere and the thermosphere in the upper atmosphere at *ca.* 85 km. *see* Fig. 16, p. 236.

mesopelagic *a. pert.* or inhabiting the ocean at depths between 200 and 1000 m.

mesophanerophyte *n.* tree from 8 to 30 m in height.

mesophil(ic) *a.* thriving at moderate temperatures, between 20 and 45 °C when *appl.* bacteria. *n.* mesophile.

mesophilous *a. appl.* plants associated with neutral soils.

mesophyte *n.* a plant thriving in a temperate climate with a normal amount of moisture.

mesoplankton *n.* (1) plankton at depths of 200 m downwards; (2) drifting organisms of medium size.

mesosaprobic *a. appl.* aquatic habitats having a decreased quantity of oxygen and substantial organic decomposition. *see* alpha-mesosaprobic, beta-mesosaprobic.

mesosaur *n.* member of the order Mesosauria, an order of lower Permian anapsid reptiles that were fish-eating and slender with long jaws.

mesoscale convective complex cluster of several thunderstorm cells. *alt.* supercell.

mesosphere *n.* the layer of atmosphere that lies between the stratosphere and the thermosphere, i.e. between the stratopause (altitude *ca.* 50 km) and the mesopause (altitude *ca.* 85 km), *see* Fig. 16, p. 236.

mesotherm *n.* plant thriving in moderate heat, at 12–19 °C, as in a warm temperate climate.

mesotidal *see* tidal range.

mesotrophic *a.* (1) having partly autotrophic and partly saprobic nutrition; (2) obtaining nourishment partly from an outside source; (3) partly parasitic; (4) providing a moderate amount of nutrition, *appl.* environment.

mesoxerophyte *n.* plant of a temperate climate that thrives in dry conditions.

Mesozoa, mesozoans *n.*, *n.plu.* a phylum or subphylum of small multicellular animals, which are parasitic on marine organisms, and have bodies composed of two layers of cells. Sometimes included in the Platyhelminthes.

Mesozoic *a. pert.* or *appl.* geological era lasting from about 248 to 65 million years ago and comprising the Triassic, Jurassic and Cretaceous periods.

messenger RNA (mRNA) type of RNA found in all cells that acts as a template for protein synthesis, each different mRNA being a copy of a single protein-coding gene (or, in bacteria, sometimes a set of adjacent genes), and that in eukaryotes is the product of extensive processing of the primary RNA transcript in the nucleus. *alt.* message. *see also* transcription.

Met methionine *q.v.*

meta- (1) prefix derived from Gk *meta*, after, signifying posterior, as in metathorax, the third and last thoracic segment in insects; (2) derived from Gk *meta*, change of, as in metamorphosis, a change in form.

metabiosis *n.* the beneficial exchange of factors (e.g. nutrients, vitamins) between species.

Metabola pterygotes *q.v.*

metabolic *a. pert.* metabolism.

metabolic activation conversion of foreign compounds into chemically reactive (often toxic or carcinogenic) forms in the body, esp. in the liver by microsomal enzymes.

metabolic pathway chain of enzyme-catalysed biochemical reactions in living cells that, e.g., convert one compound into another, or build up large macromolecules from smaller units, or break down compounds to release usable energy.

metabolic rate a measure of the rate of metabolic activity in a living organism, the rate at which an organism uses energy to sustain essential life processes such as respiration, growth, reproduction and, in animals, processes such as blood circulation, muscle tone and activity. It can be determined in numerous ways: (i) as the total heat produced over a given period; (ii) as oxygen consumption (and sometimes also carbon dioxide production) over a given period, which although easier to measure, only gives the contribution of aerobic metabolism; (iii) as the energy content of the food eaten over a given period; and (iv) by the fate of isotopically labelled water. *see* average daily metabolic rate, basal metabolic rate, energy budget, field metabolic rate, respiratory quotient, resting metabolic rate, standard metabolic rate.

metabolic scope the range of metabolic rate shown by an animal, which is the difference between the resting metabolic rate and the maximum rate of energy expenditure of which the animal is capable at maximum activity.

metabolic water water produced by oxidative processes, e.g. respiration, within the body.

metabolism *n.* integrated network of biochemical reactions in living organisms, often referring to the biochemical changes occurring in the living organism or cell as a whole. *see also* anabolism, catabolism, metabolic pathway.

metabolite *n.* any substance involved in or a product of metabolism.

metacommunication *n.* communication about the meaning of other acts of communication, such as the posture adopted by a dominant male, which signals to other males its status and likely behaviour if attacked.

metagyny protandry *q.v.*

metal *n.* (*chem.*) any of those chemical elements that, in elemental form, exhibit ductility (i.e. can be pulled to make wires), malleability (i.e. can be hammered into sheets) and high conductivity of electricity and heat. They are electropositive and tend to lose electrons to form cations. *see* Appendix 2.

metaldehyde *n.* contact and stomach-acting molluscicide used in baits to control snails and slugs.

metal fever fume fever caused by the fumes produced by certain metals when heated, e.g. nickel (Ni), lead (Pb), zinc (Zn).

metallic *a.* (1) *pert.* or *appl.* metal; (2) iridescent, *appl.* colours due to interference by fine striations or thin plates, as in insects.

metalloprotein *n.* any protein containing metal ions.

metal ore naturally occurring mineral material in which the concentration of a particular metal is high enough to make its recovery economic. This will be dependent on the demand for the metal concerned, in relation to its abundance.

metameric *a.* having the body divided into a number of segments more or less alike (metameres). *n.* metamerism.

metamorphic rocks rocks formed by the alteration of existing rock (called parent rock) by the action of extreme heat and/or pressure and/or permeating hot gases or liquids. Schists, slates and marble are examples of metamorphic rocks.

metamorphism *n.* (*geol.*) the transformation of an igneous or sedimentary rock into a metamorphic rock. This change in form is bought about by the action of heat and/or pressure and/or permeating hot gases or liquids. Metamorphic rocks may be classed as low-grade, medium-grade or high-grade depending on the degree of metamorphism that they have undergone. *see* dynamic metamorphism, metasomatic metamorphism, thermal metamorphism.

metamorphosis *n.* change in form and structure undergone by an animal from embryo to adult stage, as in insects and amphibians. In insects, incomplete metamorphosis occurs in, e.g. locusts and grasshoppers, where the larval form is relatively similar to the adult and changes gradually towards the adult form at each moult, and in which there is no non-feeding pupal stage. Complete metamorphosis occurs in, e.g. butterflies and flies, in which the larval caterpillar or maggot stage is quite unlike the adult in form and internal structure, and undergoes a radical remodelling during a non-feeding pupal stage.

Metaphyta, metaphytes *n., n.plu.* multicellular plants.

metasomatic metamorphism the formation of metamorphic rocks by the movement of hot gases and/or liquids through the parent rock. This process leads to changes in the rock's mineralogical composition as the fluids act as transport media, importing and/or exporting mineral-forming material to and from the parent rock.

Metatheria, metatherians marsupials *q.v.*

Metazoa, metazoans *n., n.plu.* multicellular animals, strictly applied only to those multicellular animals with cells organized into tissues and possessing nervous tissue.

metecdysis *n.* in arthropods, period after moult when the new cuticle hardens.

meteorites *n.plu.* solid bodies from outer space (meteors) that reach the Earth's surface without burning up completely in the atmosphere. They may be composed mainly of silicate minerals (aerolites) or of nickel-iron (siderites), or of both silicates and metals (siderolites).

meteorology *n.* scientific study of weather conditions (temperature, humidity, atmospheric pressure, etc.), esp. with a view to weather forecasting.

meteors *n.plu.* solid bodies from outer space. These may burn up in the Earth's atmosphere, when they are called shooting stars, or may reach the Earth's surface intact, when they are known as meteorites.

meth- (*chem.*) the prefix used to indicate the presence of one carbon atom in an organic compound (e.g. methane CH_4) or radical.

methabenzthiazuron *n.* soil-acting and translocated herbicide used to control blackgrass, e.g. in cereal crops such as winter wheat and winter barley.

methaemoglobin *n.* haemoglobin with the haem iron in the ferric (Fe^{III}) state, rather than ferrous (Fe^{II}) state, and thus unable to bind oxygen. It is normally found in small amounts in the blood. Increased levels are produced by the action of oxidizing agents such as nitrites and chlorates. *see* methaemoglobinaemia.

methaemoglobinaemia *n.* a severe blood disorder in infants caused by high levels of nitrate ions (NO_3^-) in drinking water. Bacterial action in the gut reduces the nitrate to nitrite ions (NO_2^-). These are absorbed into the bloodstream, where they oxidize the iron in the haemoglobin, producing methaemoglobin (*q.v.*), which reduces the oxygen-carrying capacity of the blood. At methaemoglobin concentrations above 25%, the skin and lips of the affected infant take on a bluish hue (hence 'blue-baby syndrome'). Fatalities occur in the range 60–85% methaemoglobin.

metham-sodium *n.* soil fumigant used under glass to control nematodes and a variety of soil diseases.

methanal *n.* (HCHO) the simplest aldehyde. A gas, at room temperature and pressure, with an irritating odour. *alt.* formaldehyde.

methane *n.* CH_4, the simplest hydrocarbon and the main constituent of natural gas. Methane is present in the atmosphere at a concentration of about 1.75 p.p.m. (rising at the rate of 1–2% per year). It is a greenhouse gas (*q.v.*). The anaerobic breakdown of organic material by microorganisms, e.g. in wetlands and in the intestines of ruminant animals, releases *ca.* 8.6×10^{12} to 2.9×10^{13} moles of CH_4 into the atmosphere each year. Human activity produces additional emissions, *ca.* 1.5×10^{13} to 3.6×10^{13} moles per year, primarily as the result of paddy rice production, low-temperature biomass burning, cattle rearing, waste disposal and fossil fuel extraction. Methane is removed from the troposphere by transfer to the stratosphere (where it is oxidized), absorption by soil and, most importantly, *in situ* reaction with hydroxyl (*q.v.*) free radicals or excited-state oxygen atoms. *alt.* marsh gas. *see also* biofuel, biogas.

methanogenesis *n.* the generation of the gas methane (CH_4) by living organisms, mainly bacteria. *a.* methanogenic.

methanogens *n.plu.* archaebacteria that produce the gas methane (CH_4) by the reduction of CO_2 or carbonate coupled to the oxidation of hydrogen (H_2). They are found in anoxic environments, such as marine and freshwater muds, the intestinal tracts of animals and sewage treatment plants.

methanol *n.* CH_3OH, a poisonous, flammable, colourless alcohol that is used, e.g., as antifreeze or as a solvent. In the past, it was made by destructively distilling wood but is now produced by the oxidation of methane in the presence of a catalyst. *alt.* methyl alcohol, wood spirit.

methidathion *n.* contact organophosphate insecticide and acaricide used, e.g., to control aphids on hops.

methiocarb *n.* contact and stomach-acting carbamate pesticide used, e.g., as a molluscicide to control slugs and snails in field crops.

methionine (Met, M) *n.* a sulphur-containing amino acid, constituent of proteins, essential in human diet, provides sulphur and methyl groups for metabolic reactions.

methomyl *n.* contact and systemic (entering plant via root) carbamate insecticide and nematicide.

methoxychlor *n.* wide-spectrum chlorinated hydrocarbon insecticide of relatively low toxicity whose use is still widely allowed.

methyl alcohol methanol *q.v.*

methyl bromide bromomethane *q.v.*

methyl chloroform 1,1,1-trichloroethane *q.v.*

methyl isocyanate (MIC) highly volatile, organic liquid (b.p. 39 °C), chemical formula CH_3NCO, used in the production of carbamate pesticides. *see* Bhopal disaster.

methyl mercury CH_3Hg^+, a mercury compound formed by the methylation of inorganic mercury (Hg^{2+}) by the action of microorganisms living in the sediment at the bottom of lakes and other waters. It is soluble in water and highly toxic. *alt.* monomethylmercury.

metoecious *a. appl.* parasites that are not host-specific.

metoestrus *n.* the luteal phase, period when activity subsides after oestrus.

metoxenous metoecious *q.v.*

metoxuron *n.* translocated and soil-acting herbicide used to control weeds, e.g. black-grass, in crops such as wheat and winter barley.

metre (m) *n.* the basic SI unit of length, defined (since 1983) as the length of the path travelled by light in a vacuum during a time interval of 1/299 792 458 of a second. *alt.* meter.

metribuzin *n.* translocated and soil-acting herbicide used on a variety of crops, e.g. potato and soya bean.

metric ton tonne *q.v.*

mevinphos *n.* systemic and contact organophosphate insecticide and acaricide of short environmental persistence.

Mg symbol for the chemical element magnesium *q.v.*

MHD power generation magnetohydrodynamic power generation *q.v.*

MIC methyl isocyanate *q.v.*

mica lamprophyre a fine- to medium-grained igneous rock that is generally porphyritic, rarely granular. Phenocrysts are, in the main, biotite, while some orthoclase and/or hornblende phenocrysts may also be present. The groundmass contains feldspars (sodic plagioclase or orthoclase) and amphibole or pyroxene.

mica minerals a large group of silicate minerals. They have in common structures based on sheets of $(Si,Al)O_4$ tetrahedra that are formed by the sharing of all basal oxygens (*see* Fig. 17). They have an overall stoichiometry of $A_2B_{4-6}C_8O_{20}D_4$, where, in common micas, A is K, Na or, in brittle micas, Ca, B is mostly Al, Mg or Fe, C is mostly Si or Al and D is OH or F (OH dominates in most micas).

Michaelis–Menten kinetics (Michaelis–Menten model) set of equations describing the properties of many enzyme-catalysed reactions, from which two characteristic parameters of such reactions are derived: K_m (the Michaelis constant) and V_{max}. K_m is the substrate concentration at which the reaction rate is half its maximal value, while V_{max} is the maximal rate of an enzyme-catalysed reaction under steady-state conditions.

micro- prefix derived from Gk *mikros*, small.

microaerophilic *a.* tolerating only a small amount of oxygen, *appl.* certain bacteria. *n.* microaerophile.

microalga *n.* unicellular alga visible only under a microscope. *cf.* macroalga.

microarthropod *n.* arthropod of microscopic size. *cf.* macroarthropod.

microbe *n.* any microorganism (*q.v.*), esp. a bacterium. *alt.* microorganism.

microbenthos *n.* microscopic plants, animals and microorganisms that live in or on sediments on the sea-floor or lake or river bottom.

microbial *a. pert.* or caused by microorganisms.

microbiology *n.* the science dealing with the study of microorganisms.

microbiota *n.* the population of organisms of microscopic size in any ecosystem, habitat, etc. (esp. applied to soil) and that includes bacteria, unicellular algae, fungi and protozoa. *cf.* macrobiota, mesobiota.

microbivore *n.* animal that feeds on microorganisms. *a.* microbivorous.

microbody *n.* any of a diverse class of small spherical bodies bounded by a single membrane, found in plant and animal cells (esp. liver and kidney), and including glyoxysomes (*q.v.*) and peroxisomes *q.v.*

microchiroptera *n.* small, mainly insectivorous, bats of the mammalian order Chiroptera. *cf.* megachiroptera.

microclimate *n.* (1) the climate within a very small area or in a particular habitat;

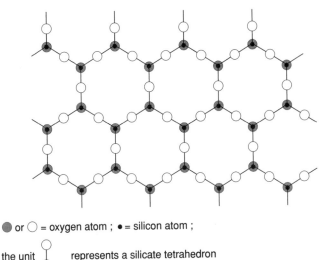

● or ○ = oxygen atom ; • = silicon atom ;

the unit ⟨tetrahedron symbol⟩ represents a silicate tetrahedron
(these are linked via shared oxygen atoms)

Note, those oxygen atoms represented thus ● reside above the silicon atoms

Fig. 17 A schematic representation of the structure of the sheets of silicate tetrahedra, as found in mica minerals.

(2) small-scale, localized climatic region such as that of an arable field. *cf.* macroclimate, mesoclimate.

microcline *n.* KAlSi$_3$O$_8$, one of the feldspar minerals *q.v.*

microconsumer decomposer *q.v.*

microcosm *n.* a world in miniature, a community that is a miniature version of a larger whole.

microcyclic *a.* (1) *appl.* rust fungi that produce only teleutospores and basidiospores during the life cycle; (2) *appl.* short and simple life cycles; (3) having a haplophase or gametophyte stage only.

microdiorite *n.* a medium-grained, generally porphyritic, igneous rock with the same mineralogy as diorite *q.v.*

microelements trace elements *q.v.*

microendemic *a.* restricted to a very small area.

microenvironment microhabitat *q.v.*

microevolution *n.* (1) evolutionary change consisting of alterations in gene frequencies, chromosome structure or chromosome numbers due to mutation and recombination and that can be noticed over a relatively short time, e.g. the acquisition of resistance to a pesticide amongst insects; (2) the relatively small-scale evolutionary change that differentiates the members of geographical races, subspecies, sibling species, etc. *cf.* macroevolution.

microfauna *n.* animals less than 200 μm long, such as protozoa, only visible under the microscope.

microflora *n.* (1) the microorganisms (bacteria, unicellular fungi and algae) living in or on an organism, or in a particular habitat or ecosystem; (2) (*bot.*) the dwarf flora of high mountains.

microfossil *n.* any microscopic fossil, as of prokaryotic and eukaryotic microorganisms, spores, pollen, microscopic animals and plants, etc.

microfungi *n.plu.* microscopic forms of fungi, the yeasts and moulds, as opposed to mushrooms and toadstools.

microgamete *n.* the smaller of two gametes in a species that produces gametes of different sizes, regarded as the male.

microgranite *n.* a rock that is like granite but with a medium grain size.

microhabitat *n.* (1) the immediate environment of an organism, esp. a small organism or microorganism; (2) a small place in the general habitat distinguished by its own set of environmental conditions. *alt.* microsite, microenvironment.

microinjection *n.* the introduction of substances into a single cell by injection with special instruments (e.g. micropipette).

micrometre (μm) *n.* unit of microscopic measurement, being 10^{-6} m, one-thousandth of a millimetre. Formerly called a micron. *alt.* micrometer.

micron micrometre *q.v.*

micronutrient elements, micronutrients *see* trace elements.

microorganism *n.* any microscopic organism, including bacteria, unicellular algae and protozoans, and microscopic fungi (yeasts and moulds). Viruses, although not strictly organisms, are also often called microorganisms. *alt.* microbe.

micropalaeontology *n.* the study of microfossils.

microparasite *n.* any parasite of microscopic size.

microphagic, microphagous *a.* (1) feeding on minute organisms or particles; (2) feeding on small prey. *n.* microphagy. *alt.* filter feeding.

microphanerophyte *n.* small tree or shrub from 2 to 8 m in height.

microphil(ic) *a.* tolerating only a narrow range of temperature, *appl.* certain bacteria.

microplankton *n.* small organisms drifting with the surrounding water, somewhat larger than those of nanoplankton.

Micropodiformes Apodiformes *q.v.*

micropopulation *n.* the population of microorganisms in a community.

micropores *see* pores.

microsaurs *n.plu.* extinct order of lepospondyl amphibians having an elongated or reptilian shape and developed limbs, each with four or fewer fingers.

microscopic *a.* able to be seen only under a microscope, *appl.* organisms or structures of very small size, i.e. less than 200 μm.

microscopy *see* dark-field microscopy, electron microscopy, light-field microscopy, optical microscopy, phase-contrast microscopy, scanning electron microscope, transmission electron microscope, tunnelling electron microscope.

microsere *n.* a successional series of plant communities in a microhabitat.

microsite microhabitat *q.v.*

microsomes *n.plu.* the smallest size particles spun down from cell homogenates in the ultracentrifuge and including broken parts of other fractions, esp. endoplasmic reticulum with ribosomes attached.

microspecies *n.* a taxonomic unit below the species level, such as a race, subspecies or variety.

microsphere *n.* structure of organic material formed by heating polypeptides, which can absorb various organic molecules from aqueous solution.

microsyenite *n.* a medium-grained igneous rock with a colour index of 0–40 and a mineralogy that is essentially alkali feldspar.

microsymbiont *n.* the smaller of two symbiotic organisms.

microtherm *n.* a plant of the cold temperate zone, which can grow below 12 °C.

microtidal *see* tidal range.

microtome *n.* machine with a sharp metal blade for slicing tissue into sections for microscopy.

microtomy *n.* the cutting of thin sections of tissues or other material, for examination by microscopy.

microwaves *n.plu.* electromagnetic radiation with wavelengths from about 1 mm to *ca.* 30 cm.

mictic *see* mixis.

middle latitudes latitudes between 20° and 50°.

middle shore zone of seashore between the average low-tide level and average high-tide level, which is usually the most extensive zone and is covered by the sea twice a day. *see* Fig. 25, p. 369.

mid-grass community grassland containing grasses of medium height, over 60 cm but under 2 m.

mid-latitude cell Ferrel cell *q.v.*

mid-latitude westerlies warm winds that blow, in each hemisphere, from the subtropical high-pressure belt (30° latitude) to the upper-mid-latitude low-pressure belt (60°), thus forming the low-altitude section of the Ferrel cell. In the northern hemisphere these global winds are known as south-westerlies. *alt.* westerlies.

mid-oceanic ridges underwater volcanic mountain ranges, which link together to form a global system over 60 000 km long. These ridges are found at the divergent plate boundaries between oceanic lithospheric plates and are associated with the production of new crustal material. *see* sea-floor spreading.

migmatite *n.* host rock, usually gneiss or schist, intimately mixed with granitic material.

migration *n.* (1) change of habitat according to season, climate, food supply, etc., that many animals undergo, often travelling very long distances along predetermined routes; (2) the movement of plants into new areas.

migratory *a.* characterized by seasonal movement from one area to another, as in many bird species.

Milankovitch model model that has been used in an attempt to explain the past sequence of glacials and interglacials. It is based on an analysis of the impact on insolation of the interplay between the natural cyclic variations in: (i) the tilt of the Earth's axis in relation to the perpendicular to the Earth's orbital plane, (ii) the shape of the Earth's elliptical orbit around the Sun and (iii) the rotation of the ellipse that is described by the Earth's orbit about the Sun.

mildews *n.plu.* fungal plant diseases typically accompanied by a powdery or downy coating on the surface of the affected part, caused by members of the oomycetes (downy mildews) and pyrenomycetes (powdery mildews).

milk fever a condition in dairy cows, characterized by lack of coordination and muscular spasms, which is caused by a low calcium content in the blood (hypocalcaemia).

milk sugar lactose *q.v.*

milk teeth first dentition of mammals, shed after or before birth. *alt.* deciduous teeth.

milkweeds *n.plu.* members of the genus *Asclepias*, well-known for containing high concentrations of cardiac glycosides.

milled peat extracted peat that is broken into granules and used as a soil substitute in horticulture or as a fuel in electricity generation. *cf.* sod peat.

milleporine *a. appl.* stony corals of the order Milleporina, which have colonies of two kinds of polyp living in pits on the surface of a massive calcareous skeleton (a corallium) and a brief medusoid stage.

millerite *n.* NiS, a nickel ore mineral of relatively minor importance.

millet *n.* may refer to various plants of the Gramineae family cultivated for their grain, including *Sorghum vulgare*, *Setaria italica*, *Pennisetum typhoideum* and *Panicum* spp.

millimicron *n.* former term for nanometre (nm), being one-thousandth of a micron (micrometre).

millipedes *n.plu.* common name for the diplopods (Diplopoda).

milpa agriculture shifting cultivation *q.v.*

milt *n.* testis or sperm of fishes.

mimesis *n.* (1) mimicry (*q.v.*); (2) the effect of the actions of one animal of a group on the activity of the others.

mimetic *a. pert.* or exhibiting mimicry.

mimetite *n.* $Pb_5(AsO_4)_3Cl$, a secondary mineral that occurs in the oxidized parts of lead ores. There is a complete solid-solution series between this mineral and pyromorphite (*q.v.*), which it closely resembles.

mimicry *n.* (1) the resemblance of one animal to an animal of a different species so that a third animal is deceived into confusing them; (2) the resemblance of an animal or plant to an inanimate object, or of an animal or part of an animal to a plant or part of a plant, usually for the purposes of camouflage. *a.* mimetic. *see also* Batesian mimicry, Müllerian mimicry.

Minamata disaster a very serious incident of mercury poisoning that occurred in Minamata, Japan, in the early 1950s. Local residents started to exhibit symptoms such as numbness of limbs and convulsions, and some babies born in the area had deformities and/or mental disorders. This collection of symptoms became known as Minamata disease. A local factory was eventually found to be responsible for contaminating Minamata Bay with mercury. This metal found its way into the human food chain via shellfish and fish, which formed a large part of the local diet. By 1975, Minamata disease had claimed over 100 lives, out of a total of 800 confirmed cases.

Minamata disease *see* Minamata disaster.

mineral fertilizers inorganic fertilizers *q.v.*

mineralization *n.* the breakdown of organic matter into inorganic components, in which form elements can be used as nutrients by plants. It occurs chiefly in the soil as a result of the activity of decomposer microorganisms. For carbon, mineralization occurs mainly through respiration, by which carbon dioxide is returned to the environment.

mineralogy *n.* the study of minerals *q.v.*

mineraloid *see* minerals.

mineral oil (1) crude oil (*q.v.*); (2) North American name for liquid paraffin (*q.v.*); (3) any oil obtained from mineral sources.

mineral reserve known deposits of a particular mineral that are recoverable under current technological and economic conditions.

mineral resource the total amount, of the mineral concerned, present on Earth, whether or not its location is known.

minerals *n.plu.* a slightly vague term generally taken to mean naturally occurring solids that are found within the Earth's crust, each of which is homogeneous, has recognizable properties and a distinct chemical composition and atomic structure. There are over 2000 known minerals. These may be classified as silicates, containing oxygen and silicon and usually other elements as well, or non-silicates, including the halides, carbonates, sulphides and sulphates. The term mineraloid is sometimes used for natural materials that do not fall fully within the definition of mineral being used. For example, the native metal mercury is a liquid and may therefore be categorized as a mineraloid by some.

mineral soil soil with a low organic matter content, variously defined as anything up to 30%.

miner's lung any pneumoconiosis (*q.v.*) found in a miner, esp. *appl.* coal miners.

minifundia *n.plu.* small farms in Central and South America.

minimal area smallest area of a community that can be sampled to find all the species present.

minimal cultivation reduced cultivation *q.v.*

minimal medium culture medium containing a basic set of nutrients only, on which normal wild-type organisms can grow, but which cannot support the growth of metabolic mutants.

minimal tillage, minimum-tillage cultivation reduced cultivation *q.v.*

minimum, law of the *see* law of the minimum.

minimum lethal dose (MLD) minimum dose of any agent sufficient to cause 100% mortality in the test population.

minimum viable population the smallest number of interbreeding individuals that can sustain a population over time. Below this number the population will eventually die out.

minor elements trace elements *q.v.*

minor gene a gene that has a small effect individually but contributes to a multifactorial phenotypic trait.

Miocene *n.* a geological epoch of the Tertiary, between Oligocene and Pliocene, lasting from about 25 to 5 million years ago.

mire *n.* bog or fen, usually referring to a peatland but also used to describe a fen developing on mineral soils, *see* blanket mire, raised mire, topogenous mire.

misfit river river that appears to be too small for its valley, possibly as a result of river capture. *alt.* underfit river.

missense mutation mutation in which one base pair is altered causing an amino acid change in the protein product of the gene.

Mississippian *a.* the name given to the Lower Carboniferous in North America.

mist *n.* (*clim.*) liquid droplets 0.1–$2 \, \mu m$ in diameter held suspended in the atmosphere.

Mistral *n.* strong, cold wind that blows down the Rhône Valley in France.

mitDNA mitochondrial DNA.

mites *n.plu.* common name for many of the Acarina *q.v.*

mitochondria *n.plu.* organelles in the cytoplasm of eukaryotic cells, having a double membrane, the inner one invaginated, and which are the sites of the tricarboxylic cycle and oxidative phosphorylation of oxidative respiration, generating ATP. They contain a small circular DNA that specifies tRNAs, rRNAs and some mitochondrial proteins. *sing.* mitochondrion.

mitogenic *a.* inducing mitosis and cell division.

mitomycin C antibiotic produced by *Streptomyces caespitosus* that inhibits nuclear division, DNA and protein synthesis in mammalian cells, and is used clinically as an antitumour agent.

mitosis *n.* the typical process of nuclear division in eukaryotic cells, in which each member of a duplicated chromosome segregates into a daughter nucleus, resulting in daughter nuclei containing identical sets of chromosomes, identical to that of the parent nucleus, and which is generally followed by cell division. *cf.* meiosis.

mitotic *a. pert.* or produced by mitosis.

mitotic index the number of dividing cells per 1000 in a population, at any given time.

mixed cloud cloud containing both water droplets and ice crystals.

mixed cropping simultaneous cultivation of two (or more) crop species on the same piece of land, often in combinations where the presence of one benefits the other(s), as between legumes and cereals. *see also* interculture, multistorey cropping. *alt.* intercropping.

mixed farming a closely integrated method of farming where a rotation system of cereals, root crops and grass–clover pastures is used to rear livestock on the farm. This was normal agricultural practice in temperate regions until the mid-1950s, when the progressive intensification of agriculture led to increasing specialization, resulting, in many cases, in the complete segregation of crop (arable) and livestock production.

mixed fertilizers compound fertilizers *q.v.*

mixed forest (1) forest containing both broad-leaved and coniferous trees; (2) broad-leaved forest in which at least 20% of the trees are of species other than the dominant species.

mixed grazing the simultaneous grazing of two (or more) different types of animals on the same land.

mixed woodland *see* mixed forest.

mixing ratio (*clim.*) ratio of the mass of water vapour to the mass of dry air in a designated volume. *alt.* water vapour mixing ratio.

mixis *n.* sexual reproduction, esp. the fusion of gametes. *a.* mictic.

mixotrophic mesotrophic *q.v.*

mixture *n.* a physical combination of more than one chemical element and/or chemical compound, the individual properties of which are still discernible. *cf.* compound.

MLD minimum lethal dose *q.v.*

Mn symbol for the chemical element manganese *q.v.*

mnemotaxis *n.* movement directed by memory, as returning to feeding place or homing.

Mo symbol for the chemical element molybdenum *q.v.*

modality *n.* the qualitative nature of a sense, stimulus, etc., e.g. taste, smell, hearing, sight are different modalities of sensory experience.

modal number the most frequently occurring chromosome number in a taxonomic group.

mode *n.* in a distribution, the most frequently occurring value.

model *n.* (1) simplified representation of a natural system or situation, which can be used to test a hypothesis; (2) proposed explanation for a natural phenomenon that can be tested by experiment.

moder *n.* type of humus intermediate between mor humus and mull humus.

moderator *n.* material used in a thermal reactor (*q.v.*) to slow down (i.e. moderate) the fast neutrons generated during nuclear fission. The slow neutrons thereby produced have appropriate kinetic energies to induce fission efficiently in the fissile fuel of the reactor. The moderators used are either light water (i.e. normal water, $_1^1H_2O$), heavy water ($_1^2H_2O$) or graphite.

mogotes *see* tower karst.

Moho, Mohorovičić discontinuity *n.* discontinuity in density that occurs at the interface between the Earth's crust and the mantle, thus allowing the thickness of the crust to be calculated. *alt.* M discontinuity.

moiety *n.* (*chem.*) part of a molecule.

moisture content amount of water (usually the mass) present in a substance, based on either wet weight or dry weight of sample, often expressed as a percentage.

moisture release curve a plot of the matric potential (*q.v.*) of a soil against its water content.

mol abbreviation for mole *q.v.*

molality *n.* one way of expressing the concentration of a chemical solution: a molal solution contains 1 mole of solute per kilogram of solvent.

molar *a.* (1) (*biol.*) adapted for grinding, as *appl.* teeth; (2) (*chem.*) most commonly, the quantity per mole. Sometimes indicates quantification per given amount, where the unit of amount is other than the mole. Exceptionally, it means quantification per unit concentration. When *appl.* a solution it indicates its concentration, e.g. a solution with a molarity of 5 mol dm^{-3} is said to be five molar (5 M), or one with a molarity of 1 mol dm^{-3} is one molar or molar (M).

molar concentration *see* molarity.

molar mass the mass, in grams, of one mole of the substance concerned. Units g mol^{-1} (frequently not stated).

molarity (M) *n.* moles of solute per litre (or cubic decimetre) of solution, i.e. mol l^{-1} or mol dm^{-3}. *alt.* molar concentration. *see also* molar.

molasses *n.* thick brown syrup produced during the processing of raw sugar. It is used, e.g., as an animal feed additive.

mold *alt.* spelling of mould *q.v.*

mole, mol *n.* the SI unit of amount of substance. It contains as many elementary units (i.e. atoms or molecules as appropriate) as there are atoms in 0.012 kg of ^{12}C. The elementary units should be specified, e.g. 1 gram-mole of a substance has a mass equal to its molecular weight expressed in grams.

molecular biology study of biological phenomena at the molecular level.

molecular clock the time elapsed since the divergence of different present-day lineages from their common ancestor can, in principle, be estimated by comparing suitable corresponding DNA sequences or protein sequences from two extant species and counting the differences that have accumulated between them. Over long periods of time the rate of certain types of unselected nucleotide change appears to be directly proportional to time elapsed, if measurements are restricted to appropriate sequences and closely related lineages. Such clocks may be calibrated in real time by comparison of sequences from species whose point of divergence is well-established from the fossil record. *alt.* evolutionary clock.

molecular evolution the changes that occur in DNA and in proteins as a result of mutation, chromosomal duplications, chromosomal rearrangements, etc., over long periods of time, which may alter function, eventually giving rise to novel genes and proteins, or which may be silent.

molecular genetics the study of the molecular structure of DNA and the information it encodes, and the biochemical basis of gene expression and its regulation.

molecular ion a molecule that carries a net overall charge, e.g. the carbonate anion (CO_3^{2-}), the ammonium cation (NH_4^+).

molecular mass sum of the atomic masses of all of the atoms in a molecule. It is generally expressed in daltons (*q.v.*). The term molecular weight, although not strictly equivalent, is often used as a synonym. *see also* relative molecular mass.

molecular phylogeny the tracing of evolutionary relationships by the comparison of DNA and protein sequences from different organisms.

molecular weight the sum of all the atomic weights of the atoms in a molecule. It now generally refers to the relative molecular mass (M_r), the ratio of the mass of one molecule of a substance to one-twelfth the mass of an atom of ^{12}C. This is a ratio and therefore dimensionless, but the term is often used in the sense of absolute molecular mass, and a value given in daltons or kilodaltons (kD).

molecule *n.* a group of atoms held together by covalent bonds.

mole drains temporary, cylindrical channels (five- to ten-year lifespan) formed by drawing a bullet-like implement (diameter *ca.* 75 cm) through the soil (a process called moling). Mole drains are typically made at a soil depth of 50–60 cm, at intervals of 2–3 m and approximately at right angles to more permanent tile drains already installed.

moling *see* mole drains.

mollic epipedon *see* mollisol.

Mollicutes *n.* a class of prokaryotes including diverse wall-less microorganisms, e.g. rickettsiae, chlamydiae and mycoplasmas.

mollisol *n.* an order of the USDA Soil Taxonomy system comprising soils characterized by a thick, dark mineral surface horizon that is >50% saturated with basic cations (mollic epipedon). The majority of mollisols have developed under grassland and are among the most fertile of the world's soils. Seven suborders are recognized, *see* alboll, aquoll, boroll, rendoll, udoll, ustoll, xeroll.

Mollusca, molluscs *n., n.plu.* a large and diverse phylum of soft-bodied, usually unsegmented, coelomate animals, many of which live enclosed in a hard shell. They include the classes Gastropoda (winkles, whelks, slugs, snails, sea slugs, etc.), Bivalvia (clams, cockles, etc.) and other smaller classes of shells, and the Cephalopoda (nautilus, squids and octopuses). The coelom is small and the main body cavity is a blood-filled haemocoel. Molluscs have well-developed sense organs and nervous system, esp. in the Cephalopoda, and a heart and blood system.

molluscicide *n.* a chemical that kills molluscs, e.g. snails.

mollusk *alt.* spelling of mollusc *q.v.*

molt *alt.* spelling of moult *q.v.*

molybdenite *n.* MoS$_2$, the most important molybdenum ore mineral.

molybdenum (Mo) *n.* metallic element (at. no. 42, r.a.m. 95.94), forming a hard silvery metal in elemental form. Found in the ore molybdenite (MoS$_2$). An essential micronutrient for plants, molybdenum is a prosthetic group in certain enzymes, including some bacterial nitrogenases.

monactinellid *a. appl.* certain sponges that bear spicules with only a single ray. *cf.* hexactinellid.

monadnock *n.* an isolated hill or mountain rising up from a flat plain.

monazite *n.* (Ce,La,Y,Th)PO$_4$, an ore mineral of thorium and cerium.

monecious *alt.* spelling of monoecious *q.v.*

monembryonic *a.* producing one embryo at a time.

Monera *n.* name given to a kingdom that includes all prokaryotes.

monestrous *alt.* spelling of monoestrus *q.v.*

Monilia *n.* large form class of deuteromycete fungi that reproduce by conidia not borne in pycnidia or acervuli, or by oidia or by budding. It includes *Penicillium* and *Aspergillum*, the false yeasts (e.g. *Cryptococcus*), and fungi that cause skin diseases in humans and animals (e.g. ringworm, *Microsporum*) and the serious human fungal pathogens *Blastomyces* and *Histoplasma*.

monkeys *see* Primates.

mono- prefix derived from Gk *monos*, single, signifying one, having one of, borne singly, etc.

monocarpic *a. appl.* plants that die after bearing fruit once. *n.* monocarp.

monoclimax theory theory, put forward in the 1920s, in which climate was viewed as the sole determinant of climax community composition. Therefore, for each particular climatic region, only one climax state was possible (climatic climax). The monoclimax theory was superseded by the polyclimax theory, which states that, while climate is the major determinant, there are a number of possible climax states depending on local conditions of topography, soil conditions and animal activity.

monocots monocotyledons *q.v.*

Monocotyledon(e)ae, Monocotyledones, monocotyledons *n., n., n.plu.* a class of angiosperm plants having an embryo with only one cotyledon, parts of the flower usually in threes, leaves with parallel veins and vascular bundles scattered throughout the stem. They include familiar bulbs, such as daffodils, snowdrops, lilies, etc., and the cereals and grasses, such as maize, wheat, rice, etc.

monocotyledonous *a. pert.* (1) monocotyledons; (2) embryo with only one cotyledon.

monocropping monoculture *q.v.*

monoculture *n.* a large area covered by a single species (or for crops, a single variety) of plant, esp. if grown year after year. *alt.* monocropping.

Monodelphia Eutheria, *see* eutherians.

monoecious *a.* (1) having male and female flowers on the same plant; (2) with male and female sex organs on same gametophyte; (3) having microsporangia and megasporangia on the same sporophyte.

monoestrous *a.* having only one period of oestrus in a sexual season. *cf.* polyoestrus.

monogamous *a.* consorting with one mate only, usually for the whole of the animal's lifetime. *n.* monogamy.

Monogenea, monogeneans *n., n.plu.* class of parasitic flatworms comprising the skin and gill flukes, which are ectoparasites mainly of fish and amphibians. They have a flattened leaf-shaped body and a simple life cycle on one host.

monogenetic *a. appl.* (1) parasites completing their life cycle in a single host; (2) origin of a new form at a single place or period. *n.* monogenesis.

monogenic *a.* (1) controlled by a single gene; (2) producing offspring all of the same sex.

monogoneutic *a.* breeding once a year.

monogyny *n.* (1) in animals generally, the tendency of each male to mate with only one female; (2) in social insects, the existence of a single functional queen in the colony.

monohybrid inheritance pattern of inheritance that results from crossing individuals identical except at one locus, one parent carrying only the dominant form of the gene and one only the recessive form of the gene. The offspring all have a dominant phenotype in the first generation, but when intercrossed, 25% of the offspring have the recessive phenotype.

monokaryon *n.* fungal mycelium whose cells carry only one haploid nucleus each.

monolinuron *n.* contact and soil-acting herbicide used to control annual weeds in crops such as potato and french beans.

monomer *n.* molecule that is the unit of a polymer, e.g. amino acids are the monomers in proteins.

monomethylmercury *see* methyl mercury.

monomorphic *a.* (1) *appl.* species in which all individuals look alike; (2) developing with no or very slight change from stage to stage, as certain protozoans and insects; (3) producing spores of one kind only. *cf.* dimorphic, polymorphic.

monomorphic loci genetic loci at which the most common homozygote has a frequency of more than 90% in a given population.

mononym *n.* (1) a designation consisting of one term only; (2) name of a monotypic genus.

monophagous *a.* subsisting on one kind of food, *appl.* insects feeding on plants of one genus only, or insects restricted to one species or variety of food plant.

monophyletic *a.* derived from a common ancestor, *appl.* taxa derived from and including a single founder species.

Monoplacophora *n.* a mainly extinct class of molluscs with a shell like a limpet, the living forms (e.g. *Neopilina*) being known only from the deep seabed.

monoploid *a.* (1) having one set of chromosomes, true haploid; (2) in a polyploid series, having the basic haploid chromosome number.

monosaccharide *n.* any of a class of simple carbohydrates, all being reducing sugars, with the general formula $(CH_2O)_n$, where n is greater than three. Examples are glyceraldehyde (a triose, $n = 3$), ribose (a pentose, $n = 5$), and glucose (a hexose, $n = 6$).

monospecific *a.* having only one species, *appl.* genus, family or other taxonomic group composed of a single species.

monosulcate *a. appl.* pollen grain with a single furrow on the surface away from that through which the pollen tube emerges.

monotaxic *a.* belonging to the same taxonomic group.

monotokous *a.* uniparous, having one offspring at birth. *alt.* monotocous.

Monotremata, monotremes *n., n.plu.* an order of primitive mammals that lay eggs, have mammary glands without nipples, and no external ears, of which the only extant species are the duck-billed platypus of Australia (*Ornithorhynchus*) and the spiny anteaters or echidnas (*Tachyglossus* and *Zaglossus*), which are found in Australia and New Guinea.

monotrophic *a.* subsisting on one kind of food.

monotropic *a.* (1) turning in one direction only; (2) visiting only one kind of flower, *appl.* insects.

monotropoid *a. appl.* mycorrhizas formed on members of the Monotropaceae, plants lacking chlorophyll that are dependent on the mycorrhiza for their carbon and energy source. An extensive root-ball of fungal and root tissue is formed, which also forms connections with the ectomycorrhizal roots of nearby green plants.

monotype *n.* single type that constitutes species or genus.

monotypic *a. appl.* (1) genera having only one species; (2) species having no subspecies.

monovalent *a.* (*chem.*) having a valency (*q.v.*) of one. *alt.* univalent.

monovoltine univoltine *q.v.*

monozygotic *a.* originating from a single fertilized ovum (zygote), as identical twins (MZ twins).

monsoon *n.* seasonal reversal in wind direction, which effects the climate of much of tropical Africa, Asia and Australasia. Typically, in summer, moistureladen winds blow onshore bringing heavy rains, while in winter, the wind direction is reversed, blowing offshore.

monsoon forest, monsoon rain forest type of rain forest that develops in tropical and subtropical regions with a high annual rainfall but marked dry and rainy seasons (monsoon rainfall), consisting of deciduous trees and shrubs that lose their leaves in the dry season. *alt.* tropical seasonal forest.

Mont Pelée (natural disaster) in May 1902, a major volcanic eruption issued from Mont Pelée on the West Indian island of Martinique. This completely devastated the nearby city of St Pierre and claimed the lives of virtually all of its 30 000 inhabitants. *see* nuée ardente.

montane *a.* (1) *pert.* mountains; (2) *appl.* the cool subalpine region just below the tree line in mountains and characterized by coniferous vegetation.

monticellite *n.* an olivine (*q.v.*) mineral of formula CaMgSiO$_4$.

monticolous *a.* inhabiting mountainous regions.

montmorillonite *n.* (Al,Mg)$_8$(Si$_4$O$_{10}$)$_3$ (OH)$_{10}$.12H$_2$O, a 2 : 1-type expanding clay mineral of the smectite group (*q.v.*). *see* clay minerals.

montmorillonite group smectite group *q.v.*

Montreal Protocol on Substances that Deplete the Ozone Layer often known as the Montreal Protocol, an international protocol that was drawn up in 1987 and came into force in January 1989, with amendments since. The main provisions were to achieve a freezing of consumption and production of the most damaging CFCs and halons at 1986 levels and to phase out their production competely by the beginning of the twenty-first century. Other substances covered by the protocol are carbon tetrachloride (CCl$_4$) and methyl chloroform (CH$_3$CCl$_3$).

moor, moorland *n.* open area of upland acid peat, with vegetation cover of heathers, sedges and certain grasses (e.g. *Molinia, Caerulea*).

moorpan *n.* hardpan formed by the accumulation of humic materials in the B-horizon.

mor, mor humus humus that forms a distinct organic horizon (the O-horizon) on the surface of an essentially inorganic soil. Its development is favoured by cold, waterlogged and/or highly acidic soil conditions such as are found under coniferous woods or heath vegetation. Three layers may be distinguished in mor humus: L-layer (undecomposed litter), F-layer (partially decomposed litter, fibrous) and H-layer (well-decomposed humus layer). *cf.* mull humus.

moraine *n.* a distinctive glacial landform produced by the deposition of till (*q.v.*), together with some stratified drift (*q.v.*). Several different types of these complex, and often transitory, landforms are recognized. Ground moraine describes the blanket of till deposited directly on the valley floor from the base of the melting glacier. Lateral moraines are ridges of material (previously carried at the edges of the now-retreating glacier) that are deposited along the valley sides. Medial moraine occupies a central position in the glacial valley and arises from the joining of two lateral moraines at the point where two glaciers meet. Push moraine may develop when the temporary advance of a previously retreating glacier shunts existing moraine into a new mound. Recessional moraine (usually formed parallel to terminal moraine) may be laid down when the retreat of a glacier is temporarily halted. Terminal moraine (also known as end moraine) is a ridge of till, deposited at right angles to the direction of ice flow, which marks the point of maximum advance of a glacier.

morbidity rate the number of cases of disease (morbidity) per 100 000 individuals.

morbilliviruses *n*. a group of RNA viruses of the paramyxovirus family, related to canine distemper virus.

more developed countries (MDCs) industrialized nations, located mainly in the northern hemisphere, characterized by low (or even zero) population growth rates, low birth and death rates, and high per capita income. *cf.* less developed countries.

Morisita's index of dispersion (I) a number that represents the spatial distribution (dispersion) of individuals within a population.

Morisita's similarity index (C_γ) an index of the similarity (*see* coefficient of community) between two communities that is weighted by the population sizes of species present in the two communities. *cf.* Gleason's index, Jaccard index, Kulezinski index, Simpson's index of floristic similarity, Sørensen similarity index.

morph *n*. one of the forms present in a polymorphic population.

morphactins *n.plu.* a group of substances derived from fluorine-9-carboxylic acids, which affect plant growth and development.

morphallaxis *n*. regeneration of a part of the body by transformation of pre-existing tissue, i.e. regeneration without growth.

morphine *n*. the chief alkaloid of opium, used clinically to relieve pain, but produces dependency with long-term use.

morphogenesis *n*. (1) the development of shape and structure; (2) origin and development of organs or parts of organisms. *alt.* morphogeny.

morphogenetic *a*. (1) *pert.* morphogenesis (*q.v.*); (2) *appl.* hormones, e.g. thyroxine, ecdysone, juvenile hormone, etc., that influence growth, development and/or metamorphosis of organisms.

morphological species morphospecies *q.v.*

morphologic index ratio expressing relation of trunk to limbs.

morphology *n*. (1) the form and structure of an organism as distinct from its physi-

ology, etc.; (2) the study of form and structure. *a*. morphological.

morphoplankton *n*. plankton organisms rendered buoyant by small size, or body shape, or structures containing oily globules, mucilage, gas, etc.

morphospecies *n*. a group of individuals that are considered to belong to the same species on grounds of morphology alone.

morphotype *n*. type specimen of one of the forms of a polymorphic species.

morph ratio cline gradual change in the frequency of different morphs in a population over its geographical range.

mortality factors environmental variables (e.g. predation) that affect the death rate in a population.

mortality rate for human populations, the number of deaths per 1000 people per year. *alt.* death rate.

mortlake oxbow lake *q.v.*

mosaic *n*. (1) disease of plants characterized by mottling of leaves, caused by various viruses, e.g. tobacco mosaic, cucumber mosaic; (2) organism whose body cells are a mixture of two or more different genotypes, e.g. human and other mammalian females, which have one of their X chromosomes inactivated at random early in development so that adult tissues contain a mixture of cells containing different active X chromosomes.

moss animals common name for the Bryozoa *q.v.*

mosses *n.plu.* common name for members of the Bryophyta (*q.v.*) a division of nonvascular, spore-bearing green plants. The mosses are divided into three classes: the 'true' mosses, the sphagnum mosses and the granite mosses (*see* individual entries). Several plants commonly called mosses belong to other groups: reindeer moss is a lichen, club mosses and Spanish moss are vascular plants, sea moss and Irish moss are algae.

moss layer the lowest horizontal ecological stratum of a plant community comprising the ground surface and plant cover, such as mosses and lichens. *alt.* ground layer.

mothballing of nuclear reactors *see* decommissioning process.

mother ship *see* factory fishing.

moths *n.plu.* the common name for many members of the order Lepidoptera, having antennae tapering to a point and not clubbed.

motile *a.* capable of spontaneous movement.

motivation *n.* internal factors controlling behaviour in an animal that lead to its achieving a goal or satisfying a need.

motivational state the combined effect of the physiological state of an animal and its perception of stimuli from the environment, which determines behaviour.

motor *a. pert.* or connected with movement, *appl.* nerves, etc.

motor spirit petrol *q.v.*

motorway *n.* road with two carriageways, each consisting of two or more lanes (plus a hard shoulder for emergency stopping) separated from each other by a central reservation and served by slip roads (*q.v.*). Particular restrictions of use are generally applicable to motorways. *alt.* autobahn, autostrada, freeway, highway.

mottles *n.plu.* (*soil sci.*) small areas of contrasting colour found within a soil, esp. *appl.* spots of red iron(III) oxide found within the predominantly blue-green/grey horizon of gley soils.

mould *n.* common name for many fungi that grow as a fluffy mycelium over a substrate. *alt.* mold.

moult *n.* the periodic shedding of outer covering, whether of feathers, hair, skin or cuticle. In crustaceans and other arthropods, it is necessary during larval growth as the exoskeleton, once hardened, cannot grow to accommodate further internal growth.

mound cultivation type of cultivation, usually practised in tropical agriculture, in which crops are grown on raised mounds of buried decomposing and/or burnt vegetation. *see also* lazy-bed cultivation.

mountain glaciers valley glaciers *q.v.*

mountain wind katabatic wind *q.v.*

mouthpart *n.* head or mouth appendage of arthropods.

m.p. melting point.

M phase the period of mitosis and cell division (cytokinesis) during the cell cycle, sometimes used to indicate mitosis alone.

mRNA messenger RNA *q.v.*

MSW municipal solid wastes *q.v.*

mtDNA mitochondrial DNA.

mucigel *n.* gelatinous material on the surface of roots in soil, comprising a mixture of plant mucilages, bacterial capsules and slime layers and colloidal soil particles.

mucilage *n.* general term for complex substances composed of various types of polysaccharides, becoming viscous and slimy when wet, widely occurring in plants, and secreted by plant roots and by bacteria (the capsule or slime layer).

mucilaginous *a. pert.*, containing or composed of mucilage.

mucin *n.* general term for various glycoproteins found in secretions such as saliva, mucus, etc.

mucivorous *a.* feeding on plant juices, *appl.* insects.

mucoid feeding feeding method used by some molluscs, which push mucus out of the mouth to trap food particles and then reingest it.

mucoids *n.plu.* glycoproteins (*q.v.*) of bone, tendon and other connective tissues.

mucopolysaccharide glycosaminoglycan *q.v.*

mucoprotein *n.* (1) glycoprotein (*q.v.*), esp. those found in mucous secretions; (2) proteoglycan *q.v.*

Mucorales *n.* class of zygomycete fungi with a well-developed mycelium and non-motile spores contained in a stalked sporangium, most living as saprophytes on dung or decaying plant and animal matter. It includes the bread moulds, e.g. *Mucor* and *Rhizopus*, and the dung fungus, *Pilobolus*.

mucous *a.* secreting, containing or *pert.* mucus.

mucous membrane any epithelial layer secreting mucus, e.g. the linings of the nasal passages, reproductive tract, gut, etc. *alt.* mucosa.

mucosa mucous membrane *q.v.*

mucus *n.* (1) slimy material rich in glycoproteins, secreted by goblet cells of mucous membranes or by mucous cells of a gland; (2) similar slimy secretion produced on the external body surface of many animals.

mudflat tidal flat *q.v.*

mudflow *see* flow movement.

mudstone *n.* a consolidated and relatively massive sedimentary rock that is made up of grains of less than 0.004 mm in diameter. *alt.* claystone. *cf.* clay (3), shale.

mud volcano a minor extrusive landform, associated with areas of declining volcanic activity, formed when mud erupts as a result of the heating of near-surface groundwater.

mulch *n.* a layer of material, e.g. wood chips, seaweed, manure or plastic, placed on the ground around crops to conserve water, protect the soil from erosion and suppress weed growth.

mull, mull humus humified organic matter that is well-mixed into a mineral soil. This type of humus occurs in soils that are moist, well-aerated and neutral or mildly acidic, such as found under grassland or deciduous woods. *cf.* mor humus.

Müllerian mimicry the resemblance of two animals to their mutual advantage, for example the yellow and black stripes of wasps and of the unpleasant-tasting cinnabar moth caterpillars, which leads to a predator that has encountered one subsequently also avoiding the other. *cf.* Batesian mimicry.

multi- prefix derived from L. *multus,* many.

multicellular *a.* (1) many-celled, *appl.* eukaryotic organisms composed of many cells specialized for different functions and organized into a cooperative structure; (2) consisting of more than one cell. *n.* multicellularity.

multidentate *a.* (1) (*biol.*) with many teeth or indentations; (2) (*chem.*) *appl.* ligands with many donor atoms.

multifactorial inheritance inheritance of phenotypic characters determined by the action of several independent genes.

multigene family a set of similar but not identical genes that encode the different members of a family of related proteins such as the interferons, the actins, the globins, etc. Multigene families are presumed to have arisen by duplication and divergence of an ancestral gene.

multimer *n.* (1) protein molecule made up of more than one polypeptide chain (protein subunit); (2) protein complex made up of several different proteins. *a.* multimeric.

multinomial *a. appl.* a name or designation composed of several terms.

multinucleate *a.* with several or many nuclei.

multiparous *a.* (1) bearing several, or more than one, offspring at a birth; (2) (*bot.*) developing several or many lateral axes.

multiple alleles a series of more than two alleles, for any given gene locus, present within a population, such loci being known as polymorphic. *alt.* genetic polymorphism.

multiple bonds (*chem.*) covalent bonds that are made by the sharing of more than one pair of electrons between a given pair of adjacent atoms. *see also* double bonds, triple bonds.

multiple cropping production of two or more crops from the same piece of land during a single growing season. *see* mixed cropping, ratooning, relay cropping, sequential cropping.

multiple cropping index (MCI) a parameter used to assess the intensity of land use in areas where multiple cropping (*q.v.*) is practised. It is the annual ratio of total crop area : total cultivated area, often expressed as a percentage.

multiple use the use of an area for a variety of purposes, e.g. the management of a forest for timber, recreation and wildlife.

multistorey cropping type of mixed cropping in which annual and perennial crops of different heights are grown together.

multituberculates *n.plu.* a class of extinct herbivorous mammals, with teeth having many small points (multituberculate), existing in the Jurassic–Eocene, with affinities to present-day monotremes.

multivariate *a.* (1) involving two or more variables; (2) *appl.* statistical analysis of several measurements made on more than one attribute of each entity under observation (e.g. height and weight).

multivoltine *a.* having more than one brood in a year, *appl.* some birds.

municipal solid waste (MSW) US term for the waste (both commercial and residential) produced in a specific municipality.

Munsell colour chart a system by which colours are classified and coded on the basis of their appearance, used, e.g., in the description of soil horizons.

mural *a.* (1) constituting or *pert.* a wall; (2) growing on a wall.

Musa genus of large tropical monocotyledonous plants that includes the banana (*M.*

sapientum) and the plantain (*M. paradisiaca*) that are toxic cultivated for food.

muscarine *n.* a toxic ptomaine base, found in the fly agaric toadstool *Amanita muscaria* and other plants.

Musci mosses *q.v.*

muscicoline *a.* living or growing among or on mosses.

muscimol *n.* hallucinogenic plant alkaloid, binding to GABA receptors in brain.

muscle *n.* contractile animal tissue involved in movement of the organism that also forms part of many internal organs. Muscle cells contain contractile protein microfibrils that contract simultaneously, usually in response to a nervous or chemical stimulus. There are three main types of muscle in vertebrates: striated or striped muscle, which forms the muscles attached to the skeleton; smooth muscle, associated with many organs, which forms the contractile layer of arteries; and the cardiac muscle of the heart.

muscology *n.* study of mosses.

Muscopsida mosses *q.v.*

muscovite *n.* $KAl_2(AlSi_3O_{10})(OH,F)_2$, one of the mica minerals *q.v.*

muscovite granite *see* granite.

mushrooms *n.plu.* common name for edible basidiomycete fungi, esp. of the genus *Agaricus*.

muskeg *n.* landscape characterized by peat bogs interspersed with areas of tussock grasses.

mustelids *n.plu.* members of the family Mustelidae: i.e. weasels, stoats, badgers, otters, polecats, martens.

mutagen *n.* any agent that can cause a mutation. *a.* mutagenic.

mutagenesis *n.* the production of mutations.

mutagenic *a.* capable of causing a mutation, as *appl.* radiation, chemicals or other extracellular agents.

mutagenize *v.* to treat with a mutagen.

mutant *n.* organism or cell carrying altered genetic material owing to which it differs from its parent or immediate precursor cell in some physical or biochemical characteristic(s). *a.* mutant.

mutate *v.* to undergo mutation.

mutation *n.* a permanent change in the amount or chemical structure of DNA,

resulting in a change in the characteristics of an organism or an individual cell as a result of alterations in, or non-production of, proteins (or RNAs) specified by the mutated DNA. Mutations occurring in body cells of multicellular organisms are called somatic mutations and are only passed on to the immediate descendants of those cells, while mutations occurring in germline cells can be inherited by the offspring. Mutations can occur spontaneously as a result of errors in normal cell processes, e.g. DNA replication, or can be induced by certain chemicals, types of radiation, etc. Alterations in DNA that do not cause any phenotypic change are also sometimes called mutations (silent mutations).

mutational load the reduction in population fitness due to the accumulation of deleterious mutations.

mutation pressure changes in gene frequencies brought about by mutational change alone.

mutation rate the rate at which mutations arise in a population. The spontaneous rate of base-pair changes (due to errors in DNA replication and environmental influences) is *ca.* 1 in 10^9 rounds of replication on experimental evidence, from which it has been calculated that a mutation will occur in a protein at a rate of *ca.* 1 per 10^6 cell generations. The mutation rate as detected by the appearance of detectably mutant organisms appears to vary between different organisms and between different genes. Mutation rate is also often expressed as the number of mutations per gamete per generation.

mutualism *n.* a special case of symbiosis in which both partners benefit from the association. *a.* mutualistic.

mutuality *n.* evolutionary strategy in regard to animal communication where both signaller and receiver benefit from the interaction.

MW (1) megawatt (*q.v.*); (2) molecular weight. *see* relative molecular mass.

MW$_e$ the electrical power output of a power station in megawatts. *see* thermal efficiency.

MW$_{th}$ the thermal power output of a power station in megawatts. *see* thermal efficiency.

MYa abbreviation for million years ago.

Mycelia Sterilia a diverse group of fungi without any currently known conidial (asexual) or sexual reproductive stages.

mycelium *n.* a network of hyphae forming the characteristic vegetative phase of many fungi, often visible as a fluffy mass or mat of hyphae.

Mycetae *n.* alternative name for the Fungi *q.v.*

mycetophage *n.* an organism that eats fungi.

myco-, myce-, mycet- prefixes derived from Gk *mykēs*, fungus.

mycobacterium *n.* bacterium of the family Mycobacteriaceae, Gram-positive non-motile rods, some spp. found in soil, others pathogenic for man and animals, e.g. *Mycobacterium tuberculosis* and *M. leprae*.

mycobiont *n.* the fungal component of a lichen or of a mycorrhiza.

mycobiota *n.* the fungi of an area or region.

mycoecotype *n.* the habitat type of mycorrhizal or parasitic fungi.

mycoflora *n.* all fungi growing in a specified area or region, or within an organism.

mycology *n.* the study of fungi.

mycophagy *n.* feeding on fungi. *a.* mycophagous.

Mycophycophyta *n.* in some classifications, the name for the lichens *q.v.*

mycoplasma-like organism (MLO) almost submicroscopic, plant-pathogenic, motile prokaryotes lacking cell walls, which are the cause of some 'yellows' diseases of plants. They are similar to spiroplasmas but are not helical.

mycoplasmas, mycoplasms *n.plu.* almost submicroscopic prokaryotic microorganisms of very simple internal structure, generally classified with the bacteria, but which lack the typical rigid bacterial cell wall and differ from bacteria in their complex life cycle. They occur in a variety of morphological forms, are obligate intracellular parasites and are responsible for several animal diseases. Formerly called the pleuropneumonia-like organisms (PPLOs). Similar microorganisms infecting plants are known as mycoplasma-like organisms. *see also* spiroplasmas.

mycorrhiza *n.* a symbiotic association between plant roots and certain fungi, in which a sheath of fungal tissue (the mantle) encloses the smallest rootlets, with fungal hyphae penetrating between the cells of the epidermis and cortex (ectomycorrhizas), or invading the cells themselves (endomycorrhizas, in which the external fungal sheath is often lacking). They are essential for optimum growth and development in many trees, shrubs and herbaceous plants. *a.* mycorrhizal.

mycorrhizoma *n.* association of fungi and a rhizome.

mycorrhizosphere *n.* region of soil immediately adjacent to a mycorrhizal surface.

mycosis *n.* animal disease caused by a fungus. *plu.* mycoses.

mycotoxin *n.* any toxin produced by a fungus.

mycotrophic *a. appl.* plants living symbiotically with fungi.

myiasis *n.* invasion of living tissue by the larvae of certain flies.

Myriapoda, myriapods *n., n.plu.* centipedes and millipedes and their relatives, terrestrial arthropods characterized by possession of a distinct head with a pair of antennae followed by numerous similar segments, each bearing legs.

Myricales *n.* an order of dicot trees and shrubs comprising the family Myriaceae (sweet gale).

myrmecioid complex one of the two major taxonomic subgroups of ants, exemplified by the subfamily Myrmeciinae.

myrmecochore *n.* an oily seed modified to attract and be spread by ants.

myrmecole *n.* an organism occupying ants' nests.

myrmecology *n.* the study of ants.

myrmecophagous *a.* ant-eating.

myrmecophil(e) *n.* (1) a guest insect in an ants' nest; (2) organism that must spend some part of life cycle with ant colonies.

myrmecophilous *a. appl.* (1) flowers, pollinated by the agency of ants; (2) fungi, serving as food for ants; (3) spiders, living with, preying on or mimicking ants.

myrmecophily *n.* beneficial relationship between ants and another organism. *see* myrmecophile, myrmecophilous, myrmecophyte.

myrmecophobic *a.* repelling ants, *appl.* plants with special glands, hairs, etc., that check ants.

myrmecophyte *n.* a plant pollinated by ants, or one that benefits from ant inhabitants and has special adaptations for housing them.

myrrh *n.* fragrant resin obtained from plants of the genus *Commiophora*.

Myrtales *n.* order of dicots, mostly shrubs and trees, including Lythraceae (loosestrife), Myrtaceae (myrtle), Onagraceae (evening primrose), Punicaceae (pomegranate), Rhizophoraceae (mangrove) and others.

Myrtiflorae Myrtales *q.v.*

Mysticeti *n.* an order of placental mammals, the baleen or whalebone whales, including the blue whale, the right whales and rorquals. *alt.* Balaenoidea.

myxamoeba *n.* in slime moulds, an amoeboid cell produced from a germinating spore.

myxinoids *n.plu.* an order of cyclostomes comprising the hagfish, as distinct from lampreys.

myxobacteria *n.plu.* group of flexible rod-shaped bacteria with a gliding movement, which aggregate into multicellular 'fruiting bodies' containing resting spores called myxospores.

myxomatosis *n.* a usually fatal disease of rabbits caused by the virus myxomatosis cuniculi. This was deliberately introduced into the rabbit population in Britain in 1953 as a form of biological control, leading to a huge decrease in numbers of rabbits.

Myxomycetes acellular slime moulds *q.v.*

Myxomycota *n.* in some classifications the name for the division of the Fungi containing all slime moulds.

Myxoviridae, myxoviruses *n.*, *n.plu.* a group of viruses containing segmented RNA genomes that includes those of influenza, mumps and measles.

Myzostomaria *n.* a class of annelids, ectoparasitic on echinoderms, and almost circular in shape.

MZ twins monozygotic (*q.v.*) twins.

N

n neutron *q.v.*

N (1) symbol for the chemical element nitrogen (*q.v.*); (2) newton (*q.v.*); (3) asparagine *q.v.*

N_A Avogadro's number *q.v.*

Na symbol for the chemical element sodium *q.v.*

NAA alpha-naphthaleneacetic acid, a synthetic auxin, used to induce roots in cuttings and prevent premature fruit drop in commercial crops.

nacre *n.* mother of pearl, iridescent inner layer of many mollusc shells, and the substance of pearls.

nacrite *n.* $Al_2Si_2O_5(OH)_4$, one of the kandite group (*q.v.*) of clay minerals (*q.v.*). A polymorph of kaolinite.

NAD nicotinamide adenine dinucleotide *q.v.*

NADH, NADH₂ reduced forms of nicotinamide adenine dinucleotide.

NADP nicotinamide adenine dinucleotide phosphate *q.v.*

NADPH, NADPH₂ reduced forms of nicotine adenine dinucleotide phosphate.

nagana *n.* a disease caused by trypanosomes (*q.v.*), spread by tsetse flies (*q.v.*). It affects hoofed animals in southern and central Africa.

naiad *n.* the aquatic nymph stage of certain insects such as dragonflies and mayflies.

naidid *a. appl.* freshwater worms of the genus *Nais*, which are often found in increased numbers in water subject to organic pollution.

Najadales *n.* order of aquatic and semiaquatic monocots including Potamogetonaceae (pondweed), Zosteraceae (eel-grass) and others.

nanism *n.* (1) dwarfism; (2) condition of unusually small size in plants or animals.

nano- (n) (1) SI prefix used to indicate that the unit it applies to is multiplied by 10^{-9},

e.g. 1 ng = 1×10^{-9} g; (2) prefix derived from Gk *nanos*, dwarf, signifying small, or smallest.

nanometre (nm) *n.* a unit of microscopic measurement and of wavelength of some electromagnetic radiation, being 10^{-9} m (one-thousandth of a micrometre, 10 Ångstrom units), formerly called millimicron.

nanophanerophyte *n.* shrub under 2 m tall.

nanoplankton *n.* microscopic floating plant and animal organisms.

NAR net assimilation rate *q.v.*

narcotic *a. appl.* drugs that can produce a state of unconsciousness, sleep or numbness.

NASA the US National Aeronautics and Space Administration.

nastic movement a plant movement caused by a diffuse non-directional stimulus. It is usually a growth movement but it may be a change in turgidity, as in the sensitive plant (*Mimosa*), which droops on contact.

natal *a. pert.* birth.

natality *n.* birth rate.

natality rate in a human population, the number of births per thousand.

natant *a.* floating on surface of water.

natatorial *a.* formed or adapted for swimming.

natatory *a.* (1) swimming habitually; (2) *pert.* swimming.

national conservation strategy (NCS) any conservation scheme adopted by a country in order to tackle specific national or regional environmental problems.

National Nature Reserve (NNR) one of several areas of Britain, deemed to be a representative example of a particular ecosystem type (e.g. open water, heath or woodland), selected and managed by English Nature, a Grade 1 Site of Special Scientific Interest (SSSI).

National Oceanic and Atmospheric Administration (NOAA) agency set up by the US Department of Commerce to bring together a number of existing services such as the National Weather Service, the Environmental Data Service and the National Environmental Satellite Service (NESS).

national park according to the IUCN definition of 1975, a large area of land that has not been significantly altered by human activities, that contains landscapes, species or ecosystems of importance and that is set aside in perpetuity to be managed in that state.

National Park and Access to the Countryside Act (1949) act covering England and Wales for the establishment and management of national parks, which in England and Wales are designated as areas of relatively unspoilt countryside of natural beauty that are to be preserved and enhanced for public enjoyment. The act also established the designation of areas of outstanding natural beauty (AONB), worthy of preservation because of their scenic value.

national parks (Britain) *see* National Park and Access to the Countryside Act.

National Radiological Protection Board (NRPB) UK statutory body concerned with all aspects of ionizing radiation.

National Rivers Authority (NRA) UK agency, established in 1989, whose remit included the policing of pollutant discharge into rivers and whose functions, as of 1996, have been subsumed within the Environment Agency *q.v.*

national vegetation classification (NVC) UK national survey of plant communities and vegetation types, which started in 1975 and which has devised a standard nomenclature for the types of plant communities found in Britain.

native *a. appl.* animals and plants that originate in the district or area in which they live.

native element (*geol.*) a mineral consisting of a chemical element found in a chemically uncombined form. Naturally occurring metallic alloys are also classed as native elements.

native species indigenous species that is normally found as part of a particular ecosystem.

natrolite *n.* $Na_2Al_2Si_3O_{10}.2H_2O$, one of the zeolite minerals *q.v.*

natural capital Earth capital *q.v.*

natural classification a classification that groups organisms or objects together on the basis of the sum total of all their characteristics, and tries to indicate evolutionary relationships. *cf.* artificial classification.

natural disasters natural events, outside human control, that kill or harm humans and/or damage property, such as earthquakes, volcanic eruptions, tornadoes, floods, etc.

natural environments areas in which the human impact on the environment is negligible or zero. *cf.* cultural environments, seminatural environments.

natural gas gas consisting mainly of methane (CH_4), usually found in association with oil reservoirs and coal seams; an important fossil fuel.

natural history the study of nature and natural objects.

natural increase a measure of the growth of a population, calculated by subtracting the number of deaths from the number of births in a given period. If the population is decreasing, a negative number will be obtained.

naturalized *a. appl.* alien species that have become successfully established.

natural levees *n.plu.* embankments formed parallel to the river channel by the deposition of the coarser river sediments during episodes of flooding. These depositional landforms may be several metres higher than the surrounding floodplain.

natural resource anything provided by the natural environment that is utilized by humans.

natural selection the process by which evolutionary change is chiefly driven according to Darwin's theory of evolution. Environmental factors, such as climate, disease, competition from other organisms, availability of certain types of food, etc., will lead to the preferential survival and reproduction of those members of a population genetically best fitted to deal with them. Continued selection will therefore lead to certain genes becoming more common in subsequent generations. Such selection, operating over very long periods of time, is believed to be able to give rise to the considerable differences now seen between different organisms.

nature *n*. general term for the living world and its environment.

nature conservation the management of the environment with a particular view to preserving and enhancing its natural wildlife.

nature reserve an area that is managed primarily to safeguard the flora, fauna and physical features it contains.

nautiloid *n*. member of the subclass Nautiloidea of the cephalopod molluscs, typified by the pearly nautilus (*Nautilus*), bearing a spiral, many-chambered shell from which the head and tentacles emerge.

Nb symbol for the chemical element niobium *q.v.*

NCS national conservation strategy *q.v.*

Nd symbol for the chemical element neodymium *q.v.*

NDOC non-dissolved organic carbon *q.v.*

Ne symbol for the chemical element neon *q.v.*

nealogy *n*. the study of young animals.

Neanderthal man a species or subspecies of *Homo*, *H. neanderthalensis*, living in the Old World during the Pleistocene.

neap tides tides generated twice every lunar month, in between the spring tides, when the Sun and Moon are at right angles with respect to the Earth. Neap tides produce the lowest high tide, the highest low tide and the smallest daily tidal range. *cf*. spring tides.

Nearctic *a. appl.* or *pert.* zoogeographical region, or subregion of the Holarctic Realm, comprising Greenland and North America, and including northern Mexico.

nearshore *n*. the zone of beach seaward of the low-water mark, i.e. seaward of the foreshore.

nebkha *n*. a small, localized dune that forms behind a clump of vegetation and tapers downwind. *alt*. shrub-coppice dune.

necroparasite *n*. parasite that kills its host organism and continues to feed on the dead remains.

necrosis *n*. localized death of cells as a result of disease or external damage, leaving small areas of dead tissue surrounded by living tissue. *a*. necrotic.

necrotroph *n*. fungus living off dead host plant tissue.

necrovirus group plant virus group containing isometric single-stranded RNA viruses that typically cause necrosis of infected tissue, type member tobacco necrosis virus.

nectar *n*. sweet liquid secreted by the nectaries of flowers and certain leaves to attract insects, and some birds, for pollination.

nectar gland, nectary *n*. (1) a group of cells secreting nectar in flowers, and in some leaves, *see* Fig. 12, p. 157; (2) the gland secreting the sweet honeydew in aphids.

nectar guides markings on petals of flowers that guide insects to the nectar, thus making cross-fertilization more likely. *alt*. honey guides.

nectariferous *a*. producing or carrying nectar.

nectarivorous, nectivorous *a*. nectar-eating. *n*. nectarivore, nectivore.

necton nekton *q.v.*

NEFA non-essential fatty acids, those fatty acids that can be synthesized *de novo* and therefore do not need to be supplied in the diet.

negative reinforcement a stimulus or series of stimuli that are unpleasant to an animal and so diminish its response to the stimulus or cause avoidance reactions.

negative tropism tendency to move or grow away from the source of the stimulus, e.g. plant shoots show negative geotropism.

nekton *n*. animals swimming actively in water, i.e. not drifting passively with the currents.

Nelumbonales *n*. order of large aquatic herbaceous dicots including the single family Nelumbonaceae (Indian lotus).

nematicide, nematocide *n*. a chemical that kills nematode worms.

Nematoda, nematodes *n*., *n.plu*. roundworms. Slender, pseudocoelomate, unsegmented worms circular in cross-section. Some (eelworms) are serious parasites of plants, others are parasitic in animals and some are free-living in soil and marine muds. Parasitic nematodes causing severe diseases in humans include the hookworms *Ancyclostoma* and *Necator*, *Trichinella* (causing trichinellosis) and *Wucheria* (the cause of elephantiasis). The soil nematode *Caenorhabditis elegans* is an important experimental organism in genetic and developmental research.

nematology *n.* the study of nematodes.

Nematomorpha *n.* a phylum of pseudocoelomate worms, sometimes known as horsehair worms, which are free-living in soil or fresh water as adults, and parasitic in arthropods when young. *alt.* threadworms.

Nemertea, nemerteans *n., n.plu.* phylum of long, slender, acoelomate, marine worms, flattened dorso-ventrally, e.g. the bootlace worm (*Lineus*). Most live on shores around the low-tide line. They have a mouth and anus, a simple blood system, and a typical muscular proboscis, which is extended to catch prey. *alt.* proboscis worms, ribbon worms.

Nemertini, nemertines Nemertea, nemerteans *q.v.*

nemoral *a.* living at the edges of woodlands, or in open woodland.

nemorose *a.* inhabiting open woodland places.

neo- prefix derived from Gk *neos*, young, signifying young or new.

neo-Darwinism the modern version of the Darwinian theory of evolution by natural selection, incorporating the principles of genetics and still placing emphasis on natural selection as a main driving force of evolution.

neodymium (Nd) *n.* a lanthanide element (at. no. 60, r.a.m. 144.24), a soft silver-coloured metal when in elemental form. It is used, e.g., to produce a purple coloration in glass.

Neogaea, Neogea *n.* zoogeographical region comprising southern Mexico, Central and South America, and the West Indies. *alt.* Neotropical Region.

Neolaurentian *a. pert.* or *appl.* early Proterozoic era.

Neolithic *a. appl.* or *pert.* the New or polished Stone Age, characterized by the use of polished stone tools and weapons and the appearance of settled cultivation.

neomorph *n.* (1) a structural variation from the type; (2) a mutant allele that produces changes in developmental processes, resulting in the appearance of a new character.

neomorphosis *n.* regeneration when the new part is unlike anything in body.

neomycin *n.* an antibiotic synthesized by *Streptomyces fradiae.*

neon (Ne) *n.* inert, colourless, odourless gas (at. no. 10, r.a.m. 20.18), one of the noble gases present in the atmosphere.

neonate *n.* newly born animal. *a.* neonatal.

neontology *n.* the study of existing organic life. In the study of evolutionary biology, neontologists are those who study evolution by comparisons between living animals and plants, while palaeontologists study evolution through the fossil record.

Neornithes *n.* subclass of birds (Aves) including all extant modern birds, the other subclass including only the extinct *Archaeopteryx.*

neoteny *n.* retention of larval characters beyond normal period and into sexually mature adult, as in some amphibians. *a.* neotenous.

Neotropical *a. appl.* (1) or *pert.* a zoogeographical region consisting of southern Mexico, Central and South America, and the West Indies; (2) a floral realm that includes the tropical and subtropical regions of America, comprising three floral regions: Central American, Pacific South American, and Parano-Amazonian Floral Regions (*see* individual entries), *alt.* Austro-Columbian.

neotype *n.* (1) a new type; (2) a new type specimen from the original locality.

Nepenthales *n.* order of herbaceous carnivorous dicots with leaves adapted for trapping small animals, and comprising the families Droseraceae (sundew) and Nepenthaceae (pitcher plants).

nepheline *n.* NaAlSiO$_4$, one of the feldspathoid minerals (*q.v.*). An end-member of a solid-solution series, the other end-member being kalsilite (KAlSiO$_4$).

nepheline syenite a rock similar to syenite (*q.v.*) but containing significant levels of nepheline (*q.v.*). Some nepheline syenites, with very low iron contents, are used in the manufacture of glasses, glazes and vitreous whiteware.

nephrite *n.* a type of jade (*q.v.*). One of the amphibole minerals *q.v.*

nephritis *n.* inflammation of the kidneys.

nepovirus group plant virus group containing nematode-transmitted isometric single-stranded RNA viruses, type member tobacco black ring virus. They are multicomponent viruses in which two genomic RNAs are encapsidated in three different virus particles, one of which lacks nucleic acid.

neptunium (Np) *n.* a transuranic element [at. no. 93, r.a.m. (of its most stable known isotope) 237], a silver-coloured metal when in elemental form. It is produced as a by-product in the synthesis of plutonium, and also occurs naturally in uranium ores.

neritic *a.* (1) *pert.* or living only in coastal waters, as distinct from oceanic; (2) *appl.* marine zone of shallow water less than 200 m deep.

neritopelagic *a. pert.* or inhabiting the sea above the continental shelf.

nerve *n.* a bundle of nerve fibres (axons) of separate neurones, which connects the central nervous system with other parts of the body.

nerve cord in invertebrates, a bundle of nerve fibres, or chain of ganglia and interconnecting nerve fibres, running the length of the body.

nervous system highly organized system of electrically active cells (nerve cells or neurones) that generate and convey signals in the form of electrical impulses. A nervous system is present in all multicellular animals except sponges, and is most highly developed in vertebrates. It receives and coordinates input from the environment and from the body through sensory receptors and conveys executive commands to muscles and glands, enabling the animal to sense and respond rapidly to external and internal stimuli. In all but the most primitive nervous systems, the nerve cells are organized into nerves and aggregates of nerve cell bodies (ganglia). The vertebrate nervous system consists of a brain and spinal cord, which constitute the central nervous system, and a peripheral nervous system, consisting of sensory cells and the peripheral nerves and their branches, which conveys signals to and from the central nervous system.

nesosilicate *n.* any silicate mineral that contains isolated SiO_4^{4-} tetrahedra, i.e. with no oxygen atoms shared between tetrahedra, e.g. andalusite.

ness *n.* a headland (UK).

NESS National Environmental Satellite Service, part of the National Oceanic and Atmospheric Administration (NOAA) in the USA.

nest epiphyte an epiphyte that builds up a store of humus around itself for growth.

nest parasitism type of parasitism in which a female of one species lays her eggs in the nest of another (host) species, which then rears the young as her own.

nest provisioning returning regularly to nests to bring food to developing offspring, as in some solitary wasps.

net above-ground production net primary production of plant material (stems, leaves, fruit, etc.) borne above the surface of the ground.

net assimilation rate (NAR) the increase in dry weight of a single plant per unit time, with reference to the total area involved in assimilation.

net below-ground production net primary production of plant material formed exclusively below the surface of the ground, e.g. roots.

net calorific value the gross calorific value (*q.v.*) from which has been subtracted the heat required to evaporate both the water produced during burning and that present in the fuel prior to combustion.

net efficiency a measure of the efficiency of an organism in converting its assimilated food to protoplasm.

net energy *see* net useful energy.

net photosynthesis photosynthesis measured as the net uptake of carbon dioxide into the leaf, equal to gross photosynthesis less respiration.

net plasmodium the kind of plasmodium found in some slime moulds, where the cells are connected by cytoplasmic strands, forming a net.

net primary production (NPP) the biomass in a plant community, which represents the difference between the increase in biomass as a result of assimilation (gross primary production) and the portion of it lost through cellular respiration. The net primary production is therefore the amount of food available for the primary consumers, the next trophic level of the food chain. *alt.* primary production. Net primary production is expressed as oven dry weight (g) of tissues per area, or energy content (kcal) of tissues per area.

net primary productivity the rate of net primary production (*q.v.*) per year.

net production the amount of food in an ecosystem available for the primary consumers, being the gross primary production minus the amount of biomass used in respiration by primary producers.

net production efficiency percentage of the total energy consumed or produced through photosynthesis that is incorporated into growth and reproduction.

net reproductive rate average number of offspring a female produces during her lifetime, symbolized by R_0.

net useful energy the total amount of useful energy available from an energy resource over its lifetime minus the amount of energy used, unavoidably wasted, and unnecessarily wasted in finding, extracting, processing and transporting it to the users.

neuartige Waldschäden (German) 'new kind of forest damage'. *see* recent forest decline.

neurobiology *n.* the study of the morphology, physiology, biochemistry and development of the brain and nervous system, and the biochemical and cell biological basis of brain function, not generally including psychology and cognitive psychology.

neuron, neurone *n.* nerve cell, basic unit of the nervous system, specialized for the conveyance and transmission of electrical impulses. Typically consists of a cell body, which contains the nucleus and other organelles, from which cytoplasmic processes project. These are the dendrites, which receive signals from other neurones, and the axon, which conducts impulses outward from the cell body.

Neuroptera *n.* order of insects with complete metamorphosis, including alder flies, lacewings and ant lions, having long antennae, biting mouthparts and two pairs of membranous wings held roof-like over the abdomen in repose.

neurotoxic *a. appl.* any toxin affecting nervous system function.

neuston *n.* organisms floating or swimming in surface water, or inhabiting the surface film.

neuter *a.* (1) sexless, neither male nor female; (2) having neither functional stamens nor pistils.

neuter *n.* (1) a non-fertile female of social insects; (2) a castrated animal.

neutral *a.* (1) neuter (*q.v.*); (2) neither acid nor alkaline, pH 7.0; (3) achromatic, as white, grey and black; (4) day-neutral *q.v.*

neutral allele neutral mutation *q.v.*

neutral fat triacylglycerol *q.v.*

neutralization *n.* process whereby an alkali is added to an acid, or vice versa, until a neutral solution (pH 7, in aqueous media) is achieved.

neutral mutation a mutation that confers no selective advantage or disadvantage on the individual.

neutral polymorphism a genetic polymorphism within a population in which the relative frequencies of the different forms are the result of chance and the action of intrinsic genetic mechanisms and are not being maintained by selection.

neutrino *n.* an elementary particle with no mass when not moving relative to the observer, i.e. it has zero rest mass, and zero electric charge. Such particles stream away from the nuclear reactors used in power generation.

neutron (n) *n.* a subatomic particle with a mass of 1.0087 a.m.u. and zero electric charge. Neutrons are found in all atomic nuclei, except those of ordinary hydrogen, ^{1}H. When not in a nucleus, the neutron is unstable ($T_{1/2}$ = 12 min), decaying into an electron, a proton and an antineutrino.

Nevado del Ruiz *see* Armero natural disaster.

névé firn *q.v.*

New Caledonian Floral Region part of the Palaeotropical Realm comprising the islands of Vanuatu (formerly New Caledonia).

newt *n.* common name for the genera *Triturus*, *Taricha* and *Notophthalamus* (in the family Salamandridae) of tailed amphibians (urodeles). They return to the water to breed and lay their jelly-coated eggs singly.

newton (N) *n.* the derived SI unit of force, defined as the force needed to give an acceleration of 1 m s^{-2} to a mass of 1 kg.

new towns towns, designed to be both economically and socially self-supporting, developed on former agricultural land, near to existing small towns or villages. These towns were primarily built to accommodate people formerly resident in overcrowded inner city areas. *alt.* overspill towns.

New Zealand flatworm *Artioposthia triangularis*, a predaceous flatworm whose recent accidental introduction into the UK has led to the destruction of huge numbers of native earthworms.

New Zealand Floral Region part of the Austral Realm comprising New Zealand and its offshore islands.

NGOs non-governmental organizations *q.v.*

Ni symbol for the chemical element nickel *q.v.*

niccolite Kupfernickel *q.v.*

niche ecological niche *q.v.*

niche breadth *see* niche width.

niche complementarity the tendency for species living in the same area to share one or more aspects of a niche (such as type of terrain) but to differ in another (such as diet). *cf.* niche differentiation.

niche differentiation, niche diversification the tendency of species sharing similar habitats to differ in at least some aspects of their ecological roles and thus not to occupy identical niches. *see also* alpha diversity.

niche glaciers the smallest of the glaciers, found in shallow hollows.

niche overlap the situation where two or more species use the same resources or the same habitat within a community and thus share the same ecological niche, leading to competition between them. Two species with identical niche requirements cannot coexist in the same community, but niche overlap may be seen where there are small and difficult-to-determine ecological differences between species.

niche packing the tendency for species living in the same area to expand their niches to the maximum available volume.

niche width the extent of the range of conditions a species requires as its niche. Species with exact requirements and very little flexibility are said to occupy narrow niches, while generalist species with wide tolerances are said to occupy wide or broad niches.

nickel (Ni) *n.* metallic element (at. no. 28, r.a.m. 58.71), forming a hard grey-white metal. Found in the ores kupfernickel, pentlandite and pyrrhotite.

nickeline Kupfernickel *q.v.*

nicotinamide adenine dinucleotide (NAD, NAD$^+$) important coenzyme, composed of nicotinamide, adenine, two riboses and two phosphate groups, found in all living cells, where it acts as a hydrogen (electron) acceptor and is reduced to NADH (+H$^+$). In this form it is an important source of reducing power in the cell, esp. as a donor of electrons to the respiratory chain that drives the synthesis of ATP.

nicotinamide adenine dinucleotide phosphate (NADP) important coenzyme, composed of NAD with an extra phosphate group attached, found in all living cells, where it acts as a hydrogen (electron) acceptor, being reduced to NADPH (+H$^+$). In this form it is an important source of reducing power in the cell, esp. in biosynthetic pathways.

nicotine *n.* alkaloid obtained from the tobacco plant *Nicotiana tabacum*, toxic to many animals because it binds to the nicotinic acetylcholine receptor and blocks the normal action of the neurotransmitter acetylcholine at neuromuscular junctions. It has been used as a contact insecticide but has been replaced because of its toxicity to humans.

nidicolous *a.* living in the nest for a time after hatching.

nidification *n.* nest building and the behaviour associated with it.

nidifugous *a.* leaving the nest soon after hatching.

night soil human ordure collected nightly from urban cesspools and used as a fertilizer, a practice that persists, e.g., in much of South-East Asia.

nimbostratus *n.* a dark grey-black, low-altitude cloud that forms continuous cover and is usually accompanied by continual precipitation.

nimby 'not in my back yard', somewhat derogatory term coined for people who object to a potentially environmentally harmful development in their locality, but by implication might not bother to protest if it were put elsewhere.

ninhydrin *n.* reagent that gives an intense blue colour with amino acids (yellow with proline).

niobium (Nb) *n.* a soft, silvery white transition metal (at. no. 41, r.a.m. 92.91), occurring naturally in the minerals columbite and pyrochlorite. It is used in special steels, high-temperature alloys and superconductors.

nitrate *n.* the NO_3^- anion, or salt or ester of nitric acid, the chief source of nitrogen for plants, and applied as ammonium nitrate, calcium nitrate and sodium nitrate inorganic fertilizers. Formed naturally in the soil from nitrite by bacterial action. Nitrates are highly soluble and are easily leached from soil to which they have been applied, causing nitrate pollution in watercourses. Excess nitrate in drinking water can cause methaemoglobinaemia (*q.v.*) in babies. *see also* nitrate sensitive areas, nitrogen cycle.

nitrate bacteria bacteria in the soil that convert nitrite to nitrate. *see* nitrifier.

nitrate radical NO_3, a trace tropospheric chemical species that is formed by the reaction between ozone (O_3) and nitrogen dioxide (NO_2) ($O_3 + NO_2 \rightarrow NO_3 + O_2$). It is a highly effective oxidant towards reduced species within the troposphere, particularly at night, when it is present at significant levels. It is much less active during day-time as its concentrations are minimal because it is readily photolysed.

nitrate sensitive areas (NSAs) areas of land designated, on the basis of their propensity to pollute fresh waters with nitrate, under a scheme launched in the UK in 1990. A range of measures were approved for use in NSAs to minimize nitrate contamination from these areas. These included avoiding the application of fertilizer or manure on fields in autumn, and planting winter cover crops.

nitratine *n.* $NaNO_3$, a mineral used as a source of nitrate. *alt.* Chile saltpetre, soda nitre.

nitric oxide (NO) *see* NO_x.

nitrification *n.* oxidation of ammonium ion (NH_4^+) to nitrite (NO_2^-), and the oxidation of nitrite to nitrate (NO_3^-), carried out chiefly by a few groups of soil bacteria (nitrifiers), mainly genera *Nitrosomonas* and *Nitrobacter*, and also by a few species of fungi. *see also* ammonification, nitrogen cycle. *see* Fig. 18.

nitrifier *n.* any of a group of autotrophic aerobic soil bacteria that can either oxidize ammonia to nitrite, e.g. *Nitrosomonas*, or nitrite to nitrate, e.g. *Nitrobacter*. *alt.* nitrifying bacteria.

nitrite *n.* the NO_2^- anion, or salt or ester of nitrous acid, formed from ammonia by soil bacteria, *see* nitrification.

nitrite bacteria bacteria in the soil that convert ammonium to nitrite. *see* nitrifier.

nitrochalk *n.* a mixture of ammonium nitrate (NH_4NO_3) and calcium carbonate ($CaCO_3$), widely used as a nitrogen fertilizer in temperate agriculture.

nitrogen (N) *n.* gaseous element (at. no. 7, r.a.m. 14.01), which in the free state is a colourless odourless unreactive gas (N_2) which makes up 78% of the Earth's atmosphere by volume. An essential macronutrient for living organisms as it is a component of organic molecules such as proteins and nucleic acids. *see* nitrogen cycle, nitrogen fixation.

nitrogen-15 naturally occurring stable isotope of nitrogen, ^{15}N, accounting for about 0.4% of atmospheric nitrogen.

nitrogen assimilation in plants, the uptake of nitrogen from the soil in the form of ammonia, nitrites and nitrates.

nitrogen balance equilibrium state of body in which nitrogen intake and excretion are equal. *alt.* nitrogen equilibrium.

nitrogen cycle the sum total of processes by which nitrogen circulates between the atmosphere and the biosphere or any subsidiary cycles within this overall process. Atmospheric elemental nitrogen (N_2) is converted by a few groups of soil and aquatic microorganisms into inorganic nitrogenous compounds in the process of nitrogen fixation. These inorganic compounds are incorporated into plants and bacteria and thence into animals, with the synthesis of complex nitrogen-containing organic molecules in their tissues. Organic nitrogen-containing compounds are subsequently broken down by bacteria and fungi (ammonification and nitrification) to generate inorganic nitrogen

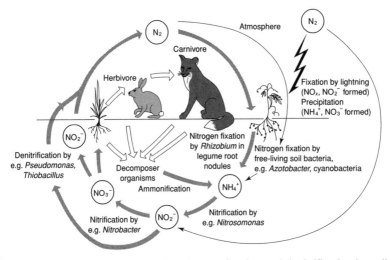

Fig. 18 The nitrogen cycle, showing nitrogen fixation and denitrification by soil microorganisms. A similar cycle occurs in water, where the primary nitrogen fixers are cyanobacteria.

compounds, i.e. ammonia, nitrites and nitrates, which may be used by plants as nutrients, or may be converted to elemental nitrogen or nitrous oxide by certain bacteria (denitrification) thus releasing nitrogen to the atmosphere. The cycle also incorporates non-biological exchanges of nitrogen between atmosphere and biosphere, as in the precipitation of inorganic nitrogen compounds in rainwater, and the fixation of atmospheric nitrogen by lightning. *see* Fig. 18.

nitrogen dioxide (NO$_2$) *see* NO$_x$.

nitrogen equilibrium nitrogen balance *q.v.*

nitrogen fertilizers inorganic fertilizers that are used to supply crop plants with the essential element nitrogen (N). *see* ammonium nitrate, ammonium sulphate, anhydrous ammonia, aqueous ammonia, calcium nitrate, sodium nitrate, urea.

nitrogen fixation the process whereby atmospheric elemental nitrogen (dinitrogen, N$_2$) is reduced to ammonia (NH$_3$), and which is carried out in the living world only by some free-living bacteria and cyanobacteria (blue-green algae) and by a few groups of bacteria in symbiotic association with plants (the *Rhizobium*–legume association and the actinomycete–non-legume asso-

ciations). The reaction is catalysed by the enzyme nitrogenase. Biological nitrogen fixation is the chief process by which atmospheric nitrogen enters the biosphere and becomes available as a nutrient to other organisms, although industrial nitrogen fixation is now of considerable significance (*see* Haber process). A smaller amount of atmospheric nitrogen is also fixed by conversion into nitrogen oxides by the action of lightning. *see also* nitrogen cycle. *see* Fig. 18.

nitrogen-fixing bacteria bacteria that are able to convert atmospheric elemental nitrogen (N$_2$) into ammonia, a form that plants and other microorganisms can use. They include free-living soil bacteria, such as *Azotobacter*, cyanobacteria, such as *Nostoc*, and members of the genus *Rhizobium*, which live symbiotically in the roots of legumes.

nitrogen monoxide (NO) *see* NO$_x$.

nitrogen oxides oxides of nitrogen *q.v.*

nitrogen peroxide (NO$_2$) *see* NO$_x$.

nitrogenous *a. pert.* or containing nitrogen.

nitrogenous wastes organic wastes with a high nitrogen content, including animal manures and urine and residues from meat processing.

nitrophile *n.* plant that grows better on nitrate than ammonium as a source of nitrogen. *a.* nitrophilous.

nitrothal-isopropyl *n.* contact fungicide usually used in mixtures to control scab and powdery mildew on apples.

nitrous oxide N_2O, an unreactive gas found throughout the troposphere at a level of about 0.3 p.p.m. but increasing at a rate of *ca.* 0.2% per year, partly as a result of the increase in intensive agriculture. The environmental consequences of this increase are not clear. N_2O is a greenhouse gas (*q.v.*). It has no tropospheric sinks and therefore eventually migrates to the stratosphere, where some of it is converted to nitric oxide (NO) and so contributes to the mechanisms that control the concentrations of ozone (O_3), *see* stratospheric ozone, NO_x. *alt.* dinitrogen oxide.

nivation hollows *see* nivation.

nivation *n.* a general term given to a combination of processes (including frost weathering, mass movement and meltwater flow) that cause localized denudation beneath snow patches and the eventual production of nivation hollows.

NMHCs non-methane hydrocarbons *q.v.*

NNR National Nature Reserve *q.v.*

No symbol for the chemical element nobelium *q.v.*

NO_x often pronounced 'knocks', the collective name for the atmospheric pollutants nitric oxide (NO) and nitrogen dioxide (NO_2). These highly reactive gases have short tropospheric residence times (typically 0.01 years, but highly variable). The main natural sources of NO_x are biomass burning (forest fires), electrical storms, *in situ* ammonia oxidation and, in the case of nitric oxide, anaerobic soil processes. A roughly comparable amount is generated anthropogenically by the burning of fossil fuels and biomass. NO_x compounds are pollutants in their own right, but the main problems they cause are associated with the secondary pollutants that they spawn. In the troposphere, they may be oxidized to nitric acid, a key component of acid deposition (*q.v.*), and their presence is a prerequisite for the generation of Los Angeles-type smog (*q.v.*). The presence of NO_x also affects levels of tropospheric ozone (*see* that entry for reactions). NO_x species are also found in the stratosphere, partly as a result of the migration of nitrous oxide across the tropopause and its subsequent oxidation to NO, and partly from emissions from high-flying aircraft, which inject principally NO directly into the stratosphere. In the stratosphere, NO_x compounds enter into the chemistry that controls the concentration of stratospheric ozone *q.v.*

NOAA National Oceanic and Atmospheric Administration *q.v.*

nobelium (No) *n.* a transuranic element [at. no. 102, r.a.m. (of its most stable known isotope) 255].

noble gases the elements helium (He), neon (Ne), argon (Ar), krypton (Kr), xenon (Xe) and radon (Rn), i.e. Group 18 of the periodic table (*see* Appendix 2). *alt.* inert gases, rare gases.

nociception *n.* sensing of painful or injurious stimuli.

nocturnal *a.* (1) seeking food and moving about only at night; (2) occurring only at night.

node *n.* (1) (*bot.*) knob or joint of a stem at which leaves arise; (2) (*phys.*) a point of zero vibration on a standing wave.

nodulated *a.* bearing nodules, in plants *appl.* esp. to roots bearing nodules containing nitrogen-fixing bacteria.

nodulation *n.* formation of nitrogen-fixing root nodules on plant roots.

nodule *n.* (1) (*geol.*) an irregularly shaped body, which may approximate to a cylinder, ellipsoid or sphere, made of one of several materials including flint, chert, siderite, pyrite, gypsum and calcite; (2) (*biol.*) *see* root nodule.

noise pollution any unwanted, disturbing or harmful sound that impairs or interferes with hearing, causes stress, affects the general quality of life, hampers concentration and work efficiency or causes accidents.

nomadism *n.* type of existence in which people (known as nomads) move around in search of more productive grazing land for their livestock. The term total or true nomadism is reserved for people who are constantly on the move while the term semi-nomadism is used for those who have a settled existence, and cultivate crops, for part of the year.

nomads *see* nomadism.

Nomarski differential-interference-contrast microscopy a type of optical microscopy that produces a high-contrast image of unstained living cells and tissue.

nomen nudum a name not valid because when it was originally published the organism to which it referred was not adequately described, defined or sketched.

nomenspecies *n.* a group of individuals bearing a binomial name, whatever its status in other respects.

non-adaptive *a. appl.* traits that tend to decrease an organism's genetic fitness.

non-allelic *a. appl.* (1) mutations that produce the same or very similar phenotypes when either is homozygous, but a normal phenotype when both are heterozygous, usually showing that the mutations are affecting different genes; (2) similar genes at two or more different loci.

non-biodegradable, non-degradable *a. appl.* material that cannot be broken down by the natural processes of decomposition by microorganisms and that therefore persists in the environment.

non-clastic rocks sedimentary rocks that are not formed from fragments derived from other rocks. *see* chemical sedimentary rocks, organic sedimentary rocks.

non-conservative properties (*oceanogr.*) *appl.* water masses (*q.v.*), these are properties of seawater that are affected by chemical or biological processes, as well as physical processes, e.g. concentrations of gases such as carbon dioxide and oxygen. *cf.* conservative properties.

non-covalent bond general term for attractive forces between atoms or molecules that do not involve the sharing of electrons. *see* hydrogen bond, ionic bond, van der Waals' forces.

non-dissolved organic carbon (NDOC) that fraction of total organic carbon in water consisting of particles >0.45 μm, which can be removed by filtration.

non-essential amino acids amino acids that can be synthesized in the body and are not required in the diet: for humans these are alanine, asparagine, aspartic acid, cysteine, glutamic acid, glutamine, glycine, proline, serine and tyrosine.

non-fissile *a. appl.* materials that do not undergo nuclear fission on bombardment with slow neutrons. *cf.* fissile.

non-frontal depressions (*clim.*) depressions (areas of low pressure) not associated with fronts. These may develop on the leeward side of mountain ranges (lee depressions) or over continental areas during summer, as a result of intense heating of the land surface (thermal depressions).

non-governmental organizations (NGOs) voluntary, non-party-political organizations, such as professional bodies, international charities, aid agencies or non-profit membership organizations, not affiliated to any government organization.

non-ionizing radiation radiation, such as radio waves, infrared light and ordinary light, that does not have enough energy to cause ionization in living tissue. *cf.* ionizing radiation.

non-material resource a resource that cannot be measured or given a monetary value, as tranquility, natural beauty, security. *see also* amenity.

non-Mendelian *a. appl.* genes or characters that are not inherited according to Mendel's laws, e.g. mitochondrial or chloroplast genes.

non-metals *n.plu.* (*chem.*) those chemical elements that, when in elemental form, do not exhibit the properties of a metal (*q.v.*). Non-metals, with the exception of the noble gases (which are inert), are electronegative and tend to lose electrons when in reaction with a metal, forming anions in the process. Bonding between non-metals tends to be covalent. *see* Appendix 2.

non-methane hydrocarbons (NMHCs) a blanket term used by atmospheric chemists to encompass all hydrocarbons (*q.v.*) except methane.

non-parametric statistics statistical analyses performed on populations not known to have a normal distribution for the variables under study.

non-persistent *a. appl.* chemicals not persisting in the environment, owing to their degradation or removal by natural processes.

non-point source of water pollution, a diffuse source of pollution with no specific point of discharge into a particular water body. Such sources include runoff from agricultural land and urban areas. *cf.* point source.

non-polar *a.* (*chem.*) *appl.* (1) molecules that do not exhibit an electric dipole; (2) covalent bonds that have an even distribution of charge. *cf.* polar.

non-protein nitrogen (*agric.*) with reference to animal feedstuffs, dietary nitrogen provided by non-protein sources, e.g. urea.

non-renewable resources environmental resources that are available in finite quantities and, once consumed, are not renewable on a human timescale, e.g. fossil fuels, minerals. *cf.* renewable resources. *alt.* exhaustible resources, fund resources.

non-selective *a. appl.* to pesticides that kill a wide range of organisms. *alt.* broadspectrum.

non-shivering thermogenesis (NST) the generation of large amounts of heat by metabolism, without shivering, of which some mammals are capable, *see* brown fat.

non-silicates *see* minerals.

non-stoichiometric *a. appl.* (1) mixtures of reactants or mixtures of products present in mole ratios that are not those dictated by the stoichiometry of the reaction; (2) compounds that exhibit variable composition (true non-stoichiometric compounds are very rare). *cf.* stoichiometric.

non-sulphur purple bacteria group of photosynthetic bacteria containing purple pigments.

non-transmissible *a. appl.* diseases that are not due to infection with a living organism and thus cannot be spread from one individual to another, e.g. cardiovascular disease, diabetes.

nontronite *n.* one of the clay minerals (*q.v.*) of the smectite group *q.v.*

non-viable *a.* incapable of surviving.

norbergite *n.* $Mg(OH,F)_2.Mg_2SiO_4$, a mineral of the humite series *q.v.*

Norfolk four-course rotation a well-known method of crop rotation in which, originally, each of the following crops were cultivated: turnips (or swedes), spring barley, red clover and winter wheat. There are now many variations of this basic rotation system.

normal distribution statistical term for a distribution in which most values fall near a central point, producing a bell-shaped curve. *alt.* Gaussian distribution.

normal fault a fault (*q.v.*) where one block of crustal rock remains stationary while the other is significantly downthrown. Normal faults arise where tensional stresses pull the crust apart. The fault scarp created indicates the angle of dip of the fault plane.

normalizing selection stabilizing selection *q.v.*

North Atlantic Drift (1) ocean current of warm water that flows polewards in the North Atlantic Ocean, part of the Gulf Stream; (2) a synonym for the entire Gulf Stream *q.v.*

north-east trades *see* trade winds.

Norther *n.* cold, strong northerly wind that occurs in winter in the central and southern USA.

northern coniferous forest boreal forest *q.v.*

northern lights *see* thermosphere.

Nor'wester *see* föhn.

nosean, noselite *n.* $Na_8(Al_6Si_6O_{24})SO_4$, one of the sodalite group (*q.v.*) of feldspathoid minerals *q.v.*

nosogenic pathogenic *q.v.*

nosology *n.* (1) branch of medicine dealing with the classification of diseases; (2) pathology *q.v.*

nothocline hybrid cline *q.v.*

Nothosauria, nothosaurs *n., n.plu.* an extinct order of streamlined fish-eating marine reptiles.

notifiable *a. appl.* certain pests and diseases, the occurrence of which must be reported to the appropriate governmental body.

no-till agriculture cultivation of crops without ploughing or turning over the soil. *alt.* direct drilling, zero tillage. *see also* reduced cultivation.

Notogaea, Notogea *n.* zoogeographical region comprising Australia, Tasmania, New Zealand, Papua-New Guinea and the islands of the Pacific Ocean.

novobiocin *n.* antibiotic synthesized by the actinomycete *Streptomyces niveus*.

Np symbol for the chemical element neptunium *q.v.*

NPK abbreviation for the elements nitrogen (N), phosphorus (P) and potassium (K), the three major plant nutrients used in fertilizers.

NPP net primary production *q.v.*

NRA National Rivers Authority *q.v.*

NRC Nuclear Regulatory Commission *q.v.*

NRPB National Radiological Protection Board *q.v.*

Nsas nitrate sensitive areas *q.v.*

NST non-shivering thermogenesis *q.v.*

N-terminus the amino-terminus of a polypeptide chain, i.e. the end that carries a free amino group.

nucivorous *a.* nut-eating.

nuclear *a. pert.* or *appl.* (1) (*phys.*) the atomic nucleus, and to the exploitation of the energy released by either splitting it or fusing two nuclei together; (2) (*biol.*) the cell nucleus.

nuclear binding energy the energy or work required to completely separate all of the nucleons (i.e. protons and neutrons) of a given atomic nucleus. It is generally expressed in units of electron volts (eV). It is so large that it can be detected as a difference in mass. The mass lost on the formation of a nucleus from its separate nucleons is related to its binding energy via Einstein's mass–energy relation, $E = mc^2$.

nuclear bomb a weapon that explodes as a consequence of either a nuclear fission reaction alone (occurring within a fissile material, e.g. ^{239}Pu or ^{235}U, under supercritical conditions) or a nuclear fusion reaction (between nuclei of, e.g., deuterium) initiated by a nuclear fission reaction. The first of these classes of nuclear bomb is called a fission bomb (*alt.* A-bomb, atom bomb, atomic bomb), while the second is termed a fusion bomb (*alt.* H-bomb, fission–fusion bomb, hydrogen bomb, thermonuclear bomb). These weapons are extremely destructive both because of their immense explosive power [equivalent to tens of kilotons of trinitrotoluene (TNT) in the case of a fission bomb or tens of megatons of TNT for a fusion bomb] and the radioactive nuclides that they spawn.

nuclear breeder reactor fast breeder reactor *q.v.*

nuclear energy the energy that is released by splitting a heavy atomic nucleus into two or more nuclei of similar mass, or by fusing two nuclei. *see* nuclear fission, nuclear fusion.

nuclear fallout the radioactive fallout (*q.v.*) resulting from a nuclear exposion or an accident at a nuclear installation that releases radioactive material into the atmosphere.

nuclear fission the splitting of a heavy atomic nucleus into two or more nuclei of similar mass, with the simultaneous ejection of neutrons and liberation of nuclear energy. *see also* induced nuclear fission, spontaneous nuclear fission.

nuclear force the extremely short-range ($\leqslant 1 \times 10^{-15}$ m) force of attraction that holds the nucleons of an atomic nucleus together.

nuclear fuel the fissile material, most commonly uranium oxides enriched in uranium-235 (*q.v.*), which is subjected to controlled nuclear fission in a nuclear reactor *q.v.*

nuclear fusion a type of nuclear reaction that occurs when two light nuclei collide with such force that their mutual electrostatic repulsion is overcome. This reaction produces a heavier nucleus and liberates large amounts of nuclear energy. For example, $^2_1\text{H} + {}^2_1\text{H} \rightarrow {}^3_2\text{He} + {}^1_0 n$ + energy. Such reactions occur at very high temperatures and are the energy source in both hydrogen bombs and the stars. Controlled, sustained nuclear fusion for the production of nuclear power has yet to be attained on Earth.

nuclear power general term for the usable power generated by harnessing the energy released by nuclear fission *q.v.*

nuclear power plant installation for generating power using the energy released by nuclear fission (*q.v.*). The heat released by the fission reaction in a nuclear reactor is used to generate steam to operate turbine generators to produce electric power. *see also* nuclear reactor.

nuclear radiation collective term for the high-speed particles (e.g. alpha particles, beta particles) and/or high-energy electromagnetic radiation (gamma rays) produced by nuclear reactions.

nuclear reaction any process that leads to a change in the number, nature or energy of the nucleons (i.e. protons and neutrons) that make up the nucleus of an atom.

nuclear reactor a device used to liberate energy and/or produce artificial elements or radioactive nuclides, using nuclear fuels under controlled conditions. *see* advanced gas-cooled reactor, boiling water reactor, candu reactor, fast breeder reactor, Magnox reactor, pressurized water reactor, steam-generating heavy-water reactor, thermal reactor.

Nuclear Regulatory Commission (NRC) federal agency in the USA responsible for licensing the construction and safe operation of nuclear reactors.

nuclear reprocessing the processing of spent nuclear fuel to extract any remaining usable material (uranium-235 and plutonium), separating it from the waste, which must then be disposed of. *see also* Thermal Oxide Reprocessing Plant.

nuclear waste *see* radioactive waste.

nuclease *n.* any of a class of enzymes that degrade nucleic acids into shorter oligonucleotides or single nucleotide subunits by hydrolysing sugar–phosphate bonds in the nucleic acid backbone.

nucleate *a.* containing a nucleus.

nuclei *plu.* of nucleus *q.v.*

nucleic acid hybridization *see* DNA hybridization, *in situ* hybridization.

nucleic acids *n.plu.* deoxyribonucleic acid (DNA) and ribonucleic acid (RNA). Very large linear polymeric molecules containing C, H, O, N and P. They are composed of chains of nucleotide subunits, RNA being composed of one chain and DNA of two. They are essential components of all living cells, where they are the carriers of genetic information (DNA and mRNA), components of ribosomes (rRNA), and involved in deciphering the genetic code (tRNA). *see also* DNA, genetic code, messenger RNA, nucleotide, polynucleotide, ribosomal RNA, RNA, transfer RNA.

nucleocapsid *n.* the nucleic acid plus protein coat of an enveloped virus.

nucleolus *n.* discrete region of the nucleus where rRNA and ribosomes are synthesized, and consisting of a fibrillar core surrounded by a granular region. *plu.* nucleoli.

nucleon *n.* each proton or neutron that makes up the nucleus of an atom.

nucleon number mass number *q.v.*

nucleoprotein *n.* any complex of protein and nucleic acid.

nucleoside *n.* any of a group of compounds consisting of a purine or pyrimidine base (commonly adenine, guanine, cytosine, thymine) linked to the sugar ribose or deoxyribose. The common nucleosides include adenosine, cytidine, uridine, thymidine, guanidine. *see also* nucleotide.

nucleosynthesis *n.* the production of elements. This occurs naturally in stars. It can also be made to occur artificially on Earth by colliding fast-moving nuclei or nucleons with the nuclei of other elements.

nucleotide *n.* phosphate ester of a nucleoside, consisting of a purine or pyrimidine base linked to a ribose or deoxyribose phosphate (up to three phosphate groups linked in series), the purine nucleotides having chiefly adenine or guanine as the base, the pyrimidine nucleotides cytosine, thymine or uracil. Nucleotides are the basic chemical subunits of DNA and RNA. Nucleotides containing deoxyribose are called deoxyribonucleotides, those containing ribose are called ribonucleotides. Nucleotides containing one phosphate group are also known as nucleoside phosphates, those containing two phosphate groups as nucleoside diphosphates, and those containing three phosphate groups as nucleoside triphosphates, such as adenosine triphosphate (ATP).

nucleotide sequence the order of the different nucleotides in RNA or DNA.

nucleus *n.* (1) (*chem./phys.*) the very small and extremely dense positively charged agglomeration of protons and neutrons (or, in the case of ordinary hydrogen, a proton only) that is found at the centre of each atom; (2) (*biol.*) a large dense organelle bounded by a double membrane, present in eukaryotic but not prokaryotic cells and that contains the chromosomes; (3) the centre of any structure, around which it grows. *plu.* nuclei.

nuclide *n.* a term for either an atom or its nucleus that has given values of both mass number (A) and atomic number (Z). The full symbol for a nuclide takes the form $^A_Z E$ (where E is the elemental symbol), frequently given in an abbreviated form, $^A E$, as Z is implied by E. For example, a nuclide with $A = 16$ and $Z = 8$ is an atom (or nucleus of that atom) of the element oxygen (symbol O). Therefore this nuclide is symbolized as either $^{16}_8 O$ or $^{16} O$.

nudibranch *a.* lacking a protective cover over the gills, *appl.* a group of shell-less marine, carnivorous gastropod molluscs (Nudibranchia), the sea slugs.

nuée ardente a glowing cloud composed of hot volcanic gases, air and fine dust that may be produced before or during a volcanic eruption. These phenomena can be highly destructive, e.g. in 1902 a nuée ardente from Mt Pelée on the Caribbean island of Martinique killed about 30 000 people in the port of St Pierre within minutes.

null allele mutant allele that results in an absence of functional gene product, the mutation itself being termed a null or amorphic mutation. Null alleles are usually recessive.

null hypothesis in planning a scientific experiment, the hypothesis that would give a certain set of experimental results under the conditions of the experiment. If the observed results depart significantly from these expected results the null hypothesis is unlikely to be true.

numerical a. (biol.) appl. hybrid of parents with different chromosome numbers.

numerical abundance the number of individuals of a species present in a given area.

numerical taxonomy classification of organisms by a quantitative assessment of their phenotypic similarities and differences, not necessarily leading to a phylogenetically based classification.

nummulitic a. containing nummulites, a type of fossil foraminiferan.

nunataks n.plu. (1) high mountain peaks that remain visible when an ice cap or ice sheet engulfs a mountain range; (2) areas on a mountain or plateau that have escaped past environmental changes, such as glaciation, and in which plants and animals of earlier floras and faunas have survived.

nuptial flight flight taken by queen bee when fertilization takes place.

nurture n. the sum total of environmental influences on a developing individual.

nut n. a dry indehiscent, one- or two-seeded, one-chambered fruit with a hard woody shell, such as an acorn.

nutation n. (1) rotational curvature of the growing tip of a plant; (2) slow rotating movement by pseudopodia, as in Amoeba.

nutrient n. (1) any substance that is actively taken up by an organism and used to maintain life; (2) any substance used or required by an organism as food. see also macronutrients, micronutrients.

nutrient budget the amounts and types of nutrients flowing into and out of an ecosystem, or circulating within the system.

nutrient cycles the exchanges of elements between the living and non-living components of an ecosystem. see biogeochemical cycle.

nutrient-holding capacity of soil, the ability to retain mineral nutrients so that they are not washed out.

nutrient loading the addition of excess levels of nutrients to an ecosystem, usually as the result of pollution, leading to effects such as the eutrophication of ponds and lakes.

nutrient stripping the removal of nutrients from sewage to prevent eutrophication.

nutrient-use efficiency the ratio of nutrients assimilated to nutrients provided.

nutrition n. the process by which an organism obtains from its environment the energy and the chemical elements and compounds it needs for its survival and growth. see autotrophic, chemotrophic, heterotrophic, phototrophic.

nutritive a. concerned with nutrition.

NVC national vegetation classification q.v.

nyctanthous a. flowering by night.

nyctinasty n. 'sleep' movements in plants, involving a change in the position of leaves, petals, etc., as they close at night or in dull weather in response to a change in the level of light and/or temperature. a. nyctinastic. alt. nyctitropism.

nyctipelagic a. rising to the surface of sea only at night.

nymph n. a juvenile form of insect without wings or with incomplete wings. They occur in insect species with incomplete metamorphosis, i.e. when the change at each moult is small and the larvae are relatively similar to the adult form. a. nymphal.

Nymphales n. order of herbaceous aquatic dicots including families Ceratophyllaceae (hornwort) and Nymphaceae (water lily).

nystatin n. an antibiotic with antifungal activity.

O

Ω ohm *q.v.*

O symbol for the chemical element oxygen *q.v.*

oak-apple *n.* type of hard spherical gall found on stems and leaves of oak and caused by larvae of a species of gall wasp.

oasis *n.* fertile place, where liquid water is found, in the middle of a desert.

oat *n. Avena sativa*, a small-grained cool temperate cereal crop, grown mainly for use as an animal feed (both grain and straw) and also to make oatmeal used in porridge, oatcakes, etc.

obligate *a.* (1) obligatory; (2) limited to one mode of life or action. *cf.* facultative.

obligate anaerobe organism that cannot tolerate oxygen and can only survive in the absence of oxygen.

obligate parasite an organism that can only live as a parasite.

obligate predator a predator that only lives off one type of prey.

obliterative coloration, obliterative shading type of coloration in which parts of an organism exposed to the brightest light are shaded more darkly, ensuring that it blends with its background more effectively.

obsidian *n.* a volcanic glass that is chemically equivalent to rhyolite. May contain infrequent phenocrysts of feldspar and quartz.

obsolescence *n.* (1) the gradual reduction and eventual disappearance of a species; (2) gradual cessation of a physiological process, or of a structure becoming disused, over evolutionary time; (3) a blurred portion of a marking on an animal.

obsolete *a.* (1) wearing out or disappearing; (2) *appl.* any character that is becoming less and less distinct in succeeding generations.

Occam's (or Ockham's) razor the principle that where several hypotheses are possible, the simplest is chosen, first proposed by William of Ockham, a medieval scholastic philosopher.

occasional species one that is found from time to time in a community but is not a regular member of it.

occluded front, occlusion (*clim.*) type of front that develops when a more rapidly moving cold front overtakes a warm front and lifts the intervening warm sector off the ground. This feature (also known as an occlusion) is frequently associated with depressions of the mid-latitudes. Two types of occlusion are recognized. *see* cold occlusion, warm occlusion.

occupational exposure limits (OELs) limits set in the workplace to restrict the exposure of individual workers to specific hazardous substances, such as benzene or carbon tetrachloride (CCl_4).

ocean *n.* geographical subdivision of the enormous body of seawater found on the surface of the Earth surrounding the continental land masses. Five oceans are recognized: Antarctic, Arctic, Atlantic, Indian and Pacific.

ocean current a body of ocean water that moves in a definite direction. Ocean currents may be closely associated with atmospheric circulation (wind-driven currents) or result from salinity and temperature differences within the oceans themselves (density currents).

ocean-floor spreading sea-floor spreading *q.v.*

oceanic *a.* inhabiting the open sea, where it is deeper than 200 m.

oceanic conveyor belt *n.* a massive oceanic circulatory system driven by convection

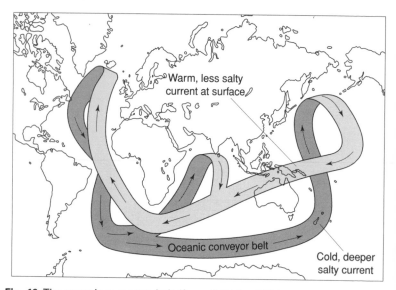

Fig. 19 The oceanic conveyor belt (from Gribbin, 1991).

(*q.v.*), which has a profound influence on climate, particularly in Europe. In the North Atlantic, cold water with a high salinity subsides and moves at depth towards the Indian and Pacific Oceans, where it pushes deep waters to the surface. Warmed by the Sun, these waters form surface currents that ultimately return to the North Atlantic, *see* Fig. 19.

oceanodromous *a.* migrating only within the ocean, *appl.* fishes. *cf.* potamodromous.

oceanography *n.* the scientific study of oceans and ocean basins.

ocean thermal energy conversion (OTEC) a technique that makes use of the naturally occurring thermal gradient of the oceans for the generation of electricity. Since the minimum difference required between the warm surface waters (the heat source) and the cold water at about 1000 m (the heat sink) is about 20 °C, the application of this technique, (currently at the research and development stage) is restricted to tropical and subtropical seas (within 25° of the Equator).

ocellate *a.* like an eye or eyes, *appl.* markings.

ochrept *n.* a suborder of the order inceptisol of the USDA Soil Taxonomy system

consisting of light-coloured, freely draining inceptisols.

ochric epipedon *see* aridisol.

octa- prefix derived from Gk *octa*, eight, signifying having eight of, arranged in eights, etc.

octahedral *a.* (*chem.*) *appl.* the geometry of the environment of an atom or ion that can be imagined to be at the centre of an octahedron, the corners of which are coincidental with the atom's nearest neighbours.

octane *n.* an alkane with the formula $CH_3(CH_2)_6CH_3$. It is a component of petroleum, as is its isomer iso-octane (2,2,4-trimethylpentane).

octane number of a fuel, an expression of the knock characteristics of a fuel destined for use in spark ignition engines. It is stated as the volume percentage of iso-octane (*q.v.*) in a normal heptane, $CH_3(CH_2)_5CH_3$: iso-octane mixture that has the same knock characteristics as the fuel concerned, under specified conditions.

octanol–water partition coefficient (K_{ow}) an equilibrium constant that relates to the distribution of a solute (So) between water (aq) and *n*-octanol (oct), represented by the general equation So(aq) $\rightleftharpoons$ So(oct).

The octanol–water partition coefficient is expressed as $K_{ow} = [\text{So(oct)}]/[\text{So(aq)}]$, where square brackets, [], represent molar concentration. This ratio is useful in assessing the potential of a substance for bioconcentration within organisms.

Octapoda *n.* an order of Cephalopoda whose members have eight tentacles and no shell, e.g. octopus.

OD optical density, *see* absorbance.

ODA (1) overseas development assistance, *see* aid; (2) Overseas Development Administration, a UK government body within the Foreign and Commonwealth Office.

odd-toed ungulates perissodactyls *q.v.*

odogen *n.* substance that stimulates the sense of smell.

Odonata *n.* order of insects including the dragonflies and damselflies, winged predatory insects with brilliant metallic colouring, whose eggs are laid in water and develop through an aquatic nymph (larval) stage that has gills.

Odontoceti *n.* the toothed whales, a suborder of the Cetacea, which are all predatory, feeding on fish and other marine animals and that include the sperm whale, killer whale, narwhal, porpoises and dolphins.

odontology *n.* the study of dental anatomy, histology, physiology and pathology.

odorimetry *n.* measurement of the strength of the sense of smell, using substances of known ability to stimulate olfaction.

OECD Organisation for Economic Co-operation and Development *q.v.*

oedema, pulmonary *see* pulmonary oedema.

OELs occupational exposure limits *q.v.*

oesophagus *n.* that part of the alimentary canal between the pharynx and stomach or part equivalent thereto. *alt.* esophagus, gullet.

oestr- *alt.* estr-.

oestrogens *n.plu.* a group of vertebrate steroid hormones, the principal female sex hormones, synthesized chiefly by the ovary and placenta in females and responsible for the development and maintenance of secondary female sexual characteristics and the growth and function of female reproductive organs. Similar compounds have been found in plants. *alt.* estrogens.

oestrous *a. pert.* oestrus.

oestrus *n.* the period of sexual heat and fertility in a female mammal when she is receptive to the male. *alt.* estrus.

oestrus cycle reproductive cycle in female mammals in the absence of pregnancy, comprising oestrus, when ovarian follicles mature and ovulation takes place, met-oestrus (*q.v.*), and pro-oestrus *q.v.*

officinal *a.* used medicinally, *appl.* plants.

offshore bar barrier island *q.v.*

Ogallala aquifer, depletion of the Ogallala aquifer is an immense groundwater reserve in the southern USA that provides water for irrigated agriculture in parts of eight states (including Texas, Kansas and Nebraska). Overexploitation of this aquifer means that, at current rates of use, it will be completely depleted within the next 40–50 years.

ohm (Ω) *n.* the derived SI unit of electric resistance, defined as the resistance between two points on a conductor when a constant potential difference of 1 V applied between them produces an electric current of 1 amp.

O-horizon the layer of fresh organic material on the surface of a soil. It comprises freshly fallen leaves and other organic debris and partially decomposed organic matter. In some situations it may be divided into distinct layers, a litter layer (L-layer), a fermentation layer (F-layer) and a humified layer (H-layer). *see* Fig. 28, p. 385. *see also* mor.

oil *n.* (1) naturally occurring liquid hydrocarbon fossil fuel, which is formed from organic material within the Earth's crust, *see* crude oil; (2) (*biol.*) a lipid that is a liquid under normal conditions, esp. those glycerides and esters of fatty acids that are liquid at 20 °C, the fatty acids in oils being in general less saturated than those in fats. *cf.* fats.

oil bodies, oil storage bodies lipid bodies *q.v.*

oil field area of commercial oil production, e.g. Gulf of Texas, USA.

oil gland (1) any gland secreting oils; (2) in birds, a gland in the skin that secretes oil used in preening the feathers.

oil recovery extrication of crude oil from underground oil deposits, a process that can involve up to three recovery stages. In primary recovery, the oil comes to the surface naturally due to the release of pressure exerted by natural gas and water present in the oil deposit. In secondary recovery, techniques such as pumping or injecting water or air into the well are used. In tertiary recovery (also known as enhanced recovery), more complex techniques such as steam flooding may be employed to increase the percentage of oil extracted further.

oil sand tar sand *q.v.*

oil-seed rape *Brassica napus* forma *oleifera*, a member of the family Cruciferae, grown for its oil-rich seeds. The oil is extracted for use, e.g., in margarine, while the residual cake can be used as animal fodder.

oil shale *n.* shale (fine-grained sedimentary rock) containing appreciable levels of kerogen *q.v.*

oil slick layer of oil floating on the surface of a water body, as a result of either deliberate discharge, natural extrusion or accidental spillage. Marine oil slicks may cover thousands of square kilometres, causing extensive ecological damage to both marine and coastal habitats. For example, *see Exxon Valdez* incident.

oil spill accidental release of oil into the environment, esp. into water.

oil terminal temporary storage point for crude oil, brought in by pipeline from oil fields, prior to its further transportation, esp. by ships.

oil traps geological structures in which porous rocks containing crude oil (termed reservoir rocks) are capped, or otherwise sealed, by impermeable rock layers.

old field term used in USA for a meadow *q.v.*

old-growth forest uncut virgin forest containing trees that are hundreds and sometimes thousands of years old, as in the forests of Douglas fir, western hemlock, giant sequoia and redwoods in the western USA.

Olea europea the olive, a tree of great economic importance for its oil-yielding fruits.

oleaginous *a.* containing or producing oil.

Oleales *n.* order of dicot trees, shrubs and climbers comprising the family Oleaceae (olive, privet).

olefins *n.plu.* (*chem.*) an old term for the alkenes *q.v.*

olfaction *n.* (1) the sense of smell; (2) the process of smelling.

olig- prefix derived from Gk *oligos*, few, signifying having few, having little of, etc.

Oligocene *n.* a geological epoch in the Tertiary, between the Eocene and Miocene, lasting from about 38 to 25 million years ago.

Oligochaeta, oligochaetes *n., n.plu.* class of mainly freshwater and terrestrial annelid worms characterized by possession of few bristles (setae or chaetae) on each segment and no parapodia and that includes the earthworms.

oligoclase *n.* one of the plagioclase (*q.v.*) minerals.

oligogenic *a. appl.* characters controlled by a few genes responsible for major heritable changes.

oligohalabous *a. appl.* plankton that live in brackish water (with a salinity of less than 5 parts per thousand).

oligohaline *a. appl.* brackish water with salinity of 0.5–5 parts per thousand.

oligomer *n.* a molecule composed of only a few monomer units.

oligomycin *n.* antibiotic that inhibits ATP synthesis in mitochondria by interacting with one of the proteins in the mitochondrial coupling factor.

oligonucleotide *n.* short chain of nucleotides (*q.v.*). *see also* polynucleotide.

oligophagous *a.* restricted to a single order, family or genus of food plants, *appl.* insects.

oligophyletic *a.* derived from a few different lines of descent.

oligosaccharide *n.* a molecule composed of only a few (*ca.* four to 20) monosaccharide units.

oligosaprobic *a. appl.* (1) aquatic environment with a high dissolved oxygen content and little organic decomposition; (2) category in the saprobic classification of river organisms comprising those that can only live in water unpolluted by organic pollutants, e.g. brown trout (*Salmo trutta*) and stonefly nymphs (plecopterans). *cf.* alpha-mesosaprobic, beta-mesosaprobic, polysaprobic.

oligothermic *a.* tolerating relatively low temperatures.

oligotokous *a.* bearing few young. *alt.* oligotocous.

oligotrophic *a.* (1) providing or *pert.* inadequate nutrition; *appl.* (2) waters relatively low in nutrients, such as the open oceans compared with the continental shelves, and such as some lakes whose waters are low in dissolved minerals and which cannot support much plant life; (3) microorganism that thrives and predominates in a nutrient-poor environment, *n.* oligotroph. *cf.* eutrophic.

oligotrophophyte *n.* plant that will grow on poor soil.

oligotrophy *n.* the ability to live in a nutrient-poor environment, as, e.g., many soil actinomycetes.

oligotropic *a.* visiting only a few allied species of flowers, *appl.* insects.

oligoxenous *a. appl.* parasites adapted for life in only a few species of hosts.

olive *see Olea europea.*

olivine *n.* a group of rock-forming isostructural orthosilicate minerals with the formula M_2SiO_4, where M is a divalent metal, most commonly magnesium and/or iron. *see also* fayalite, forsterite.

olivine basalt basalt that contains olivine in addition to the essential minerals of a basalt *q.v.*

olivine gabbro *see* gabbro.

ombrogenous *a. appl.* wet habitats arising from precipitation rather than from water in the ground.

ombrophile *a.* adapted to living in a rainy place, *appl.* plants, leaves.

ombrophobe *n.* a plant that does not thrive under conditions of heavy rainfall.

ombrophyte *n.* a plant adapted to rainy conditions.

omni- prefix derived from L. *omnis*, all.

omnicolous *a.* capable of growing on different substrates, *appl.* lichens.

omnivore *n.* animal that eats both plant and animal food. *a.* omnivorous.

onchocerciasis *n.* a tropical disease that causes impaired vision and eventual blindness caused by infection with the parasitic roundworm *Onchocerca volvulus*, which is transmitted to humans by blackflies. *alt.* river blindness.

oncoviruses *n.plu.* a subfamily of retroviruses, comprising the RNA tumour viruses.

Divided into types C, B and D. Formerly known as oncornaviruses.

one gene–one enzyme hypothesis the original form of the idea that each gene specifies one polypeptide chain, developed in the 1930s and 1940s from the study of biochemical mutants.

onion-skin weathering exfoliation *q.v.*

ontogeny *n.* the history of development and growth of an individual. *a.* ontogenetic. *cf.* phylogeny.

Onychophora, onychophorans *n., n.plu.* class of primitive worm-like terrestrial arthropods with a soft flexible cuticle, often treated as a separate phylum, which live in damp habitats in warm climates, They possess a pair of antennae and a pair of stiff jaw appendages, followed by a number of pairs of unjointed hollow walking legs. There is only one extant genus, *Peripatus. alt.* velvet worms.

onyx *n.* SiO_2, varieties of the mineral chalcedony (*q.v.*) that contain straight, parallel bands that differ in colour. May be considered to be a variety of agate *q.v.*

oocyte *n.* female ovarian cell in which meiosis occurs to form the egg. Cells undergoing the first meiotic division are often termed primary oocytes, after which they become secondary oocytes, which undergo the second meiotic division to become mature eggs.

oogamy *n.* the union of unlike gametes, usually a large non-motile female gamete and a small motile male gamete. *a.* oogamous.

oogenesis *n.* formation, development and maturation of the female gamete or ovum.

ooliths *n.plu.* ellipsoidal or spheroidal structures found in some types of rocks, <2 mm in diameter and formed of concentric layers. *see* oolitic.

oolitic *a. pert.* ooliths (*q.v.*), *appl.* rocks containing ooliths, e.g. oolitic limestone, oolitic ironstone.

Oomycota, oomycetes *n., n.plu.* phylum of simple non-photosynthetic, saprobic or parasitic, unicellular or filamentous protists sometimes classified as fungi. Unlike most fungi, their cell walls contain cellulose. Sexual reproduction is oogamous and they reproduce asexually by motile zoospores. They include the water moulds, and the causative organisms of several important plant diseases, e.g. downy mildew of grapes (*Plasmopora*) and potato blight (*Phytoph-thora infestans*).

oosphere *n.* a female gamete or egg, esp. as produced in an oogonium by algae and oomycetes.

ootocoid *a.* giving birth to young at a very early stage and then carrying them in a pouch, such as marsupials.

ootokous *a.* egg-laying. *alt.* ootocous.

ooze *n.* (1) a deposit containing skeletal parts of minute organisms and covering large areas of the ocean floor; (2) soft mud.

OP osmotic pressure *q.v.*

opal *n.* $SiO_2.nH_2O$, an amorphous silica mineral (*q.v.*) that contains *ca.* 6–10 wt % water. Some varieties of opal are used as gemstones.

OPEC Organisation of Petroleum Exporting Countries *q.v.*

open *a. appl.* plant community that does not completely cover the ground but leaves bare areas that can be colonized.

open canopy, open forest, open woodland type of wooded area with trees sufficiently widely spaced not to touch each other, and in which the crowns shade <20% of the ground surface. *cf.* closed forest.

opencast mining, open cut mining, open pit mining type of surface mining in which the overburden is removed from an extensive area to expose the underlying mineral deposit (e.g. of coal or iron ore), which is subsequently removed.

open system in thermodynamic analysis, denotes a system (*q.v.*) that can exchange both matter and energy with its surroundings.

operant behaviour spontaneous animal behaviour that occurs without any apparent stimulus.

operant conditioning type of procedure for studying animal behaviour in which rewards and punishments are used to select, strengthen or weaken behaviour patterns.

operational taxonomic unit (OTU) any group such as genus, species, etc., evaluated on taxonometric methods.

ophidians *n.plu.* snakes *q.v.*

ophitic *a. appl.* a rock texture (*q.v.*) in which small grains of plagioclase are enclosed in larger grains of pyroxene. Commonly seen in diabases (*q.v.*) and gabbros *q.v.*

ophiurans Ophiuroidea *q.v.*

Ophiuroidea *n.* a class of echinoderms commonly known as brittle stars, an individual having a star-shaped body with the arms clearly marked off from the disc.

Opiliones *see* harvestmen.

opisthobranch *n.* member of the mollusc subclass Opisthobranchia (e.g. sea slugs, sea hares), which are marine, and in which the shell is much reduced or absent.

opium *n.* addictive drug obtained from the opium poppy, *Papaver somniferum*, consisting of the dried milky juice from the slit poppy capsules, which acts as a stimulant, narcotic and hallucinogen, formerly widely used to ease pain but now replaced by its derivative alkaloids, such as morphine.

opophilous *a.* feeding on sap.

opportunistic *a. appl.* (1) microorganisms that are normally non-pathogenic but can cause disease in certain conditions, e.g. in immunosuppressed or otherwise debilitated individuals; (2) species specialized to exploit newly opened habitats.

OPs organophosphates *q.v.*

optical density (OD) absorbance *q.v.*

optical isomer either of two optically active isomeric forms of a compound, one of which rotates a beam of plane-polarized light to the right (dextrorotatory, *d*) and the other to the left (laevorotatory, *l*).

optical microscopy type of microscopy that uses light to create the image. *alt.* light microscopy. *cf.* electron microscopy.

optimal *a.* (1) the most efficient, the most cost-effective; (2) *appl.* animal behaviour, as optimal foraging (*q.v.*), optimal reproductive strategy (*q.v.*), etc.

optimal foraging behaviour that enables an animal to collect as much energy as possible in a certain time.

optimal reproductive strategy behaviour that results in an animal leaving the largest possible number of viable offspring.

optimal yield the highest rate of increase that a population can sustain in a given environment.

optimum *n.* (1) the most suitable degree of any environmental factor or set of factors for full development of organism concerned; (2) point at which best response is obtained in any system.

optimum population population that produces the maximum sustainable yield.

optimum yield *see* maximum sustainable yield.

Opuntiales Cactales *q.v.*

OR orientation response *q.v.*

orb web the most familiar type of spider's web, with a spiral centre supported on radiating threads anchored to a roughly triangular frame.

Orchidales *n.* an order of monocot herbs, with tubers or stems swollen into pseudo-bulbs, often epiphytic or sometimes saprophytic, often with showy flowers, and comprising the family Orchidaceae (orchids).

orchid mycorrhizas mycorrhizas formed by basidiomycete fungi on the embryos of orchids (family Orchidaceae), which are necessary for the orchid's successful development.

order *n.* taxonomic group of related organisms ranking between family and class. *see* classification.

order of magnitude rough way of expressing the number of something in terms of the closest power of 10, as tens, hundreds, thousands, etc.

order of reaction *see* rate law.

ordination *n.* method of graphical analysis of the distribution of communities according to some factor such as similarity in species or relative position along an environmental gradient. The data gathered from different communities is plotted in such a way that the most similar communities appear together.

Ordovician *a. pert.* or *appl.* geological period lasting from about 500 to 440 million years ago.

ores *n.plu.* naturally occurring mineral deposits from which one or more metallic elements can be extracted. *see* metal ores.

orebodies *n.plu.* deposits of ores which are sufficiently large to be commercially worked. They may be formed by a number of enrichment processes, e.g. *see* contact metasomatism, hydrothermal deposition, pegmatite formation, sedimentary deposition.

organ *n.* any part or structure of an organism adapted for a special function or functions, e.g. heart, stomach, kidney, etc.

organelle *n.* a structure within a eukaryotic cell in which certain functions and processes are localized, e.g. nucleus, mitochondrion.

organic *a.* (1) *pert.*, derived from, or showing the properties of, a living organism; (2) *appl.* molecules containing carbon.

organic chemistry that branch of chemistry concerned with the synthesis and characterization of carbon-containing compounds.

organic compounds compounds that contain carbon and hydrogen covalently bonded together in molecules. They may also contain atoms of other elements, most commonly oxygen, nitrogen, sulphur, phosphorus or one or more of the halogens *q.v.*

organic farming farming without the use of industrially produced (artificial) fertilizers and pesticides.

organic manures *n.plu.* bulky organic materials, such as farmyard manure, compost and night soil, used to supply essential nutrients for crop growth. *alt.* biofertilizers.

organic matter material derived from living organisms, usually refers to the decaying or decayed plant and animal matter in soil.

organic sedimentary rocks non-clastic rocks formed by the accumulation of material of organic origin, e.g. chalk.

organic soil soil with a high organic matter content (definitions vary but typically taken to be >30%).

organic weathering the disintegration of rocks *in situ* as a result of plant, animal and/or microbial activity. *alt.* biological weathering.

organism *n.* any living thing.

organismal, organismic *a. pert.* (1) an organism as a whole; (2) or *appl.* factors or processes involved in maintaining the integrity and life of an individual.

Organisation for Economic Co-operation and Development (OECD) an international organization established in 1961 to promote trade. Member countries are: Australia, Austria, Belgium, Canada, Denmark, Finland, France, Germany, Greece, Iceland, Republic of Ireland,

Italy, Japan, the Netherlands, New Zealand, Norway, Portugal, Spain, Sweden, Switzerland, Turkey, UK and USA.

Organisation of Petroleum Exporting Countries (OPEC) an organization established in 1961 to oversee a common sales policy for the export of crude oil from member countries, i.e. Algeria, Ecuador, Gabon, Indonesia, Iran, Iraq, Kuwait, Libya, Nigeria, Qatar, Saudi Arabia, the United Arab Emirates and Venezuela.

organochlorines *n.plu.* organic compounds containing chlorine. These are mainly manufactured compounds that have very few counterparts in nature. Organochlorines, such as polychlorinated biphenyls (*q.v.*) and the pesticide DDT (*q.v.*), are persistent in the environment. They are able to enter food chains and undergo biomagnification, eventually posing a risk to top predators, including humans.

organomercurials *n.plu.* toxic organic compounds containing mercury that have been used as fungicides, esp. as seed dressings.

organophosphates (OPs), organophosphorus compounds *n.plu.* (*agric.*) pesticides, e.g. malathion and diazinon, similar in structure to some compounds active as nerve gases. These were developed as more selective and less persistent alternatives to organochlorine pesticides such as DDT.

organotin *n.* any compound in which tin is directly bonded to a carbon atom of an organic moiety. In general, such compounds are toxic and some, esp. tributyltins (*q.v.*), are used as biocides, e.g. in marine antifouling paints (*q.v.*). Organotins are also used as stabilizers in PVC plastics.

Oriental Region zoogeographical region, which is part of Arctogaea, and which comprises the Indian subcontinent south of the Himalayas, South-East Asia and Indonesia north and west of Wallace's line, which runs between Borneo and Sulawesi and the Indonesian islands of Bali and Lombok.

Oriental sore leishmaniasis *q.v.*

orientation response (OR) the physiological response of an organism to a sudden change, e.g. a sound or a novel sight, in its environment.

origin of life life is generally thought to have originated on the early Earth, some 4000 million years ago, from simple carbon-containing compounds, beginning with the evolution of some form of self-replicating molecule.

orimulsion *n.* fuel consisting of 70% bitumen and 30% water, which may be used as a substitute for oil in modified power stations.

ornis avifauna *q.v.*

ornithic *a. pert.* birds.

ornithine cycle urea cycle *q.v.*

Ornithischia, ornithischians *n., n.plu.* an order of Mesozoic dinosaurs, commonly called the bird-hipped dinosaurs, individuals having a pelvis resembling that of a modern bird. They were all herbivorous and included both bipedal (ornithopods) and quadrupedal (ceratopians, stegosaurs and ankylosaurs) members.

Ornithogaea *n.* the zoogeographical region that includes New Zealand and Polynesia.

ornithology *n.* the study of birds.

ornithophilous *a. appl.* flowers pollinated through the agency of birds. *n.* ornithophily.

orogenesis, orogeny *n.* the creation of mountain belts by tectonic activity.

orogenic belt a belt of mountains that is usually curved or linear when viewed from above. *alt.* orogenic mountain belt.

orographic rain rain formed when warm, moist maritime air is forced to rise when it meets an area of high relief such as a coastal mountain range. Most of the rain falls on the windward side, while the leeward side receives little precipitation. *alt.* relief rain.

orthent *n.* a suborder of the order entisol of the USDA Soil Taxonomy system consisting of entisols that develop at high altitudes over hard rock.

orthids *see* aridisol.

orthite allanite *q.v.*

ortho- prefix derived from Gk *orthos*, straight.

orthoclase *n.* $KAlSi_3O_8$, one of the feldspar minerals *q.v.*

orthod *n.* a suborder of the order spodosol of the USDA Soil Taxonomy system, consisting of spodosols with an illuvial horizon rich in humus, aluminium oxides and iron oxides.

orthogenesis *n.* evolution along some apparently predetermined line independent of natural selection or other external forces.

orthograde *a.* walking upright, with the body in a vertical position.

orthokinesis *n.* (1) movement in which the organism changes its speed when it meets an unfavourable environment, the speed depending on the intensity of the stimulus, resulting in the dispersal or aggregation of organisms; (2) variation in linear velocity with intensity of stimulation.

orthologous *a. appl.* genes in different species that are homologous because they are derived from a common ancestral gene, e.g. alpha-globin genes from humans and horses.

Orthomyxoviridae, orthomyxoviruses *n., n.plu.* family of large, enveloped RNA viruses with single-stranded segmented genomes, and including influenza.

Orthoptera, orthopterans *n., n.plu.* order of insects that includes crickets, locusts and grasshoppers (in some classifications also cockroaches and mantises), with long antennae, biting mouthparts, long narrow tough fore wings and broad membranous hind wings, usually also having enlarged hindlegs for jumping and stridulating organs, which produce the characteristic sounds of these insects.

orthopyroxenes *n.plu.* pyroxene minerals (*q.v.*) that crystallize in the orthorhombic system. *cf.* clinopyroxenes.

orthoquartzite *n.* sandstone (*q.v.*) that contains very little except quartz grains cemented with silica.

orthotype *n.* genotype originally designated.

Oryza sativa rice *q.v.*

Os symbol for the chemical element osmium *q.v.*

oscillatory waves (*oceanogr.*) deep-water waves in which individual water particles move in an essentially circular orbit. The size of the orbit decreases as distance from the surface increases until, at a depth of approximately half the wavelength *q.v.*, movement is negligible. This point is known as the wave base. In this type of wave there is little net movement of water, because it is the wave form that moves rather than the water. *alt.* waves of oscillation. *cf.* translatory waves.

osmium (Os) *n.* metallic element (at. no. 76, r.a.m. 190.2), found naturally in osmiridium and as osmium sulphide.

osmoconformer *n.* organism that does not regulate the osmotic concentration of its internal fluids, which therefore vary with the osmotic concentration of the external environment, as some estuarine invertebrates.

osmolality *n.* the osmotic concentration of a solution, usually expressed in osmoles (1 osmole = 1 molal, *see* molality).

osmomorphosis *n.* change in shape or structure due to changes in osmotic pressure, such as due to changes in salinity.

osmophore *n.* the group of atoms responsible for the odour of a compound, which combines with the chemoreceptors on olfactory neurones in the nasal epithelium or other sensory organ.

osmoregulation *n.* in animals, regulation of the osmotic pressure of body fluids by controlling the amount of water and/or salts in the body.

osmoregulator *n.* organism that actively regulates the osmotic concentration of its internal fluids.

osmosis *n.* diffusion of a solvent, usually water, through a semipermeable membrane from a region of high solvent concentration to a region of low solvent concentration, i.e. from a dilute solution to a more concentrated one or from the pure solvent to a solution.

osmotaxis *n.* a taxis in response to changes in osmotic pressure.

osmotic *a. pert.* osmosis.

osmotic pressure (OP) a measure of the osmotic activity of a solution, defined as the minimum pressure that must be exerted to prevent the passage of pure solvent into the solution when the two are separated by a semipermeable membrane. Osmotic pressure is proportional to solute concentration.

osmotroph *n.* any heterotrophic organism that absorbs organic substances in solution, such as some bacteria and fungi. *cf.* phagotroph.

ossification *n.* formation of bone, replacement of cartilage by bone.

ossify *v.* to change to bone.

Ostariophysi *n.* a superorder of teleost fishes that includes the carps, minnows, catfishes and loaches, characterized by the presence of Weberian ossicles.

Osteichthyes *n.* the bony fishes, a class comprising all fish except for the Agnatha and the Selachii, which have a bony skeleton, usually an air bladder (swimbladder) or lung and a cover (operculum) over the gill openings.

osteolepid *a.* having a skin armoured with bony scales, as in the Crossopterygii.

osteology *n.* the study of the structure, nature and development of bones.

Ostracoda, ostracods *n., n.plu.* a group of small aquatic crustaceans having a bivalved carapace enclosing the head and body, and reduced trunk and abdominal limbs.

ostracoderms *n.plu.* extinct Palaeozoic jawless fishes (Agnatha) that were armoured with an exoskeleton of dermal bone.

OTEC ocean thermal energy conversion *q.v.*

Otto engine a four-stroke spark ignition internal combustion engine.

OTU operational taxonomic unit *q.v.*

outbreeding *n.* the mating of individuals who are not closely related. *a.* outbred. *alt.* outcrossing. *cf.* inbreeding, cross-fertilization.

outcrop *n.* an area of exposed rock.

outcrossing outbreeding *q.v.*

outfield *see* infield–outfield system.

outlier *n.* (1) (*geol.*) area of younger rocks completely surrounded by older rocks. *cf.* inliers; (2) *see* gross error.

outwash deposit stratified drift *q.v.*

outwash plain *n.* large plain composed of sorted glacial drift deposited by several meltwater streams issuing from the margin of a glacier. *alt.* sandur (*plu.* sandar) (Icelandic).

ova *plu.* of ovum.

ovary *n.* in plants and animals, the reproductive organ in which female gametes or egg cells are produced. In flowering plants it comprises the enlarged portion of the carpel(s) containing the ovules (*see* Fig. 12, p. 157) and after fertilization develops into the fruit containing the seeds.

oven-dried soil soil dried to a constant weight at 105 °C.

overburden *n.* the soil and rock overlying a mineral deposit such as a coal seam, which is removed before surface mining begins.

overcompensating density dependence *see* density overcompensation.

overdominance *n.* the condition where a heterozygote has a more extreme phenotype than either of the corresponding homozygotes.

overexploitation *n.* the use of renewable resources at a level that the system cannot sustain, e.g. overfishing, overgrazing, thus causing long-term damage to the ecosystem and the disappearance of species and, in some cases, physical damage such as erosion or desertification.

overfishing *n.* the harvesting of fish at such a rate that, if continued, will cause fish stocks to become severely depleted.

overflow *a. appl.* behaviour in which an inappropriate response occurs to a certain stimulus in order to satisfy certain drives, such as a dog displaying maternal care to a bone.

overgrazing *n.* the removal of vegetation cover by the activities of browsing animals, thus exposing the underlying soil to the forces of erosion.

overgrowth competition competition due to individuals of plants or sessile animals physically growing over another, limiting their access to resources such as light or nutrients.

overharvesting *n.* taking out of a system more than it can replace each year.

overhunting *n.* taking out so many members of an animal population that reproduction is hampered, leading to a decline in numbers and even to eventual extinction.

overland flow surface runoff *q.v.*

overpopulation *n.* situation where there are more people in a particular area than can be adequately supported by the resources and technology available locally. *cf.* underpopulation.

overseas development assistance (ODA) *see* aid.

overshoot *n.* situation in which the population size of a species temporarily exceeds the carrying capacity of its habitat, leading to a sharp reduction in the population.

overspill towns new towns *q.v.*

overstorey *n.* topmost layer of vegetation in the forest, formed of the leaves and branches of the tallest trees. *cf.* understorey.

overturned folds *see* anticlines.

oviparous *a.* egg-laying. *n.* oviparity.

oviposition *n.* the deposition of eggs on a surface, esp. in insects and fish.

ovoviviparous *a. pert.* organisms that produce an egg with a persistent membrane, but which hatches within the maternal body. *n.* ovoviviparity.

ovule *n.* in seed plants, the structure consisting of the megagametophyte and megaspore, surrounded by the nucellus enclosed in an integument, and which develops into a seed after fertilization.

ovum *n.* a female gamete (*q.v.*). *alt.* egg, egg cell. *plu.* ova. *see also* oocyte.

oxamyl *n.* contact and systemic carbamate pesticide used as a nematicide on onions, potatoes and sugar-beet, and as an insecticide.

oxbow lake cresent-shaped segment of river left isolated when the river cuts through a narrow neck of land embraced by a pronounced meander, to carve a new and shorter channel. *alt.* cutoff, mortlake.

oxic horizon *see* oxisol.

oxidant oxidizing agent *q.v.*

oxidase *n.* any enzyme that catalyses oxidation–reduction reactions using molecular oxygen as the electron acceptor. *cf.* reductase, dehydrogenase.

oxidation *n.* the addition of oxygen to, or the loss of hydrogen or loss of electrons from, a compound, atom or ion. The substance is then said to be oxidized. *cf.* reduction.

oxidation catalyst-based catalytic converter a device for facilitating the oxidation of unburnt hydrocarbons (RH) and carbon monoxide (CO) within the exhaust gases of a petrol (i.e. gasoline), engine, running on unleaded fuel. Now used as part of a three-way catalytic converter (*q.v.*) but formerly used alone. When used alone, the engine is made to run rich (suppressing NO_x formation), air is admitted into the exhaust gases, which then pass into a chamber containing the catalyst (e.g. platinum or palladium) supported on a porous matrix. The oxygen present converts the RH and CO into carbon dioxide and water. *see also* reduction catalyst-based catalytic converter. *alt.* two-way catalytic converter.

oxidation pond large, shallow tank used in the secondary treatment of sewage in warm climates. Settled sewage slowly passes through the tanks and is oxidized by microorganisms. Oxygen for this process comes from the photosynthetic activity of algae, which are themselves sustained by the nutrient species released by microbial action. An advantage of these ponds is that the algae can be harvested and used as animal feed or a source of biomass energy. *alt.* stabilization pond.

oxidation state (*chem.*) the actual, or a notional, charge on the atoms of an element in a given chemical species. The oxidation state of an element in its elemental form (i.e. not in a compound) is zero. For a monoatomic ion, e.g. Cl⁻, the oxidation state is the actual charge on the ion (−1 in this case). The oxidation state of elements within compounds or molecular ions can be assessed on the basis of the information given in Appendix 3, in conjunction with the following formula: $\Sigma o = c$, where Σo is the sum of the oxidation states of all of the atoms present in the chemical species concerned and c is the species' overall charge. Oxidation states are usually quoted in roman numerals, either in parentheses after the name of the element concerned or superscripted to the right-hand side of its elemental symbol, e.g. iron(III) or Fe^{III}. *alt.* oxidation number.

oxidative phosphorylation process in aerobic organisms in which ATP is formed from ADP and orthophosphate (P_i), the process being driven by electron flow along an electron transport chain with O_2 as the final acceptor, the electrons being derived from the oxidation of fuel molecules during the anaerobic phase of respiration.

oxides of carbon a collective term for carbon monoxide (CO) and carbon dioxide (CO_2).

oxides of nitrogen compounds consisting of nitrogen and oxygen. Three of these are significant primary atmospheric pollutants: nitrous oxide (N_2O), nitric oxide (NO) and nitrogen dioxide (NO_2). NO and NO_2 are collectively known as NO$_x$ (q.v.). alt. nitrogen oxides.

oxidizing agent chemical species that readily accepts electrons, alt. oxidant.

oxisol n. an order of the USDA Soil Taxonomy system, consisting of highly weathered soils, with practically no humus content, found only in tropical regions with a high rainfall. They are characterized by a deep, red-coloured, subsurface horizon dominated by iron and aluminium oxides (oxic horizon). Oxisols have a high (non-sticky) clay content and very deep profiles. They form under savanna grassland and tropical rain forest and are generally of low fertility. Five suborders are recognized, see aquox, perox, torrox, udox, ustox.

oxydemeton-methyl n. systemic and contact organophosphate insecticide and acaricide used on many horticultural and agricultural crops to control aphids and red spider mites.

oxygen (O) n. gaseous element (at. no. 8, r.a.m. 15.99), in the free state most commonly a colourless, odourless gas (O$_2$), but can also form ozone (O$_3$). As O$_2$ it forms 21% of the atmosphere by volume. It is highly reactive, and is the most abundant element on Earth, occurring also in the Earth's crust and in many types of organic molecules. An essential element for living organisms. Many organisms also require molecular oxygen (O$_2$) to carry out respiration. see also oxygen cycle, ozone.

oxygen cycle the circulation of oxygen through the biosphere and the atmosphere. Oxygen is released by green plants and some photosynthetic bacteria, e.g. the cyanobacteria, during photosynthesis and is taken in by aerobic organisms, including green plants, for respiration.

oxygen debt a deficit in stored chemical energy that builds up when a normally aerobic tissue, such as muscle, is working with an inadequate oxygen supply. It then consumes oxygen above the normal rate for some time until energy supplies are restored by respiration.

oxygen-demanding wastes organic wastes, esp. sewage effluent, that, when released into waterbodies, have the effect of depleting the dissolved oxygen content of the water. This depletion is caused by the respiration of aerobic bacteria that utilize the organic substrate. In extreme cases, all the dissolved oxygen present in the water is used up and the aquatic fauna perishes. The aerobic bacteria are then replaced by anaerobic bacteria, which produce foul-smelling toxic products, such as hydrogen sulphide and ammonia. alt. oxygen-demanding effluent.

oxygen dissociation curve (1) a graph of the percentage saturation of haemoglobin with oxygen against concentration of oxygen, which gives information about the dissociation of oxyhaemoglobin under different environmental conditions or in different animals; (2) any graph showing the dissociation of oxygen from a substance.

oxygenic a. oxygen-producing.

oxygenic photosynthesis type of photosynthesis in which oxygen is produced as a waste product, which is the type of photosynthesis carried out by green plants, algae and cyanobacteria.

oxygenotaxis oxytaxis (q.v.). a. oxygenotactic.

oxygen quotient (Q$_{O_2}$) (biol.) the volume of oxygen (O$_2$) in microlitres gas at normal temperature and pressure taken in per hour per milligram dry weight.

oxygen sag curve curve that is obtained when the amount of dissolved oxygen in a watercourse receiving a source of organic pollution is plotted against distance downstream of the discharge. The curve shows a sharp decrease in dissolved oxygen immediately downstream of the discharge followed by a gradual increase as one goes downstream.

oxyphilous a. tolerating only acid soils and substrates.

oxyphobe a. unable to tolerate acid soils, appl. plants.

Oxyphotobacteria n. class of photosynthetic bacteria comprising the cyanobacteria, i.e. those bacteria that produce oxygen as a byproduct of photosynthesis.

oxyphyte n. a plant thriving on acid soils. alt. calcifuge.

oxytaxis n. a taxis in response to the stimulus of oxygen. a. oxytactic.

oxytropism n. tendency of organisms or organs to be attracted by oxygen.

ozone *n.* O_3, a pungent-smelling, triatomic molecular form of oxygen, occurring in trace levels in the atmosphere. It is a powerful oxidant, used as a bleaching agent and to disinfect water. In the stratosphere it forms the ozone layer, which absorbs considerable solar ultraviolet radiation and shields the Earth's surface from the harmful effects of this radiation. Ozone is also formed as a pollutant in the lower atmosphere from, e.g., nitrogen oxides. It is damaging to herbaceous plants at levels greater than 100 parts per 10^9. *see* Chapman reaction, stratospheric ozone, tropospheric ozone.

ozone hole *see* stratospheric ozone.
ozone layer *see* stratospheric ozone.
ozone shield *see* stratospheric ozone.
ozonosphere *n.* region within the stratosphere where the concentration of ozone in the Earth's atmosphere is greatest. *see* stratospheric ozone.

P

p proton *q.v.*

P (1) symbol for the chemical element phosphorus (*q.v.*); (2) proline *q.v.*

P₁, P₂ parental and grandparental generations, respectively, in genetic crosses.

Pa (1) symbol for the chemical element protactinium (*q.v.*); (2) pascal *q.v.*

pachyderms *n.plu.* any of various nonruminant hoofed mammals, e.g. an elephant, or rhinoceros, that are generally large, and have thick tough skins.

Pacific North American Floral Region part of the Holarctic Realm comprising North America west of the Rocky Mountains from southern Alaska south to the Mexican border.

Pacific Ring of Fire Ring of Fire *q.v.*

Pacific South American Floral Region part of the Neotropical Realm comprising the Andes and the coastal strip to their west from Ecuador south to central Chile.

pack ice floating sea ice.

paddock grazing *n.* type of rotational grazing, used for dairy cows, in which the grazing areas consist of small enclosures (paddocks), each of which provides the herd with one day's grazing. These are used in turn on a 21- or 28-day grazing cycle. *cf.* set stocking.

paddy rice *see* rice.

paedogamy pedogamy *q.v.*

paedogenesis *n.* (*biol.*) reproduction in young or larval forms, as in axolotl.

paedomorphosis pedomorphosis *q.v.*

Paeoniales *n.* order of herbaceous dicots, some of which are shrubs, with large deeply cut leaves and large showy flowers. Comprises the family Paeoniaceae (peony).

PAGE polyacrylamide gel electrophoresis *q.v.*

PAH polycyclic aromatic hydrocarbon *q.v.*

pahoehoe *n.* a lava flow with a characteristic smooth, wavy appearance. This occurs in very fluid lavas when the slightly hardened surface is dragged into folds by the movement of lava underneath.

Palaearctic *a. appl.* or *pert.* zoogeographical region, or subregion of the Holarctic Realm, including Europe, North Africa, Western Asia, Siberia, northern China and Japan.

palaeo- prefix derived from Gk *palaios* ancient. *alt.* paleo-.

palaeobiology *n.* the study of the biology of extinct plants, animals and microorganisms.

palaeobotany *n.* study of fossil plants and plant impressions.

Palaeocene *n.* the earliest epoch of the Tertiary period, before the Eocene and lasting from around 65 to 55 million years ago. *alt.* Paleocene.

palaeoclimatology *n.* the study of past climatic conditions from tree-ring data, geological and other evidence.

palaeodendrology *n.* the study of fossil trees and tree impressions.

palaeoecology *n.* the study of the relationship between past organisms and the environment in which they lived.

palaeoflora *n.* fossilized plants from a particular area or a particular geological time.

Palaeogaea *n.* the area comprising the Palaearctic, Ethiopian, Indian and Australian zoogeographical regions.

palaeogenetics *n.* the application of the principles of genetics to the interpretation of the fossil record and the evolution of now extinct species.

palaeogeography *n.* the study of the positions of continents and oceans and of the geography of major landmasses in prehistoric times. *alt.* paleogeography.

Palaeognathae *n.* the ratites, flightless birds of the subclass Neornithes, such as the kiwi, cassowary and ostrich, which are secondarily flightless.

Palaeolaurentian *a. pert.* or *appl.* Archaen.

Palaeolithic *a. appl.* or *pert.* the Old Stone Age, characterized by a hunter–gatherer economy and chipped stone tools. *alt.* Paleolithic.

Palaeoniscoidei, Palaeonisciformes, palaeoniscids *n.*, *n.*, *n.plu.* a group of actinopterygian fishes, most now extinct, existing from the Devonian to the present day and including the bichars of the Nile. They are carnivorous with large sharp teeth and usually a very heterocercal tail.

palaeontology *n.* the study of past life on Earth from fossils and fossil impressions.

palaeosere *n.* the development of vegetation throughout the Palaeozoic.

palaeosol *n.* fossil soil.

palaeospecies *n.* a group of extinct organisms that are assumed to have been capable of interbreeding and so are placed in the same species.

Palaeotropical kingdom *see* Palaeotropical Realm.

Palaeotropical Realm floristic area comprising Africa except for the northern part around the Mediterranean Sea and the southern tip (but including Madagascar), Arabia and Asia, including the Indian subcontinent and the islands of the Indian Ocean south from the Himalayas to New Guinea, and some Pacific islands. It is made up of 14 floral regions: East African, Ethiopian, Fijian, Hawaiian, Indian, Macaronesian, Madagascan, Malaysian-Papuan, New Caledonian, Polynesian, Saharo-Arabican, South-East Asian, Sudanian-Sindian and West African (*see* individual entries). *alt.* Palaeotropical kingdom.

Palaeozoic *a. pert.* or *appl.* geological era lasting from about 590 to 248 million years ago and comprising the Cambrian, Ordovician, Silurian, Devonian, Carboniferous and Permian periods. *alt.* Paleozoic.

palaeozoology *n.* the study of the biology of extinct animals from fossils and fossil impressions.

Paleocene Palaeocene *q.v.*

Palaeozoic Paleozoic *q.v.*

palisade tissue the layer or layers of photosynthetic cells beneath the epidermis of many foliage leaves.

palladium (Pd) silvery white metallic element (at. no. 46, r.a.m. 106.4).

Palmaceae, palms *n.*, *n.plu.* tropical and subtropical family of monocot plants typically with large leathery fan-like leaves. They include climbing and tree species. The fruit is a berry or drupe. The palms are of great economic importance and cultivated species include the date palm (*Phoenix dactylifera*), oil palms (*Elaeis* spp.) and the coconut palm (*Cocos nucifera*).

palmigrade *a.* walking on the sole of the foot.

palmoid *a. appl.* palms and palm-like forms.

palsa *n.* a periglacial landform found in bogs. It is similar to a pingo (*q.v.*) but composed largely of peat with a core of discrete ice lenses.

paludal *a.* (1) marshy; (2) *pert.*, or growing in, marshes or swamps.

paludicole *a.* living in marshes.

palustral, palustrine *a.* growing in marshes or swamps.

palynology *n.* (1) the study of pollen preserved in, e.g., peat and its distribution, *alt.* pollen analysis; (2) the study of spores.

pampas *n.plu.* vast grassland plains of South America, south of the Amazon.

pan- suffix meaning all or throughout.

pan *n.* (1) (*soil sci.*) relatively hard layer or horizon within a soil caused by compaction, cementation or the presence of a very high clay content; (2) (*geog.*) a large enclosed depression. Such features are found in arid and semi-arid regions. Deflation (*q.v.*) is thought to play a key role in the formation and maintenance of these basins.

PAN (1) the peroxyacyl nitrates (*q.v.*); (2) peroxyacetyl nitrate *q.v.*

panclimax *n.* two or more related climax communities of similar compositions that exist under similar climatic conditions.

Pandanales *n.* order of monocots, mainly sea coast or marsh plants with tall stems supported by aerial roots, and leaves running in spirals, comprising the family Pandanaceae (screwpine).

pandemic *a.* very widely distributed. *n.* epidemic disease with a world-wide distribution.

Pangaea *n.* the supercontinent made up of all the present continents fitted together before their separation by continental drift.

pangamic *a. appl.* indiscriminate or random mating. *n.* pangamy.

panicle *n.* a branched flowerhead, strictly one in which the branches are alternate and side-branches also branch alternately.

panicoid *a. appl.* millets of the genus *Panicum.*

panmictic *a.* characterized by, or resulting from, random matings.

panmixia *n.* indiscriminate interbreeding. *alt.* panmixis.

panphytotic *n.* a pandemic affecting plants.

panspermia *n.* a theory popular in the nineteenth century and that enjoys periodic revivals, that life did not originate on Earth but arrived in the form of bacterial spores or viruses from an extraterrestrial source.

Panthalassa *n.* the ocean surrounding the supercontinent Pangaea *q.v.*

panthalassic *a.* living in both coastal and oceanic waters.

pantodonts *n.plu.* group of extinct North American and Asian herbivorous placental mammals living from the Paleocene to Oligocene.

pantotheres *n.plu.* an order of mammals from the Jurassic, possible ancestors of living therians, having molar teeth showing the basic pattern found in living forms.

pantropical *a.* distributed throughout the tropics, *appl.* species.

Papaverales *n.* order of mainly herbaceous dicots, some shrubs and small trees, and including the families Fumariaceae (fumitory), Hypecoaceae and Papaveraceae (poppy).

paper chromatography separation technique carried out with paper as the support on which the substances to be separated migrate differentially in a solvent.

papillomaviruses *see* Papovaviridae.

Papovaviridae, papovaviruses *n., n.plu.* family of small, non-enveloped, double-stranded DNA viruses that includes polyoma and simian virus 40 (SV40), and the papillomaviruses that cause common warts. Polyoma and SV40 are oncogenic

in the cells of certain species, and some papillomaviruses may be involved in human cervical cancer.

Papuan *a. appl.* subregion of Australian zoogeographical region comprising New Guinea and islands westwards to Wallace's line.

PAR photosynthetically active radiation *q.v.*

para- prefix derived from Gk *para*, beside, signifying situated near, or surrounding.

parabiosis *n.* amicable use of the same nest by different species of ant, which, however, keep their broods separate. *a.* parabiotic.

parabolic dune a cresent-shaped dune commonly found in semi-arid and coastal regions where vegetation cover restricts sand movement. These are distinguishable from barchans (*q.v.*) in that their horns point upwind and not downwind. Parabolic dunes often develop from blow-outs.

Paracanthopterygii, paracanthopterygians *n., n.plu.* group of advanced teleost fish including the cod and its relatives.

paracme *n.* (1) the evolutionary decline of a taxon after reaching the highest point of development; (2) the declining or senescent period in the life history of an individual.

paraffin oil liquid containing C11–C12 hydrocarbons (b.p. 150–300 °C) produced by the fractional distillation of crude oil, and used in e.g. domestic heaters and paraffin lamps. *alt.* kerosene, kerosine.

paraffins *n.plu.* (*chem.*) an old term for the alkanes *q.v.*

paraffin wax white translucent solid consisting of a mixture of higher alkanes (C23–C29, m.p. 50–60 °C), produced by the fractional distillation of crude oil, and used, e.g., for polishes and candles.

paragenesis *n.* (1) condition in which an interspecific hybrid is fertile with the parental species but not with other similar hybrids; (2) a subsidiary mode of reproduction.

paraheliotropism paraphototropism *q.v.*

paralectotype *n.* specimen of a series used to designate a species, which is later designated as a paratype.

paralimnic *a. pert.* or inhabiting the lake shore.

parallel descent, parallel evolution (1) evolution in similar direction in different groups; (2) the independent acquisition of similar traits in two related species.

parallel drainage *see* drainage pattern.

parallel flow the flow of two fluids in the same direction.

paralogous *a. appl.* (1) similarities in anatomy that are not related to common descent or similar function; (2) two genes in a genome that are similar because they derive from a gene duplication, e.g. alpha- and beta-globin. *n.* paralogy. *cf.* orthologous.

paramo *n.* (1) any treeless plateau found high in the tropical Andes Mountains of South America below the permanent snow line; (2) term used in northern Andes for the vegetation of high tropical mountains. This type of vegetation is referred to as puna in the central Andes region.

paramorph *n.* (1) any variant form or variety; (2) a form induced by environmental factors without underlying genetic change.

paramutualism facultative symbiosis *q.v.*

Paramyxoviridae, paramyxoviruses *n., n.plu.* family of large, enveloped, single-stranded RNA viruses including measles and Newcastle disease virus.

Parano-Amazonian Floral Region part of the Neotropical Realm comprising Brazil and Bolivia.

Paranthropus genus of fossil hominids from southern Africa, subsequently renamed *Australopithecus robustus*.

parapatric *a. appl.* distribution of species or other taxa that meet in a very narrow zone of overlap. *n.* parapatry.

parapatric speciation the separation of a group into different neighbouring species, with a limited amount of genetic exchange where the two species overlap.

paraphototropism *n.* tendency of plants to turn the edges of their leaves towards intense illumination, thus protecting the surfaces.

paraphyletic *a. appl.* groups such as the reptiles that have evolved from and include a single ancestral species (known or hypothetical) but that do not contain all the descendants of that ancestor. In the case of the reptiles, the birds also have the same common ancestor. *n.* paraphyly.

paraquat *n.* contact herbicide used in general weed control.

parasematic *a. appl.* markings, structures or behaviour tending to mislead or deflect attack by an enemy. *cf.* aposematic, episematic.

parasexual cycle a cycle of plasmogamy, karyogamy and haploidization in some fungi that superficially resembles true sexual reproduction but that may take place at any time in the life cycle.

parasite chain a food chain passing from large to small organisms.

parasite *n.* an organism that for all or some part of its life derives its food from a living organism of another species (the host), usually living in or on the body or cells of the host, which is usually harmed to some extent by the association. *a.* parasitic.

parasitic castration (1) castration caused by the presence of a parasite, as in male crabs infested by the barnacle *Sacculina*; (2) sterility in various other plants or animals caused by a parasite attacking sex organs.

parasitic male a dwarf male that is parasitic on its female and has a reduced body in all but the sex organs, as in some deep-sea fish.

parasitism *n.* a special case of symbiosis in which one partner (the parasite) receives advantage to the detriment of the other (the host).

parasitocoenosis *n.* the whole complex of parasites living in or on any one host.

parasitoid *n.* (1) an organism alternately parasitic and free-living; (2) animal, generally an insect, in which the adults are free-living but that lay their eggs in the body of another animal in which the larvae develop, usually only killing the host on hatching and emergence.

parasitology *n.* the study of parasites, esp. animal parasites.

parasocial *a. appl.* social group in which some features of complete eusociality are absent.

parasymbiosis *n.* the living together of organisms without mutual harm or benefit.

parately *n.* evolution from material unrelated to that of type, but resulting in superficial resemblance.

parathion *n.* highly toxic organophosphorus insecticide, no longer used in some countries.

paratrophic *a. appl.* method of nutrition of obligate parasites.

paratype *n.* specimen described at same time as the one regarded as type specimen of a new genus or species.

paratyphoid fever a bacterial disease of humans caused by *Salmonella paratyphi*, resembling typhoid fever (*q.v.*) but of a milder nature.

Parazoa *n.* a term sometimes used for those multicellular animals, such as sponges, having a loose organization of cells and not forming distinct tissues or organs.

parental rock parent rock *q.v.*

parental generation the parent individuals in a genetic cross, designated P₁.

parental investment any behaviour towards offspring that increases the chances of the offspring's survival at the cost of the parent's ability to invest in other offspring.

parent material (*soil sci.*) material from which a soil develops.

parent nucleus an atomic nucleus that undergoes radioactive decay to yield a daughter nucleus.

parent rock the original rock, igneous or sedimentary, that is altered by the action of extreme heat and/or pressure and/or permeating hot gases or liquids to form a metamorphic rock. *alt.* parental rock.

parity *n.* the number of times a female has given birth, regardless of the number of offspring produced at any one birth.

park-and-ride system transport system whereby people can park their cars on the outskirts of a city and finish their journey into the centre by public transport.

parr *n.* young of some species of fish.

parsimony principle in molecular taxonomy, the principle that organisms that are closely related, i.e. that diverged more recently, will have fewer differences in their DNA than those that diverged longer ago.

parthenogenesis *n.* reproduction from a female gamete without fertilization by a male gamete.

parthenogenetic *a. appl.* (1) organisms produced by parthenogenesis; (2) agents that can activate an unfertilized ovum.

partial dominance codominance *q.v.*

particulate *a.* consisting of individual particles, esp. *appl.* finely divided solids or liquid droplets suspended in air or water. *see also* Aitken nuclei, dust, grit, haze, mist, smoke.

particulate organic matter small particles of organic material in water, produced from dead and decaying organisms.

partition chromatography chromatographic technique where the substances to be separated are partitioned between two solvent phases.

partition coefficient (K) an equilibrium constant that relates to the distribution of a solute (So) between two immiscible solvents (solv₁ and solv₂), a process that may be represented by the general equation $So(solv_1) \rightleftharpoons So(solv_2)$. The partition coefficient is expressed as $K = [So(solv_2)]/[So(solv_1)]$, where square brackets, [], represent molar concentration.

parts per billion (p.p.b.) the amount of a substance in every billion (10^9) parts of another substance, often used to describe the amount of a pollutant in water or air.

parts per million (p.p.m.) for trace concentrations of a substance in a solid, p.p.m. stands for grams of substance per million grams of sample (this is equivalent to mg kg^{-1} or μg g^{-1}), while in solution, p.p.m. refers to grams of solute per million millilitres of solution (equivalent to mg l^{-1} or μg ml^{-1}). In the gas phase, p.p.m. refers to litres of substance per million litres of gas (equivalent to μl l^{-1}, and is sometimes written p.p.m.v. (parts per million, by volume). For ultra trace levels, parts per billion (p.p.b.) may be used, where, in each of the above applications, 1 p.p.m. = 1000 p.p.b.

parts per thousand (p.p.t., ‰) the amount of a substance in every thousand parts of another substance.

parturition *n.* the act of giving birth.

Parvoviridae, parvoviruses *n., n.plu.* family of small, non-enveloped, single-stranded DNA viruses that includes Aleutian disease of mink, adeno-associated viruses and human parvoviruses.

PAS periodic acid-Schiff reagent, a dye used for staining proteins rich in carbohydrate side chains.

pascal (Pa) *n.* derived SI unit of pressure, equal to one newton per square metre (N m^{-2}).

Passeriformes, passerines *n.*, *n.plu.* large order of birds, which includes small and medium-sized perching birds and songbirds such as crows, tits, warblers, thrushes and finches.

Passiflorales *n.* order of dicot shrubs, herbs, climbers and small trees including the families Caricaceae (pawpaw) and Passifloraceae (passion flower).

passive dispersal distribution of seeds or spores or dispersal stages of animals by purely passive means such as wind or ocean currents.

passive solar heating systems *see* solar heating systems.

pasteurization *n.* method of partial sterilization, used for milk, wine and other beverages, by heating at 62 °C for 30 min or at 72 °C for 15 s, followed by rapid cooling.

pastoral farming a type of extensive agriculture in which domestic livestock are reared in areas of natural or seminatural vegetation known in this context as rangeland. *alt.* pastoralism, rough grazing.

pastoralism pastoral farming *q.v.*

pasture *n.* grassland also containing a mixture of low-growing herbaceous plants, managed for grazing.

patabiont *n.* animal that spends all its life in the litter on the forest floor.

patacole *n.* animal that lives temporarily in the litter on the forest floor.

patchy habitat an environment in which there are marked variations in quality and size of a particular habitat.

pathogen *n.* any disease-causing microorganism.

pathogenesis *n.* the origin or cause of the pathological symptoms of a disease.

pathogenic *a.* causing disease, *appl.* a parasite (esp. a microorganism) in relation to a particular host.

pathology *n.* (1) science dealing with disease or dysfunction; (2) the characteristic symptoms and signs of a disease.

patoxene *n.* animal that occurs accidentally in the litter on the forest floor.

patriclinal, patriclinous *a.* with hereditary characteristics more paternal than maternal. *alt.* patroclinal, patroclinous.

patrilineal *a.* (1) *pert.* paternal line; (2) passed from male parent.

patristic *a. appl.* plants, *pert.* similarity due to common ancestry.

patterned ground ground covered with a repeated regular arrangement of sorted rock and soil debris, found in periglacial environments. Patterns include stone polygons (borders of frost-shattered stones surrounding elevated centres of moist mud), stone nets (intersecting stone borders of stone polygons) and, on gently sloping ground, stone stripes (stone lines separated by broader bands of fine material).

Pavlovian conditioning classical conditioning *q.v.*

Pb symbol for the chemical element lead *q.v.*

P/B ratio the ratio of annual net primary productivity to total biomass in a community.

PCBs polychlorinated biphenyls *q.v.*

PCDDs polychlorinated dibenzo-*p*-dioxins *q.v.*

PCDFs polychlorinated dibenzofurans *q.v.*

PCE tetrachloroethene *q.v.*

PCR cycle (*mol. biol.*) one complete round of DNA replication in the polymerase chain reaction *q.v.*

PCR cycle (*bot.*) photosynthetic carbon reduction cycle. *see* Calvin cycle.

Pd symbol for the chemical element palladium *q.v.*

PD$_{50}$ a measure of activity for certain viruses, the dose at which 50% of test animals show paralysis.

pea *n. Pisum sativum*, a seed legume (family Leguminosae) grown for food and fodder.

pea and bean weevil *Sitona* spp., an insect pest of peas, beans and other legumes. The larvae feed on the roots and the adults feed on the leaves.

pea enation mosaic virus group plant virus group with a single member, pea enation mosic virus, an isometric single-stranded RNA virus. It is a multicomponent virus in which two genomic species of RNA are encapsidated in different particles.

peanut groundnut *q.v.*

pearl *n.* the abnormal growth formed around a minute grain of sand or other foreign matter that gets inside the shell of certain bivalve molluscs and that consists of many layers of nacre.

pearl millet *Panicum miliceum*, a tropical cereal crop cultivated mainly in Africa and Asia to provide a staple food (also used for animal fodder).

peasouper London-type smog *q.v.*

peat *n*. dark brown organic deposit consisting of partly decomposed plant material, associated with wetland areas. Peat accumulates when the rate of production of vegetation exceeds the rate at which it is decomposed by microorganisms. Their action is inhibited primarily by a lack of oxygen, but other factors, such as a lack of nutrients, low temperatures or low pH, can contribute. *see* blanket peat, fen peat.

peat bog *see* blanket peat, bog.

peat mosses *see* sphagnum mosses.

peck order a social hierarchy, esp. in birds, ranging from the most dominant and aggressive animal down to the most submissive.

pecorans *n.plu.* giraffes, deer and cattle.

pectolite *n*. $Ca_2NaHSi_3O_9$, one of the pyroxenoid minerals *q.v.*

ped *see* peds.

pedalfer *n*. any of a group of soils, in humid regions, usually characterized by the presence of aluminium and iron compounds and the absence of carbonates.

pedicel *n*. short stalk, as of flower, fruit, etc. *see* Fig. 12, p. 157.

pedigree *n*. in genetics, a diagram showing the ancestral history of a group of related individuals.

pediment *n*. a gently-sloping rock platform, with an incline of less than 7°, found at the foot of mountains or steep-sided hills. These landforms are characteristic of arid and semi-arid regions. They are thought to be formed either by sheet erosion (*q.v.*) during earlier wetter periods (pluvials) or by parallel retreat of the slope due to weathering.

pediplain *n*. an extensive, nearly level erosional plain formed by the coalescence of many rock pediments (*q.v.*), over a long period of time. This process is known as pediplanation.

pediplanation *see* pediplain.

pedocal *n*. any of a group of soils, of arid and semi-arid regions, characterized by the presence of carbonate of lime.

pedogenic *a. pert.* the formation of soil.

pedogenesis *n*. (1) formation of soil; (2) paedogenesis *q.v.*

pedology soil science *q.v.*

pedomorphosis *n*. retention of juvenile traits in adults. *a.* pedomorphic.

pedon *n*. smallest sample of a soil that exhibits all of the properties of the soil sampled, including the arrangement of soil horizons in the soil profile. A group of several pedons constitutes a polypedon.

pedonic *a. appl.* organisms of freshwater lake bottoms.

pedoturbation *n*. agitation of soil caused by biological activity and physical processes.

peds *n.plu.* relatively robust aggregates of soil solids, capable of withstanding several cycles of wetting and drying. They are clearly visible and easily identifiable, being separated from one another by voids or lines of weakness.

peduncle *n*. (1) (*bot.*) the stalk of a flowerhead; (*zool.*) (2) the stalk of sedentary protozoans, crinoids, brachiopods and barnacles; (3) link between thorax and abdomen in arthropods.

pegmatite *see* pegmatite formation.

pegmatite formation pegmatites are coarse-grained rocks that are formed during the latter stages of the solidification of intrusions. In these stages, a water-rich immiscible fluid may form within the remaining melt. Pegmatites result if this fluid is forced into cracks in the newly formed igneous and/or surrounding rocks, where it solidifies. Pegmatites are most commonly derived from acidic magmas. However, they may also occur as a result of basic intrusions. While pegmatites have essentially the same mineralogy as the rest of the intrusion, they may contain significant concentrations of minerals rich in less abundant elements with large ionic radii, including in some cases uranium and thorium.

pegmatites, granite *see* granite pegmatites.

pegmatitic *a. pert.* pegmatite, *appl.* the coarsest-grained igneous rocks. *see* pegmatite formation.

Pekin(g) man an extinct fossil hominid found near Peking and at first called *Sinanthropus pekinensis*, then *Pithecan-*

thropus pekinensis, and now classified as *Homo erectus*.

pelagic *a*. living in the sea or ocean at middle or surface levels.

pelasgic *a*. moving from place to place.

Pelecaniformes *n*. an order of large aquatic, fish-eating birds with all four toes webbed, including the pelicans, cormorants, boobies and gannet.

pelitic rocks sedimentary rocks composed of very small-sized particles, e.g. shale, mudstone.

pellagra *n*. deficiency disease caused by insufficient tryptophan and nicotinamide (niacin) in the diet. It can be treated with niacin.

pelleted fuel *see* waste-derived fuel.

pellet fusion inertial confinement fusion *q.v.*

pelophilous *a*. growing on clay.

Pelycosauria pelycosaurs *n., n.plu.* order of aberrant and primitive mammal-like reptiles of the Carboniferous–Permian, having a primitive sprawling gait and including the sail-back lizards.

peneplain *n*. nearly flat landscape produced by extensive erosion over millions of years.

penetrance *n*. the percentage of individuals possessing a particular genotype who show the associated phenotype, i.e. in whom the trait is expressed. Complete penetrance is shown by a trait if it is expressed in all persons who carry it; incomplete penetrance is shown if it is not expressed at all in some individuals who carry it.

penicillin *n*. any of various antibiotics, based on a beta-lactam ring structure, produced by the mould *Penicillium notatum* and related species, and that inhibit bacterial cell wall synthesis leading to osmotic lysis.

pennate *a*. (1) divided in a feathery manner; (2) feathered; (3) having a wing; (4) in the shape of a wing; (5) *appl*. diatoms that are bilaterally symmetrical in valve view.

Pennsylvanian *a. appl.* (1) and *pert.* an epoch of the Carboniferous period, lasting from 320 to 286 million years ago; (2) Coal Measures in North America.

pent- (*chem.*) the prefix used to indicate the presence of five carbon atoms in a chain in an organic compound (e.g. pentane $CH_3CH_2CH_2CH_2CH_3$) or radical.

penta- prefix derived from Gk *pente*, five, signifying having five of, arranged in fives, etc.

pentadactyl limb limb with five digits, the basic limb form characteristic of extant amphibians, reptiles, birds and mammals, although in some groups, digits have been lost, reduced or modified.

Pentastomida *n*. a phylum of worm-like parasites that live in the nasal passages of predatory mammals with larvae, found mainly in herbivorous mammals, resembling parasitic mites. *alt.* tongue worms.

pentose *n*. any monosaccharide having the formula $(CH_2O)_5$, e.g. arabinose, xylose and ribose.

peppered moth *Biston betularia. see* industrial melanism.

peptidase *n*. enzyme that splits off one or two amino acid units from the end of a polypeptide chain.

peptide *n*. a chain of a small number (up to *ca.* 20) of amino acids linked by peptide bonds.

peralkaline granite rock very similar to granite (*q.v.*) but characteristically containing alkali-rich amphibole and/or alkalirich pyroxene.

per capita per head of population, as in income *per capita*.

percentage area method technique for determining the density and distribution of a species by measuring the percentage area covered by it in a series of quadrats.

percentage successive mortality apparent mortality *q.v.*

perception *n*. the mental interpretation of physical sensations produced by stimuli from the outside world.

perched aquifer an aquifer (*q.v.*) that develops above the level of the regional water table due to the local presence of an impermeable rock layer. These pockets of groundwater are common features of karst terrain.

perched blocks *see* erratics.

perched water table the upper boundary of a perched aquifer *q.v.*

perching birds a name for the Passeriformes *q.v.*

percolating filter trickling filter *q.v.*

percolation *n.* the downward movement of liquid water through the unsaturated layers of the soil under the influence of gravity.

perdominant *n.* a species present in almost all the associations of a given type.

perennation *n.* (1) of a plant, survival from year to year; (2) survival for a number of years. *a.* perennating, *appl.* roots, buds, etc.

perennial *n.* plant that persists for several years.

perfect flower flower that has functional male and female organs.

perfect stage in ascomycete fungi, the sexual reproductive phase in their life history, in which they produce asci.

peri- prefix derived from Gk *peri*, around, and signifying surrounding, or situated around.

perianth *n.* (1) the outer whorl of floral leaves of a flower when not clearly divided into calyx and corolla; (2) collectively, the calyx and corolla; (3) cover or sheath surrounding the archegonia in some mosses; (4) tubular sheath surrounding developing sporophyte in leafy liverworts.

pericarp *n.* the tissues of a fruit that develop from the ovary wall, comprising an outer layer, which is sometimes fleshy, and an inner layer (endocarp), which is often fibrous or hard, as in the 'stone' enclosing the seed in plums, cherries, etc.

peridotite *n.* a medium- to coarse-grained ultramafic igneous rock (colour index >90). It contains olivine (as an essential mineral) and pyroxene. The amphibole hornblende is commonly present. Types of peridotite include lherzolite (the dominant type), wehrlite and harzburgite.

periglacial *a.* *pert.* or *appl.* to conditions, processes and landforms found at the periphery of glaciers and ice sheets (both existing and Pleistocene). However, its usage has become widened to include other cold climate areas where processes of frost weathering are operative.

period *n.* in geological time, a subdivision of an era, e.g. the Jurassic is a period of the Mesozoic era.

periodic plankton organisms that are only found in the plankton at certain times.

periodic table a listing of the chemical elements in such a way that elements with similar properties are placed together in vertical columns (called groups). It is constructed by listing the elements in order of increasing atomic number, a new row being started whenever a new shell of electrons is occupied for the first time. *see* Appendix 2.

periods *n.plu.* (*chem.*) the horizontal rows of the periodic table, e.g. the second period consists of lithium (Li), beryllium (Be), boron (B), carbon (C), nitrogen (N), oxygen (O), fluorine (F) and neon (Ne). *see* Appendix 2.

peripheral nervous system the nervous system of vertebrates other than the brain and spinal chord, comprising sensory receptors of trunk, limbs and internal organs and nerves other than the cranial nerves.

periphyton *n.* the community of plants and animals adhering to parts of rooted aquatic plants.

Perissodactyla, perissodactyls *n., n.plu.* order comprising the odd-toed ungulate mammals, in which the weight is borne mainly on the third toe, and which includes the horse, tapir and rhinoceros. *cf.* Artiodactyla.

Peritrichia, peritrichans *n., n.plu.* a group of ciliate protozoans that are usually fixed permanently to the substratum and have few cilia, e.g. *Vorticella*.

permafrost *n.* a layer, usually subsurface, of permanently frozen ground that may be hundreds of metres thick. 'Permanent' is defined, in this context, as a period of time spanning at least two consecutive winters during which temperatures within the layer are perpetually <0 °C. Three types of permafrost are recognized. *see* continuous permafrost, discontinuous permafrost, sporadic permafrost. *see also* active layer.

permafrost table the upper surface of the permafrost layer.

permanent crop perennial crop-tree or shrub cultivated for its fruit, leaves or sap, which is harvested on a regular basis, over a period of years.

permanent cropping (1) the cultivation of a permanent crop (*q.v.*); (2) practice of cultivating crops continuously on an area of land, possibly including fallow on a seasonal basis only.

permanent grassland cultivated grassland, more than five years old, whose quality is determined mainly by the percentage of perennial ryegrass present in the sward. *alt.* permanent pasture.

permanent hardness *see* hard water.

permanent pasture permanent grassland *q.v.*

permanent plot a piece of land set aside for long-term ecological studies.

permanent teeth, permanent dentition a set of teeth developed after milk (i.e. deciduous) teeth, the second set of most, third set of some mammals. Some mammals do not develop a second set of teeth.

permanent wilting point point at which the water content of the soil is so low that plants growing in it are permanently wilted. *alt.* wilting point.

permeability *n.* (1) of a membrane, its capacity to allow the free passage of specified substances across it; (2) of rock, the capacity of a rock to transmit water. This is dependent on the nature of the pore spaces in the rock and whether or not fractures, such as joints, are present. A permeable rock therefore is one that permits the passage of water either through its pores (known as a porous rock) or along its joints (known as a pervious rock). An impermeable rock is one that will not absorb water or permit it to pass through.

permeable rock *see* permeability (2).

permeants *n.plu.* animals that move freely from one community or habitat to another.

permethrin *n.* contact insecticide used on fruit and vegetable crops to control caterpillars.

Permian *a. pert.* or *appl.* geological period lasting from about 286 to 248 million years ago.

permineralization *n.* type of fossilization in which minerals are deposited within spaces within the tissue.

perox *n.* a suborder of the order oxisol of the USDA Soil Taxonomy system, consisting of moist oxisols rich in humus.

peroxisome *n.* a small membrane-bounded organelle found in animal cells and containing catalase and peroxidases. In liver cells, it is believed to be important in detoxification reactions, e.g. of ethanol. *see also* glyoxysome. *alt.* microbody.

peroxyacetyl nitrate (PAN) $CH_3CO.O_2.NO_2$, one of the peroxyacyl nitrates (*q.v.*) found in Los Angeles-type smogs.

peroxyacyl nitrates (PANs) a class of organic compounds with the general formula $RCO.O_2.NO_2$. They are formed in Los Angeles-type smogs. They are respiratory irritants and lachrymators in humans and are highly damaging to plants, which are sensitive to concentrations of around 0.05 p.p.m. Known by either of the abbreviations PAN or PANs, the first of which is also used for a specific peroxyacyl nitrate, namely peroxyacetyl nitrate $(CH_3CO.O_2.NO_2)$.

perpetual resource resource, such as solar energy, that can be considered inexhaustible on a human timescale.

perseveration *n.* the persistence of a response after the original stimulus has ceased.

persistent *a. appl.* pesticides or other chemicals that are not degraded in the environment.

Personatae Scrophulariaceae *q.v.*

perthophyte *n.* a parasitic fungus that obtains nourishment from host tissues after having killed them by a poisonous secretion.

pervious rock *see* permeability (2).

pest *n.* any organism that reduces agricultural productivity via loss of soil fertility and/or direct damage to livestock or crops.

pesticide *n.* general term for any chemical agent that kills unwanted weeds, fungi or animal pests.

pesticide tolerance permissible legal level of pesticides in human or animal foods.

pest resurgence rapid regrowth of pest populations in the wake of pesticide application. If broad-spectrum pesticides are used, this resurgence may be facilitated by reduced competition and/or reduced predation.

petal *n.* part of a flower (*q.v.*), a modified sterile leaf, often brightly coloured and with others forming the corolla or the inner series of perianth segments, depending on the type of flower. *see* Fig. 12, p. 157.

petalite *n.* $LiAlSi_4O_{10}$, a lithium ore mineral that is a tectosilicate *q.v.*

petiole *n.* (1) (*bot.*) the stalk of a leaf; (2) (*zool.*) slender stalk connecting thorax and abdomen in certain insects, such as wasps; (3) any slender stalk-like structure.

Petri dish shallow circular glass or plastic dish used for growing microorganisms or cultured cells, etc., in a suitable medium. *alt.* plate.

petrifaction, petrification *n.* fossilization through replacement of organic material by minerals from solution, subsequently turning to stone.

petrochemicals *n.plu.* chemicals derived from crude oil (petroleum) or natural gas.

petrol *n.* a complex mixture that is used extensively as a transport fuel. It is a largely hydrocarbon (C5–C8) derivative of petroleum, with which other fuels and additives, e.g. tetraalkyl lead antiknock agents, are often blended. *alt.* gasoline, motor spirit, petroleum spirit.

petrolatum *n.* mixture of hydrocarbons in semisolid form (white or yellow in colour) produced by the fractional distillation of crude oil. *alt.* petroleum jelly.

petroleum crude oil *q.v.*

petroleum jelly petrolatum *q.v.*

petroleum spirit petrol *q.v.*

petrology *n.* the scientific study of the structure, composition and origin of rocks.

petrophyte *n.* a rock plant.

PFA pulverized fuel ash *q.v.*

PFJS polar front jet stream *q.v.*

PGRs plant growth regulators *q.v.*

pH a measure of the acidity of a solution, the negative $\log_{10}$ of the hydrogen ion (H^+) concentration (strictly the hydronium ion (H_3O^+) concentration). A neutral solution (at 25 °C) has a hydrogen ion concentration of 10^{-7} mol dm^{-3} and thus the pH of a neutral solution is 7. That of acid solutions is less than 7 and of alkaline solutions greater than 7.

Phaeophyta *n.* the brown algae or brown seaweeds. Mainly multicellular and marine, they can reach large sizes and some groups possess complex internal structure. Their green chlorophyll is masked by the brown pigment fucoxanthin so that they appear brownish. Carbohydrate reserves are in the form of laminarin.

phaeoplankton *n.* plankton living in the top 30 m of water.

phage bacteriophage *q.v.*

phagotroph *n.* any heterotrophic organism that ingests nutrients as solid particles. *a.* phagotrophic. *cf.* osmotroph.

phalloidin *n.* toxic alkaloid obtained from the toadstool *Amanita phalloides* that binds to actin filaments and prevents cell movement.

phanerogams *n.plu.* (1) all seed-bearing plants; (2) formerly used for plants with conspicuous flowers. *a.* phanerogamic.

phanerophyte *n.* tree or shrub.

Phanerozoic *n.* eon comprising the Paleozoic, Mesozoic and Cenozoic eras.

phaoplankton *n.* surface plankton living at depths at which light penetrates.

pharotaxis *n.* the movement of an animal towards a definite place, the stimulus for which is acquired by conditioning or learning.

phase-contrast microscopy type of optical microscopy that enables unstained living cells and tissue to be studied by using the way different parts of the cell diffract light to give a high-contrast image.

Phasmida *n.* order of insects including the stick insects, which are long slender insects with long legs, and the leaf insects, which have flattened bodies with leaf-like flaps on their limbs, both being excellently camouflaged in the bushes and trees in which they live.

Phe phenylalanine *q.v.*

phene *n.* a phenotypic character that is genetically determined.

phenetic *a. appl.* classification based purely on similarities in phenotypic characters, not necessarily reflecting relationships by evolutionary descent.

Phenoclor a trade name under which polychlorinated biphenyls (PCBs) were sold.

phenocontour *n.* (1) a contour line on a map showing the distribution of a certain phenotype; (2) a line connecting all places within a region at which a biological phenomenon, e.g. flowering of a plant, occurs at the same time; (3) a contour line delimiting an area corresponding to a given frequency of a variant form.

phenocopy *n.* a modification produced by environmental factors that simulates a genetically determined change.

phenocrysts *n.plu.* the large crystals within porphyritic (*q.v.*) rocks. *alt.* insets.

phenogram *n.* a tree-like diagram classifying organisms according to the conclusions of numerical taxonomy (*q.v.*). *cf.* cladogram. *see* Fig. 7, p. 79.

phenol *n.* C_6H_5OH, that one of the phenols (*q.v.*) that is derived from benzene. It is a poisonous solid (m.p. 41 °C) with a characteristic smell. *alt.* carbolic acid.

phenological *a.* (1) *pert.* phenology; (2) *appl.* isolation of species owing to differences in flowering or breeding season.

phenology *n.* the recording and study of periodic biological events, such as flowering, breeding, migration, etc., in relation to climate and other environmental factors.

phenols *n.plu.* a group of organic chemicals, all of which have at least one -OH group attached to a carbon of a benzene ring.

phenome *n.* all the phenotypic characteristics of an organism.

phenon *n.* group of organisms placed together by numerical taxonomy.

phenotype *n.* (1) the visible or otherwise measurable physical and biochemical characteristics of an organism, a result of the interaction of genotype and environment; (2) a group of individuals exhibiting the same phenotypic characters. *a.* phenotypic. *cf.* genotype, genotypic.

phenotypic plasticity the range of variability shown by the phenotype in response to environmental fluctuations.

phenyl *n.* the -C_6H_5 aryl radical derived from benzene.

phenylalanine (Phe, F) *n.* amino acid with an aromatic side chain, constituent of protein, essential in human diet.

pheromone *n.* a chemical released, usually in minute amounts, by one organism that is detected and acts as a signal to another member of the same species, such as the volatile sexual attractants released by some female insects that can attract males from a distance. Some pheromones act as alarm signals.

philopatry *n.* tendency of an organism to stay in or return to its home area.

philoprogenitive *a.* having many offspring.

phloem *n.* the principal food-conducting tissue of vascular plants, extending throughout the plant body. It is composed of elongated conducting vessels, sieve tubes (in angiosperms) or sieve cells (in ferns and gymnosperms), both containing clusters of pores (sieve areas) in the walls, through which the protoplasts of adjacent cells communicate. Sugars and amino acids are the main nutrients transported via the phloem. Parenchymatous companion cells (or albuminous cells in gymnosperms) closely associated with the conducting elements are involved in the delivery to and uptake of material from the phloem. Phloem also contains supporting fibres (bast). *see also* vascular bundle, xylem.

phlogopite *n.* $KMg_3(AlSi_3O_{10})(OH,F)_2$, one of the mica minerals *q.v.*

phocids *n.plu.* members of the Phocidae, comprising the seals.

Phoenicopteriformes *n.* an order of birds in some classifications, including the flamingoes.

pholadophyte *n.* a plant living in hollows, shunning bright light.

Pholidota *n.* an order of placental mammals known from the Pleistocene or possibly Oligocene, the only extant member being the pangolin (scaly anteater), having no teeth, and the body covered with overlapping scales.

phonation *n.* production of sounds, e.g., by insects.

phonolite *n.* a fine-grained, generally porphyritic, rock with a composition equivalent to that of nepheline syenite *q.v.*

phorate *n.* systemic organophosphate insecticide used, e.g., in the control of caterpillars, aphids, frit flies and wireworms.

phoresia, phoresy *n.* the carrying of one organism by another, without parasitism, as in certain insects.

Phoronida *n.* a small phylum (only 15 species) of marine worm-like coelomate animals that secrete chitinous tubes in which they live. The mouth is surrounded by a horseshoe-shaped crown of tentacles (a lophophore) that projects from the tube.

phosalone *n.* contact organophosphate insecticide and acaricide used, e.g., to control aphids and red spider mites on apples, and also as a sheep dip.

phosphamidon *n.* systemic organophosphate insecticide used to control sap-feeding insects such as aphids and rice stem-borers.

phosphate *n.* the anion PO_4^{3-} or a salt of phosphoric acid, H_3PO_4. They are essential to the metabolism of living organisms because inorganic phosphate is required for the synthesis of ATP. Plants and microorganisms take up phosphorus mainly in the form of phosphates, and various phosphates are used as fertilizers. Excess phosphate washed into streams and lakes contributes to eutrophication and the formation of algal blooms. *see also* phosphorus cycle.

phosphate rock rock, whether igneous or sedimentary, that contains a high proportion of phosphate minerals.

phospholipid *n.* any of a group of amphipathic lipids (*q.v.*) with either a glycerol or a sphingosine backbone, fatty acid side chains and a phosphorylated alcohol headgroup, which form the lipid bilayer in all biological membranes. They include the glycerolipids phosphatidylcholine, phosphatidylethanolamine, phosphatidylinositol and phosphatidylserine, and the sphingolipid sphingomyelin.

phosphorescence *n.* the luminescence of marine protozoans, copepods and the majority of deep-sea animals, which is produced without accompanying heat.

phosphorus (P) *n.* non-metallic element (at. no. 15, r.a.m. 30.97) that exists in nature only in the combined state, mainly as phosphates in minerals, e.g. apatite, and in organic matter. It is an essential nutrient for living organisms, and is included in combined fertilizers. *see also* organophosphorus compounds, phosphate, phosphorus cycle.

phosphorus cycle the movement of phosphorus within the biosphere and between the biosphere and the inorganic environment. Phosphorus released to the soil in the form of phosphates through the weathering of minerals is taken up by plants and microorganisms and from them passes into animals. Phosphorus is released back to the soil in animal wastes and by the decomposition of organic matter.

phosphorus fertilizers straight inorganic fertilizers that are used to supply crop plants with the essential element phosphorus (P). *see* basic slag, dicalcium phosphate, rock phosphate, super-

phosphate, superphosphoric acid, triple superphosphate.

photic zone that part of any water column that is illuminated by sufficient sunlight to produce a net growth of phytoplankton as a consequence of photosynthesis.

photo- prefix derived from Gk *phōs*, light, indicating response to, sensitivity to or causation by light.

photoassimilate *n.* the carbon-containing compounds produced as a result of photosynthesis.

photoautotroph *n.* an organism that uses light as an energy source and carbon dioxide as the main source of carbon, such as green plants and some bacteria. *alt.* photolithotroph.

photobiology *n.* the field of biology dealing with the effects of light on organisms.

photochemical *a. appl.* and *pert.* chemical changes brought about by light.

photochemical smog Los Angeles-type smog *q.v.*

photodynamics *n.* the study of the effects of light on plants.

photofission *n.* nuclear fission (*q.v.*) in response to bombardment with gamma-ray photons.

photogenic *a.* (1) light-producing; (2) luminescent.

photoheterotroph *n.* an organism that uses light as a source of energy, but derives much of its carbon from organic compounds, such as the photosynthetic nonsulphur purple bacteria.

photokinesis *n.* a kinesis (random movement) in response to light.

photolithotroph photoautotroph *q.v.*

photolysis *n.* the splitting of a compound or molecule by the action of light, as the splitting of water into hydrogen and oxygen.

photomorphogenesis *n.* any effect on plant growth produced by light.

photon *n.* the particulate unit of electromagnetic radiation (e.g. light) carrying a quantum of energy. 1 mol photons (or 1 mol quanta) is the number of photons corresponding to Avogadro's number of particles (6.023×10^{23}) and is the number of photons required to convert 1 mol of a substance to another form with 100% efficiency if captured in a single step.

Photon number incident on a surface normal to the beam in a given time is the photon flux (often called photon flux density, but this is not recommended) and is measured in mol per metre squared per second.

photonasty *n*. response of plants to diffuse light stimuli, or to variations in illumination.

photoperiod *n*. (1) duration of daily exposure to light; (2) the length of day favouring optimum functioning of an organism.

photoperiodicity, photoperiodism *n*. the response of an organism to the relative duration of day and night, such as the flowering of many plants and mating of many animals, which is triggered by the lengthening or shortening of the days as the seasons change.

photophilous *A*. seeking, and thriving in, strong light.

photophobic *A*. not tolerating light, shunning light.

photophosphorylation *n*. the formation of ATP using energy from light during photosynthesis. *cf*. oxidative phosphorylation.

photoplagiotropy *n*. tendency to take up a position transverse to the incident light.

photoregulation *n*. regulation by light.

photorespiration *n*. type of 'wasteful' respiration occurring in green plants in the light, different from normal mitochondrial respiration, consuming oxygen and evolving carbon dioxide using chiefly glycolate derived from the primary photosynthate as substrate, occurring to a much greater extent in C3 plants (*q.v.*) than in C4 plants *q.v.*

photosynthate *n*. product(s) of photosynthesis.

photosynthesis *n*. (1) in green plants, the synthesis of carbohydrate in chloroplasts from carbon dioxide as a carbon source and water as a hydrogen donor with the release of oxygen as a waste product, using light energy trapped by the green pigment chlorophyll to drive the synthesis of ATP and NADPH (the light reaction). These energy carriers are subsequently used as energy sources in carbohydrate synthesis utilizing atmospheric CO_2 (the dark reaction); (2) in bacteria, a similar process but sometimes using hydrogen donors other

than water and producing waste products other than oxygen. *a*. photosynthetic. *see also* Calvin cycle.

photosynthetically active radiation (PAR) radiation capable of driving the light reactions of photosynthesis, wavelength 380–710 nm.

photosynthetic carbon reduction cycle (PCR cycle) Calvin cycle *q.v.*

photosynthetic efficiency the conversion factor of the energy falling per unit area on a photosynthetic tissue and the energy value of the biochemical compounds produced.

photosynthetic quotient ratio between the volume of oxygen produced and the volume of carbon dioxide used in photosynthesis.

photosynthetic reaction centre *see* reaction centre.

photosynthetic unit proposed functional unit composed of several hundred chlorophyll molecules, a reaction centre and accessory pigments, which is required to generate one oxygen molecule in photosynthesis.

photosynthetic zone of sea or lakes, the vertical zone in which photosynthesis can take place, between surface and compensation point.

photosystem I (PSI), photosystem II (PSII) the multimolecular light-capturing and electron-transporting complexes present in the thylakoid membranes of chloroplasts and involved in the light reactions of photosynthesis.

phototaxis *n*. movement in response to light. Positive phototaxis is movement towards a light source while negative phototaxis is movement away from a light source. *a*. phototactic.

phototroph *n*. organism using sunlight as a source of energy. *a*. phototrophic.

phototropism *n*. growth movement of plants in response to stimulus of light. Where the stimulus is sunlight, sometimes called heliotropism.

photovoltaic cell device used to capture light energy, esp. solar radiation, and directly convert it into electricity. In the context of solar energy technology, these devices are also called solar cells.

photovoltaic conversion the direct conversion of solar radiation into electricity.

Phragmobasidiomycetes, Phragmobasidiomycetidae *n.* group of basidiomycete fungi that form basidiospores on a septate basidium, and that include the rust and smut fungi.

phragmosis *n.* the use of part of the body to close a burrow (in reptiles or amphibians).

phreatic eruptions gigantic volcanic explosions caused when water (seawater, fresh water or groundwater) enters a superheated magma chamber.

phreaticolous *a. appl.* organisms living in underground fresh water.

phreatic zone zone of saturation *q.v.*

phreatophyte *n.* plant with very long roots that reach the water table.

phthisis *see* tuberculosis.

phyco- prefix derived from Gk *phykos*, seaweed, signifying to do with algae.

phycobiliproteins *n.* protein pigments with phycobilin chromophores, e.g. phycoerythrin and phycocyanin, found in some algae and cyanobacteria, where they act as accessory photosynthetic pigments.

phycobiont *n.* the algal partner in a symbiosis, e.g., in lichens and in certain marine invertebrates.

phycology *n.* the study of the algae.

Phycomycetes *n.plu.* a general name for the simple, mainly aquatic fungi of the classes Chytridiomycetes, Hyphochytridiomycetes, Plasmodiophoromycetes and Oomycetes, now generally classified as protists.

phycophaein *n.* a brown pigment in brown algae, thought to be an oxidation product of fucosan.

Phycophyta *n.* in some classifications the name for the algae.

phyla *plu.* of phylum *q.v.*

phylacobiosis *n.* mutual or unilateral protective behaviour, as of certain ants. *a.* phylacobiotic.

Phylactolaemata *n.* in some classifications a class of freshwater Bryozoa.

phyletic *a.* (1) *pert.* a phylum or a major branch of an evolutionary lineage; (2) *appl.* a group of species related to each other by common descent; (3) *pert.* a line of direct descent.

phyletic evolution sequence of evolutionary changes leading to a sequence of species or forms arising through time in a single line of descent.

phyletic gradualism the idea that evolutionary change is built up in small steps, the change in any single generation being extremely small. Each stage must have a selective advantage, how ever marginal, eventually giving rise to new forms, organs and functions by a cumulative effect. The term phyletic gradualism is sometimes also used to refer to the view that evolution proceeds at a steady rate by such imperceptibly small changes, and in this usage is often contrasted with the idea of punctuated equilibria and a non-uniform rate of evolution.

phyllite *n.* a fine- to medium-grained rock produced by low-grade regional metamorphism (*q.v.*) from pelitic rocks (*q.v.*). It has pronounced schistosity (*q.v.*), and cleavage surfaces have a lustrous sheen. It is typically greenish or greyish in colour, due to the presence of chlorite and/or muscovite.

phyllode *n.* winged leaf-stalk with flattened surfaces placed laterally to stem, functioning as a leaf.

phyllophagous *a.* feeding on leaves.

phylloplane *n.* the leaf surface.

phyllosilicate *n.* any silicate mineral that contains silica tetrahedra that are joined to each other via the sharing of three oxygens per tetrahedron, thereby forming sheets of silica tetrahedra.

phyllotaxis, phyllotaxy *n.* the arrangement of leaves on an axis or stem, which for spirally arranged leaves may be expressed as the number of circuits of the stem that have to be made and the number of leaves that have to be passed to progress from the point of attachment of one leaf to that immediately above it, e.g. 1/2 for leaves positioned 180° apart.

phylogenetic *a.* (1) *pert.* the evolutionary history and line of descent of a species or higher taxonomic group; (2) *appl.* tree, a diagram showing the evolutionary relationship of taxa to each other.

phylogenetics *n.* (1) the line of descent of a species or higher taxon; (2) approach to classification that attempts to reconstruct evolutionary genealogies and the historical course of speciation.

phylogeny *n.* the evolutionary history and line of descent of a species or higher taxonomic group. *cf.* ontogeny.

phylum *n.* in classification, the highest taxonomic grouping within the animal kingdom, a primary grouping consisting of animals constructed on a similar general plan, and thought to be evolutionarily related. In plants, the similar category is called a division. Examples: Cnidaria (sea anemones, jellyfish, corals, etc.), Porifera (sponges), Platyhelminthes (flatworms, flukes and tapeworms), Mollusca (molluscs), Arthropoda (spiders, insects, crustaceans) and Chordata (includes the vertebrates). *plu.* phyla. *see* Appendix 7.

physical containment in genetic engineering and in microbiology generally, the level of physical security and safety required in laboratory procedures, different levels being recommended for work involving microorganisms of differing degrees of pathogenicity.

physical quality of life index (PQLI) parameter based on a mean value of three indices (each spanning 0–100), one for each of life expectancy, child mortality and literacy. The PQLI is used to indicate the physical well-being of a country's population, with a value of 77 taken as the benchmark below which the basic human requirements are not being met.

physical weathering the disintegration of rocks *in situ* by mechanical stresses imposed by physical changes. The process of physical weathering always incorporates changes of volume, either in the rock mass itself (*see* insolation weathering) or of exogenous material that has entered fissures in the rock (*see* frost weathering and salt weathering). *alt* mechanical weathering.

physiographic *a. pert.* geographical features of landforms.

physiographic succession plant succession influenced mainly by topography and local climate.

physiological longevity maximum expected life span in the absence of a fatal disease or accident.

physiological races outwardly similar races within a species that differ in their physiology as a result of genetically determined factors, as exemplified by some plant pathogenic fungi that differ in their virulence towards different varieties of the host species.

physiology *n.* that part of biology dealing with the functions and activities of organisms, as opposed to their structure. *a.* physiological.

physisorption *see* adsorption.

physostigmine *n.* an alkaloid derived from the Calabar bean, an inhibitor of the enzyme acetylcholinesterase. *alt.* eserine.

phytal zone shallow lake bottom and its rooted vegetation.

phyto- prefix derived from Gk *phyton*, plant.

phytoactive *a.* stimulating plant growth.

phytoalexin *n.* any of a group of substances produced by plant cells in response to wounding or attack by parasitic fungi or bacteria and that are involved in resistance to infection and in limiting the damage caused by accidental wounding.

phytobiotic *a.* living within plants, *appl.* certain protozoans.

phytochemistry *n.* the chemistry of plants.

phytochory *n.* dissemination of pathogens through the agency of plants.

phytochrome *n.* a light-sensitive protein pigment in plants. It exists in two forms: P_r, which is sensitive to red light, which converts it into P_{fr}, which is sensitive to far-red light. Far-red light converts P_{fr} back to P_r. P_{fr} is active in stimulating developmental processes such as flowering in short-day plants and germination in some seeds, whereas P_r is inactive.

phytocoenology *n.* the study of plant communities.

phytocoenosis *n.* the assemblage of plants living in a particular locality.

phytogenesis *n.* the evolution, or development, of plants.

phytogenetics *n.* plant genetics.

phytogenous *a.* of vegetable origin, produced by plants.

phytogeocoenosis *n.* the vegetative parts of an ecosystem.

phytogeographical kingdoms one way in which the world has been divided into major floral geographical areas. The kingdoms are: Antarctic, Australian, Neotropical, Boreal, Palaeotropical and South African (*see* individual entries). *see also* floral realm.

phytogeography *n.* the study of the geographical distribution of plant species.

phytomass *n.* plant biomass *q.v.*

Phytomastigophora *n.* flagellates having many plant characteristics, including the presence of chloroplasts.

phytome *n.* plants considered as an ecological unit.

phytoparasite *n.* any parasitic plant.

phytopathology *n.* the study of plant diseases.

phytophage *n.* animal that feeds on plants, usually used for the smaller sap-sucking and leaf-eating insects, etc., rather than the larger herbivores. *a.* phytophagous.

Phytophthora infestans fungus responsible for potato blight *q.v.*

phytophysiology *n.* plant physiology.

phytoplankton *n.* all photosynthetic plankton, including, e.g., unicellular algae and cyanobacteria.

phytoreovirus group one of the two genera of plant viruses of the family Reoviridae, the other being the fijiviruses, containing isometric double-stranded RNA viruses that are considered to be insect viruses that are becoming adapted to plants. The genome is composed of 12 separate RNAs.

Phytoseiulus persimilis a predatory mite used in the control of glasshouse red spider mite (*Tetranychus urticae*).

phytosociology *n.* the study of all aspects of the ecology of plants and the influences on them.

phytosuccivorous *a.* living on plant juices.

phytotoxic *a.* toxic to plants.

phytotoxin *n.* any toxin originating in a plant.

phytotron *n.* a large chamber in which a controlled environment can be created to study plant growth and development.

phytotrophic autotrophic *q.v.*

phytotype *n.* representative type of plant.

P_i inorganic phosphate, orthophosphate *q.v.*

pica *n.* in humans, an abnormal craving to eat substances not suitable as food, e.g. soil, chalk and paint.

Piciformes *n.* an order of birds including the woodpeckers.

pick-your-own (PYO) type of horticulture, particularly suited to soft fruits, such as raspberries and strawberries, in which members of the public pick the produce, which they then purchase, directly from plants grown by a producer.

Picornaviridae, picornaviruses *n., n.plu.* family of small, non-enveloped, single-stranded RNA viruses including poliomyelitis, the rhinoviruses, which cause common colds, and the aphthoviruses, which include foot and mouth disease.

piedmont *n.* lowland plain at the base of a mountain range.

piedmont glaciers glaciers that form on lowlands or at the foot of mountains when several valley glaciers (*q.v.*) emerge and coalesce.

piedmontite, piemontite *n.* $Ca_2(AlFeMn)_3$-$Si_3O_{12}(OH)$, a rare mineral found in some manganese ores and low-grade schists.

piezoelectric *a.* becoming electrically polarized when subjected to mechanical stress.

pigment *n.* colouring matter in plants and animals.

pigmentation *n.* disposition of colouring matter in an organ or organism.

pillars *see* speleothems.

pillow lava balloon-like masses of extrusive igneous rock formed when molten lava enters water, characteristic of submarine eruptions.

Pinaceae, pines *n., n.plu.* coniferous trees and shrubs that bear their needles in bunches.

'pinch' fusion an approach to controlled nuclear fusion (*q.v.*) in which, effectively, two concentric plasmas are established for a brief split second by the passage of a large current through them both. The current in the outer plasma generates a magnetic field that 'pinches' the inner plasma, while the massive current through the inner plasma makes its temperature rise rapidly, causing ignition.

pincushion gall reddish thread-like growth found on roses and caused by larvae of the gall wasp *Diplolepis rosae*.

pinging, pinking knocking *q.v.*

pingo *n.* an isolated hill composed of silt, clay or gravel with a single ice mass at its core, found in periglacial environments.

pinnate *a.* (1) divided in a feathery manner; (2) having lateral processes; (3) (*bot.*) *appl.* a compound leaf having leaflets on each side of an axis or midrib.

Pinnipedia, pinnipeds *n.*, *n.plu.* order of mammals including the seals, walruses and sealions.

Pinophyta the gymnosperms *q.v.*

pioneer community the organisms that establish themselves on bare ground at the start of a primary succession.

pioneer species species that are the first to colonize bare ground, usually algae, lichens and mosses.

pipe drains tile drains *q.v.*

Piperales *n.* order of woody and herbaceous dicots comprising the families Piperaceae (the spice pepper) and Saururaceae (lizard's tail).

piperidine *n.* an alkaloid obtained from pepper, *Piper nigrum.*

pirimicarb *n.* contact carbamate insecticide of short environmental persistence used to control aphids on numerous agricultural and horticultural crops.

pirimiphos-methyl *n.* contact organophosphate pesticide used, both as an insecticide and an acaricide, against field and stored-product pests.

piroplasms, piroplasmas *n.plu.* parasitic protozoa of the class Piroplasmea, which infect red blood cells. *Babesia* is the causal agent of red water fever of cattle. Piroplasms are characterized by an intracellular stage in which the parasite is not contained in a vacuole, but has its cell membrane in direct contact with the cytoplasm of the host cell.

Pisces *n.* the fishes *q.v.*

piscicolous *a.* living in fish.

pisciculture *n.* fish farming.

piscivorous *a.* fish-eating, living entirely or mainly on fish.

pistil *n.* the female reproductive organ of a flower, which may be the carpels collectively when fused into a single structure or each carpel with its stigma and style in flowers with carpels separate.

pistillate *a.* (1) bearing pistils; (2) *appl.* a flower bearing pistils but no stamens. *cf.* staminate.

pitchblende *n.* the massive form of the major uranium ore mineral uraninite (UO_2).

pitchstone *n.* a volcanic glass that is chemically equivalent to rhyolite. Contains abundant phenocrysts of feldspar and quartz.

pithecanthropines *n.plu.* fossil hominids formerly placed in the genus *Pithecanthropus* and now considered to be members of the species *Homo erectus.*

placental mammals mammals that develop a persistent placenta, as in eutherians. *cf.* marsupials.

placer deposits economically valuable deposits of dense minerals, e.g. gold or cassiterite (a tin ore, SnO_2), formed when the minerals concerned are released from their source rocks by weathering and become concentrated under the influence of gravity, generally by the action of moving water (rivers and/or waves).

Placodermi, placoderms *n.*, *n.plu.* an extinct class of Early Devonian–Early Carboniferous primitive jawed fish, with archaic jaw suspension, crushing dental plates and bony dermal plates on the head and thorax.

Placodontia, placodonts *n.*, *n.plu.* extinct order of fully aquatic Triassic marine reptiles, having a short armoured body, some with a turtle-like carapace.

Placozoa *n.* phylum of extremely simple metazoans consisting of a single known species, *Trichoplax adhaerens*, a flattened sac-like organism with a fluid-filled internal cavity, cilia covering the body surface, and lacking differentiated tissues, organs, or any discernible head, tail or bilateral symmetry.

plaggen epipedon plaggen soil *q.v.*

plaggen soil man-made surface soil layer, greater than 50 cm thick, formed by the prolonged deposition of manure and other organic material, such as seaweed, and found in association with old settlements. *alt.* plaggen epipedon.

plaggept *n.* a suborder of the order inceptisol of the USDA Soil Taxonomy system consisting of inceptisols with an artificial surface layer (>50 cm thick), formed, e.g., by repeated manuring.

plagioclase *n.* $NaAlSi_3O_8$–$CaAl_2Si_2O_8$, a solid-solution series that constitutes part of the feldspar minerals (*q.v.*). Within this series, there is a continuous progression from albite ($NaAlSi_3O_8$), via oligoclase, andesine, labradorite and bytownite to anorthite ($CaAl_2Si_2O_8$).

plagioclimax *n.* subclimax stage in plant succession that persists as a result of animal or human interference, e.g. heather moorlands. *alt.* biotic climax.

plagiosaurs *n.plu.* an advanced order of amphibians, existing from Permian to Triassic times, individuals having very wide flat bodies in advanced forms, and a body armour of interlocking plates.

plagiosere *n.* an ecological succession deviating from its natural course as a result of continuous human intervention.

plagiotropism *n.* growth tending to incline a structure from the vertical plane to the oblique or horizontal as in lateral roots and branches. *a.* plagiotropic.

plague *n.* a disease caused by the bacterium *Yersinia* (*Pasteurella*) *pestis*, which occurs in bubonic and pneumonic forms, and is spread to humans by infected fleas of black rats. Between 1348 and 1350, about 25% of the population of Europe is thought to have been killed by bubonic and pneumonic plague. *alt.* Black Death.

planarian *n.* any member of the order Tricladida, free-living flatworms (platyhelminths) living in streams, ponds, lakes and the sea, having a broad, flattened body, well-developed sense organs at the anterior end and an intestine with three main branches.

planation surface area of relatively flat relief denuded as a result of fluvial or marine erosion. *alt.* erosion surface.

Planck's constant the fundamental constant, h, that relates the energy of a photon (E) to its frequency (v), thus: $E = hv$. The Planck constant has a value of $6.626\ 196 \times 10^{-34}$ J s.

plankter *n.* an individual planktonic organism.

planktivorous *a.* feeding on plankton.

planktohyponeuston *n.* aquatic organisms that gather near the surface at night but spend their days in the main water mass.

plankton *n.* the usually small marine or freshwater plants (phytoplankton) and animals (zooplankton) drifting with the surrounding water.

planktotrophic *a.* feeding on plankton.

Plantae *n.* the plant kingdom, in most modern classifications comprising the algae, bryophytes (mosses and liverworts), seedless vascular plants (ferns, club mosses, horsetails) and seed plants (gymnosperms and angiosperms). In older classifications, the fungi and even bacteria are also included, but these groups are now placed in separate kingdoms. The algae are sometimes placed in the Protista. *see* Appendices 5 and 8.

plantain *see Musa.*

plantation *n.* (1) a large estate in the subtropics or tropics where tree species are cultivated to produce cash crops such as timber, rubber or cocoa; (2) stand of planted forest.

plant-available *a. appl.* nutrients in soil that are in a form readily available for plant use.

plant genetic resources the genetic diversity present within plants, usually *appl.* crop plants and their wild relatives, and that is of potential use in plant breeding.

plant growth regulators (PGRs) *n.plu.* natural or synthetic substances used in the cultivation of ornamental crops, horticultural crops, and, more latterly, arable crops, to control certain aspects of plant growth. Chlormequat, for example, is widely used on cereal crops to encourage the development of short, strong stems.

plant hormone a substance produced by a plant that regulates its growth and development. Plant hormones include auxins, ethylene and gibberellins.

plantigrade *a.* walking with the whole sole of the foot touching the ground.

plant pathogen any agent (bacterium, fungus, virus or mycoplasma) causing disease in plants.

plant pathology the study of the diseases of plants.

plant rhabdovirus group single-stranded RNA plant viruses of the family Rhabdoviridae, e.g. sonchus yellow net virus, resembling the animal and insect members of the family, having bullet-shaped or rod-shaped particles with a lipid outer coat in which glycoproteins are embedded. They are considered to be insect viruses that are becoming adapted to plants.

plant viruses viruses that infect plants.

plasma *n.* (1) (*biol.*) the liquid part of body fluids, such as blood, milk or lymph, as opposed to suspended material, such as cells and fat globules; (2) (*phys.*) *see* states of matter.

plasmalemma plasma membrane *q.v.*

plasma membrane membrane (*q.v.*) bounding the surface of all living cells, formed of a fluid lipid bilayer in which proteins carrying out the functions of enzymes, ion pumps, transport proteins, receptors for hormones, etc., are embedded. It regulates the entry and exit of most solutes and ions, few substances being able to diffuse through unaided. *alt.* cell membrane, plasmalemma.

plasmid *n.* small circular DNA replicating independently of the chromosome in bacteria and unicellular eukaryotes such as yeasts, which is maintained at a characteristic stable number from generation to generation. Plasmids typically carry genes for antibiotic resistance, colicin production, the breakdown of unusual compounds, etc. They are widely used in genetic engineering as vectors into which foreign genes are inserted for subsequent cloning or expression in bacterial cells.

plasmodial slime moulds acellular slime moulds *q.v.*

Plasmodiophoromycetes *n.* a class of simple fungi, obligate endoparasites of vascular plants, algae and fungi, and including *Plasmodiophora brassicae*, the cause of clubroot in cabbages and other brassicas. They grow as a plasmodium inside the host cell and reproduce by motile spores and gametes.

plasmodium *n.* (1) in the plasmodial slime moulds, a multinucleate, fan-shaped mass of streaming protoplasm without a cell wall that may cover several square metres and that forms the non-reproductive stage of the organism; (2) parasitic protozoan of the genus *Plasmodium*, e.g. the causal agent of malaria *Plasmodium falciparum*. *a.* plasmodial.

plasmolysis *n.* the withdrawal of water from a plant cell by osmosis if placed in a strong salt or sugar solution, resulting in contraction of cytoplasm away from the cell walls.

plaster of Paris powdered $CaSO_4.O.5H_2O$ made by heating gypsum (*q.v.*). It sets on mixing with water.

plastid *n.* a cellular organelle containing pigment, esp. in plants, e.g. chloroplast.

plate boundaries (*geol.*) boundaries where lithospheric plates meet. *see* convergent boundary, divergent boundary, transform boundary.

plate count method of measuring the number of bacteria or other microorganisms in a sample by counting the number of colonies appearing on a nutrient plate spread with a small amount of diluted sample.

plate tectonics a widely accepted theory that the Earth's surface is divided into a series of essentially rigid plates of lithosphere and that the plastic nature of the underlying asthenosphere allows these plates to move relative to one another. *see also* continental drift.

plateau *n.* extensive, elevated tract of land with an essentially horizontal surface. *plu.* plateaus, plateaux. *alt.* tableland.

plateau lava a lava flow that forms a plateau (*q.v.*) of large area (>1000 km² to *ca.* 300 000 km²).

platinum (Pt) *n.* metallic element (at no. 78, r.a.m. 195.09), forming a silvery-white metal. It occurs naturally as the native element and in heavy metal sulphide ores.

platinum group metals a collective term for the chemical elements ruthenium (Ru), rhodium (Rh), palladium (Pd), osmium (Os), iridium (Ir) and platinum (Pt), which share similar properties. *alt.* platinum metals.

platy *a. appl.* soil crumbs in which the vertical axis is shorter than the horizontal.

Platyhelminthes, platyhelminths *n., n.plu.* a phylum of multicellular, acoelomate animals, commonly called flatworms, which are flattened dorso-ventrally, are bilaterally symmetrical, and have the epidermis (ectoderm) and gut separated by a solid mass of tissue. They include the free-living Turbellaria and the parasitic Monogea (skin and gill flukes), Trematoda (gut, liver and blood flukes) and Cestoda (tapeworms).

platypus *n.* the duck-billed platypus of Australia, *see* Monotremata.

platyrrhines *n.plu.* the New World monkeys.

play *n.* behaviour exhibited esp. by young animals in which they explore the environment and learn by trial and error during the time when life is fairly easy for them.

playa *n.* an area of salt-encrusted ground representing the dry phase of an ephemeral playa lake. *alt.* salt flat.

playa lake a shallow ephemeral saline lake. *alt.* alkali lake.

Plecoptera *n.* order of insects commonly called stoneflies, similar in many respects to mayflies (order Ephemeroptera), but with two 'tails' and hind wings larger than fore wings.

Plectomycetes *n.* a group of ascomycete fungi, commonly called the blue, green and black moulds from the colour of their asexual spores (conidia), and generally bearing asci in closed ascocarps (cleistothecia). The asexual (imperfect) stages of many plectomycetes are similar to those of *Aspergillus* and *Penicillium*.

pleiomorphic pleomorphic *q.v.*

pleioxenous *a.* living on more than one host during life cycle, *appl.* parasites.

Pleistocene *n.* the glacial and post-glacial epoch following the Tertiary, lasting from around 2 million to 10 000 years ago.

pleomorphic *a.* (1) being able to change shape; (2) existing in different shapes at different stages of the life cycle. *see also* polymorphic.

plesiobiotic *a.* (1) living in close proximity, *appl.* colonies of ants of different species; (2) building contiguous nests, *appl.* ants and termites. *n.* plesiobiosis.

plesiomorphic *a.* in cladistics, *appl.* the original pre-existing member of a pair of homologous characters.

plesiomorphous *a.* having a similar form.

Plesiosauria, plesiosaurs *n., n.plu.* order of Mesozoic reptiles that were fully aquatic with barrel-shaped bodies and paddle-shaped limbs.

pleurilignosa *n.* rain forest.

pleurodont *a. appl.* teeth that are attached to the inside surface of the jaw, as opposed to the outer edge (acrodont) or in sockets (thecodont).

Pliocene *n.* the geological epoch that followed the Miocene and preceded the Pleistocene, lasting from around 5 to 2 million years ago.

ploidy *n.* the number of sets of chromosomes in a cell, e.g. haploid, diploid, triploid, tetraploid.

plotophyte *n.* a plant adapted for floating.

plough layer Ap-horizon *q.v.*

plough pan compacted soil layer formed beneath the plough layer as a result of the use of heavy agricultural machinery, esp. in wet weather.

plucking *n.* type of glacial erosion in which chunks of bedrock are removed by the glacier as it flows forwards. *alt.* quarrying.

Plumbaginales *n.* order of dicot herbs or small shrubs that comprises the family Plumbaginaceae (e.g. leadworts sea lavender, thrifts), which are xerophytes or halophytes of steppes, semideserts and sea coasts.

plume *n.* a stream of fluid waste, after emission but prior to its dispersion in the environment, while it still has discernible boundaries, e.g. smoke freshly emitted from a stack. Hot water from a power station immediately after discharge into a river is known as a thermal plume,

plunge line (*oceanogr.*) the point at which waves break. This occurs when wave height and water depth are roughly equal.

pluton *n.* a body of plutonic rock *q.v.*

plutonic rocks *n.plu.* intrusive igneous rocks formed at great depths. These are characterized by exceptionally large crystals formed as a result of the very slow rate at which the magma (*q.v.*) cools. *alt.* abyssal rocks. *see* batholith.

plutonium (Pu) *n.* radioactive element (at. no. 94, r.a.m. 244), synthesized from ^{238}U in nuclear reactors and used as a nuclear fuel. All isotopes are radioactive. A critical mass of ^{239}Pu can initiate a nuclear explosion.

plutonium-239 an isotope of plutonium. It is an alpha emitter with a radioactive half-life of 24 390 years. It is considered to be one of the most toxic materials known. The maximum tolerated dose is estimated to be $<1 \times 10^{-6}$ g per person.

pluvials *n.plu.* time periods during which land-based moisture levels (including lakes and rivers) were heightened due to decreased evapotranspiration and/or increased precipitation levels. Pluvials used to be considered to be necessarily simultaneous with glacials (*q.v.*) but there is evidence that they also occurred during interglacials, e.g. during the Holocene interglacial.

Pm symbol for the chemical element promethium *q.v.*

pneumatolytic deposition hydrothermal deposition (*q.v.*) that occurs above *ca.* 400 °C.

pneumoconiosis *n.* any disease of the bronchi or lungs resulting from the inhalation of airborne particles of minerals or metals, and characterized by coughing and the inflammation and fibrosis of the respiratory tissues. *see also* asbestosis, silicosis.

pneumotaxis *n.* reaction to stimulus consisting of a gas, esp. carbon dioxide in solution.

Po symbol for the chemical element polonium *q.v.*

poaching *n.* (1) illegal commercial hunting or fishing; (2) (*agric.*) the creation of potholes in the sward in wet weather by the trampling action of livestock.

poad *n.* a meadow plant.

Poales *n.* order of herbaceous monocots, the grasses, comprising the family Poaceae (Graminae).

pod *n.* (1) legume (*q.v.*); (2) a husk; (3) the cocoon in which eggs are laid in locusts; (4) a school of fish in which the bodies of the individuals actually touch; (5) a group of whales.

Podicipediformes *n.* an order of small water birds with unwebbed feet, including the grebes.

Podostemales *n.* an order of dicots living in fast-flowing water and on rocks in rivers, with often filamentous or ribbon-like leaves and stems. Comprises the family Podostemaceae.

podzol, podsol *n.* grey forest soil, found in cold temperate regions, and formed on heathlands and under coniferous forest by podzolization (*q.v.*). A podzol is characterized by the presence of mor humus (*q.v.*), a pale-coloured leached E-horizon (eluvial horizon low in organic matter) and an accumulation of iron, aluminium and humic material in the B-horizon. *see* Fig. 28, p. 385.

podzolization, podsolization *n.* the leaching of soluble complexes (formed by chelation) of aluminium and iron(II) from the A-horizon and the subsequent deposition of these metals, together with organic matter, in the B-horizon. This process leads to the formation of a soil called a podzol *q.v.*

Pogonophora, pogonophorans *n.*, *n.plu.* phylum of marine sessile worm-like invertebrates, with similarities to hemichordates, which live in chitin tubes. *alt.* beard worms. *see also* Vestimentifera.

poikilitic *a. appl.* an igneous rock texture (*q.v.*) in which small mineral crystals are enclosed within larger crystals of another mineral. *see also* ophitic.

poikiloblastic *a. appl.* the texture of those metamorphic rocks that have a relatively fine-grained matrix within which are embedded mineral crystals (called poikiloblasts) that themselves contain many grains of another mineral or minerals.

poikiloblasts *see* poikiloblastic.

poikilohydrous *a. appl.* (1) plants that cannot regulate their rate of water loss and thus contain much the same amount of water as their environment; (2) plants becoming dormant in the dry season after losing most of their water.

poikilosmotic *a.* having internal osmotic pressure varying with that of the surrounding medium.

poikilothermic *a. appl.* animals whose body temperature varies with that of the surrounding medium. Although most poikilotherms are also ectothermic (*q.v.*), the two terms are not synonyms and describe different aspects of thermoregulation. *n.* poikilothermy. *cf.* homoiothermic.

point bar a curved deposit laid down on the convex inside of a river channel bend.

point source of water pollution, an identifiable source of concentrated pollution discharge. Such sources include sewage outfalls and waste pipes from factories. *alt.* discrete source. *cf.* non-point source.

polar *a.* (1) *pert.* or *appl.* the North or South Poles and the regions surrounding them; (2) (*chem.*) *appl.* covalent bonds and molecules that have an uneven distribution of charge, *cf.* non-polar.

polar cell cell of primary atmospheric circulation system that operates in each hemisphere between the upper mid-latitude low-pressure belt (60°) and the pole itself (area of high pressure). *see* Fig. 32, p. 419.

polar continental air masses cold, dry, stable air masses that develop over Siberia, northern Canada, the Arctic and the Antarctic during the winter.

polar easterlies cold winds that blow, in each hemisphere, from the high pressure area over the pole itself towards the upper mid-latitude low-pressure belt (60°). During transit, these winds become easterly under the influence of the Coriolis force. They constitute the low-altitude section of the polar cell (*q.v.*). *alt.* easterlies. *see* Fig. 32, p. 419.

polar front (1) a major front, which occurs in each hemisphere, where the warm mid-latitude westerlies meet the cold polar easterlies at the junction between the Ferrel and polar cells, *see* Fig. 32, p. 419; (2) sometimes refers specifically to the polar front of the northern hemisphere.

polar front jet stream (PFJS) jet stream located at the boundary between the Ferrel cell and the polar cell. With a strong (discontinous) westerly air flow, this jet stream (which occurs in both hemispheres, between 40° and 60° latitude) is associated with the generation and movement of mid-latitude depressions.

polar high large area of high pressure located over each pole.

polar maritime air masses cool, moist, unstable air masses that develop at high latitudes over the Atlantic and Pacific Oceans.

polar molecule a molecule with an overall electric dipole. For example, hydrogen chloride is made up of polar molecules, each one, i.e. H–Cl, carries a partial positive charge on the H atom and an equal but opposite partial negative charge on the Cl atom.

polar orbit *see* polar satellite.

polar satellite artificial satellite placed at an altitude of between 500 and 1500 km, which orbits the Earth longitudinally about once every 90 min. During each orbit, the satellite crosses the equatorial plane at *ca.* 90° and passes directly over or close to both poles. This type of orbit is known as a polar orbit.

polar stratospheric clouds (PSCs) clouds of both nitric acid trihydrate ($HNO_3.3H_2O$) and more dilute nitric acid,

formed in the stratosphere at temperatures lower than −80 °C.

polder *n.* an area of land formerly submerged beneath sea, lake or river water that has been reclaimed by artificial drainage and is protected by dikes. Polders are generally used for agricultural purposes.

Polemoniales *n.* an order of dicot trees, shrubs, vines and herbs, including the families Boraginaceae (borage), Polemoniaceae (phlox), Convolvulaceae (morning glory), Cuscutaceae (dodder) and others.

poliomyelitis *n.* viral infection of the spinal cord and brain caused by poliovirus, an enterovirus, usually transmitted by personal contact. *alt.* polio.

polje *n.* a large flat-floored depression, which can cover in excess of 200 km^2, found in karst regions. The alluvium-covered floors of these depressions provide fertile agricultural land. Streams that flow onto the polje floor often disappear into sinks called ponors.

pollakanthic *a.* having several flowering periods.

pollarding *n.* a method of tree management in which the young tree is lopped at some distance above the ground (usually less than 3 m), and the branches then regularly pruned back, producing a crown of long straight shoots on a short single trunk. *cf.* coppicing.

pollen *n.* fine (usually yellow) powder produced by anthers and male cones of seed plants, composed of pollen grains that each enclose a male gamete. Pollen grains are durable and their size, shape and surface ornamentation is characteristic of the species that produced them. They provide a means of identifying previous vegetation on a site, *see* pollen analysis.

pollen analysis quantitative and qualitative determination of pollen grains preserved in deposits such as peat, from which the former vegetation of the area can be reconstructed.

pollen profile the vertical distribution of preserved pollen grains in a deposit such as peat, giving an indication of the former vegetation at different periods. *alt.* pollen record.

pollen record pollen profile *q.v.*

pollen spectrum the relative numerical distribution or percentage of pollen grains of different species preserved in a deposit.

pollination *n.* transfer of pollen from anther (in angiosperms) or male cone (gymnosperms) to stigma or female cone, respectively.

pollinator *n.* any insect or other animal whose activities affect pollination.

pollutant *n.* substance or effect, e.g. heat and noise, that causes pollution of air, soil or water.

polluter-must-pay principle concept that industries responsible for environmental pollution should bear the cost both of remedial action and future prevention of environmental contamination.

pollution *n.* any harmful or undesirable change in the physical, chemical or biological quality of air, water or soil as a result of the release of, e.g., chemicals, radioactivity, heat or large amounts of organic matter (as in sewage). Usually *appl.* changes arising from human activity although natural pollutants, e.g. volcanic dust, sea salt, are known.

polonium (Po) *n.* highly radioactive element (at. no. 84, r.a.m. 210) occurring in trace amounts in uranium ores. The most stable isotope, [209]Po, has a half-life of 103 years.

poly- prefix derived from Gk *polys*, many.

polyandry *n.* having more than one male mate at a time. *a* polyandrous.

Polychaeta, polychaetes *n.*, *n.plu.* the bristle worms, a class of mainly marine annelid worms, comprising the ragworms, lugworms, etc., an individual being characterized by possession of parapodia bearing numerous chaetae, which are used for crawling, and a pronounced head bearing tentacles, palps and often eyes.

polychlorinated biphenyls (PCBs) a large group of toxic synthetic lipid-soluble chlorinated hydrocarbons that are used in various industrial processes and that have become persistent and ubiquitous environmental contaminants that can be concentrated in food chains.

polychlorinated dibenzo-*p*-dioxins (PCDDs) family of organochlorine compounds made up of 75 congeners (*see* Fig. 20), including the extremely toxic 2,3,7,8-tetrachlorodibenzo-*p*-dioxin (TCDD). These widespread environmental contaminants are of anthropogenic origin. They are manufactured, not deliberately but as impurit-

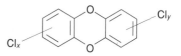

Fig. 20 The chemical structure of PCDDs.

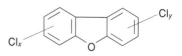

Fig. 21 The chemical structure of PCDFs.

ies, in a number of industrial processes, e.g. the bleaching of paper using chlorine, and through the incomplete incineration of chlorine-containing materials, e.g. PCBs and the plastic PVC. *alt.* dioxins.

polychlorinated dibenzofurans (PCDFs) family of organochlorine compounds made up of 135 congeners (*see* Fig. 21). These widespread environmental contaminants are of anthropogenic origin. They are manufactured, not deliberately but as impurities, in a number of industrial processes, and through the incomplete incineration of chlorine-containing materials, e.g. PCBs and the plastic PVC. *alt.* furans.

polychloroethene polyvinyl chloride *q.v.*

polyclad *n.* member of the order Polycladida, marine turbellarian flatworms with a very broad flattened leaf-shaped body, and large numbers of eyes at the anterior end.

polyclimax *n.* a climax community consisting of several different climax associations, none of which shows a tendency to give way to any other.

polyclimax theory *see* monoclimax theory.

polyculture *n.* the culture of a number of different species together in the same area, *appl.* cultivation of mixed crops and, in freshwater aquaculture, the rearing of several different fish species within the same pond.

polycyclic aromatic hydrocarbons (PAHs) multi-ringed aromatic compounds that are found in, e.g., soot, coal tar, cigarette smoke and barbecued meat. Examples include pyrene and benzo(a)pyrene (*see* Fig. 22). Some of these compounds are known carcinogens *q.v.*

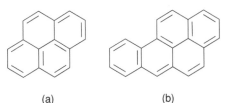

Fig. 22 The structures of (a) pyrene and (b) benzo(a)pyrene.

Polydnaviridae *n.* family of enveloped insect DNA viruses. Each particle contains many double-stranded DNAs of variable molecular weight.

polyethism *n.* division of labour among members of an animal society, as in the social insects.

polyethylene polythene *q.v.*

Polygales *n.* order of dicot herbs, shrubs and small trees including the families Polygalaceae (milkwort) and others.

polygamy *n.* (1) having more than one mate at a time; (2) (*bot.*) having male, female or hermaphrodite flowers on the same plant. *a.* polygamous.

polygenes *n.plu.* genes that each have a small effect and that collectively produce a multifactorial or polygenic phenotypic trait.

polygenesis *n.* (1) derivation from more than one source; (2) origin of a new type at more than one place or time.

polygenetic *a.* derived from more than one source or line of descent. *alt.* polyphyletic. *n.* polygenesis.

polygenic *a. appl.* (1) phenotypic characters (such as height, eye colour, etc., in humans) that are determined by the collective effects of a number of different genes; (2) inheritance, the inheritance of such traits, *alt.* multifactorial inheritance.

Polygonales *n.* order of herbaceous dicots, rarely trees, and comprising the family Polygonaceae (buckwheat).

polygoneutic *a.* rearing more than one brood in a season.

polygyny *n.* (1) having more than one female mate at a time; (2) (*bot.*) having numerous styles. *a.* polygynous.

polyhaline *a. appl.* brackish water of salinity 18–30 parts per thousand, approaching that of seawater.

polyhalite *n.* $K_2Ca_2Mg(SO_4)_4 \cdot 2H_2O$, a mineral used as a source of potassium.

polyhybrid *n.* a hybrid heterozygous for many genes.

polymer *n.* large organic molecule made up of repeating identical, or similar, subunits.

polymerase chain reaction (PCR) technique for selectively replicating a particular stretch of DNA *in vitro*, using DNA polymerase and the appropriate primers, to produce a large amount of a particular gene. This technique can make use of uncloned genomic DNA as the starting material, thus avoiding the need for DNA cloning in microorganisms.

polymerization *n.* the formation of a polymer from smaller subunits.

polymorph *n.* (1) (*geol./chem.*) any one of the several crystal types exhibited by a particular chemical of a given composition. Those chemicals or minerals that have polymorphs are said to exhibit polymorphism.

polymorphic *a.* (1) existing in two or more different forms, *appl.* minerals, chemicals and also *appl.* genes, characters, morphological forms, etc., within a biological species or population; (2) showing a marked degree of variation in body form during the life cycle or within the species; (3) *pert.* or containing variously shaped units, i.e. cells, or individuals in a colony.

polymorphic loci genetic loci with two or more alleles, conventionally defined as loci at which the most common homozygote has a frequency of less than 90% in a given population.

polymorphism *n.* (1) the existence within a species or a population of different forms of individuals, *see also* enzyme polymorphism, genetic polymorphism; (2) occurrence of different forms of, or different forms of organs in, the same individual at different periods of life; (3) (*geol./chem.*) *see* polymorph.

Polynesian Floral Region part of the Palaeotropical Realm comprising the Pacific islands east of Indonesia, except for the Fijian and Hawaiian Islands.

polynucleotide *n.* unbranched chain of nucleotides (*q.v.*) linked through an alternating sugar–phosphate backbone.

polynya *n.* large area of ice-free water in an otherwise ice-covered sea around the Antarctic.

polyoestrous *a.* having a succession of periods of oestrus in one sexual season. *cf.* monoestrous.

polyp *n.* (1) a sedentary individual or zooid of a colonial animal; (2) in coelenterates, an individual having a tubular body, usually with a mouth and ring of tentacles on top, like a miniature sea anemone.

polypedon *see* pedon.

polypeptide *n.* a chain of amino acids linked together by covalent bonds. Proteins are composed of one or more polypeptide chains. The polypeptide chains of proteins are synthesized on the ribosomes, using messenger RNA as a template. *see also* genetic code, protein.

polyphagous *a.* (1) eating various kinds of food; (2) of insects, using many different food plants.

polyphenism *n.* the occurrence in a population of several phenotypes that are not genetically controlled.

polyphosphate *n.* a word denoting the presence of $(PO_4)_n^{x-}$ anionic chains and/or rings. Polyphosphates are salts of polyphosphoric acids, $H_x(PO_4)_n$. Polyphosphates are added to detergents to act as a water softener, pH buffer and dirt particle dispersant. Disposal of polyphosphate-containing detergents to inland waterways can be responsible for phosphate-induced artificial eutrophication.

polyphyletic *a. appl.* a taxonomic group having origin in several different lines of descent.

polyphyodont *a.* having many successive sets of teeth.

Polyplacophora *n.* a class of molluscs that includes the chitons *q.v.*

polyploid *a.* having more than two chromosome sets, as triploid (three), tetraploid (four), etc. *n.* an organism with more than two chromosome sets per somatic cell.

polyploidy *n.* the polyploid condition, which may be the normal state of the somatic tissues of the whole organism as in some plants and (more rarely) animals, or a reduplication of chromosome number found in only some tissues or cells, and that can be induced artificially by chemicals such as colchicine.

polypores *n.plu.* group of basidiomycete fungi, including the bracket fungi, coral fungi and the cantharelles, in which the basidocarp is usually leathery, papery or woody and the hymenium may be smooth, ridged, warty or spiny or form the lining of tubes or gills on the underside of the basidiocarp. *cf.* agarics.

polysaccharide *n.* any of a diverse class of high-molecular-weight carbohydrates formed by the linking together by condensation of monosaccharide, or monosaccharide derivative, units into linear or branched chains. They include homopolysaccharides (composed of one type of monosaccharide only) and heteropolysaccharides (composed of a mixture of different monosaccharides). Found as storage products, e.g. starch and glycogen, and structural components of cell walls, e.g. cellulose, xylans and arabinans, and as components of glycoconjugates, *see* proteoglycans. *alt.* glycan.

polysaprobic *a.* in the saprobic classification of river organisms *appl.* category of organisms that can live in water heavily polluted with organic pollutants, in which decomposition is mainly anaerobic, e.g. sewage fungus, bloodworms and the rat-tailed maggot (*Eristalis tenax*). *cf.* alpha-mesosaprobic, beta-mesosaprobic, oligosaprobic.

polytetrafluoroethene (PTFE) *n.* plastic (trade names Teflon, Fluon) used, e.g., in electrical insulation, formed by the polymerization of tetrafluoroethene (CF_2CF_2). *alt.* polytetrafluoroethylene.

polythene *n.* tough, flexible thermoplastic material produced by the polymerization of ethene (C_2H_4), used, e.g., for packaging and insulation purposes. *alt.* polyethylene.

polythermic *a.* tolerating relatively high temperatures.

polythetic *a. appl.* a classification based on many characteristics, not all of which are necessarily shown by every member of the group.

polytocous, polytokous *a.* (1) producing several young at birth; (2) fruiting repeatedly.

polytopic *a.* occurring or originating in several places.

polytypic *a.* (1) having or *pert.* many types; *appl.* (2) species having geographical subspecies; (3) genus having several species.

polyunsaturated *a. appl.* fatty acids with more than one C=C double bond in their hydrocarbon chains.

polyvarietal cultivation the planting of several different varieties of the same crop in the same area of land.

polyvinyl chloride (PVC) colourless thermoplastic material formed by the polymerization of vinyl chloride (CH_2CHCl). *alt.* polychloroethene.

polyvoltine *a.* producing several broods in one season.

polyxenous *a.* adapted to life in many different hosts, *appl.* parasites.

Polyzoa Bryozoa *q.v.*

pome *n.* fruit derived from a compound inferior ovary in which the fleshy portion is largely the enlarged base of the perianth or receptacle, e.g. apples, pears.

ponderal *a.* (1) *pert.* weight; (2) *appl.* growth by increase in mass.

pongid *n.* any anthropoid ape other than the gibbons or siamang, i.e. chimpanzee, gorilla and orang-utan.

ponors *see* polje.

pooid *a. appl.* grasses of the genus *Poa*, e.g. *Poa annua* (meadow-grass).

pool and riffle sequence pattern of alternating deep and shallow sections, known as pools and riffles, respectively, found along the course of a river channel.

POPs persistent organic pollutants.

population *n.* a group of individuals of a species living in a certain area.

population crash sharp reduction in the population of a species when its numbers

exceed the carrying capacity of the habitat. *alt.* dieback.

population cycle pattern of regularly repeated changes in population numbers, with repeating increases followed by decreases.

population density number of individuals, usually with reference to a given species, living in a specified area.

population dispersion the distribution of the members of a population throughout its habitat.

population distribution the variation in population density over a given area.

population dynamics the changes in the structure of a population over time, i.e. the changes in the relative numbers of individuals of particular ages, different sexes or different forms.

population ecology the study of factors influencing the numbers and structure of a given population.

population explosion rapid increase in the population size of a given biological species, e.g. as a result of its introduction into a new geographical location.

population genetics the study of how genetic principles apply to groups of interbreeding individuals (a population) as a whole.

population momentum the potential for population growth as a large group of young individuals reaches reproductive age.

population profile analysis of the different sizes of age groups within a population.

population pyramid diagrammatic representation of the age distribution within a population, with the youngest age group at the base.

population trajectory the graphical tracking of changes in population numbers in response to an outside intervention such as predation or human exploitation.

population vulnerability analysis (PVA) an evaluation of a population's likelihood of extinction.

pores *n.plu.* (*soil sci.*) spherical or cylindrical voids present between soil particles, made, e.g., by earthworms or former roots. Those larger than 0.05 mm are termed macropores while those less than 0.05 mm in size are termed micropores. *cf.* fissures.

pore space in soil, the spaces between particles of soil collectively.

pore space ratio volume of soil pores divided by the total soil volume.

Porifera *n.* a phylum of simple multicellular animals, commonly called sponges, with a simple body enclosing a single central cavity (in the simple sponges) or penetrated by numerous interconnected cavities. The body wall consists of an outer layer of epithelium separated from an inner layer of ciliated choanocytes (feeding cells) by a mesogloeal layer. There are no nerve or muscle cells. Water is drawn into the internal cavities through pores (ostia), food particles are taken up by the choanocytes, and the water flows out through a large pore (the osculum). There are three classes: the Calcarea, the calcareous sponges (e.g. *Leucosolenia*), which have spicules of calcium carbonate embedded in the mesogloea and projecting to the outside; the Hexactinellida, the glass sponges (e.g. *Euplectella*, Venus's flower basket), with silica spicules; and the Demospongia, which includes some species with silica spicules and some species without, and which often have the body wall strengthened by a tangled mass of fibres (e.g. the bath sponge *Spongia*).

porosity *n.* (1) of rock, the ratio of the volume occupied by pore spaces (voids) to the total volume of the rock sample; (2) of soil, *see* soil porosity *q.v.*

porous rock *see* permeability (of rock).

porphyritic *a. appl.* (1) the texture of igneous rocks that have large crystals (phenocrysts) embedded in a finer-grained or glassy matrix (the groundmass); (2) rocks with porphyritic textures, e.g. porphyritic basalt.

porphyritic granite *n.* any granite that contains phenocrysts (of feldspar).

porphyritic microgranite a medium-grained intrusive igneous rock containing phenocrysts of feldspar and quartz. It is found in sills, dykes and veins and has essentially the same mineralogy as granite *q.v.*

porphyroblastic *a. appl.* the texture of metamorphic rocks that have a relatively fine-grained matrix within which are embedded well-shaped mineral crystals (called porphyroblasts) that grew *in situ* during metamorphism.

porphyroblasts *see* porphyroblastic.

porpoise *n.* member of the family Phocoenidae of the suborder Odontoceti (toothed whales) of the Cetacea (*q.v.*). Porpoises are smaller and dumpier than dolphins and lack the typical dolphin 'beak'. They also cannot leap completely out of the water like dolphins.

positional goods non-material goods, such as natural beauty, clean air, tranquillity, or the enjoyment of art. *alt.* amenity, public goods.

positive feedback type of regulation of a system in which the response to a stimulus causes an enhanced response to that stimulus.

positive reinforcement a stimulus, or series of stimuli, which is pleasant to an animal and increases its response.

positive taxis, positive tropism tendency to move (taxis) or grow (tropism) towards the source of the stimulus.

positron *n.* the positive electron, a particle with the mass of an electron but of opposite charge, symbol $^0_{+1}e$ or β^+. Its mass and the magnitude of its charge are the same as those of the negative electron, but its charge is of opposite sign. When a positron and a negative electron meet, they are both annihilated.

positron emission a type of radioactive decay in which a fast-moving positron (*q.v.*) is emitted from certain nuclei. This process decreases the atomic number of the nucleus concerned by one unit.

post- prefix derived from L. *post*, after, signifying situated behind, the hindmost part of an organ or structure, or occurring after.

post-climax *a. appl.* a stable plant community whose composition reflects previous climatic conditions that were more favourable (e.g. moister, cooler) than usual for the region.

post-clisere *n.* a series of vegetative formations (*q.v.*) that arise when the climate becomes wetter.

posterior *a.* (1) situated behind; (2) nearer the tail end; (3) dorsal in human anatomy; (4) behind the axis; (5) superior or next to the axis. *cf.* anterior.

post-glacial *a. appl.* Holocene *q.v.*

post-oestrus metoestrus *q.v.*

potable *a.* drinkable, *appl.* water of a standard acceptable for human consumption.

potamobenthos *n.* the bottom-living organisms in a river or other freshwater body.

potamodromous *a.* migrating only in fresh water.

Potamogetonales Najadales *q.v.*

potamoplankton *n.* the plankton of rivers, streams and their backwaters.

potash *n.* potassium carbonate, KCO_3.

potash alum *see* alum.

potash feldspar *n.* feldspar minerals (*q.v.*) with the formula $KAlSi_3O_8$, e.g. orthoclase, sanidine and microcline. *alt.* potassic feldspar, potassium feldspar.

potassium (K) *n.* highly reactive metallic element (at. no. 19, r.a.m. 39.10), forming a soft silvery solid that reacts violently with water. It occurs in nature in a wide range of minerals. Essential for all life because of the participation of the ion, K^+, in many cellular processes. Potassium compounds are included in combined fertilizers.

potassium–argon dating a geological dating technique in which the ratio of radioactive potassium-40 (which occurs naturally in rock) to argon-40 (the stable isotope formed by its decay) present in a sample is used to estimate its age. It is used on rocks over 100 000 years old.

potassium chloride *n.* KCl, (*agric.*) inorganic potassium fertilizer with a potassium content equivalent to 63% by weight K_2O when pure, supplied as granules, the most widely used potassium fertilizer.

potassium feldspar potash feldspar (*q.v.*). *alt.* potassic feldspar.

potassium fertilizers *n.plu.* straight inorganic fertilizers that supply the essential element potassium (K). *see* kainit, potassium chloride, potassium nitrate, potassium sulphate.

potassium nitrate *n.* KNO_3, (*agric.*) inorganic potassium fertilizer with a potassium content equivalent to 46.5% by weight K_2O, when pure, supplied as a powder.

potassium sulphate *n.* K_2SO_4, (*agric.*) this material, in an impure form, is used as an inorganic potassium fertilizer with a potassium content equivalent to 48–50% by weight K_2O, supplied as a powder.

potato *n. Solanum tuberosum* (family Solananceae), a crop plant originating in South America and now grown widely in the cooler, wetter parts of the world, whose tubers are eaten as food. *cf.* sweet potato.

potato blight a very serious disease of potatoes, caused by the fungus *Phytophthora infestans*. It was responsible for the Irish potato famine in the mid-nineteenth century.

potential energy energy stored in an object by virtue of that object's position or the position of its parts, as in a coiled spring.

potential evapotranspiration the theoretical maximum amount of water that could be lost by the combination of transpiration from vegetation and evaporation from the ground, given an unlimited supply of water.

potential temperature the temperature that a fluid would become if it were to be adiabatically brought to a pressure of 1 bar (= 1000 mbar = 1×10^5 N m^{-2} = *ca.* 1 atm at sea-level).

potexvirus group group of single-stranded RNA plant viruses with flexuous rod-shaped particles, type member potato virus X.

pothole *n.* (*geog.*) (1) deep hole formed by abrasion in the solid rock of a river bed, *alt.* swirlhole; (2) in limestone regions, a vertical shaft leading to an underground cave system.

potyvirus group group of single-stranded RNA plant viruses with flexuous rod-shaped particles, type member potato virus Y.

powdery mildews parasitic fungi of the order Erysiphales in the Pyrenomycetes (*q.v.*), the powdery appearance being due to the large numbers of conidia (spores) formed on the surface of the host tissue.

power *n.* (*phys.*) the rate at which work is done, i.e. work per unit time, expressed, in the SI system, using the derived unit, the watt.

power station an industrial plant designed to generate electricity.

Poxviridae, poxviruses *n., n.plu.* family of large, double-stranded DNA viruses that includes vaccinia and smallpox, fowl pox, sheep pox and myxoma.

PP primary production *q.v.*

PP_i pyrophosphate *q.v.*

p.p.b. parts per billion (10^9).

p.p.m. parts per million *q.v.*

p.p.m.v. parts per million, by volume. *see* parts per million.

ppt chemical precipitate.

p.p.t. parts per thousand *q.v.*

PQLI physical quality of life index *q.v.*

Pr symbol for the chemical element praseodymium *q.v.*

prairie *n.* in North America, the natural grassland covering the semi-arid middle of the continent in the mid-latitudes and that consists of tall-grass prairie in the cooler moister areas, most of which has now been converted into agricultural land, and short-grass prairie.

praseodymium (Pr) *n.* a lanthanide element (at. no. 59, r.a.m. 140.91), a soft silver-coloured metal when in elemental form. Praseodymium is obtained, e.g., from monazite and used to produce a yellow coloration in ceramics and glasses.

pratal *a.* (1) *pert.* meadows; (2) *appl.* flora of rich humid grasslands.

pre- prefix derived from L. *prae*, before, signifying situated before or occurring before.

preadaptation *n.* any previously existing anatomical structure, physiological process or behaviour pattern that makes new forms of evolutionary adaptation more likely.

prebiotic *a.* before life appeared on Earth.

Precambrian *a. pert.* or *appl.* time before the Cambrian, reckoned generally as the time from the earliest formation of rocks until around 590 million years ago, and divided into two eons, the Proterozoic and the earlier Archaean. The Precambrian saw the origin of life, the evolution of living cells and the evolution of the eukaryotic cell. The first multicellular animals arose towards the end of the Precambrian.

precautionary principle a way of assessing activities that may have a damaging effect on people and/or the environment generally that takes the 'better safe than sorry' approach. It emphasizes the responsibility of those who seek to alter things to show that they will not cause harm. It advocates the ideas that preventive or other action should be taken in advance of rigorous scientific proof if the damage that can be foreseen would be serious, life-threatening or irreversible, that a resource should not be automatically extracted even though it may seem desirable in the short term if the long-term effects cannot be assessed, that hazardous activities, esp. novel activities, should be introduced and managed with the greatest care and caution, and that the burden of proof that an activity is not damaging should be shifted from the victim to the perpetrator. *cf.* risk–benefit analysis.

precipitate *see* precipitation.

precipitation *n.* (1) (*clim.*) collective term for rain, hail, snow, etc., *alt.* hydrometeors; (2) (*geol.*) the deposition of minerals from solution as a result of evaporation; (*chem.*) (3) chemical reaction involving the formation of an insoluble material (precipitate) from reactants in solution; (4) the formation of a solid (precipitate) from its solution in a solvent, following supersaturation, as a consequence of either evaporation of the solvent or a change in temperature.

precipitation scavenging wet deposition *q.v.*

precipitator ash ash that has been removed from the flue gases of combustion processes by the use of electrostatic precipitators.

precision *n.* (*chem.*) the degree of agreement between quantitative analytical data, each of which has been obtained by exactly the same process. High precision equates with close agreement. *cf.* accuracy.

preclimacteric *a.* a period before the climacteric, i.e. the time of ripening.

preclimax *n.* (1) the plant community immediately preceding the climax community; (2) a plant community reflecting a warmer, drier previous climate.

precocial *a. appl.* young that are able to move around and forage at a very early stage, esp. in birds. *cf.* altricial.

predaceous, predacious *a.* preying upon, *appl.* fungi of the family Zoopagaceae, which trap and feed on protozoans and nematode worms.

predation *n.* the catching and killing of other organisms for food.

predator *n.* any organism that catches and kills other organisms for food. *a.* predatory.

predator chain food chain that starts from plants and passes from herbivores to carnivores. *cf.* parasite chain, saprophyte chain.

predator–prey relationship interaction between two organisms of different species in which one (the predator) captures and feeds on the other (the prey).

preferential species species that are present in several different communities, but are more common or thriving in one particular community.

prehensile *a.* adapted for grasping and holding.

preliminary treatment of sewage *see* sewage treatment.

premolars *n.plu.* teeth located between canines and molars in mammalian dentition.

prenatal *a.* before birth, *appl.* tests for genetic defects performed on a foetus in the womb.

prescribed burning controlled burning of woodland to prevent build-up of dead wood or woody undergrowth that could cause more destructive fires in the future.

presentation time minimum duration of continuous stimulus necessary for production of a response.

preservation *n.* maintenance of the environment as it is, without change or use.

pressure *n.* force per unit area. Units are the pascal (Pa), newtons per square metre (N m^{-2}), millimetres of mercury (mm Hg), atmosphere or bar.

pressure gradient (*clim.*) change in atmospheric pressure per unit distance, measured in the horizontal plane. *alt.* barometric gradient.

pressure gradient force (*clim.*) the force that causes air to be accelerated from a region of relatively high pressure to one of lower pressure. This has both horizontal and vertical components. The latter of these is essentially balanced by the force of gravity.

pressure melting point the term applied to the melting point of the basal ice of a glacier. This is lower than 0 °C owing to the increased pressure exerted by the weight of the overlying ice layers.

pressure vessel the container within which the core of a nuclear reactor resides. It retains the coolant. In most designs, the pressure vessel is made of steel and is within, and physically separate from, the biological shield (*q.v.*). There are reactors, however, in which the pressure vessel and biological shield are one and the same steel-lined prestressed concrete container.

pressure–volume work work done when a system contracts or expands against an external force.

pressurized water reactor (PWR) a type of fission nuclear reactor that utilizes Zircaloy-clad fuel pellets of enriched uranium oxide. It uses light water as the coolant and moderator.

prey *n.* organism that is captured and used as food by another organism (the predator).

Priapulida, priapulids *n.*, *n.plu.* phylum of burrowing and marine worm-like pseudocoelomate animals with a warty and superficially ringed body, with spines around the mouth.

prickly pear cactus *Opuntia* spp., a type of cactus, native to North and South America, which became a serious weed problem in Australia after its introduction there in *ca.*1840. The subsequent introduction of the Argentinian moth *Cactoblastis cactorum* has kept this weed largely under control.

prills *n.plu.* (*agric.*) spherical granules of any solid inorganic fertilizer, such as ammonium nitrate or urea.

primary *a.* (1) first; (2) principal; (3) original.

primary amines *see* amines.

primary atmospheric circulation global system of large-scale circulatory air movements within the troposphere.

primary consumer herbivore *q.v.*

primary decomposers soil organisms, chiefly saprophytic bacteria and fungi, that act on undecomposed organic material in the soil, breaking it down into simpler compounds and assimilating its carbon into their own mass. *see also* secondary decomposers.

primary ecological succession primary succession *q.v.*

primary effluent *see* sewage treatment.

primary energy the energy content of a fuel as it exists in the ground or as growing biomass. *see also* delivered energy, useful energy.

primary energy consumption energy conversion, by human endeavour, from a form found in nature. The burning of coal to generate electricity is therefore primary energy consumption, the use of the electricity generated in the heating of a building is not. Often cited in tonnes of oil equivalent for a given year.

primary forest forest in its original natural state, untouched by man. *alt.* virgin forest.

primary host host in which a parasite lives for much of its life cycle and in which it becomes sexually mature.

primary landforms *see* landform.

primary minerals minerals present in unweathered rock.

primary pollutants environmentally damaging substances that are present in the chemical form in which they were released into the environment. Primary pollutants may arise as the result of either human activity or natural processes. *cf.* secondary pollutants.

primary producer autotroph *q.v.*

primary production (PP) the assimilation and fixation of inorganic carbon and other inorganic nutrients into organic matter by autotrophs, which are therefore called primary producers.

primary productivity the amount of organic matter fixed by the autotrophic organisms in an ecosystem per unit time.

primary recovery *see* oil recovery.

primary sere natural succession of plant communities from the pioneer community, which colonizes bare ground, through to the climax community. *alt.* prisere.

primary sewage treatment *see* sewage treatment.

primary sexual characters differences between the sexes relating to the reproductive organs and gametes.

primary sludge *see* sewage treatment.

primary standard (*chem.*) a compound that is highly pure that is used to find the concentration of an analyte of unknown concentration by a suitable means.

primary succession sequence of different plant communities developing over time, initiated when new uncolonized habitats are created. Such habitats may be created either through human activities, e.g. waste heaps from mining operations, or by natural phenomena, e.g. cooled lava from fresh volcanic eruptions. *cf.* secondary succession.

primary tillage *see* tillage.

Primates *n.* an order of mammals known from the Paleocene and including tree shrews, lemurs, monkeys, apes and man. They are largely arboreal with limbs modified for climbing, leaping or brachiating (swinging), large brains in relation to body size, a shortening of the snout and elaboration of the visual apparatus, often with stereoscopic vision.

prime meridian meridian on which the Royal Observatory at Greenwich in London, UK, lies, defined as having a longitude of 0°. *alt.* standard meridian.

prime movers the ultimate factors that determine the direction of evolutionary change. They are of two kinds: basic genetic mechanisms, preadaptations and constraints imposed by an organism's existing developmental programme on the one hand, and the set of all environmental influences that constitute the agents of natural selection on the other.

primitive *a.* (1) of earliest origin; (2) not differentiated or specialized; (3) *appl.* traits that appeared first in evolution and that give rise to other, more advanced, traits. They are often, but not always, less complex than the advanced ones.

primordial *a.* (1) primitive *q.v.*; (2) original; (3) first begun; (4) first formed; (5) *appl.*, e.g., to embryonic cells that will develop into particular cell types or tissues, e.g. primordial germ cell.

primordium *n.* (1) original form; (2) a developing structure at the stage at which it starts to assume a form, *alt.* anlage; (3) (*bot.*) group of immature cells that will form a particular structure, such as a leaf or flower.

Primulales *n.* order of dicot herbs, shrubs and trees comprising the woody tropical families Myrsinaceae and Theophrastaceae and the temperate family Primulaceae (primrose).

Principes Arecales (*q.v.*), the palms.

principle *see* competitive exclusion principle, parsimony principle, polluter-must-pay principle, precautionary principle.

prion *n.* proteinaceous infectious particle, a protein complex lacking nucleic acid, which has been implicated in the transmission of transmissible spongiform encephalopathies such as scrapie in sheep, bovine spongiform encephalopathy (BSE) in cattle and Kreutzfeldt–Jakob disease in humans, but whose status is still uncertain.

prisere primary sere *q.v.*

prismatic *a.* (*soil sci.*) *appl.* soil peds (*q.v.*) that are vertical pillars with level tops. *cf.* columnar.

private good an economic good that is owned and enjoyed on a private and exclusive basis and that can be divided and sold. *cf.* commons, public good.

private-property resource any resource, esp. land, owned by a private individual or group of individuals other than local or national government. *cf.* commons, public-property resource. *see also* property rights.

pro- prefix derived from Gk *pro*, before, denoting previous to, in front of, the precursor of, or from L. *pro*, forward, for.

Pro proline *q.v.*

proangiosperm *n.* a fossil type of angiosperm.

probe *n.* (1) well-defined, usually radioactively labelled, fragment of DNA or RNA used to find and identify corresponding sequences in nucleic acids by selectively hybridizing with them, *see also* DNA hybridization; (2) labelled antibody used to detect and identify proteins.

Proboscidea *n.* an order of herbivorous placental mammals, known from the Eocene to present, including the elephants and the extinct mammoths and mastodons. They are of great size, having a massive skeleton, stout legs, an elongated trunk and incisors modified as tusks.

proboscis worms a common name for the Nemertea *q.v.*

procaryote prokaryote *q.v.*

Procellariiformes *n.* an order of ocean birds with external tubular nostrils and hooked beaks, including the albatrosses, shearwaters and petrels. *alt.* tubenoses.

process *n.* (1) a biochemical reaction or a procedure; (2) an elongated portion of a cell, such as the axon and dendrites of nerve cells; (3) an elongated projection from any structure.

prochirality *n.* property of molecules lacking handedness in their chemical structure,

i.e. their mirror images can be superimposed on each other, and that are optically inactive. *a.* prochiral. *cf.* chirality.

Prochlorophyta, prochlorophytes *n.*, *n.plu.* photosynthetic prokaryotes containing chlorophyll *a* and chlorophyll *b* but lacking phycobiliproteins, and therefore resembling plant chloroplasts rather than cyanobacteria. They include both ectosymbiotic and free-living species.

proclimax *n.* stage in a sere appearing instead of usual climatic climax and not determined by climate. *alt.* subclimax.

procrypsis *n.* (1) shape, pattern, colour or behaviour tending to make animals less conspicuous in their normal environment; (2) camouflage. *a.* procryptic.

producer *n.* an autotrophic organism, usually a photosynthetic green plant or photosynthetic microorganism, which synthesizes organic matter from inorganic materials and is an early stage in a food chain. *alt.* primary producer.

product, mathematical the result of multiplying two, or more, quantities together.

production *n.* (1) (*ecol.*), the assimilation of nutrients into biomass. *see* net primary production; (2) primary production *q.v.*

production efficiency the amount of biomass stored in a given system compared with the amount actually produced or taken in.

productivity *n.* the rate of production, i.e. the amount of organic matter fixed by an ecosystem per unit time. *see* primary productivity.

products *n.plu.* (*chem.*) the chemical species formed during a chemical reaction. *cf.* reactants. *see also* chemical equations.

proecdysis *n.* in arthropods, the period of preparing for moulting with the laying down of new cuticle and the detachment of the older one from it.

profile transect (1) a profile of vegetation, drawn to scale and intended to show the heights of plant shoots, *alt.* stratum transect; (2) a line transect that also shows the changing level of the ground, *see* transect sampling.

profundal zone the zone of a lake lying below the compensation point, comprising the deep water and the lake bottom.

progeotropism *n.* positive geotropism.

progressive provisioning the feeding of a larva in repeated meals. *cf.* mass provisioning.

Progymnophyta, progymnophytes *n.*, *n.plu.* division of extinct spore-bearing woody plants, with secondary xylem similar to that of gymnosperms, that are believed to be possible ancestors of the gymnosperms.

prohydrotropism *n.* positive hydrotropism.

Prokaryotae *n.* kingdom of living organisms comprising all prokaryotes (*q.v.*). Now often considered to consist of two superkingdoms, the Bacteria (*q.v.*) and the Archaea (*q.v.*). *alt.* Monera. *see* Appendix 9.

prokaryotes *n.plu.* unicellular organisms (e.g. bacteria, mycoplasmas, cyanobacteria) whose small, simple cells lack a membrane-bounded nucleus, mitochondria, chloroplasts and other membrane-bounded organelles typical of plant, animal, fungal, protozoan or algal cells. Their DNA is in the form of a single circular molecule not complexed with histones. In modern classifications they are placed in a separate kingdom, Monera or Prokaryotae, or into two separate superkingdoms, the Archaea (*q.v.*) and the Bacteria (*q.v.*). *a.* prokaryotic. *cf.* eukaryotes.

prolamines *n.plu.* simple proteins found in the seeds of cereals, soluble in ethanol and including gliadin from wheat, zein from maize, hordein from barley.

proliferation *n.* (1) increase by frequent and repeated reproduction; (2) increase by cell division.

proline (Pro, P) *n.* a cyclic amino acid (more properly an imino acid) with a hydrocarbon side chain, a constituent of proteins. *see also* hydroxyproline.

promethium (Pm) *n.* a radioactive element that is one of the lanthanides [at. no. 61, r.a.m. (for its most stable known isotope) 145], produced during the fission of uranium in nuclear reactors.

promiscuous *a.* mating with several different individuals over a breeding season.

promontory headland *q.v.*

promoter *n.* in carcinogenesis, any agent that hastens the process of carcinogenesis while not being a carcinogen on its own. *alt.* cocarcinogen.

pro-oestrus *n.* (1) the phase before oestrus or heat; (2) period of preparation for pregnancy.

prop- (*chem.*) the prefix used to indicate the presence of three carbon atoms in a chain in an organic compound (e.g. propane $CH_3CH_2CH_3$) or radical.

propachlor *n.* soil-acting herbicide used on vegetable crops to control germinating annual weeds.

propagation *n.* reproduction of a plant or animal.

propagative *a.* reproductive, *appl.* a cell, a phase in life cycle, an individual in a colonial organism.

propagator *n.* a box with a transparent lid used for rooting cuttings or germinating seeds. The provision of bottom heat, although not essential, encourages both of these processes.

propagule *n.* any spore, seed, fruit or other part of a plant or microorganism capable of producing a new plant and used as a means of dispersal. *alt.* diaspore.

propellant, aerosol *see* aerosol.

property rights rights of ownership or control over land or a resource. This may refer to conventional legally enforceable rights over private or public property, but also to rights that are not necessarily legally recognized or enforceable or pertain to individual ownership. For example, indigenous peoples could be considered to have property rights in the area they inhabit even though they may not be its legal owners, or the global community could be considered to have communal property rights in threatened habitat of international importance, wherever it may be, or in the non-pollution of the atmosphere. *see also* commons, private good, private-property resource, public good, public-property resource.

propham *n.* soil-acting carbamate herbicide used to control germinating weeds in crops such as sugar-beet and peas.

prophylactic *a.* disease-preventing.

propiconazole *n.* fungicide used to control fungus diseases of barley and wheat.

propineb *n.* fungicide used, e.g., to control apple scab and potato blight.

propolis *n.* resinous substance from buds of certain trees, used by worker bees to fasten comb portions and fill up crevices.

propoxur *n.* contact carbamate insecticide used, e.g., to control aphids on hops.

prop roots adventitious aerial roots growing downwards from the stem, as in mangrove and maize, and helping to support the stem.

propyzamide *n.* soil-acting herbicide used for weed control in orchards and forests.

prosimian *n.* any primate, such as lemurs and tarsiers, belonging to the primitive suborder Prosimii.

prosobranch *n.* member of the mollusc subclass Prosobranchia (e.g. abalone, winkles and whelks), which is aquatic and has a robust spiral shell that can be closed with an operculum, and a mantle cavity with one or two ctenidia.

protactinium (Pa) *n.* radioactive metallic element of the actinide series [at. no. 91, r.a.m. (of its most stable known isotope) 231]. It occurs naturally in uranium ores and its isotope ^{233}Pa is formed by the beta decay of thorium-233, which is itself produced by the neutron bombardment of thorium-232 in fast breeder reactors.

protandry *n.* condition of hermaphrodite plants and animals where male gametes mature and are shed before female gametes mature. *a.* protandrous.

Proteales *n.* an order of xerophytic shrubs and trees with either entire or much divided leaves covered with a thick cuticle and hairs, and showy flowers, comprising the family Proteaceae (protea).

protease proteinase *q.v.*

protein *n.* one of the chief constituents of living matter, any one of a vast group of large polymeric organic molecules containing chiefly C, H, O, N, S. Proteins are essential in living organisms as enzymes, structural components of cells and tissues, and in the control of gene expression. An individual protein molecule consists of one or more unbranched polypeptide chains constructed from amino acids linked covalently together by peptide bonds. The chains are folded into three-dimensional structures that differ from one type of protein to another. The polypeptide chains are sometimes associated with non-protein compounds (e.g. haem, flavin) termed pros-

thetic groups. Each protein chain has a unique, genetically determined, amino acid sequence, which dictates its three-dimensional structure and thus its function. Polypeptide chains are synthesized by translation of mRNA at the ribosomes.

proteinaceous *a. pert.* or composed of protein.

proteinase *n.* any enzyme that degrades proteins by splitting internal bonds between amino acids to produce peptides. *alt.* protease.

protein engineering the alteration of the structure of a protein by deliberate modification of the gene that encodes it.

protein family group of proteins of related sequence and function that arise from the duplication and divergence of their genes from a single ancestral gene.

protein quality the nutritional value of a protein, which is determined both by its digestibility and by whether it contains adequate amounts of the essential amino acids that animals cannot synthesize for themselves.

protein superfamily group of proteins descended from a common ancestral protein but that have subsequently diverged considerably in sequence and/or structure and have acquired different functions.

protein synthesis synthesis of a protein at the ribosomes using messenger RNA as a template. *see also* genetic code, translation, transfer RNA.

proteoglycan *n.* any of a class of compounds consisting of polysaccharide (95%) and protein (5%) units, forming the ground substance of connective tissue, important in determining the viscoelastic properties of joints, etc.

proteolysis *n.* breakdown of proteins and peptides into their constituent amino acids by enzymatic or chemical hydrolysis. *a.* proteolytic.

Proterozoic *n.* a geological eon of the Precambrian, before the Cambrian, lasting from around 2500 million years to 590 million years ago, and whose rocks contain few fossils, mainly blue-green algae (cyanobacteria) and soft-bodied animals of problematical affinities.

prothallus *n.* (1) the hyphae of lichens during the initial growth stages; (2) a small haploid gametophyte, as in algae, ferns and some gymnosperms, bearing antheridia or archegonia or both, and developing from a spore.

protherians *n.plu.* egg-laying mammals, including the extinct triconodonts and multituberculates, and the extant monotremes.

Protista, protists *n.*, *n.plu.* in modern classifications a kingdom comprised of eukaryotic unicellular, colonial and simple multicellular organisms that do not fall easily into either the plant or animal kingdoms. The Protista are usually held to comprise the algae, including the multicellular seaweeds and other macroalgae, diatoms, protozoans, the water moulds and the cellular and acellular slime moulds. *alt.* Protoctista.

proto- prefix derived from Gk *prōtos*, first.

Protoavis a possible fossil bird from the late Triassic, some 75 million years earlier than *Archaeopteryx*, but whose identification as a bird is still controversial.

protochordates *n.plu.* group of animals comprising the hemichordates, urochordates and cephalochordates, having gill slits, a dorsal hollow central nervous system, a persistent notochord and a post-anal tail.

protoepiphyte *n.* a plant that is an epiphyte all its life and does not start life rooted to the ground or come to root in the ground later.

protogyny *n.* the condition of hermaphrodite plants and animals in which female gametes mature and are shed before maturation of male gametes. *a.* protogynous.

protologue *n.* the printed matter accompanying the first description of a name.

proton (p) *n.* (1) (*phys.*) a subatomic particle with a mass of 1.0078 a.m.u. and an electric charge designated +1 (equal to 1.602×10^{-19} C). Protons are found in the nuclei of all atoms; (2) (*chem.*) an alternative name for the hydrogen ion, H^+.

protoplasm *n.* (1) living matter; (2) the total substance of a living cell, which in the case of a eukaryotic cell consists of cytoplasm and nucleoplasm.

protoplast *n.* (1) plant cell with cell wall removed; (2) the living component of a cell, i.e. the protoplasm not including any cell wall.

protostomes *n.plu.* collectively all animals with a true coelom and spiral cleavage

of the egg, and in which the blastopore becomes the mouth (i.e. molluscs, annelids, arthropods, phoronids, bryozoans and brachiopods). *cf.* deuterosomes.

Prototheria, prototherians *n.*, *n.plu.* subclass of primitive mammals that includes the orders Triconodonta, Multituberculata and Monotremata, of which only the monotremes, e.g. duck-billed platypus, are extant.

prototroph *n.* nutritionally independent, wild-type strain of bacterium or fungus that has no special nutritional requirements. *cf.* auxotroph.

prototrophic *a.* (1) nourished from one supply or in one manner only; (2) feeding on inorganic matter, *appl.* iron-, sulphur- and nitrifying bacteria and green plants.

prototype *n.* (1) an original type species or example; (2) an ancestral form.

Protozoa, protozoans *n.*, *n.plu.* a group of diverse unicellular, heterotrophic, generally non-photosynthetic, aquatic eukaryotes, lacking cell walls, formerly classified as unicellular animals. They include the Mastigophora (the flagellates, including the photosynthetic 'plant' flagellates, such as *Chlamydomonas*), the Sarcodina (amoebas and foraminiferans, and radiolarians and heliozoans), the Ciliophora (the ciliates), the Sporozoa (parasitic protozoans such as *Eimeria*, which causes coccidiosis, *Plasmodium*, the malaria parasite, and the piroplasms, such as *Babesia*), and the Cnidospora, which cause disease in fish and other animals. Protozoans are often now classified along with algae and other simple unicellular eukaryotes in a separate kingdom, Protista. *a.* protozoan. *see* Appendix 8.

protozoology *n.* that branch of biology dealing with protozoans.

protozoon *n.* individual protozoan cell.

Protura *n.* order of insects including the bark lice, minute insects with 12 segments in the abdomen, no antenna or compound eyes and very small legs. Found under the bark of trees, in turf and in soil.

proustite *n.* Ag_3AsS_3, a silver ore mineral. *alt.* light ruby silver.

provenance *n.* the original region in which a species was found.

provirus *n.* virus DNA that has become integrated into a host cell's chromosome and is carried from one cell generation to the next in the chromosome, not producing infective virus particles.

provisioning *n*. providing food for young, as in mass provisioning (*q.v.*), nest provisioning (*q.v.*), progressive provisioning, *q.v.*

proximal *a*. nearest to the body, or to centre, or to place of attachment. *cf*. distal.

proximate *a*. (1) nearest to, next to; (2) *appl*. cause, direct immediate cause.

proximate analysis of a food a rough estimate of the nutritive value of a food made by first determining the total nitrogen and multiplying by 6.25 to get a rough value for total protein, then determining the fat content by ether extraction, and finally determining the carbohydrate content by the difference between the above two values added together and the total dry weight of the sample.

proximate analysis of fuels the establishment of the percentage of each of the following components of a solid fuel: ash, fixed carbon, free moisture and volatile matter.

proximo-distal axis axis running from point of attachment of a limb to the body to the tip of the limb.

Prymnesiophyta, prymnesiophytes *n*., *n.plu*. mainly marine division of algae containing the coccolithophoroids, unicellular flagellate microorganisms armoured in calcareous 'scales' (coccoliths), and that are important in the marine phytoplankton.

psamment *n*. suborder of the order entisol of the USDA Soil Taxonomy system consisting of entisols found in areas of sand, e.g. Sahara Desert.

psammon *n*. the organisms living between sand grains, as of freshwater and marine shores.

psammophyte *n*. a plant that grows in sandy or gravelly ground.

psammosere *n*. a plant succession originating in a sandy area, as on dunes.

PSCs polar stratospheric clouds *q.v.*

pseud- prefix derived from Gk *pseudes*, false.

pseudannual *n*. a plant that completes its growth in one year but provides a bulb or other means of surviving the winter.

pseudaposematic *a*. *appl*. harmless species imitating warning coloration or other

protective features of harmful or distasteful animals, i.e. showing Batesian mimicry.

pseudepisematic *a*. having false coloration or markings, as in protective mimicry or for allurement or aggressive purposes.

pseudoallelic *a*. *appl*. two or more mutations that behave as alleles of the same locus in a complementation test but that can be separated by crossing over and that indicate the presence of a complex locus *q.v.*

pseudoaquatic *a*. thriving in wet ground.

pseudobulb *n*. a thickened internode of orchids and some other plants, for storage of water and food reserves.

pseudodominance *n*. expression of a recessive allele in the absence of the dominant allele.

pseudofossil *n*. natural inorganic feature in rock that can be mistaken for a fossil.

pseudogamy *n*. (1) union of hyphae from different thalli; (2) activation of ovum by sperm that plays no part in further development; (3) pseudomixis *q.v.*

pseudomixis *n*. sexual reproduction by fusion of vegetative cells instead of gametes, leading to zygote formation.

pseudomonads *n.plu*. bacteria of the family Pseudomonadaceae, widely distributed in soil and water, typically aerobic or facultatively anaerobic heterotrophs, Gram-negative polarly flagellated rods, some spp. containing blue or green fluorescent pigments.

pseudomycorrhiza *n*. mild pathological fungal infection of plant roots, superficially resembling mycorrhiza.

pseudoparasitism *n*. accidental entry of a free-living organism into the body and its survival there.

Pseudoscorpiones, psedoscorpions *n*., *n.plu*. order of small arachnids, commonly called false scorpions, which resemble scorpions but whose opisthosoma is not divided into two regions.

pseudosematic pseudepisematic *q.v.*

pseudotrophic *a*. *appl*. mycorrhiza when the fungus is parasitic.

PSI photosystem I *q.v.*

PSII photosystem II *q.v.*

Psilophyta *n*. one of the four major divisions of extant seedless vascular plants, represented by only two living genera. They are tropical plants of simple structure, having a rootless sporophyte, dichotomously branching rhizomes and aerial branches with small scale-like appendages (*Psilotum*) or larger bract-like outgrowths (*Tmesipteris*). *alt.* Psilopsida.

Psittaciformes *n.* order of birds including the parrots.

Psocoptera, psocids *n.*, *n.plu.* order of small insects, commonly called book lice and bark lice, an individual having incomplete metamorphosis, a globular abdomen and often no wings.

psychrometer *n.* an instrument consisting of two thermometers, one of which has its bulb covered in a wet cloth (wet-bulb thermometer) and one that does not (dry-bulb thermometer). The wet- and dry-bulb temperatures are used to calculate specific humidity, relative humidity and the mixing ratio.

psychrophil(ic) *a.* thriving at relatively low temperatures. For bacteria, these are below 20 °C.

Pt symbol for the chemical element platinum *q.v.*

pteridine *n.* organic compound composed of two fused six-membered rings of nitrogen and carbon with various substituents, which is a constituent of many natural compounds such as pterins (e.g. leucopterin, xanthopterin) and folic acid.

pteridology *n.* the branch of botany dealing with ferns.

Pteridophyta, pteridophytes *n.*, *n.plu.* major group of spore-bearing vascular plants: the ferns, club mosses, horsetails and the Psilophyta, sometimes treated as a division, Pteridophyta. *see* Lycophyta, Psilophyta, Pterophyta, Sphenophyta.

Pteridospermophyta, pteridosperms *n.*, *n.plu.* extinct division of seed-bearing vascular plants, the seed ferns *q.v.*

pterin *n.* any of a group of pigments, derivatives of pteridine, widespread in insects as eye pigments and in wings, which are also found in vertebrates and plants.

Pterobranchia, pterobranchs *n.*, *n.plu.* class of colonial hemichordates, living in secreted tubes, the individuals each possessing a crown of tentacles similar to the lophophore of bryozoans.

pterodactyls *n.plu.* the common name for the pterosaurs *q.v.*

pteropaedes *n.plu.* birds able to fly when newly hatched.

Pterophyta *n.* one of the four major divisions of the spore-bearing vascular plants, commonly called the ferns. The sporophyte has roots, stems and large leaves (fronds) that bear the sporangia. The gametophyte is a tiny thallus-like plant that bears the sex organs, i.e. the female archegonia and the male antheridia.

pteropods *n.plu.* group of marine gastropod molluscs with wing-like extensions to the foot, commonly called sea butterflies.

Pteropsida *n.* plant classification that has been used in different ways, as an alternative to Filicophyta (ferns), or for a larger grouping containing the ferns and seed plants.

Pterosauria, pterosaurs *n.*, *n.plu.* order of Jurassic and Cretaceous archosaurs, flying reptiles commonly called pterodactyls, which have membranous wings supported by greatly elongated fourth fingers.

PTFE polytetrafluoroethene *q.v.*

Pu symbol for the chemical element plutonium *q.v.*

puberty *n.* the beginning of sexual maturation.

pubescent *a.* covered with soft hair or down.

public good a good that cannot be divided and sold in units, is owned by no-one in particular, and can be consumed by all, such as clean air, clean water, natural beauty and wildlife. Public goods have the characteristics that the consumption of the good by one person does not diminish the amount of that good consumed by another person and that one person cannot prevent another from consuming the resource. *alt.* commons. *cf.* private good. *see also* amenity.

public-property resource any resource, esp. land, that is owned jointly by all citizens but is managed on their behalf by a government agency, e.g. public parks, national parks in many countries. *cf.* private-property resource.

public trust doctrine the commitment to put back into the environment at least the equivalent of what is being removed in any particular development. *see also* precautionary principle.

puffball *n.* common name for gasteromycete fungi of the order Lycoperdales *q.v.*

pullet *n.* a young hen between first lay and first moult.

pulmonary oedema condition, similar in its effect to drowning, in which water from the body fluids enters the lungs, caused, e.g., by altitude or exposure to high concentrations of corrosive or irritant pollutants, e.g. methylisocyanate (MIC). *see* Bhopal disaster.

pulmonates *n.plu.* molluscs of the subclass Pulmonata, the snails and slugs, characterized by lack of ctenidia and in which the mantle cavity is used as a lung.

pulp *n.* (1) the fibrous raw material used for paper making, obtained by the mechanical and chemical treatment of coniferous softwoods (and other cellulose-rich sources); (*biol.*) (2) any soft fleshy tissue in animals and plants, as interior of fruit; (3) internal cavity of vertebrate tooth, containing connective tissue, nerves and blood vessels; (4) (*mining*) crushed ore, esp. when combined with water.

pulse *n.* the seed of a legume, e.g. peas, beans, lentils, etc.

pulse crops seed legumes *q.v.*

pulsed-field gel electrophoresis electrophoretic technique for separating large pieces of DNA, e.g. chromosomes, in which an electric field is applied first in one direction and then in a direction at an angle to the first.

pulverized fuel ash (PFA) ash collected from the flue gases of power stations fuelled by finely ground coal. *cf.* furnace bottom ash.

pumice *n.* a variety of rhyolite (*q.v.*) that is highly vesicular.

pumped storage scheme scheme designed to generate electricity when demand is high. Water is pumped from a low-level reservoir to a high-level one during periods of low electricity demand. When demand increases, water from the high-level reservoir is allowed to return to the low-level one, driving electricity-generating turbines in the process.

puna *n.* (1) plateau high in the Andes of Peru; (2) term used in central Andes of South America for the vegetation of high tropical mountains. This type of vegetation is referred to as paramo in the northern Andes region.

punctuated equilibrium the view that the course of evolution has been marked by long periods of little or no evolutionary change (stasis) punctuated by short periods of rapid evolution. The view is based on an interpretation of the fossil record, which in some cases appears to show such a pattern. *cf.* phyletic gradualism.

punishment negative reinforcement *q.v.*

Punnett square a conventional representation used to calculate the proportions of different genotypes in the progeny of a genetic cross, e.g. for parents *Aa* and *aa*:

	A	*a*
a	*Aa*	*aa*
a	*Aa*	*aa*

pupa *n.* in insects with complete metamorphosis, a resting stage in the life cycle where the larval insect is enclosed in a protective case, within which tissues are reorganized and metamorphosis into a new form, usually the adult, occurs. *a.* pupal. *v.* pupate.

pure-breeding *a. appl.* line or variety in which a given heritable trait or traits appear in all the progeny of every generation.

pure line organisms originating from a single homozygous ancestor or identical homozygous ancestors, which are therefore themselves homozygous for a given heritable trait or traits, and thus breed true for those traits. *alt.* inbred line.

purine *n.* a type of nitrogenous organic base, of which adenine and guanine are most common in living cells, occurring in nucleic acids, where they pair with pyrimidines. Forms a nucleotide when linked to ribose or deoxyribose phosphates. Purine nucleotides, esp. those of adenine, are important cofactors and enerygy-rich compounds in metabolism.

puromycin *n.* antibiotic that becomes incorporated into a polypeptide chain as the chain is being synthesized, causing the release of the incomplete chain from the ribosome.

purple bacteria group of photoautotrophic bacteria, e.g. *Rhodopseudomonas*, which contain bacteriochlorophyll and the purple protein bacteriorhodopsin.

purple sulphur bacteria photoautotrophic bacteria, mainly aquatic, which oxidize sulphide to sulphur.

purposive behaviour goal-related behaviour.

push moraine *see* moraine.

push waves *see* body waves.

putrefaction *n.* decomposition of organic material, esp. the usually anaerobic breakdown of proteins by microorganisms, resulting in incompletely oxidized, ill-smelling compounds such as mercaptans, alkaloids and polyamines.

putrescible *a.* liable to rot, to undergo putrefaction (*q.v.*), *appl.*, e.g., to sewage sludge, animal and vegetable wastes.

PVC polyvinyl chloride *q.v.*

P waves *see* body waves.

PWR pressurized water reactor *q.v.*

pycnocline *n.* the sizable step in the density profile of a typical body of seawater.

Pycnogonida, pycnogonids *n., n.plu.* class of chelicerate marine arthropods (*q.v.*) commonly known as sea spiders, individuals having a long slender body consisting of an anterior cephalon, a trunk with four pairs of long walking legs and a short segmented abdomen. Some species bear chelicerae and feelers, others have neither.

pygmy male a purely male form, usually small, found living close to the ordinary hermaphrodite form in certain animals, as in some polychaete worms and barnacles.

PYO pick-your-own *q.v.*

pyramid of biomass a representation of the total biomass at each level of a food chain, which forms a pyramid, the biomass at lower levels (e.g. primary producers) being greater than that at higher levels (e.g. carnivores).

pyramid of energy a representation of the energy available per unit time at each trophic level in an ecosystem, usually expressed in kilocalories per square metre per year. A pyramid of energy can also represent the energy flow through the ecosystem, i.e. that proportion of energy

that is passed on to each trophic level. *see* Fig. 11, p. 127.

pyramid of numbers a representation of the numbers of organisms at different levels of a food chain, which forms a pyramid, greater numbers of organisms being present at the lower levels (e.g. primary producers) than at higher levels (e.g. carnivores). *alt.* Eltonian pyramid.

pyrargyrite *n.* Ag_3SbS_3, a silver ore mineral. *alt.* dark ruby silver.

pyrazon chloridazon *q.v.*

pyrazophos *n.* systemic organophosphate fungicide used to control powdery mildews on hops and apples.

Pyrenomycetes *n.* group of ascomycete fungi, commonly called the flask fungi, in which the generally club-shaped asci are usually borne in a hymenial layer in flask-shaped or spherical ascocarps (perithecia) that open in a terminal pore. They include the plant parasitic powdery mildews, the saprophytic pink bread mould *Neurospora*, the agents of several plant cankers (including the coral-spot fungus), anthracnoses and leaf spot diseases.

pyrethrin *n.* either of two structurally related insecticides, one of which (pyrethrin I) has the formula $C_{21}H_{28}O_3$ whereas the other (pyrethrin II) has the formula $C_{22}H_{28}O_5$.

pyrethroid *n.* any highly selective insecticide akin in its properties to the natural insecticide pyrethrum. Pyrethroids have low toxicity to mammals and are used, e.g., in household fly sprays. *alt.* synthetic pyrethroid.

pyrethrum *n.* a highly selective natural insecticide made from the flowers of a number of species of chrysanthemum, esp. *Chrysanthemum roseum*.

pyrimidine *n.* a type of nitrogenous organic base, of which cytosine, uracil and thymine are most common in living cells, occurring in nucleic acids, where they pair with purines. Forms a nucleotide when linked to ribose or deoxyribose phosphates. Some pyrimidine nucleotides also act as phosphate donors and energy-rich compounds in metabolism.

pyrimidine dimer structure produced in DNA by ultraviolet light in which

● or ○ = oxygen atom ; ● = silicon atom ;

the unit represents a silicate tetrahedron
(these are linked via shared oxygen atoms)

Note, those oxygen atoms represented thus ● reside above the silicon atoms

Fig. 23 A schematic representation of the structure of the chains of silicate tetrahedra, as found in pyroxene minerals.

adjacent pyrimidines on the same strand become covalently linked, blocking DNA replication and transcription, and causing mutation.

pyrite iron pyrites *q.v.*

pyrites *n.* sulphide minerals of certain metals, e.g. iron pyrites *q.v.*

pyroclastic rock rock composed of an accumulation of pyroclasts (*q.v.*). Once formed into a rock, this material is referred to as tephra.

pyroclasts *n.plu.* fragments of rock thrown into the air during the violent eruption of volcanoes. These are classified according to size: ash (<2 mm across), lapilli (2–64 mm across) and volcanic bombs (>64 mm across).

pyroclimax community climax community maintained by the periodic occurrence of fire, e.g. pine forests of the southern and western USA.

pyrolusite *n.* MnO$_2$, the principal manganese ore mineral.

pyrolysis *n.* chemical decomposition by the action of heat.

pyrometallurgy *n.* type of chemical processing involving the use of high temperatures in the extraction of metals from their ores.

pyrometasomatic deposits *see* skarns.

pyrometasomatism contact metasomatism *q.v.*

pyromorphite *n.* Pb$_5$(PO$_4$)$_3$Cl, a secondary mineral that occurs in the oxidized parts of lead veins. There is a complete solid-

solution series between pyromorphite and mimetite *q.v.*

pyrope *n.* one of the garnet minerals *q.v.*

pyrotheres *n.* a group of South American placental mammals of Eocene–Oligocene, somewhat resembling elephants, with tusk-like teeth and tending to large size.

pyroxene minerals an important group of rock-forming silicates that are structurally related. They are all based on chains of silicate tetrahedra (*q.v.*, *see* Fig. 23), with each chain held to the neighbouring chain by cationic counterions. In each chain, each tetrahedron shares two oxygens, one with each of its two neighbouring tetrahedra. Within any one chain, the 'bases' of the tetrahedra are essentially co-planar, as are their apices. These planes are parallel to the axis of the chain and approximately parallel to one another. The tetrahedra alternate in their orientation about the axis of the chain, such that the unshared basal oxygen of any one tetrahedron is on the opposite side of the chain axis to that of either of its nearest neighbour tetrahedra. Pyroxenes share the general formula $A_{1-x}(B,C)_{1+x}D_2O_6$, where, in common pyroxenes, A = Na or Ca, B = Mg or divalent Fe, C = Al or trivalent Fe and D = Si or Al, allowing for several solid-solution series among these minerals. *see also* clinopyroxenes, orthopyroxenes, pyroxenoid minerals.

pyroxenite *n.* a medium- to coarse-grained ultramafic igneous rock (colour index >90) that is mainly orthopyroxene or clinopyroxene in composition.

pyroxenoid minerals a group of silicate minerals, which differ structurally from the pyroxene minerals (*q.v.*) but that have the same 1 : 3 Si : O ratio.

Pyrrophyta *n.* the dinoflagellates, a group of largely unicellular biflagellated organisms sometimes known as whirling whips, considered either as protists or as part of the plant kingdom. They include both photosynthetic and heterotrophic forms and are important members of both marine and freshwater plankton. A feature of many dinoflagellates is the plates of cellulose immediately under the plasma membrane, which form a sculptured wall (theca) around the cell.

pyruvic acid *n.* three-carbon organic acid ($CH_3COCOOH$) produced during glycolysis, converted to acetyl CoA, in which form it is the starting point of the tricarboxylic acid cycle of aerobic respiration. Often referred to as the ionized form, pyruvate.

Q

Q (1) glutamine (*q.v.*); (2) ubiquinone *q.v.*

Q_{CO_2} the volume of carbon dioxide in microlitres of gas at normal temperature and pressure given out per hour per milligram dry weight.

Q_{O_2} oxygen quotient *q.v.*

Q_{10} temperature coefficient *q.v.*

quadrat *n.* a sample area enclosed within a frame, usually a square, within which a plant community, or sometimes an animal community, is analysed.

quadrivoltine *a.* having four broods in a year.

quadrumanous *a.* having hind feet as well as fore feet constructed as hands, as in most Primates except man.

quadrupedal *a.* walking on four legs.

quaking bog type of bog covered with a floating mass of vegetation (often sphagnum moss) that moves when walked on.

qualitative *a.* concerned only with the nature of organisms or substances under investigation. *cf.* quantitative.

qualitative inheritance the inheritance of phenotypic characters that occur in two or more distinct states within a population and do not grade into each other, the states representing combinations of different alleles at a single locus. *alt.* simple Mendelian inheritance. *cf.* quantitative inheritance *see also* polygenic inheritance.

quangos *n.plu.* quasi non-governmental organizations. In the UK, agencies wholly or partly funded from public funds, but not within the Civil Service, which manage many areas of government responsibility.

quantitative *a.* concerned with the amounts, as well as the nature, of organisms or substances under investigation. *cf.* qualitative.

quantitative inheritance the inheritance of characters determined by many different genes acting independently that appear as continuously variable characters within a population. *cf.* qualitative inheritance.

quantitative trait in genetics, a phenotypic character determined by the effects of many genes, which shows a continuously graded spectrum of variation within a population that can only be measured quantitatively, such as height, weight, etc.

quantitative variation continuous variation *q.v.*

quantum *n.* a unit of light. *plu.* quanta. *a.* quantal. *see* photon.

quarantine *n.* period of isolation in which an individual is kept in order to prevent the spreading of a disease.

quarrying *n.* (1) the excavation of rock, esp. for building material, from surface deposits by cutting and/or blasting; (2) plucking *q.v.*

quartz *n.* SiO_2, a crystalline silica mineral (*q.v.*) with many varieties, e.g. rock crystal (*q.v.*), amethyst (*q.v.*), citrine *q.v.*

quartz gabbro *see* gabbro.

quartz syenite syenite (*q.v.*) that contains relatively high levels of quartz ($\leqslant 10\%$ v/v) but not enough to be classed as granite *q.v.*

quartzite *n.* a very hard metamorphic rock transformed from quartz sandstone by either contact metamorphism or regional metamorphism.

quasisocial insects those social insects in which there is cooperative care of the brood but in which each female still lays eggs at some time.

Quaternary *a. pert.* or *appl.* geological period lasting from about 2 million years ago to present, comprising the Pleistocene and Holocene epochs.

quaternary structure the structural arrangement of the various subunits in a protein composed of several polypeptide chains (protein subunits).

queen *n.* a member of the reproductive caste in eusocial and semisocial insects, sometimes, but not always, morphologically different from the workers.

queen substance the set of pheromones by which a queen honey bee attracts workers and controls their reproductive activities, generally denotes trans-9-keto-2-decenoic acid, the most powerful of the components.

quicklime *see* lime.

quiescence *n.* temporary cessation of development or other activity, owing to an unfavourable environment.

quillworts *n.plu.* an order, the Isoetales, of vascular, non-seed-bearing plants of the division Lycophyta, having linear leaves and a 'corm' with complex secondary thickening.

quinine *n.* alkaloid extracted from bark of the South American tree *Cinchona officinalis* that has been used medicinally as an antimalarial and anti-fever drug.

R

r roentgen *q.v.*

r (1) coefficient of relationship (*q.v.*); (2) correlation coefficient (*q.v.*); (3) intrinsic rate of increase *q.v.*

r_max *see* biotic potential.

R (1) arginine (*q.v.*); (2) roentgen *q.v.*

R₀ net reproductive rate *q.v.*

Ra symbol for the chemical element radium *q.v.*

rabbit *see* Lagomorpha.

rabbit-fishes a small group of marine fishes, class Holocephalii, with long slender tails and large pectoral fins.

rabies *n.* acute, life-threatening, viral disease spread by the saliva of infected animals, e.g. from dogs or monkeys to humans via bites. It affects the nervous system, causing, e.g., convulsions and aversion to water. *alt.* hydrophobia.

race *n.* (1) group of individuals within a species that forms a permanent and distinguishable variety; (2) a rhizome, as of ginger.

racemate *n.* a mixture of two optical isomers, dextrorotatory (*d*) and laevorotatory (*l*), whose steric formulae are mirror images of each other and not superimposable.

raceme *n.* flowerhead having a single axis bearing stalked flowers arranged spirally around it, the bottom flowers opening first, as in hyacinth.

rachis *n.* (1) the shaft of a feather; (2) a stalk or axis.

rad (1) radian (*q.v.*); (2) unit formerly used to measure the amount of ionizing radiation absorbed by living tissue, 1 rad being equal to 100 erg per gram tissue. It has been replaced by the gray (Gy), with 1 rad = 10^{-2} Gy.

radial *a.* (1) *pert.* the radius; (2) growing out like rays from a centre; (3) *pert.* ray of

an echinoderm; (4) *appl.* leaves or flowers growing out like rays from a centre.

radial *n.* (1) cross-vein of an insect wing; (2) supporting skeleton of a fin-ray.

radial drainage *see* drainage pattern.

radial growth growth in the thickness of a stem, produced by concentric rings of new cells formed from the cambium.

radial symmetry having a plane of symmetry about each radius or diameter, as in many flowers and some animals, e.g. sea anemones and starfish.

radian (rad) *n.* the supplementary SI unit of plane angle. One radian, by definition, is the angle, at a circle's centre, that is subtended by an arc that is equal in length to the radius of the circle of which the arc is part. One radian = 57.296° (to five significant figures), 2π rad = 360°.

radiant *a.* (*phys.*) (1) emitting rays; (2) *pert.* radiation (*q.v.*); (3) (*biol.*) *pert.* ecological or evolutionary radiation.

radiant *n.* an organism or group of organisms dispersed from an original geographical location.

radiate *a.* (*biol.*) (1) radially symmetrical; (2) diverging or spreading out from a centre; (3) stellate.

radiate *v.* (1) to diverge or spread from a central point; (2) to emit rays.

radiation *n.* (1) (*phys.*) a general term for the action or condition of emitting electromagnetic rays, waves or subatomic particles, e.g. alpha particles, *see also* electromagnetic radiation, nuclear radiation; (2) (*biol.*) the relatively rapid increase in numbers of new species of a particular type of animal or plant and their diversification and spread into many new habitats, e.g. the mammalian radiation that occurred after the end of the Cretaceous period, when most present-day types of mammals arose. *alt.* adaptive radiation.

radiation biology the study of the effects of potentially damaging radiation on living organisms.

radiation cooling (*clim.*) rapid nocturnal cooling of the ground as a result of terrestrial radiation, and the resultant cooling of the lower air layers through conduction. This type of atmospheric cooling can lead to the formation of dew, hoar frost or radiation fog. *alt.* contact cooling.

radiation dose equivalent dose equivalent *q.v.*

radiation ecology the study of radiation and radioactive elements in the environment, and their effects on living organisms.

radiation fog type of fog formed when the land surface cools rapidly at night, lowering the temperature of the air above to below its dew-point and causing condensation of atmospheric water vapour.

radiation sickness illness caused by exposure to radiation. Short-term effects include loss of appetite, nausea and vomiting, while long-term exposure can produce sterility, suppress bone marrow function and production of blood cells, and cause decreased resistance to infection and cancers.

radiative cooling lowering of the temperature of a system as a consequence of the emission of electromagnetic radiation.

radiative forcing the difference between the solar energy absorbed by the Earth and that radiated back into space as terrestrial radiation.

radical *a.* (*bot.*) arising from root close to ground, *appl.* basal leaves and flower stems.

radical *n.* (1) molecule, or esp. a fragment of a molecule, that contains one or more unpaired electrons, *alt.* free radical; (2) group of atoms that is common to more than one compound, which is rarely, if ever, found to exist independently but which survives intact after reactions that alter other parts of a compound within which it is found. Examples are -OH, -NH$_2$, -C$_6$H$_5$, etc.

radicivorous *a.* root-eating.

radicle *n.* embryonic plant root, developing at the lower end of the hypocotyl.

radicolous *a.* living in or on roots.

radii *plu.* of radius.

radio- a prefix used instead of the adjective radioactive, e.g. radionuclide is the same as radioactive nuclide, radiocarbon is the same as a radioactive isotope of carbon.

radioactive *a. pert.* or possessing radioactivity *q.v.*

radioactive decay the spontaneous disintegration of any unstable nuclide with the emission of ionizing particles (i.e. alpha particles or beta particles) and electromagnetic radiation (as gamma radiation). The nuclides of all elements with an atomic number >83 are subject to this process. *alt.* decay.

radioactive fallout *see* fallout.

radioactive family radioactive series *q.v.*

radioactive half-life *see* half-life.

radioactive isotope *see* isotope, radioisotope.

radioactive series any succession of nuclides formed by the stepwise radioactive decay of an unstable isotope. With the exception of the first isotope, each nuclide is a daughter of the previous one in the series. All such series terminate when a stable isotope (often of lead) is made. *alt.* radioactive family.

radioactive waste waste that is contaminated with radionuclides (*q.v.*), produced from nuclear installations, laboratories and hospitals. *see* high-level radioactive waste, intermediate-level radioactive waste, low-level radioactive waste.

radioactivity *n.* the disposition of some elements to undergo spontaneous disintegration of their nuclei associated with the emission of ionizing particles (alpha particles or beta particles) or electromagnetic radiation (as gamma radiation).

radioautography autoradiography *q.v.*

radiobiology *n.* the study of the effects of radiation, esp. potentially harmful ionizing radiation such as X-rays, on living cells and organisms.

radiocarbon *n.* radioactive isotope of carbon, usually referring to ^{14}C, occurring naturally in small amounts in the atmosphere, used in biochemical and physiological research and as an indicator for dating in archaeology. *alt.* carbon-14.

radiocarbon dating the use of the uptake of the rare radioactive isotope of carbon, ^{14}C, during carbon fixation by plants to date the remains of organic material in archaeology. The difference between the proportion of ^{14}C in the material that would be expected if the organic material were newly synthesized and the actual proportion of ^{14}C reflects the time since the plant died and over which the ^{14}C has decayed. The radiocarbon method can be used to date material between 3000 and 40 000 years old.

radioecology radiation ecology *q.v.*

radioimmunoassay (RIA) *n.* a very sensitive method for the detection and measurement of substances using radioactively labelled antibodies or antigens.

radioiodine *n.* radioactive isotope of iodine, ^{131}I (half-life 8.6 days), used for studying the thyroid and in treatment of thyroid cancers.

radioisotope *n.* radioactive isotope of an element, such as tritium (^{3}H), phosphorus (^{32}P), radiocarbon (^{14}C) and radioiodine (^{131}I). The examples given are widely used in experimental biology to label tracer compounds, biological molecules, etc.

radiolarians *n.plu.* group of marine planktonic protists of the phylum Actinopoda (*q.v.*) (formerly classified as protozoans of the class Sarcodina); individuals are characterized by symmetrical skeletons of silicaceous spicules.

radiometer *n.* instrument used for measuring the radiation emitted by an object.

radiometric dating methods of dating of rocks or organic remains that depend on measuring the amount of a radioactive isotope in the sample and comparing it with the amount of the stable isotope it decays into. From this, the age of the material can be calculated. *see* potassium–argon dating, radiocarbon dating,

radiometry *n.* measurement of the radiation emitted by an object, e.g. infrared radiometry from satellites is used to measure the temperature of the oceans.

radiomimetic *a.* resembling the effects of radiation, *appl.* chemicals causing mutations.

radionuclide *n.* an unstable atomic nucleus, which undergoes spontaneous radioactive decay, emitting radiation and usually eventually changing from one element into another.

radioresistant *a.* offering a relatively high resistance to the effects of radiation, esp. ionizing radiation such as X-rays.

radiosensitive *a.* sensitive to the effects of radiation, esp. ionizing radiation such as X-rays.

radiosonde *n.* (*clim.*) an instrument package, equipped with radio, which is carried up through the atmosphere by balloon to relay information about humidity, temperature and pressure during its ascent. A more refined variation of the radiosonde, known as a rawinsonde, is also able to provide information about wind speed and direction.

radiosymmetrical *a.* having similar parts similarly arranged around a central axis.

radiowaves *n.plu.* electromagnetic radiation with frequencies ranging from 10 kilohertz (kHz) to 3×10^5 megahertz (MHz).

radium (Ra) *n.* luminescent radioactive metallic element (at. no. 88, r.a.m. 226.03) found in trace amounts in uranium ores. The most stable isotope, ^{226}Ra, has a half-life of 1620 years before decaying to radon.

radon (Rn) *n.* radioactive noble gas (at. no. 86, r.a.m. 222), occurring naturally in groundwater, and in some soils and rocks, such as granite, and in minute amounts in the atmosphere. The most stable isotope, ^{222}Rn, has a half-life of 3.825 days. Radon is a carcinogen and in houses built of granite and similar rocks, or built on granite, it can accumulate to hazardous levels.

radon daughters the decay products of radon: polonium-218, lead-214, bismuth-214 and polonium-214, which are all short-lived radioactive metals.

raffinose *n.* a trisaccharide found in sugar beet, cereals and some fungi, giving glucose, fructose and galactose on hydrolysis.

Rafflesiales *n.* order of plant parasitic dicots in which the body is reduced to a simple thallus and that comprises two families, Hydnoraceae and Rafflesiaceae.

rain *n.* precipitation in the form of water droplets of 0.5–2.5 mm diameter. Rain is produced by the condensation of water vapour in ascending air masses. Three different types are recognized depending on the causative mechanisms of uplift. *see* convectional rain, cyclonic rain, orographic rain. *see also* acid precipitation.

rain day period of 24 h during which 0.2 mm or more of precipitation falls.

raindrop impact erosion splash erosion *q.v.*

rainfall *n.* amount of precipitation that falls at a given location per unit time, as measured with a rain gauge (*q.v.*), usually expressed in millimetres or centimetres per year.

rain forest forest biomes that develop in areas with an annual rainfall of more than 200 cm, evenly distributed throughout the year. *see* monsoon rain forest, temperate rain forest, tropical rain forest.

rain gauge simple instrument used for the direct measurement of rainfall (*q.v.*), essentially consisting of a receptacle with a horizontal opening of known size exposed just above the Earth's surface.

rainout *see* wet deposition.

rainshadow *n.* area on the leeward side of a mountain range that receives little precipitation. *see* orographic rain.

rainsplash erosion splash erosion *q.v.*

raised beach a shore platform (*q.v.*) left stranded above the existing sea-level as a result of a fall in relative sea-level.

raised bog, raised mire convex lens-shaped acid peatland developed in fen basins or river floodplains in wet climates.

ram tup *q.v.*

r.a.m. relative atomic mass *q.v.*

ramapithecids, ramapithecines *n.plu.* a group of Miocene ape-like fossils from Asia and Africa, including the genus *Ramapithecus*, which show hominid-like features in the teeth.

ramet *n.* an individual member of a clone, such as an offshoot of a plant reproducing by stolons, etc.

Ramsar Convention Convention on Wetlands of International Importance Especially as Waterfowl Habitats, an international convention adopted in 1971, concerned with the protection and preservation of wetland habitats.

ranches *see* ranching.

ranching *n.* type of commercial pastoral farming (*q.v.*) in which large numbers of sheep or cattle are allowed to graze over extensive areas of natural/seminatural vegetation. Such farms/estates are known as ranches or, in Australia, stations.

random drift (1) random accumulation of changes in nucleotide sequence occurring over long periods of time that do not appear to be due to any selective forces; (2) the random changes in gene frequency that can occur in a small population over time as a result of sampling of gametes in each generation. *alt.* genetic drift.

random error error that alters the spread, i.e. precision (*q.v.*), of the data. *alt.* indeterminate error.

random sampling a sampling technique in which all points within the study area, volume or population have an equal chance of being sampled.

range *n.* (1) the series of measured values between a highest and a lowest value; (2) (*ecol.*) the area within which an animal or group of animals seeks food; (3) (*agric.*) an area of unenclosed, unintensively managed grassland on which livestock are allowed to graze freely. *alt.* rangeland.

rangeland range (3) *q.v.*

ranivorous *a.* feeding on frogs.

rank of coal *see* high-rank coal, low-rank coal.

rank–abundance diagram *see* dominance–diversity curve.

ranunculaceous *a. pert.* a member of the dicot flower family Ranunculaceae, the buttercups and their relatives.

Ranunculales *n.* order of herbaceous and woody plants, climbers or shrubs, and including the families Berberidaceae (barberry), Ranunculaceae (buttercup) and others.

rape *see* oil-seed rape.

rapids *n.* section of a river where flow is faster and more turbulent than other reaches, characterized by a steeper gradient and the presence of rock exposures.

raptatory *a.* preying.

raptorial *a.* adapted for snatching or robbing, *appl.* birds of prey.

raptors *n.plu.* birds of prey, e.g. hawks, eagles, owls, etc.

rare *a.* (1) not commonly found; (2) unusual and occurring in only a few cases; (3) IUCN definition *appl.* species or larger taxa that have small populations, and, although not at present considered endangered or vulnerable, are at risk, e.g., because of their highly restricted distribution within a habitat, or because they are thinly spread over a very large area, as are some large carnivores, *see also* endangered, rarity, vulnerable.

rare earths lanthanides *q.v.*

rare gases noble gases *q.v.*

rarity *n.* categories of rarity of plant and animal species that have been defined by the International Union for Conservation of Nature and Natural Resources (IUCN). *see* endangered, rare, vulnerable.

rasorial *a.* adapted for scratching or scraping the ground, as in fowls.

rate *n.* quantity, amount or degree of something measured in relation to the quantity or amount of something else (most generally time), as in uptake of CO_2 per area of leaf tissue per unit time (photosynthetic rate).

rate constant *see* rate law.

rate law (*chem.*) an experimentally derived equation that relates the rate of a given reaction to the concentration of its reactants and products. For the general reaction, represented by the equation $aA + bB \rightleftharpoons cC + dD$, the rate law takes the form: rate $= k \times [A]^e \times [B]^f \times [C]^g \times [D]^h$, where the powers e, f, g and h are commonly 0, 1 or 2, although fractions and negative numbers are also found. The proportionality constant, k, is called the rate constant. Different reactions have different values of k. In reactions that essentially go to completion, or reactions in their initial stages, when very little product has been produced, the rate law generally simplifies to rate $= k \times [A]^e \times [B]^f$. The powers define the order of reaction. If $e = 1$, the reaction is first-order with respect to A, if $e = 2$, then it is second-order with respect to A, etc. The sum of all of the powers in the rate law is the overall order of reaction, e.g. if both e and f are one, overall the reaction is second-order.

rate of natural change the difference between the crude birth rate and the death rate for a given population, showing the speed at which the population is increasing or decreasing.

rate of population growth the increase in a given population (in terms of birth rate minus death rate plus migration into the population) divided by the size of the population.

rate of reaction (*chem.*) the rate, with respect to time (t), of the disappearance of the reactants or appearance of the products of a given reaction, each divided by their coefficient of stoichiometry. Thus, for the general reaction represented by the equation $aA + bB \rightleftharpoons cC + dD$,

$$\text{rate} = -\frac{d[A]}{a \times dt} = -\frac{d[B]}{b \times dt}$$
$$= +\frac{d[C]}{c \times dt} = +\frac{d[D]}{d \times dt}$$

where d stands for 'infinitesimal change in', the square brackets, [ ], represent the concentration of the species; and a, b, c and d are the coefficients of stoichiometry in the above chemical equation. A minus sign indicates loss of reactant, while a plus sign represents a gain of product.

rating *n.* the thermal power produced within a nuclear fission reactor per unit mass of fuel.

ratites *n.plu.* a group of flightless birds comprising the ostriches, emus, rheas, cassowaries and kiwis. They have rudimentary wings, a breast bone without a keel and fluffy feathers with no barbs.

ratooning *n.* type of multiple cropping in which a second crop is grown from the roots of a harvested crop, e.g. in sugar cane.

Raunkiaer's life forms a classification of plants by the type of perennating organs they possess and their position in relation to soil or water level. *see* chamaeophyte, cryptophyte, geophyte, helophyte, hemicryptophyte, hydrophyte, phanerophyte, therophyte.

raw sewage untreated human bodily wastes.

raw sludge semisolid sewage before it undergoes primary treatment.

raw wastewater any domestic or industrial wastewater that has not been purified.

rawinsonde *see* radiosonde.

ray-finned fishes the common name for the Actinopterygii *q.v.*

Rb symbol for the chemical element rubidium *q.v.*

rDNA (1) DNA specifying ribosomal RNA (*q.v.*); (2) recombinant DNA *q.v.*

re- prefix derived from L. *re*, again.

Re symbol for the chemical element rhenium *q.v.*

reach *n.* an open stretch of flowing water, esp. *appl.* rivers.

reactants *n.plu.* the chemical species consumed during a chemical reaction. *cf.* products. *see also* chemical equations.

reaction *see* chain reaction, nuclear reaction, rate of reaction.

reaction centre the protein–chlorophyll complex in the photosystems of chloroplasts and other photosynthetic membranes in which chlorophyll electrons excited by light are transferred to the electron-transport chain.

reaction time (RT) the time interval between completion of presentation of a stimulus and the beginning of the response. *alt.* latent period.

reaction wood wood modified by bending of stem or branches, apparently in an attempt to restore the original position, and including compression wood in conifers and tension wood in dicotyledons.

realgar *n.* a red mineral with the empirical formula AsS which is naturally occurring tetraarsenic tetrasulphide, As_4S_4.

realized niche the actual place and role in an ecosystem that an organism or species occupies, as opposed to its niche under ideal conditions.

recalcitrant *a.* non-biodegradable, *appl.* organic, usually man-made, compounds in the soil.

receiving water any body of water, e.g. sea, lake or river, into which treated or untreated wastewater is discharged.

recent forest decline the as yet unexplained widespread damage to forest trees observed over large areas of Europe, eastern Canada and north-east USA. This phenomenon was first noticed in the early 1970s in Germany, but tree-ring studies suggest that, in some areas at least, the problem may date back to the 1950s. Atmospheric pollution is most commonly cited as the principal causative agent. *alt.* neuartige waldschäden, tree dieback, waldsterben.

Recent *n.* the geological epoch following the Pleistocene and lasting until the present day. *alt.* Holocene.

receptacle *n.* (*bot.*) of a flower, the point from which floral organs such as ovary, anthers, petals, etc., arise. *see* Fig. 12, p. 157. *alt.* floral axis.

receptor *n.* (1) specialized tissue or cell sensitive to a specific stimulus; (*neurobiol.*) (2) sensory organ; (3) sensory nerve ending; (4) (*cell biol.*) any site on or in a cell to which a neurotransmitter, hormone, drug, metabolite, virus, etc., binds specifically, in some cases to activate a specific cellular response, in others to gain access to the cell. Such a site is composed of a specific protein, glycoprotein or polysaccharide.

recessional moraine *see* moraine.

recessive *a. appl.* (1) alleles or mutations that are not reflected in the phenotype when present as one member of a heterozygous pair, only determining the phenotype when present in the homozygous state; (2) phenotypic characters expressed only in the homozygous state. *n.* recessivity.

recharge zone the area through which water is able to infiltrate into an aquifer.

reciprocal altruism social behaviour in which altruistic acts by one individual towards another are reciprocated, rare in most animals, but seen, e.g., in some monkeys and anthropoid apes, which will band together to aid each other in disputes against other members of the troop.

reciprocal crosses two crosses between the same pair of genotypes or phenotypes in which the sources of the gametes are reversed in one cross.

reciprocal feeding trophallaxis *q.v.*

reciprocal hybrids two hybrids, such as the mule and the hinny, one descended from a cross between a male of one species and female of the other, the other from a cross between a female of the first species and a male of the second.

reclamation *n.* the rehabilitation of a previously polluted or contaminated site by, e.g., soil clearance, filling-in and replanting. Also applied to the rehabilitation of a site destroyed by a natural disaster.

recombinant *a. appl.* (1) genotypes, phenotypes, gametes, cells or organisms produced as a result of natural genetic recombination (*q.v.*); (2) DNA, *see* recombinant DNA; (3) proteins, e.g. recombinant insulin, recombinant growth hormone: proteins produced from cells containing recombinant DNA directing their synthesis; (4) somatic mammalian cells, organisms, bacteria, yeasts, viruses, into which recombinant DNA has been introduced, or whose genomes have been modified *in vitro* by recombinant DNA techniques.

recombinant *n.* any chromosome, cell or organism that is the result of recombination, either natural (genetic recombination) or artificial (*see* recombinant DNA).

recombinant DNA (1) DNA produced by joining together, *in vitro*, genes from different sources. Also used to describe DNA that has in any way been modified *in vitro* to introduce novel genetic information; (2) DNA produced as a result of natural genetic recombination.

recombinant DNA technology collectively the techniques for the production of artificially genetically modified organisms, cells and microorganisms, and their applications in research and biotechnology.

recombinant fraction, frequency or value proportion of recombinant gametes produced by an individual (with respect to two genetic loci), calculated as the number of recombinant gametes divided by the total number of gametes, used to calculate the distance apart of two loci on the chromosome and relative positions of loci on the chromosome.

recombination *see* genetic recombination, recombinant DNA.

recruitment *n.* entry of new individuals into a population by reproduction or immigration.

rectangular drainage *see* drainage pattern.

recti- prefix derived from L. *rectus*, straight.

recumbent folds *see* anticlines.

recycling *n.* the recovery and re-use of materials from domestic refuse and industrial wastes. *see* direct recycling, indirect recycling, re-use.

red algae common name for the Rhodophyta *q.v.*

red copper ore natural Cu_2O. *alt.* cuprite.

Red Data books books produced by the International Union for the Conservation of Nature and Natural Resources (IUCN) to provide information, e.g. population size, geographical distribution, on threatened plant and animal species.

red drop the phenomenon that the quantum yield of photosynthesis falls sharply when the wavelength of light is greater than 680 nm. This is due to the fact that only photosystem I can be driven by light of longer wavelength.

redevelopment *n.* the major reconstruction of buildings, roads, etc., in run-down, built-up areas.

red light light of wavelength 620–680 nm.

Red Queen hypothesis idea that each evolutionary advance by one species is detrimental to other species so that all species must evolve as fast as possible simply to survive.

red spider mite a pest esp. of orchard and glasshouse crops. The glasshouse red spider mite *Tetranychus urticae* can be biologically controlled by another mite, *Phytoseiulus persimilis*.

red tide a bloom of red dinoflagellates, e.g. *Gonyaulax polyedra*, which colours the sea red. Toxins contained in some of these microorganisms are concentrated in the shellfish that feed on them and can cause fatal poisoning in humans who eat the shellfish.

reduced *a.* (1) in an anatomical context *appl.* structures that are smaller than in ancestral forms; (2) (*chem.*) *appl.* molecule that has had oxygen removed or hydrogen or electrons added, or an atom that has had electrons added.

reduced cultivation, reduced tillage an alternative method of cultivation in which the number and intensity of tillage operations is reduced, compared with conventional cultivation (*q.v.*). This may, for example, include the use of cultivators that disturb the soil to the minimum depth needed for drilling. *alt.* minimal cultivation, minimal tillage. *see also* no-till agriculture.

reducer organism decomposer *q.v.*

reducing agent chemical species that readily gives up electrons, *alt.* reductant.

reducing power a general term for the presence in cells of compounds such as $NADH_2$ and NADPH, which are hydrogen and electron donors in metabolic reduction reactions.

reductant reducing agent *q.v.*

reductase *n.* any enzyme that catalyses reduction of a compound, used esp. where hydrogen transfer from the donor is not readily demonstrable.

reduction *n.* decreasing the oxygen content or increasing the proportion of hydrogen in a molecule, or adding an electron to an atom or ion. *cf.* oxidation.

reduction catalyst-based catalytic converter a device fitted to the exhaust system of a petrol (i.e. gasoline) engine. Its purpose is to facilitate the reaction of any nitric oxide (NO) present with the reducing agents (e.g. carbon monoxide, unburnt hydrocarbons) found in the waste gases, thus producing carbon dioxide, water and molecular nitrogen. These devices use rhodium as the catalyst, supported on an inert matrix with a high surface area. They may be used alone (coupled to a lean-running engine), or as part of a modern three-way catalytic converter (*q.v.*). Unleaded fuel must be used. *see also* oxidation catalyst-based catalytic converter.

reduction division first meiotic division, sometimes used for meiosis as a whole.

reductionist *a. appl.* strategy of breaking down a complex system into its components and examining the components independently to resolve how the system as a whole works. In biology, refers to the idea that complex phenomena such as embryonic development, inheritance, mental processes, etc., are in principle completely explicable in terms of the basic principles of biochemistry, molecular genetics, etc. The opposing anti-reductionist view maintains that at higher levels of organization novel properties emerge that have no direct correspondence to lower-level processes and therefore cannot be wholly explained in their terms. *n.* reductionism. *cf.* holistic.

reduction potential, standard (E^0, E^o) an electrochemical measure (in volts) which, for any substance that can exist in an oxidized or reduced form, gives its affinity for electrons relative to hydrogen in standard conditions (the redox potential of the H^+/H_2 couple is defined as 0 V). A highly negative standard reduction potential indicates a strong reducing agent, and a highly positive standard reduction potential a strong oxidizing agent. *alt.* standard electrode potential.

reductive pentose phosphate pathway Calvin cycle *q.v.*

reed-bed system a water treatment system in which the macrophytes used for purification are reeds. *see* aquatic macrophyte treatment system.

reef *n.* (1) (*oceanogr.*) narrow chain of rocks lying at or near the surface of the water; (2) (*geol.*) in Australia, vein of gold-bearing quartz.

refection *n.* reingestion of incompletely digested food by some animals, such as eating faecal pellets, or in rumination. *see also* coprophagy.

reforestation *n.* the replanting of trees in areas previously cleared of their forest cover.

refrigerant *n.* (1) fluid, e.g. ammonia or freon, used as a heat-transfer medium in refrigerators; (2) any material, e.g. solid carbon dioxide or liquid nitrogen, that is used to cool another.

refuge, refugium *n.* (1) an area that has remained unaffected by environmental changes to the surrounding area, such as a mountain area that was not covered with ice during the Pleistocene, and in which the previous flora and fauna has survived; (2) area designated for the preservation of wildlife and which is left undisturbed.

refuse *n.* solid waste.

refuse-derived fuel waste-derived fuel *q.v.*

reg *n.* a stone-covered desert plain with little or no vegetation. Erosion by wind and water is responsible for the selective removal of the finer surficial particles.

regelation *n.* a type of flow involved in the basal sliding (*q.v.*) of glaciers. This occurs when a small obstacle is encountered by the glacier. Higher pressure on the up-glacier side of the obstacle causes the ice to melt. This refreezes on the down-glacier side of the obstacle, where pressure is reduced.

regional metamorphism the formation of metamorphic rocks by compaction and/or recrystallization facilitated by the action of the increasing pressure/temperature associated with progressive burial and mountain building.

regolith *n.* layer of unconsolidated material that covers the surfaces of virtually all rocks. Regolith may either be formed *in situ* by chemical and physical weathering of the bedrock (known in this context as saprolite) or be brought in from outside by the action of wind, moving water or ice.

regosols *n.plu.* soils that are developed on fairly deep unconsolidated parent material such as dune sands or volcanic ash.

regression *n.* (1) reversal in the apparent direction of evolutionary change, with simpler forms appearing; (2) the replacement of a climax ecosystem with a previous stage in the succession, e.g. the replacement of forest by grassland after felling.

regression analysis analysis of the relationship between two variables when one is dependent upon another independent variable.

regular *a. appl.* any structure, such as flower, organism, showing radial symmetry. *alt.* actinomorphic *q.v.*

regular bedding (*geol.*) bedding characterized by plane bedding surfaces separating beds with parallel sides.

regulator *n.* (*biol.*) an animal that maintains its internal environment in a state that is largely independent of external conditions. *cf.* conformer.

regulatory genes genes that direct the production of proteins that regulate the activity of other genes, or that represent control sites in DNA at which gene expression is regulated.

reinforcement *n.* an event that alters an animal's response to a stimulus, positive reinforcement being reward and increasing its response, negative reinforcement being disagreeable or painful and suppressing its response.

reinforcing selection operation of selection pressures on two or more levels of organization, such as population, family and individual, in such a way that certain genes are favoured at all levels and their spread through the population is accelerated.

rejuvenation *n.* (*geog.*) a renewal of vertical erosion by a river. This may be caused by an increase in the river's load discharge (static rejuvenation), or by a fall in the river's base level or uplift of the surrounding land masses (dynamic rejuvenation). River rejuvenation is associated with the formation of river terraces and knickpoints.

relationship, coefficient of *see* coefficient of relationship.

relative abundance rough estimate of the population density of a species in a community, arrived at by counting the individuals of a particular species seen over a given time or in a given place and dividing by the total of all species. Usually expressed as a percentage.

relative age, relative date chronology based on stratigraphy and fossil composition rather than on physical dating methods. *cf.* absolute age.

relative atomic mass (r.a.m.) the ratio of the mass of one atom of a substance to one-twelfth the mass of an atom of ^{12}C. It is a ratio and therefore dimensionless. *alt.* atomic weight.

relative basal area the basal area (*q.v.*) of a species divided by the total basal area of all species in the community, usually expressed as a percentage.

relative dating techniques *see* dating techniques.

relative density the density of a species divided by the total density of all species in the community, usually expressed as a percentage.

relative error (*chem.*) the absolute error (*q.v.*) expressed as a percentage, or as parts per thousand (p.p.t.), of the true, or accepted, value.

relative frequency the number of samples a species occurs in divided by the total number of samples, usually expressed as a percentage.

relative humidity amount of water vapour present in air relative to that present in saturated air (*q.v.*) under the same conditions of temperature and pressure (expressed as a percentage).

relative molecular mass (M_r) the ratio of the mass of one molecule of a substance to one-twelfth the mass of an atom of ^{12}C. It is a ratio and therefore dimensionless. *alt.* molecular weight. *see also* molecular mass.

relative resource scarcity localized or temporary shortage of a resource, which may be due to numerous causes. *cf.* absolute resource scarcity.

relative transpiration the rate at which water is lost from vegetation by transpiration, divided by the rate at which water would be lost by evaporation from the surface of open water under the same conditions.

relay cropping type of multiple cropping in which seedlings of a second crop are planted among maturing plants (annual or biennial) of a first crop.

release *see* competitive release.

releaser *n.* a stimulus or group of stimuli that activates an inborn tendency or pattern of behaviour, as of species-specific behaviour.

relict *a.* (1) not now functional, but originally adaptive, *appl.* structures; (2) surviving in an area isolated from the main area of distribution owing to intervention of environmental events such as glaciation, *appl.* species, populations. *alt.* relic. (3) (*geol./geog.*) features that survive significant changes to the conditions that formed them, e.g. *appl.* structures present in metamorphic rocks that have been inherited from the parent rock.

relict dunes dunes covered with vegetation and which, as a result, are no longer active. These wind-formed landforms are commonly found on the periphery of existing arid regions.

relief *n.* a way of describing a particular area with respect to the magnitude of the height difference between the highest and lowest elevations within the area. Thus a plain is an area of low relief while a mountain range with high peaks and deep valleys is described as an area of high relief.

relief rain orographic rain *q.v.*

rem *n.* abbreviation for roentgen equivalent man, the unit dose of ionizing radiation that gives the same biological effect as that due to 1 roentgen of X-rays. The rem has been replaced by the sievert, with 1 sievert = 100 rem.

remote sensing the collection of data without direct contact with the object or area observed, as the collection of information on land and oceans by, e.g., radar, aerial photography and infrared photography from aircraft and satellites.

rendoll *n.* suborder of the order mollisol of the USDA Soil Taxonomy system consisting of mollisols of humid climatic regions, developed from calcium-rich parent material.

rendzina *n.* any of a group of rich, dark, greyish-brown, limy soils of humid or subhumid grasslands, having a brown upper layer and yellowish-grey lower layers.

renewable energy useful energy derived from resources that are not depleted as a result. Hydropower, solar, wind, wave, tidal, ocean thermal, biomass and geothermal energy are all usually included in this category.

renewable resources environmental resources that are permanently available in continuous supply, e.g. plants and animals (if not overexploited), solar energy and water power. *cf.* non-renewable resources. *alt.* flow resources.

Renner complex a group of chromosomes that passes from generation to generation as a unit, as in the evening primrose, *Oenothera.*

Reoviridae, reoviruses *n., n.plu.* family of icosahedral, non-enveloped, double-stranded RNA viruses of animals and plants, including human rotaviruses that cause diarrhoea in children.

repeat *a. appl.* two or more physical or biological samples, of the same size, that have been treated in an identical fashion, sequentially (i.e. at different times). *cf.* replicate.

replacement capacity (1) the ability of a biological resource to regenerate its original state after harvesting or other use; (2) net reproductive rate.

replacement-level fertility birth rate that keeps a population constant, exactly replacing deaths by births. *alt.* zero population growth.

replacement name scientific name adopted as substitute for one found invalid under the rules of the International Codes of Nomenclature.

replica plating production of an exact replica of a plate containing bacterial colonies by transfer of bacteria to a new plate by 'blotting' with a velvet pad, filter paper, etc., which retains the exact positions of the colonies relative to each other. Destructive identification procedures can be carried out on the replica plate leaving the master plate as an untouched source of bacteria, DNA, etc.

replicate *a. appl.* two or more physical or biological samples, of the same size, that have been treated in an identical fashion, simultaneously. *cf.* repeat.

replication *n.* (1) duplication, as of DNA, by making a new copy of an existing molecule; (2) duplication of organelles, such as mitochondria, chloroplasts and nuclei, and of cells.

reporter gene a marker gene inserted in a recombinant DNA vector, etc., whose activity can be easily tracked and the distribution of the introduced DNA assessed.

reproduction *n.* the formation of new individuals by sexual or non-sexual means.

reproduction curve a plot giving the relationship between the number of individuals at a particular stage in one generation and the numbers at that stage in a previous generation.

reproductive age the age range within which an organism is sexually mature and capable of reproduction.

reproductive allocation the proportion of an organism's resources that are devoted to reproduction in a given time period.

reproductive cost the lowering of the survival rate or growth rate, and the subsequent decrease in its reproductive potential, due to an organism devoting more of its resources to reproduction.

reproductive effort the total number of seeds and vegetative offshoots produced by a plant.

reproductive isolation the inability of two populations to interbreed because they are geographically isolated, or isolated from each other by differences of behaviour, mating time (or in plants maturation times of male and female sex organs), or genital morphology. This is a phase in the development of new species.

reproductive potential (1) the number of offspring a given population of a species could produce in ideal conditions; (2) the number of offspring produced by a single female in a population.

reproductive rate the number of offspring produced over a given time period.

reproductive success the number of offspring of an individual surviving at a given time.

reproductive value the expected number of female offspring remaining to be born to each female of age x, symbolized by v_x.

Reptilia, reptiles *n., n.plu.* a class of amniote, air-breathing, poikilothermic ('cold-blooded') tetrapod vertebrates, mostly terrestrial, an individual having a dry horny skin with scales, plates or scutes, functional lungs throughout life, one occipital condyle and a four-chambered heart. Most reptiles lay eggs with a leathery shell but some are ovoviviparous. They include the tortoises and turtles, the tuatara, lizards and snakes, crocodiles, and many extinct forms, such as dinosaurs, pterosaurs, etc.

RER rough endoplasmic reticulum, *see* endoplasmic reticulum.

reserve *see* nature reserve.

reserve cellulose cellulose found in plant storage tissue and subsequently used for nutrition after germination.

reservoir *n.* (1) as *appl.* biogeochemical cycles, a portion of the material that constitutes the contents of the cycle in question that is separate from the remainder of the contents of the cycle by virtue of physical location, physical state and/or chemical speciation. The size of a given reservoir is usually expressed in kilograms or moles; (2) an artificial lake created by damming and flooding a valley. The stored water may be used for domestic supply or to provide a head of water for the generation of hydroelectric power (*q.v.*); (3) host that carries a pathogen but is itself unharmed and acts as a source of infection to others, *alt.* carrier.

reservoir rocks *see* oil traps.

residence time (τ) as *appl.* biogeochemical cycles, the ratio of the amount of material in a reservoir to the total flux in or out of it. *alt.* average transit time, lifetime, turnover time.

residual deposits deposits of economically valuable minerals formed *in situ* when the bulk of the more soluble material originally present has been removed by severe leaching, e.g. bauxite in leached tropical soils.

residual reproductive value (RRV) probable contribution of individuals in specific age groups to produce offspring in the future, excluding the progeny they already have.

residual volume volume of air remaining in lungs after strongest possible breathing out.

residue *n.* (1) pesticides and other chemicals remaining in the environment after they have carried out their function; (2) (*chem./biochem.*) a compound such as a monosaccharide, nucleotide or amino acid when it is part of a larger molecule.

resilience *n.* (*biol.*) ability of an ecosystem to restore itself to its original condition after being disturbed. *cf.* resistance.

resin *n.* any of various high-molecular-weight substances, including resin acids, esters and terpenes, which are found in mixtures in plants and often exuded from wounds, where they may protect against insect and fungal attack as they harden to glassy amorphous solids.

resistance *n.* (1) (*ecol.*) ability of an ecosystem to resist being displaced by perturbations, a measure of how effectively it is buffered against change, *cf.* resilience; (2) the degree to which an individual or a species is unable to be infected with a particular disease, or is unaffected by a particular poison; (3) (*phys.*) electrical resistance, the opposition to the flow of electricity through an object. It is the reciprocal of conductance and is expressed in units of volts per ampere (ohms, Ω).

resistance transfer factor (RTF) that part of a transmissible drug-resistance plasmid that mediates conjugation and plasmid transfer to another bacterium.

resolution *n.* the size of an object that can be viewed clearly in a given microscope, *alt.* resolving power.

resource *n.* anything provided by the environment to satisfy the requirements of a living organism, e.g. food, living space.

resource allocation (1) resource partitioning (*q.v.*); (2) the way in which an individual organism uses the limited energy available to it for different purposes, e.g. growth, reproduction, foraging, etc.

resource competition the result of density-dependent competition within or between species for limited resources. *cf.* interference competition.

resource depletion zone the area around an organism where it has partially used up its food resources.

resource-holding potential the ability of an animal to gain and maintain possession of essential resources by fighting.

resource partitioning the division of scarce resources in an ecosystem so that species with similar requirements use the same resources at different times, in different ways, or in different places.

resource switching the ability of some organisms to change from one food source to another if the second source becomes more abundant.

respiration *n.* any or all of the processes used by organisms to generate metabolically usable energy, chiefly in the form of ATP, from the oxidative breakdown of foodstuffs. Refers to processes ranging from the exchange of oxygen and carbon dioxide between an organism and the environment to the biochemical processes generating ATP at the cellular level. *a.* respiratory, *pert.* or involved in respiration. *see also* aerobic respiration, anaerobic respiration, glycolysis, oxidative phosphorylation.

respiratory index the amount of carbon dioxide produced per unit of dry weight per hour by a respiring organism.

respiratory pigments (1) pigments such as haemoglobin and other haem proteins that form an association with oxygen and carry it from the respiratory surfaces to tissue cells; (2) pigments concerned with cellular respiration, components of the respiratory chain, e.g. cytochromes.

respiratory quotient (RQ) ratio of volume of carbon dioxide produced to volume of oxygen used in respiration.

respiratory substrate any substance that can be broken down by living organisms during respiration to yield energy.

respiratory surface the surface at which gas exchange occurs between the environment and the body, such as gill lamellae, alveoli of lungs, etc.

respondent behaviour animal behaviour performed in response to an obvious stimulus.

response *n.* (1) the activity of a cell or organism in terms of movement, secretion of hormones or other substances, enzyme production, changes in gene expression, etc., as a result of a stimulus; (2) the behaviour of an organism as a result of fluctuations in the environment.

response latency the time interval between a stimulus and the response.

resting metabolic rate (RMR) the metabolic rate, esp. of an animal, measured at rest. *cf.* basal metabolic rate.

Restionales *n.* order of xeromorphic tufted or climbing monocots including the family Restionaceae and others.

restoration ecology the study of ecology with a view to the re-establishment of vegetation on derelict land.

resurgence *n.* (*geog.*) the reappearance of an underground river at the surface. This frequently occurs when a permeable rock, such as limestone, meets an impermeable rock layer.

reticulate evolution evolution caused by recombination and splitting into groups among several interbreeding populations.

retinol vitamin A$_1$.

retro- prefix derived from L. *retro*, backwards.

retrogressive *n. appl.* evolutionary trends towards more primitive rather than more complex forms.

Retroviridae, retroviruses *n., n.plu.* family of enveloped, single-stranded RNA viruses including the RNA tumour viruses and the human immunodeficiency virus (HIV) as well as many apparently harmless, non-oncogenic viruses. They have a unique life history, copying their RNA into DNA by means of the viral enzyme reverse transcriptase. The DNA then enters a host cell chromosome, where it may continue to direct the production of virus particles, or may remain quiescent for many cell generations. The integrated DNA (the provirus) is passed on to all the cell's progeny. Vertebrates appear to carry a number of so-called endogenous proviruses permanently in their genomes without any ill effects.

retting *n.* part of the processing of flax into linen in which the flax stems are soaked in water and softened by the action of bacteria. After retting the fibres can be separated.

re-use *n.* type of recycling in which items are re-used with minimal reprocessing, e.g. returnable milk bottles, which are cleaned and refilled.

reverse fault a fault (*q.v.*) where one block of crustal rock is forced to ride over another along a fault plane. This is caused by compression of the crustal rocks into a smaller horizontal space. The overriding block is referred to as upthrown while the block forced down is termed downthrown. The upthrown block initially forms a fault scarp (*q.v.*) that overhangs the downthrown block.

reverse osmosis process in which pressure is applied in order to reverse the normal process of osmosis (*q.v.*). Reverse osmosis can be used in desalination. In this instance, the application of pressure to a solution of salt separated from pure water by a semipermeable membrane forces water to move from the brine into the pure water.

reversing dune a dune with two periodically opposing slip-faces (*q.v.*), which develops in response to a strong seasonal wind pattern. Reversing dunes may be regarded as a type of transverse dune *q.v.*

reversing tidal current (*oceanogr.*) a tidal current whose direction is reversed during part of the tidal cycle. These occur in coastal rivers and harbours for example.

reversion *n.* a mutation that reverses the effects of a previous mutation in the same gene. *alt.* back mutation.

reward positive reinforcement *q.v.*

Rh symbol for the chemical element rhodium *q.v.*

rhabdocoel *n.* member of the order Rhabdocoela, small turbellarian flatworms whose intestines are simple and sac-like.

Rhabdoviridae, rhabdoviruses *n., n.plu.* family of bullet-shaped, enveloped, single-stranded RNA viruses including rabies and vesicular stomatitis virus.

Rhaetic, Rhaetian *a. appl.* (1) fossils found in marls, shales and limestone; (2) stage that occurs between the Triassic and Lias.

Rhamnales *n.* order of trees, shrubs or woody climbers including Rhamnaceae (buckthorn) and Vitaceae (grape).

rhamnose *n.* six-carbon (hexose) sugar found in the lipopolysaccharide outer membrane of some Gram-negative bacteria, and in plant cell wall polysaccharides.

Rheiformes *n.* an order of flightless birds, including the rheas.

rhenium (Re) *n.* a rare-earth silvery-grey metallic element (at. no. 75, r.a.m. 186.2).

rheophile, rheophilic *a.* preferring to live in running water. *n.* rheophily.

rheophyte *n.* plant that lives in running water.

rheoplankton *n.* the plankton of running waters.

rheotaxis *n.* a taxis in response to the stimulus of a current, usually a water current.

rheotropism *n.* a growth curvature in response to a water or air current. *a.* rheotropic.

rhinitis, allergic hayfever *q.v.*

rhinoviruses *n.plu.* a numerous group of RNA viruses of the family Picornaviridae, the cause of the common cold and similar minor respiratory ailments in humans.

rhipidistians *n.plu.* group of extinct crossopterygian fish existing from Devonian to Permian times, and believed to include the ancestors of land vertebrates.

rhiz-, rhizo- prefix derived from Gk *rhiza*, a root.

rhizobacteria *n.plu.* soil bacteria associated with root surfaces.

rhizobia *n.plu.* soil bacteria of the genus *Rhizobium* and related genera, Gram-negative rods that form nodules on the roots of leguminous plants, in which they carry out symbiotic nitrogen fixation.

rhizoid *a.* root-like.

rhizoid *n.* a filamentous outgrowth from algal or bryophyte prothallus that functions like a root.

rhizome *n.* a thick horizontal stem usually underground, bearing buds and scale leaves, sending out shoots above and roots below.

rhizomorph *n.* a root-like or bootlace-like structure formed from interwoven hyphae in some basidiomycete fungi such as honey fungus (*Armillaria*).

rhizophagous *a.* root-eating.

rhizoplane *n.* part of the rhizosphere immediately adjacent to the root surface, comprising a layer *ca.* 1 μm thick.

Rhizopodea, rhizopods *n., n.plu.* class of Protozoa, of the superclass Sarcodina, mainly free-living, found in freshwater and marine habitats, including the amoebae and foraminiferans, characterized by pseudopodia and no flagella. In some groups, such as the foraminiferans, the body is surrounded by a casing or test, sometimes calcified.

rhizosphere *n.* area of soil immediately surrounding and influenced by plant roots.

rhodium (Rh) *n.* shiny silvery-white metallic element (at. no. 45, r.a.m. 102.91), found naturally with platinum group metals and in cupro-nickel deposits.

rhodochrosite *n.* $MnCO_3$, a manganese ore mineral.

rhodonite *n.* $(Mn,Fe,Ca)SiO_3$, one of the pyroxenoid minerals (*q.v.*). Found in deposits of manganese ores. Its pink–brown colour renders it valuable as a decorative stone.

Rhodophyta, Rhodophyceae *n.* the red algae, a group of largely multicellular, structurally complex photosynthetic organisms classified either in the plant kingdom or as a division of the Protista. They are composed of close-packed filaments. The red colour is due to water-soluble phycobilin pigments. The storage carbohydrate is floridean starch, resembling amylopectin. The red algae are largely marine and there are many tropical species, and they usually grow attached to rocks or other substrates. Their cells have no flagella at any stage.

rhomb porphyry a variety of microsyenite *q.v.*

R-horizon the unweathered parent rock, or bedrock, at the base of a soil. *see* Fig. 28, p. 385. *alt.* D-horizon.

Rhyncocephalia, rhynchocephalians *n., n.plu.* order of mainly extinct reptiles, with one living member, the tuatara (*Sphenodon punctatus*), a lizard-like animal confined to a few islands off New Zealand. *alt.* Sphenodonta.

rhyolite *n.* a very fine- to fine-grained volcanic rock with the same mineralogy as granite (*q.v.*). May be porphyritic and may contain amygdales or vesicles.

ria *n.* a river valley, usually perpendicular to the coast, submerged by a relative rise in sea-level. This is the fluvial equivalent of the glacial feature called a fjord.

RIA radioimmunoassay *q.v.*

ribbon lakes finger lakes *q.v.*

ribbon worms Nemertea *q.v.*

ribitol *n.* sugar alcohol derived from ribose, a constituent of the teichoic acids of bacterial cell walls.

riboflavin *n.* vitamin B$_2$, water-soluble vitamin consisting of ribose linked to the nitrogenous base dimethylisoalloxazine, synthesized by all green plants and most microorganisms, occurring free in milk and in some tissues of higher organisms and green plants, and in all living cells as a component of the coenzymes flavin adenine dinucleotide (FAD) and flavin mononucleotide (FMN). Liver, yeast and green vegetables are particularly rich sources. Deficiency causes skin cracking and lesions (ariboflavinosis).

ribonuclease (RNase) *n.* any of various enzymes that cleave RNA into shorter oligonucleotides or degrade it completely into its constituent ribonucleotide subunits. *alt.* nuclease.

ribonucleic acid *see* RNA.

ribonucleoprotein (RNP) *n.* any complex of RNA and protein.

ribonucleoside *n.* a nucleoside (*q.v.*) containing the sugar ribose.

ribonucleotide *n.* a nucleotide (*q.v.*) containing the sugar ribose.

ribose *n.* a pentose sugar, present in RNA and also an intermediate in the Calvin cycle of photosynthesis.

ribosomal DNA (rDNA), ribosomal genes the DNA encoding ribosomal RNAs, which in many eukaryotes are present in many copies and clustered in chromosomal regions that form the nucleolar organizers.

ribosomal RNA (rRNA) major component of ribosomes and the most abundant RNA species in cells. In eukaryotes it is synthesized in the nucleolus from rRNA genes serially repeated many times in the chromosomes. Several different types known, denoted by their sedimentation coefficients, e.g. 23S, 16S and 5S RNAs in eukaryotic ribosomes.

ribosome *n.* small particle found in large numbers in all cells both free in cytoplasm and attached to the endoplasmic reticulum in eukaryotic cells. It is composed of RNA and protein, and is the site at which translation of messenger RNA and protein synthesis takes place.

ribozyme *n.* an RNA molecule with enzymatic acitivity.

ribulose (Ru) *n.* five-carbon ketose sugar that as phosphate and bisphosphate is involved in carbon dioxide fixation in photosynthesis and in other metabolic pathways such as the pentose phosphate pathway.

ribulose 1,5-bisphosphate (RuBP) five-carbon sugar phosphate that is the primary carbon dioxide acceptor in photosynthesis. *alt.* (formerly) ribulose 1,5-diphosphate (RuDP).

ribulosebisphosphate carboxylase/oxygenase (RuBPc/o, Rubisco) enzyme found in chloroplasts of all green plants and in photosynthetic bacteria that catalyses the fixation of carbon dioxide into carbohydrate via ribulose 1,5-bisphosphate as acceptor, also having oxygenase activity, which is involved in photorespiration, estimated to be one of the most abundant enzymes on Earth.

rice *n. Oryza sativa*, a small-grained cereal crop with a long history of cultivation, grown predominantly in South-East Asia. Rice provides staple food for about 50% of the world's population. The vast majority of rice is cultivated under wetland conditions (known as paddy rice or wetland rice) but some is grown on land not subjected to flooding (known as dryland rice or upland rice).

rich *a. appl.* mixtures of fuel and air in which the amount of air is less than that required for complete combustion of the fuel. *cf.* lean. Tuning an internal combustion engine from a lean to a rich mixture tends to suppress its fuel efficiency, while increasing the amounts of unburnt hydrocarbons and carbon monoxide produced and decreasing the quantity of NO$_x$ generated. Rich fuel mixtures were used in conjunction with oxidation catalyst-based catalytic converters (*q.v.*) before three-way catalytic converters (*q.v.*) were introduced.

Richter scale a logarithmic scale used to assess the magnitude of earthquakes, based on the amplitude of the ground motion produced during earthquakes. A magnitude of 2 or less means a barely perceptible tremor, a magnitude of 5 means a potentially destructive earthquake, and a magnitude of 8 or more will be among the largest earthquakes known.

rickets *n.* inadequate calcification of bone in children, caused by deficiency of vitamin D.

rickettsiae *n.plu.* small prokaryotic obligate parasites, causing typhus and similar fevers in humans and animals and transmitted by ticks, mites and lice. Their simple cells lack cell walls and are obligate intracellular parasites of mammalian cells. Diseases caused by louse-borne rickettsiae include typhus fever and trench fever. Tick-borne rickettsiae cause Rocky Mountain spotted fever and other tick-borne typhuses, while scrub typhus is transmitted by mites. *sing.* rickettsia.

ridge *n.* (*clim.*) elongated region of high pressure sandwiched between two regions of lower pressure.

riebeckite *n.* $Na_2(Mg,Fe,Al)_5Si_8O_{22}(OH)_2$ or $Na_2Fe_5Si_8O_{22}(OH)_2$, one of the amphibole minerals (*q.v.*). Crocidolite (*q.v.*), also called blue asbestos, is a fibrous variety of riebeckite.

rifampicin *n.* semisynthetic derivative of the antibiotic rifamycin.

rifamycin *n.* antibiotic from *Streptomyces* spp. that specifically inhibits the initiation of RNA synthesis in bacterial cells.

riffle *n.* shallow broken water in a stream running over a stony bed. *see also* pool and riffle sequence.

rift *n.* a major, extended down-faulted block, i.e. graben. These form rift valleys, e.g. the East African rift valley.

rift valley *see* rift.

right of way (1) legal right (established by long usage or grant) of a person to traverse land belonging to another; (2) also *appl.* footpath, bridleway or road that a person can lawfully use, esp. one that crosses another's land.

rigor *n.* contraction and loss of irritability of muscle on heating or after death, or in some states such as shock, fever, etc.

rill *n.* small channel (only a few centimetres deep and wide), usually transitory in nature, which forms naturally in arid and semi-arid regions as a result of heavy rainfall. Rills can also develop in more humid regions where the natural vegetation cover has been removed.

rill erosion the erosion of slopes by rills *q.v.*

rime *n.* deposit of ice found on the windward side of solid objects such as trees and telegraph poles, formed when super-cooled droplets of water freeze upon contact.

ring barking girdling *q.v.*

ringed worms common name for the annelids *q.v.*

Ring of Fire circum-Pacific zone of earthquakes and volcanoes. *alt.* Pacific Ring of Fire.

ring-porous *appl.* (1) wood in which the vessels tend to be larger and have thinner walls than in diffuse-porous wood; (2) wood in which the vessels formed early in the season are clearly larger than those formed later, producing a clear ring in cross-section.

ring road road or roads along which through traffic is directed in order to avoid a city or town centre.

ring species two species that overlap in range and behave as true species with no interbreeding, but are connected by a series (the ring) of interbreeding subspecies so that no true specific separation can be made.

Rio Conference, Rio Summit the United Nations Conference on Environment and Development (*q.v.*), held in 1992.

riparian *a.* frequenting, growing on, or living on the banks of streams or rivers.

rip current (*oceanogr.*) a narrow current that flows seaward from a beach, carrying sediment. Rip currents tend to occur at regular intervals along the shore and are fed by longshore currents.

ripple marks features originally created by the action of wind or waves on ancient sands that have been preserved during the formation of sedimentary rocks.

risk analysis the identification of hazards and the evaluation of their economic and social costs. *alt.* risk assessment.

risk–benefit analysis the evaluation of long- and short-term risks associated with a project or process and their offsetting against the economic and social benefits of the project.

river blindness onchocerciasis *q.v.*

river capture capture of the discharge of one river by another river via the process of headwater erosion (*q.v.*). The river whose headwaters have been captured is known as the beheaded stream and the place of capture is often marked by a sharp bend (known as the elbow of capture). *alt.* abstraction, beheading, stream piracy.

river erosion *see* fluvial erosion.

riverine *a.* (1) of rivers; (2) living in rivers.

river profile longitudinal section of the course of a river, from source to mouth. *alt.* thalweg.

river terrace a flat bench of land, formerly the floodplain of a river. River terraces are usually formed as a result of vertical erosion by the river and are separated from the river's new floodplain by steep slopes.

RMR resting metabolic rate *q.v.*

r.m.s. *see* standard deviation *q.v.*

Rn symbol for the chemical element radon *q.v.*

RNA *n.* ribonucleic acid, a nucleic acid composed of a single chain of ribonucleotide subunits of variable sequence, containing the bases uracil, guanine, cytosine and adenine. Found in all cells as transfer RNA (*q.v.*), ribosomal RNA (*q.v.*) and messenger RNA (*q.v.*), all cellular RNAs being synthesized by transcription (*q.v.*) of chromosomal DNA. It is the primary genetic material in some viruses. *cf.* DNA.

RNA viruses *n. plu.* viruses having RNA as their genetic material, and including the Reoviridae, Togaviridae, Coronaviridae, Picornaviridae, Caliciviridae, Rhabdoviridae, Paramyxoviridae, Orthomyxoviridae, Arenaviridae, Bunyaviridae, and Retroviridae among vertebrate viruses, and most plant viruses. *cf.* DNA viruses. *see* Appendix 10.

RNP ribonucleoprotein *q.v.*

Roaring Forties (*clim.*) term coined by early sailors to describe the belt of strong and stormy westerlies encountered around latitude 40° in the southern hemisphere. At latitudes 50° S and 60° S, these westerly winds were known as the Furious Fifties and Screaming Sixties, respectively.

robust *a.* heavily built, *appl.* australopithecines, *Australopithecus robustus.*

roche moutonnée an asymmetrical mound shaped by glacial processes. The smoothed up-glacier, i.e. upstream, side is the result of abrasion, while the craggy down-glacier side has been produced by plucking (*q.v.*). Larger forms are known as flyggbergs.

rock *n.* naturally occurring non-living solid material that makes up the Earth's crust.

Although rocks may be composed of only one mineral (*q.v.*), they are usually an assemblage of several minerals. Rocks need not be hard; the term encompasses such materials as clay and sand. *see also* igneous rocks, metamorphic rocks, sedimentary rocks.

rock crystal *n.* SiO_2, colourless quartz.

rock cycle the cycling of rock as illustrated in Fig. 24.

rockfall *n.* fall movement (*q.v.*) of rock fragments.

rock flour fine particles of silt and clay formed in glaciers by the grinding up of debris.

rock gypsum an evaporite rock that is essentially gypsum ($CaSO_4.2H_2O$) but that frequently also contains anhydrite, calcite, halite, dolomite, iron oxides and clay minerals.

rocking stones *see* erratics.

rock phosphate (*agric.*) inorganic phosphorus fertilizer, of variable chemical composition, with a phosphorus content equivalent to 25–30% P_2O_5, supplied in powder form (often highly insoluble).

rock salt an evaporite rock that is essentially halite. Also used as a synonym for halite *q.v.*

rock sea felsenmeer *q.v.*

rockslide *see* slide movement.

rock varnish desert varnish *q.v.*

Rodentia, rodents *n., n.plu.* the largest order of placental mammals, known from the Paleocene and including rats, mice, voles, hamsters, porcupines, beavers and squirrels. They are omnivorous or herbivorous, and have continously growing chisel-like incisors adapted for gnawing and no canines.

rodenticide *n.* chemical that kills rodents such as rats and mice.

roding *n.* the patrolling flight of birds defending territory.

Rodolia cardinalis an Australian ladybird beetle used successfully in the biological control of cottony-cushion scale *q.v.*

roentgen (r, R) unit of ionizing radiation corresponding to an amount of ionizing radiation sufficient to produce two ionizations per cubic micrometre of water or living tissue.

Röntgen rays X-rays *q.v.*

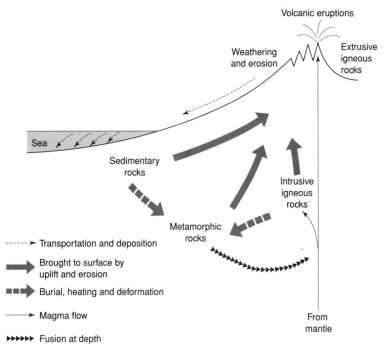

Volcanic eruptions

Weathering
and erosion

Extrusive
igneous
rocks

Sea

Sedimentary
rocks

Intrusive
igneous
rocks

Metamorphic
rocks

- - - -> Transportation and deposition

Brought to surface by
uplift and erosion

Burial, heating and deformation

——→ Magma flow

From
mantle

▶▶▶▶▶ Fusion at depth

Fig. 24 The rock cycle.

root *n*. (1) (*bot*.) descending portion of plant, fixing it in soil, and absorbing water and minerals, and having a characteristic arrangement of vascular tissues; (2) (*zool*.) embedded portion of tooth, hair, nail or other structure.

root cap a protective cap of tissue at the extreme tip of a plant root.

root hairs unicellular outgrowths from epidermal cells of plant roots, by which the plant takes up water and solutes from soil.

root-mean-square value standard deviation *q.v.*

root nodule structure formed on the roots of leguminous and some non-leguminous plants that contains nitrogen-fixing bacteria.

root parasitism condition shown by semiparasitic plants, the roots of which penetrate the roots of neighbouring plants and draw nutrients from them.

root sheath *n*. a protective sheath surrounding the developing radicle of some flowering plants such as grasses, *alt*. coleorhiza.

rootstock *n*. (1) more-or-less underground part of stem; (2) rhizome *q.v.*

root tubers swollen food-storing roots of certain plants such as lesser celandine and orchids.

rosaceous *a*. (1) resembling a rose; (2) *pert*. the Rosaceae family of flowering dicot plants, which includes, as well as the roses, many members cultivated for their edible fruits, such as strawberries, cherries, apples, plums, etc.

Rosales *n*. order of mainly woody dicots, including the families Chrysobalanaceae (coco plum), Neuradaceae and Rosaceae (rose, apple, etc.)

rosin *n*. a resin obtained from pine.

Rossby waves horizontal wave-like motions in the mid-latitude westerly circumpolar upper tropospheric air flow. Rossby waves vary with time in their number, position and amplitude, taking as little as one or two months to form and decay.

rostral *a*. (1) towards the anterior end of the body; (2) *pert*. a beak or rostrum.

rotary current (*oceanogr*.) a tidal current that perpetually changes direction. Rotary currents occur in the open ocean, where there are no obstructions.

rotation (*agric*.) *see* crop rotation.

rotational bush fallow bush fallow *q.v.*

rotational grazing livestock grazing system in which a series of grazing areas are used in rotation, thus allowing the regeneration of pasture during the intervening rest periods. *see also* paddock grazing.

rotenone *n*. derris ($C_{23}H_{22}O_6$), a natural contact insecticide and acaricide extracted from the root of the *Derris* tree. It is highly toxic to fish but relatively safe for mammals and birds and is rapidly degraded. In common use as a household and garden insecticide.

Rotifera, rotifers *n*., *n.plu*. phylum of microscopic, multicellular, pseudocoelomate animals, living mostly in fresh water. They are generally cone-shaped with a crown of cilia at the widest end surrounding the mouth. The epidermis is separated from the internal organs by a fluid-filled space, the pseudocoel. Formerly called wheel animals, as the crown of beating cilia looks as though it is rotating.

roughage *n*. food rich in indigestible fibrous material.

rough endoplasmic reticulum *see* endoplasmic reticulum.

rough grazing pastoral farming *q.v.*

round dance type of repeated circular dance in bees that alerts other bees to the existence of a food source near the hive.

roundworms *n.plu*. common name for the Nematoda *q.v.*

royal jelly material supplied by worker bees to female larvae in royal cells, which is necessary for the transformation of larvae into queens.

R plasmids plasmids carrying genes for resistance to commonly used antibiotics, found in many disease-causing bacteria. Some R plasmids are transmissible to other bacteria of the same and other related species. *alt*. drug-resistance plasmid.

RQ respiratory quotient *q.v.*

rRNA ribosomal RNA *q.v.*

RRV residual reproductive value *q.v.*

r-selected species, r-strategists species typical of variable or unpredictable environments, characterized by small body size and rapid rate of increase. They are able to colonize early in a succession. *alt*. opportunist. *cf*. K-selected species.

r selection selection favouring rapid rates of population increase, esp. prominent in species that colonize short-lived environments or undergo large fluctuations in population size. *cf*. K selection.

RT reaction time *q.v.*

RTF resistance transfer factor *q.v.*

Ru (1) symbol for the chemical element ruthenium (*q.v.*); (2) ribulose *q.v.*

rubber *n*. the coagulated latex of several trees, mainly *Hevea* spp., being long-chain polymers of isoprene and hydrocarbons.

rubellite *n*. a variety of the mineral tourmaline *q.v.*

rubiaceous *a*. *pert*. a member of the dicot family Rubiaceae (the blackberries, raspberries, etc.).

rubidium (Rb) *n*. highly reactive silvery white metal (at. no. 37, r.a.m. 85.47), which is used in photoelectric cells.

rubifaction *n*. soil-forming process in which the soil becomes reddened as a result of hematite (Fe_2O_3) formation.

Rubisco ribulose bisphosphate carboxylase/oxygenase *q.v.*

RuBP ribulose 1,5-bisphosphate *q.v.*

ruby *n*. a variety of the mineral corundum (*q.v.*) that is red due to the presence of trace levels of chromium. Used as a gem.

ruby silver proustite (*q.v.*) and pyrargyrite (*q.v.*) are known as the ruby silver ores.

rudaceous rocks *see* clastic rocks.

ruderal *a*. (1) growing among rubbish or debris; (2) growing by the roadside or in disused fields.

ruling reptiles archosaurs *q.v.*

rumen *n*. in ruminants (cud-chewing mammals) the first stomach, in which food is digested by bacteria and from which it can be regurgitated into the mouth for further chewing.

ruminants *n.plu*. herbivorous mammals, such as cows, sheep, goats, deer, antelopes and giraffes, that chew the cud and have complex, usually four-chambered, stomachs containing microorganisms that break down the cellulose in plant material.

runner *n.* a specialized stolon consisting of a prostrate stem rooting at the node and forming a new plant that eventually becomes detached from the parent, as in the strawberry.

runoff *n.* the drainage of water into streams and rivers from hard surfaces or from soil.

rupestrine, rupicoline, rupicolous *a.* growing or living among rocks.

rural *a. pert.* countryside, country lifestyles, farming, *appl.* e.g. planning, development, housing, population. *cf.* urban.

rush *see* Juncaceae.

rust fungi common name for a group of basidiomycete fungi, the Uredinales, many of which are serious and widespread plant pathogens of numerous important crops, causing rust disease. They are characterized by the production of thick-walled teleutospores that germinate to produce a basidium and basidiospores, which give rise to a mycelium and, in many rusts, one or more other kind of binucleate spore. They have complex life histories on two different hosts, e.g. the black stem rust of cereals (*Puccinia graminis*) uses the barberry (*Berberis vulgaris*) as an intermediate host.

rut *n.* period of sexual heat in male animals, when they often fight for females, defend territory, etc., before mating.

Rutales *n.* order of dicot trees and shrubs, rarely herbs, with leaves often dotted with glands, and including the families Anacardiaceae (cashew), Meliaceae (mahogany), Rutaceae (citrus, rue), Simaroubaceae and others.

ruthenium (Ru) *n.* brittle shiny grey element (at. no. 44, r.a.m. 101.07) found in small amounts in some platinum ores.

rutile *n.* TiO_2, a titanium ore mineral.

rye *n. Secale cereale*, a small-grained temperate cereal crop, grown particularly in Eastern Europe, where it is used to make 'black bread'.

S

σ sigma, (1) symbol for standard deviation (*q.v.*); (2) occasionally used as a symbol for 0.001 seconds.

s second (of time) *q.v.*

S (1) symbol for the chemical element sulphur (*q.v.*); (2) serine (*q.v.*); (3) Svedberg unit *q.v.*

S₁ first selfing generation, the offspring of a self-cross.

sabuline *a.* (1) sandy; (2) growing in sand, especially coarse sand.

saccharomycetes yeasts *q.v.*

Saccharum officinarum sugar-cane *q.v.*

saddle col *q.v.*

safe concentration the maximum concentration of a toxic substance that has no observable effect on a species after long-term exposure over one or more generations.

safety rods control rods *q.v.*

sagittal *a. appl.* section or division in median longitudinal plane.

sago palm *see* Cycadophyta.

Saharo-Arabian Floral Region part of the Palaeotropical Realm comprising Arabia and an area on the eastern side of the Persian Gulf.

Sahel *n.* the semidesert region that lies between the Sahara Desert and the tropical coast of West Africa.

Saint Anthony's fire historical name for ergotism. *see* ergot.

Saint John's wort Klamath weed *q.v.*

salamanders *see* Urodela.

salcrete *n.* duricrust (*q.v.*) formed by the deposition of sodium chloride (NaCl). *alt.* halite crust.

Salicales *n.* order of dicot trees and shrubs comprising the family Salicaceae (willow).

Salientia *n.* in some classifications, the name given to the order of amphibians comprising the frogs and toads. *alt.* Anura.

salination, salinization *n.* the deposition of excessive amounts of soluble mineral salts in the soil, making it unfit for cultivation, caused by high surface evaporation often exacerbated by artificial irrigation over long periods. *see also* upward translocation.

salinity *n.* the concentration of total dissolved salts present in a given water body, e.g. river, lake or sea. In the case of seawater, this is generally assessed by measuring its electrical conductivity and is expressed in grams of total dissolved salts per kilogram of water.

salmonellae *n.plu.* bacteria of the genus *Salmonella* that include species causing food poisoning and the causal agent of typhoid fever, *S. typhi. sing.* salmonella.

Salmonidae, salmonids *n., n.plu.* a family of the Salmoniformes that includes the genus *Salmo* (including salmon, rainbow and brown trout).

Salmoniformes *n.* large order of fairly primitive marine and freshwater teleost fishes, including trout, salmon, pike, etc.

salps Thaliacea *q.v.*

SALR saturated adiabatic lapse rate *q.v.*

salsuginous *a.* growing in soil impregnated with salts, as in a salt marsh.

salt *n.* (1) sodium chloride (NaCl), *alt.* common salt; (2) the ionic product of a reaction between an acid and a base.

saltation *n.* (1) a jumping movement; (2) in a river, a transportation process whereby sand- and gravel-sized particles are bounced along the stream bed; (3) due to wind, the transportation process whereby particles are bounced along the ground surface by the wind; (4) (*evol.*) the idea that major evolutionary changes can take place within a single generation through 'macromutations' – mutations of large effect. In its most extreme form this idea is no longer held by most modern evolutionary biologists. *cf.* phyletic gradualism.

saltatorial *a.* adapted for, or used in, leaping, *appl.* limbs of jumping insects.

saltatorians *n.plu.* crickets and grasshoppers, members of the insect order Orthoptera (called Saltatoria in some classifications).

salt bond, salt bridge electrostatic bond *q.v.*

salt flat playa *q.v.*

saltigrade *a.* moving by leaps, as some insects and spiders.

salt linkage electrostatic bond *q.v.*

salt marsh an intertidal mud flat that forms, e.g., behind a spit or in a sheltered bay on temperate coasts and is colonized by salt-tolerant plants, supporting characteristic plant and animal communities.

saltwater water containing a considerable proportion of salt (>3% or 30 parts per thousand).

saltwater intrusion incursion of seawater (*q.v.*) into freshwater aquifers in coastal regions. This occurs when abstraction from such freshwater reserves exceeds their replenishment.

salt weathering type of physical weathering that occurs in coastal areas and hot desert regions, caused by the evaporation of slightly saline water within rocks and the subsequent formation of salt crystals.

samara *n.* winged fruit, typical of elm and ash, composed of a single winged achene.

samarium (Sm) *n.* a lanthanide element (at. no. 62, r.a.m. 150.35), a soft silver-coloured metal when in elemental form. It is used, e.g., to absorb neutrons in nuclear reactors.

sample *n.* part of a whole taken in order to carry out chemical, physical, biological and/or statistical analysis.

sampling, population selecting a small portion, i.e. the sample, of a population as representative of the population as a whole. *see* random sampling, transect sampling.

Samun *see* föhn.

San Andreas fault transform boundary (*q.v.*) between the North American and Pacific lithospheric plates, located along the Californian coast. This fault is associated with earthquake activity.

sand *n.* (1) a soil textural class, *see* texture; (2) those mineral grains present in a soil's fine earth with diameters in the range 0.02–2 mm (in the International or Atterberg system), 0.05–2 mm (in the USDA system) or 0.06–2 mm (in the system used by the Soil Survey of England and Wales, British Standards and the Massachusetts Institute of Technology).

sand dollars common name for sea urchins of the order Clypeasteroidea, having round flattened tests.

sand sea erg *q.v.*

sandspit spit *q.v.*

sandstone *n.* a clastic sedimentary rock in which one-quarter or more by volume of the grains are sand-sized, i.e. 0.0625–2 mm.

sandstorm *n.* a cloud of sand-sized particles, typically extending only a few metres above ground level, blown into the air by strong winds and that restricts horizontal visibility to <1000 m. Sandstorms are usually found only in arid areas.

sandur outwash plain *q.v.*

sand waves submerged features, with a wave length of 200–300 m, formed offshore in coastal zones with a tidal range of less than 2 m.

sanguicolous *a.* living in blood.

sanguivorous *a.* feeding on blood.

sanidine *n.* $KAlSi_3O_8$, one of the feldspar minerals *q.v.*

sanitary landfill controlled tipping *q.v.*

sanitation *n.* (1) in relation to public hygiene or public health, the practice of hygienic methods of waste disposal, water provision, etc., to prevent disease; (2) the removal of dead plant tissues to remove sites of entry for pathogens.

Santales *n.* an order of woody dicots, often parasitic on other angiosperms or rarely on gymnosperms and including the families Santalaceae (sandalwood), Viscaceae (mistletoe) and others.

Santotherm a trade name under which polychlorinated biphenyls (PCBs) were sold.

sap *n.* (1) sugary fluid carried by phloem in plants; (2) cell sap *q.v.*

Sapindales *n.* order of dicot trees, shrubs, lianas and rarely herbs, including the Aceraceae (maple), Hippocastanaceae (horse chestnut), Sapindaceae (soapberry) and others.

saponin *n.* any of various steroid glycosides present in many plants, such as soapwort and soapbark, which produce a soapy solution in water.

sapphire *n.* a crystalline, transparent variety of the mineral corundum (*q.v.*) that is blue due to the presence of trace levels of cobalt or other metals. Used as a gem.

sapping *n.* (*geog.*) process by which the base of a cliff is undermined, e.g. by groundwater outflow (spring sapping) or wave action, causing collapse of the cliff face.

sapric *a. appl.* organic soil material in the most advanced stage of decomposition.

saprist *see* histosol.

saprobe *n.* a heterotrophic organism (usually *appl.* microorganisms) that lives by breaking down or decomposing dead organic matter. *alt.* saprobiont, saprophage, saprotroph. *a.* saprobic.

saprobic *a.* living on dead and decaying matter, breaking it down in the process. *alt.* saprophagous, saprophytic (*appl.* plant, fungi and bacteria), saprozoic (*appl.* animals). *n.* saprobe.

saprobic classification, Saprobien system a biotic index (*q.v.*), used esp. in continental Europe, for assessing the degree of organic pollution of a body of water, which is based on recognition of four stages in the oxidation of organic matter, each characterized by the presence and relative abundance of certain groupings (saprobic groupings) of indicator species, *see* alpha-mesosaprobic, beta-mesosaprobic, oligosaprobic, polysaprobic.

saprobiont saprobe *q.v.*

saprogenic, saprogenous *a.* (1) causing decay; (2) resulting from decay.

saprolite *n.* material produced *in situ* by deep weathering of the bedrock, esp. of crystalline rocks such as gneiss and granite. *see also* regolith.

sapropelic *a.* living among the debris of bottom ooze.

saprophage saprobe *q.v.*

saprophagous *a.* feeding on dead and decaying organic matter, *appl.* microorganisms.

saprophyte *n.* plant, fungus or bacterium that gains its nourishment directly from dead or decaying organic matter. *a.* saprophytic.

saprophyte chain a food chain starting with dead organic matter and passing to saprophytic microrganisms.

saprotroph *n.* any organism that feeds on dead organic matter. *a.* saprotrophic.

saprozoic *a. appl.* animal that lives on dead or decaying organic matter. *n.* saprozoite.

sapwood *n.* the more superficial, younger, paler, softer wood of trees, which is water-conducting and contains living cells. *alt.* alburnum, splintwood. *cf.* heartwood.

Sarcodina *n.* class or superclass of Protozoa containing the Rhizopodea (amoebae and foraminiferans) and Actinopodea (heliozoans and radiolarians), which have pseudopodia or actinopodia, little differentiation of the body and no flagella at any stage.

Sarcomastigophora *n.* in some classifications a subphylum of the Protozoa containing the flagellates, amoebas, foraminiferans, radiolarians and heliozoans.

Sarcopterygii, sarcopterygians *n., n.plu.* a group of mostly extinct bony fishes having fleshy fins and nostrils opening into the mouth, comprising the lungfishes, Dipnoi (*q.v.*), and the crossopterygians, Crossopterygii *q.v.*

sard *n.* SiO_2, brown varieties of chalcedony (*q.v.*). *cf.* carnelian.

Sargasso Sea the calm central region enclosed by the currents of the large gyre in the North Atlantic Ocean. It is characterized by large floating masses of sargasso weed, the brown seaweed *Sargassum*.

Sarraceniales *n.* order of carnivorous herbaceous dicots, typically with pitcher-like leaves for trapping insects. It comprises a single family, the Sarraceniaceae (pitcher plants).

sastrugi *n.plu.* small dunes of hard-packed snow formed parallel to the prevailing wind direction. These wind-formed surface features are found on the Antarctic and Greenland ice sheets.

satellite RNA encapsidated small plant pathogenic RNAs that require co-infection with a specific virus, known as a helper virus, for replication and encapsidation. *alt.* satellite virus, virusoid.

satellite virus satellite RNA *q.v.*

satin spar a fibrous variety of gypsum *q.v.*

saturated *a. appl.* fatty acids with a fully hydrogenated carbon backbone, and to fats in which they predominate. *cf.* unsaturated.

saturated adiabatic lapse rate (SALR) rate of temperature decrease of an ascending air mass, cooling adiabatically, that is saturated with water vapour. The SALR is influenced by the temperature of the air mass and hence the amount of water vapour present; values range between 0.4 and 0.9 °C of decline per 100 m ascent.

saturated air (*clim.*) air containing water vapour that is in dynamic equilibrium with its condensed phase, i.e. liquid water or ice. The vapour content of saturated air varies with both temperature and pressure. Saturated air is said to have a relative humidity of 100%.

saturated hydrocarbons (*chem.*) hydrocarbons that do not contain multiple bonds between adjacent carbon atoms. *cf.* unsaturated hydrocarbons.

saturated solution any solution in which solute in solution is in dynamic equilibrium with undissolved solute.

saturated vapour any vapour that is in dynamic equilibrium with either its liquid or its solid phase.

saturated vapour pressure the pressure that a saturated vapour exerts. This is independent of the volume occupied by the vapour, but dependent on temperature. Boiling occurs when the atmospheric pressure is equalled by the saturated vapour pressure of the liquid.

saturation level dew-point *q.v.*

saurian *a. appl., pert.* or resembling a lizard.

Saurischia *n.* a large order of Mesozoic archosaurs, commonly called lizard-hipped dinosaurs, including both bipedal carnivores (the theropods) and very large quadrupedal herbivores, e.g. the sauropods.

sauropods *n.plu.* a group of gigantic, herbivorous lizard-hipped dinosaurs that included *Diplodocus*, *Apatosaurus* (*Brontosaurus*) and *Brachiosaurus*.

savanna *n.* (1) subtropical or tropical dry grassland with drought-resistant vegetation and scattered trees; (2) transitional zone between dry grassland or semidesert and tropical rain forest. *alt.* savannah.

sawflies *see* Symphyta.

saxatile, saxicoline, saxicolous *a.* living in, on or among rocks.

Saxifragales *n.* order of dicot trees, shrubs, lianas and herbs, including the families Crassulaceae (orpine), Escalloniaceae (escallonia), Grossulariaceae (gooseberry), Hydrangeaceae (hydrangea), Saxifragaceae (saxifrage) and others.

Sc symbol for the chemical element scandium *q.v.*

scale (*chem.*) fur *q.v.*

scale insect any member of the family Coccidae of the order Hemiptera (bugs), feeding on plants, in which the wingless females remain fixed to the food plant and are covered with a waxy covering, the 'scale'. Some are serious plant parasites, others yield shellac and cochineal. *alt.* mealy bug.

scandium (Sc) *n.* metallic element (at. no. 21, r.a.m. 44.96) occurring in some lanthanide ores.

scanning electron microscope (SEM) electron microscope that produces an image of the surface of the specimen from electrons reflected from that surface.

scansorial *a.* characterized by, or adapted to, climbing.

Scaphopoda *n.* a class of marine molluscs, commonly called tusk shells or elephant-tooth shells, an individual having a tubular shell, a reduced foot and no ctenidia.

scapolite group $(Na,Ca,K)_4Al_3(Al,Si)_3$-$Si_6O_{24}(Cl,F,OH,CO_3,SO_4)$ a group of tecto-silicate minerals found in metamorphic rocks. Includes the solid-solution series that has meionite, $Ca_4Al_6Si_6O_{24}(SO_4,CO_3)$, and marialite, $Na_4Al_3Si_9O_{24}Cl$, as its end-members, with intermediates between these extremes known as wernerite.

scarp and vale topography chalk landscape in which chalk escarpments alternate with clay vales, found in south-east England, UK.

scarp slope *see* cuesta.

scatophagous *a.* dung-eating.

scavenger *n.* an animal that feeds on organic refuse or on animals that have been killed by other predators or have died naturally.

schattenseite slope ubac slope *q.v.*

scheelite *n.* $CaWO_4$, a major tungsten ore mineral.

Schisandrales Illicidales *q.v.*

schistosity *n.* (*geol.*) a fabric (*q.v.*) caused by (and characteristic of) regional metamorphism that consists of foliation characterized by the preferred orientation of mineral grains that are tabular in shape and large enough to be seen by the naked eye.

schistosome *n.* parasitic digenean blood fluke (class Trematoda) infesting mammals, such as *Schistosoma mansoni*, which causes schistosomiasis (bilharzia) in humans in tropical regions. The larvae develop in certain freshwater snails.

schistosomiasis *n.* disease that affects over 200 million people, mainly in the tropics, caused by parasitic digenean blood flukes of the genus *Schistosoma* (class Trematoda), which have their larval stage in certain freshwater snails. Schistosomiasis, although not often fatal, causes chronic fatigue, anaemia and haemorrhaging. *alt.* bilharzia, snail fever.

schist *n.* any rock that exhibits schistosity *q.v.*

school *n.* a group of fish or marine mammals, such as porpoises, that swim together in an organized fashion.

schorl *n.* a variety of the mineral tourmaline *q.v.*

schwingmoor *n.* vegetation that grows out from land over water to form a floating mat on the surface.

scientific method the rational formulation of hypotheses, collection of data and testing of hypotheses against observations or experimental results that is the basis of the scientific approach to explaining natural phenomena.

scientific name the binomial (*q.v.*) Latin name of an organism.

scion *n.* a part removed from a plant and used for grafting.

sciophyll skiophyll *q.v.*

Scitamineae Zingiberales *q.v.*

scleractinians *n.plu.* an order (the Scleractinia or Madreporina) of usually colonial Zooantharia, known as true corals, an individual having a compact calcareous skeleton and polyps with no siphonoglyph.

sclereid *n.* a type of plant cell with a thick lignified wall, making up some seed coats, nutshells, the stone of stone fruits like plums, cherries, etc., and that gives the flesh of pears its gritty texture.

sclerenchyma *n.* (1) (*bot.*) plant tissue with thickened, usually lignified, cell walls, which acts as a supporting tissue; (2) (*zool.*) hard tissue of coral.

sclerophyll *n.* (1) a plant with tough evergreen leaves; (2) one of the leaves of such a plant.

sclerophyllous *a.* hard-leaved, *appl.* leaves that are resistant to drought through having a thick cuticle, much sclerenchymatous tissue and reduced intercellular air spaces.

sclerotium *n.* a resting or dormant stage of some fungi in which they become a mass of hardened or mummified tissue. *plu.* sclerotia.

scolite *n.* a fossil worm burrow.

scombrids *n.plu.* fish of the mackerel and tuna family (Scombridae).

scorpion *see* Scorpiones.

Scorpiones, Scorpionoidea *n.* order of arachnids including the scorpions, an individual having a dorsal carapace on the prosoma, an opisthosoma divided into two regions, with the posterior segments forming a flexible tail bearing a terminal poisonous sting that is used to paralyse prey. They are viviparous.

scorpion flies common name for the Mecoptera *q.v.*

Scotobacteria *n.* the large class of bacteria containing all heterotrophic Gram-negative bacteria and some other heterotrophic groups, the name signifying 'bacteria indifferent to light'.

scotoplankton *n.* plankton that live in poorly illuminated environments.

Scottish Environment Protection Agency government agency in Scotland, UK, which has taken over the former duties of the river purification boards, Her Majesty's Industrial Pollution Inspectorate and responsibilities for regulation of waste and air pollution.

Scottish Natural Heritage (SNH) government agency responsible for nature conservation and countryside policy in Scotland, UK.

scouring *n.* (*agric.*) diarrhoea in domestic livestock, esp. calves.

SCP single-cell protein *q.v.*

scramble competition situation where a resource is shared proportionately between competitors, depending on their density.

scraper *n.* organism that feeds by scraping microorganisms, such as algae, off surfaces.

scrapie *n.* a neurodegenerative disease of sheep, one of the transmissible spongiform encephalopathies, caused by an agent, probably a prion, not yet fully characterized.

Screaming Sixties *see* Roaring Forties.

scree talus *q.v.*

scree slope talus slope *q.v.*

screening *n.* the analysis of a large number of chemicals, environmental samples, individuals, etc., to detect the presence of some particular property, chemical, etc.

Scrophulariales *n.* order of dicot trees, shrubs, herbs and vines, including the families Acanthaceae (acanthus), Bignoniaceae (jacaranda), Buddleiaceae (buddleia), Solanaceae (nightshade, potato) and others.

scrounger *n.* in animal behaviour, an animal that waits for another animal of a different species to catch its prey, and then takes the prey from it. *see also* kleptoparasite.

scrub *n.* a plant community dominated by shrubs or small or stunted trees.

scrubber *n.* general term applied to any device used to clean flue gases, prior to their emission, either through the removal of particulate matter and/or gaseous pollutants, e.g. sulphur dioxide.

scurvy *n.* deficiency disease caused by a lack of vitamin C (ascorbic acid), which, among other symptoms, prevents formation of effective collagen fibres, leading to skin lesions and blood vessel fragility.

Scyphozoa, scyphozoans *n.*, *n.plu.* the jellyfish, a class of marine coelenterates of the phylum Cnidaria, with a dominant medusa stage that is free-swimming or attached by an aboral stalk, and no velum.

s.d. standard deviation *q.v.*

Se symbol for the chemical element selenium *q.v.*

sea anemones common name for an order (Actiniaria) of coelenterates of the Zoantharia, which are generally solitary and have no skeleton. An individual has a hollow cylindrical body, often anchored to rocks, opening at one end in a small mouth surrounded by a ring of tentacles often numbering multiples of six.

sea arch a small island or headland perforated at its base to form an arch.

sea breeze gentle wind that blows onshore during the day, i.e. from the sea to the land. This airflow is created by a decrease in air pressure over the land due to its more rapid heating during the day, compared with that of the adjacent sea. *cf.* land breeze.

sea butterflies pteropods *q.v.*

sea combs common name for the Ctenophora *q.v.*

sea cows the Sirenia, including the dugong and manatee, marine placental mammals highly specialized for an aquatic life with naked bodies and front limbs modified as paddles.

sea cucumbers a common name for the Holothuroidea *q.v.*

sea devils *see* whirlwind.

sea fans gorgonians *q.v.*

sea-floor spreading the production of new oceanic crustal material, due to the upwelling of magma within the mid-oceanic ridges, and the spreading of this material outwards across the sea-floor. *alt.* ocean-floor spreading.

sea gooseberries common name for the Ctenophora *q.v.*

sea lilies common name for the Crinoidea *q.v.*

sea loch Scottish (UK) term for a narrow arm of the sea.

seamounts *n.plu.* completely submerged volcanic mountains that can rise 3000 m above the surrounding abyssal plain. Seamounts with tops worn flat by marine erosion are known as guyots.

sea mouse common name for the polychaete *Aphrodite*, which has a broad stout oval shape.

sea pens group of corals of the subclass Alcyonaria, which form stalked colonies markedly resembling quill pens, composed of two different kinds of polyp.

search(ing) image a transitory filtering of external visual stimuli that enables an animal to focus its attention on finding, e.g., a prey item of a particular colour or shape.

seashore *n.* the ground bordering the sea, between the highest high-water and lowest low-water marks, also including the splash zone above high-water mark. *see also* backshore, beach, foreshore, intertidal,

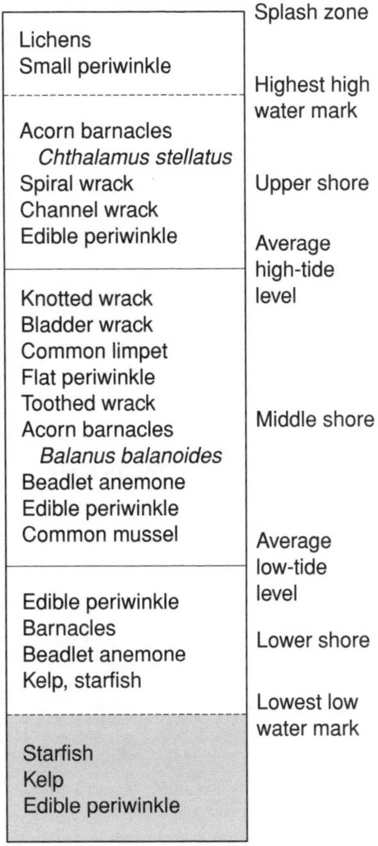

	Splash zone
Lichens Small periwinkle	
	Highest high water mark
Acorn barnacles *Chthalamus stellatus* Spiral wrack Channel wrack Edible periwinkle	Upper shore
	Average high-tide level
Knotted wrack Bladder wrack Common limpet Flat periwinkle Toothed wrack Acorn barnacles *Balanus balanoides* Beadlet anemone Edible periwinkle Common mussel	Middle shore
	Average low-tide level
Edible periwinkle Barnacles Beadlet anemone Kelp, starfish	Lower shore
	Lowest low water mark
Starfish Kelp Edible periwinkle	

Fig. 25 Typical zonation for a sheltered Atlantic European rocky shore.

littoral zone, lower shore, middle shore, splash zone, sublittoral, upper shore. *see* Fig. 25 for an example of shore zonation.

sea slugs shell-less marine molluscs of the class Gastropoda, subclass Opisthobranchia.

sea spiders Pycnogonida *q.v.*

sea squirts a common name for the Ascidiacea *q.v.*

sea stars a common name for the Asteroidea *q.v.*

sea urchins a common name for the Echinoidea *q.v.*

seawater *n.* the typical water of the sea, it has a salinity *ca.* 35 000 p.p.m. total dissolved solids, *ca.* 85% of which is common salt, sodium chloride (NaCl).

sea waves (*oceanogr.*) chaotic waves, with different wave periods (*q.v.*), generated by local winds. As these move away from the zone of generation, they separate into more regular, long-period waves known as swell waves.

seaweed *n.* marine multicellular algae belonging to various groups. *see* Chlorophyta, Phaeophyta, Rhodophyta.

Secchi depth *see* Secchi disc.

Secchi disc simple device for the assessment of water clarity. It consists of a white, or black and white, flat circular plate, 10–14 cm in radius. During use, it is kept horizontal while being lowered into the water. The depth at which it just ceases to be visible from the surface is known as the Secchi depth.

second (of time) (s) (1) the basic SI unit of time, defined as the duration of 9 192 631 770 periods of radiation corresponding to the transition between two hyperfine levels of the ground state of caesium-133; (2) one-sixtieth of a minute of time.

secondary amines *see* amines.

secondary atmospheric circulation regional system of atmospheric circulation, such as the Asian monsoon.

secondary compounds, secondary metabolites compounds produced by plants and microbes, e.g. antibiotics, plant alkaloids, flower pigments, that are not essential to the growth of the organism.

secondary consumer carnivore that eats herbivores.

secondary decomposers microorganisms and small invertebrates, e.g. nematodes, mites, protozoa, that feed on the primary decomposers, i.e. bacteria and fungi, in the soil.

secondary ecological succession secondary succession *q.v.*

secondary effluent effluent produced as a result of the secondary treatment of sewage, *see* sewage treatment.

secondary forest, secondary woodland forest or woodland that has developed as a result of secondary succession or replanting after complete clearance of pre-existing forest, or which has been planted.

369

secondary growth in plants, growth bringing about an increase in the thickness of stem and root, as opposed to extension of plant body at the apices of shoots and roots, and which is most marked in trees and shrubs. It is initiated at lateral meristems which are the vascular cambium, a layer of tissue encircling root and stem between phloem and xylem, producing new xylem and phloem (secondary xylem and phloem) and the cork cambium that contributes to the bark.

secondary landforms *see* landform.

secondary minerals minerals formed from primary minerals by weathering reactions. The majority of secondary minerals are either hydrated metal oxides (mostly of iron and aluminium) or one of the many silicate minerals known as clay minerals *q.v.*

secondary pest outbreak increase in the population of a normally non-pest species to pest status, which may be produced when the normal predators of that species are suddenly killed off.

secondary poisoning poisoning of non-target species by poison put out for another species.

secondary pollutant any environmentally damaging substance that is not released directly into the environment but forms as the result of the chemical alteration of, or interactions between, primary pollutants. Examples are ozone and peroxyacyl nitrates.

secondary producers in an ecosystem, the primary consumers (herbivores).

secondary production in an ecosystem, the yield due to the primary consumers (herbivores).

secondary productivity the rate of biomass formation by the secondary producers (herbivores) in an ecosystem.

secondary recovery *see* oil recovery.

secondary sexual characteristics features characteristic of a particular sex other than the gonads and genitalia, usually developing under the influence of androgens and oestrogens.

secondary sludge sludge produced as a result of the secondary treatment of sewage. *see* trickling filter, activated sludge tank.

secondary standard (*chem.*) a compound or solution, the purity or concentration of which must be established by quantitative reaction with a primary standard prior to its own use as a standard in analytical chemistry.

secondary succession sequence of different plant communities developing over time that takes place when the existing vegetation is abruptly removed by natural disasters or by human activities such as deforestation or strip mining. *cf.* primary succession.

secondary tillage *see* tillage.

secondary treatment of sewage *see* sewage treatment.

second filial generation (F$_2$) the result of crossing two individuals from the first generation (F$_1$) of an experimental cross.

second-growth forest secondary forest *q.v.*

second law of thermodynamics *see* thermodynamics, laws of.

Second World, the term used for industrialized socialist countries with centrally planned economies, such as China.

secrete *v.* (1) to hide; (2) to release material or fluid from a cell or glandular tissue.

secretion *n.* material or fluid that is produced and released from a cell or gland.

section *n.* (1) thin slice of tissue prepared for microscopy; (2) a taxonomic group, often used as a subdivision of a genus, but used in different ways by different authors and never precisely defined.

secular *a.* long-term, over a long period of time.

secured landfill controlled tipping *q.v.*

sedentaria *n.plu.* sessile or sedentary organisms.

sedentary *a.* not free-living, *appl.* animals attached by a base to some substratum.

sediment *n.* (1) soil and rock particles transported and deposited by agents such as winds, rivers or glaciers; (2) particulate material that has settled at the bottom of a waterbody under the force of gravity.

sedimentary deposition the formation of rocks, called sedimentary rocks, by the deposition of the products of weathering and erosion under large waterbodies such as lakes or oceans. Most rocks formed by this process are not orebodies, but there are significant exceptions. Both evaporite formation and chemical precipitation have produced deposits of commercial significance, e.g. of iron ore.

sedimentary rocks rocks produced by the compression, compaction and cementation of sediments, a process known as lithification. Sedimentary rocks may be classified as being either clastic, i.e. formed from fragments of other rocks, or nonclastic.

sedimentation *n.* (*geol.*) the process of sediment (*q.v.*) deposition.

sedoheptulose *n.* seven-carbon ketose sugar, involved esp. in carbon dioxide fixation in photosynthesis.

seed *n.* propagative unit of gymnosperm and angiosperm plants, formed from a fertilized ovule and consisting of an embryo, food store and protective coat.

seed *v.* to introduce microorganisms into a culture medium.

seed bank (1) the viable seeds contained in the top layers of soil; (2) a conservation collection of seeds of wild plant species and local cultivated varieties, usually of important crop plants, kept in long-term storage, usually freeze-dried under liquid nitrogen. Seeds can be germinated as required to provide material for study and for plant breeding programmes.

seed ferns a division of fossil seed-bearing vascular plants that had fern-like leaves bearing seeds.

seed leaf cotyledon *q.v.*

seed legumes leguminous plants, such as peas and beans, grown for their seeds, which are then harvested for human consumption or for animal feed. *alt.* grain legumes, pulse crops.

seed plants all seed-bearing plants, i.e. the gymnosperms and angiosperms, sometimes collectively termed Spermatophyta.

seed storage proteins simple proteins produced in large quantities within seeds, where they act as nitrogen storage compounds, and that are broken down into amino acids then utilized during germination and seedling growth.

seed tree tree left when others are cut, to reseed the surrounding area.

seed-tree cutting type of forestry practice in which nearly all the trees on one site are harvested in one cutting, leaving a few seed-producing, wind-resistant trees

uniformly distributed as a source of seed to regenerate.

segregation of alleles the first of Mendel's laws, which describes the fact that alleles of a gene segregate unchanged by passing into different gametes at the formation of the next generation.

seif dune a type of linear dune (*q.v.*) that is sinuous in form and may extend for over 100 km.

seine netting type of marine fishing in which a seine net, i.e. one designed to hang vertically in the water between floats and weights, is brought slowly together to enclose pelagic fish. *see also* drift netting.

seismic sea wave tsunami *q.v.*

Seismic Sea Wave Warning System system set up in the Pacific to detect earthquake activity and alert relevant authorities to the speed and direction of any resulting seismic sea wave (tsunami).

seismic waves shock waves generated by earthquakes. *see* body waves, surface waves.

seismograph *n.* an instrument used to record the tremors and shocks generated by earthquakes in the Earth's crust.

seismology *n.* the scientific study of seismic waves generated within the Earth by earthquakes or man-made explosions. Seismology not only yields information about the Earth's internal structure but is also invaluable both in the prediction of future earthquakes and in the monitoring of nuclear test treaties.

seismonasty *n.* plant movement in response to mechanical shock or vibration. *a.* seismonastic.

seismotaxis *n.* a taxis in response to mechanical vibrations.

Selachii, selachians *n.*, *n.plu.* a class (or in some classifications an order) of cartilaginous fishes containing the sharks, dogfishes, skates and rays, having claspers and fins with a constricted base, existing from the Devonian to the present day.

selectable marker any characteristic by which a cell with a particular property can be selected during an experiment. One commonly used type of selectable marker is a gene, e.g. for antibiotic resistance, that is placed on a vector so that any cell receiving a recombinant DNA molecule can be selected by survival in a medium containing antibiotics.

selection *n.* (1) the non-random survival and reproduction of one organism compared with another, due either to natural processes or to human intervention, which results in the preferential reproduction of certain genotypes in the conditions prevailing; (2) natural selection (*q.v.*). *see also* directional selection, disruptive selection, sexual selection, stabilizing selection.

selection pressure the effect of any feature of the environment that contributes to natural selection, e.g. food shortage, predator activity, competition from members of the same or other species, etc.

selection, coefficient of *see* coefficient of selection.

selective advantage *pert.* any character that gives an organism a greater chance of surviving to reproductive age, breeding and rearing viable offspring.

selective breeding breeding from selected individuals, e.g. of crop plant or livestock animal, in order to improve attributes such as productivity or disease resistance within the population.

selective cutting the forestry practice of selecting and removing only the most mature trees in a given area. *cf.* clearcutting.

selective herbicide chemical that kills only specific plant species.

selenite *n.* a colourless, transparent variety of gypsum *q.v.*

selenium (Se) *n.* metalloid element (at. no. 34, r.a.m. 78.96) occurring as selenide impurities in sulphide ores and in clansthalite. It occurs in several allotropes, including red and grey solids and a glassy form. It is an essential micronutrient for living organisms but when it occurs in high levels in soils it is toxic to non-native plants and the animals that graze them.

selenophyte *n.* plant tolerating quite high levels of selenium in the soil.

selenotropism *n.* tendency to turn towards the Moon's rays.

self-compatible self-fertile *q.v.*

self-fertile *a.* capable of being fertilized by its own male gametes, *appl.* hermaphrodite animals and flowers.

self-fertilization the fusion of male and female gametes from the same individual.

self-incompatible self-sterile *q.v.*

selfing *n.* self-pollination or self-fertilization.

selfishness *n.* in sociobiology, behaviour that benefits the individual in terms of genetic fitness at the expense of the genetic fitness of other members of the same species.

self-pollination transfer of pollen grains from the anthers to the stigma of the same flower.

self-purification of rivers, the natural process of bacterially mediated decomposition that leads to the eventual recovery of a river from organic pollution.

self-sterile *a.* incapable of being fertilized by its own male gametes, *appl.* hermaphrodite animals and flowers.

self-thinning the natural decrease in density of plants in an area as they grow and compete with each other.

selva *n.* the tropical rain forest.

SEM scanning electron microscope *q.v.*

sematic *a.* functioning as a danger signal, as warning colours or odours, *appl.* warning and recognition markings. *see also* episematic, parasematic.

semelparity *n.* the production of offspring by an organism all at one time. *a.* semelparous. *cf.* iteroparous.

semi- prefix derived from L. *semi*, half, signifying half or partly.

semi-arid regions dry regions with sufficient rainfall (280–400 mm per annum) to support steppe or savanna grassland, and some agriculture. *alt.* semideserts.

semideserts *see* semi-arid regions.

semidominant codominant *q.v.*

semidiurnal tides a tidal regime with two high tides and two low tides approximately every 24 h. These are common on the Atlantic coasts of Europe and the USA.

semilethal *a.* (1) not wholly lethal; (2) *appl.* genes, causing a mortality of more than 50% or permitting survival until reproduction has been effected.

seminatural environments environments in which the basic type of vegetation cover has been altered by human activities but in which there has been no cultivation or intentional alteration of species composition. Examples would be the conversion of land to grassland or heathland after deforestation. *cf.* cultural environments.

seminomadism *see* nomadism.

semiotics *n.* the study of communication.

semioviparous *a.* between oviparous and viviparous, as in marsupials, whose young are imperfectly developed when born.

semiparasite *n.* a plant that only derives part of its nutrients from the host and has some photosynthetic capacity.

semipermeable *a.* (1) partially permeable; (2) *appl.* membrane, permeable to a solvent, esp. water, but not to solutes.

semisaprophyte *n.* a plant partially saprophytic.

semisocial insects those social insects in which there is cooperative care of the brood and a separate sterile worker caste but in which there is no overlap of generations caring for the brood. *cf.* eusocial insects.

semispecies *n.* a taxonomic group intermediate between a species and subspecies, esp. formed as a result of geographical isolation.

semolina *see* durum wheat.

senescence *n.* (1) advancing age; (2) the complex of ageing processes that eventually lead to death. *a.* senescent.

senility *n.* degeneration due to old age.

sensitive *a.* (1) capable of receiving impressions from external objects; (2) reacting to a stimulus.

sensitivity *n.* the degree to which a land system or ecosystem undergoes change due to natural forces after human interference.

sensu lato in a broad sense.

sensu stricto in a restricted sense.

sentient *a.* (1) capable of perceiving through the senses; (2) conscious.

sepal *n.* modified sterile leaf, often green and with others forming the calyx, or outer series of perianth segments, of a flower (*q.v.*). (*see* Fig. 12, p. 157). Sometimes the same colour as and resembling the petals.

sepia *n.* the brown 'ink' released by the cuttlefish *Sepia* to distract predators when threatened.

sepicolous *a.* living in hedges.

sepiolite–attapulgite series *n.* a series of clay minerals (*q.v.*) with structures reminiscent of those of the amphibole minerals *q.v.*

septal *a. pert.* (1) a septum; (2) hedgerows, *appl.* flora.

septarian *a.* (*geol.*) *appl.* a nodule (*q.v.*) or concretion (*q.v.*) that contains irregular veins or cracks.

septic tank a tank used for the treatment of domestic wastewater where households are not connected to mains sewerage. A septic tank acts as a combined sedimentation tank and anaerobic digester. The liquid effluent produced is allowed to soak away into the soil, while the sludge is periodically removed from the tank to be treated in a conventional sewage treatment plant.

septoria diseases *see* glume blotch, leaf blotch.

sequential cropping type of multiple cropping in which one crop is followed directly by another, without an intervening period of fallow. *alt.* successional cropping.

Ser serine *q.v.*

SER smooth endoplasmic reticulum. *see* endoplasmic reticulum.

seral *a.* (1) *pert.* sere (*q.v.*); (2) *appl.* a plant community before reaching equilibrium or climax; (3) *pert.* blood serum.

seration *n.* a series of communities within a biome.

serclimax *n.* a stable community formed as the result of the interruption of a succession by some repeated natural disturbance such as fire.

sere *n.* (1) a successional series of plant communities; (2) a stage in a succession.

serial endosymbiosis theory the idea that mitochondria and chloroplasts, and possibly some other organelles of eukaryotic cells, originated as endosymbiotic microorganisms, mitochondria being acquired first, chloroplasts being acquired later only in the line or lines that led to algae and plants.

series, radioactive *see* radioactive series.

serine (Ser, S) *n.* beta-hydroxyalanine, an amino acid with an aliphatic hydroxyl side chain, non-essential in the human diet, constituent of protein, which is also an important intermediate in phosphatide synthesis.

serology *n.* the study of immune sera and the use of antisera to characterize pathogens, antigens, cells, etc.

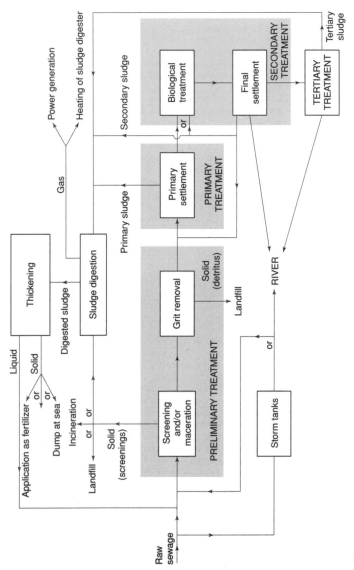

Fig. 26 The sewage treatment process.

serotinal, serotinous *a.* (1) appearing or blooming late in the season; (2) *pert.* later summer; (3) flying late in the evening, *appl.* bats.

serotinous cones cones that remain tightly closed on the tree for years until they open to release the seeds in response to a stimulus such as the heat of a forest fire.

serotype *n.* a subdivision of a species, esp. of bacteria or viruses, characterized by its antigenic character.

Serpentes *n.* suborder of reptiles comprising the snakes.

serpentine *n.* $Mg_3Si_2O_5(OH)_4$, a group of phyllosilicate minerals that includes the polymorphs chrysotile, antigorite and lizardite. The first of these is the most common and is fibrous and, when highly so, is a form of asbestos *q.v.*

serpentinite *n.* a medium- to coarse-grained rock that is composed mainly of serpentine minerals and that has a colour index of >90. It is formed by the alteration (serpentinization) of ultramafic rocks including peridotite (the principal source of serpentinite), dunite and pyroxenite.

serpentinization *n.* alteration that produces serpentine (*q.v.*) minerals.

serule *n.* a succession within a microhabitat.

sesquioxide *n.* oxide of a trivalent element, e.g. aluminium oxide (Al_2O_3), iron(III) oxide (Fe_2O_3).

sessile *a.* (1) sitting directly on base without a stalk or pedicel, *appl.* flowers, leaves, etc; (2) attached or stationary, as opposed to free-living or motile, *appl.* animals, protozoans, etc.

seston *n.* (1) microplankton (*q.v.*); (2) all bodies, living and non-living, floating or swimming in water.

set-aside *n.* (1) formerly productive agricultural land that is deliberately left fallow or used for non-agricultural purposes (such as recreation or forestry) and on which a subsidy is paid to the farmer; (2) the policy of encouraging farmers to establish set-aside areas by the payment of subsidies. This operates in many of the developed nations, with the aim of cutting excess food production.

set of chromosomes the basic haploid set of chromosomes of any organism.

set stocking a grazing system whereby livestock are kept on the same piece of pasture for a long period of time. *cf.* rotational grazing. *alt.* continuous stocking.

settleable solids suspended solids present in an effluent that will settle out within a period of 2 h when the effluent is held motionless.

settlement *see* subsidence.

Seveso incident an explosion at the ICMESA works at Seveso, Italy on 10 July 1976 that allowed a cloud of toxic chemicals, including 2,4,5-trichlorophenol and TCDD (dioxin), to escape into the atmosphere. This caused serious contamination of the surrounding area.

sewage *n.* water within the sewerage system of a community. It consists of the outflow from domestic and industrial premises and, in some cases, the runoff from roads. Only a very small fraction of sewage is waste material, the rest being water. However, despite its apparently low waste content, the discharge of untreated sewage into surface waters can lead to gross pollution. *see* oxygen-demanding wastes.

sewage fungus pale slimy growth that often occurs in water with heavy organic pollution, either as a slime or a fluffy fungoid growth with streamers, attached to the bed of the river or pond. It consists of a characteristic community of microorganisms dominated by filamentous and zoogloeal bacteria, but also including fungi and protozoa.

sewage sludge semisolid mixture of organic and inorganic material produced during the treatment of sewage. The safe disposal of this type of sludge, with its burden of toxic metals, synthetic organic chemicals and microorganisms, poses major environmental problems. Hence, sludge is generally biodigested anaerobically and then disposed of by incineration, landfill, dumping at sea or spreading on agricultural land.

sewage treatment treatment of raw sewage designed primarily to lower its pathogen content but also to decrease its solids content and biochemical oxygen demand. Treatment may consist of up to four separate stages (*see* Fig. 26). In *preliminary* *treatment*, large objects such as bottles are removed by screens, and grit is separated by gravity. In the next stage, *primary treatment*, the sewage is allowed to slowly traverse a tank. During this process, about one-half of the suspended solids fall to the bottom forming primary sludge, while the remaining settled sewage (also known as primary effluent) then enters the next stage, *secondary treatment*. In this stage, the settled sewage is oxidized by a suspension of microorganisms (*see* activated sludge tank, oxidation pond, trickling filter). If a very high-quality effluent is required, this is followed by *tertiary treatment* in which a variety of techniques are used to reduce the biochemical oxygen demand and/or concentrations of suspended solids, nutrients, toxicants (such as heavy metals) and/or pathogens further.

Sewall Wright effect genetic drift *q.v.*

sex *n.* the sum characteristics, structures, features and functions by which a plant or animal is classed as male or female. In some animals sex is entirely genetically determined, in others the sex may change in response to environmental conditions.

sex attractants natural compounds secreted by females of some insect species to attract males. *see also* pheromone.

sex chromosomes chromosomes, such as X and Y in humans, that form non-homologous pairs in one of the sexes, and whose presence, absence, or particular form may determine sex; called W and Z in other groups of vertebrates.

sex-conditioned trait sex-influenced trait *q.v.*

sex-controlled genes genes present and expressed in both sexes, but manifesting themselves differently in males and females.

sex hormones in mammals chiefly the oestrogens, androgens and gonadotropins.

sex-influenced trait phenotypic character whose expression is influenced by the sex of the individual, such as the simple Mendelian trait of baldness in humans, in which the allele is dominant in males but rarely shows an effect in females. *alt.* sex-conditioned trait.

sex-limited trait phenotypic character expressed only in one sex.

sex-linked *a. appl.* (1) genes carried on only one of the sex chromosomes and that therefore show a different pattern of inheritance in crosses where the male carries the gene from those in which the female is the carrier; (2) an inherited trait that is manifest in only one sex.

sex ratio the ratio of males to females in a population, which may be given as a proportion of male births, number of males per 100 females or per 100 births, or as the percentage of males in the population.

sexual *a. pert.* (1) sex; (2) reproduction, any kind of reproduction that involves the fusion of gametes to form a zygote.

sexual coloration colours displayed during the breeding season but not at other times, often different in the two sexes.

sexual cycle (1) menstrual cycle (*q.v.*); (2) oestrus cycle *q.v.*

sexual dimorphism marked differences, in shape, size, morphology, colour, etc., between the male and female of a species.

sexual imprinting imprinting (*q.v.*) influencing mating preference as an adult.

sexual reproduction reproduction involving the formation and fusion of two different kinds of gametes to form a zygote, usually resulting in progeny with a somewhat different genetic constitution from either parental type and from each other.

sexual selection the difference in the ability of individuals of different genetic types to acquire mates, and therefore the differential transmission of certain characteristics to the next generation. It is made up of the choices made between males and females on the basis of some outward characteristic such as bright plumage, length of tail, etc., in birds, and competition between members of the same sex.

SGHWR steam generating heavy water reactor *q.v.*

shade-intolerance inability of some species of plants to grow in the shade.

shade-tolerance ability of some species of plants to thrive in shady situations.

shake waves *see* body waves.

shale *n.* a clastic sedimentary rock, formed from compacted mud, which splits readily into thin layers. The individual grains of clay minerals that make up the shale are too small to be detected with the naked

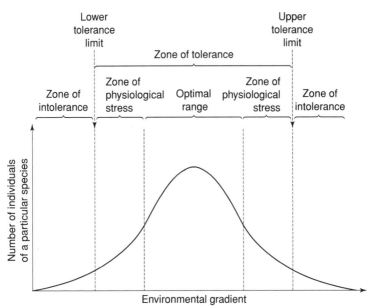

Fig. 27 Tolerance limits.

eye (<0.0039 mm). *cf.* clay (*geol.* definition), mudstone.

Shannon–Weiner diversity measure of equitability (*q.v.*) that incorporates the sum of the proportional contribution of a species to the total population, total biomass, or productivity of the community. Minimum values mean that one species is strongly dominant in the community; maximum values mean that all species are equally dominant.

sharecropping *n.* arrangement whereby a tenant farmer pays the landlord for use of the land with a share of the crop, rather than with money.

shear boundary transform boundary *q.v.*

shearling *n.* a sheep after its first shearing.

shear waves *see* body waves.

sheet erosion the erosion of fine-grained surficial particles over a wide area by sheetflow *q.v.*

sheetflow *n.* the unconfined flow of surface water down a slope. This occurs, e.g., in desert regions that experience sudden heavy rainfall. *alt.* sheetwash.

sheetwash sheetflow *q.v.*

Shelford's law of tolerance concept attributed to the American ecologist Victor Shelford in 1913. This law may be stated thus: individual environmental factors have both minimum and maximum levels (known as tolerance limits) outside of which a particular species either cannot reproduce or cannot survive (known as zones of intolerance). Between these limits is the zone of tolerance, itself made up of a central optimal range where the species is abundant, flanked by zones of physiological stress, where its population is smaller. *see* Fig. 27.

shell *n.* (1) the hard outer covering of an animal, fruit or some eggs; (2) a calcareous, siliceous, bony, horny or chitinous covering; (3) (*chem.*) *pert.* atomic structure, any group of electrons that share the same value for the principal quantum number.

shelterbelt *n.* a tract of trees and shrubs established to protect crops, soil and/or buildings from the effects of wind. *alt.* windbreak.

shelterwood cutting the removal of all mature trees in an area in a series of cuttings, typically over a period of ten years.

shield volcano an extensive, low-domed volcano built up from successive depositions of fluid basic lava. The Pacific island of Hawaii is composed of a group of shield volcanoes. One of these, Mauna Loa, rises 9000 m from the ocean floor (4000 m above sea-level) and has a base greater than 200 km across.

shifting cultivation a type of extensive agriculture traditionally practised in the humid tropical rain forests of Africa, South America and South-East Asia. Small tribal or family groups clear patches of the native forest by felling and burning the trees. The resultant clearing is used to grow a mixture of food crops until, after about three years, the decline in soil fertility forces the people to move on and clear another patch of the forest. *alt.* milpa agriculture, slash and burn cultivation, swidden agriculture.

shigellosis *n.* acute diarrhoea caused by bacteria of the genus *Shigella*, transmitted by the consumption of food or water contaminated with sewage.

shingle beach a beach (*q.v.*) composed of gravel and/or pebbles.

shipworm *n.* bivalve mollusc that is wormlike in form, with a much-reduced shell at the foot and two chalky plates closing the siphon, and that bores into wood.

shivering *n.* a mechanism of heat production found in many animals, used by birds and mammals as an emergency protection to maintain body temperature in cold conditions.

shoal *n.* a group of fish.

shock waves *see* seismic waves.

shoddy *n.* wool waste (containing 2–15% nitrogen) used as a slow-release nitrogen fertilizer.

shoot *n.* (1) the part of a vascular plant derived from the plumule, being the stem and usually the leaves; (2) a sprouted part, branch or offshoot of a plant.

shooting stars *see* meteors.

shore platform an essentially horizontal rock bench formed by the retreat of coastal cliffs. Three different types of shore platform are recognized. *see* high-tide platform, intertidal platform, low-tide platform.

alt. marine abrasion platform, wave-cut platform.

shore *see* beach, seashore.

short-day plants plants that will only flower if the daily period of light is shorter than some critical length: they usually flower in early spring or autumn. The critical factor is in fact the period of continuous darkness they are exposed to. *cf.* long-day plants, day-neutral plants.

shortest day *see* solstice.

short-grass community type of grassland that develops on poorer soils in dry regions, e.g. short-grass prairie, short-grass steppe, and that consists of grasses no more than 60 cm tall and small herbaceous plants.

short-horned flies Brachycera *q.v.*

short-horned grasshoppers Acrididae *q.v.*

short-range forecast *see* weather forecasting.

short-term exposure limit *see* threshold limit value.

short-wave radiation (*clim.*) in certain contexts used as a synonym for solar radiation *q.v.*

shredder *n.* an aquatic animal that eats coarse particles of organic matter, which it shreds into smaller pieces.

shrub *n.* low-growing (less than 6 m high) woody plant that does not have a main trunk and that branches from the base.

shrub–coppice dune nebkha *q.v.*

shrub layer bush layer *q.v.*

Si symbol for the chemical element silicon *q.v.*

SI units units that conform to the metric Système International d'Unités. This system has seven basic units, i.e. metre (m), kilogram (kg), second (s), mole (mol), kelvin (K), ampere (A) and candela (cd), two supplementary units, i.e. radian (rad) and steradian (sr), and 18 named derived units including volt (V), watt (W) coulomb (C), ohm (Ω), joule (J), newton (N), pascal (Pa), becquerel (Bq), sievert (Sv) and gray (Gy). *see* individual entries.

sial *n.* the predominantly granite, nonsedimentary rocks of the Earth's continental crust. These rocks are very old (mainly over 1500 million years) and generally light in colour. The term sial is derived from *s*ilica and *a*luminium, two major components of the minerals of these rocks. *alt.* sialic rocks. *cf.* sima.

sibling species true species that do not interbreed but are difficult to separate on morphological grounds alone.

siblings *n.plu.* offspring of the same parents. *alt.* sibs.

sibmating *n.* mating between siblings.

sibs siblings *q.v.*

sibship *n.* collectively, the siblings of one family.

siccicolous *a.* drought-resistant.

sick building syndrome collection of symptoms, e.g. nausea, headaches and skin rashes, suffered by occupants of a building and perceived to be related to conditions therein.

sickle dance a dance of bees where the bee performs a sickle-like semicircular movement and the axis of the semicircle indicates the direction of the food source.

siderite *n.* FeCO$_3$, a mineral that is occasionally important as an iron ore mineral. *alt.* chalybite.

siderites *see* meteorites.

siderolites *see* meteorites.

siderophile *a. appl.* species that thrives in an iron-rich environment.

sierozem *n.* grey soil, containing little humus, of middle-latitude continental desert regions.

sievert (Sv) *n.* the SI unit of radiation dose equivalent, equal to 1 J of energy per kilogram of absorbing tissue, and that replaces the rem, with 1 Sv = 100 rem.

sigma (σ) symbol for standard deviation *q.v.*

sigmoid growth curve S-shaped growth curve *q.v.*

sign stimulus an environmental stimulus that acts as a releaser of species-specific behaviour.

silage *n.* a winter feedstuff made from green crops, such as grass or forage maize, harvested and then conserved by fermentation in a succulent condition.

silent mutation a mutation that does not affect function or production of gene product and therefore has no effect on phenotype. They are usually point mutations that change one codon into another specifying the same amino acid or a substitute amino acid that does not affect protein function

(*alt.* neutral mutations), or mutations in a DNA region that has no genetic function.

silica minerals a group of minerals with the formula SiO$_2$. They may contain impurities and are often coloured as a consequence. They may be transparent, translucent or opaque. The degree of crystallinity varies from fully crystalline (e.g. quartz), through cryptocrystalline to amorphous.

silica tetrahedron the basic structural unit of all silicate minerals; it can be pictured as a tetrahedron with oxygen atoms centred on each of its four corners and a silicon atom at its centre. Such tetrahedra may be joined together by shared oxygen atoms, thereby forming rings, chains, sheets or three-dimensional networks. For each unshared oxygen atom in such structures there is a single negative charge. Such negative charges must be balanced by cationic counterions within the overall silicate structure.

silicate clays clay minerals *q.v.*

silicates *see* minerals.

siliceous *a.* consisting of or containing silica.

silicification *n.* type of fossilization in which organic material is replaced by silica (i.e. silicon dioxide, SiO$_2$).

silicole *n.* a plant thriving in markedly siliceous (silica-rich) soil.

silicon (Si) *n.* black-brown metalloid element (at. no. 14, r.a.m. 28.09), whose atoms form large molecular structures rather like diamond. It is the second most abundant element on Earth after oxygen, and occurs in a wide range of minerals, e.g. many gemstones and semiprecious stones, quartz, talc, clay and mica. It is the main constituent of many rocks and soils.

silicosis *n.* a type of pneumoconiosis caused by the inhalation of minute particles of any one of several minerals/rocks, e.g. slate or silica.

silicula *n.* broad flat capsule divided into two by a false septum and found in members of the Cruciferae, such as honesty (*Lunularia*). *cf.* siliqua.

siliqua *n.* long thin capsule divided into two by a false septum, found in members of the Cruciferae such as the wallflower. *cf.* silicula.

silk *n.* very strong fine protein fibre produced by various insects and spiders. It is extruded as a fluid from specialized glands and hardens on contact with the air. It is composed of the fibrous protein fibroin and other proteins such as sericin, and is used to make, e.g., webs, and egg and pupal cocoons. Commercial silk is produced by the silkworm (*Bombyx mori*).

silkworm *n.* the larva of the silk-moth *Bombyx mori*, which spins a cocoon of silk around itself when it pupates, from which silk fibre is gathered commercially.

sill *n.* a minor, essentially horizontal, intrusion formed when magma intrudes between the existing rock strata.

sillimanite *n.* Al_2SiO_5, a nesosilicate mineral of aluminous metamorphic rocks, typically of gneisses and schists generated by high-grade regional metamorphism. *alt.* fibrolite.

silo *n.* a container or chamber used for the storage of grain or the preservation of green crops such as silage.

silt *n.* (1) a soil textural class recognized by the USDA; (2) those mineral grains present in a soil's fine earth with diameters in the range 0.002–0.02 mm (in the International or Atterberg system), 0.002–0.05 mm (in the USDA system) or 0.002–0.06 mm (in the system used by the Soil Survey of England and Wales, British Standards and Massachusetts Institute of Technology).

siltstone *n.* any sedimentary rock made up of grains between 0.004 and 0.0625 mm in diameter.

Silurian *a. pert.* or *appl.* geological period lasting from about 438 to 408 million years ago, and in which the first land plants arose.

silva *alt.* spelling of sylva *q.v.*

silver (Ag) *n.* metallic element (at. no. 47, r.a.m. 107.87), a shiny white malleable metal in elemental form. It occurs widely in nature as veins of the free metal or as sulphide ores. It is found as a native element (*q.v.*), a sulphide (silver glance), a chloride (horn silver) and other compounds. It is used as a decorative metal and in the electronics and photographic industries.

silverfish *n.plu.* common name for many members of the Thysanura *q.v.*

silvicolous *a.* inhabiting or growing in woodlands.

silviculture *n.* the cultivation of trees and management of forests and woodland for timber.

sima *n.* the rocks that form the oceanic crust of the Earth. These rocks are very young (usually less than 200 million years) and dark in colour and are mainly basalts. The term sima is derived from *si*lica and *mag*nesium, two major components of the minerals of these rocks. *alt.* simatic rocks. *cf.* sial.

simazine *n.* soil-acting herbicide used to control germinating weeds, e.g., in orchards.

simian *a.* possessing characteristics of, or *pert.*, the anthropoid apes.

similarity index *see* coefficient of community.

simple Mendelian trait genetic trait determined by a single gene.

Simpson's index of floristic resemblance an index of the similarity between two communities that is calculated by dividing the number of species common to the two communities by the total number of species in the smaller of the two communities. *cf.* Gleason's index, Jaccard index, Kulezinski index, Morisita's similarity index, Sørensen similarity index.

Simpson diversity index (D) a measure of the species diversity in a community. It is based on the probability of randomly picking two individuals of different species from the community and is calculated from the formula $D = [N(N − 1)]/\sum n(n − 1)$, where N is the total number of individuals of all species counted, and n is the number of individuals of individual species. The higher the value of D the more diverse the site.

Simpson dominance index (C) a measure of the dominance in a community based on the probability of randomly picking two individuals of the same species from a community.

Sinanthropus *see* Pekin man.

single bond a covalent bond (*q.v.*) that consists of one pair of electrons shared between the bonded atoms. Diagrammatically, this is represented as a single line between the bonded atoms, e.g. the single bond in H_2 may be shown as H–H.

single-cell protein (SCP) protein derived from unicellular microorganisms, e.g. bacteria or fungi, grown on a feedstock of crude oil, carbohydrate raw materials, or wastes from food processing as a carbon source, which may be used for human or animal consumption.

singleton *n.* a single offspring.

sinistral *A.* on or *pert.* the left.

sink *n.* (1) any process, chemical or physical in nature, that results in the loss of a pollutant from a given part of the environment. The operation of a sink mechanism may not remove the elements of the polluting compound from the part of the environment concerned. For example, atmospheric hydrogen sulphide (H_2S) is oxidized in the air to sulphur dioxide (SO_2). During this process the sulphur is transformed from one gaseous species to another, and so remains in the atmosphere; (2) (*biol.*) any cell, tissue or organism that is a net importer and end-user of a metabolite or other resource. A storage root of a plant is, for example, a sink for sugars synthesized in the leaves, converting them into storage polysaccharide.

sinkhole *see* doline.

Sino-Japanese Floral Region part of the Holarctic Realm comprising Japan, northern and eastern China, and the northern Himalayas.

sinuosity *n.* a quantification of the irregularity of a river channel that is observable from above. It is expressed as the ratio of river channel length to valley length. Thus a straight river channel would have a sinuosity of 1, whereas a river channel following a highly convoluted route would have a sinuosity of around 3.

Siphonaptera *n.* order of wingless insects with large hind legs, commonly called fleas, which as adults are ectoparasites of the skin of birds and mammals. They drink blood through piercing mouthparts, have a worm-like larval stage and a pupa.

siphonophores *n.plu.* group of pelagic hydrozoans that form colonies consisting of both polyps and medusoid forms, some individuals of the colony modified as a float for swimming as in the Portuguese Man o' War, sometimes mistakenly called jellyfish.

Siphonopoda Cephalopoda *q.v.*

Siphunculata *n.* the Anoplura (*q.v.*), the sucking lice.

Sipuncula, sipunculids *n., n.plu.* phylum of marine unsegmented coelomate worms. An individual has the anterior end of the body introverted and used as a proboscis and tentacles round the mouth.

Sirenia, sirenians *n., n.plu.* order of placental mammals commonly called sea cows, including the dugong and manatee, which are highly modified for aquatic life, with naked bodies and front limbs modified as paddles.

sirens *n.plu.* family of eel-like amphibians that live in muddy pools in south-east USA and northern Mexico.

sisal *n. Agave sisalana*, a Mexican plant grown for the fibres (also known as sisal) contained in its leaves, which are used in rope-making.

sister taxa two taxa (species or other type of grouping) that have arisen from a single evolutionary change and thus are closely related to each other.

SIT sterile insect technique *q.v.*

site *n.* (1) the combination of non-biological environmental conditions at a particular location that will affect the type of vegetation that will develop, including soil type, drainage, slope and climate; (2) location.

Site of Special Scientific Interest (SSSI) a site in the UK deemed worthy of conservation by English Nature (*q.v.*) on the basis of such features as its fauna, flora and/or geology.

site-specific mutation a general term covering various techniques by which specific changes can be introduced into isolated DNA *in vitro*.

skarns *n.plu.* rocks formed within contact aureoles by the metasomatic influx of material into carbonate-rich sediments, esp. limestones, during contact metamorphism. Some skarns (pyrometasomatic deposits) constitute orebodies, as they are rich in one or more valuable commodity, e.g. Fe, Cu, W, graphite, Zn, Pb, Mo, Sn, U.

skeleton *n.* hard framework, internal or external, that supports and protects softer parts of a plant, animal or unicellular organism, and to which muscles usually attach in animals. *see also* endoskeleton, exoskeleton.

skid trail *n.* a forest track along which logs are hauled. *alt.* snig track.

skin *n.* the outermost covering of an animal, plant, fruit or seed. *see also* epidermis. In mammals, the skin is composed of two layers, an outer epidermis derived from ectoderm, which becomes cornified at the surface and is continuously being renewed, and an inner dermis, derived from mesoderm, composed of connective, vascular and muscular tissues.

skiophyll *n.* a leaf modified to function in shady conditions.

skiophyllous *a.* shade-loving, growing in shade.

skiophyte *n.* a shade plant.

skotoplankton *n.* plankton living at depths at which light does not penetrate.

skototaxis *n.* a positive taxis towards darkness, not a negative phototaxis.

skutterudite *n.* a cobalt and nickel ore mineral with a composition that approximates to $(Co,Ni)As_3$. *see also* chloanthite, smaltite.

slag *n.* waste material (mainly non-metallic) produced when metal ores undergo smelting.

slaked lime *see* lime.

slash and burn cultivation shifting cultivation *q.v.*

slate *n.* a fine-grained metamorphic rock transformed from shales and mudstones by low-grade regional metamorphism. Like the original shale, slate cleaves evenly (i.e. splits) along parallel planes.

slavery dulosis *q.v.*

sleeping sickness *see* trypanosome.

sleep movements change in position of leaves, petals, etc., at night that may be brought about by external stimuli of light and temperature changes, or may be an endogenous circadian rhythm.

sleet *n.* a term used in the UK to denote precipitation consisting of either partly melted snowflakes or a mixture of snow and rain.

slide movement a type of mass movement faster than flow movement (*q.v.*) and less reliant on the presence of water. Landslides (composed of soil) and rockslides may be initiated by heavy rains, earthquakes or by human activities, e.g. road construction in mountainous areas.

slime *n.* viscous substance secreted by plants, fungi and animals and rich in glycans (in fungal and plant slimes) or glycoproteins (in most animal slimes).

slime bacteria myxobacteria *q.v.*

slime moulds *n.plu.* common name for the acellular slime moulds (*q.v.*) and cellular slime moulds *q.v.*

slime nets Labyrinthulomycota *q.v.*

slip-face the steep lee slope of a dune, where sand accumulates at the angle of stability of dry sand (between 30° and 35° from the horizontal).

sliproad *n.* roadway allowing access to or egress from a carriageway of a motorway *q.v.*

slow neutrons neutrons each with a kinetic energy of less than *ca.* 10 eV. Such neutrons are of an appropriate kinetic energy to induce fission efficiently in the fissile fuel of a nuclear reactor. They are obtained by allowing the fast neutrons generated during nuclear fission to pass through a moderator *q.v.*

slow viruses viruses that only show effects a long time after infection.

sludge *n.* (1) solid or semisolid waste; (2) *see* sewage sludge.

slug *n.* (1) shell-less terrestrial gastropod mollusc; (2) migrating pseudoplasmodium of a cellular slime mould.

slump bedding (*geol.*) bedding formed on the sea-floor when newly deposited wet, incoherent beds of sediment slide or slump down a slope. It consists of a highly contorted version of the original bedding pattern with former layers broken and intermixed. *alt.* slump structure.

slump structure slump bedding *q.v.*

slurry *n.* (1) suspension of particles of solid in a liquid; (2) (*agric.*) *appl.* dung and urine mixed with water, a form of livestock excrement that can be directly applied to fields. *alt.* liquid organic manure.

Sm symbol for the chemical element samarium *q.v.*

smallpox *n.* an acute contagious viral disease of humans, characterized by fever and the formation of pustules on the skin, which was officially declared eradicated in 1979, after a vaccination programme. *alt.* variola.

smaltite *n.* a mineral that approximates to cobalt arsenide. It is very closely related to skutterudite (*q.v.*), and some consider the terms smaltite and skutterudite to be synonymous.

smectite group a group of 2 : 1-type clay minerals (*q.v.*) that includes montmorillonite (*q.v.*), beidellite (*q.v.*), nontronite, saponite, hectorite and sauconite. The first three of these are the most important smectites in soils. *see also* bentonite, Fuller's earth. *alt.* montmorillonite group.

smelting *n.* metal extraction process in which the ore is fused, i.e. melted, and the metal reduced to its elemental form. This usually involves a reducing agent, most commonly carbon (in the form of charcoal or coke), carbon monoxide (CO) or hydrogen (H_2).

smithsonite *n.* $ZnCO_3$, a zinc ore mineral. *alt.* calamine (a name also applied to hemimorphite *q.v.*).

smog *n.* a mixture of *sm*oke and f*og*. *see* London-type smog, Los Angeles-type smog.

smoke *n.* (*clim.*) air-suspended particles, each <2 μm in diameter, derived from combustion.

smoke control zone smokeless zone *q.v.*

smokeless fuel a term usually reserved for solid fuel formed by heat-treating coal in essentially anoxic conditions, at *ca.* 600 °C.

smokeless zone a designated urban area in which the amount of smoke emitted from chimneys is highly restricted by law. *alt.* smoke control zone.

smooth endoplasmic reticulum (SER) *see* endoplasmic reticulum.

smut *n.* an airborne disease affecting several types of cereal crop, e.g. wheat, barley, oats and maize, caused by species of fungi belonging to the genus *Ustilago*. The exception is stinking smut of wheat, which is caused by *Tilletia caries*.

smut fungi common name for members of the Ustilaginales, a group of parasitic heterobasidiomycetes some of which are serious pathogens of important crops, causing smut diseases that form black, dusty masses of thick-walled teleutospores, resembling soot or smut.

Sn symbol for the chemical element tin *q.v.*

snail fever schistosomiasis *q.v.*

snails *n.plu.* generally refers to members of the large group of terrestrial and freshwater gastropod molluscs of the subclass Pulmonata, which have a helically coiled shell and no ctenidia, and in which the mantle cavity is used as a lung. *see also* sea snails.

snakes *n.plu.* members of the reptilian suborder Ophidia (or Serpentes) of the class Squamata, being highly specialized legless carnivorous reptiles with elongated sinuous bodies covered with scales, which move by undulations of the body. The bite of many snakes is poisonous.

SNH Scottish Natural Heritage *q.v.*

snig track skid trail *q.v.*

snoods *see* line fishing.

snow *n.* precipitation in the form of ice crystals, which may be loosely aggregated to form snowflakes. Snow develops when dew-point temperatures are below 0 °C, causing the direct conversion of atmospheric water vapour into ice.

snowfield *n.* an area of permanent snow and ice found at high altitudes and/or high latitudes, above the snow line.

snowflakes *see* snow.

snow line level above which snow lies on the ground throughout the year. *alt.* firn line.

snow mould a disease mainly affecting wheat and rye, caused by exceptionally low temperatures and/or fungal attack. Infected plants are stunted, with shrivelled grain and weak roots.

snow rot a soil-borne disease mainly affecting winter cereals, esp. barley, caused by the fungus *Typhula incarnata*. Infected plants are yellowish with poor tiller and root development.

soapstone *n.* the massive form of the mineral talc (*q.v.*). *alt.* steatite.

sobemovirus group plant virus group containing isometric single-stranded RNA viruses, type member southern bean mosaic virus.

social facilitation the initiation or increase in an ordinary behaviour pattern by the presence or actions of another animal.

social forestry agroforestry *q.v.*

social group group of animals that live and travel together and depend on each other to some degree.

social insects the species of insect that live in organized groups and colonies, with division of labour between different morphological forms and that include many bees, ants and wasps (order Hymenoptera) and the termites (order Isoptera).

social mimicry evolution of a similar form or behaviour in unrelated species so as to facilitate interaction, such as the evolution of orchid flowers that resemble bees to attract bees for pollination.

social parasite (1) parasite that exploits the normal behaviour of another species for its own advantage, such as brood parasites like the cuckoo; (2) a parasite of social insects.

social pathology disturbances of behaviour and physiology that result from overcrowding.

social selection selection of traits that are promoted by social behaviour.

sociation *n.* a minor unit of vegetation.

socies *n.* an association of plants representing a stage in the process of succession.

society *n.* (1) a number of organisms forming a community; (2) in animals, a true society is characterized by cooperative behaviour between individuals, a division of labour and, in the social insects, morphological distinctions between members carrying out different functions; (3) a community of plants other than dominants, within an association or consociation.

sociobiology *n.* the study of the biological and genetic basis of social organization and social behaviour and their evolution in animals, a field of study that has caused controversy when applied to human social behaviour and organization.

sodalite *n.* $Na_8(Al_6Si_6O_{24})Cl_2$, one of the sodalite group (*q.v.*) of feldspathoid minerals *q.v.*

sodalite group *n.* a group of the feldspathoid minerals (*q.v.*). Includes sodalite, $Na_8(Al_6Si_6O_{24})Cl_2$, nosean, $Na_8(Al_6Si_6O_{24})SO_4$, and haüyne, $(Na,Ca)_{4-8}(Al_6Si_6O_{24})(SO_4)_{1-2}$.

soda nitre nitratine *q.v.*

sodification alkalinization *q.v.*

sodium (Na) *n.* metallic element (at. no. 11, r.a.m. 22.99), forming a soft silvery-white solid that reacts violently with water. It occurs naturally as sodium chloride (NaCl) in seawater and in salt deposits and as other compounds such as sodium borate (borax). As the ion, Na^+, it is essential for cellular activity in living organisms, and is an essential nutrient.

sodium nitrate *n.* $NaNO_3$, (*agric.*) inorganic nitrogen fertilizer containing 16.5% by weight N, supplied in granular form, but now rarely used.

sod peat lumps of peat cut from a vertical section, commonly used as a fuel in small boilers. *cf.* milled peat.

soft corals group of colonial coelenterates, typified by the sponge-like 'dead men's fingers' (*Alcyonarium*), in which the gastric cavities of individual polyps are connected by fine tubes and the bulk of the colony is made up of mesogloea.

soft detergents *see* detergent swans.

soft water water that readily forms a lather with soap. Such water has, of necessity, a low concentration of both calcium and magnesium cations. *cf.* hard water.

softwoods *n.plu.* conifers and their woods.

soil *n.* the thin surface layer of the Earth's crust between the lithosphere and the atmosphere. It constitutes part of the biosphere as it supports plant, animal and microbial life. It is composed of humus, mineral matter, soil water and soil atmosphere, in various proportions.

soil-acting herbicide *n.* herbicide that is applied to the soil, from where it acts on the target plants through their roots.

soil aeration the dynamic exchange of gases between the soil atmosphere and the air of the troposphere.

soil air, soil atmosphere the mixture of gases that occupies those voids within the soil that are not filled with soil water.

soil conservation any method used to protect the soil from erosion, deterioration or depletion as a result of natural forces and/or human activities. *see* contour ploughing, shelterbelt, strip cropping, stubble mulch cultivation.

soil consistence the sum of those attributes of a soil that determine its behaviour when subjected to forces that may crush or mould. For example, a soil with a friable consistence is one that readily crumbles. *alt.* soil consistency.

soil erosion a natural process, mediated by water and/or wind, whereby the rate of removal of soil exceeds the rate at which it is produced by weathering of the underlying bedrock. Accelerated soil erosion occurs in many parts of the world (especially in dry tropical regions) as a result of human activities, primarily those involving inappropriate agricultural techniques.

soil fertility the condition of a soil with regard to its suitability for crop growth. This includes such factors as its texture, drainage and water-holding characteristics, and its content of available chemical elements that are essential for plant growth.

soil fissures *see* fissures.

soil fragment recognizable macroscopic structural unit of soils made by breaking peds (*q.v.*) across natural lines of weakness.

soil horizon a horizontal zone of soil of distinct composition and texture. A mature soil is made up of several distinct soil horizons. *see* Fig. 28. *see also* soil profile.

soiling zero grazing *q.v.*

soiling crop a crop cultivated for harvesting and immediate feeding to livestock. *see* zero grazing.

soil map map showing the distribution of different soil types in an area, used, e.g., in land use classification.

soil moisture deficit the amount of water necessary to restore the soil to field capacity *q.v.*

soil monolith soil sample taken vertically down a soil profile, which is subsequently mounted intact.

soil morphology the biological, chemical and physical attributes of those soil horizons that constitute a given soil profile.

Soil Order in the USDA Soil Taxonomy system, the broadest category of soils. Eleven orders are currently recognized: entisol, histosol, vertisol, inceptisol, aridisol, mollisol, alfisol, spodosol, ultisol, oxisol and andisol. *see* individual entries.

soil organic matter the fraction of the soil that is made up of partially decayed plant and animal remains, together with material synthesized by soil microorganisms.

soil pit hole dug to expose a soil profile (*q.v.*) for examination.

soil pores *see* pores.

soil porosity (θ) volume of soil pores divided by the total soil volume. This may be expressed in cubic centimetres of pores per cubic centimetre of soil, as a percentage (volume/volume) or, as it is a dimensionless ratio, as a figure without units.

soil profile vertical section through a soil showing the different horizons from the surface to underlying parent material. *see* Fig. 28.

soil science the scientific study of soils including their formation, properties, classification and distribution. *alt.* pedology.

soil separate any one of the three recognized fractions of a soil's fine earth, based on particle size, i.e. clay, silt or sand.

soil solution soil water together with the dissolved salts it contains.

soil structure the arrangement of voids, individual soil particles and aggregates of these particles within a soil.

soil survey the systematic analysis, classification and mapping of the soils of a given locality.

Soil Taxonomy widely used system of soil classification devised by the USDA and formerly known (pre-1975) as the Comprehensive Soil Classification System (CSCS).

soil texture *see* texture. *see also* coarse-textured soil, fine-textured soil, medium-textured soil.

soil water (1) water present in the soil voids and rock pores above the level of the water table, *cf.* groundwater; (2) water that is lost from a soil sample maintained at 105 °C for at least 24 h.

soil water potential amount of water held in a soil, usually expressed as a percentage of the maximum amount the soil can hold (the water-holding capacity).

solanaceous *a. appl.* member of the Solanaceae, the nightshade family, which also includes important crop plants and ornamentals such as potato (*Solanum tuberosum*), capsicums (*Capsicum* spp.), tomato (*Solanum lycopersicon*), tobacco (*Nicotiana tabacum*) and petunia.

solar cells *see* photovoltaic cell.

solar constant the total solar energy that crosses a plane that is perpendicular to the

Soil horizon		Description	
O		Litter layer (L-layer); loose fragmented organic material in which structures such as twigs and leaves are discernible.	Layers of fresh organic material. In some conditions these form a distinct mor humus (*q.v.*) layer.
		Fermentation layer (F-layer); partially decomposed organic material.	
		Humified layer (H-layer); much transformed organic material with no discernible plant remains.	
A		Dark-coloured horizon, containing humus and minerals. In cultivated soils, an Ap layer (plough layer) is developed in this layer.	Eluvial layer.
	E	Light-coloured horizon, with low organic content due to high eluviation. Develops in some soils as a result of podzolization (*q.v.*).	
B		Dark layer of accumulated silica and clay minerals, iron oxides, and organic matter, leached or transported from the upper layers. This layer shows the clearest differentiation into, e.g., blocky or prismatic soil peds.	Illuvial layer. In some soils contains a gley layer or layers of accumulated calcium carbonate or sulphate.
C		Weathered parent material with little or no structure.	
D (or R)		Underlying parent rock.	

Increasing Depth

Note that not all horizons/layers are developed in all soil types, and their precise composition will differ between soil types.

Fig. 28 General soil horizon nomenclature.

incoming radiation per unit time per unit area at the boundary between the atmosphere and space. Estimates of this constant vary, but a figure of 1370 W m^{-2} is probably the best yet.

solar energy energy from the Sun. This energy is utilized by green plants to drive the process of photosynthesis (*q.v.*) and is therefore of fundamental importance to the existence of life on Earth. Man has been able to make direct use of only a tiny fraction of this superabundant energy source, e.g. in solar heating systems (*q.v.*) and in solar cells *q.v.*

solar heating systems heating systems, passive or active, that make direct use of solar energy. Passive solar heating systems do not involve any moving parts but rely on design features, such as thick stone walls, to trap solar energy and subsequently release it as heat. In active solar heating systems, pumps and/or motors are used to move heat-absorbing fluids from the collection area to the place of use.

solar pond a type of low-temperature solar thermal technology used to produce electricity. In these ponds, a gradient of salt concentration is maintained to provide a dense saline bottom layer overlain by an upper layer of almost fresh water. The bottom layer heats up (to *ca.* 80 °C) and is drawn off to boil an organic fluid. The vapour produced drives a turbine for the generation of electricity.

solar radiation radiation emitted by the Sun, often referred to as short-wave radiation.

solfatara *n.* a type of fumarole in which the escaping gases are predominantly sulphurous.

solid *n.* matter in a form that, if placed into a container, will rest on the bottom of the container without moulding itself to the inner surfaces of that container. *cf.* liquid, gas.

solid waste any unwanted and discarded material that is not a liquid or a gas.

solifluction *see* flow movement.

soligenous *a. appl.* wet habitats that are supplied by groundwater rather than by precipitation, e.g. fens.

solonchak *n.* light-coloured alkali soil with a high salt content found in semidesert regions. They are infertile and low in organic matter.

solonetz *n.* dark-coloured alkali soil found in semidesert regions where the more soluble salts have been leached out. They support some vegetation but are relatively infertile.

solpugids *n.plu.* order of arachnids (the Solpugida or Solifuga), commonly called false spiders or sun spiders, having a very hairy body, large chelicerae, and segmented prosoma and opisthosoma.

solstice *n.* one of the two days during each year when the Sun's rays, at noon, strike the Earth vertically at either the Tropic of Cancer (summer solstice in the northern hemisphere) or the Tropic of Capricorn (winter solstice in the northern hemisphere). In the northern hemisphere, the winter solstice (shortest day) falls on *ca.* 22 December and the summer solstice (longest day) on *ca.* 21 June. This situation is reversed in the southern hemisphere.

solubility product (K_{sp}) an equilibrium constant that relates to the dissolution of poorly soluble salts (a process that may be represented by the general equation

$$A_xB_{y(s)} \rightleftharpoons xA^{y+} + yB^{x-})$$

The solubility product is expressed as K_{sp} = $\{A^{y+}\}^x\{B^{x-}\}^y$, where curly brackets, $\{\}$, represent activity based on molar concentration. In dilute solutions, the activity of a solute approximates to its molar concentration.

soluble *a.* capable of being dissolved.

solum *n.* in soils, the collective term used to denote the eluvial and illuvial soil layers, i.e. the A- and B-horizons.

solute *n.* substance (solid, liquid or gas) that is dissolved in another substance, usually a liquid (known as the solvent) to form a solution.

solution *n.* dissolved substance (solute) plus the substance, usually a liquid, in which it is dissolved (solvent).

solution dolines *see* doline.

solution mining a technique used to recover metals from spoil heaps and low-grade ore deposits in which the metal is leached directly from the rock and subsequently recovered from the leachate, e.g. dilute sulphuric acid is used in this way to extract copper.

solvent *n.* a liquid, e.g. water, trichloroethylene or benzene, capable of dissolving other substances in it.

solvent extraction a separation technique that utilizes two immiscible solvents. In one of these, a mixture containing a desired material is dissolved, while the other is used to selectively remove the desired component from the dissolved mixture, used, e.g., in copper ore processing.

soma *n.* the animal or plant body as a whole with the exception of the sex cells. *a.* somatic.

somaclonal variation genetic variation arising from mutations in somatic plant cells undergoing regeneration in culture.

somatic *a.* (1) *pert.* soma (*q.v.*); *appl.* (2) cell, body cell as opposed to cells of the germ line; (3) mutation, mutation occurring in a body cell; (4) number, number of chromosomes in somatic cells.

song birds Passeriformes *q.v.*

sonnenseite slope adret slope *q.v.*

Sonoran *a. appl.* or *pert.* a zoogeographical region of southern North America, including northern Mexico, between Nearctic and Neotropical regions.

soot *n.* the black particulate material that is formed when either fossil fuels or biomass are incompletely burned. It is an impure form of elemental carbon, associated with which are many organic compounds, including polycyclic aromatic hydrocarbons *q.v.*

sorbitol *n.* a faintly sweet alcohol isomeric with mannitol.

Sørensen similarity index a measure of the similarity of species composition between two plant communities. It is calculated by doubling the number of species common to the two communities and dividing by the sum of the total number of species in the two communities. *see also* Gleason's index, Jaccard index, Kulezinski index, Morisita's similarity index, Simpson's index of floristic resemblance.

sorghum *n. Sorghum* spp., a large-grained tropical cereal crop of African origin. It is grown as a staple food in several parts of Africa and also as a fodder crop in the south-western USA.

sorption *n.* retention of material at surface by adsorption or absorption.

sour crude crude oil in which the presence of hydrogen sulphide (H_2S) is detectable. *cf.* sweet crude.

sourveld *n.* areas of veld (*q.v.*) that have been degraded by overgrazing.

South African Floral Region part of the Austral Realm comprising South Africa, Botswana and Namibia. *alt.* South African kingdom.

South-East Asian Floral Region part of the Palaeotropical Realm comprising southern Burma, northern and central Malaysia, Cambodia, Laos and Vietnam.

south-east trades *see* trade winds.

Southerly Buster cold, dry wind, blowing from the south, that occurs in south-eastern Australia.

southern lights *see* thermosphere.

Southern Realm Austral Realm *q.v.*

South Oceanic Floral Region part of the Austral Realm comprising the islands of the South Atlantic and Indian Oceans south of latitude 50° S, e.g. South Georgia, and the South Sandwich Islands.

south-westerlies *see* mid-latitude westerlies.

sovkhoz *n.* state-owned farm in the former USSR, where workers were paid a weekly wage.

soya bean, soybean *n. Glycine max*, a pulse crop grown for the high oil and protein content of its seeds. It is used in the manufacture of edible oils, industrial oils and a high-protein flour used to make meat substitutes. *alt.* soya.

sp. species *q.v.*

SP suction pressure *q.v.*

spadix *n.* a branched inflorescence with an elongated axis, sessile flowers and an enveloping green or coloured leaf-like structure (spathe), as in wild arum.

spangle gall small reddish gall found on the underside of oak leaves and caused by larvae of the gall wasp *Neuroterus quercusbaccarum.*

Spanish cooking oil tragedy a public health disaster that began in central Spain in 1981 and continued for several years. Hundreds of people died (*ca.* 600 by March 1987) and many thousands more became ill as a result of using a particular brand of cooking oil. This had been sold as fit for human consumption when, in fact, the (rape-seed) oil was only suitable for industrial use.

spat *n.* the spawn or young of bivalve molluscs.

spathe *n.* a large enveloping leaf-like structure, green or coloured, that protects a spadix *q.v.*

Spathiflorae Arales *q.v.*

spawn *n.* (1) collection of eggs deposited by bivalve molluscs, fishes, frogs, etc.; (2) mycelium of certain fungi. *v.* to deposit eggs, as by fishes, etc.

specialist, specialist species species that can survive and thrive only within a narrow range of habitat and/or climatic conditions, or that can use only a very limited range of food, and is therefore usually less able to adapt to changing environmental conditions.

specialization *n.* adaptation to a particular mode of life or habitat in the course of evolution.

speciation *n.* (1) (*biol.*) the evolution of new species; (2) (*chem.*) chemical speciation *q.v.*

species *n.* (1) (*biol.*) in sexually reproducing organisms, a group of interbreeding individuals not normally able to interbreed with other groups, being a taxonomic unit having two names in binomial nomenclature (e.g. *Homo sapiens*), the generic name and specific epithet, both of which are italicized. Similar and related species are grouped into genera. Species can be subdivided into subspecies, geographic races, varieties and, for cultivated plants, named cultivars and, for domesticated animals, breeds or strains. Species in asexually reproducing organisms such as bacteria are to a large extent based on morphological, genetic and biochemical characteristics, habitat and host range; (2) (*chem.*) chemical species *q.v.*

species–abundance curve graph illustrating the relationship between the number of species (plotted on the *y*-axis) and the number of individuals per species (on the *x*-axis).

species aggregate a group of very closely related species that have more in common with each other than with other species of the genus.

species–area curve graph illustrating the number of species (plotted on the *y*-axis) found in a given area (increase in area plotted on the *x*-axis). The shape of the curve provides information on the species diversity and species richness of an area and is helpful in determining the most efficient size of plot for sampling.

species–area effect the general rule that the number of species present in a given area of a particular habitat is less than the total number of species that are supported by the habitat as a whole. Thus fragmentation of even a species-rich habitat will lead to loss of species.

species, chemical *see* chemical species.

species composition the different species in a given area or ecosystem.

species diversity the relative abundance of the different species within a given area

or ecosystem. An environment with a high species diversity has roughly equal numbers of the different species present; one with a low species diversity will have a large number of just one or two species and small numbers of the others. *cf.* species richness.

species flock group of numerous species, endemic to a particular small area and ecologically diverse, such as the cichlid fish of some African lakes, which are thought to have evolved from a single ancestral species.

species monitoring general term for repeated surveys intended to determine the species present and/or their abundance and distribution in a given area over time.

species packing the number of niches that can be filled with different species within a community.

species pair sibling species *q.v.*

species richness the number of different species within a given community or area.

species selection the view that selection can act at the level of the species as well as the individual, which is advanced as an explanation of some long-term evolutionary trends.

species-specific behaviour behaviour patterns that are inborn in a species and are performed by all members under the same conditions and that are not modified by learning.

specific *a.* (1) peculiar to; (2) *pert.* a species; (3) *appl.* characteristics distinguishing a species; (4) restricted to interaction with a particular substrate, *appl.* enzymes, antibodies, etc.

specific enthalpy of combustion the enthalpy (*q.v.*) of combustion per unit mass of a fuel, measured in kilojoules per gram or megajoules per kilogram. Unlike most enthalpies, specific enthalpies of combustion are expressed as positive values for exothermic processes. *alt.* calorific value, heating value, specific enthalpy.

specific epithet the second name in a Latin binomial.

specific humidity ratio of the mass (weight) of water vapour in the air to the mass of moist air.

specific hunger the preference for certain foods containing an essential mineral or vitamin shown by animals on a diet deficient in the substance concerned.

specificity *n.* (1) the condition of being specific; (2) being limited to a species; (3) restriction of parasites, bacteria and viruses to particular hosts; (4) restriction of enzymes to certain substrates, restriction of antibody to interactions with particular antigens, etc.

specific surface area the surface area per unit mass of a given solid, measurable in square metres per kilogram (SI units) and also may be given in square centimetres per gram.

spectrum, electromagnetic *n.* the frequency, or wavelength, range of electromagnetic radiation. The main subdivisions of this spectrum are, in order of increasing wavelength (decreasing frequency): gamma rays, X-rays, ultraviolet radiation, visible light, infrared radiation, microwaves and radiowaves.

speleology *n.* the study of caves and cave life.

speleothems *n.plu.* calcite deposits precipitated in limestone caves in a variety of interesting forms. These include stalagmites, which grow upwards, stalactites, which grow downwards, and columns (*alt.* pillars). Each column is created when a stalagmite and a stalactite meet and merge. Speleothems can be dated to yield important information on the development of karst landscapes.

sperm *n.* (1) a spermatozoon (*q.v.*); (2) a male gamete.

spermaceti *n.* oil found in the heads of sperm whales and porpoises, which used to be processed into a fine white wax for commercial use.

spermatogenesis *n.* formation, development and maturation of the male gametes, the sperm.

Spermatophyta, spermatophytes *n.*, *n.plu.* in some classifications, a major division of plants containing all seed-bearing plants, i.e. gymnosperms (e.g. cycads, conifers, *Ginkgo, Ephedra* and quillworts) and angiosperms (flowering plants).

spermatozoon *n.* mature motile male gamete in animals, typically consisting of a head containing the nucleus, and a tail consisting of a single flagellum by which the spermatozoon moves. *alt.* sperm. *plu.* spermatozoa.

sperm competition type of sexual competition in animals in which there is competition not for access to females but for fertilization.

spessartine *n.* one of the garnet minerals *q.v.*

Sphaeropsida *n.* in some classifications, a morphological class of deuteromycete fungi that reproduce by conidia borne in pycnidia, and that includes numerous plant pathogens (e.g. *Phyllosticta, Dendrophoma, Septoria*).

sphagnicolous *a.* inhabiting sphagnum peat moss.

sphagniherbosa *n.* plant community on peat, and containing large amounts of *Sphagnum* moss.

sphagnum mosses mosses of the class Sphagnidae, also called peat mosses or bog mosses, having the gametophore with branches in whorls, many dead water-absorbing cells, and comprising one genus, *Sphagnum*. They are distinguished from other mosses by leaves lacking midribs and a plant body with no rhizoids when mature. The spore capsule lacks a peristome and the protonema germinating from the spore is plate-like rather than filamentous as in the true mosses.

sphalerite *n.* the most common zinc mineral. It is an important zinc ore mineral and has the composition ZnS. *alt.* Black Jack, blende, zinc blende.

sphecology *n.* the study of wasps.

sphene titanite *q.v.*

Sphenisciformes *n.* order of birds including the penguins.

Sphenodonta *see* Rhynchocephalia.

Sphenophyta *n.* one of the four main divisions of extant seedless vascular plants, commonly called horsetails and represented by a single living genus, *Equisetum*. They have a sporophyte with roots, jointed stems, and leaves in whorls, and have strobili of reflexed sporangia borne on sporangiophores.

spheroidal weathering exfoliation *q.v.*

spheroplast *n.* bacterial or yeast cell from which the wall has been removed, leaving a naked protoplast.

spherulite *n.* an essentially spherical body (generally *ca.* 1–20 mm across) that may be found in silicic volcanic rock. Spherulites are partly or wholly comprised of needle-shaped crystals that are radially oriented.

spherulitic *a. appl.* texture of a rock characterized by the presence of abundant spherulites.

sphingolipid *n.* any of a group of complex lipids, glycolipids and phospholipids, which contain the amino alcohol sphingosine and not glycerol, and including sphingomyelin, cerebrosides, gangliosides and ceramide.

spiders *n.plu.* an order of arachnids, the Araneida, individuals having spinning glands on the opisthosoma, from which they produce 'spider silk' for webs, and poison glands on the chelicerae.

spike *n.* a flowerhead with stalkless flowers or secondary small spikes (spikelets) of flowers borne alternately along a single axis.

spinel *n.* (1) specifically, the mineral $MgAl_2O_4$; (2) in general, minerals with the formula AB_2O_4 (where A is a divalent metal and B is a trivalent metal) that are isostructural with $MgAl_2O_4$. Spinels may be of gem quality.

spinneret *n.* (1) organ perforated with tubes connecting to the silk glands in spiders, and from which the liquid silk is released to form webs; (2) similar organ from which a cocoon is spun in some insects.

spinning glands glands that secrete silky material in arthropods, for webs in spiders and cocoons in insect caterpillars.

spiny-headed worms Acanthocephala *q.v.*

spiracle *n.* (1) hole in the sides of thoracic and abodminal segments of insects and myriopods, through which the tracheal respiratory system connects with the exterior, and that can be opened and closed; (2) small round opening, like a vestigial gill slit, immediately anterior to the hyomandibular cartilage of elasmobranch fishes; (3) various other exterior openings connected with breathing or respiration in other animals.

spirillum *n.* any of a group of bacteria with helically curved or twisted thread-like cells. Some are flagellate. Some are free-living in freshwater or marine environments, others saprophytic or parasitic, including some human pathogens such as *Spirillum minor*, the cause of rat-bite fever. *plu.* spirilla.

spirochaete *n.* member of a group of slender, helically coiled bacteria with flexible body and no rigid cell wall. They include free-living, commensal and parasitic forms, including some human pathogens such as *Treponema pallidum*, the causal organism of syphilis. *alt.* spirochete.

spiroplasmas, spiroplasms *n.plu.* almost submicroscopic spiral-shaped prokaryotic pathogenic microorganisms found in plants. *see also* mycoplasmas.

Spirotrichia, spirotrichans *n.*, *n.plu.* a group of ciliate protozoans having a well-defined gullet surrounded by a ring of composite cilia called the undulating membrane, e.g. *Stentor*.

spit *n.* (*geog.*) a narrow tongue of sediment extending from the beach into open water, formed by the process of long-shore drift (*q.v.*). Spits usually form where the coast abruptly changes direction. *alt.* sandspit.

spite *n.* in animal behaviour, the name given to a behaviour by which an animal reduces its own fitness in the process of harming another animal.

splash erosion displacement of soil particles through the impact of raindrops. *alt.* raindrop impact erosion, rainsplash erosion.

splash zone zone of a seashore above the high-tide mark which may be wetted by sea spray at high tide.

splintwood sapwood *q.v.*

spodic horizon *see* spodosol.

spodosol *n.* an order of the USDA Soil Taxonomy system consisting of mineral soils with an illuvial horizon rich in organic matter, aluminium oxides and iron oxides (spodic horizon). The overlying eluvial horizon is typically ash-grey as a result of intense leaching of most of its minerals (albic horizon). Spodosols occur only in humid regions (especially those of cold and temperate latitudes), mainly under coniferous forests. Four suborders of spodosols are recognized, *see* aquod, ferrod, humod, orthod.

spodumene *n.* $LiAlSi_2O_6$, one of the pyroxene minerals (*q.v.*) of the clinopyroxene subgroup. Found in some granite pegmatites. Valued as a source of lithium. The transparent varieties hiddenite (green) and kunzite (lilac) are prized as gemstones.

spoil heaps mounds of waste material generated during mining operations.

sponges *n.plu.* common name for the Porifera *q.v.*

spongicolus *a.* living in sponges.

spontaneous *a. appl.* processes that, once started, will continue without intervention from outside the system, e.g. the burning of a fuel in air. For a spontaneous process to occur, the entropy (disorder) of the universe must increase.

spontaneous mutations those changes in DNA sequence occurring as a result of normal cellular processes and random interaction with the environment. *cf.* induced mutations.

spontaneous nuclear fission nuclear fission (*q.v.*) that does not require external stimulation. *cf.* induced nuclear fission.

spoon worms Echiura *q.v.*

sporadic *a. appl.* (1) plants confined to limited localities; (2) scattered individual cases of a disease.

sporadic permafrost a type of permafrost (*q.v.*) that characteristically occurs as isolated pockets of frozen ground. It is found at high altitudes in lower latitudes. *alt.* alpine permafrost.

sporangium *n.* a cell or multicellular structure in which asexual non-motile spores are produced in fungi, algae, mosses and ferns. *plu.* sporangia.

spore *n.* a small, usually unicellular, reproductive body from which a new organism arises, produced by some plants (ferns and mosses), fungi, bacteria and protozoa.

sporo-, -spore, -sporous word elements from Gk *sporos*, a seed, *pert.* spores and the structures that produce them.

sporophyte *n.* the diploid or asexual phase in the alternation of generations in plants, which in some plants produces spores. *cf.* gametophyte.

Sporozoa, sporozoans *n., n.plu.* a subphylum of parasitic protozoans containing many that cause disease in humans and domestic animals, and in other vertebrates and invertebrates. They include *Plasmodium*, the causal agent of human malaria, and *Eimeria*, the agent of coccidiosis in cattle, sheep and poultry. They

usually have no feeding or locomotory organelles. Sporozoans have a complex life cycle, with asexual and sexual generations, sometimes in two hosts. The stage that infects new cells is a haploid sporozoite.

SPOT Système Pour l'Observation de la Terre *q.v.*

spotted slate a rock formed by the contact metamorphism of pelitic rocks (*q.v.*). It has much in common with slate *(q.v.)* but is distinguished by the presence of spots, spherical, ovoid or rectangular in shape, that are < 4 mm across.

spp. abbreviation for *plu.* of species *q.v.*

spray irrigation sprinkler irrigation *q.v.*

spring *n.* (1) the season between winter and summer; (2) in the UK, the months February, March and April, or the months March, April and May; (3) astronomically, the season that commences at the spring equinox and terminates at the summer solstice, i.e. in the northern hemisphere, between 21 March and 21 June; (4) a place where water issues from the ground. Springs are often formed when the progress of water percolating through permeable rock is impeded by the presence of an impermeable rock layer.

spring equinox *see* equinox.

spring sapping *see* sapping.

springtails *n.plu.* common name for the Collembola *q.v.*

spring tides tides generated twice every lunar month when the Sun and Moon are in alignment. Spring tides produce the highest high tide, the lowest low tide and the greatest daily tidal range. *cf.* neap tides.

sprinkler irrigation overhead method of irrigation in which water, under pressure, is transported by fixed or movable pipes and delivered via nozzles in the form of a fine spray. *alt.* spray irrigation.

spruce budworm *Choristoneura fumiferana*, an economically important insect pest of northern temperate and boreal forests. The larvae cause extensive damage by defoliating the trees.

squall line line of thunderstorm cells, which can stretch for hundreds of kilometres, arranged perpendicular to the direction of motion.

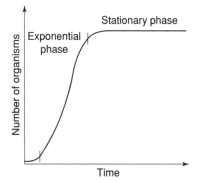

Fig. 29 S-shaped growth curve.

Squamata *n.* order of reptiles comprising the snakes, lizards and amphisbaenians. *alt.* squamates.

squids *n.plu.* common name for a group of cephalopods, an individual having a tubular body and a mouth surrounded by two arms for catching prey and eight arms with sucking discs. They swim by jet propulsion and contain only a vestigial shell.

sr steradian *q.v.*

Sr symbol for the chemical element strontium *q.v.*

SSEW Soil Survey of England and Wales.

S-shaped growth curve type of growth curve (*q.v.*) in which the rate of population growth (after an initial lag phase, where there is little or no increase in numbers) is exponential and then slows down as conditions for growth become progressively less favourable (usually as a result of diminishing food supplies). Eventually an equilibrium is reached where the birth rate equals the death rate (stationary phase). In some circumstances this may be followed by a stage where the death rate exceeds the birth rate as conditions continue to deteriorate (decline phase). *see* Fig. 29. *cf.* J-shaped growth curve. *alt.* sigmoid growth curve.

ssp. subspecies *q.v.*

SSSI Site of Special Scientific Interest *q.v.*

stability *n.* (1) ability of a community or ecosystem to withstand or recover from changes or stresses imposed from outside, *see* constancy, inertia, resilience; (2) (*clim.*)

characteristic of air that, when forced to rise (or fall), tends to resume its original position once the motivating force halts. *cf.* instability.

stability constant (K_{stab}) an equilibrium constant that relates to the formation of complexes, i.e. species with dative bonds (a process that may be represented by the general equation $M^{n+} + xL^{y-} \rightleftharpoons ML_x^{(n-y)+}$). The stability constant is expressed as $K_{stab} = \{ML_x^{(n-y)+}\}/\{M^{n+}\}\{L^{y-}\}^x$, where curly brackets, { }, represent activity based on molar concentration. In dilute solutions, the activity of a solute approximates to its molar concentration. *alt.* formation constant.

stability–time hypothesis the hypothesis that the greatest biological species diversity is found in stable environments, i.e. in communities that have existed for a long time.

stabilization pond oxidation pond *q.v.*

stabilizing factors those factors that tend to stabilize an ecosystem against disturbance, returning populations rapidly to their original sizes.

stabilizing selection selection that operates against the extremes of variation in a population and therefore tends to stabilize the population around the mean for any particular character or group of characters.

stable age distribution maintenance of relatively constant proportions of each age class within a population, which tends to occur in populations that have existed for a long time in a stable environment.

stable population a population in which the numbers remain much the same over time, where the number of births plus immigrations is equal to the number of deaths.

stachyose *n.* a tetrasaccharide present in certain plant roots and other sources, made up of two galactose residues, one glucose and one fructose.

stack *n.* (1) chimney; (2) (*geog.*) a remnant of an eroded headland in the form of a column of rock.

stack gas, stack gases the gaseous waste carried by a chimney-stack.

stade *n.* (1) a stage in development or life history of a plant or animal; (2) interval between two successive moults. *alt.* stadium.

staggers *n.* a condition related to milk fever (*q.v.*), which is common in grazing cows and is caused by a low magnesium content in the blood (hypomagnesaemia). *alt.* grass staggers.

stagnicolous *a.* living or growing in stagnant water.

stagnogleying *see* gleying.

stalactites *see* speleothems.

stalagmites *see* speleothems.

stamen *n.* male reproductive organ (microsporophyll) of a flower (*q.v.*), consisting of a stalk or filament bearing an anther in which pollen is produced. *see* Fig. 12, p. 157.

stand *n.* (1) aggregation of plants of uniform species and age, distinguishable from the adjacent vegetation; (2) group of trees.

standard *n.* (1) a unit of measurement; (2) (*chem.*) a compound or solution with a known composition that is used to establish the concentration of an analyte (*q.v.*). *see also* primary standard, secondary standard; (3) (*biol.*) petal standing up at the back of a papilionaceous flower; (4) upstanding petal of iris flower; (5) free-standing tree or shrub not supported by a wall.

standard *a.* (*chem.*) *appl.* solutions of primary standards (*q.v.*) of accurately known concentration.

standard (sea-level) air pressure atmospheric pressure at sea-level. This has a value of 1013.25 millibars and gives a reading of 760 mm on the mercury barometer.

standard addition a method used in quantitative instrumental chemical analysis. In this technique, known amounts of standard are added to known amounts of sample, prior to the application of the instrumental measurement. The analyte concentration is then found by extrapolation. This technique is useful in the minimization of certain types of matrix effect *q.v.*

standard deviation (σ, s.d., *s*) statistical measure of the variation around the mean, $\bar{x}$, in a given set of *n* numbers. It is equal to the square root of the variance. In a normal distribution (*q.v.*) curve, a distance of one standard deviation on either side of the mean will include 68% of the cases.

standard electrode potential *see* reduction potential, standard.

standard enthalpy of combustion the enthalpy (*q.v.*) change observed on the complete combustion of a substance in its standard state (*q.v.*) to yield the products also in their standard states. It is referenced to a particular temperature (usually 298 K) and is expressed in kilojoules per mole of the substance that is burned. Symbol ΔH_c°.

standard enthalpy of formation the standard enthalpy of formation of any element in its most stable form at 1 atm pressure and at a stated temperature (usually 298 K) is defined to be zero. The standard enthalpy of formation of any compound, at a given temperature, is the standard reaction enthalpy (*q.v.*) associated with its synthesis from its elements in their most stable form at a pressure of 1 atm., at the given temperature. Symbol ΔH_f°.

standard error the standard deviation of the distribution of the means of random samples taken from a frequency distribution. It is calculated by dividing the standard deviation of the frequency distribution by the square root of the number of observations (*n*) in the sample minus one (*n* − 1), or just by *n* if the sample is large. *alt.* standard error of the mean (s.e.m.).

standardization *n.* (*chem.*) the act of determining the concentration of a secondary standard. *v.* standardize.

standardized *a.* (*chem.*) *appl.* solutions of secondary standards, the concentrations of which have been established by use of a primary standard.

standardized birth rate *see* birth rate.

standard meridian prime meridian *q.v.*

standard metabolic rate the metabolic rate measured under a set of given conditions.

standard nutritional unit the unit expressing the energy available at a certain trophic level for the next level in the food chain, usually expressed in kilocalories ($\times 10^6$) per hectare per year.

standard reaction enthalpy the enthalpy (*q.v.*) change that accompanies the conversion of reactants in their standard states to the products in their standard states. Such enthalpies may be cited for any temperature; however, this is commonly 298 K. Symbol ΔH°.

standard reduction potential *see* reduction potential, standard.

standard state (*chem.*) any substance is in its standard state when it is pure and at 1 atm. pressure.

standard temperature and pressure (STP) standard conditions characterized by a temperature of 273.15 K (0 °C) and a pressure of 1 atm. (equal to 760 mm Hg).

standing biomass, standing crop the biomass of a particular area, ecosystem, etc., at any specified time.

St Anthony's fire *see* Saint Anthony's fire.

staphylococci *n.plu.* Gram-positive bacteria of the genus *Staphylococcus*, which are small spherical cells (cocci), often arranged in irregular clusters. Pathogenic strains cause skin lesions and wound infections, and occasionally more disseminated disease. *sing.* staphylococcus.

staple food food that forms a major dietary constituent, usually a cereal or root crop, e.g. rice in Asia and potatoes in UK.

starch *n.* polysaccharide made up of a long chain of glucose units joined by alpha-1,4 linkages, either unbranched (amylose) or branched (amylopectin) at an alpha-1,6 linkage. It is the storage carbohydrate in plants, occurring as starch granules in amyloplasts, and is hydrolysed by animals during digestion by amylases, maltase and dextrinases to glucose via dextrins and maltose.

star dune a pyramid-shaped dune with a number of slip-faces, separated by elongated arms of sand. This type of dune is created by strong winds blowing in succession from a number of different directions.

starfish common name for the Asteroidea *q.v.*

star navigation learned method of navigation apparently used by many migrating song birds, which have been shown to orient themselves according to the stars.

stasipatric speciation formation of new species that results from chromosomal rearrangement, producing individuals who are reproductively isolated from the original species.

stasis *n.* (1) in evolution, the apparent lack of major evolutionary change over long periods of time in any given lineage, as

seen in the fossil record. *see* punctuated equilibrium, stabilizing selection; (2) stoppage or retardation in growth or development, or of movement of animal fluids.

state farm government-owned farm run with paid labour.

state function function of state *q.v.*

state property function of state *q.v.*

states of matter matter may be in one of three states, gas, liquid or solid, depending on the temperature, pressure regime and the nature of the matter concerned. Plasma (hot ionized gas) is sometimes considered to be the fourth state of matter.

static rejuvenation *see* rejuvenation.

stationary front front that marks the boundary between two stationary air masses.

stationary orbit geostationary orbit *q.v.*

stationary phase of population growth *see* S-shaped growth curve.

stations *see* ranching.

statis life table life table showing the age structure of a population at a given instant of time.

statutory limit upper (or lower) limit that it is illegal to exceed (or fall below).

steady state, dynamic *see* dynamic steady state.

steam fog type of fog formed when cold air passes over a warmer waterbody, causing evaporating water to condense. Ice fog may form if air temperatures are low enough to freeze the water vapour.

steam generating heavy water reactor (SGHWR) a type of fission nuclear reactor that utilizes Zircaloy-clad fuel pellets of enriched uranium oxide. It has light water as the coolant and heavy water as the moderator.

steam turbine a turbine (*q.v.*) that uses steam as the working fluid.

steatite *n.* the massive form of the mineral talc (*q.v.*). *alt.* soapstone.

stecklings *n.plu.* young sugar beet plants cultivated for the production of seed.

steers *n.plu.* male cattle that have been castrated.

STEL short-term exposure limit. *see* threshold limit value.

stem *n.* the main axis of a vascular plant, bearing buds and leaves or scale leaves, and reproductive structures, e.g. flowers, usually borne above ground (but *see* rhizome), and having a characteristic arrangement of vascular tissue.

stem and bulb nematode *Ditylenchis dipsaci*, a minute eelworm, which attacks oats. Affected plants become twisted and swollen and fail to produce ears.

stem rust an airborne disease affecting wheat, caused by the fungus *Puccinia graminis*. *alt.* black rust, black stem rust. *see also* rust fungi.

stenobaric *a. appl.* animals adaptable only to small differences in pressure or altitude.

stenobathic *a.* having a narrow vertical range of distribution.

stenobenthic *a. pert.*, or living, within a narrow range of depth on the sea bottom. *cf.* eurybenthic.

stenochoric *a.* having a narrow range of distribution.

stenohaline *a. appl.* organisms adaptable to a narrow range of salinity only.

stenohygric *a. appl.* organisms adaptable to a narrow variation in atmospheric humidity.

Stenolaemata *n.* a class of Bryozoa *q.v.*

stenothermal, stenothermic *a. appl.* organisms adaptable to only slight variations in temperature.

stenotopic *a.* (1) restricted to living in one habitat; (2) having a restricted range of geographical distribution.

steppe *n.* dry and treeless temperate grassland covering extensive areas of southeastern Europe and Asia.

steradian (sr) *n.* the supplementary SI unit of solid angle. One steradian is the angle at the apex of a cone that, when coincidental with the centre of a sphere, will delimit at the sphere's surface an area that equals the square of the length of the radius.

stereoisomer *n.* any of two or more compounds with the same atomic composition but differing in their structural configuration.

stereokinesis *n.* movement, or inhibition of movement, in response to contact stimuli.

stereoscopic vision binocular vision, in which images from both eyes are integrated, allowing the seeing of objects in three dimensions.

stereotaxis, stereotaxy *n.* the response of an organism to the stimulus of contact with a solid, such as the tendency of some organisms to attach themselves to solid objects or to live in crannies or tunnels.

stereotropism *n.* growth movement in plants associated with contact with a solid object, as the tendency for stems and tendrils of climbers to twine around a support.

stereotyped behaviour *see* fixation.

sterile *a.* (1) incapable of reproduction; (2) incapable of conveying infection, *alt.* aseptic; (3) axenic *q.v.*

sterile insect technique (SIT) an insect pest control technique in which large populations of the pest species are raised in the laboratory and then sterilized by irradiation, prior to release into the wild. The mating of these sterile insects with fertile individuals from the wild population can reduce the pest population to acceptable levels or eradicate it altogether.

sterile male technique method of biological pest control in which large numbers of sterile males are released that outnumber the normal males in the competition for mates and thus lower the rate of reproduction in the pest population.

sterilize *v.* (1) in animals to render incapable of reproduction; (2) to render material containing microorganisms incapable of conveying infection, usually by heat or chemical treatment.

steroid *n.* any of a large group of complex polycyclic lipids with a hydrocarbon nucleus and various substituents, synthesized from acetyl CoA via isoprene, squalene and cholesterol, and that include bile acids, sterols, various hormones, cardiac glycosides and saponins.

steroid hormones a group of steroids that act as hormones in animals, e.g. oestrogens, testosterone and its derivatives, glucocorticoids, mineralocorticoids, and the insect hormone ecdysterone.

sterol *n.* any steroid alcohol. They are ubiquitous in plants, animals and fungi, as components of the cell membranes, and including ergosterol (a typical fungal sterol), cholesterol (in animal cells), and phytosterol (in plants).

Stevenson screen a specially constructed wooden box on legs used to house meteorological instruments. The design of this shelter, white-painted and double-louvred, protects the instruments from direct radiation while allowing air to flow freely past the thermometers.

stibnite *n.* Sb_2S_3, the most common mineral of antimony and the major ore mineral of this metal. *alt.* antimonite, antimony glance.

stick insects common name for many of the Phasmida *q.v.*

stigma *n.* (1) (*bot.*) the upper portion of the carpel that receives the pollen and that is usually connected to the ovary by an elongated structure, the style, *see* Fig. 12, p. 157; (2) (*zool.*) various pigmented spots and markings, such as the coloured wing spot of some butterflies and other insects; (3) a pigmented spot near base of flagellum in photosynthetic euglenoids, involved in photoreception and phototaxis.

stilbite *n.* $NaCa_2Al_5Si_{13}O_{36}.14H_2O$, one of the zeolite minerals *q.v.*

stimulus *n.* an agent that causes a reaction or change in an organism or any of its parts.

stinkhorns *n.plu.* common name for fungi of the Phallaels, an order of Gasteromycetes having a foetid odour to the gleba and a mature fruit body resembling a phallus.

stipe *n.* (1) stalk, esp. of mushrooms, toadstools and other stalked fungi; (2) stalk of seaweeds; (3) stem of a fern frond.

stipule *n.* leaf-like or bract-like outgrowth at the base of a leaf-stalk.

St John's wort *see* Saint John's wort.

STJS subtropical jet stream *q.v.*

stochastic *a. appl.* a process in which there is an element of chance or randomness.

stock *n.* (1) the population in a given area (as in fish stocks); (2) true-breeding strain of an organism kept for breeding; (3) one or a group of individuals initiating a line of descent; (*bot.*) (4) stem of tree or bush receiving bud or scion in grafting; (5) the perennial part of a herbaceous plant; (*zool.*) (6) an asexual zooid that produces sexual zooids of one sex by gemmation, as in polychaetes; (7) livestock; (8) (*geol.*) a major intrusion, similar to but much smaller than a batholith *q.v.*

Stockholm Conference the United Nations Conference on the Human Environment, held in Stockholm in 1972, the first major international conference on the global environment held by the UN, which was attended by most UN member governments and resulted in the formation of the United Nations Environment Programme (UNEP).

stocking density with reference to grazing livestock, the number of a particular kind of animal per unit area of land at a given point in time.

stocking rate with reference to grazing livestock, the number of a particular kind of animal per unit area of land over a given period of time.

stock-recruitment models mathematical models by which the expected number of offspring added to the population in a given time can be calculated from the size of the population.

stock resources mineral resources that are valued and used for human needs, including metal ores, aggregate materials (*q.v.*) and fossil fuels.

stoichiometric *a.* (1) *pert.* stoichiometry; *appl.* (2) mixtures of reactants or mixtures of products in which the mole ratios are those dictated by the stoichiometry of the reaction; (3) compounds of fixed composition. *cf.* non-stoichiometric.

stoichiometry *n.* (*chem.*) (1) the study of the composition of chemical species and the mole ratios in which reactants and products are consumed and produced; (2) a statement of mole ratios as embodied in (i) chemical formulae, e.g methane (CH_4) has a $1 : 4$ C : H ratio, and (ii) balanced chemical equations, e.g. hydrogen (H_2) and oxygen (O_2) react with a $2 : 1$ stoichiometry as shown by: $2H_2 + O_2 \rightarrow 2H_2O$.

stoichiometry, coefficients of the numbers placed in front of the formulae of each reactant and product in a chemical equation. They indicate the mole ratios in which the reactants must be present if they are all to be consumed fully if the reaction were to go to completion, and the mole ratios in which the products are produced.

stolon *n.* (1) a creeping plant stem or runner capable of developing rootlets and stem and ultimately forming a new individual; (2) hypha connecting two bunches of rhizoids in fungi; (3) similar structures in other organisms.

stomach pesticide *n.* pesticide that acts as a poison via the stomach, after ingestion by the pest either in the form of poisoned bait or via treated foliage.

stomata *n.plu.* (1) minute openings in the epidermis of aerial parts of plants, esp. on undersides of leaves, through which air and water vapour enter the intercellular spaces, and through which carbon dioxide and water vapour from respiration are released. Stomata can be opened or closed by changes in turgor of the two guard cells that surround the central pore; (2) any small openings or pores in various structures. *a.* stomatal. *sing.* stoma.

stomatal index the ratio between number of stomata and number of epidermal cells per unit area.

stoneflies *n.plu.* common name for the Plecoptera *q.v.*

stone lines rows of stones placed along the contours of a slope in order to reduce surface runoff, a traditional local technique used in Burkina Faso.

stone nets *see* patterned ground.

stone pavement desert pavement *q.v.*

stone polygons *see* patterned ground.

stones *n.plu.* according to the Soil Survey of England and Wales, British Standards and the Massachusetts Institute of Technology, that part of air-dry soil that, once it has been gently ground, will not pass through a sieve with holes of 2 mm diameter. Synonymous with the word gravel used by the International, or Atterberg, and USDA systems. *cf.* fine earth.

stone stripes *see* patterned ground.

stoneworts *n.plu.* common name for the Charophyta *q.v.*

stony corals a group of colonial coelenterates, typified by the reef-building corals, in which individual polyps are embedded in a matrix of calcium carbonate and connected by living tissue.

store animals cattle or sheep reared in such a way that their fatty tissue is undeveloped and their muscular tissue is underdeveloped, but their skeletal development is normal. At this point, they are sold on for fattening, and eventual slaughter.

storey *n.* layer of a given height in a plant community.

storm surge abnormally high wall of water driven ashore by high winds associated with the approach of a hurricane.

stotting *n.* warning behaviour in some gazelles, which bound away with a stiff-legged gait and tails raised, displaying a white rump.

STP standard temperature and pressure *q.v.*

straight fertilizers fertilizers with a single active constituent, i.e. nitrogen (N), phosphorus (P) or potassium (K), generally in the form of an inorganic compound(s). *cf.* compound fertilizers.

strain *n.* (1) pure-breeding variant line in domesticated animals and cultivated plants; (2) (*microbiol.*) in bacteria and other microorganisms, a population of genetically identical individuals with some genetically determined characteristic differentiating them from other strains of the same species; (3) *appl.* bodies undergoing deformation (e.g. rocks), it is the ratio of the amount of change of a given dimension (i.e. length, area or volume) to the corresponding dimension in the undeformed body.

strata *plu.* of stratum *q.v.*

strategic mineral a fuel or other mineral essential to the economy or defence of a country and that is typically stockpiled.

stratification *n.* (1) vertical arrangement in layers, e.g. of soil layers within a soil profile, sediments within sedimentary rock, distinct temperature bands within a waterbody (lake or sea), vegetation types within an undisturbed ecosystem; (2) (*hort.*) process of exposing seeds to below freezing temperatures, which is required to break dormancy in some seeds.

stratified drift sorted glacial drift that has been deposited some distance from the glacier by meltwater streams. *alt.* fluvioglacial deposit, outwash deposit.

stratigraphical column geological timescale *q.v.*

stratocumulus *n.* a white-grey, patchy, low-altitude cloud that appears in rolls or in long rows, accompanied by occasional showers.

stratopause *n.* the upper boundary of the stratosphere, which separates it from the

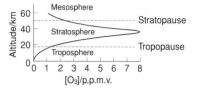

Fig. 30 Ozone (O₃) levels in the atmosphere.

mesosphere, located at an altitude of *ca.* 50 km. *see* Fig. 30.

stratosphere *n.* the layer of atmosphere that occurs between the troposphere and mesosphere. It extends from the tropopause (altitude *ca.* 8–16 km) to the stratopause (altitude *ca.* 50 km) (*see* Fig. 16, p. 236, Fig. 30). The majority of atmospheric ozone is found in this layer.

stratospheric ozone the triatomic oxygen molecule, ozone (O_3), when found in the stratosphere (*q.v.*). Ozone concentrations are maximal (*ca.* 8–10 p.p.m.v. at an altitude of *ca.* 22 km) within this region of the atmosphere (*see* Fig. 30). This band of relatively concentrated ozone has become known as the ozone layer, although this is a misnomer given the low absolute concentration of O_3. Stratospheric ozone absorbs ultraviolet light, principally in the wavelengths *ca.* 230–320 nm (roughly equating to ultraviolet-B), thus shielding living organisms from its damaging effects. Emissions of man-made ozone-destroying pollutants, such as chlorofluorocarbons (CFCs), pose a significant threat to stratospheric ozone levels, with a massive decline in the springtime concentrations of ozone – the ozone hole – having been recorded over the Antarctic. This phenomenon, first recorded in 1985, has continued to be observed, with indications that it is becoming more severe with time. A less intense, but significant, thinning of the ozone layer at high northern latitudes has also been observed. Halogenated hydrocarbons, particularly CFCs, their brominated analogues (i.e. halons) and bromomethane (CH_3Br) undergo photolysis in the stratosphere to yield chlorine and/ or bromine atoms, which enter into cata-

lytic cycles that destroy ozone. An international agreement to reduce emissions of the most damaging ozone-destroying pollutants is now in force, *see* Montreal Protocol. *see also* bromomethane, carbon tetrachloride, chlorinated hydrocarbons, chlorofluorocarbons, halons, hydrofluorocarbons, nitrous oxide, NO_x, 1,1,1-trichloroethane. *alt.* ozone shield, ozonosphere.

strato-volcanoes composite volcanoes *q.v.*

stratum *n.* (1) a layer, as of cells or tissue; (2) a group of organisms inhabiting a vertical division of an area; (3) vegetation of similar height in a plant community, such as trees; (4) (*geol.*) a layer of sedimentary rock >1 cm thick. *plu.* strata.

stratum transect a profile of vegetation, drawn to scale and intended to show the heights of plant shoots. *alt.* profile transect.

stratus *n.* a white-grey low-altitude cloud that forms a continuous cover and is often accompanied by drizzle.

straw *n.* the stalks of cereal crops such as wheat, barley, rye and oats.

stream capacity the maximum sediment load that a stream with a given discharge is able to carry.

stream competence the maximum particle size that a river is capable of transporting at a specific flow velocity.

stream erosion *see* fluvial erosion.

stream piracy river capture *q.v.*

Strepsiptera *n.* an order of small insects with incomplete metamorphosis, commonly called stylopids, whose larvae and females are parasites of other insects and whose males are free-living. Fore wings are halteres, hind wings are fan-shaped.

streptococci *n.plu.* Gram-positive bacteria of the genus *Streptococcus*, which are small spherical cells (cocci) forming long chains. Many are harmless commensals living in the throat and gut, but some are human and animal pathogens, causing tonsilitis, scarlet fever and tissue destruction. *sing.* streptococcus.

streptolydigins *n.plu.* class of antibiotics that inhibit bacterial transcription.

Streptomycetaceae, streptomycetes *n., n.plu.* family of filamentous prokaryotic microorganisms widespread in soil, characterized by formation of a permanent mycelium and reproduction by means of conidia, some spp. producing antibiotics including streptomycin, chloramphenicol and the tetracyclines.

streptomycin *n.* trisaccharide antibiotic synthesized by the streptomycete *Streptomyces griseus* which inhibits bacterial protein synthesis. It is used to treat tuberculosis in humans, and downy mildew on hops.

Streptoneura Prosobranchia *q.v.*

streptonigrin *n.* antibiotic synthesized by the streptomycete *Streptomyces flocculus* which causes chromosome breakage.

stress *n.* negative environmental factors that have deleterious effects on an individual or on a population, such as disease, drought, unaccustomed cold or heat, and pollution.

stress responses the various responses made by cells and whole organisms to heat shock, cold shock and other stresses.

stridulating organs special structures on various parts of the body of certain insects, such as grasshoppers, crickets and cicadas, that produce the characteristic 'song' of these insects. In crickets and grasshoppers, the mechanism resembles that of a scraper being run along a toothed file as two parts of the body are rubbed against each other. In cicadas, membranes on either side of the abdomen are made to vibrate by muscular action.

stridulation *n.* the characteristic sound made by grasshoppers, crickets and cicadas.

Strigiformes *n.* the owls, an order of mainly nocturnal, short-necked, large-headed birds of prey.

strike *n.* (*geol.*) the azimuthal orientation, measured with a compass, of an imaginary horizontal line that is on the surface of an inclined geological plane (e.g. faults, bedding, foliations) and that is at right angles to the direction of true dip.

strike-slip fault transcurrent fault *q.v.*

strip cropping, strip farming the practice of growing two or more crops in alternating narrow bands in order to check soil erosion. When these bands lie perpendicular to the incline of a hill, this method is termed contour strip cropping.

strip grazing grazing system in which livestock are given access (by repositioning electrified fences) to narrow strips of fresh pasture or root crops on a daily or twice-daily basis.

strip mining type of open pit mining in which the mined material is removed in successive, parallel strips.

strobilus *n.* (*bot.*) (1) cone-shaped assemblage of sporophylls in horsetails, club mosses and conifers, *alt.* cone; (2) in flowering plants, a spike formed by persistent membranous bracts, each having a pistillate flower; (*zool.*) (3) stage in development of some jellyfish, a sessile polyplike form that separates off a succession of disc-like embryos by segmentation; (4) chain of proglottids in tapeworms.

stroma *n.* (1) (*mycol.*) tissue composed of hyphae, or of fungal cells and host tissue, in or upon which a spore-bearing structure may be produced; (2) (*bot.*) the colourless material inside chloroplasts in which the green-pigmented grana are embedded and in which carbon dioxide fixation takes place.

stromatolites *n.plu.* layered structures, sometimes of considerable size, formed in certain warm shallow waters by mats of cyanobacteria (blue-green algae). Fossils of similar structure have been found in Precambrian rocks, indicating the presence of life at that time.

strong *a.* (*chem.*) *appl.* (1) an acid or a base that completely ionizes in solution. An example of a strong acid is HNO_3 in water; an example of a strong base is NaOH in water; (2) electrolytes that completely or almost completely dissociate into ions when in solution.

strontianite *n.* $SrCO_3$, a mineral utilized as a source of strontium and used in fireworks to produce a crimson flare.

strontium (Sr) *n.* metallic element (at. no. 38, r.a.m. 87.62), forming a silvery-white malleable metal in elemental form. Found in many different soils as strontianite ($SrCO_3$) and celestine ($SrSO_4$). The radioactive isotope ^{90}Sr (half-life 28 years) is present in nuclear fallout and represents a danger to health because strontium is chemically similar to calcium and is therefore incorporated in small amounts into bone.

strophanthin *n*. a glycoside with effects on the nervous system, obtained from various plants of the family Apocynaceae and used as a tropical arrow poison.

strophotaxis *n*. twisting movement or tendency, in response to an external stimulus.

structural colours colours of fish's skin, insect's wings, etc., that are not due to pigment but to surface structure, e.g. reflecting layers, plates of guanine crystals, etc.

structural diversity variations in the physical characteristics of a site that increase the number of species that can live there.

structure *n*. (1) (*geol.*) rock features that can be defined geometrically and are larger in scale than the attributes generated by the size, shape and arrangement of individual mineral grains. Typically, structural features are evident on a scale that ranges from centimetres to tens, if not hundreds, of metres and are therefore best seen in the field. *cf.* texture; (2) (*soil sci.*) the arrangement of voids, individual soil particles and aggregates of these particles in a soil. This has both macroscopic and microscopic manifestations. The former can be seen with the naked eye or with the aid of a hand lens, while the latter requires the application of optical or electron microscopy to become evident. *cf.* texture.

Struthioniformes *n*. an order of flightless birds including the ostriches.

strychnine *n*. an alkaloid produced from seeds of *Strychnos* spp. and some other plants that is a mammalian poison because of its effects on the nervous system.

stubble *n*. the above-ground remains of harvested crops, esp. cereals.

stubble mulch cultivation a method of cultivation in which the stubble from the previous cereal crop is left in place and the subsequent crop is planted in a seedbed prepared below the surface with minimal disturbance. This technique, used, e.g., in the North American wheat belt, helps to conserve water and to prevent wind and water erosion. *alt.* trash mulch cultivation.

style *n*. in a flower (*q.v.*), the portion of the carpel connecting stigma and ovary, often slender and elongated. *see* Fig. 12, p. 157.

stylopids *n.plu.* common name for the Strepsiptera *q.v.*

sub- prefix from L. *sub*, under, signifying beneath, below (as in anatomical terms), less than (as in subthreshold), not quite, nearly, somewhat (esp. in descriptions of plant and animal parts, as in subdentate, slightly toothed (of leaves), subcarinate, somewhat keel-shaped). In classification it indicates a group just below the status of the taxon above it, as in subclass, etc.

subalpine *a. appl.* (1) ecological zone immediately below the alpine zone (i.e. just below the timberline) on high mountains, at around 1300–1800 m; (2) to the plants and animals that live there.

subangular blocky *appl.* soil peds (*q.v.*) that are block-like in appearance with concave and/or convex surfaces and rounded corners. *cf.* blocky.

subatomic particles particles that make up atoms or that are fragments of atoms.

sub-bituminous coal *n*. soft, brown-black coal, intermediate in rank between lignite and bituminous coal. *see* coal series. *alt.* brown coal.

subclass *n*. taxonomic grouping between class and order, e.g. Theria (mammals that have live-born young) in the class Mammalia.

subclimax *n*. stage in plant succession preceding the climax, which persists because of some arresting factor such as fire, human activity, etc.

subcritical *a. appl.* nuclear fission chain reactions in which the number of neutrons produced by one generation is smaller than in the previous generation. A subcritical condition will result in the eventual cessation of the fission reaction. *cf.* critical, supercritical.

subdominant *n*. species that may seem more abundant than the true dominant species in a climax plant community at particular times of the year or in particular places.

subduction *n*. the process that takes place when two lithospheric plates converge and one plate is forced down (subducted) beneath the other. This occurs when an oceanic plate is forced under either another oceanic plate or a continental plate. These areas are known as subduction zones and are characterized by earthquakes and volcanic activity.

subduction zone the area where the process of subduction (*q.v.*) occurs.

subglacial debris glacial debris transported at the base of the glacier.

sublethal *a.* not causing death directly, but having cumulative deleterious effects.

sublimation *n.* process by which a solid is converted directly into a gas, also applied to the reverse process.

subliminal *a. appl.* stimuli that are not strong enough to evoke a sensation.

sublittoral *a.* (1) below littoral, *appl.* the shallow water zone of the sea from the extreme low-tide level to a depth of around 200 m; (2) zone of a lake too deep for rooted plants to grow.

submarine canyon large, steep-sided trench that cuts across the continental margin.

submersed *a. appl.* plants growing entirely under water.

submission *n.* the behaviour of a losing animal in a conflict where it takes up a submissive posture to prevent further attack.

subnatural environments habitats in which human influence has caused some change to the natural vegetation cover, but the type and structure of the vegetation is still essentially the same, e.g. many existing temperate forests. *cf.* natural environments, seminatural environments, cultural environments.

subphylum *n.* taxonomic grouping between phylum and class, e.g. Vertebrata (in the phylum Chordata).

subsere *n.* (1) plant succession on a denuded area; (2) secondary succession *q.v.*

subsidence *n.* the process by which the ground surface is lowered, frequently as a result of human activities. Subsidence may involve the sudden collapse of material into a cavity (cavity collapse) or it may be a gradual process (settlement) bought about by the lowering of the water table (*q.v.*), through, e.g., depletion of groundwater reserves for human use.

subsistence economy economy where the aim is to produce enough goods to meet the basic requirements for survival, with little or no surplus for trade.

subsistence farming *n.* type of farming in which crops are cultivated and livestock reared to supply the family or local community with food, with little or no surplus for exchange. *cf.* commercial farming.

subsoil *n.* soil found beneath the A-horizon (*q.v.*). *see also* B-horizon.

subsoiling *n.* agricultural practice of mechanically breaking up a compacted subsoil (without turning it over) to improve drainage.

subsong *n.* the first attempts at song by a young bird, which resembles the adult song but is imprecise and lacks some elements, with ill-defined phrasing and lack of tonal purity.

subspecies (ssp.) *n.* a taxonomic term usually meaning a group consisting of individuals within a species having certain heritable distinguishing characteristics separating them from other members of the species and forming a breeding group, but which can still interbreed with other members of the species. *alt.* variety.

substrate *n.* (1) substance on which an enzyme acts in biochemical reactions; (2) respiratory substrate, substance undergoing oxididation during respiration; (3) any material used by microorganisms as a source of food; (4) inert substance containing or receiving a nutrient solution on which microorganisms grow; (5) the base to which a sedentary animal or plant is fixed, *alt.* substratum.

substratum *see* substrate.

subtidal *a. pert.* area of shore that is below the mean low-water level of the spring tides.

subtropical *a. appl.* regions, their climate and their flora and fauna, that border the tropics on either side.

subtropical jet stream (STJS) jet stream located at the boundary between the Ferrel cell and the Hadley cell. With a strong (continuous) westerly airflow, this jet stream, which occurs in both hemispheres between 25° and 30° latitude, is associated with the development of monsoons and hurricanes.

subtypical *a.* deviating slightly from type.

suburb *n.* the peripheral part of a town or city that is largely residential in nature. *a.* suburban.

suburbanized village commuter village *q.v.*

succession *n.* (1) a geological, ecological or seasonal sequence of species; (2) the sequence of different communities developing over time in the same area, leading to a dynamic steady state or climax community (used esp. of plant or microbial communities), *see also* mechanisms of succession, primary succession, secondary succession; (3) the occurrence of different species over time in a given area.

successional cropping sequential cropping *q.v.*

successive percentage mortality apparent mortality *q.v.*

succulent *a.* (1) full of juice or sap; *appl.* (2) fruit having a fleshy pericarp, such as berries; (3) plants adapted to dry and desert conditions, with swollen water-storing stems and leaves.

sucker *n.* (1) (*bot.*) a branch of stem, at first running underground and then emerging, which may eventually form an independent plant; (2) haustorium (*q.v.*); (3) (*zool.*) an organ adapted to attach to a surface by creating a vacuum. In some animals, it is used for the purpose of feeding, in others to assist locomotion or attachment.

sucking lice common name for the Anoplura *q.v.*

sucrose *n.* a non-reducing disaccharide present in many green plants and hydrolysed by the enzymes invertase or sucrase or by dilute acids to glucose and fructose. *alt.* beet sugar, cane sugar, invert sugar, saccharose, saccharobiose.

suction pressure (SP) the capacity of a plant cell to take up water by osmosis, being the difference between the osmotic pressure of the cell sap causing water to enter and the back pressure exerted by the cell wall (turgor pressure). When a cell is fully turgid it has no suction pressure. *alt.* water potential.

Suctoria, suctorians *n.*, *n.plu.* group of predatory ciliate protozoans, which usually lose their cilia as adults and possess one or more tentacles, by which they catch their prey.

Sudanian-Sindian Floral Region part of the Palaeotropical Realm comprising the Sahel region of Africa, Sudan and northwest India.

suffrescent *a.* slightly shrubby, *appl.* plants that are woody at the base but herbaceous above and that do not die back to ground level in winter.

sugar *n.* (1) the general name for any mono-, di- or trisaccharide; (2) sucrose *q.v.*

sugar-beet *n.* *Beta vulgaris* (family Chenopodiaceae), a root crop grown in temperate regions for the production of sugar and for fodder.

sugar-cane *n.* *Saccharum officinarum*, a tropical plant of the family Gramineae grown for the sugar content of its stems.

suids *n.plu.* members of the mammalian family Suidae, the pigs.

suines *n.plu.* members of the mammalian suborder Suina, which includes the nonruminant artiodactyls, comprising the hippopotamuses, pigs, peccaries and a number of extinct groups.

sulfur, sulfide, sulfate *see* sulphur, sulphide, sulphate.

sulphate *n.* the SO_4^{2-} anion, or a salt or ester containing this anion. A salt of sulphuric acid, H_2SO_4. *alt.* sulfate.

sulphate-reducing bacteria *see* sulphur bacteria.

sulphide *n.* the S^{2-} anion, or a salt containing this anion. *alt.* sulfide.

sulphite *n.* the SO_3^{2-} anion, or a salt or ester containing this anion. A salt of sulphurous acid, H_2SO_3. *alt.* sulfite.

sulphur (S) *n.* non-metallic element (at. no. 16, r.a.m. 32.06), forming a flammable yellow solid. It occurs in nature as deposits of the native element and as sulphides and sulphates and in organic matter. It is an essential element for living organisms as many proteins contain sulphur. *alt.* sulfur. *see also* sulphur bacteria, sulphur cycle.

sulphur bacteria a group of unrelated bacteria that can variously utilize sulphur or sulphide as a respiratory substrate or electron acceptor in photosynthesis or reduce sulphate to sulphide. They comprise the photosynthetic green sulphur bacteria and purple suphur bacteria, which oxidize sulphide to sulphur (and can also fix nitrogen), the colourless non-photosynthetic sulphur bacteria (e.g. *Thiobacillus*), which oxidize sulphide to sulphur and sulphate, and the heterotrophic sulphate-reducing bacteria (e.g. *Desulfovibrio*), which reduce sulphate to sulphide.

sulphur cycle a cycle of biological processes by which sulphur circulates within the biosphere, which includes assimilation of sulphur by plants as soil sulphate, incorporation into plant and animal protein, putrefaction of dead organic matter by bacteria that release sulphide, which can be converted to elemental sulphur, or to sulphate and back to sulphide by a heterogeneous group of sulphur bacteria.

sulphur dioxide SO_2, a gaseous pollutant of the atmosphere. Its tropospheric concentrations range from <1 p.p.b. in locations remote from industrial activity to 2 p.p.m. in highly polluted areas. It has both natural sources (totalling *ca.* 2.3×10^{12} mol SO_2 per year) and anthropogenic sources (totalling 1.6×10^{12} mol SO_2 per year, esp. from the burning of sulphur-containing fossil fuels (particularly coal). It has a variable but very short residence time in the troposphere (typically <0.02 years). Sulphur dioxide is a respiratory irritant (esp. in conjunction with particulates), can cause damage to plants and is oxidized in the troposphere to sulphur trioxide (SO_3) and sulphuric acid (H_2SO_4), both of which are important in the formation of acid deposition (*q.v.*). Some SO_2 reaches the stratosphere as a result of direct injection during major volcanic activity and the *in situ* photochemical oxidation of carbonyl sulphide (OCS) of marine origin. Once in the stratosphere, the SO_2 is oxidized, forming droplets of sulphuric acid. These have been implicated in the natural process of stratospheric ozone loss.

sulphurous smog London-type smog *q.v.*

summer *n.* (1) the season between spring and autumn, frequently taken to be, in the northern hemisphere, the period from mid-May to mid-August; (2) astronomically, the season that commences at the summer solstice and terminates at the autumnal equinox, i.e. in the northern hemisphere, between 21 June and 23 September.

summer solstice *see* solstice.

summer wood late wood *q.v.*

sun-basking *n.* behaviour shown by many poikilothermic animals to control body temperature.

sun spiders common name for the Solpugida. *see* solpugids.

super- prefix derived from L. *super*, over. In classification, a group just above the status of the taxon following it, as in superclass.

supercell mesoscale convective complex *q.v.*

supercity megalopolis *q.v.*

superclass *n.* taxonomic grouping between subphylum and class.

supercritical *a. appl.* nuclear fission chain reactions in which the number of neutrons produced by one generation is greater than in the previous generation. Under these conditions, the rate of reaction will rapidly increase, possibly reaching explosive rates, as in a nuclear bomb. *cf.* critical, subcritical.

superdominance overdominance *q.v.*

superfamily *n.* (1) taxonomic group ranking above a family but below an order.

superfluent *n.* an animal species of the same importance in an ecosystem as a sub-dominant plant species in a succession.

supergene enrichment (*geol.*) the natural enhancement of the grade of an ore as a consequence of the leaching of metalliferous material through the orebody to a point where it is concentrated by a change in the chemistry of the orebody, esp. at the water table.

supergene family gene superfamily *q.v.*

superheated steam gaseous water at a temperature >100 °C.

superimposed drainage a drainage pattern that appears to be unrelated to the existing rock surface. This arises when the original rock surface on which the drainage developed has been removed by denudation *q.v.*

superkingdoms *n.plu.* three primary kingdoms proposed for the classification of all living organisms: the Bacteria, the Archaea (archaebacteria) and the Eukarya (eukaryotes).

supernumerary chromosomes extra heterochromatic chromosomes present in some plants above the normal number for the species, such as B chromosomes.

superorganism *n.* any society, such as a colony of a eusocial insect species, possessing features of organization analogous to the properties of a single organism.

superovulation *n.* the production of an unusually large number of eggs at any one time.

superphosphate *n.* a mixture of $Ca(H_2PO_4)_2$ and $CaSO_4$, used as an inorganic phosphorus fertilizer, with a phosphorus content equivalent to 18–21% P_2O_5, supplied in powder form.

superphosphoric acid *n.* a mixture of H_3PO_4 and polyphosphoric acids, e.g. $H_4P_2O_7$, used as an inorganic phosphorus fertilizer, with a phosphorus content equivalent to 60% P_2O_5, supplied as a liquid.

supersaturated air air cooled below its dew-point that, in the absence of suitable nucleation points (e.g. dust), attains a relative humidity in excess of 100%.

superspecies *n.* a group of closely related species having many morphological resemblances.

supertanker *n.* a fast ship designed to carry liquid cargo (esp. oil) with a capacity in excess of 275 000 tonnes.

supplemental air volume of air that can be expelled from the lungs after normal breathing out.

supporting tissue (1) in plants, tissue made of cells with thickened walls, such as collenchyma and sclerenchyma, adding strength to the body of the plant; (2) in animals, skeletal tissue forming endo- or exoskeleton.

supra- prefix derived from L. *supra*, above, signifying situated above.

supraglacial debris glacial debris transported on the glacier surface.

supralittoral *a. pert.* seashore above high-water mark, or splash zone.

supratidal *a.* above high-water mark, *appl.* the splash zone and organisms living there.

surf *n.* the turbulent water caused by the breaking of waves on the seashore.

surface active compound surfactant *q.v.*

surface creep the process whereby large particles (>0.25 mm in diameter) are pushed forward along the surface of the ground by the wind.

surface fire fire that burns the undergrowth but does not reach the crowns of trees.

surface mining surface extraction of mineral materials, e.g. metal ores, coal and phosphates, from near-surface deposits. *see* open pit mining, strip mining. *cf.* deep mining.

surface runoff (*geog.*) moving surface water (*q.v.*). *alt.* overland flow.

surface soil the uppermost layer of soil, which is disturbed during tillage, i.e. Ap-horizon, or its equivalent in uncultivated soils. *alt.* topsoil.

surface water water from rain and other precipitation that has not yet infiltrated the soil, or evaporated, or entered channels such as streams, rivers and lakes.

surface water gleying *see* gleying.

surface waves (*geog./geol.*) seismic waves that travel along the surface of the Earth's crust. *cf.* body waves. *alt.* L waves.

surfactant *n.* a chemical species that, when present in liquid, is adsorbed at the boundary between the liquid and its surroundings and that alters the surface properties of that liquid, e.g. surface tension. Surfactants are used as wetting agents in detergents, facilitating better contact between water and dirt. *alt.* surface active compound.

surplus yield model a model for harvesting a natural resource that balances harvesting and recruitment rates by harvesting no more than the maximum sustainable yield.

surroundings *n.* in thermodynamic analysis, the surroundings are those parts of the universe that are outside the system.

survival curve curve produced by plotting the proportion of individuals surviving (from a starting population of known size), on a logarithmic scale, against age. *alt.* survivorship curve.

survival of the fittest the principle that those organisms most suited to their environment will be the most likely to survive and reproduce, which underlies the idea of natural selection. *see* fitness.

survivorship *n.* (1) the proportion of a population that reaches a given age; (2) the number of individuals in a particular age group counted at the beginning of a period that were surviving at the end of that period, often denoted by l_x.

survivorship schedule demographic data giving the number of individuals surviving to each particular age in a population.

survivorship curve survival curve *q.v.*

susceptibility *n.* the tendency of an individual or a species to succumb to infection by a particular disease, or to be affected by a particular poison. *cf.* resistance.

suspended load material, e.g. sand and silt, carried in suspension by a stream or river.

suspension feeder aquatic animal that feeds by straining particulate organic matter or plankton from water. *alt.* filter feeder.

sustainability *n.* the situation in which any particular process or way of life is carried out in such a way, or at such a level, that it could be continued indefinitely without harming the environment's life-support systems, or reducing its ability to assimilate wastes or depleting renewable or non-renewable resources in the foreseeable future. The term is often used interchangeably with the term sustainable development (*q.v.*), although sustainability *per se* has no implication of economic development or growth.

sustainable development, sustainable economic development an ill-defined but widely used concept in environmental political discussion, esp. on a global scale, and one of the central themes of the 1992 United Nations Conference on Environment and Development and Agenda 21. As defined in the Brundtland Report (*q.v.*), sustainable development is 'development that meets the needs of the present without compromising the ability of future generations to meet their own needs.' As thus defined, it recognizes the right of people, esp. the poorest people, to an increase in their standard of living and quality of life, while at the same time avoiding uncompensated and future environmental costs for future generations. The term is often used as a synonym for sustainability *q.v.*

sustainable economic growth sustainable development *q.v.*

sustainable management the management of any area, ecosystem or natural resource in such a way that it is maintained in its present state, or even improved, and not depleted or harmed.

sustainable yield, sustained yield highest rate at which a renewable resource can be used without reducing its supply. *alt.* maximum sustainable yield.

Sv seivert *q.v.*

Svedberg unit (S) unit in which the rate of sedimentation of a particle in the ultracentrifuge is expressed ($1\ S = 10^{-13}\ s$ under standard conditions) and which is an indirect measurement of size and molecular weight.

swallet swallowhole *q.v.*

swallowhole *n.* an underground shaft down which a surface stream disappears in limestone regions. *alt.* swallet.

swamp *n.* wet ground, saturated or periodically flooded, dominated by woody plants and with no surface accumulation of peat.

sward stick an implement used in grassland management to measure the height of the grass and therefore assess its suitability for different types of grazing animals.

swash *n.* the thin sheet of water that rushes up the beach when waves break. The return flow of water, under the influence of gravity, is known as backwash.

S waves *see* body waves.

swayback *n.* a disease of lambs caused by copper deficiency, in which nerve degeneration leads to a lack of limb coordination.

sweat glands specialized glands in the skin of some mammals, through which water and salts are exuded to aid evaporative cooling of the body.

sweating *n.* exudation of water from the body's surface through sweat glands, which is used as a cooling mechanism by some mammals, with the evaporation of the water from the surface having a cooling effect.

sweepstakes dispersal chance dispersal of organisms across water on floating objects.

sweet crude crude oil in which the presence of hydrogen sulphide (H_2S) is not detectable. Oil refineries prefer this low-sulphur crude to sour crude *q.v.*

swell waves *see* sea waves.

swidden agriculture shifting cultivation *q.v.*

swimmeret *n.* an abdominal appendage in crustaceans used for swimming.

swirlhole pothole *q.v.*

switch plant a xerophyte that produces normal leaves when young, then sheds them, and photosynthesis is taken over by another structure such as a cladode or phyllode.

syenite *n.* a coarse-grained or pegmatitic igneous rock with a colour index of 0–40. It contains alkali feldspar (which dominates the mineralogy) with or without sodium-rich plagioclase. Minor amounts of quartz (quartz syenite, grading into granite) or nepheline (grading into nepheline syenite) may be present, as may be pyroxene, amphibole or biotite.

sylva *n.* (1) forest of a region; (2) forest trees collectively.

sylvine, sylvite *n.* KCl, an evaporite mineral that is exploited as a source of potassium compounds.

sym-, syn- prefixes from the Gk *syn*, with.

symbiont *n.* one of the organisms in a symbiosis.

symbiosis *n.* (1) a close and usually obligatory association of two organisms of different species living together, not necessarily to their mutual benefit; (2) often used exclusively for an association in which both partners benefit, which is more properly called mutualism. *a.* symbiotic.

symbiotic nitrogen-fixing bacteria the rhizobia and some actinomycetes, which only fix nitrogen when living within the roots of their host plants.

symmetrodonts *n.plu.* an order of Mesozoic trituberculate mammals having molar teeth with three or more cusps in a triangle.

symmetry *n.* regularity of form. *see* bilateral symmetry, radial symmetry.

symparasitism *n.* the development of several competing species of parasites within or on one host.

sympatric *a. appl.* distinct species inhabiting the same or overlapping geographic areas that do not interbreed. *cf.* allopatric.

sympatric speciation the evolution of distinct species from a population living in the same area.

symphily *n.* the situation where one species of insect lives as a guest (symphile) in the nest of a social insect, which feeds and protects it in return for its secretions, which are used as food, e.g. certain beetles in the nests of ants and termites.

Symphyta *n.* the sawflies, a suborder of hymenopteran insects, having no well-defined 'waist', and considered more primitive than members of the suborder Apocrita (bees, wasps, ants, ichneumon flies). The ovipositor is serrated like a saw.

symplast *n.* the interconnected protoplasm of plant tissue, comprising the protoplasm of individual cells connected by plasmodesmata through the cell walls. *a.* symplastic.

symplesiomorphy *n.* in the cladistic method of classification, the case where a homologous character state shared between two or more taxa is believed to have originated as a novelty in a common ancestor earlier than the most recent common ancestor. *a.* symplesiomorphic, symplesiomorphous. *cf.* homoplasy, synapomorphy.

Synanthae Cyclanthales *q.v.*

synanthropic *a.* associated with humans or their dwellings.

synapomorphy *n.* in cladistic classifications denotes a homologous character common to two or more taxa and thought to have originated in their most recent common ancestor. *a.* synapomorphous. *cf.* apomorphous.

synaposematic *a.* having warning colours in common, *appl.* mimicry of a more powerful or dangerous species as a means of defence.

synapsid *a.* having a skull with one ventral temporal fenestra on each side.

Synapsida, synapsids *n., n.plu.* subclass of reptiles living from the Carboniferous to the Triassic, the mammal-like reptiles, with synapsid skulls, some forms of which gave rise to the mammals, and including the pelycosaurs and therapsids.

synchorology *n.* (1) study of the distribution of plant and animal associations; (2) geographical distribution of communities.

synchronic *a.* (1) contemporary; (2) existing at the same time, *appl.* species, etc. *cf.* allochronic.

synchronizer *n.* some environmental factor, such as light or temperature, that interacts with an endogenous circadian rhythm, causing it to be synchronized (entrained) precisely to a 24-h cycle rather than free-running. *alt.* zeitgeber.

synchronous orbit geostationary orbit *q.v.*

synclines *n.plu.* trough-like downfolds in rock strata, produced by folding. *cf.* anti-clines. *see* Fig. 2, p. 25.

syncryptic *a. appl.* animals that appear alike, although unrelated, due to common protective resemblance to the same surroundings.

syncytium *n.* a multinucleate mass of protoplasm that is not divided into separate cells. *a.* syncytial. *cf.* coenocyte.

synecology *n.* the ecology of plant or animal communities. *cf.* autecology.

synergic, synergistic *a.* acting together, often to produce an effect greater than the sum of the two agents acting separately. *n.* synergism, synergy.

synfuel *n.* liquid fuel derived from coal, tar sands or oil shales. *alt.* synthetic fuel.

syngameon *n.* group of species within which hybridization can occur.

syngamodeme *n.* a population unit made up of coenogamodemes whose members can form sterile hybrids with each other.

syngamy *n.* sexual reproduction in unicellular organisms where the two cells that fuse are morphologically similar.

syngas *n.* a mixture of hydrogen (H_2) and carbon monoxide (CO) generated, along with methane (CH_4), during coal gasification (*q.v.*). *alt.* synthetic gas.

synonym *n.* in classification, an alternative Latin name.

synoptic chart weather map *q.v.*

syntenic *a. appl.* genetic loci that lie in the same order on the same chromosome in different species.

synthetic fuel synfuel *q.v.*

synthetic gas syngas *q.v.*

synthetic pyrethroid pyrethroid *q.v.*

synthetic theory of evolution *see* neo-Darwinism.

syntopic *a.* sharing the same habitat within the same geographical range, *appl.* different species or to phenotypic variants within a species.

syntype *n.* any one specimen of a series used to designate a species when holotype and paratypes have not been selected. *alt.* cotype.

synusia *n.* a plant community of relatively uniform composition, living in a particular environment and forming part of the larger community of that environment.

syrinx *n.* sound-producing organ in birds, situated at junction of windpipe with bronchi. It consists of two patches of thin membrane (tympaniform membrane) in the wall of each bronchus. Sound is produced when the tympanic membranes are pushed inwards partially blocking the bronchi. Air pushed out of the lungs makes the membranes vibrate and produces a sound. *plu.* syringes, syrinxes.

system *n.* in thermodynamic analysis, the system is the portion of the universe that is under consideration.

systematic error error that alters the accuracy (*q.v.*) of the outcome of a quantitative analytical procedure. *alt.* determinate error.

systematics *n.* (1) the classification of living things with regard to their natural relationships; (2) taxonomy (*q.v.*); (3) the identification, classification and nomenclature of organisms.

Système International d'Unités *see* SI units.

Système Pour l'Observation de la Terre (SPOT) series of orbiting satellites designed for Earth observation, commencing in 1986 with SPOT-1. The French equivalent of the Landsat of the USA.

systemic *a.* (1) throughout the body, involving the whole body; *appl.* (2) herbicides that are taken up by a plant through its roots and translocated to other parts of the plant, where they exert their effects; (3) pesticides that are taken up into plants through the roots and transferred via the sap to the insect pests feeding on that plant. They are useful for ornamental plants but cannot usually be used on food crops.

T

τ tau, symbol for residence time *q.v.*

θ theta, symbol for soil porosity *q.v.*

T (1) tritium (*q.v.*); (2) threonine (*q.v.*); (3) thymine *q.v.*

2,4,5-T 2,4,5-trichlorophenoxyacetic acid, a translocated herbicide that acts as a defoliant and can be used to clear woody plants in grasslands. It is a cause of environmental concern because of the presence of TCDD (*q.v.*). *see also* Agent Orange.

Ta symbol for the chemical element tantalum *q.v.*

tableland plateau *q.v.*

table mountain a term sometimes used for a large mesa *q.v.*

tactic *a. pert.* a taxis.

taiga *n.* northern coniferous forest zone, esp. in Siberia, adjacent to the tundra.

tailings *n.* in mining operations, the waste material remaining after ore processing.

take-all *n.* a soil-borne disease affecting wheat and barley, esp. on calcareous soils, caused by the fungus *Ophiobolus graminis*. Infected plants produce bleached ears that contain little or no grain (a result of premature ripening). *alt.* whiteheads.

talc *n.* $Mg_3Si_4O_{10}(OH)_2$, a phyllosilicate mineral. Occurs as an alteration product of magnesium silicates. When massive, called soapstone or steatite.

talik *n.* the unfrozen ground found beneath the permafrost (*q.v.*). Also *appl.* pockets of unfrozen water trapped within the permafrost or between the permafrost table (*q.v.*) and the frozen active layer *q.v.*

tall-grass community type of grassland with grasses up to 2 m or more high that develops in parts of the tropics (tall-grass savanna) and in temperate regions (tall-grass prairie, tall-grass steppe).

talus *n.* coarse angular rock fragments (produced by the physical weathering of rock faces) that accumulate at the base of steep slopes. *alt.* scree.

talus cone steep-sided accumulation of talus *q.v.*

talus creep slow movement of talus (*q.v.*) downhill.

talus slope slope covered with talus (*q.v.*), formed by the convergence of several talus cones. *alt.* scree slope.

Tamaricales *n.* order of dicot trees, shrubs or, rarely, herbs, with small scale-like or heather-like leaves. They include the families Fouquieriaceae (ocotilla), Frankeniaceae (sea heath) and Tamaricaceae (tamarisk).

tannins *n.plu.* complex aromatic compounds some of which are glucosides, occurring in the bark of various trees, possibly giving protection or concerned with pigment formation, and used in tanning and dyeing.

tantalite *n.* a mineral with a composition $(Fe,Mn)(Nb,Ta)_2O_6$, in which tantalum (Ta) predominates over niobium (Nb). A niobium and tantalum ore mineral. *see also* columbite.

tantalum (Ta) *n.* metallic element (at. no. 73, r.a.m. 180.95), forming a heavy grey malleable metal in elemental form. It occurs with niobium in the mineral tantalite.

tanzanite *n.* a blue variety of the mineral zoisite (*q.v.*), prized as a gemstone.

tapeworms *n.plu.* common name for the Cestoda *q.v.*

taphocoenoses *n.plu.* burial communities, assemblages of fossils of widely differing types that result from conditions of burial and sediment transport and do not usually represent the original inhabitants of that site.

taphonomy *n.* the study of the conditions of burial of fossils.

taphrophyte *n.* a ditch-dwelling plant.

tapioca *see* cassava.

taproot *n.* long straight main root formed from radicle of embryo in gymnosperms and dicotyledons. In plants such as carrot and turnip it forms a swollen food-storage organ.

Tardigrada, tardigrades *n., n.plu.* (1) a phylum of small animals, commonly called water bears, that have some features similar to arthropods and are sometimes included in the Arachnida; (2) an infraorder of Edentata, including the tree sloths.

target organism organism that is intended to be killed by a pesticide.

tarn *n.* a small circular lake that develops post-glacially in a cirque *q.v.*

tar-oil spray *n.* spray used in winter on deciduous fruit trees and bushes to kill eggs of pests such as aphids and scale insects.

tar sand sandstone impregnated with bitumen (a very heavy, viscous oil), a potentially commercial source of oil. *alt.* oil sand.

tarsus *n.* (1) ankle and heels bones, collectively; (2) segment of insect leg beyond the tibia, which bears the terminal claw; (3) fibrous connective tissue plate of eyelid. *a.* tarsal.

tassel-finned fishes a common name for the crossopterygians *q.v.*

tau (τ) residence time *q.v.*

tautomerism *n.* the simultaneous presence of two isomeric forms of a compound (referred to as tautomers) that are in a state of dynamic equilibrium, as in, e.g., Keto-entol tautomerism, in which both $-CH_2-c(o)-CH_2-$ (i.e. the Keto-form) and $-C=C(oH)-Ch_2-$ (i.e. the enol-form) forms of the compound are present.

tautonyn *n.* the same name given to a genus and one of its species or subspecies.

taxa *plu.* of taxon *q.v.*

Taxales *n.* order of gymnosperms, being evergreen shrubs or small trees with small linear leaves, and ovules solitary and surrounded by an aril, such as the yews (family Taxaceae).

taxes *plu.* of taxis *q.v.*

taxis *n.* (1) a movement of a freely motile cell, such as a bacterium or protozoan, or of part of an organism, towards (positive) or away from (negative) a source of stimulation, such as light (phototaxis) or chemicals (chemotaxis); (2) orientation behaviour related to a directional stimulus. *a.* tactic. *plu.* taxes.

taxol *n.* anticancer drug obtained from the Pacific yew tree, which inhibits cell division by binding to microtubules and preventing their normal assembly and disassembly.

taxon *n.* the members of any particular taxonomic group, e.g. a particular species, genus, family, etc. *plu.* taxa.

taxonometrics numerical taxonomy *q.v.*

taxonomic category a category used in the classification of living organisms, e.g. phylum, class, order, family, genus or species.

taxonomy *n.* (1) the science of classification, esp. of living organisms, which arranges organisms into hierarchical groupings on the basis of anatomy, morphology, and genetic and biochemical characteristics; (2) the analysis of an organism's characteristics for the purpose of classification. *see also* systematics.

taxospecies, taxonomic species species defined by similarities in morphological characters only and that do not necessarily correspond to biological species.

Tb symbol for the chemical element terbium *q.v.*

TB tuberculosis *q.v.*

2,3,6-TBA translocated herbicide used to control numerous broad-leaved annual and perennial weeds.

TBTs tributyltins *q.v.*

Tc symbol for the chemical element technetium *q.v.*

TC₅₀ a measure of infectivity for viruses, the dose at which 50% of tissue cultures treated with the virus become infected and show degeneration.

TCA (1) trichloracetic acid. Used as a soil-acting herbicide to control weeds such as wild oats and couch grass; (2) 1,1,1-trichloroethane, an industrial and dry-cleaning solvent.

TCA cycle tricarboxylic acid cycle *q.v.*

TCDD 2,3,7,8-tetrachlorodibenzo-*p*-dioxin *q.v.*

TCE trichloroethene, a toxic and potentially carcinogenic solvent, which has been used as a degreasing agent and dry-cleaning agent.

TDN total digestible nutrients *q.v.*

TDS total dissolved solids *q.v.*

Te symbol for the chemical element tellurium *q.v.*

tea *see Camellia.*

tear fault transcurrent fault *q.v.*

technetium (Tc) *n.* radioactive metallic element (at. no. 43, r.a.m. 98.91) that is a fission product of uranium and can be produced by neutron bombardment of molybdenum. There are no known terrestrial sources.

tectonic (*geol.*) *a.* (1) *pert.* those major processes within the crust of the Earth that lead to its structural deformation; (2) *appl.* rock masses, landforms, etc., associated with tectonic activity.

tectonic plates lithospheric plates *q.v.*

tectonism diastrophism *q.v.*

tectosilicate *n.* any silicate mineral that contains silica tetrahedra in a three-dimensional network such that each tetrahedron shares all four of its oxygens with neighbouring tetrahedra.

tedding *n.* (*agric.*) spreading out the new-mown grass during haymaking to accelerate the drying process.

teeth *n.plu.* (*zool.*) (1) hard, bony outgrowths from jaws of mammals, each tooth composed of a core of soft tissue (pulp) supplied with nerves and blood vessels, surrounded by a layer of dentine and covered with an outer layer of enamel. The adult set generally comprises molars or grinding teeth, premo-lars, canines and incisors or biting teeth; (2) similar structures in jaws, throat, etc., of reptiles and fish; (3) any similarly shaped structures in invertebrates used for rasping, seizing or grinding food and that are generally composed of chitin or keratin; (4) (*bot.*) the pointed outgrowths on edges of leaves, calyx, etc. *sing.* tooth.

Teflon trade name for polytetrafluoroethene (PTFE).

teleological *a. appl.* explanations for the evolution of particular functions or structures that suppose a purpose or design to evolution.

teleology *n.* (1) the doctrine of final causes, the invalid view that evolutionary developments are due to the purpose or design that is served by them; (2) similar type of explanation applied to biological or cellular process, or animal behaviour, which presupposes an impossible awareness of a particular goal.

teleonomy *n.* the idea that if a structure or process exists in an organism it must have conferred an evolutionary advantage. *a.* teleonomic.

Teleostei, teleosts *n.*, *n.plu.* group of fish including all modern bony fishes except lungfishes, holosteans and crossopterygians, with thin bony scales covered by an epidermis, a homocercal tail, a hydrostatic air bladder (swimbladder), no spiracle and no spiral valve in the gut.

teleplanic larvae larvae that become dispersed over immense distances in the ocean.

tellurium (Te) *n.* semiconducting metalloid element (at. no. 52, r.a.m. 127.60), whose compounds are toxic. It is present in ores (tellurides) with other metals such as gold and silver.

telotaxis *n.* movement along line between animal and source of stimulus.

TEM transmission electron microscope *q.v.*

Temnospondyli, temnospondyls *n.*, *n.plu.* extinct order of labyrinthodont amphibians, found in the Carboniferous.

temperate *a. appl.* (1) moderate climate with distinct seasons, having warm to hot summers and cool to cold winters; (2) the regions between latitudes 23° 27′ and 66° 33′ N and 23° 27′ and 66° 33′ S of the Equator, in which temperate climates are found, and, e.g., to the ecosystems, biomes, etc., found within these regions.

temperate deciduous forest the major biome in temperate regions when annual rainfall is sufficient to support the growth of trees (75–150 cm), being composed of deciduous broad-leaved trees, such as oaks, ash, beech, maples and chestnut, the particular species composition varying with different continents and regions.

temperate evergreen forest type of forest found in warm-temperate coastal climates of Japan and elsewhere with high levels of rainfall, characterized by broad-leaved evergreens such as rhododendron, live oak, magnolias and hollies.

temperate karst type of limestone landscape, such as found in former Yugoslavia, characterized by solution depressions, underground cave networks and craggy rock masses.

temperate rain forest type of forest dominated by conifers that develops in some temperate coastal areas with high annual rainfall (75–375 cm) and continual moisture from ocean fogs, such as the coastal redwood forests of western North America. *alt.* moist coniferous forest.

temperature coefficient (Q_{10}) the increase in the rate of a process for a 10 °C rise in temperature, calculated as the quotient of the reaction rates at the lower and higher temperature.

temperature inversion an atmospheric condition in which relatively cold, dense air lies below warmer air. This feature exists in the stratosphere and, under certain climatic conditions, may form temporarily within the troposphere. A localized temperature inversion may occur in the troposphere when night-time radiative cooling (*q.v.*) lowers the temperature of the air near the Earth's surface. Inversions of this type are usually broken down during the day time, when sunlight warms the ground. Tropospheric temperature inversions contain essentially static air. They can therefore allow significant levels of pollution to build up over industrialized centres, leading to the formation of either London-type smogs (*q.v.*) or Los Angeles-type smogs (*q.v.*). *alt.* inversion.

temporal *a.* (1) *pert.* time; (2) in region of temples; (3) *appl.* compound bone on side of mammalian skull whose formation includes fusion of petrosal and squamosal bones.

temporal isolation prevention of interbreeding between species of plants or animals as a result of differences in the timing of reproductive events such as shedding of pollen or mating.

temporary hardness *see* hard water.

tendril *n.* specialized twining stem, leaf, petiole or inflorescence by which climbing plants support themselves.

teneral *a.* (1) immature; (2) *appl.* stage of some insects on emergence from the nymphal covering.

Tenericutes *n.* one of the main divisions of the prokaryotic kingdom in some classifications, which includes the various wall-less prokaryotes, such as rickettsiae, mycoplasmas and chlamydiae.

tennantite *n.* $Cu_{12}As_4S_{13}$ or $(Cu,Fe)_{12}As_4S_{13}$, a copper ore mineral. It is the end-member of a continuous series, with tetrahedrite $[Cu_{12}Sb_4S_{13}$ or $(Cu,Fe)_{12}Sb_4S_{13}]$ at the other end formed as antimony (Sb) substitutes for arsenic (As).

ten per cent law the generalization that 90% of energy at one trophic level in a food chain is lost as respiration when being transformed into the energy of the next trophic level.

tension wood reaction wood of dicotyledons, having little lignification and many gelatinous fibres, and produced on the upper side of bent branches.

tenuivirus group group of single-stranded RNA plant viruses with filamentous particles, occasionally branched.

teosinte *n. Zea mexicana*, a member of the Gramineae that occurs in Central America and is thought to be an ancestor of cultivated maize.

tepal *n.* a petal-like perianth segment that is not differentiated into petal or sepal, as in magnolias.

tephra *n.* the material that constitutes pyroclastic rock *q.v.*

teratogen *n.* any agent that can cause malformations during embryonic development, e.g. the drug thalidomide.

teratogenic *a.* capable of causing birth defects.

terbium (Tb) *n.* a lanthanide element (at. no. 65, r.a.m 158.92), a silver-coloured metal when in elemental form.

terminal *a.* (1) *pert.*, or situated at, the end, as the terminal bud at end of twig; (2) final.

terminal moraine *see* moraine.

termitarium *n.* elaborately constructed nest of a termite colony.

termites *n.plu.* common name for the Isoptera *q.v.*

termiticole *n.* an organism that lives in a termite nest, e.g. some fungi and insects.

terpene, terpenoid *n.* any of a large class of natural plant products based on the isoprenoid unit, and including beta-carotene, lycopene, violaxanthin and other xanthophylls, and gibberellins.

terracettes *n.plu.* small steps, each usually less than 0.5 m deep and 0.5 m wide, which typically occur in groups along the contours of a slope, caused by soil instability.

terracing *n.* the conversion of a hillside into a series of level shelves of earth. These man-made terraces, capable of retaining both soil and water, are then used for the intensive cultivation of crops, e.g. rice.

terra rosa red clay soil formed on limestones, most common in areas with a mediterranean climate.

terrestrial *a.* (1) *pert.*, living, or found on land, as opposed to in water or in the atmosphere; (2) *pert.* Earth.

terrestrial radiation radiation emitted from the Earth, often referred to as long-wave radiation.

terricolous *a.* inhabiting the soil.

terriherbosa *n.* terrestrial herbaceous vegetation.

territoriality *n.* a social system in which an animal establishes a territory, which it defends against other members of the same species.

territory *n.* (1) area defended by an animal or group of animals, mainly against other members of the same species; (2) an area sufficient for the food requirements of an animal or aggregation of animals; (3) foraging area.

Tertiary *a. pert.* or *appl.* geological period (or sometimes considered as an era) lasting from about 65 to 2 million years ago.

tertiary amines *see* amines.

tertiary consumer a carnivore that eats other carnivores.

tertiary effluent effluent produced as a result of the tertiary treatment of sewage. *see* sewage treatment.

tertiary parasite an organism parasitic on a hyperparasite.

tertiary recovery *see* oil recovery.

tertiary sludge sludge produced as a result of the tertiary treatment of sewage. *see* sewage treatment.

tertiary treatment of sewage *see* sewage treatment.

test, testa *n.* (1) shell or hard outer covering; (2) (*bot.*) seed coat.

test cross the mating of an organism to a double recessive for a given character in order to determine whether it is homozygous or heterozygous for the character under consideration.

Tethys (Tethyian Sea) the sea that existed between Laurasia and Gondwanaland during the Mesozoic.

tetra- prefix derived from Gk *tetras*, four, signifying having four of, arranged in fours, divided into four parts.

tetraalkyl lead a class of compounds with the general formula PbR_4, where R is an alkyl group. Those with R = methyl (CH_3) and/or ethyl (CH_2CH_3) are used as antiknock additives in petrol. The extensive use of these additives has resulted in the widespread contamination of the environment with lead.

2,3,7,8-tetrachlorodibenzo-*p*-dioxin (TCDD) extremely toxic congener of the dioxin family. It is found as a trace contaminant in a number of manufactured products, including certain disinfectants, some herbicides (such as Agent Orange *q.v.*) and paper bleached with chlorine. It is also formed, in trace amounts, when organochlorine compounds, including PCBs and the plastic PVC, are burned in an uncontrolled manner. *see also* polychlorinated dibenzo-*p*-dioxins, Seveso incident.

tetrachloroethene *n.* a non-flammable organochlorine that has been used as a dry-cleaning solvent and degreaser, but which is carcinogenic. *alt.* tetrachloroethylene.

tetrachloromethane carbon tetrachloride *q.v.*

tetrachlorvinphos *n.* contact organophosphate insecticide used to control caterpillars and dipterous pests of crops such as cotton and rice.

tetracycline *n.* antibiotic produced by a *Streptomyces* sp. which inhibits the binding of tRNA to bacterial ribosomes and can be used against both Gram-negative and Gram-positive bacterial infections.

tetradifon *n.* contact organochlorine acaricide used to control red spider mites under glass and on fruit crops.

tetraethyl lead *n.* $Pb(CH_2CH_3)_4$, a tetraalkyl lead compound that is used as an antiknock additive in petrol.

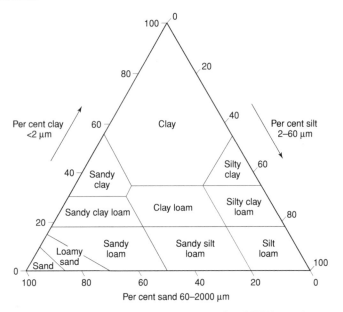

Fig. 31 Soil textural classes (from White, 1979, after SSEW, 1974).

tetrahedrite *n.* $Cu_{12}Sb_4S_{13}$ or $(Cu,Fe)_{12}$-Sb_4S_{13}, a copper ore mineral. It is the end-member of a continuous series, with tennantite [$Cu_{12}As_4S_{13}$ or $(Cu,Fe)_{12}As_4S_{13}$] at the other end, formed as arsenic (As) substitutes for antimony (Sb).

tetraploid *a.* having four times the haploid number of chromosomes, *appl.* cells, nuclei, organisms. *n.* an organism with four sets of chromosomes per somatic cell.

tetrapod *n.* a four-legged vertebrate.

tetrahedral *a. appl.* the geometry of the environment of an atom or ion that can be imagined to be at the centre of a tetrahedron, the corners of which are coincidental with the atom's nearest neighbours.

texture *n.* (1) (*geol.*) the size, shape and arrangement of the constituent grains of a rock; (2) (*soil sci.*) the particle size distribution of the mineral component of a soil's fine earth (*q.v.*). A soil may be allocated to a textural class on the basis of the proportion of sand, silt and clay-sized particles within its fine earth. Unfortunately, there is more than one system for doing

this in common use: e.g. the system used by the Soil Survey of England and Wales (*see* Fig. 31) differs from that of the USDA.

textured vegetable protein soya beans processed to resemble meat in texture and used as a meat substitute or meat extender.

Th symbol for the chemical element thorium *q.v.*

thalassoid *a. pert.* freshwater organisms resembling, or originally, marine forms.

thalassoplankton *n.* marine plankton.

Thaliacea *n.* class of free-swimming tunicates (urochordates), commonly called salps, having a gelatinous test and two distinct phases in the life history, one reproducing sexually, the other asexually.

thallium (Tl) *n.* soft white metallic element (at. no. 81, r.a.m. 204.37), occurring naturally in crookesite. Its compounds are extremely toxic and are used as pesticides.

Thallobacteria *n.* a class of prokaryotes containing the actinomycetes and related forms.

thallophyte *n.* any member of the plant kingdom in which the plant body is not divided into root, stem and leaves, e.g. an alga.

thallus *n.* a simple plant body not differentiated into leaf and stem, as of lichens, multicellular algae and some liverworts.

thalweg river profile *q.v.*

thanatocoenosis *n.* fossil assemblage composed of organisms brought together after their death, but which did not necessarily live together.

thanatosis *n.* (1) habit or act of feigning death; (2) death of a part.

thaumatin *n.* a protein with an intensely sweet taste.

thaw lakes small, shallow lakes formed in thermokarst depressions, a common feature of periglacial environments.

Theales *n.* order of dicot trees, shrubs and woody climbers, rarely herbaceous, with evergreen leaves. Includes many families, e.g. Dipterocarpaceae, Hypericaceae (St John's wort) and Theaceae (tea).

theave *n.* a female sheep between its first and second shearing.

thecodonts *n.plu.* group of primitive reptiles of Permian–Triassic age, including bipedal or crocodile-like forms, and thought to be ancestral to several groups, with thecodont teeth (embedded in sockets within the bone).

Theobroma cacao tree whose seeds are cocoa beans, from which chocolate is made.

theobromine *n.* 3,7-dimethylxanthine, a purine similar to caffeine, found in coffee, tea and chocolate, which is a stimulant of the central nervous system and a diuretic.

theophylline *n.* plant alkaloid closely related to theobromine and obtained originally from tea leaves, which inhibits the hydrolysis of cyclic AMP by phosphodiesterase and is also a diuretic and heart stimulant.

Therapsida, therapsids *n.*, *n.plu.* an order of extinct mammal-like reptiles, living from the Permian to the Triassic, believed to be ancestral to mammals.

Theria, therians *n.*, *n.plu.* a subclass of mammals, including all living mammals except monotremes, with molar teeth bearing a triangle of cusps (tribosphenoid), cervical ribs fused to vertebrae and a spiral cochlea.

therm *n.* a unit used for quantifying heat. 1 therm = 100 000 British thermal units *q.v.*

thermal conduction *see* conduction, thermal.

thermal conductivity rate at which a body can conduct heat.

thermal depressions *see* non-frontal depressions.

thermal efficiency an expression of the efficiency of any one heat-to-other-energy-form conversion (including the heat-to-work transformations of heat engines, *q.v.*). It equals (useful energy output)/(energy input) and may be expressed as a percentage. For power stations, thermal efficiency is taken to be the ratio of electrical power output (MW_e) to thermal power output (MW_{th}).

thermal enrichment a positive effect on an aquatic ecosystem as a result of a small rise in the temperature of the water.

thermal metamorphism the formation of metamorphic rocks by the action of heat alone. These rocks are often characterized by the presence of new minerals formed when high temperatures cause the parent rock to melt and recrystallize.

Thermal Oxide Reprocessing Plant (THORP) the plant at Sellafield, Cumbria, UK, run by the government-owned company British Nuclear Fuels plc (BNFL), which reprocesses spent fuel from nuclear reactors. Operations began in 1994.

thermal panting method of body cooling employed by some birds and mammals, and a few reptiles, in which heat is lost from the body by panting, thus increasing the evaporation of water from the moist surface tissues of the respiratory tract.

thermal plume *see* plume.

thermal pollution raising of the temperature of the environment as a result of discharge of heat into it from industrial processes, e.g. the release of cooling water from a power station into a river, which may affect living organisms both by the rise in temperature itself and by the decreased amount of dissolved oxygen that the warmer water can hold.

thermal processing waste disposal (or waste conversion) involving pyrolysis or incineration. *see also* Thermal Oxide Reprocessing Plant *q.v.*

thermal reactors nuclear reactors that utilize neutrons that are in thermal equilibrium with the atoms of the reactor. These neutrons cause fission of fissile material (generally ^{235}U, although ^{233}U and ^{239}Pu can also be used), liberating energy, fission fragments and more neutrons. These liberated neutrons are fast-moving. They are slowed down, i.e. moderated, until they are in thermal equilibrium with the atoms of the reactor, after which they can initiate further fission reactions. During these reactions the fissile material is consumed or 'burned', hence these reactors are also called burner reactors.

thermal shock a very rapid change in water temperature resulting in mass kills of fish and other aquatic organisms.

thermal springs hot springs *q.v.*

thermium *n.* plant community in warm or hot springs.

thermoacidophilic bacteria group of archaebacteria (e.g. *Sulfolobus, Thermococcus, Thermoproteus*) adapted for life in acidic hot springs, thriving at temperatures of 70–75 °C and pH range 1–3. *alt.* thermoacidophiles.

thermobiology *n.* the study of the effects of thermal energy on all types of living organisms and biological molecules.

thermocleistogamy *n.* self-pollination of flowers when unopened owing to unfavourable temperature.

thermocline *n.* (1) the upper layer of water of rapidly changing temperature found in lakes and seas in summer; (2) zone between the warm surface water and colder deep water, in which the temperature gradually decreases, as in a lake, reservoir or ocean.

thermoduric *a.* resistant to relatively high temperatures (70–80 °C), *appl.* microorganisms.

thermodynamic temperature absolute temperature *q.v.*

thermodynamics *n.* the study of energy transformations.

thermodynamics, the laws of these may be stated thus: *Zeroth law*: consider three bodies, A, B and C. If A and B are in thermal equilibrium with C, then A and B must also be in thermal equilibrium. A, B and C must therefore share a common temperature. *First law*: the energy content of a system of constant mass can be neither increased nor decreased. *Second law*: the entropy (disorder) of an isolated system can either remain constant or increase. *Third law*: at a temperature of absolute zero (0 K, –273.15 °C), a perfect crystalline material has zero entropy.

thermogenesis *n.* heat generation, e.g. in hibernating mammals to maintain body temperature, carried out by brown adipose tissue.

thermohaline circulation (*oceanogr.*) the deep-ocean circulation. This is propelled chiefly by differences in seawater density, brought about by variations in temperature and salinity.

thermokarst *n.* a periglacial landscape pitted with surface depressions formed by the melting of ground ice. It is named thermokarst because of its resemblance to true karst (*q.v.*). The development of thermokarst may be triggered by a number of human activities, including drilling for oil and the removal of the tundra vegetation.

thermonasty *n.* nastic movement in plants in response to variations of temperature. *a.* thermonastic.

thermoneutral zone for an endothermic animal, the temperature range it can tolerate without change in its metabolic rate.

thermonuclear bomb *see* nuclear bomb.

thermonuclear plasma a plasma within which a thermonuclear reaction (*q.v.*) occurs.

thermonuclear reaction any nuclear fusion reaction that occurs within a very high-temperature plasma. The temperatures are sufficiently high to imbue the atomic nuclei present with enough kinetic energy to overcome, on collision, the mutual electrostatic repulsion that results from their like charges. Once this high-energy collision has occurred, the short-range nuclear forces may allow the nuclei to coalesce, generating a new, heavier, nucleus. This process may be accompanied by the liberation of subatomic particles. When light nuclei fuse, large amounts of energy are released. The temperatures required for a thermonuclear reaction are typically $>20 \times 10^6$ °C for a hydrogen bomb, or 5×10^8 to 5×10^9 °C for a controlled reaction in tritium and deuterium gas.

thermoperiodicity, thermoperiodism *n.* the response of living organisms to regular changes of temperature, either with day or night or season to season.

thermophil(ic) *a.* thriving at relatively high temperatures, above 45 °C, *appl.* certain bacteria. *n.* thermophil.

thermophilous *a.* heat-loving, *appl.* plants.

thermophobic *a.* able to live or thrive only at relatively low temperatures.

thermophyte *n.* a heat-tolerant plant.

thermoplastic *n.* any plastic, such as polyvinyl chloride (PVC) or polythene, that can be made soft by heating. During recycling, some such plastics can be reformed into useful products by heating and remoulding. *cf.* thermoset plastics.

thermoregulation *n.* the control of body temperature, either by metabolic or behavioural means, so that it maintains a more or less constant temperature.

thermoset plastic plastic that cannot be softened by heating. Unlike thermoplastics (*q.v.*), discarded thermoset plastics *cannot* be readily reformed into useful products by heating and remoulding.

thermosphere *n.* the layer of the atmosphere found above the mesosphere. The thermosphere extends upwards from *ca.* 85 to *ca.* 500 km (*see* Fig. 16, p. 236). In this layer, ionization of atmospheric gases by ultraviolet radiation, cosmic radiation and solar X-rays can produce a vivid show of coloured lights visible from Earth at higher latitudes. This phenomenon is called the aurora australis or southern lights (at latitudes of >70° S), or the aurora borealis or northern lights (at latitudes of >70° N).

thermotactic *a.* (1) *pert.* thermotaxis; (2) *appl.* optimum, the range of temperature preferred by an organism.

thermotaxis *n.* movement in response to temperature stimulus.

thermotropism *n.* curvature in plants in response to temperature stimulus.

theromorphs *n.plu.* an order of aberrant and primitive mammal-like reptiles of the Carboniferous–Permian, having a primitive sprawling gait and including the sailback lizards.

therophyllous *a.* (1) having leaves in summer; (2) with deciduous leaves.

theropod *a. appl.* a group of small carnivorous bipedal dinosaurs, a suborder of the Saurischia, the lizard-hipped dinosaurs.

theta (θ) symbol for soil porosity *q.v.*

thiabendazole *n.* systemic fungicide used to control storage rots of potatoes.

thiamin(e) *n.* a water-soluble vitamin, a member of the vitamin B complex, found esp. in seed embryos and yeast, its absence causing beri-beri in humans, or polyneuritis, and which is a precursor of the coenzyme thiamine pyrophosphate required for carbohydrate metabolism. *alt.* vitamin B_1.

thigmotropism *n.* (1) growth curvature in response to a contact stimulus found in clinging plant organs, such as stems, tendrils; (2) response of sessile organisms to stimulus of contact. *a.* thigmotropic.

thinophyte *n.* plant of sand dunes.

thiobacilli *n.plu.* non-filamentous, chemoautotrophic bacteria of the genus *Thiobacillus*, characterized by their use of elemental sulphur or other inorganic sulphur compounds.

thiofanox *n.* systemic insecticide used to control aphids on sugar-beet and potatoes.

thiogenic *a.* sulphur-producing, *appl.* bacteria utilizing sulphur compounds.

thiols *n.plu.* class of foul-smelling organic compounds containing sulphur in the form of the –SH functional group directly bonded to carbon. Thiols are associated with the decomposition of organic matter. Formerly known as mercaptans.

thiophanate-methyl *n.* systemic carbamate fungicide used, e.g., on apples to treat scab, mildew and storage rots.

thiophil(ic) *a. appl.* an organism thriving in the presence of sulphur compounds, e.g. certain bacteria.

Thiopneutes *n.* in some classifications the name for the sulphate-reducing bacteria, e.g. *Desulfovibrio, Desulfuromonas.*

thiram *n.* fungicide used, e.g., to control *Botrytis* on a number of horticultural crops.

third body (*chem.*) this is M in reactions of the type: A + B + M → M + C where A and B are reactants and C is the product. In these reactions M (which may, e.g., be a molecule or a solid surface) is needed to absorb energy given out during new bond formation, so stabilizing the product.

third law of thermodynamics *see* thermodynamics, the laws of.

Third World, the general term used for the less developed countries, found mainly in Latin America, Africa and Asia. *cf.* First World, Second World.

thomsonite *n.* $NaCa_2Al_5Si_5O_{20}.6H_2O$, one of the zeolite minerals *q.v.*

thorax *n.* (1) in higher vertebrates, that part of the trunk between neck and abdomen, the chest in humans, containing the heart, lungs, etc.; (2) the region behind the head in other animals; (3) in insects, the first three segments behind the head, bearing legs and wings.

thorite *n.* $ThSiO_4$, a radioactive mineral found in coarse granite. This mineral is exploited as a source of the element thorium.

thorium (Th) *n.* a radioactive element of the actinide series (at. no. 90, r.a.m. 232.04), a soft silvery-white metal when in elemental form. It occurs naturally in monazite and thorite. Its isotope ^{233}Th may be used as a fertile nuclide in fast breeder reactors.

thorny-headed worms Acanthocephala *q.v.*

THORP Thermal Oxide Reprocessing Plant *q.v.*

Thr threonine *q.v.*

threadworms (1) Nematomorpha (*q.v.*); (2) small nematode worms, commonly also called pin worms, common inhabitants of the human bowel.

threatened *a. appl.* a wild species that is still abundant in its natural range but is likely to become endangered because of sharply declining numbers in parts of its range. *see also* endangered, rare, vulnerable.

Three Mile Island incident a nuclear accident that occurred at Three Mile Island, near Harrisburg, Pennsylvania, USA on 28 March 1979. The core of one of the plant's two nuclear reactors became overheated, causing the release of radioactive water into the Susquehanna River and radioactive steam into the atmosphere. At one stage, there was grave danger of a complete meltdown of the reactor's core but in the event, only a partial meltdown occurred. There were no injuries or deaths

attributed to this incident nor were any individuals seriously overexposed to radioactivity. However, this civilian nuclear accident seriously undermined public confidence in nuclear power at the time.

three-cell model a global model of primary atmospheric circulation consisting of three circulatory cells in each hemisphere (*see* Hadley cell, Ferrel cell, polar cell) that is used to explain the primary circulation of the atmosphere. *see* Fig. 32. *alt.* tricellular model.

three-way catalytic converter a two-stage device fitted to the exhaust system of a petrol (i.e. gasoline) engine to convert NO_x, unburnt hydrocarbons (RH) and carbon monoxide (CO) to molecular nitrogen, carbon dioxide and water. The exhaust gases from an engine, running on a near stoichiometric fuel : air mix, pass into a chamber fitted with a reduction catalyst-based catalytic converter (*q.v.*) that converts the NO_x to N_2. Air is then admitted and the gas stream passes into another chamber, this time occupied by an oxidation catalyst-based catalytic converter (*q.v.*) This oxidizes any ammonia formed in the first chamber and the RH and CO present. Unleaded fuel must be used.

threonine (Thr, T) *n.* an amino acid, aminohydroxybutyric acid, constituent of protein, essential in human diet.

thresh *v.* to separate the grain part of harvested cereal crops from the straw and husks by beating or rubbing, either by machine or using hand implements.

threshold *n.* (1) minimal stimulus required to produce sensation; (2) level or value that must be reached before an event occurs.

threshold effect the harmful effect of a small change in environment that exceeds the limit of tolerance of an organism or population and that becomes evident, e.g., as a sudden and dramatic decrease in population size.

threshold limit value (TLV) maximum concentration of a particular airborne pollutant to which it is believed the majority of workers may be exposed day after day, without ill effects (derived by the American Conference of Government Industrial Hygienists, ACGIH). Several categories of TLVs are recognized. The most common

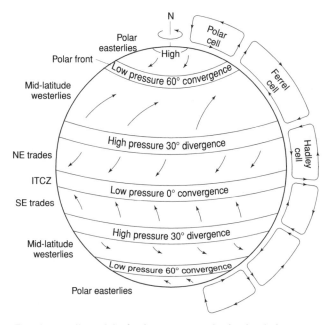

Fig. 32 The three-cell model of primary atmospheric circulation.

is the TLV–time-weighted average (TLV–TWA) based on exposure for five 8-h periods per week. The TLV–short-term exposure limit (TLV–STEL), set at a higher limit than the TLV–TWA, may also be used. Workers may be exposed to the TLV–STEL for a maximum of four 15-min periods in the course of an 8-h working day, with the proviso that the TLV–TWA for that day is not passed.

thrips *n.plu.* common name for the insect order Thysanoptera *q.v.*

throughfall *n.* snow and rainfall that comes down to the ground through a plant canopy and carries substances washed off the plant surfaces.

throughflow *n.* downslope flow of water through the soil, roughly parallel to the surface of the ground.

thrust fault a fault (*q.v.*) formed by compressional forces in which the angle of dip is very low. The upthrown block moves almost horizontally over the downthrown block, covering a greater proportion of it

compared with the situation in a reverse fault.

thulium (Tm) *n.* a lanthanide element (at. no. 69, r.a.m. 168.93), a soft grey metal when in elemental form.

thunder *n.* loud noise (ranging from a sharp crash to a prolonged rumbling) that accompanies a lightning flash during a thunderstorm.

thunderhead cumulonimbus *q.v.*

thunderstorm *n.* storm characterized by the formation of massive cumulonimbus clouds, heavy precipitation, thunder and lightning, and caused by the development of strong convection cells. *see* convection.

Thymelaeales *n.* order of dicot shrubs comprising the family Thymelaeaceae (mezereon).

thymidine *n.* a nucleoside made up of the pyrimidine base thymine linked to deoxyribose. *alt.* deoxythymidine.

thymidine triphosphate (TTP, dTTP) thymidine nucleotide containing a triphosphate group, one of the four deoxyribonucleotides needed for DNA synthesis.

thymine (T) *n*. a pyrimidine base, constituent of DNA, and which is the base in the nucleoside thymidine.

thymine dimer mutagenic structure formed in DNA by the chemical reaction between two adjacent thymine bases, commonly occurring as a result of ultraviolet radiation.

thyroxin(e) *n*. iodine-containing thyroid hormone, tetraiodothyronine, derived from the amino acid tyrosine, which initiates metamorphosis in tadpoles, and is essential for normal growth and development in mammals. *alt*. tetraiodothyronine.

Thysanoptera *n*. order of small slender sap-sucking insects, commonly called thrips, with piercing mouthparts and no or very narrow wings fringed with long setae, which can become pests. They have an incomplete metamorphosis.

Thysanura, thysanurans *n*., *n.plu*. order of insects containing the silverfish and the firebrat, small primitive wingless insects, an individual having small leg-like appendages on the abdomen and three long processes at the posterior end.

Ti symbol for the chemical element titanium *q.v.*

ticks *n.plu*. common name for many of the Acari *q.v.*

tidal *a. pert*. tides (*q.v.*), *appl*. e.g. the kinetic energy of waters moving with the tide, electrical power generated using this motion.

tidal barrage barrier, equipped with sluice gates and turbines, built across a narrow neck of water of suitable tidal range (>5 m), as part of a tidal power plant. Depending on design, electricity may be generated on the ebb tide only, or on both the ebb and flood tides (two-way generation).

tidal bore a wall of water that can develop in advance of the incoming tidal flow in estuaries with a large tidal range, e.g. the Severn bore in the Severn Estuary, UK.

tidal current a horizontal water movement set in motion by the tides.

tidal flat a large expanse of muddy land exposed at low tide, found, e.g., in sheltered estuaries or lagoons. *alt*. mudflat.

tidal inlet a channel that breaches a barrier island *q.v.*

tidal range the mean vertical difference between the level of the sea at high tide and low tide. Tidal ranges may be classified as microtidal (<2 m), mesotidal (2–4 m) or macrotidal (>4 m).

tidal rhythms endogenous physiological or behavioural rhythms governed by the bimonthly tidal cycle, seen in many marine and shore animals, which use it to synchronize their daily behavioural rhythms with the changes in high- and low-tide levels that occur throughout the cycle.

tides *n.plu*. the periodic rise and fall in the level of the sea, observable at any one point. This is caused by the gravitational attraction of the Moon and, to a lesser extent, the Sun on the oceanic waters of the Earth.

tied accommodation accommodation conditional on employment by a particular employer, a situation often found on farms in the UK.

tiger worms *Eisenia foetida*. *see* wormery.

tile drains underground porous pipes installed to improve soil drainage. *alt*. pipe drains.

till *n*. unsorted glacial drift that is deposited directly from the ice. *alt*. boulder clay, glacial till.

till *v*. to cultivate the soil.

tillage *n*. the physical preparation of the soil prior to planting or sowing. This may be divided into primary tillage (the initial breaking up of the soil, usually by ploughing) and secondary tillage (subsequent operations, e.g. harrowing, to achieve a finer tilth). *see also* reduced cultivation.

tiller *n*. flowering stem of a grass.

tillite *n*. fossil till (*q.v.*). Tillite, unlike till, is consolidated.

tilth *n*. the physical condition of the soil with respect to the conditions needed for plant growth.

timber line tree line *q.v.*

time and energy budget in behavioural ecology, the quantitative evaluation of the amount of time and energy an animal expends on various tasks.

time-delay population model model for population growth that describes the lag time of the increase and decrease of the predator population behind the population oscillations of the prey population.

time lag length of time between an event, such as an environmental disturbance, and its manifestation.

Times Beach incident an incident in which the entire town of Times Beach, Missouri, USA, became contaminated with TCDD (dioxin) and eventually had to be abandoned. This came about when waste oil contaminated with TCDD (from an industrial plant manufacturing 2,4,5-trichlorophenol) was mixed with other oils and used as a dust-control spray on local roads and other surfaces.

time-weighted average, threshold limit value see threshold limit value.

tin (Sn) *n.* metallic element (at. no. 50, r.a.m. 118.69) forming a malleable white metal in elemental form. It resists corrosion and is used to plate other metals, such as steel, and in alloys such as bronze and pewter. Extracted from the ore cassiterite (SnO_2).

Tinamiformes *n.* an order of birds including the tinamous.

tinctorial *a.* producing dyestuff.

tinstone cassiterite *q.v.*

Ti plasmid a plasmid carried by the crown-gall bacterium *Agrobacterium tumefaciens*, part of which (T-DNA) becomes integrated into the chromosomes of infected tissue. It normally encodes unusual amino acids that are used by the bacterium as a food source. Foreign genes artificially spliced into T-DNA are also stably integrated into the plant cell chromosomes and *Agrobacterium* and its T-DNA is one of the most important vectors in present use for introducing foreign DNA into plant cells to produce transgenic plants. Modified T-DNA that retains its capacity for integration but does not cause tumours is used.

tipophyte *n.* a pond plant.

tissue *n.* an organized aggregate of cells of a particular type or types, e.g. nervous tissue, connective tissue.

tissue culture the growth in the laboratory of cells taken from an animal or plant. *see also* cell culture.

tissue growth efficiency ratio of energy contained in biomass production to the energy assimilated by the organism (or trophic level as a whole). *alt.* net growth efficiency.

titanite *n.* $CaTiO(SiO_4)$, a common accessory mineral in various igneous rocks (esp. coarse-grained varieties) and metamorphic rocks (notably schists, gneisses and some metamorphosed limestones). A titanium ore mineral. *alt.* sphene.

titanium (Ti) *n.* metallic element (at. no. 22, r.a.m. 47.90) occuring in rutile (TiO_2) and ilmenite ($FeTiO_2$) and in some organic matter. Titanium compounds used as white pigments are less toxic than lead pigments.

titer *alt.* spelling of titre *q.v.*

titrant *see* titration.

titration *n.* method of analysis in which measured amounts of one reagent, known as the titrant, are added to a specific amount of another reagent, of unknown concentration, until the end point of the reaction (signalled, e.g., by the appearance of a precipitate or a colour change) is reached.

titre *n.* the concentration of specific antibodies, antigens, virus particles, etc., in a sample, often presented in terms of the maximum dilution at which the sample still gives a reaction in an immunological test with a standard preparation of an appropriate antigen or antibody, or in other biological or biochemical assays. *alt.* titer.

Tl symbol for the chemical element thallium *q.v.*

TLV threshold limit value *q.v.*

TLV–STEL threshold limit value–short-term exposure limit. *see* threshold limit value.

TLV–TWA threshold limit value–time-weighted average. *see* threshold limit value.

Tm symbol for the chemical element thulium *q.v.*

toad *n.* common name for those members of the amphibian order Anura that are stout-bodied, with a warty skin, and live in damp places, returning only to the water to spawn. There is no biological or evolutionary distinction between frogs and toads and many anuran families contain species of both types. The term frog is generally used to cover all anurans.

toadstool *n.* common name for agaric fungi other than edible mushrooms of the genus *Agaricus*, but which has no biological significance.

tobamovirus group group of single-stranded RNA plant viruses with rigid rod-shaped particles, type member tobacco mosaic virus, which has been the subject of much work on the molecular biology of virus structure.

tobravirus group group of rigid rod-shaped single-stranded RNA viruses, type member tobacco rattle virus. They are multicomponent viruses in which two genomic RNAs are encapsidated in two different virus particles.

TOC total organic carbon *q.v.*

Togaviridae, togaviruses *n.*, *n.plu*, family of small, enveloped, single-stranded RNA viruses that includes Sindbis virus and those that cause yellow fever and German measles (rubella). Many are arthropod-borne viruses (arboviruses), such as Semliki Forest virus, Eastern equine encephalomyelitis and Ross river virus.

Tokamak *n.* a torus-shaped device used to produce controlled nuclear fusion via the magnetic confinement of a thermonuclear plasma. This technology is still at the development stage; it has not yet succeeded in generating sustainable, useful fusion power.

token stimulus a stimulus that operates indirectly by having become linked through experience with an action or object. For example, colour is the token stimulus attracting bees to some flowers, although they actually want to eat the nectar.

tolerance *n.* the ability to survive and grow in the presence of a normally toxic substance, e.g. heavy metals.

tolerance limits *see* Shelford's law of tolerance.

tolerance theory a theory of plant succession. This hypothesizes that any species can colonize initially but some are competitively superior and come to dominate the mature community. May be viewed as intermediate between facilitation theory (*q.v.*) and inhibition theory *q.v.*

toll *n.* a charge made for the right of passage, e.g., along a road or over a bridge.

tomato spotted wilt virus group group of single-stranded RNA plant viruses with spherical particles comprising a lipid envelope surrounding a ribonucleoprotein core. The genome is made up of three different linear RNAs.

tombolo *n.* a spit of sand that links one island to another, or to the mainland.

tombusvirus group group of single-stranded RNA plant viruses with isometric particles, type member tomato bushy stunt virus.

ton, British (UK) 2240 pounds or 1.016 tonnes.

tongue worms Pentastomida *q.v.*

tonne (t) *n.* 1000 kg or 0.9842 British (UK) ton. *alt.* metric ton.

tonotaxis *n.* a taxis in response to a change in density.

tool use the manipulation of an object by an animal to achieve some end, which it could not achieve without it.

toothed whales Odontoceti *q.v.*

tooth shells Scaphopoda *q.v.*

topaz *n.* $Al_2SiO_4(OH,F)_2$, a nesosilicate mineral of pegmatites and granites. Found also as worn particles in alluvial deposits. Prized as a gemstone.

top carnivore the predator at the top of a food web or chain.

topocline *n.* a geographical variation, not always related to an ecological gradient, but to other factors such as topography climate.

topodeme *n.* a deme occupying a particular geographical area.

topogamodeme *n.* individuals occupying a precise locality that form a reproductive or breeding unit.

topogenous mire a mire that develops in places where there is a permanently high water table.

toponym *n.* (1) the name of a place or region; (2) a name designating the place of origin of a plant or animal.

toposequence catena *q.v.*

topotaxis *n.* movement induced by spatial differences in stimulation intensity and orientation in relation to sources of stimuli.

topotropism *n.* orientation towards the source of a stimulus. *a.* topotropic.

topotype *n.* (1) a specimen from the locality of the original type species; (2) a population that has become different from other populations of the same species because of adaptations to local geographical features.

topsoil surface soil (*q.v.*). *see also* A-horizon.

tor *n.* an exposed mass of block-jointed rock, e.g. granite tors of Dartmoor, Devon, UK.

tornado *n*. vortex of air (100–500 m in diameter) formed around a core of very low pressure, which descends to the ground from the base of cumulonimbus clouds. Tornadoes are characterized by very high winds (50–200 m s⁻¹) and heavy precipitation and are responsible for extensive damage to property and loss of human life. *alt*. twister.

torpor *n*. state of complete inactivity accompanied by decreased body temperature and greatly reduced metabolic rate shown by some animals, which may occur on a daily basis, or seasonally, when it is more usually known as hibernation.

torrand *see* andisol.

torrert *see* vertisol.

Torrey Canyon incident the first major oil tanker incident. Extensive pollution of the coastlines of Brittany, France and Cornwall, UK, occurred when 117 000 tonnes of oil were lost from the stricken ship *Torrey Canyon* in the English Channel in March 1967.

torrox *n*. a suborder of the order oxisol of the USDA Soil Taxonomy system, consisting of oxisols of arid regions, possibly formed when different climatic conditions prevailed.

tortoises *see* Chelonia.

total digestible nutrients (TDN) a measure of the nutrient value of a food, which combines chemical analysis with estimates of digestibility obtained from analysis of the composition of the faeces.

total dissolved solids (TDS) amount of solid material remaining after a sample of effluent or water has been evaporated to dryness, measured in milligrams per litre.

total exchange capacity cation exchange capacity *q.v.*

total fertility rate estimate of the number of living offspring a female will produce in her lifetime.

total nomadism *see* nomadism.

total organic carbon (TOC) an alternative method to the biochemical oxygen demand (BOD) (*q.v.*) for estimating the potential of organic oxygen-demanding wastes (*q.v.*) to pollute waterbodies. In this technique, the total organic carbon of a sample is evaluated by fully oxidizing it (often in a small furnace) causing the organic carbon present to be evolved as carbon dioxide (CO_2). This gas is readily quantified, allowing the TOC content of the original sample to be expressed as parts per million of carbon. The total organic carbon contains non-dissolved organic carbon (NDOC), which is organic particles of >0.45 μm diameter, and dissolved organic carbon (DOC), which is what remains when the non-dissolved carbon is removed by filtration.

total range the entire area covered by an individual animal in its lifetime.

tourmaline *n*. $(Na,K,Ca)(Li,Mg,Fe,Mn,Al)_3$-$(Al,Fe,Cr,V)_6(BO_3)_3(Si_6O_{18})(O,OH,F)_4$, a cyclosilicate mineral commonly found in granite pegmatites, in granites that have been altered metasomatically by boron-rich fluids and in sediments newly derived from these rocks. Also occurs in gneisses and schists as an accessory mineral. It has several named varieties, some of which may be distinguished by colour, e.g. rubellite (pink), schorl (black) and dravite (brown).

tower karst a type of karst (*q.v.*) landscape found in the humid tropics and subtropics, e.g. in Belize and southern China. This limestone terrain is dominated by flat-topped towers (sometimes referred to as mogotes), which may rise to heights of more than 100 m above the surrounding plain.

town gas coal gas *q.v.*

toxic *a*. (1) *pert*., caused by, or of the nature of, a poison; (2) poisonous.

toxicity *n*. (1) the nature of a poison; (2) the strength of a poison. *see* LD_{50}.

toxicology *n*. the study of poisons and their effects on living organisms.

toxic waste form of hazardous waste that causes death or serious injury to those that come in contact with it.

toxin *n*. any poison derived from a plant, animal or microorganism. *see also* mycotoxin, phytotoxin, zootoxin.

TP turgor pressure *q.v.*

trace elements (1) (*biol*.) elements such as copper (Cu), zinc (Zn) and manganese (Mn) that are present in minute amounts in living organisms. These are called micronutrient elements only if they are known to be bioelements, i.e. essential for proper growth and development; (2) (*geol.*) elements that are present in a mineral at <1% and that are considered to be non-essential.

trace fossil fossilized tracks or burrows of ancient organisms found in sedimentary rock.

tracer *n.* a molecule or atom that has been labelled chemically or radioactively and can thus be followed in a system (mechanical, biological or chemical).

Tracheophyta, tracheophytes vascular plants *q.v.*

trachyte *n.* a fine-grained, generally porphyritic, rock with a composition equivalent to that of syenite *q.v.*

traction *n.* (1) by river, transportation process whereby rocks, pebbles and boulders are rolled or slid along the river bed; (2) by wind, transportation process whereby soil and rock debris is rolled or slid along the ground surface.

traction load bedload *q.v.*

trades trade winds *q.v.*

trade winds global winds that converge on the intertropical convergence zone (equatorial low-pressure belt) from the horse latitudes (subtropical high-pressure belts). These winds blow from the north-east in the northern hemisphere (north-east trades) and from the south-east in the southern hemisphere (south-east trades).

tragedy of the commons degradation or depletion of a common resource, e.g. fish stocks or clean water, as a result of individuals or groups making maximal use of the resource without assuming sufficient responsibility for its conservation. *see also* commons, public good.

trait *n.* a heritable characteristic that is particular to an organism and can vary from individual to individual, such as eye colour or flower colour.

tramlines *n.plu.* wheel-tracks spaced at regular intervals in cereal fields. These are usually established during seed drilling and re-used by machines during subsequent operations such as fertilizer application.

transalpine *a.* situated to the north of the Alps.

transcription *n.* the copying of any DNA strand, nucleotide by nucleotide, into RNA, following the base-pairing rules. Transcription is carried out by the enzyme RNA polymerase and produces an RNA complementary in nucleotide sequence to the DNA template.

transcurrent fault a fault (*q.v.*) where the crustal blocks move horizontally along the fault plane. There is no vertical displacement and therefore no upthrown or downthrown blocks. *alt.* strike-slip fault, tear fault, wrench fault.

transect *n.* a line (line transect), strip (belt transect), or profile (profile transect), chosen to sample the organisms present within a particular habitat and to gain an idea of their distribution. The simplest type of transect is a line transect, which is made by laying down a straight line and recording the organisms touched by the line. A belt transect consists of recording within a strip, typically up to 1 m wide, along such a line. A profile transect also records changes in ground level along the line, or (for vegetation) the heights of the vegetation along the line, drawn to scale. *see also* quadrat, transect sampling.

transection *n.* section across a longitudinal axis. *alt.* cross-section, transverse section.

transect sampling sampling technique in which samples are taken at regular intervals along a line (line transect) or within a strip between two parallel transect lines (belt transect). Transect sampling is used, e.g., to chart changes in soil type and vegetation, e.g. on land bordering motorways or land contaminated with industrial pollutants.

transfection *n.* the genetic modification of cultured eukaryotic cells by DNA added to the culture medium, which enters the cells and in some cases is stably incorporated into the genome. *v.* transfect.

transfer efficiency the percentage of energy transferred from one trophic level to another in a food chain or food web.

transfer RNA (tRNA) type of small RNA, found in all cells, that carries amino acids to the ribosomes for protein synthesis. Each different tRNA is specific for one kind of amino acid only, e.g. $tRNA^{Cys}$, $tRNA^{Lys}$, etc. The tRNAs line up amino acids in the correct sequence by specific base pairing of a three-base anticodon on each tRNA with the complementary codon on messenger RNA.

transformation *n.* (1) metamorphosis, or a change in form; (2) (*genet.*) the genetic modification of a bacterium (or other type of cell) by the uptake of free DNA by the cell. In bacteria it can occur naturally, during mixed infections. Transformation is also used in experimental bacterial genetics and to produce bacteria containing artificially produced recombinant DNA.

transformation series *see* evolutionary transformation series.

transform boundary (*geol.*) type of plate boundary that occurs when two lithospheric plates move, one past the other, without either moving apart or colliding. *alt.* conservative boundary, shear boundary.

transform fault a special type of transcurrent fault that marks the plate boundary between two lithospheric plates.

transgene *n.* any gene introduced into an animal or plant artificially by the techniques of genetic engineering, such organisms being known as transgenic animals or plants.

transgenic *a. appl.* animals and plants into which genes from another species have been deliberately introduced by genetic engineering. In mammals, this involves introducing the gene into an ovum, fertilized egg or blastocyst taken from the animal, fertilizing it *in vitro* if necessary and then replacing it in a pseudopregnant foster mother. In plants, the bacterium *Agrobacterium tumefaciens* carrying the required gene is often used to infect and transmit the gene to cultured plant tissue, from which a transgenic plant can then be regenerated.

transgressive *a. appl.* a species that overlaps two adjacent communities.

transhumance *n.* the practice of moving livestock on a seasonal basis to areas of more favourable pasture.

transient *a.* (1) passing; (2) of short duration.

transient polymorphism the existence of two or more distinct types of individuals in the same breeding population only for a short while, one type then replacing the other.

transition zone *see* biosphere reserve.

translation *n.* process by which the genetic information encoded in messenger RNA (mRNA) directs the synthesis of proteins.

Each polypeptide chain is synthesized on a ribosome, using mRNA as the template, and matching the codons in mRNA to the complementary anticodons on transfer RNAs carrying individual amino acids. Thus the sequence of amino acids in the protein chain follows that of the codons in RNA (and thus ultimately in DNA). *see also* genetic code, transcription.

translatory waves (*oceanogr.*) shallow-water waves that result from the transformation of oscillatory waves (*q.v.*). This transformation occurs where the depth of water is less than that of the wave base and the motion of the individual water particles becomes progressively more elliptical due to interference by the sea-floor. As a result, there is net forward movement of water. *alt.* waves of translation.

translocated herbicide herbicide that has the ability to move through the plant, prior to acting on the growth processes.

translocation *n.* (1) movement or removal to a different place or habitat; (2) movement of material in solution within an organism, esp. in phloem of plant; (3) (*genet.*) chromosomal rearrangement in which part of a chromosome breaks off and is rejoined to a non-homologous chromosome. *see also* reciprocal translocation.

transmissible *a. appl.* diseases that can be transmitted from one individual to another by contact, or by air, water, food or insect vectors, i.e. diseases caused by bacteria, viruses, fungi and other parasites, *cf.* non-transmissible.

transmissible spongiform encephalopathies a group of neurodegenerative diseases of humans and animals in which the brain has a characteristic sponge-like appearance after death and which includes scrapie in sheep, bovine spongiform encephalopathy (BSE), and Kreutzfeld–Jakob disease in humans. The transmissible agent, sometimes called a prion, and the nature of the transmission are not yet fully identified and understood.

transmission electron microscope (TEM) electron microscope that produces an image from the diffraction of electrons after passage through the specimen.

transmutation *n.* the transformation of the nuclei of one element into those of another. This change always involves a nuclear reaction (*q.v.*) of one type or another.

transpiration *n.* the evaporation of water through stomata of plant leaves and stem.

transpirational pull the continuous loss of water through leaves by transpiration, causing the flow of water through xylem from roots to leaves.

transpiration efficiency the amount of plant biomass produced (in grams) per 1000 g of water transpired.

transpiration ratio the amount of water transpired per unit weight of plant biomass produced.

transpiration stream the movement of water and inorganic solutes upwards from the roots to the leaves through the xylem.

transudate *n.* any substance that has oozed out through a membrane or pores.

transuranic elements those elements with atomic numbers greater than that of uranium (atomic number 92). They are produced by bombarding a heavy nucleus with a light nucleus and all are radioactive. They are not found in nature.

transverse dune a dune whose long axis is at right angles to the prevailing wind direction. Transverse ridges, barchanoid ridges and barchans (*q.v.*) are the commonest types of transverse dune.

transverse ridges simple, straight dunes formed in parallel, perpendicular to the predominant wind direction.

trash mulch cultivation stubble mulch cultivation *q.v.*

traumatic *a. pert.* or caused by, a wound or other injury.

traumatonasty *n.* a movement in response to wounding.

traumatotropism *n.* a growth curvature in plants in response to wounds.

trawling *n.* type of marine fishing in which a strong sock-shaped net is dragged along the bottom of the sea to catch fish.

tree *n.* a woody perennial plant that has a single main trunk at least 7.5 cm in diameter at 1.3 m height, a definitely formed crown of foliage and a height of at least 4 m.

tree dieback *see* recent forest decline.

tree farm a deliberately planted and intensively managed forest consisting of one or a very few different commercial species.

tree ferns a group of tropical and subtropical ferns (e.g. *Cyathea*) with aerial stems several metres high.

tree layer the highest horizontal layer in a plant community, comprising the tree canopy.

tree line line marking the altitude above which trees are unable to grow, also *appl.* to the northern and southern latitudinal limits of tree growth. *alt.* timber line.

tree rings growth rings *q.v.*

tree-ring dating dendrochronology *q.v.*

trehalose *n.* a disaccharide composed of two glucose units, esp. abundant in some lichens and in insects, in which it is one of the principal fuels for the flight muscles.

trellis drainage *see* drainage pattern.

Trematoda, trematodes *n., n.plu.* class of parasitic flatworms, including the digenean gut, liver and blood flukes, such as *Fasciola*, the liver fluke of sheep and cattle, and *Schistosoma* spp., which cause the serious disease schistosomiasis in humans, with tissue damage and inflammation.

tremolite *n.* $Ca_2Mg_5Si_8O_{22}(OH)_2$, one of the amphibole minerals (*q.v.*), the magnesian end-member of the tremolite–actinolite series *q.v.*

tremolite–actinolite series *n.* a solid-solution series of amphibole minerals with the formula $Ca_2(Mg,Fe)_5Si_8O_{22}(OH)_2$. *see also* actinolite, tremolite.

triacylglycerol *n.* any of a class of lipids common in living organisms, uncharged fatty acid esters of the three-carbon sugar alcohol glycerol, storage form of fatty acids and important components of plant and animal fats and oils. *alt.* neutral fat, triglyceride.

triademefon *n.* systemic fungicide used, e.g., to control mildew on oats and yellow rust on wheat.

trial-and-error conditioning or learning a kind of learning in which a random and spontaneous response becomes associated with a particular stimulus, because that response has always produced a reward whereas other responses have not done so. *alt.* instrumental conditioning, operant conditioning.

tri-allate *n.* soil-acting herbicide used, e.g., to control wild oats and blackgrass in wheat and barley.

Triassic *a. pert.* or *appl.* geological period lasting from about 248 to 213 million years ago. *n.* Triassic, sometimes abbreviated to Trias.

triazophos *n.* broad-spectrum contact organophosphate insecticide and acaricide.

tribe *n.* a subdivision of a family in plant taxonomy, differing in minor characters from other tribes. Tribal names generally have the ending -eae.

tributaries *n.plu.* streams that flow into the main (trunk) river.

tributyltins (TBTs) *n.plu.* $(C_4H_9)_3SnX$ (where X is typically F) or $((C_4H_9)_3Sn)_2O$, compounds that belong to a group known as the organotins. TBTs are manufactured compounds that have no counterparts in nature. They are extremely toxic over a broad spectrum and are used, e.g., as biocides in marine antifouling paints *q.v.*

tricarboxylic acid cycle (TCA cycle) a key series of metabolic reactions in aerobic cellular respiration, occurring in the mitochondria of animals and plants, and in which acetyl CoA formed from pyruvate produced during glycolysis is completely oxidized to CO_2 via interconversions of various carboxylic acids (oxaloacetate, citrate, ketoglutarate, succinate, fumarate, malate). It results in the reduction of NAD and FAD to NADH and $FADH_2$, whose reducing power is then used indirectly in the synthesis of ATP by oxidative phosphorylation. The TCA cycle also produces substrates for many other metabolic pathways. *alt.* citric acid cycle, Krebs cycle.

tricarboxylic acid organic acid bearing three -COOH groups, as in citric acid and aconitic acid.

tricellular model of primary atmospheric circulation. *see* three-cell model.

1,1,1-trichloroethane (TCA) *n.* methyl chloroform (CCl_3CH_3), a non-flammable industrial solvent that is much less toxic than carbon tetrachloride, but which is also an ozone-destroying chemical and is covered by the Montreal Protocol *q.v.*

trichloroethene (TCE) *n.* $CHClCCl_2$, colourless non-flammable liquid organochlorine solvent used, e.g., as a solvent in dry cleaning. A potential carcinogen. *alt.* trichloroethylene.

trichlorofluoromethane *see* chlorofluorocarbons.

2,4,5-trichlorophenoxyacetic acid 2,4,5-T *q.v.*

trichlorphon *n.* contact and stomach organophosphate insecticide used, e.g., to control warble fly on cattle.

Trichomycetes *n.* class of endo- and exoparasitic fungi living in or on arthropods, having a simple or branched mycelium, and reproducing by non-motile spores.

Trichoptera *n.* order of insects commonly known as caddis flies, whose aquatic larvae (caddis worms) often build elaborate protective cases incorporating pieces of sand, leaf fragments, etc., or nets of silk in which food is trapped. They have a complete metamorphosis. The adults have two pairs of long slender wings and much reduced mouthparts and rarely feed.

trickle irrigation drip irrigation *q.v.*

trickling filter type of reactor commonly used in the secondary treatment of sewage. It consists of a tank filled with inert solid particles (*ca.* 3–5 cm across) that are covered with a mixed, essentially microbial, community, mainly developed from the sewage. Settled sewage is sprinkled over the top of the tank via moving booms. It percolates down the filter, where its organic content is largely oxidized by organisms established on the solid support. Oxygen (O_2) for this process is provided by air that is passively drawn into the tank. The secondary sludge produced is then removed downstream in final settlement tanks. *alt.* biological filter, percolating filter.

triclads *n.plu.* order of elongated turbellarians, the Tricladida, commonly called planarians, having an intestine with three main branches and well-developed sense organs.

Tricoccae Euphorbiales *q.v.*

Triconodonta, triconodonts *n., n.plu.* order containing the earliest known mammals, living from the late Triassic to the Jurassic, somewhat resembling shrews, and with triconodont molar teeth (with three cusps in a straight line).

tridemorph *n.* systemic fungicide used, e.g., on cereals to control mildew.

trietazine *n.* soil-acting herbicide used in mixtures to control annual broad-leaved weeds and grasses in field crops such as peas and potatoes.

trifluralin *n.* soil-acting herbicide used to control germinating weeds before planting or sowing vegetable crops.

triforine *n.* systemic fungicide used, e.g., to control a number of fungus diseases of apples.

triglyceride triacylglycerol *q.v.*

trigoneutic *a.* producing three broods in each breeding season.

trihybrid *n.* a cross whose parents differ in three distinct characters. *a.* heterozygous for three pairs of alleles.

Trilobita, trilobites *n., n.plu.* group of fossil arthropods (*q.v.*), known from the Cambrian to Permian, with a body divided into three regions, usually considered as a phylum. An individual had a single pair of antennae and numerous similar biramous appendages.

Trimerophyta *n.* group of primitive vascular plants, known from the mid-Devonian and now extinct, thought to represent the ancestors of the ferns and progymnosperms. The main axis was branched, with some branches bearing sporangia, and they lacked leaves.

2,2,4-trimethylpentane *n.* ($CH_3C(CH_3)_2$-$CH_2CH(CH_3)CH_3$), an isomer of octane. *alt.* iso-octane.

trimonoecious *a.* with male, female and hermaphrodite flowers on the same plant.

trimorphism *n.* occurrence of three distinct forms or forms of organs in one life cycle or one species.

trinomial *a.* consisting of three names, as names of subspecies.

trioecious *a.* producing male, female and hermaphrodite flowers on different plants.

trionym *see* trinomial.

triose *n.* any monosaccharide having the formula ($CH_2O)_3$, such as glyceraldehyde, dihydroxyacetone phosphate.

triple bond a covalent bond (*q.v.*) that consists of three pairs of electrons shared between the bonded atoms. Diagrammatically, this is represented as a triple line between the bonded atoms, e.g. the triple bond in N_2 may be shown thus N≡N.

triple superphosphate $Ca(H_2PO_4)_2$, inorganic phosphorus fertilizer, with a phosphorus content equivalent to 47% P_2O_5, supplied in powder form.

triploblastic *a. appl.* embryos with three primary germ layers, i.e. endoderm, mesoderm and ectoderm. *cf.* diploblastic.

triploid *a.* having three times the haploid number of chromosomes, *appl.* cells, nuclei, organisms. *n.* an organism with three sets of chromosomes per somatic cell.

tripolite diatomaceous earth *q.v.*

tripton *n.* non-living material drifting in the plankton.

trisaccharide *n.* a carbohydrate made up of three monosaccharide units, e.g. raffinose.

tritium *n.* the isotope of hydrogen that has a mass number of three; it is symbolized ^3_1H, ^3H or T. It is radioactive with a half-life of 12.5 years.

trituberculates *n.plu.* a group of Mesozoic therian mammals known mainly from remains of jaws and teeth, the molar teeth having the characteristic triangle of cusps, and that are forerunners of the living therians. *alt.* tribosphaenids.

Triuridales *n.* order of saprophytic monocots with scale leaves and comprising the family Triuridaceae.

trivalent *a.* having a valency (*q.v.*) of three.

trivoltine *a.* having three broods in a year.

trixenous *a.* of a parasite, having three hosts.

tRNA transfer RNA *q.v.*

Trochodendrales *n.* order of dicot shrubs and trees including the families Tetracentraceae and Trochodendraceae.

trochophore *n.* free-swimming top-shaped pelagic larval stage of marine annelids, bryozoans and some molluscs, forming part of the zooplankton.

troctolite *n.* a coarse-grained igneous rock with a colour index of 30–90, essentially consisting of plagioclase feldspar (labradorite, bytownite, anorthite) and olivine/serpentine. *alt.* troutstone.

troglobiont *n.* any organism living only in caves.

Trogoniformes *n.* order of very soft-plumaged birds including the trogons.

trombe wall interior wall designed for heat absorption.

tropept *n.* suborder of the order inceptisol of the USDA Soil Taxonomy system comprising reddish/brownish, freely draining tropical inceptisols.

trophallaxis *n.* in social insects, the exchange of alimentary liquid among colony members and guest organisms, either mutually or unilaterally.

trophic *a. pert.* or connected with nutrition and feeding.

trophic level (1) a level in a food chain defined by the method of obtaining food, and in which all organisms are the same number of energy transfers away from the original source of the energy (e.g. photosynthesis) entering the ecosystem. *see* herbivore, primary producer, secondary consumer, tertiary consumer; (2) the nutrient status of a body of water. *see* eutrophic, oligotrophic.

trophic level efficiency the ratio of the production of one trophic level to that of the level immediately below it.

trophobiont *n.* an organism that lives in a symbiosis where each partner feeds the other (called a trophobiosis), e.g. ants and aphids.

trophotaxis *n.* response to stimulation by an agent that may serve as food.

trophotropism *n.* tendency of a plant organ to turn towards food, or of an organism to turn towards food supply.

tropic *a.* (1) *pert.* tropism, *appl.* movement or curvature in response to a directional or unilateral stimulus; (2) tropical, *appl.* regions.

tropical *a. appl.* climate characterized by high temperature, humidity and rainfall, found in a belt between latitudes 23° 27′ N and 23° 27′ S, and to the flora and fauna of these regions.

tropical continental air masses hot, dry, unstable air masses that develop in the subtropical high-pressure belts of the horse latitudes, over land regions, e.g., the Sahara.

tropical cyclone hurricane *q.v.*

tropical karst type of limestone landscape characterized by large solution features and steep-sided cone-shaped hills, often clothed in vegetation.

tropical maritime air masses hot, humid, unstable air masses that develop in the subtropical high-pressure belts of the horse latitudes over the oceans, e.g. Atlantic and Pacific Oceans.

tropical rain forest evergreen broadleaf forest that develops in areas near the Equator with a climate of high temperature, humidity and rainfall and no marked seasons, and that is characterized by a high biological diversity and productivity. Tropical rain forest is found in the Amazon Basin, parts of Central America, central West Africa, parts of the south-eastern African coast and Madagascar, South-East Asia and Indonesia, Papua-New Guinea and the northern tip of Australia. *see also* rain forest.

tropical savanna *see* savanna.

tropical seasonal forest *see* monsoon forest.

Tropic of Cancer most northerly latitude (23° 28′ N) where the Sun's noontime rays strike the surface of the Earth vertically. This occurs on the day of the northern hemisphere's summer solstice (21 June).

Tropic of Capricorn most southerly latitude (23° 28′ S) where the Sun's noontime rays strike the surface of the Earth vertically. This occurs on the day of the southern hemisphere's summer solstice (December 21).

tropics, the *n.* geographical region straddling the Equator, located between the Tropic of Cancer (latitude 23° 28′ N) and the Tropic of Capricorn (latitude 23° 28′ S).

tropism *n.* a plant or sessile animal growth movement, usually curvature towards (positive) or away from (negative) the source of stimulus.

tropopause *n.* the upper boundary of the troposphere, which separates it from the stratosphere. *see* Fig. 16, p. 236.

tropophil(ous) *a.* (1) tolerating alternating periods of cold and warmth, or of moisture or dryness; (2) adapted to seasonal changes.

tropophyte *n.* (1) a plant that adapts to the changing seasons, being more or less hygrophilous in summer and xerophilous in winter; (2) a plant growing in the tropics.

troposphere *n.* the lowermost layer of the atmosphere (in which most of the weather occurs). The altitude of the upper boundary of this zone (the tropopause) varies between *ca.* 8 km at the poles and *ca.* 16 km at the Equator. *see* Fig. 16, p. 236.

tropospheric ozone the term used for the triatomic oxygen molecule, ozone (O_3), when found in the troposphere (*q.v.*), where it is considered to be a pollutant. There is a close interplay between the levels of tropospheric ozone and NO_x. Tropospheric ozone can oxidize nitric oxide, thus:

$$NO + O_3 \rightarrow O_2 + NO_2 \qquad (1)$$

This reaction is rapid. However, it does not result in the total depletion of either of the reactants. Significantly, NO is regenerated photolytically from nitrogen dioxide:

$$NO_2 + hv \ (\lambda < 430 \ nm) \rightarrow NO + O \qquad (2)$$

The oxygen atoms formed can then react with molecular oxygen (O_2), regenerating ozone:

$$O + O_2 + M \rightarrow O_3 + M \qquad (3)$$

where M is a third body, such as a molecule or solid surface, which is needed to absorb energy given out during new bond formation, so stabilizing the product. The process represented by equations 2 and 3 is the only verified *in situ* source of tropospheric ozone and is responsible for the rise in the levels of this pollutant during episodes of Los Angeles-type smog (*q.v.*). Ozone also enters the troposphere by downward movement from the stratosphere, where it is formed photochemically. *see* stratospheric ozone.

tropotaxis *n.* a taxis in which an animal orients itself in relation to the source of a stimulus by a simultaneously comparing the amount of stimulus on either side of it by symmetrically placed sense organs.

trough *n.* (*clim.*) elongated region of low pressure sandwiched between two regions of higher pressure.

troutstone troctolite *q.v.*

Trp tryptophan *q.v.*

true-breeding *a. appl.* organisms that are homozygous for any given genotype and therefore pass it on to all their progeny in a cross with an identical homozygote.

true corals common name for the Scleractinia, an order of mainly colonial hydrozoans, having a compact calcareous skeleton and no siphonoglyph.

true cost the cost of a good when its internal costs and its external costs are included in its market price.

true dip *see* dip, true.

true flies Diptera *q.v.*

true mosses the Bryophyta *q.v.*

true nomadism *see* nomadism.

true predator an organism that kills its prey soon after attacking. *cf.* parasite.

truncated spurs (*geog.*) previously interlocking spurs (*q.v.*) whose tips have been removed as glacial erosion widened the river valley.

trypanosome *n.* member of a genus of parasitic flagellate protozoans, *Trypanosoma*, which includes the organisms causing trypanosomiasis, or sleeping sickness in Africa (*T. brucei*) and Chagas disease in South America (*T. cruzi*), both transmitted by blood-sucking insects.

trypanosomiasis *see* trypanosome.

tryptophan (Trp, W) *n.* beta-indolealanine, an amino acid with an aromatic side chain, constituent of protein, essential in human diet and a precursor of the auxin indoleacetic acid in plants.

tsetse flies *n.plu.* any of the blood-sucking dipterous insects that belong to the genus *Glossina*. These African parasites are important as they transmit various diseases, including nagana in cattle.

tsunami *n.* long-wavelength sea wave generated by sudden movement of the ocean floor. This may be caused by volcanic activity or earthquakes. Although initially of low amplitude, a tsunami can reach a wave height of 10–20 m in shallow coastal waters, often with devastating results. *plu.* tsunami. *alt.* seismic sea waves.

TTP the nucleotide thymidine triphosphate (*q.v.*). Unlike the abbreviations ATP, CTP and GTP, TTP usually refers to the deoxyribose form of the nucleotide (sometimes written as dTTP), because thymine is not a constituent of RNA.

tuatara *see* Rhyncocephalia.

tubenoses Procellariiformes *q.v.*

tuber *n.* a thickened, fleshy, food-storing underground root, or similar underground stem with surface buds, e.g. potato.

tuberculosis *n.* a serious disease caused by the tubercle bacillus *Mycobacterium tuberculosis*. In humans, it can infect most parts of the body, often starting in the lungs (where it is called consumption). *alt.* TB.

tubeworms *see* Annelida, Echiura, Phoronida, Pogonophora, Pterobranchia, Vestimentifera.

tubicolous *a.* inhabiting a tube.

tubificid worms freshwater worms of the genus *Tubifex*, which are tolerant of heavy organic pollution.

Tubulidentata *n.* order of placental mammals known from the Miocene, or possibly the Eocene, whose only living member is the African anteater or aardvark (*Orycteropus*), which possesses unique peg-like teeth with tubular canals in the dentine, is ant-eating and has powerful digging forelimbs.

tuff *n.* the pyroclastic igneous rock formed from consolidated volcanic ash (particle size < 2 mm).

tumour *n.* a growth resulting from the abnormal proliferation of cells that may be self-limiting or non-invasive, when it is called a benign tumour. Some tumours continue proliferating indefinitely and also can invade underlying tissues and be carried in the blood stream to other parts of the body (metastasis). These are known as malignant tumours, or cancers.

tumour viruses *see* DNA tumour viruses, RNA tumour viruses.

tundra *n.* treeless polar region with permanently frozen subsoil, which may support grasses, mosses, lichens, herbaceous plants and dwarf shrubs. *see also* alpine tundra.

tungsten (W) *n.* metallic element (at. no. 74, r.a.m. 183.85) forming a lustrous silvery white metal. It occurs in wolframite, $(Fe,Mn)WO_4$, stolzite, $PbWO_4$, and scheelite, $CaWO_4$.

tunicamycin *n.* antibiotic that inhibits the glycosylation of glycoproteins in eukaryotic cells.

Tunicata, tunicates *n., n.plu.* subphylum of chordates containing the classes Ascidiacea (sea squirts), Larvacea and Thaliacea (salps), in which chordate features are found only in the larva and are generally lost in the adult.

tunnelling electron microscope type of electron microscope that gives atomic-level resolution of certain types of structures or surfaces, including biological macromolecules.

tup *n.* a male sheep kept for breeding, *alt.* ram. *v.* to serve a ewe.

Turbellaria, turbellarians *n., n.plu.* class of free-living flatworms with a leaf-like shape and a ciliated epithelium.

turbidity current (1) submarine mudflow, triggered, e.g., by earthquake activity; (2) any rapidly moving body of water containing suspended material caused, e.g., by a river in spate or storm.

turbine *n.* any motor designed to utilize the kinetic energy of a stream of fluid (e.g. steam, water, air) to rotate a shaft via the reaction and/or impact of the fluid's current with blades mounted around the shaft.

turbulent flow a fluid flow pattern in which the flow is broken up, and motion varies both spatially and temporally in terms of magnitude and direction. *cf.* laminar flow.

turgor *n.* (1) the swelling of a plant cell due to the internal pressure of its vacuolar contents; (2) distension of any living tissue due to internal pressures. *a.* turgid, turgescent.

turgor pressure (TP) the pressure set up inside a plant cell due to the hydrostatic pressure of the vacuole contents pressing on the rigid cell wall. It provides mechanical support to non-woody plant stems, and changes in turgor pressure due to osmosis are responsible for the opening and closing of stomata and some seismonastic movements.

turnover *n.* (1) in an ecosystem, the ratio of productive energy flow to biomass; (2) the fraction of a population that is exchanged per unit time through loss by death or emigration and replacement by reproduction and immigration.

turnover time (1) time taken to complete a biological cycle; (2) time from birth to death of an organism; (3) residence time *q.v.*

turquoise *n.* a blue, green or blue-green coloured mineral with the formula $CuAl_6(PO_4)_4(OH)_8.nH_2O$, where $n = 4$ or 5, which is prized as a gemstone.

turtles *see* Chelonia.

tusk shells common name for the Scaphopoda *q.v.*

TWA time-weighted average. *see* threshold limit value.

twinning *n.* the property exhibited by any grain of crystalline material (e.g. a mineral grain) that consists of two or more

431

intergrowing crystals. Grains that exhibit this property are said to be twinned.

twister tornado *q.v.*

twitch grass couch grass *q.v.*

two-dimensional gel electrophoresis technique for the separation of proteins, in which the mixture is first separated by isoelectric focusing or electrophoresis in one direction, and then subjected to electrophoresis at a direction at right-angles to the first.

two-way catalytic converter oxidation catalyst-based catalytic converter *q.v.*

two-way generation of electricity *see* tidal barrage.

tycholimnetic *a.* temporarily attached to bed of lake and at other times floating.

tychoplankton *n.* (1) drifting or floating organisms that have been detached from their previous habitat, as in plankton of the Saragasso Sea; (2) inshore plankton.

tychopotamic *a.* thriving only in backwaters.

tylopods *n.plu.* members of the mammalian suborder Tylopoda, of the order Artiodactyla, which includes the camel family and a number of extinct groups.

tymovirus group group of single-stranded RNA plant viruses with isometric particles, type member turnip yellow mosaic virus.

type *n.* (1) sum of characteristics common to a large number of individuals, e.g. of a species, and serving as the basis for classification; (2) a primary model, the actual specimen described as the original of a new genus or species. *alt.* holotype.

type locality locality in which the holotype or other type used for designation of a species was found.

type number the most frequently occurring chromosome number in a taxonomic group.

type specimen the single specimen chosen for the designation and description of a new species.

Typhales *n.* order of marsh or aquatic monocots with rhizomes and linear leaves, comprising the families Sparaginiaceae (bur reed) and Typhaceae (cat-tail).

typhoid fever a serious bacterial disease of humans caused by *Salmonella typhi*, transmitted by faecally contaminated food or water. Symptoms include diarrhoea, vomiting and high temperature. *alt.* enteric fever.

typhoon hurricane *q.v.*

typical *a.* (1) *appl.* specimen conforming to type or primary example; (2) exhibiting in marked degree the characteristics of species or genus.

typogenesis *n.* (1) phase of rapid type formation in phylogenesis; (2) quantitative or explosive evolution.

typonym *n.* a name designating or based on type specimen or type species.

Tyr tyrosine *q.v.*

tyramine *n.* a phenolic amine formed by decarboxylation of tyrosine, produced in small amounts in animal liver, also produced by bacterial action on tyrosine-rich substrates, secreted by cephalopods, and found in various plants such as mistletoe and in ergot of rye, and that causes a rise in arterial blood pressure.

tyrocidin *n.* cyclic decapeptide antibiotic produced by *Bacillus brevis*.

tyrosine (Tyr, Y) *n.* amino acid with aromatic side chain, a constituent of protein, and also important as a precursor of, e.g., adrenaline and noradrenaline, thyroxine and melanin.

tyuyamunite *n.* $Ca(UO_2)_2(VO_4)_2.5–10H_2O$, a uranium ore mineral.

U

u atomic mass unit, dalton *q.v.*

U (1) symbol for the chemical element uranium (*q.v.*); (2) uracil *q.v.*

U internal energy *q.v.*

ubac slope shaded (north-facing) side of an Alpine valley. *alt.* schattenseite slope. *cf.* adret slope.

UCR unconditional response or reflex *q.v.*

UCS unconditional stimulus *q.v.*

udalf *n.* suborder of the order alfisol of the USDA Soil Taxonomy system consisting of alfisols of temperate or warm, and moist climatic regions.

udand *see* andisol.

udert *see* vertisol.

udoll *n.* a suborder of the order mollisol of the USDA Soil Taxonomy system consisting of mollisols of temperate or warm, and moist climatic regions.

udox *n.* suborder of the order oxisol of the USDA Soil Taxonomy system consisting of oxisols of tropical climatic regions with little or no dry season.

udult *n.* suborder of the order ultisol of the USDA Soil Taxonomy system consisting of ultisols with low organic matter contents found in temperate or warm, and moist climatic regions.

ultimate *a. appl.* factor thought to be the fundamental cause of a given biological phenomenon. *cf.* proximate.

ultisol *n.* order of the USDA Soil Taxonomy system consisting of moist soils of warm/tropical climatic regions, characterized by the presence of an illuvial horizon rich in clay (argillic horizon), less than 35% saturated with basic cations. Ultisols develop under savanna grassland or forest. Five suborders are recognized (all favourable for cultivation). *see* aquult, humult, udult, ustult, xerult.

ultra-abyssal hadral *q.v.*

ultrabasic *a.* (*geol.*) *appl.* igneous rocks that contain <45% w/w silica (SiO_2).

ultracentrifuge *n.* instrument in which extracts of broken cells can be separated into their different components by spinning at various speeds (up to 150 000 *g*), the different organelles sedimenting at different rates; also used to separate large molecules of different molecular weights.

ultradian rhythm biological rhythm with a periodicity greater than 24 h.

ultramafic *a. appl.* rocks that are largely (>90%) made up of Fe–Mg minerals.

ultramicroscopic *a. appl.* structures or organisms too small to be visible under the light microscope but that can be seen under the electron microscope.

ultramicrotome *n.* machine with fine glass or diamond knife-blade for slicing ultra-thin tissue sections for electron microscopy.

ultrastructure *n.* the fine structure of cells as seen under the electron microscope.

ultraviolet light electromagnetic radiation of wavelengths between those of the violet end of the visible light spectrum and X-rays. Although invisible to the human eye, it can be captured on photographic film. The ultraviolet spectrum is subdivided by wavelength into A (400–320 nm), B (320–280 nm) and C (280–10 nm) bands, of which UV-B is the most energetic and most harmful to living organisms. Much of the solar ultraviolet radiation, esp. the shorter wavelengths, is absorbed by the strato-spheric ozone layer before reaching the Earth's surface.

umbel *n.* a flower-head in which each flower or cluster of flowers arises from a common centre, forming a flat-topped or rounded cluster.

Umbellales Cornales *q.v.*

umbellifer *n.* member of the large dicot family Umbelliferae, typified by cow parsley, carrot, etc., having small flowers borne in umbels, and much-divided leaves. They include many plants grown for food, but also some highly poisonous species, e.g. hemlock.

umbraticolous *a.* growing in a shaded habitat.

umbrept *n.* suborder of the order inceptisol of the USDA Soil Taxonomy system comprising reddish/brownish, free-draining, acidic inceptisols rich in organic matter.

UN *see* United Nations.

unburnt hydrocarbons uncombusted hydrocarbons (organic compounds composed only of hydrogen and carbon) present in the waste gases expelled from flues and exhausts.

UNCED United Nations Conference on Environment and Development (*q.v.*), held in Rio de Janeiro in June 1992. *alt.* Rio Conference, Earth Summit.

UNCLOS United Nations Conference on the Law of the Sea *q.v.*

unconditional reflex (UCR) an inborn reflex, produced involuntarily in response to a stimulus. *alt.* unconditioned reflex.

unconditional response (UCR) *see* classical conditioning. *alt.* unconditioned response.

unconditional stimulus (UCS) a stimulus that produces a simple reflex response. *alt.* unconditioned stimulus.

unconfined aquifer *see* aquifer.

unconformity *n.* a phenomenon sometimes observable in sedimentary rocks where an eroded rock bed is in direct contact with a more recently deposited one. This represents a time interval during which deposition was temporarily interrupted by a period of erosion. *cf.* disconformity.

UNCSD United Nations Commission on Sustainable Development *q.v.*

undercompensating density dependence type of density-dependent response in which the change in population growth rate is less than the original decrease in density that stimulated the response, resulting in a smaller increase over the original density than expected.

underfit river misfit river *q.v.*

underground *n.* rail network that runs largely underground, served by stations located at the surface, e.g. the Paris metro.

underground drainage river system that runs below ground level, characteristic of chalk and karst landscapes.

undergrowth *n.* general term for the vegetation (shrubs, tree seedlings and herbaceous plants) growing under trees in a wood or forest.

undernourished *a. appl.* humans or animals receiving <90% of their minimum dietary requirements over a long period of time.

underpopulation *n.* situation where the resources available in a particular area far exceed those necessary to maintain a reasonable living standard for those individuals that make up its current human population.

understorey *n.* vegetation layer between tree canopy and the ground cover in a forest or wood. It is composed of shrubs and small trees.

undetectable concentrations levels of pollutants that are too low to be measured by current equipment.

undifferentiated *a.* (1) not differentiated, in immature state, *appl.* embryonic cells that have not yet acquired a specialized structure and function; (2) *appl.* meristematic and stem cells.

UNEP United Nations Environment Programme *q.v.*

UNESCO United Nations Educational, Scientific and Cultural Organisation *q.v.*

uneven-aged stand group of trees of different ages, often of different species.

ungulates *n.plu.* hoofed mammals. *see* Artiodactyla, Perissodactyla.

unguligrade *a.* walking upon hoofs, which are formed from the tips of the digits.

uni- prefix from L. *unus*, one, generally meaning having one of.

unicellular *a.* having only one cell, or consisting of one cell.

unidentate *a.* (*chem.*) *appl.* ligands with one donor atom.

uniform spatial distribution distribution of individual organisms throughout an ecosystem at roughly equal distances in all directions from each other. *alt.* even distribution.

uniformity *n.* in ecology, the tendency of the component species of an association to be uniformly distributed within it.

unimodal *a.* having only one mode, *appl.* frequency distribution with a single maximum.

uniparous *a.* producing one offspring at a birth.

Uniramia *n.* in some classifications, a phylum of arthropods containing the insects and myriapods.

unisexual *a.* (1) of one or other sex, distinctly male or female; (2) (*zool.*) sometimes *appl.* animal producing both sperm and eggs; (3) (*bot.*) *appl.* plants and flowers having stamens and carpels in separate flowers.

unit leaf rate the increase in dry weight produced by a plant over a given time period, in relation to its leaf area. *alt.* net assimilation rate.

United Nations Commission on Sustainable Development (UNCSD) commission established after the Rio Conference to implement Agenda 21 and to provide guidance on national strategies for sustainable development.

United Nations Conference on Environment and Development (UNCED) international conference held in Rio de Janeiro in 1992, attended by UN member governments and a wide range of non-governmental organizations and special interest groups, and sometimes referred to as the Earth Summit. It produced Agenda 21, a blueprint for pursuing sustainable development (*q.v.*) in the twenty-first century, and two global conventions were signed. *see* United Nations Convention on Biological Diversity, United Nations Convention on Climate Change.

United Nations Conference on the Law of the Sea (UNCLOS) a series of conferences held between 1972 and 1984 that resulted in the development of new marine laws regulating the exploitation of marine resources. The resulting Law of the Sea Convention established exclusive economic zones extending 320 km from a country's shoreline.

United Nations Convention on Biological Diversity framework convention opened for signature at the United Nations

Conference on Environment and Development at Rio in 1992 and ratified in December 1993, which aimed to prevent the global loss of biological diversity. Its three main objectives were to conserve global biological diversity through national plans for monitoring, conserving and sustaining national biological diversity, to use biological resources in a sustainable manner, and to share equitably the benefits accruing from the exploitation of genetic resources. *alt.* Biodiversity Convention.

United Nations Convention on Climate Change framework convention opened for signature at the United Nations Conference on Environment and Development at Rio in 1992 and ratified in 1994, which aimed to stabilize greenhouse gas concentrations and prevent further atmospheric warming.

United Nations Educational, Scientific and Cultural Organisation (UNESCO) a specialized agency (established in 1945) of the UN to encourage international collaboration through educational, scientific and cultural links, thus promoting international peace and security.

United Nations Environment Programme (UNEP) a programme, established in 1972, whose purpose was to stimulate and coordinate environmental action primarily within the system of the UN.

univalent *a.* (1) (*chem.*) having a valency (*q.v.*) of one, *alt.* monovalent; (2) (*biol.*) *appl.* a single unpaired chromosome at meiosis.

universe *n.* all of the space, matter and energy that is believed to exist. In thermodynamic analysis, the universe equals the system plus its surroundings.

univoltine *a.* producing only one brood in a year or breeding season.

unleaded petrol petrol to which tetraalkyl lead antiknock agents have *not* been added.

unsaturated *a.* *appl.* fatty acids with one or more double bonds, C=C, in their hydrocarbon chain, and to fats and oils in which they predominate.

unsaturated hydrocarbons (*chem.*) hydrocarbons that contain one or more double and/or triple bonds between adjacent carbon atoms. *cf.* saturated hydrocarbons.

unsaturated zone zone of aeration *q.v.*

upland rice *see* rice.

upper atmosphere (1) the part of the atmosphere found beyond the stratopause, i.e. above *ca.* 50 km, and that encompasses the mesosphere, thermosphere and exosphere (*see* Fig. 16, p. 236); (2) the part of the atmosphere that cannot be directly observed by the use of balloons, i.e. atmosphere above *ca.* 30 km.

upper shore zone of seashore between the average high-tide level and the highest high-water mark, which supports only a few species. *see* Fig. 25, p. 369.

upthrown *a. appl.* a block of rock, bounded by a fault, that moves upwards as a result of tectonic activity.

upward translocation (*soil sci.*) the net upward movement of soluble and sparingly soluble material through the soil profile. This process is mediated by the soil solution moving upwards under capillary action and occurs in arid soils where the rate of evapotranspiration exceeds rainfall. Where sodium compounds dominate the material translocated, which they frequently do, the process is known as salinization.

upwelling *n.* (*oceanogr.*) large-scale upward movement of seawater, bringing cold water from the depths of an ocean to its surface.

upwelling zones (*oceanogr.*) marine zones where upwelling (*q.v.*) brings nutrient-rich bottom sediments into the photic zone, thus stimulating primary, and hence secondary, productivity.

uracil (U) *n.* a pyrimidine base, constituent of RNA and the base in the nucleoside uridine.

uraninite *n.* UO_2, the major uranium ore mineral, called pitchblende when massive.

uranium (U) *n.* dense white radioactive element (at. no. 92, r.a.m. 238.03) occurring naturally in uraninite (pitchblende), carnotite and coffinite. Naturally occurring uranium consists of three isotopes: ^{234}U, ^{235}U (approx 0.7%) and ^{238}U (>99%), which decay over a long time (a half-life of 4.5×10^9 years for ^{235}U) to produce lead. ^{235}U is the only fissile isotope: it can be split by neutron bombardment to form smaller nuclei and release energy, producing more neutrons that can initiate fission in other nuclei, thus starting a chain reaction.

uranium-235 an isotope of the element uranium, symbol $^{235}_{92}U$ or ^{235}U. It is the most widely used fissile material 'burned' in nuclear fission reactors. The natural abundance level of this isotope is 0.715% of total uranium.

urate (uric acid) *n.* 2,6,8-trioxypurine, an almost insoluble end-product of purine metabolism in certain mammals, the product of the breakdown of nucleic acids and proteins, excreted in urine in primates, main nitrogenous excretion product in birds, reptiles and some invertebrates, esp. insects; also produced in plants.

urban *a. pert.* city or town, *appl.* e.g. planning, development, housing, population. *cf.* rural.

urban climate microclimate that develops over a city. Such microclimates form under the modifying influence of the city's topography and the presence of waste heat and air pollutants, and are frequently characterized by the development of an urban heat island (*q.v.*). *alt.* city climate.

urban heat island distinct zone of elevated air temperatures associated with urban areas, caused by the release of waste heat.

urbanization *n.* the process of urban development.

urea *n.* carbamide, NH_2CONH_2, a soluble waste product of the breakdown of proteins and amino acids in mammals and some other animals, chief nitrogenous constituent of the urine, and also found in some fungi and higher plants. It is used in agriculture as a nitrogen fertilizer containing 46.6% by weight N, supplied as prills (i.e. spherical granules); may cause leaf scorch.

urea cycle metabolic cycle principally involving arginine, citrulline and ornithine, and found in all terrestrial vertebrates except reptiles and birds, in which the ammonium ions formed during amino acid breakdown are converted to urea for excretion. *alt.* arginine–urea cycle, Krebs–Henseleit cycle, ornithine cycle.

ureotelic *a.* excreting nitrogen as urea, as in adult amphibians, elasmobranch fishes, mammals. *cf.* ammonotelic, uricotelic.

uric acid *see* urate.

uricotelic *a.* excreting nitrogen as uric acid, as in insects, birds, reptiles. *cf.* ureotelic.

uridine *n.* a nucleoside made up of the pyrimidine base uracil linked to ribose.

uridine triphosphate (UTP) uridine nucleotide containing a triphosphate group, one of the four ribonucleotides needed for synthesis of RNA, takes part in many metabolic reactions in a manner analogous to ATP, esp. forming activated UDP-sugars, which are the activated forms of sugars used in the synthesis of polysaccharides *in vivo.*

urine *n.* (1) fluid excretion from the kidneys in mammals, containing waste nitrogen chiefly as urea; (2) solid or semisolid excretion in birds and reptiles, containing waste nitrogen chiefly as uric acid.

Urochordata, urochordates tunicates *q.v.*

Urodela, urodeles *n., n.plu.* one of the three orders of extant amphibians, containing the newts and salamanders, amphibians with well-developed tails and two pairs of legs of more or less equal length. Called the Caudata in some classifications. *cf.* Anura.

Urticales *n.* order of dicots including herbs, vines, shrubs and trees, and including the families Cannabaceae (hemp), Moraceae (mulberry), Ulmaceae (elm) and Urticaceae (nettle).

USDA United States Department of Agriculture.

useful energy energy in the form it is required by the user, i.e. heat, light, motive power. It differs from delivered energy as it accounts for the inefficiency of energy conversion by the end-use appliance. *see also* delivered energy, primary energy.

useful work the work interaction between a system and its surroundings that is not pressure–volume work *q.v.*

U-shaped valley *see* glacial trough.

ustalf *n.* suborder of the order alfisol of the USDA Soil Taxonomy system consisting of alfisols of warm climatic regions, intermittently dry for long periods of time.

ustand *see* andisol.

ustert *see* vertisol.

ustoll *n.* a suborder of the order mollisol of the USDA Soil Taxonomy system consisting of mollisols of monsoon climatic regions.

ustox *n.* suborder of the order oxisol of the USDA Soil Taxonomy system consisting of oxisols found in tropical climatic regions that experience a long dry season.

ustult *n.* suborder of the order ultisol of the USDA Soil Taxonomy system consisting of ultisols of warm regions with high rainfall but a distinct dry season.

utilization efficiency ratio of energy assimilated by a given trophic level to the energy produced by the trophic level immediately below it.

UTP uridine triphosphate *q.v.*

UV ultraviolet radiation *q.v.*

UV-A the region of the spectrum of ultraviolet light that is not normally injurious to living organisms. UV-A has wavelengths from 320 to 400 nm (in some classifications, radiation with wavelengths of 315–320 nm is also included). *see also* UV-B, UV-C.

UV-B the region of the spectrum of ultraviolet light that is most likely to cause injury to living organisms. UV-B has wavelengths from 290 to 315 nm (in some classifications, light with wavelengths of 280–290 and 315–320 nm is also included). Much of the solar radiation within this region does not reach the biosphere as it is absorbed by stratospheric ozone. *see also* UV-A, UV-C.

UV-C the region of the spectrum of ultraviolet light that is potentially most injurious to biological systems. UV-C has wavelengths from 10 to 280 nm (in some classifications, light with wavelengths of 280–290 nm is also included). Solar radiation within this region does not pose a threat to living organisms, because it does not reach the biosphere as it is completely absorbed by the atmosphere. *see also* UV-A, UV-B.

uvala *n.* a large complex depression, sometimes >1.5 km in diameter, found in karst (*q.v.*) terrain, which is frequently formed by the merger of dolines *q.v.*

uvarovite *n.* one of the garnet minerals *q.v.*

V

V (1) symbol for the chemical element vanadium (*q.v.*); (2) volt (*q.v.*); (3) valine *q.v.*

vaccine *n.* a preparation of microorganisms or their antigenic components that can induce protective immunity against a pathogenic bacterium or virus but that does not itself cause disease. Vaccines may be composed of killed pathogenic microorganisms (killed vaccines), or live nonpathogenic strains of virus or bacterium (live vaccines), or may be composed of isolated protein antigens (subunit vaccines).

vacuole *n.* a large permanent or temporary fluid-filled membrane-bounded cavity in the cytoplasm of eukaryotic cells, found esp. in plants, algae and protozoa, which may contain cell sap (in plant and algal cells) or water or partially digested food (in protozoans).

vacuum activity a fixed-action pattern of activity carried out by an animal in the absence of any external stimulus.

vadose zone zone of aeration *q.v.*

vagile *a.* (1) freely motile; (2) able to migrate. *n.* vagility.

vaginicolous *a.* building and inhabiting a sheath or case.

valence electrons electrons held within the valence shell of an atom.

valence shell (1) the outermost shell of electrons of an atom; (2) the atom's electrons that are involved in the formation of chemical bonds.

valency *n.* (1) (*chem.*) the combining power of a radical or atom (or the atoms of an element). In a compound, the valency of all radicals or atoms must be satisfied. Thus a trivalent element (e.g. aluminium, Al) will combine with a divalent one (e.g. oxygen, O) in a ratio of 2 : 3 (Al_2O_3). However, this useful concept is limited in its applicability as many elements exhibit variable valency; (2) (*biol.*) of an antibody, the number of antigen-binding sites per molecule.

valine (Val, V) *n.* amino acid with a hydrocarbon side chain, constituent of protein, essential in the diet of man and some animals.

valley *see* U-shaped valley, V-shaped valley.

valley glaciers glaciers that move downhill within the confines of steep-sided valleys. *alt.* alpine glaciers, mountain glaciers.

valley train alluvium derived from glacial moraine that is transported by meltwater and deposited in a glacial trough, in the wake of a receding glacier.

valley wind anabatic wind *q.v.*

vamidothion *n.* systemic organophosphate acaricide and insecticide used to control pests, such as red spider mites and woolly aphids, on fruit crops.

van der Waals' forces (1) the weak non-covalent attractive forces that operate between neutral molecules and/or atoms; (2) a synonym for the London force *q.v.*

vanadium (V) *n.* metallic element (at. no. 23, r.a.m. 50.94) forming a bright white metal. It occurs naturally in the minerals magnetite, vanadite (V_2S_5) and vanadinite [$Pb_5(VO_4)Cl$] and in crude oil. Vanadium is a prosthetic group in some bacterial nitrogenases.

vaporization evaporation *q.v.*

vapour *n.* a gas that can be liquefied, without decreasing its temperature, by the application of pressure.

variance (σ^2) the variation of a set of values around the mean, which is given by the square of the standard deviation (*q.v.*), which is the mean of the squares of the deviations of each value from the overall mean.

variance-to-mean ratio a measurement of the dispersion of a population in which numbers of individuals are defined using quadrats, and the variance of this set of numbers is then divided by the mean. If the dispersion is random as described by the Poisson distribution, the variance divided by the mean is 1. Values of <1 indicate uniform spatial distribution, while values of >1 indicate a clumped distribution.

variation *n.* (*biol.*) (1) divergence from type in certain characteristics; (2) the phenotypic differences that exist between individuals of the same species (other than those due to age) and that reflect both genetic differences and the influence of the environment. *see also* continuous variation, discontinuous variation, genetic variation; (3) (*stat.*) deviation from the mean. *see* variance.

variety *n.* a taxonomic group below the species level. The term is used in different senses by different specialists.

variola smallpox *q.v.*

varves *n.plu.* sediment layers deposited annually in lakes by meltwater from glaciers. Each varve consists of lighter coloured, coarser sand overlain by darker coloured, finer silt.

vascular bundle a discrete strand of xylem and phloem cells, sometimes separated by a strip of cambium, in the stems of some plants.

vascular plants common name for all plants containing xylem and phloem. They include the club mosses, ferns, cycads, gymnosperms and angiosperms.

vascular system (1) (*bot.*) the tissues that carry water and solutes around the plant, i.e. the xylem and phloem; (2) (*zool.*) the blood and/or lymphatic systems.

vector *n.* (*biol.*) (1) any agent, living or inanimate, that acts as an intermediate carrier or alternative host for a pathogenic organism and transmits it to a susceptible host; (2) in genetic engineering, phage, plasmid or virus DNA into which another DNA is inserted for introduction into bacterial or other cells for amplification (DNA cloning) or studies of gene expression.

vedalia a ladybird beetle. *see* cottony-cushion scale.

veering *a. appl.* winds, a change in direction clockwise in the northern hemisphere,

and anticlockwise in the southern hemisphere. *cf.* backing.

vegetation *n.* general term for the plant cover of an area.

vegetative *a. appl.* (1) stage of growth in plants and fungi in which biomass is being added and reproduction has not yet occurred; (2) foliage shoots on which flowers are not formed; (3) reproduction by bud formation or other asexual method in plants and animals.

vegetative apomixis asexual reproduction in plants by, e.g., rhizomes, stolons or bulbils.

vegetative propagation the production of new plants from parts of the parent plant (such as roots, stems or leaves) other than its seed.

veil *n.* in fungi, the sheet of fine tissue stretching from the stipe over the cap in some basidiomycete fungi and that is ruptured as the fruit body develops, remaining as the ring on a stalk and sometimes as patches on the cap. *alt.* velum.

veins *n.plu.* (*zool.*) (1) branched vessels that convey blood to heart; (2) of insect wing, fine extensions of the tracheal system that support the wing; (3) (*bot.*) of leaves, branching strands of vascular tissue; (4) (*geol.*) sheet-like mineralized bodies that have formed within fractures in a pre-existing rock. Such fissures are typically irregular and veins commonly branch, sometimes forming complex networks. In some locations, vein systems occur in which the veins show some orientational regularities. Veins that are rich in metallic sulphides and/or other minerals of industrial importance may represent commercially significant deposits. *alt.* lodes.

veld *n.* open temperate grasslands of southern Africa. *alt.* veldt. *see also* sourveld.

veliger *n.* second larval stage in some molluscs, developing from the trochophore.

velvet *n.* soft vascular skin that covers the antlers of deer during their growth and is rubbed off as the antlers mature.

velvet worms common name for members of the Onychophora *q.v.*

vent community *see* hydrothermal vent community.

ventifacts *n.plu.* pebbles with many smooth sides, shaped mainly by wind abrasion and characteristic of stony deserts. Pyramid-shaped ventifacts with three wind-worn faces are known as dreikanters.

ventral *a*. (1) *pert*. or nearer the belly or underside of an animal or under surface of leaf, wing, etc.; (2) *pert*. or designating that surface of a petal, etc., that faces centre or axis of flower. *cf*. dorsal.

vermiculite *n*. (Mg,Ca)$_{0.3}$(Mg,Fe,Al)$_3$(Al,Si)$_4$-O$_{10}$(OH)$_2$.nH$_2$O, a phyllosilicate mineral generated by the alteration of magnesian micas, especially biotite (*q.v.*). Found as a 2 : 1-type expanding clay mineral *q.v.*

vermin *n*. any animals, esp. rodents or insects, of an objectionable or noxious kind that, either directly or indirectly, have a deleterious effect on humans, their environment or activities.

vernal *a*. *pert*. or appearing in mid or late spring.

vernal equinox *see* equinox.

vernalization *n*. the exposure of certain plants or their seeds to a period of cold that is necessary either to cause them to flower at all or to make them flower earlier than usual, and is used esp. on cereals, such as winter varieties of wheat, oats and rye.

Vertebrata, vertebrates *n*., *n.plu*. subphylum of the Chordata, animals characterized by the possession of a brain enclosed in a skull, teeth, ears, kidneys and other organs, and a well-formed bony or cartilaginous vertebral column or backbone enclosing the spinal cord. The Vertebrata includes the classes Agnatha (lampreys and hagfish), Holocephalii (rabbit fish), Selachii (sharks, dogfishes and rays), Osteichthyes (bony fish), Amphibia, Reptilia, Aves (birds) and Mammalia.

vertical temperature gradient the rate of change of temperature of the atmosphere with altitude above sea-level. *see also* lapse rate.

vertisol *n*. order of the USDA Soil Taxonomy system consisting of mineral soils with a high clay content (>35%). Vertisols swell during the wet season and shrink during the dry season to develop deep, wide cracks. Four suborders, closely related to climate, are recognized: torrert (arid climate), udert (humid climate), ustert (monsoon climate) and xerert (mediterranean climate).

vesicles *n.plu*. (1) small spherical air spaces in tissues; (2) any small cavities or sacs usually containing fluid; (3) the three primary cavities in the human brain; (4) (*mycol*.) hyphal swellings as in mycorrhiza; (5) (*zool*.) hollow prominences on shell or coral; (6) (*geol*.) voids present in some igneous rocks. These cavities are formed by the release of volcanic gases formerly dissolved in magma. Rocks that contain vesicles are described as vesicular.

vesicular *a*. (1) composed of or marked by the presence of vesicle-like cavities; (2) bladder-like.

vesicular–arbuscular mycorrhiza (v–a mycorrhiza) common type of endomycorrhiza characterized by the occurrence of vesicles (swellings on invading hyphae) and arbuscules (discrete masses of branched hyphae) in infected tissues, the lack of a fungal sheath around the roots, and hyphae ramifying within and between the cells of the root cortex.

vespertine *a*. blossoming or active in the evening.

vespoid *a*. wasp-like.

vessel element one type of water-conducting cell found in the xylem of angiosperms, with heavily lignified secondary cell walls and large perforations through the cell wall, especially the end walls. Joined end to end with similar cells to form a long hollow tube or xylem vessel.

vestigial *a*. (1) of smaller and simpler structure than corresponding part in an ancestral species; (2) small and imperfectly developed.

Vestimentifera, vestimentiferans *n*., *n.plu*. phylum proposed to include certain genera of the Pogonophora (e.g. *Riftia*, *Lamellibrachia*), sessile deep-sea worms that produce fixed chitin tubes in bottom sediments or on decaying wood on the sea-floor, which carry symbiotic sulphide-oxidizing chemoautotrophic bacteria contained within a structure called a trophosome, and whose haemoglobin can carry hydrogen sulphide as well as oxygen.

vestiture *n*. a body covering, as of scales, hair, feathers, etc.

viable *a*. (1) capable of living; (2) capable of developing and surviving parturition, *appl*. foetus.

viatical *a. appl.* plants growing by the roadside.

vibratile *a.* oscillating, *appl.* antennae of insects.

vibrio *n.* any of a group of bacteria with short curved cells, appearing comma-shaped under the microscope, esp. the cholera bacillus, *Vibrio cholerae*, and related organisms.

vibrissa *n.* (1) stiff hair growing on the nostril or face of an animal, e.g. the whiskers of a cat or mouse, often acting as a tactile organ; (2) feather at base of bill or around eye; (3) one of the sensitive hairs of an insectivorous plant, as of *Dionaea*, Venus' fly trap. *plu.* vibrissae.

vicariance, vicariation *n.* the separate occurrence of corresponding species, e.g. reindeer and caribou, in corresponding but separate environments, divided by a natural barrier, such species being known as vicariants. *a.* vicarious.

vicilin *n.* a seed storage protein of legumes.

vicinism *n.* tendency to variation due to proximity of related forms.

vinblastine *n.* plant alkaloid derived from *Vinca* spp. and used as an anticancer drug. It inhibits microtubule formation and kill rapidly dividing cells by disrupting the mitotic spindle.

vincristine *n.* plant alkaloid derived from *Vinca* spp. that has similar anticancer activity to vinblastine *q.v.*

vinclozolin *n.* contact fungicide used, e.g., to control *Botrytis* on strawberries.

vines *n.plu.* members of the Vitaceae, the grape family, mainly woody climbing plants that climb by tendrils and whose fruit is a berry. Includes the grapevine (*Vitis vinifera*) and Virginia creeper (*Parthenocissus*).

vinyl chloride CH_2CHCl, colourless gas that is polymerized to form polyvinyl chloride (PVC). *alt.* chloroethene.

Violales *n.* order of dicot shrubs or small trees, less often herbs, including the families Cistaceae (rock rose), Violaceae (violet) and others.

viral *a. pert.*, belonging to, consisting of, or due to, a virus.

virescence *n.* production of green colouring in petals instead of usual colour.

virgin forest primary forest *q.v.*

virgin land land that has never been used for cultivation.

virgin site greenfield site *q.v.*

virilization *n.* (1) masculinization of genetic females caused by disturbances of sex hormone metabolism, due to various causes; (2) precocious sexual development in genetic males.

virion *n.* mature infectious virus particle consisting of nucleic acid core and protein coat (and lipid outer envelope if applicable).

viroids *n.plu.* small circles of RNA that cause various diseases in plants, replicated entirely by host cell enzymes and not coding for any proteins. Transmitted from plant to plant by insect vectors.

virology *n.* the study of viruses.

virulence *n.* the ability to cause disease. *a.* virulent, *appl.* bacteria, viruses, etc.

virulence gene gene in a pathogenic microorganism that is responsible for its ability to cause disease.

virus *n.* minute, intracellular obligate parasite, visible only under the electron microscope. A virus particle consists of a core of nucleic acid, which may be DNA or RNA, surrounded by a protein coat, and in some viruses a further lipid/glycoprotein envelope. Viruses are unable to multiply or express their genes outside a host cell as they require host cell enzymes to aid DNA replication, transcription and translation. Viruses cause many diseases of man, animals and plants. Viruses infecting bacteria are called bacteriophages. *see also* DNA viruses, RNA viruses. *see* Appendix 10.

virusoid *n.* small encapsidated plant pathogenic RNA that requires co-infection with a specific virus (a helper virus) for replication and encapsidation. *alt.* satellite RNA.

viscera *n.plu.* the internal organs collectively. *a.* visceral.

viscid silk a highly extensible and sticky type of silk produced by spiders, which forms the spiral of a typical orb web. *see also* frame silk.

viscous *a. appl.* fluids, having a high viscosity, i.e. a high internal flow resistance, e.g. coal-tar.

visible light electromagnetic radiation with wavelengths from *ca.* 380 to *ca.* 780 nm.

This is the part of the electromagnetic spectrum that can be observed by the human eye.

vitamin *n.* any of various organic compounds needed in minute amounts for various metabolic processes in animals. They are synthesized by many microorganisms, by plants and by some lower animals, but must be supplied in the diet of higher animals. The lack of the appropriate vitamin causes a deficiency disease.

viticulture *n.* the cultivation of grapevines.

Vitis genus of dicot climbing plants, of which the grapevine *V. vinifera* and related species are cultivated for their edible fruits, which are used to make wine, and are also dried to produce raisins, currants and sultanas (vine fruits).

vitrand *see* andisol.

vitrification *n.* process by which a suitable material is made into a glassy solid via heating to an appropriate temperature. This process may be used to immobilize high-hazard wastes (including high-level radioactive waste) prior to disposal.

viverrids *n.plu.* members of the Viverridae, a family of carnivores including the genet.

viviparous *a.* (1) producing young alive rather than laying eggs, i.e. all mammals except monotremes, and some animals in other groups; (2) *appl.* plants, having seeds that germinate while still attached to the parent plant, e.g. mangrove. *n.* viviparity.

volant *a.* adapted for flying or gliding.

volatile *a.* readily vaporized, *appl.* any material with a high vapour pressure.

volatilization *n.* conversion into a gaseous form.

volatilize *v.* to transform (or cause to transform) a condensed phase, i.e. a liquid or a solid, into a gaseous one.

volcanic ash *see* pyroclasts.

volcanic bombs *see* pyroclasts.

volcanic cone cone-shaped hill or mountain formed by the accumulation of lava and pyroclasts from repeated volcanic eruptions.

volcanic dome a small volcanic mound formed when acid lava is slowly emitted from a volcanic vent, in the absence of pyroclastic activity. Such domes may occur within the craters of recently erupted volcanoes, e.g. as seen in Mount St Helens in the aftermath of the 1980 eruption.

volcanic eruption the eruption of magma through an orifice in the Earth's crust. This can involve both the surface flow of lava and the violent expulsion of rock fragments (known as pyroclasts) into the air. *see* central eruption, fissure eruption.

volcanic neck volcanic plug *q.v.*

volcanic pipe the conduit found immediately beneath a volcanic vent.

volcanic plug a mass of solidified lava that blocks up a volcanic pipe. *alt.* volcanic neck.

volcanic rocks extrusive rocks *q.v.*

volcanic tuff *see* tuff.

volcanic vent a single pipe-like opening at the surface of the Earth through which liquid magma, hot gases and rock fragments escape. *see also* hydrothermal vent community.

volt (V) *n.* the derived SI unit of electric potential, defined as the potential difference between two points on a conducting wire carrying a current of 1 ampere, when the power dissipated between the points is 1 watt.

voltine *a. pert.* number of broods in a year.

volva *n.* cup-shaped remnant of the universal veil that remains around the base of the stalk in some fungi such as *Amanita*. *plu.* volvae.

V-shaped valley river valley whose typical V-shaped cross-section is formed as a result of fluvial erosion, characteristic of the upper course of a river.

vulnerable *a.* IUCN definition *appl.* species or larger taxa (i) thought likely to move into the endangered category in the near future if circumstances do not change, as most or all of their populations are decreasing through, e.g., loss of habitat or over-exploitation; (ii) whose populations are seriously depleted and whose security is not assured; or (iii) that are at present abundant but are threatened by major adverse factors throughout their range. *see also* endangered, rare, rarity.

vultures *n.plu.* large birds of prey belonging to two distinct families: the Accipitridae, which contains the Old World vultures (as well as eagles, hawks, etc.), and the Cathartidae, the New World vultures (the condors and turkey vultures), which lack the strong beak and talons of the Old World vultures.

W

W (1) symbol for the chemical element tungsten (*q.v.*); (2) tryptophan *q.v.*

waders *n.plu.* shore and inland birds that wade in shallow water to feed, e.g. sandpipers, herons, snipe, cranes.

wadi *see* arroyo.

waggle dance the sequence of movements by which honey bees communicate the location and distance of food sources and new nest sites. It comprises a repeated figure of eight movement, made up of a straight run, a loop back to right (or left), then another straight run, then a loop back in the opposite direction and so on, the straight run containing information on the direction and distance away of the target.

waldsterben *n.* extensive decline and death of trees that has occurred in Central Europe since the 1970s. *see* recent forest decline.

wall pressure (WP) turgor pressure *q.v.*

Wallace's line an imaginary line separating the Australian and Oriental zoogeographical regions, running between Bali and Lombok, between Sulawesi and Borneo, and then eastwards of the Philippines.

warble fly hairy dipterous fly belonging to the genus *Hypoderma* (or related genera), the larvae of which infest cattle, producing small abscesses beneath the skin (known as warbles).

ware potatoes the best potatoes (in terms of size, shape, condition, etc.) in a crop, and that are destined for human consumption.

warfarin *n.* anticoagulant rodenticide used to control rats and mice.

warm-based glacier a glacier whose basal ice layer is at its pressure melting point (*q.v.*). This type of glacier occurs in both temperate and polar regions.

warm-blooded *a.* colloquial but meaningless term *appl.* animals whose body heat is primarily generated and regulated internally. The terms endothermic (*q.v.*) or homoiothermic (*q.v.*) should be used to describe such animals.

warm front borderline between an advancing warm air mass forced to rise over colder, denser air, found to the fore of the warm sector in a depression.

warm occlusion type of occluded front (*q.v.*) where the overtaking cold front is warmer than the air beyond the warm front.

warm-season plant plant that grows most during the warmer seasons of the year.

warm sector wedge-shaped section of warm air that separates the warm front from the cold front in a depression.

warning coloration bright and distinctive coloration, such as the yellow and black stripes of a wasp, that warns a potential predator that the plant or animal is unpleasant-tasting or dangerous. *alt.* aposematic coloration.

washout *see* wet deposition.

wasp *n.* insect of the superfamily Vespoidea (true wasps) of the order Hymenoptera, generally with a smooth shiny body and a well-defined waist between thorax and abdomen. Some species are social and some solitary. All feed their young on small insects, larvae, etc. Social wasps live in colonies with a queen, males and workers, in a cellular nest made from paper produced from chewed wood pulp. *see also* digger wasps.

waste *n.* any moveable material that is perceived to be of no further use and is permanently discarded.

waste-derived fuel (WDF) fuel derived from municipal solid waste or agricultural wastes, such as sawdust and straw, used, e.g., to supplement coal. WDF may be compressed into pellets before use (pelleted fuel). *alt.* refuse-derived fuel.

waste disposal, waste management *see* concentrate and contain, dilute and disperse, incineration, thermal processing. *see also* radioactive waste.

waste heat heat that is generated but not utilized during industrial, commercial or domestic processes, the cause of thermal pollution.

waste load amount of waste, both municipal and domestic, produced per person per year within a given area.

waste-pickers *n.plu.* people who make at least part of their livelihood from scavenging useful or saleable items from refuse tips. This practice is particularly prevalent in the developing world.

waste tip site where waste is disposed of.

wastewater *n.* general term for the effluent from sewage.

water balance the balance between the water intake of an organism directly, in food and as metabolic water (*q.v.*), and the water lost by excretion and evaporation.

water bears the common name for the tardigrades *q.v.*

waterbody any area of water, such as a pond, stream, lake, sea or ocean.

water budget of soil difference between input from precipitation and potential losses through evapotranspiration. A water budget is described as positive when the former exceeds the latter and negative when the reverse is true.

water consumption the use of water withdrawn from a particular source such that it is not returned to the surface water or groundwater from which it was taken.

water cycle the process by which the Earth's supply of water is converted from one physical state to another by evaporation and condensation, is moved between the land, oceans and atmosphere, and is passed through the bodies of plants and animals.

waterfall *n.* any portion of a stream or river in which natural flow is essentially vertical.

water fleas *see* Branchiopoda.

water-holding capacity of soil the maximum amount of water that can be held in the soil.

water hyacinth usually refers to *Eichhornia crassipes*, which has become a noxious weed in the USA, Australia and Africa, choking waterways.

water-layer weathering a general term used for weathering processes that involve the wetting and drying of surfaces during the tidal cycle.

waterlogged *a. appl.* soil saturated with water, in which the water table is at or near the surface and all the soil spaces are filled with water.

water masses large portions of the oceans, each of which may be distinguished from the others on the basis of characteristic features, particularly their salinity and temperature signatures.

water meadow grassland bordering a river that regularly floods in winter.

water of crystallization hydrate water incorporated within a crystal lattice.

water parting watershed *q.v.*

water potential (WP) suction pressure *q.v.*

watershed *n.* an imagined line, often along a ridge of high land, that separates adjacent drainage basins. *alt.* divide, drainage divide, water parting.

water slater small freshwater isopod crustacean, resembling the terrestrial woodlouse, with no carapace, and which carries young in a brood pouch under the hind part of the body.

water softening any process, e.g. ion exchange, by which the ions that cause hardness in water, i.e. $Ca^{2+}_{(aq)}$ and $Mg^{2+}_{(aq)}$, are removed from that water.

waterspout *n.* (*clim.*) term used for a tornado that moves over water.

water table upper surface of the zone of saturation, below which all available pores in soil and rock are filled with water.

water-use efficiency the amount of dry matter (net primary production) produced per given amount of water consumed by a plant, which is effectively the ratio of photosynthesis to transpiration.

water vapour mixing ratio mixing ratio *q.v.*

watt (W) *n.* derived SI unit of power (*q.v.*), defined as 1 J s^{-1}.

wattle *n.* (1) common name for members of the plant genus *Acacia*; (2) fleshy excrescence under throat of cock or turkey or of some reptiles.

wave *n.* any periodic change in either a physical property (e.g., pressure as in sound waves, electromagnetic field strength as in electromagnetic radiation, e.g. light) or the location of particles of matter about a mean (as in ocean waves).

wave base *see* oscillatory waves.

wave-cut notch a nick cut into the base of a coastal cliff by wave action.

wave-cut platform shore platform *q.v.*

wave energy kinetic energy stored in ocean waves.

wave height (*oceanogr.*) the vertical distance between the crest (highest point) and the trough (lowest point) of a wave, estimated in deep water.

wavelength (λ) *n.* the distance between any two adjacent points that are located at equivalent positions on a wave's cycle (e.g. in ocean waves, the distance from one crest to the next). *alt.* wave length.

wave period (*oceanogr.*) the time interval, usually expressed in seconds, between successive wave crests passing a fixed point.

wave power a term used to encompass the technology that is used to convert the energy in ocean waves into useful work, usually in the form of electricity.

wave refraction (*oceanogr.*) the process whereby incoming wave crests are bent until they are roughly parallel to submarine contours. This results in a concentration of wave energy (and hence erosion) on headlands, where waves converge, and a dissipation of wave energy in bays, where waves diverge.

wave steepness (*oceanogr.*) the ratio of the height of a wave to its wavelength.

wave velocity (*oceanogr.*) the rate of forward movement of a wave crest.

waves of oscillation oscillatory waves *q.v.*

waves of translation translatory waves *q.v.*

waxes *n.plu.* esters of fatty acids with long-chain monohydric alcohols, insoluble in water and difficult to hydrolyse, found as protective waterproof coatings on leaves, stems, fruits, animal fur and integument of insects, etc., and including beeswax and lanolin.

W chromosome the X chromosome in animals in which the female is the heterogametic sex.

WCMC World Conservation Monitoring Centre *q.v.*

WCS World Conservation Strategy *q.v.*

WCU World Conservation Union *q.v.*

WDF waste-derived fuel *q.v.*

weak *a.* (*chem.*) *appl.* (1) an acid or a base that does not completely ionize in solution. An example of a weak acid is acetic acid, i.e. ethanoic acid (CH_3COOH), in water, and an example of a weak base is ammonia (NH_3) in water; (2) electrolytes that slightly dissociate into ions when in solution.

wean *v.* to remove a young mammal from its mother's milk supply.

weather *n.* the physical condition of the Earth's atmosphere (particularly the troposphere) at a specific time and place with regard to wind, temperature, cloud cover, fog and precipitation.

weather chart weather map *q.v.*

weather forecasting prognosis of future weather, either for the next month (long-range forecast), the next four days (medium-range forecast) or the next 24 h (short-range forecast). The last of these includes an indication of the likely weather patterns up to three days from the time of the forecast.

weathering *n.* the decomposition and disintegration of rocks *in situ* by the action of external factors, such as rain, frost, snow, sun or wind. *see* biological weathering, chemical weathering, physical weathering.

weathering front the interface between the regolith (layer of weathered material) and the unweathered bedrock below.

weathering profile the term applied to a regolith when distinct horizons can be identified.

weather map map that gives meteorological information, e.g. air pressure, air temperature, precipitation, wind speed, wind direction, at a particular time. *alt.* synoptic chart, weather chart.

web *n.* (1) membrane stretching from toe to toe as in frogs and swimming birds; (2) network of threads spun by spiders. *see* orb web. *see also* food web.

Weber's line imaginary line separating islands with a preponderant Indo-Malayan fauna from those with a preponderant Papuan fauna.

WECS wind energy conversion system, a term used for wind turbines used for electricity production where turbine and generator are combined.

weed *n.* a plant growing where it is not wanted.

wehrlite *see* peridotite.

Weil's disease *see* leptospirosis.

Welwitschiales *n.* an order of Gnetophyta including the single extant genus *Welwitschia*, having a mainly subterranean stem and two thick long leaves surviving throughout the plant's life.

wernerite *n.* part of the scapolite group (*q.v.*) of minerals.

West African Floral Region part of the Palaeotropical Realm comprising the southern coastal region of West Africa and Africa east to Lake Tanganyika and south to central Angola.

westerlies mid-latitude westerlies *q.v.*

wet-bulb temperature *see* psychrometer.

wet-bulb thermometer *see* psychrometer.

wet deposition the removal of material (dust, gases, chemical pollutants, etc.) from the atmosphere by its incorporation into rain or other precipitation, both within clouds (known as rainout) and/or beneath clouds (known as washout). *alt.* precipitation scavenging.

wether *n.* a castrated male sheep.

wetland rice *see* rice.

wetland, wetlands *n., n.plu.* area or areas habitually saturated with water, which may be partly or wholly covered permanently, occasionally or periodically by fresh or salt water up to a depth of 6 m and that includes bogs, fens, water meadows, marshland and salt marshes, shallow ponds, river estuaries and intertidal mud flats, but excludes rivers, streams, lakes and oceans.

wet meadow *alt.* name for water meadow *q.v.*

whales *see* cetaceans.

whaling *n.* deliberate killing of wild cetaceans, such as the minke whale and sperm whale, usually for commercial purposes.

wheat bulb fly *n. Leptohylemyia coarctata*, an insect pest of winter wheat and barley. The whitish-grey larvae attack crop plants in late spring, causing the central shoot to turn yellow and die.

wheat *n. Triticum aestivum*, a small-grained cereal crop grown for fodder and for food, esp. bread-making. Wheat is one of the oldest cultivated grains, along with maize and rice, and is one of the world's major food crops, both in terms of area occupied and world production. *alt.* corn (UK usage). *see also* durum wheat, einkorn, emmer, gluten.

wheel animals *n.plu.* common name for the Rotifera *q.v.*

wheel-ore bournonite *q.v.*

whey *n.* the watery part of milk left after the curds have been removed during cheese-making. This byproduct is widely used in food manufacture. However, if it is released untreated into the environment, it may cause serious water pollution.

whip scorpions common name for members of the Uropygi, an order of arachnids having the last segment bearing a long jointed flagellum.

whirlwind *n.* vortex of air formed around a core of very low pressure, similar to, but smaller and less intense than, a tornado. Whirlwinds may be termed land devils or sea devils depending on the type of surface over which they pass.

whisk ferns common name for the Psilophyta *q.v.*

white asbestos *see* chrysotile.

whitefly *n. Trialeurodes vaporariorum*, a pest of glasshouse crops. *see also Encarsia formosa*.

whiteheads take-all *q.v.*

whiteout *n.* condition in which surface features, and the horizon, are very hard to distinguish due to the multiple scattering of sunlight between snow-covered ground and the base of a low cloud.

WHO World Health Organisation *q.v.*

whole-tree clearcut method of logging in which the entire above-ground part of the tree is removed from the site.

whorl *n.* (1) (*bot.*) circle of flowers, parts of a flower, or leaves arising from one point; (*zool.*) (2) the spiral turn of a univalve shell; (3) the concentric arrangement of ridges of skin on fingers.

wilderness *n.* land that has never been permanently occupied by humans, or exploited by mining, agriculture, logging, etc., in any way.

wildlife *n.* collective term for undomesticated plants and animals living independently of humans, esp. when in their natural habitats.

wildlife corridor narrow continuous area of favourable habitat in otherwise unfavorable surroundings, which allows the movement of animals, birds and plants along it. *alt.* green corridor, habitat corridor.

wildlife park, wildlife reserve an area designated as a refuge for wildlife.

wild oat common wild oat *q.v.*

willemite *n.* Zn_2SiO_4, a zinc ore mineral.

willy-willy In Australia, a hurricane *q.v.*

wilting *n.* loss of turgidity in plant cells, due to inadequate water absorption.

wilting coefficient percentage of moisture in soil when wilting takes place.

wilting point the moisture content of a soil at the stage when all available water has been removed and plants lose turgor and consequently wilt. *alt.* permanent wilting point.

wind *n.* horizontal flow of air from a region of higher air pressure to one of lower pressure ('the wind doth blow from high to low').

windbreak shelterbelt *q.v.*

wind chill index index of how cold humans would feel under given conditions of wind speed and air temperature. The wind cools the body by removing sensible and latent heat, thus making the temperature feel lower than it actually is.

wind deflation *see* deflation.

wind-driven currents *see* ocean currents.

wind energy the kinetic energy of moving air.

wind farm collection of wind turbines (sometimes thousands, e.g. in California, USA) used for electricity generation.

wind gap a valley formed by a river that is no longer present.

wind power a term used to encompass the technology that is used to convert wind energy into useful work, either in the form of mechanical work, e.g. for milling corn or pumping water, or in the form of electricity.

windrows (1) organic material spread into long rows, to facilitate large-scale composting; (2) debris floating on a lake in long lines parallel to the prevailing wind direction.

wind turbine a turbine (*q.v.*) used to transform the kinetic energy of moving air into useful work, often for the purpose of electricity generation.

windward side side of a mountain (or other area of high relief) facing towards the oncoming wind. *cf.* leeward side.

wing *n.* (*zool.*) (1) fore limb modified for flying in pterodactyls, birds, bats; (2) epidermal structure modified for flying, in insects; (3) large lateral process on sphenoid bone; (4) (*bot.*) one of two lateral petals in a papilionaceous flower; (5) (*bot.*) lateral expansion on many fruits and seeds; (6) any broad membranous expansion.

winged insects common name for the Pterygota *q.v.*

winged stem stem having expansions of photosynthetic tissue, as in some vetches.

wingless insects common name for the Apterygota *q.v.*

winnow *v.* to expose to wind or air currents so that lighter particles are removed, *appl.* cereal grain, also *appl.* rock surfaces eroded by wind. *see* loess.

winter *n.* (1) the season between autumn and spring, in the northern hemisphere commonly taken to be the months December, January and February; (2) astronomically, the season that commences at the winter solstice and terminates at the spring equinox, i.e. in the northern hemisphere, between 22 December and 21 March.

winter bud dormant bud, protected by hard scales during winter.

winter egg egg of many freshwater animals, provided with a thick shell that preserves it as it lies quiescent over the winter and that hatches in spring.

winter solstice *see* solstice.

wireworms *n.plu.* larvae of the click-beetle (*Agriotes* spp.). These slender, tough-skinned larvae attack the roots of cereals and vegetable crops including potatoes and sugar beet.

witches' broom twiggy growth occurring on some trees, caused by infection with fungi or mites.

withdrawn water water that is returned directly to its source of supply after use. *cf.* consumed water.

within-habitat comparison the study and comparison of species diversity in two different areas with similar habitats.

WMO World Meteorological Organisation *q.v.*

wolframite *n.* (Fe,Mn)WO$_4$, the main tungsten ore mineral.

wollastonite *n.* CaSiO$_3$, one of the pyroxenoid minerals *q.v.*

wood *n.* (1) secondary xylem (*q.v.*); (2) the hard, generally non-living part of a tree.

woodland *n.* type of tree cover where trees are numerous but are generally not growing together as closely as in a forest, and not forming a completely continuous canopy.

woodlice *see* Isopoda.

wood spirit methanol *q.v.*

woody plant any perennial plant, e.g. tree or shrub, having secondary lignified xylem in its stem.

work *n.* any interaction (*q.v.*) between two bodies, or any system and its surroundings, that is neither heat nor the transfer of matter. All work interactions must, in theory, have the capacity to be entirely transformed into mechanical work (*q.v.*). The unit in the SI system is the joule (J).

worker *n.* a member of the labouring, non-reproductive caste of semisocial and eusocial insect species.

World Bank colloquial term for the International Bank for Reconstruction and Development (IBRD) (*q.v.*). *see also* World Bank Group.

World Bank Group collective term for the International Bank for Reconstruction and Development (*q.v.*) and the International Development Association *q.v.*

World Commission on Environment and Development also known as the Brundtland Commission, after its chairwoman, a commission formed under UN auspices in 1983, charged with identifying and promoting the cause of 'sustainable development'. It produced the Brundtland Report in 1987, which led to the convening of the 1992 United Nations Conference on Environment and Development *q.v.*

World Conservation Monitoring Centre (WCMC) an independent charity established by three international conservation agencies, the IUCN (WCU), WWF and UNEP, to provide an authoritative source of data on global conservation issues. The WCMC is located in Cambridge, UK.

World Conservation Strategy (WCS) major report published collectively by the IUCN, UNEP and WWF in 1980 that called upon countries to prepare national conservation strategies based on sustainable policies. Succeeded by *Caring for the Earth: A Strategy for Sustainable Living* published in 1991 (IUCN, UNEP and WWF).

World Conservation Union (WCU) formerly the International Union for Conservation of Nature and Natural Resources (IUCN), an international conservation agency, established in 1948, with more than 400 member organizations (governments, non-governmental organizations and voluntary commissions) from over 100 countries. *see also* Red Data Books.

World Food Programme a joint FAO–UN body established in 1961, whose mandate included the provision of long-term food aid to countries in need and emergency relief in disaster situations.

World Health Organisation (WHO) a UN agency, founded in 1948 to promote cooperation between nations in disease control and in improving the health of populations throughout the world, especially in developing countries.

World Heritage Convention the Convention Concerning the Protection of the World Cultural and Natural Heritage, adopted by UNESCO in Paris in 1972. Under this convention, the World Heritage List was established to identify and protect world heritage sites of outstanding universal value, e.g. Hadrian's Wall, UK.

World Heritage List *see* World Heritage Convention.

World Meteorological Organisation (WMO) an organization established in

1873 to facilitate cooperation between nations regarding information of a meteorological or hydrological nature.

Worldwatch Institute (WWI) US not-for-profit research organization founded in 1974 and concerned with identifying, analysing and publicizing emerging global environmental problems.

World Wide Fund for Nature (WWF) an international non-governmental organization established in 1961 to raise funds for conservation projects, e.g. protection of endangered species such as the giant panda. Formerly the World Wildlife Fund.

World Wildlife Fund *see* World Wide Fund for Nature.

worm bin wormery *q.v.*

wormery *n.* a specially designed container (typical capacity 90 litres) in which tiger worms (*Eisenia foetida*, also known as brandling worms) or red worms (*Eisenia andrei*) are used to actively transform organic kitchen waste into compost. *alt.* worm bin.

wound cambium cambium forming protective tissue at the site of injury in plants.

wound hormones substances produced by wounded plant cells, which stimulate renewed growth of tissue near the wound.

wound response *see* plant defence response.

wrench fault transcurrent fault *q.v.*

WWF World Wide Fund for Nature (*q.v.*), formerly the World Wildlife Fund.

WWI Worldwatch Institute *q.v.*

X

X the female sex chromosome in mammals.

xanthein *n.* a water-soluble yellow pigment in cell sap.

xanthin *n.* a yellow carotenoid pigment in flowers.

xanthine *n.* a purine, 2,6-dioxypurine, found especially in animal tissues, such as muscle, liver, pancreas and spleen, and in urine, and also in certain plants. It is a breakdown product of AMP and guanine, and is oxidized to urate (uric acid).

xanthism *n.* colour variation in which the normal colour is replaced almost entirely by yellow.

Xanthophyceae the yellow-green algae, *alt.* Xanthophyta.

xanthophyll *n.* any of a group of widely distributed yellow or brown carotenoid pigments, oxygenated derivatives of carotenes, and including lutein, violaxanthin, neoxanthin, cryptoxanthin. *alt.* phylloxanthin.

Xanthophyta *n.* the yellow-green algae, a phylum of yellow-green photosynthetic protists, which have two unequal flagella, the longer hairy and the shorter smooth. They contain the chlorophylls *a*, *c*, c_2 and *e* and have xanthin pigments. Storage products are oils. Common in fresh water, many xanthophytes have multicellular and syncytial forms. Considered as members of the Chrysophyta (*q.v.*) in some classifications. *alt.* Xanthophyceae.

xanthopterin *n.* a yellow pigment, a pterin, found esp. in the wings of yellow butterflies and in the yellow bands of wasps and other insects, and also in mammalian urine, which can be oxidized to leucopterin and converted to folic acid by microorganisms.

X chromosome (1) the female sex chromosome in mammals, two copies of which are present in each somatic cell of females with one copy being permanently inactivated, and one (active) copy being present in males; (2) in general, the sex chromosome present in two copies in the homogametic sex, and in one copy in the heterogametic sex.

Xe the symbol for the chemical element xenon *q.v.*

xenic *a. pert.* a culture containing one or more unidentified microorganisms.

xeno- prefix derived from Gk *xenos*, strange.

xenobiosis *n.* the condition where colonies of one species of social insect live in the nests of another species and move freely among them, obtaining food from them by various means but keeping their broods separate.

xenobiotic *a.* (1) not found in nature, i.e. entirely man-made; (2) foreign to a living organism, *appl.* foreign substances such as drugs, etc.; (3) *appl.* microorganisms able to degrade foreign organic compounds, e.g. man-made organic pesticides in soil.

xenodeme *n.* a deme of parasites differing from others in host specificity.

xenoecic *a.* living in the empty shell of another organism.

xenogenous *a.* (1) originating outside the organism; (2) caused by external stimuli.

xenolith *n.* a piece of a foreign rock embedded in an igneous rock. *cf.* autolith.

xenon (Xe) *n.* noble gas (at no. 54, r.a.m. 131.30) that is present in the atmosphere in one part in 20 million.

xerad xerophyte *q.v.*

xeralf *n.* suborder of the order alfisol of the USDA Soil Taxonomy system consisting of alfisols of warm climatic regions that have long, dry summers and moist winters (mediterranean climates).

xerand *see* andisol.

xerarch *a. appl.* seres progressing from xeric towards mesic conditions.

xerert *see* vertisol.

xeric *a.* (1) dry, arid; (2) tolerating, or adapted to, arid conditions.

xero- prefix derived from the Gk *xeros*, dry.

xeroderma pigmentosum rare inheritable skin disease in humans, in which skin is abnormally sensitive to ultraviolet or sunlight, producing parchment skin, ulceration of the cornea and a predisposition to skin cancer. It is caused by a defect in a DNA repair enzyme responsible for excising pyrimidine dimers formed in DNA on exposure to ultraviolet or sunlight.

xeroll *n.* suborder of the order mollisol of the USDA Soil Taxonomy system consisting of mollisols of mediterranean climatic regions.

xeromorphic *a.* structurally modified so as to withstand drought, *appl.* desert plants such as cacti, etc. *n.* xeromorphy.

xerophilous *a.* able to withstand drought, *appl.* plants adapted to a limited water supply. *n.* xerophil.

xerophyte *n.* a plant adapted to arid conditions, either having xeromorphic characteristics, or being a mesophyte growing only in a wet period.

xerosere *n.* a plant succession originating on dry soil.

xerothermic *a. appl.* organisms thriving in hot, dry conditions.

xerult *n.* suborder of the order ultisol of the USDA Soil Taxonomy system consisting of ultisols of mediterranean climatic regions.

Xiphosura *n.* order of aquatic arthropods, commonly called king or horseshoe crabs, in the class Merostomata, an individual having a heavily chitinized body with the prosoma covered with a horseshoe-shaped carapace. *Limulus* is a living example, but the group is known from the Palaeozoic.

X-linked gene any gene carried on the X chromosome. *alt.* sex-linked gene.

XP xeroderma pigmentosum *q.v.*

X-ray crystallography technique for determining the three-dimensional atomic structures of molecules that can be crystallized, from the diffraction patterns of X-rays passing through the crystal.

X-rays electromagnetic radiation with wavelengths from *ca.* 5×10^{-9} to *ca.* 6×10^{-12} m.

xylan *n.* any of a group of polysaccharides composed of a central chain of linked xylose residues, with other monosaccharides attached as single units or side chains, found in the cell walls of many angiosperms.

xylem *n.* the main water-conducting tissue in vascular plants, which extends throughout the body of the plant and is also involved in transport of mineral matter, food storage and support. Primary xylem is derived from the procambium, secondary xylem, e.g. the wood of trees and shrubs, from the vascular cambium. Xylem is composed of tracheary elements: tracheids and (in angio-sperms) vessel elements. Both are elongated hollow cells, with thickened, usually heavily lignified walls, and lacking proto-plasts when mature. They are joined end to end to form a continuous conducting tube. *see also* phloem, vascular bundle.

xylocarp *n.* a hard woody fruit.

xyloma *n.* (1) hardened mass of mycelium that gives rise to spore-bearing structures in certain fungi; (2) a tumour of woody plants.

xylophagous *a.* wood-eating, as are certain termites and beetles.

xylophilous *a.* (1) preferring wood; (2) growing on wood.

xylophyte *n.* a woody plant.

xylose *n.* a five-carbon aldose sugar, a constituent of polysaccharides, found esp. in the cell walls of some plants.

xylotomous *a.* able to bore or cut wood.

xylulose *n.* a five-carbon ketose sugar. As xylulose 5-phosphate, it is involved esp. in carbon dioxide fixation in photosynthesis.

Y

Y (1) symbol for the chemical element yttrium (*q.v.*); (2) tyrosine (*q.v.*); (3) male sex chromosome in mammals.

Y-linked gene any gene carried on the Y chromosome.

yardangs *n.plu.* low ridges separated by troughs, which occur in very arid desert regions. These aeolian features, which lie parallel to the prevailing wind direction, are produced mainly by wind abrasion cutting into weakly consolidated rock.

yarovization vernalization *q.v.*

Yb symbol for the chemical element ytterbium *q.v.*

Y chromosome (1) in mammals, the male sex chromosome, which is smaller than and non-homologous with the X chromosome, being absent from cells of females and present in one copy in the somatic cells of males; (2) in general, the chromosome that pairs with the X chromosome in the heterogametic sex.

yeasts *n.plu.* unicellular ascomycete fungi of the family Saccharomycetaceae, which include the bread and brewing yeast *Saccharomyces cerevisiae*. Yeasts are ubiquitous inhabitants of the soil and plant surfaces, esp. on sugary substrates, which they ferment when growing anaerobically, typically producing alcohols and carbon dioxide.

yellow cake crude uranium oxide (UO_2).

yellow fever an acute viral subtropical/ tropical disease, transmitted to humans by the bite of female *Aedes aegypti* mosquitoes. Symptoms include fever, jaundice and vomiting blood. *alt.* black vomit.

yellow-green algae *see* Xanthophyta.

yellow rust a disease affecting wheat and barley, caused by the fungus *Puccinia striiformis*. Symptoms include bright yellow pustules on the leaves and spores on the grain. Yields are greatly reduced if the attack is severe. *see also* rust fungi.

yolk *n.* nutrient material rich in protein and fats forming a large part of the ova of many egg-laying animals, e.g. amphibians, reptiles and birds, and which nourishes the developing embryo.

ytterbium (Yb) *n.* a lanthanide element (at. no. 70, r.a.m. 173.04), a soft silver-coloured metal when in elemental form. It is used as a constituent of steel alloys.

yttrium (Y) *n.* metallic element (at. no. 39, r.a.m. 88.91).

Yu-Cheng incident the poisoning of *ca.* 2000 people in Taiwan, in 1978, through the ingestion of rice oil (used for cooking) contaminated with polychlorinated biphenyls (*q.v.*). This incident was very similar to the Yusho incident (*q.v.*), which occurred a decade earlier.

Yusho incident the poisoning of *ca.* 1800 people in Japan, in 1968, through the ingestion of rice oil contaminated with polychlorinated biphenyls (*q.v.*). This incident appeared to be linked to a rise in the rates of miscarriage and birth defects. It is now believed that the toxicity of the contaminated rice oil was greatly elevated by the presence of polychlorinated dibenzofurans (*q.v.*), which were formed when the PCBs were heated in the presence of air.

Z

Z either glutamine or glutamic acid in the single-letter code for amino acids.

Z atomic number *q.v.*

Zadok's Scale (*agric.*) a system used to identify, and give a numerical code to, the growth stages (and therefore the development) of small-grain cereals.

Z chromosome the Y chromosome when the female is the heterogametic sex.

zeatin *n.* a natural cytokinin isolated from maize (*Zea mays*), a derivative of adenine.

zeaxanthin *n.* yellow carotenoid pigment found in many plant cells, including those of maize grains, in some classes of algae and in egg yolk and that is an isomer of lutein.

zein *n.* a simple protein in seeds of maize, which lacks tryptophan and lysine.

zeitgeber *n.* a general term for a synchronizing agency, such as environmental cues responsible for keeping circadian rhythms of plants in tune with the daily 24-h light–dark cycle.

zeolite minerals a group of framework aluminosilicate minerals that share the formula $M_{a/b}[(AlO_2)_a(SiO_2)_c].nH_2O$, where M represents counterions with a charge of b^+, e.g. Na^+, Ca^{2+}. They contain interlinked submicroscopic cavities, within which both the water and counterions (M^{b+}) reside. The accessible nature of these cavities means that these minerals can undergo both reversible dehydration and cation exchange. There are about 40 naturally occurring and >130 entirely synthetic zeolites known and characterized. They have a wide range of industrial applications as they can be used as cation exchangers (e.g. in washing powders, where they may replace polyphosphate as an ingredient), drying agents, absorbents and selective catalysts.

zero grazing the practice of cutting fresh grass, on a daily basis, to feed to livestock, e.g. dairy cows, kept in continuous confinement. *alt.* green soiling, soiling.

zero population growth (ZPG) situation in which the number of births in a population is equal to the number of deaths, thus keeping the population level constant. *alt.* replacement-level fertility.

zeroth law of thermodynamics *see* thermodynamics, the laws of.

zero tillage direct drilling *q.v.*

zeuge *n.* a tabular mass of resistant rock underlain by a pinnacle of less resistant rock. This desert landform is produced by differential wind erosion.

zinc (Zn) *n.* metallic element (at. no. 30, r.a.m. 65.38), forming a hard brittle bluish-white metal that is resistant to corrosion. It occurs naturally in ores such as zinc blende (ZnS), smithsonite ($ZnCO_3$) and hemimorphite ($ZnSiO_4.H_2O$). Zinc has many industrial uses and is an essential micronutrient for living organisms.

zinc blende sphalerite *q.v.*

zincite *n.* ZnO, a zinc ore mineral.

zineb *n.* fungicide used, e.g., to control rust fungi on chrysanthemums and blight on potatoes.

Zingiberales *n.* order of tropical monocot herbs with rhizomes, including Cannaceae (canna), Musaceae (banana), Strelitziaceae (bird of paradise flower), Zingiberaceae (ginger) and others.

Zinjanthropus *see* australopithecines.

Zircaloy *n.* alloy of zirconium and tin used for cladding nuclear fuel rods in some reactor designs, e.g. boiling water, candu and pressurized water reactors.

zircon *n.* $ZrSiO_4$, a common accessory mineral in igneous rocks. Also occurs in metamorphic rocks. It is dense and durable and therefore tends to be concentrated in river and beach sands. It is used as a gemstone when transparent. A source of zirconium.

zirconium (Zr) *n.* metallic element (at. no. 40, r.a.m. 91.22) present in minerals in acid igneous rocks and a constituent of the gemstone zircon ($ZrSiO_4$). It is used in photoflash bulbs and in alloys that are used in the cladding of fuel elements in nuclear reactors.

Zn symbol for the chemical element zinc *q.v.*

Zoantharia n. subclass of Anthozoa, including the stony corals and sea anemones, which are solitary or colonial, have paired mesenteries usually in multiples of six, and have skeletons, if present, external and not made of spicules.

zoic *a.* (1) containing remains of organisms and their products; (2) *pert.* animals or animal life.

zoisite *n.* $Ca_2Al_3Si_3O_{12}(OH)$, a mineral of metamorphic rocks. Tanzanite, a blue variety of zoisite, is prized as a gemstone.

zonation *n.* arrangement or distribution in zones.

Zonda *see* föhn.

zone *n.* (1) an area characterized by similar conditions or by similar fauna or flora throughout; (2) a belt or area to which certain species are limited; (3) stratum or set of beds characterized by a typical fossil or set of fossils; (4) an area or region of the body. *a.* zonal.

zone fossil index fossil *q.v.*

zone of ablation the lower part of a glacier, where outputs, in the form of direct evaporation and meltwater streams, exceed inputs, from avalanches and snowfall. *alt.* zone of melting.

zone of accumulation the upper part of a glacier, where inputs, from avalanches and snowfall, exceed outputs, in the form of direct evaporation and meltwater streams.

zone of aeration the subsurface zone between the ground surface and the water table in which not all of the available pores in the soil and rock are filled with water. *alt.* unsaturated zone, vadose zone.

zone of equilibrium the zone in a glacial system where the rate of accumulation equals the rate of melting (ablation). This corresponds to the snow line *q.v.*

zone of intolerance *see* Shelford's law of tolerance.

zone of leaching *see* A-horizon.

zone of melting zone of ablation *q.v.*

zone of physiological stress *see* Shelford's law of tolerance.

zone of saturation the subsurface zone found beneath the zone of aeration in which all the available pores in the soil and rock are completely filled with water. The upper boundary of this zone constitutes the water table. *alt.* phreatic zone.

zone of tolerance *see* Shelford's law of tolerance.

zonolimnetic *a.* of or *pert.* a certain zone in depth.

zooanthellae *n.plu.* cryptomonads symbiotic with certain marine protozoans.

zoobenthos *n.* the fauna of the sea bottom, or of the bottom of inland waters.

zoobiotic *a.* parasitic on, or living on an animal.

zoochlorellae *n.plu.* symbiotic green algae living in the cells of various animals.

zoochoric *a. appl.* plants dispersed by animals. *n.* zoochory.

zoocoenosis *n.* an animal community.

zooerythrin *n.* red pigment found in the plumage of various birds.

zoofulvin *n.* yellow pigment found in the plumage of various birds.

zoogenesis *n.* the origin of animals, *appl.* usually to phylogenetic origin.

zoogenic *a.* arising from the activity of animals.

zoogenous *a.* produced or caused by animals.

zoogeographical regions large areas of the world with distinct natural faunas. At the highest level are three large regions or zoogeographical realms: Arctogaea, Neogaea and Notogaea. Arctogaea comprises the Palaearctic (Europe, North Africa, Asia south to the Himalayas), the Nearctic (Greenland, North America south to central Mexico), the Ethiopian (Africa south of the Sahara), and the Oriental (India, Indochina, Malaysia, the Philippines and Indonesian islands west of Wallace's line) regions. The Palaearctic and Nearctic regions are sometimes considered together as the Holarctic Realm. Neogaea or the Neotropical Region comprises South America, Central America and southern Mexico, and the West Indies. Notogaea comprises Australia, New Zealand, most of the Pacific islands and the Indonesian islands east of Wallace's Line.

zoogeography *n.* the geographical distribution of animal species.

zoogloea *n.* a mass of bacteria embedded in a mucilaginous matrix and frequently forming an iridescent film on the surface of water. *a.* zoogloeal.

zooid *n.* an individual in a colonial animal, such as an individual polyp in a coral.

zoology *n.* the science dealing with the structure, functions, behaviour, history, classification and distribution of animals.

Zoomastigina *n.* heterogenous phylum of non-photosynthetic, mainly unicellular protists bearing from one to many thousands of flagella, commonly known as the animal flagellates, zooflagellates or zoomastigotes. Some, e.g. *Naegleria*, can change from a flagellated to an amoeboid form. They include intestinal parasites of fish and amphibians (e.g. the opalinids), the kinetoplastids, typified by the genus *Trypanosoma*, which causes sleeping sickness in humans, wood-digesting protozoa that live in the gut of termites (e.g. *Staurojenia*), the choanoflagellates, which resemble the body cells of sponges, and several other groups. They correspond to the Zoomastigophora or Protomonadina of older classifications.

Zoomastigomorpha *n.* in older classifications, a class of protozoans including non-photosynthetic flagellates.

zoomorphic *a.* having the form of an animal.

zoonosis *n.* a disease of animals that can be transmitted to man. *plu.* zoonoses.

Zoopagales *n.* class (or in some classifications an order) of zygomycete fungi parasitic on small soil animals. They are known as the predaceous fungi or animal traps, and capture soil amoebae, rhizopods and nematodes by attaching to them and growing within or on them.

zooparasite *n.* any parasitic animal.

zoophilous *a. appl.* plants adapted for pollination by animals other than insects.

zoophobic *a. appl.* plants shunned by animals because they are protected by spines, hairs, etc.

zoophyte *n.* an animal resembling a plant in appearance, e.g. some colonial hydrozoans.

zooplankton *n.* animal plankton.

zoosis *n.* any disease caused by animals.

zoospore *n.* motile, flagellated asexual reproductive cell in protozoans, algae and fungi.

zootic climax any stable climax community dependent for its maintenance on an animal activity such as grazing.

zootoxin *n.* any toxin produced by animals.

zootrophic heterotrophic *q.v.*

zootype *n.* representative type of animal.

zooxanthellae *n.plu.* parasitic or symbiotic yellow or brown algae living in various animals.

zooxanthin *n.* yellow pigment found in the plumage of certain birds.

ZPG zero population growth *q.v.*

Zr symbol for the chemical element zirconium *q.v.*

zwitterion *n.* an ionized molecule with both positive and negative charges, as all amino acids.

zygomorphic *a. appl.* flowers and other structures that are bilaterally, rather than radially, symmetrical.

Zygomycota, zygomycetes *n., n.plu.* division of terrestrial fungi including the bread moulds, fly fungi and animal traps (predaceous fungi) characterized by sexual reproduction by fusion of gametangia, the production of a resting sexual spore (zygospore) and asexual reproduction by non-motile spores. *alt.* Zygomycotina (when considered as a class).

zygopterans *n.plu.* damsel flies, members of the suborder Zygoptera, of the order Odonata.

zygote *n.* a cell formed from the union of two gametes or reproductive cells. *alt.* fertilized egg.

zygotic *a.* (1) *pert.* a zygote; *appl.* (2) a mutation occurring immediately after fertilization; (3) number, the diploid or somatic number of chromosomes; (4) gene activity, activity of an embryo's own genes, as opposed to the activity of gene products laid down in the egg by the mother.

zymogen *n.* functionally inactive precursor of certain enzymes, the active form being produced by specific cleavage of the polypeptide chain. *alt.* proenzyme.

zymogenic *a. pert.* or causing fermentation.

zymogenous *a. appl.* microflora in soil normally present in the resting state and only becoming active when a fresh supply of organic material is added.

Appendix 1

CHEMICAL ELEMENTS, NAMES AND SYMBOLS

Name	Symbol	Name	Symbol	Name	Symbol	Name	Symbol
Actinium	Ac	Erbium	Er	Neodymium	Nd	Sodium	Na
Aluminium	Al	Europium	Eu	Neon	Ne	Strontium	Sr
or aluminum		Fermium	Fm	Neptunium	Np	Sulphur or sulfur	S
Americium	Am	Fluorine	F	Nickel	Ni	Tantalum	Ta
Antimony	Sb	Francium	Fr	Niobium	Nb	Technetium	Tc
Argon	Ar	Gadolinium	Gd	Nitrogen	N	Tellurium	Te
Arsenic	As	Gallium	Ga	Nobelium	No	Terbium	Tb
Astatine	At	Germanium	Ge	Oxygen	O	Thallium	Tl
Barium	Ba	Gold	Au	Palladium	Pd	Thorium	Th
Berkelium	Bk	Hafnium	Hf	Phosphorus	P	Thulium	Tm
Beryllium	Be	Helium	He	Platinum	Pt	Tin	Sn
Bismuth	Bi	Holmium	Ho	Plutonium	Pu	Titanium	Ti
Boron	B	Hydrogen	H	Polonium	Po	Tungsten	W
Bromine	Br	Indium	In	Potassium	K	Unnilennium	Une
Cadmium	Cd	Iodine	I	Praseodymium	Pr	Unnilhexium	Unh
Calcium	Ca	Iridium	Ir	Promethium	Pm	Unniloctium	Uno
Californium	Cf	Iron	Fe	Protactinium	Pa	Unnilpentium	Unp
Carbon	C	Krypton	Kr	Radium	Ra	Unnilquadium	Unq
Cerium	Ce	Lanthanum	La	Radon	Rn	Unnilseptium	Uns
Cesium	Cs	Lawrencium	Lr	Rhenium	Re	Uranium	U
or caesium		Lead	Pb	Rhodium	Rh	Vanadium	V
Chlorine	Cl	Lithium	Li	Rubidium	Rb	Xenon	Xe
Chromium	Cr	Lutetium	Lu	Ruthenium	Ru	Ytterbium	Yb
Cobalt	Co	Magnesium	Mg	Samarium	Sm	Yttrium	Y
Copper	Cu	Manganese	Mn	Scandium	Sc	Zinc	Zn
Curium	Cm	Mendelevium	Md	Selenium	Se	Zirconium	Zr
Dysprosium	Dy	Mercury	Hg	Silicon	Si		
Einsteinium	Es	Molybdenum	Mo	Silver	Ag		

Appendix 2

THE PERIODIC TABLE OF THE ELEMENTS

Key:

1
H
1.008

← Atomic number
← Relative atomic mass (i.e. molar mass)

Group numbers (shown in bold italics)

Non-metals

Metals

Period	1	2	3	4	5	6	7	8	9	10	11	12	13	14	15	16	17	18
1	1 H 1.008																	2 He 4.00
2	3 Li 6.94	4 Be 9.01											5 B 10.81	6 C 12.01	7 N 14.01	8 O 16.00	9 F 19.00	10 Ne 20.18
3	11 Na 22.99	12 Mg 24.31											13 Al 26.98	14 Si 28.09	15 P 30.98	16 S 32.06	17 Cl 35.45	18 Ar 39.95
4	19 K 39.10	20 Ca 40.08	21 Sc 44.96	22 Ti 47.90	23 V 50.94	24 Cr 52.00	25 Mn 54.94	26 Fe 55.85	27 Co 58.93	28 Ni 58.71	29 Cu 63.54	30 Zn 65.37	31 Ga 69.72	32 Ge 72.59	33 As 74.92	34 Se 78.96	35 Br 79.91	36 Kr 83.80
5	37 Rb 85.47	38 Sr 87.62	39 Y 88.91	40 Zr 91.22	41 Nb 92.91	42 Mo 95.94	43 Tc 98.91	44 Ru 101.07	45 Rh 102.91	46 Pd 106.4	47 Ag 107.87	48 Cd 112.40	49 In 114.82	50 Sn 118.69	51 Sb 121.75	52 Te 127.60	53 I 126.90	54 Xe 131.30
6	55 Cs 132.91	56 Ba 137.34	57 La 138.91	72 Hf 178.49	73 Ta 180.95	74 W 183.85	75 Re 186.2	76 Os 190.2	77 Ir 192.2	78 Pt 195.09	79 Au 196.97	80 Hg 200.59	81 Tl 204.37	82 Pb 207.19	83 Bi 208.98	84 Po 210	85 At 210	86 Rn 222
7	87 Fr 223	88 Ra 226.03	89 Ac 227.03															

Lanthanides

58 Ce 140.12	59 Pr 140.91	60 Nd 144.24	61 Pm 146.92	62 Sm 150.35	63 Eu 151.96	64 Gd 157.25	65 Tb 158.92	66 Dy 162.50	67 Ho 164.93	68 Er 167.26	69 Tm 168.93	70 Yb 173.04	71 Lu 174.97

Actinides

90 Th 232.04	91 Pa 231.04	92 U 238.03	93 Np 237.05	94 Pu 239.05	95 Am 241.06	96 Cm 247.07	97 Bk 249.08	98 Cf 251.08	99 Es 254.09	100 Fm 257.10	101 Md 258.10	102 No 255	103 Lr 257

Appendix 3

THE ELEMENTS WITH ESSENTIALLY INVARIANT OXIDATION STATES IN THEIR COMPOUNDS

Element	Oxidation state
Elements of group 1[a]	+1
Elements of group 2[a]	+2
Hydrogen in compounds with non-metals	+1
Hydrogen in compounds with metals	−1
Fluorine	−1
Oxygen	−2[b] unless combined with F
	or −1 in peroxides (O_2^{2-})
	or −1/2 in superoxides (O_2^-)
	or −1/3 in ozonides (O_3^-)

[a] See Appendix 2 for a listing of the elements in these groups.
[b] −2 is by far the most common oxidation state for oxygen in compounds.

Appendix 4

THE GEOLOGICAL TIMESCALE (NOT DRAWN TO SCALE) IN MILLIONS OF YEARS BEFORE PRESENT

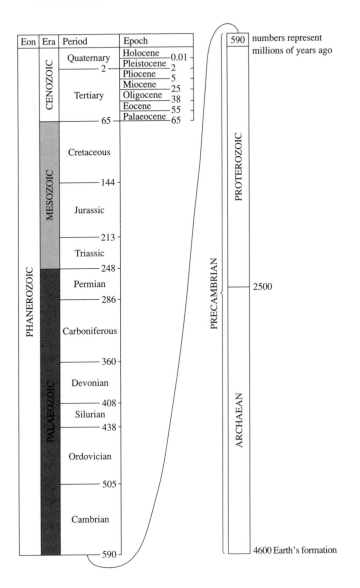

Appendix 5

OUTLINE OF THE PLANT KINGDOM

The following outline is intended as a brief summary of the different types of plants that exist and does not represent a rigorous taxonomic classification. The names of divisions, classes and orders, and the rank assigned to the various groups, differ considerably from authority to authority, and no attempt has been made to give all alternatives. The nomenclature of the main groups largely follows that used in P. H. Raven, R. F. Evert and H. Curtis, *Biology of Plants*, 5th edn (Worth, New York, 1992) and L. Margulis and K. V. Schwartz, *Five Kingdoms*, 2nd edn (Freeman, New York, 1988). The nomenclature and arrangement of the flowering plants follow that proposed by Takhatajan and Cronquist rather than the older systems of Engler and Bentham and Hooker (see S. Holmes, *An Outline of Plant Classification*, Longman, Harlow, 1983). Fungi and lichens have been classified within the plant kingdom, but the quite different nature and evolutionary origins of the fungi are now recognized by placing them in a separate kingdom, and they are considered as such in Appendix 6. The lichens are also included there. The blue-green algae (cyanobacteria) are also sometimes considered as honorary plants for historical reasons, but are prokaryotes and are included in the kingdom Prokaryotae (Appendix 9). The algae, both unicellular and multicellular, are considered as members of the Protista (Appendix 8), but those divisions traditionally included in the plant kingdom are also listed here.

All divisions and classes named here, and some orders and common names, have an entry in the body of the dictionary.

Algae

DIVISION CHRYSOPHYTA (diatoms and golden algae)
DIVISION PYRROPHYTA (dinoflagellates)
DIVISION EUGLENOPHYTA (euglenoids)
DIVISION RHODOPHYTA (red algae)
DIVISION PHAEOPHYTA (brown algae)
DIVISION CHLOROPHYTA (green algae)

Bryophytes

DIVISION HEPATOPHYTA (liverworts)
DIVISION ANTHOCEROPHYTA (hornworts)
DIVISION BRYOPHYTA
       class Sphagnidae (sphagnum or peat mosses)
       class Andreaeidae (granite or rock mosses)
       class Bryidae (true mosses)

Early vascular plants, now extinct

DIVISION RHYNIOPHYTA
DIVISION ZOSTEROPHYLLOPHYTA
DIVISION TRIMEROPHYTA (possibly the progenitor of the ferns, horsetails and progymnophytes)

Pteridophytes: seedless vascular plants

DIVISION PSILOPHYTA (PSILOTOPHYTA) (whisk ferns, only two living genera, *Psilotum* and *Tmesipteris*)

DIVISION LYCOPHYTA (lycophytes)

    order Lycopodiales (lycopods or club mosses)

      Lepidodendrales (tree lycophytes, extinct)

      Selaginellales (one living genus, *Selaginella*)

      Isoetales (quillworts)

      and other orders (extinct)

DIVISION SPHENOPHYTA (horsetails)

    order Equisetales (one living genus, *Equisetum*)

      and other orders (extinct)

DIVISION PTEROPHYTA (FILICOPHYTA) (ferns)

    order Marattiales (giant ferns)

      Ophioglossales (e.g. *Ophioglossum*, adder's tongue)

      Osmundales (e.g. *Osmunda*)

      Filicales (most living ferns, e.g. maidenhair fern, filmy fern, hart's tongue, bracken)

      Marsileales (water ferns, e.g. *Pilularia*, pillwort)

      Salviniales (water ferns, e.g. *Azolla*)

DIVISION PROGYMNOPHYTA (progymnosperms, extinct, e.g. *Archaeopteris*)

Spermatophytes: seed plants

Gymnosperms

DIVISION PTERIDOSPERMOPHYTA (seed ferns, extinct)

DIVISION CYCADEOIDOPHYTA (BENNETTITALES) (cycadeoids, extinct)

DIVISION CYCADOPHYTA (cycads)

DIVISION GINKGOPHYTA (one species, *Ginkgo*)

DIVISION CONIFEROPHYTA (conifers and their allies)

    order Coniferales (most living conifers, e.g. firs, monkey puzzle, cedars, cypresses, junipers, pines, redwoods)

      Cordaitales (extinct)

      Voltziales (extinct)

      Taxales (yews, *Torreya*)

DIVISION GNETOPHYTA

    order Welwitschiales (one species, *Welwitschia*)

      Ephedrales (one genus, *Ephedra*)

      Gnetales (one genus, *Gnetum*)

Angiosperms

DIVISION ANTHOPHYTA (MAGNOLIOPHYTA) (the common names in brackets after orders refer to families)

  class Dicotyledones (Magnoliopsida)

   subclass Magnoliidae

    order Magnoliales (e.g. magnolia, nutmeg)

      Laurales (e.g. laurel, calycanthus)

Piperales (e.g. pepper, the spice)
Aristolochiales (birthwort)
Rafflesiales (rafflesia)
Nymphaeales (e.g. water lily)
subclass Ranunculidae
order Illiciales (e.g. star anise)
Nelumbonales (Indian lotus)
Ranunculales (e.g. buttercup, barberry)
Papaverales (poppy, fumitory)
Sarraceniales (pitcher plant, family Sarraceniaceae)
subclass Hamamelididae
order Trochodendrales
Hamamelidales (e.g. witch hazel, plane)
Eucommiales
Urticales (e.g. elm, nettle, mulberry)
Casuarinales (she oak)
Fagales [beech (includes oaks), birch, hazel]
Myricales (sweet gale)
Leitneriales
Juglandales (e.g. walnut)
subclass Caryophyllidae
order Caryophyllales (e.g. pink, amaranth, goosefoot)
Cactaceae (cacti)
Polygonales (polygonum)
and other orders
subclass Dilleniidae
order Paeoniales (peonies)
Theales (e.g. tea, St John's wort)
Violales (e.g. violet, rock rose)
Passiflores (passion flower, pawpaw)
Cucurbitales (cucurbits: e.g. marrow, squash, gourd)
Datiscales (e.g. begonia)
Capparales [e.g. crucifers (brassicas), caper, mignonette]
Tamaricales (e.g. tamarisk)
Salicales (willow)
Ericales (e.g. heather, wintergreen)
Primulales (e.g. primrose)
Malvales (e.g. mallow, cocoa, lime-tree)
Euphorbiales (e.g. spurge, box-tree, jojoba)
Thymelaeales [daphne (mezereon)]
subclass Rosidae
order Saxifragales (e.g. gooseberry, saxifrage, hydrangea)
Rosales [e.g. rose (rose, apple, hawthorn, etc.), coco plum]
Fabales (Leguminosae) (e.g. peas, beans, etc., mimosa)
Nepenthales [sundews, pitcher plant (family Nepenthaceae)]
Myrtiflorae (e.g. myrtle, mangrove, pomegranate, evening
primrose, loosestrife)

Hippuridales (mare's tail, gunnera)
Rutales (e.g. rue, citrus fruits, mahogany)
Sapindales (Acerales) (e.g. maple, horse chestnut)
Geraniales (e.g. balsam, geranium, nasturtium)
Cornales (e.g. dogwood, umbellifers, *Davidia*, ginseng)
Celastrales (e.g. holly)
Rhamnales (e.g. grape, buckthorn)
Oleales (Ligustrales) (olive, privet)
Elaeagnales (oleaster)
and other orders
subclass Asteridae
order Dipsacales (e.g. honeysuckles, valerian)
Gentianales (e.g. gentian, bog bean)
Polemoniales (e.g. borage, convolvulus, dodder, phlox)
Scrophulariales (Personatae) (e.g. acanthus, buddleia, African violet, nightshade)
Lamiales (e.g. verbena, mint)
Campanulales (e.g. bellflower, lobelia)
Asterales [composites (e.g. daisy, cornflower, thistle, dandelion)]
class Monocotyledones (Liliopsida, Liliatae)
subclass Alismidae
order Alismales (e.g. flowering rush)
Potamogetonales (e.g. pondweed, eel-grass)
subclass Liliidae
order Liliales (e.g. agave, lily, daffodil, onion, yam)
Iridales (e.g. iris)
Zingiberales (e.g. canna, banana, ginger)
Orchidales (orchid)
subclass Commelinidae
order Juncales (e.g. rush)
Cyperales (sedge)
Bromeliales (pineapple)
Commelinales (e.g. tradescantia, xyris)
Restionales
Poales (grasses)
subclass Arecidae
order Arecales (palms)
Arales (e.g. arum)
Pandanales (screw pine)
Typhales (bur reed, cat-tail)

Appendix 6

OUTLINE OF THE FUNGI (KINGDOM MYCETAE)

This outline follows that given in L. Margulis and K. V. Schwartz *Five Kingdoms*, 2nd edn (Freeman, New York, 1988) and C. J. Alexopolous and C. W. Mims, *Introductory Mycology*, 3rd edn (Wiley, New York, 1979) but includes only those groups regarded as the 'true fungi' or Eumycota. The slime moulds, chytrids, hypochytrids, oomycetes and hyphomycetes are now considered as protists and are included in Appendix 8.

True fungi (Eumycota)

DIVISION ZYGOMYCOTA

 class Mucorales (e.g. *Mucor, Rhizopus, Pilobolus*)

 class Entomophthorales (fly fungi)

 class Zoopagales (predaceous fungi or animal traps)

DIVISION ASCOMYCOTA

 class Hemiascomycetae [yeasts (e.g. *Saccharomyces cerevisiae*), leaf-curl fungi]

 class Euascomycetae [black moulds, blue moulds, pyrenomycetes (flask fungi, e.g. powdery mildews), discomycetes (cup fungi, e.g. *Monilinia* brown rot of peach, *Rhytisma acerinum* tar spot of maples), morels, truffles]

 class Laboulbeniomycetae (parasitic on insects)

 class Loculoascomycetae [e.g. *Elsinoe* spp. (citrus scab, grape anthracnose, raspberry anthracnose)]

DIVISION BASIDIOMYCOTA

 class Homobasidiomycetae

 subclass Hymenomycetes

  order Aphyllophorales (chanterelles, coral fungi)

   Stereales

   Thelephorales

   Polyporales (bracket fungi)

   Corticales

   Boletales (ceps and other boletes)

   Russulales (russulas and lactarias)

   Trichlomatales (e.g. tricholomas)

   Pluteales (e.g. *Pluteus*)

   Cortinariales (cortinarias)

   Agaricales (e.g. *Agaricus, Amanita)*

   and other orders

 subclass Gasteromycetes

  order Hymenogastrales

   Lycoperdales (puffballs, earthstars)

   Sclerodermales (earthballs)

   Phallales (stinkhorns)

   Nidulariales (bird's nest fungi)

class Heterobasidiomycetae
  subclass Teliomycetes (rusts and smuts)
  subclass Dacrymycetes (jelly fungi)
DIVISION DEUTEROMYCOTA (FUNGI IMPERFECTI) (an artificial grouping
  of fungi with no known sexual stage)
    form class Sphaeropsida (pycnidial fungi)
    form class Melanconia (conidia)
    form class Monilia (e.g. *Penicillium*)
    form class Mycelia Sterilia (e.g. *Rhizoctonia*)

Lichens

DIVISION MYCOPHYCOPHYTA (Lichens)
    class Ascolichenes (lichens in which an ascomycete is the fungal partner)
    class Basidiolichenes (lichens in which a basidiomycete is the fungal partner)
    class Deuterolichenes (lichens in which a deuteromycete is the fungal partner)

Appendix 7

OUTLINE OF THE ANIMAL KINGDOM

This outline is intended simply to give an overall view of the different types of animals that exist, and is not a rigorous or comprehensive taxonomic classification. Only extant phyla are included, but some extinct groups within phyla are indicated. The names of phyla, classes and orders and the rank given to the various groups differ from authority to authority, and no attempt has been made to give all the alternatives. The arrangement and nomenclature of phyla used here generally follow L. Margulis and K. V. Schwartz *Five Kingdoms*, 2nd edn (Freeman, New York, 1988). A traditional zoological classification of the protozoans is included here for historical reasons; a more modern treatment of the protozoa, in which the group is divided into numerous separate phyla, is included in Appendix 8. See entries in the body of the dictionary for more information.

PHYLUM PROTOZOA (unicellular organisms)
  subphylum Sarcomastigophora
    superclass Mastigophora
      class Phytomastigophora (photosynthetic flagellates, e.g. *Chlamydomonas, Euglena, Gymnodinium*)
      class Zoomastigophora [non-photosynthetic flagellates, e.g. *Trypanosoma* (parasitic), choanoflagellates]
    superclass Opalinata (multiflagellate protozoans inhabiting the gut of some amphibians)
    superclass Sarcodina
      class Rhizopodea [amoebas (including the parasitic amoebas) and foraminiferans]
      class Actinopodea (radiolarians and heliozoans)
  subphylum Ciliophora (the ciliates)
    class Ciliatea
      subclass Holotrichia (e.g. *Paramecium, Tetrahyena*)
      subclass Peritrichia (e.g. *Vorticella*)
      subclass Suctoria (e.g. *Podophyra*)
      subclass Spirotrichia (e.g. *Stentor*)
  subphylum Sporozoa (exclusively parasitic protozoa)
    class Telosporea
      subclass Gregarinia (gregarines)
      subclass Coccidia (e.g. *Eimeria, Plasmodium*)
    class Piroplasmea (e.g. *Babesia*)
  subphylum Cnidospora (exclusively parasitic protozoa)
    class Myxosporidea (e.g. *Myxobolus*)
    class Microsporidea

In modern classifications protozoa are included in the separate kingdom Protista along with other groups of eukaryotic microorganisms of uncertain affinity (*see* Appendix 8).

Invertebrates

PHYLUM PLACOZOA one species known, *Trichoplax adhaerens*
PHYLUM PORIFERA (sponges)
      class Calcarea (calcareous sponges)
      class Hexactinellidea (glass sponges, e.g. *Euplectella*, Venus's flower basket)
      class Demospongia (e.g. *Spongia*, the bath sponge)
PHYLUM CNIDARIA[*]
      class Anthozoa (corals and sea anemones)
        subclass Alcyonaria (soft corals, sea fans, sea pens)
        subclass Zoantharia (sea anemones, stony corals)
      class Hydrozoa (milleporine corals, solitary hydroids, e.g. *Hydra*, and colonial hydroids, e.g. *Bougainvillea*, and the siphonophores, e.g. *Physalia*, the Portuguese man o' war)
      class Scyphozoa (true jellyfishes)
      class Cubozoa (box-jellies)
PHYLUM CTENOPHORA[*] (sea gooseberries or comb jellies)
      class Tentacula
      class Nuda
PHYLUM MESOZOA
      class Rhombozoa (dicyemids, heterocyemids)
      class Orthonectidea (orthonectids)
PHYLUM PLATYHELMINTHES (flatworms)
      class Turbellaria (free-living flatworms)
          order Acoela
              Rhabdocoela
              Tricladida
              Polycladida
              and other orders
      class Monogenea (flukes)
          order Monopisthocotylea (skin flukes)
              Polyopisthocotylea (gill flukes)
      class Trematoda (flukes)
          order Aspidobothrea
              Digenea (gut, liver and blood flukes)
      class Cestoda (tapeworms)
PHYLUM NEMERTINA (NEMERTEA, RHYNCHOCOELA) (nemertine worms, ribbon worms)
      class Anopla
      class Enopla
PHYLUM GNATHOSTOMULIDA
PHYLUM GASTROTRICHA
PHYLUM ROTIFERA (rotifers)
      class Seisonacea
      class Bdelloidea
      class Monogononta

* The Cnidaria and Ctenophora collectively are the coelenterates.

PHYLUM KINORHYNCHA
PHYLUM LORICIFERA (only ten species known)
PHYLUM ACANTHOCEPHALA (thorny-headed worms)
PHYLUM ENTOPROCTA* (moss animals)
PHYLUM NEMATODA (nematodes or roundworms)
  class Aphasmidia (Adenophorea)
  class Phasmidia (Secernentea)
PHYLUM NEMATOMORPHA (gordian worms, horse-hair worms)
PHYLUM ECTOPROCTA* (BRYOZOA) (moss animals)
PHYLUM PHORONIDA (tubeworms, *ca.* 15 species known)
PHYLUM BRACHIOPODA (lamp shells)
  class Inarticulata
  class Articulata
PHYLUM MOLLUSCA (molluscs)
  class Monoplacophora (*Vema, Neopilina*)
  class Aplacophora (solenogasters)
  class Caudofoveata
  class Polyplacophora (chitons)
  class Pelecypoda (Bivalvia) (clams, etc.)
   subclass Protobranchia (*Nucula*)
   subclass Lamellibranchia (most other genera)
  class Gastropoda
   subclass Prosobranchia (winkles, etc.)
   subclass Opisthobranchia (sea slugs, sea hares)
   subclass Pulmonata (whelks, snails and land slugs)
  class Scaphopoda (tusk shells)
  class Cephalopoda
   subclass Nautiloidea (pearly nautilus)
   subclass Ammonoidea (ammonites, extinct)
   subclass Coleoidea (octopus, squid, cuttlefish)
PHYLUM PRIAPULIDA
PHYLUM SIPUNCULA (peanut worms)
PHYLUM ECHIURA (spoon worms)
PHYLUM ANNELIDA (ringed worms)
  class Polychaeta (e.g. ragworms, lugworms, myzostomarians)
  class Oligochaeta (earthworms, etc.)
  class Hirudinea (leeches)
PHYLUM TARDIGRADA (water bears)
PHYLUM PENTASTOMIDA (PENTASTOMA) (tongue worms)
PHYLUM ONYCHOPHORA (velvet worms)
PHYLUM ARTHROPODA (arthropods)
 subphylum Trilobitomorpha (extinct)
  class Trilobita
 subphylum Chelicerata
  class Merostomata (horseshoe crabs)
  class Arachnida

* The Entoprocta and Endoprocta were formerly grouped together as the Bryozoa.

order Scorpiones (scorpions)
Pseudoscorpiones (false scorpions)
Araneae (Araneida) (spiders)
Palpigrada (palpigrades)
Solifuga (Solpugida) (solifugids)
Opiliones (harvestmen)
Acari (Acarina) (ticks and mites)
class Pycnogonida (sea spiders)
subphylum Crustacea
class Cephalocarida
class Branchiopoda (water fleas, etc.)
class Ostracoda (ostracods)
class Copepoda (copepods)
class Mystacocarida
class Branchiura (fish lice)
class Cirripedia (barnacles)
class Malacostraca (crabs, lobsters, shrimps, woodlice)
subphylum Atelocerata
class Diplopoda (millipedes)
class Chilopoda (centipedes)
class Pauropoda
class Symphyla
class Insecta
order Diplura
Thysanura (silverfish)
Collembola (springtails)
Protura (bark lice)
Odonata (dragonflies)
Ephemeroptera (mayflies)
Plecoptera (stoneflies)
Dictyoptera (cockroaches)
Dermaptera (earwigs)
Embioptera (web-spinners)
Isoptera (termites)
Psocoptera (book lice and their allies)
Anoplura (biting and sucking lice)
Orthoptera (locusts, grasshoppers and crickets)
Thysanoptera (thrips)
Hemiptera (bugs)
Neuroptera (lacewings, ant lions)
Mecoptera (scorpion flies)
Trichoptera (caddis flies)
Lepidoptera (moths and butterflies)
Coleoptera (beetles)
Diptera (houseflies, mosquito, tsetse fly, craneflies)
Siphonaptera (fleas)
Strepsiptera (stylopids)

Hymenoptera (ants, wasps, bees, sawflies, ichneumon flies, gall wasps)

PHYLUM POGONOPHORA (beard worms, tubeworms)

PHYLUM VESTIMENTIFERA (beard worms, tubeworms)

PHYLUM ECHINODERMATA (echinoderms)

  subphylum Pelmatozoa

     class Crinoidea (sea lilies and feather stars)

  subphylum Eleutherozoa

     class Stelleroidea [starfish, brittle stars: Asteroidea (starfish) and Ophiuroidea (brittle stars)]

     class Echinoidea (sea urchins, sand dollars)

     class Holothuroidea (sea cucumbers)

       and many extinct classes

PHYLUM CHAETOGNATHA (arrow worms)

PHYLUM HEMICHORDATA (hemichordates)

     class Enteropneusta (acorn worms, tongue worms)

     class Pterobranchia (pterobranchs)

     class Planktosphaeroidea (known from a larval form only)

     class Graptolita (graptolites, extinct)

PHYLUM CHORDATA (the chordates, including tunicates (urochordates), cephalochordates and vertebrates)

  subphylum Tunicata (Urochordata)

     class Ascidiacea (sea squirts)

     class Larvacea

     class Thaliacea (salps)

  subphylum Cephalocordata (lancelets)

     class Leptocardii

Vertebrates

  subphylum Agnatha (vertebrates without jaws)

     class Cyclostomata (lampreys, hagfishes, slime eels)

       and the extinct class Ostracodermi (ostracoderms: cephalaspids, anaspids, pteraspids and thelodonts)

  subphylum Gnathostomata (jawed vertebrates)

   superclass Pisces

    class Chondrichthyes (cartilaginous fishes)

      subclass Holocephali (rabbit fishes, chimaeras)

      subclass Elasmobranchii

         order Heterodontiformes (the Port Jackson shark)

           Hexanchiformes

           Lamniformes (most dogfishes and sharks)

           Raiiformes (rays and skates)

           Torpediniformes (electric rays)

           and extinct orders

    class Osteichthyes (bony fishes)

      subclass Actinopterygii

        infraclass Palaeoniscoidei (extinct except *Polypterus* and *Erpetoichthys*)

        infraclass Chondrostei (sturgeons)
        infraclass Holostei (extinct except *Amia* and *Lepisosteus*)
        infraclass Teleostei (most living bony fishes)
          superorder Elopomorpha (eels, tarpons, etc.)
          superorder Clupeomorpha (herrings, etc.)
          superorder Osteoglossomorpha (mooneyes, knife-fish and bony
                    tongues)
          superorder Protacanthopterygii
            order Salmoniformes (salmon)
            order Gonorhynchiformes (millifishes and deep-sea lantern fishes)
              and extinct orders
          superorder Ostariophysi
            order Cypriniformes (characins, American knife-fishes, carps and
              minnows)
              Siluriformes (catfishes)
          superorder Paracanthopterygii
            order Gadiformes (cod, etc.)
               and other orders
          superorder Acanthopterygii
            order Atheriniformes (tooth-carps, etc.)
              Perciformes (perches, mackerel, tuna, etc.)
              Pleuronectiformes (flatfishes)
              and other orders
    class Sarcopterygii
      subclass Crossopterygii
          order Rhipidistia (extinct)
              Actinistia (extinct except for the coelacanth *Latimeria*)
      subclass Dipnoi (lungfishes)
    and extinct classes acanthodians and placoderms
  superclass Tetrapoda
    class Amphibia
      subclass Labyrinthodontia (extinct)
      subclass Lepospondyli (extinct)
      subclass Lissamphibia (includes all living amphibians)
          order Anura (Salientia) (frogs and toads)
              Urodela (newts and salamanders)
              Apoda (caecilians)
    class Reptilia
      subclass Anapsida
          order Cotylosauria (Captorhinida) (extinct)
              Mesosauria (extinct)
      subclass Testudinata
          order Chelonia (tortoises and turtles)
      subclass Lepidosauria
          order Sphenodonta (extinct except the tuatara, *Sphenodon*)
              Eosuchia (extinct)
              Squamata

suborder Lacertilia (lizards)
     Amphisbaenia (amphisbaenids)
     Serpentes (snakes)
subclass Archosauria
   order Thecodontia (extinct)
     Saurischia (lizard-hipped dinosaurs, extinct)
     Ornithischia (bird-hipped dinosaurs, extinct)
     Pterosauria (pterosaurs, extinct)
     Crocodylia (crocodiles and alligators)
     and other extinct orders
subclass uncertain
   order Nothosauria (extinct)
     Placodontia (extinct)
     Plesiosauria (extinct)
     Ichthyosauria (extinct)
     and other extinct orders
subclass Synapsida (mammal-like reptiles) (extinct)
   order Pelycosauria
     Therapsida
class Aves (birds)
  subclass Archaeornithes (*Archaeopteryx* only, extinct)
  subclass Odontornithes (extinct)
   order Hesperornithiformes Ichthyorniformes
  subclass Neornithes (all other birds)
   order Tinamiformes (tinamous)
     Rheiformes (rheas)
     Struthioniformes (ostriches)
     Casuariiformes (cassowaries and emus)
     Aepyornithiformes (e.g. *Aepyornis*, extinct)
     Dinornithiformes (moas and kiwis)
     Podicipediformes (Colymbiformes) (grebes)
     Procellariformes (albatrosses, shearwaters and petrels)
     Sphenisciformes (penguins)
     Pelecaniformes (pelicans and allies)
     Anseriformes (waterfowl, ducks, geese, swans)
     Phoenicopteriformes (flamingoes and allies)
     Ciconiiformes (herons and allies)
     Falconiformes (falcons, hawks, eagles, buzzards and other birds of prey)
     Galliformes (grouse, pheasants, partridges, etc.)
     Gruiformes (hemipodes, cranes, bustards, rails, coots, etc.)
     Charadriiformes (shorebirds, waders, gulls, auks)
     Gaviiformes (divers)
     Columbiformes (doves, pigeons, sand grouse)
     Psittaciformes (parrots, etc.)
     Cuculiformes (cuckoos and others)
     Strigiformes (owls)

Caprimulgiformes (nightjars)
Apodiformes (swifts)
Coliiformes (colies)
Trogoniformes (trogons)
Coraciiformes (kingfishers, bee-eaters, hoopoes)
Piciformes (woodpeckers)
Passeriformes (perching birds and songbirds, a very large
order, including finches, crows, warblers, sparrows and
weavers, etc.)

class Mammalia
subclass Prototheria
order Monotremata (duck-billed playtpus)
and several extinct orders
subclass Theria
infraclass Metatheria
order Marsupalia
suborder Polyprotodonta (opossums, etc.)
Diprotodonta (kangaroos, etc.)
infraclass Eutheria (placental mammals)
order Insectivora (shrews, moles, etc.)
Chiroptera (bats)
Dermoptera (flying lemurs)
Fissipedia (dogs, cats, bears, mustelids and other specialized carnivores)
Pinnipedia (seals)
Cetacea (whales and dolphins)
suborder Odontoceti (toothed whales and dolphins)
Mysticeti (baleen whales)
order Rodentia (rodents)
Lagomorpha (rabbits, hares, etc.)
Artiodactyla (even-toed ungulates)
suborder Suina (pigs, hippopotamus, etc.)
Tylopoda (camels, etc.)
Ruminantia (deer, antelope, etc.)
order Perissodactyla (odd-toed ungulates)
suborder Hippomorpha (horses)
Ceratomorpha (rhinoceroses and tapirs)
Proboscidea (elephants)
order Sirenia (dugongs, manatees)
Primates
suborder Prosimii (tree shrews, lemurs, etc.)
Anthropoidea (monkeys, apes and humans)
and many extinct groups

Appendix 8

OUTLINE OF THE KINGDOM PROTISTA

The kingdom Protista (or Protoctista) comprises a diverse assemblage of unicellular and simple multicellular eukaryotic organisms, which form groups each representing a separate evolutionary line that split off early in eukaryote evolution, and which do not sit happily in the animal, plant or fungal kingdoms. The classification here follows L. Margulis and K. V. Schwartz *Five Kingdoms*, 2nd edn (Freeman, New York, 1988). See entries in the body of the dictionary for more information.

Commonly known as algae

PHYLUM DINOFLAGELLATA (DINOMASTIGOTA or PYRROPHYTA)
  (dinoflagellates, e.g. *Gymnodinium, Gonyaulax*)
PHYLUM CHRYSOPHYTA (golden algae, e.g. *Ochromonas*)
PHYLUM HAPTOPHYTA (PRYMNESIOPHYTA) (haptomonads,
  coccolithophoroids, e.g. *Prymnesium*)
PHYLUM EUGLENOPHYTA (euglenoids, e.g. *Euglena*)
PHYLUM CRYPTOPHYTA (cryptomonads)
PHYLUM XANTHOPHYTA (yellow-green algae, e.g. *Vaucheria*)
PHYLUM EUSTIGMATOPHYTA (e.g. *Pleurochloris*)
PHYLUM BACILLARIOPHYTA (diatoms, e.g. *Asterionella, Navicula,
  Thalassiosira*)
PHYLUM PHAEOPHYTA (brown algae, e.g. *Fucus, Laminaria*)
PHYLUM RHODOPHYTA (red algae, e.g. *Porphyra, Corallina, Chondrus*)
PHYLUM GAMOPHYTA (conjugating green algae, e.g. *Mougeotia, Spirogyra*)
PHYLUM CHLOROPHYTA (green algae, e.g. *Chlamydomonas, Chlorococcus,
  Chlorella, Volvox, Ulva, Oedogonium, Stigeoclonium, Chaetomorpha,
  Acetabularia, Nitella, Platymonas*)

Commonly known as protozoa

PHYLUM CARYOBLASTEA (one species, *Pelomyxa palustris*)
PHYLUM RHIZOPODA (amoebas, e.g. *Acanthamoeba, Amoeba, Difflugia*)
PHYLUM ZOOMASTIGINA (zooflagellates, e.g. *Crithidia, Giardia, Naegleria,
  Trypanosoma*)
PHYLUM ACTINOPODA (radiolarians, acantharians)
PHYLUM FORAMINIFERA (e.g. *Globigerina*)
PHYLUM CILIOPHORA (ciliates, e.g. *Colpoda, Didinium, Stentor, Vorticella*)
PHYLUM APICOMPLEXA (sporozoans: gregarines, coccidians, e.g. *Eimeria,
  hemosporidians, e.g. *Plasmodium*, piroplasms, e.g. *Babesia*)
PHYLUM CNIDOSPORIDIA (microsporidians or myxosporidians, e.g. *Glugea*)

Formerly classified in the Fungi

PHYLUM LABYRINTHULOMYCOTA (slime nets, e.g. *Labyrinthula*)
PHYLUM ACRASIOMYCOTA (acrasiomycetes or cellular slime moulds, e.g.
  Dictyostelium)

PHYLUM MYXOMYCOTA (myxomycetes or plasmodial slime moulds, e.g. *Echinostelium*, *Physarum*, *Stemonitis*)

PHYLUM PLASMODIOPHOROMYCOTA (plasmodiophorans, e.g. *Plasmodiophora brassicae*, club-root)

PHYLUM HYPHOCHYTRIDIOMYCOTA (water moulds or hyphochytrids, e.g. *Rhizidiomyces*)

PHYLUM CHYTRIDIOMYCOTA (water moulds or chytrids, e.g. *Allomyces*, *Blastocladiella*)

PHYLUM OOMYCOTA (oomycetes: e.g. *Phytophthora infestans*, potato blight, downy mildews, e.g. *Peronospora*, parasitic water moulds)

Appendix 9

OUTLINE OF THE PROKARYOTES (SUPERKINGDOMS BACTERIA AND ARCHAEA)

Until recent years there have been few means to classify prokaryotes in a way that reflects their evolutionary relationships. The bacteria have traditionally been divided into some dozen large groups on the basis of shape, Gram-staining (which reflects the structure and composition of the cell wall) and other biochemical and physiological properties. The brief outline given below largely follows that proposed in *Bergey's Manual of Systematic Bacteriology*, 4 vols, J. G. Holt (ed.) (Williams & Wilkins, Baltimore, Md, 1984–1989). The division of the prokaryotes into two Superkingdoms, the Bacteria and the Archaea, has been proposed on the evidence of molecular phylogeny and is also indicated here.

Superkingdom Bacteria (formerly the Eubacteria)

DIVISION GRACILICUTES

    class Scotophobia 'bacteria indifferent to light': includes all heterotrophic Gram-negative bacteria and some other groups, e.g.

        Spirochaetes (e.g. *Treponema, Borrelia*)

        Sulphate-reducing bacteria (e.g. *Desulfovibrio*)

        Nitrogen-fixing aerobic bacteria (e.g. *Azotobacter, Rhizobium*)

        Pseudomonads (e.g. *Pseudomonas, Xanthomonas*)

        Enterobacteria (e.g. *Escherichia, Salmonella, Klebsiella*)

        Vibrios (e.g. *Vibrio cholerae*)

        Gram-negative cocci (e.g. *Aeromonas*)

        Spirilla (e.g. *Spirillum*)

        Stalked bacteria (e.g. *Caulobacteria*)

        Budding bacteria (e.g. *Hyphomicrobium*)

        Aggregated bacteria (e.g. *Spherotilus*)

        Sheathed bacteria

        Chemoautotrophic bacteria (the nitrifying and denitrifying bacteria, e.g. *Nitrobacter, Nitrosomonas*, the sulphur-oxidizing bacteria, e.g. *Thiobacillus*, and the methane-ozidizing bacteria, e.g. *Methylomonas*)

        Myxobacteria (the gliding bacteria, e.g. *Beggiatoa*, and the myxobacteria, e.g. *Stigmatella*).

    class Anoxyphotobacteria (anaerobic phototrophic bacteria: the green and purple photosynthetic sulphur bacteria, and the purple non-sulphur bacteria, e.g. *Chlorobium, Rhodospirillum, Rhodomicrobium*)

    class Oxyphotobacteria (the cyanobacteria, e.g. *Anabaena, Nostoc, Microcystis*)

    class Prochlorophyta (the prochlorophytes, *Prochloron, Prochlorothrix*)

DIVISION FIRMICUTES

    class Firmibacteria (Gram-positive rods and cocci)

        Aerobic cocci (e.g. *Micrococcus, Staphylococcus*)

        Anaerobic fermenting cocci (e.g. *Streptococcus*)

Fermenting rods (e.g. *Bacteroides*)
Aerobic endospore-forming rods (e.g. *Bacillus* spp.)

class Thallobacteria
Actinobacteria (the actinomycetes, e.g. *Streptomyces* and *Actinomyces*, and the mycobacteria, e.g. *Mycobacterium tuberculosis*)
Corynebacteria (e.g. *Corynebacterium*)

DIVISION TENERICUTES
class Mollicutes (diverse wall-less prokaryotes of doubtful origins, e.g. rickettsiae, chlamydiae, mycoplasmas)

Superkingdom Archaea (formerly the Archaebacteria)

DIVISION MENDOSICUTES
class Archaebacteria (prokaryotes with cell walls lacking peptidoglycan, and having ether-linked membrane lipids, and distinctive ribosomes and rRNA sequences)
Methanogens (e.g. *Methanobacterium, Methanococcus*)
Extreme halophiles (e.g. *Halobacterium*)
Thermoacidophiles (e.g. *Thermococcus, Thermoproteus*)

Appendix 10

VIRUS FAMILIES

DNA virus families

Double-stranded DNA

Bacteriophages
  Corticoviridae (e.g. PM2)
  Lipothrixviridae (e.g. TTV1)
  Myoviridae (T4 and the T-even phages)
  Plasmaviridae (e.g. MVL2)
  Podoviridae (T7 and the T-odd phages)
  Siphoviridae (e.g. lambda and P22)
  SSV1 group
  Tectoviridae (e.g. PRD1)
Plant viruses
  Caulimoviruses
Animal viruses
  Adenoviridae (adenovirus)
  Baculoviridae (e.g. insect baculovirus)
  Hepadnaviridae (e.g. hepatitis B)
  Herpesviridae (e.g. herpesviruses, chickenpox, Epstein–Barr virus (glandular fever))
  Iridoviridae (e.g. insect iridescent viruses)
  Papovaviridae (e.g. papillomaviruses, SV40)
  Polydnaviridae
  Poxviridae (e.g. vaccinia, mumps, measles, smallpox)

Single-stranded DNA

Bacteriophages
  Microviridae (e.g. ΦX174, G4)
  Inoviridae (fd)
Plant viruses
  Geminiviruses
Animal viruses
  Parvoviridae (e.g. canine distemper virus)

RNA virus families

Double-stranded RNA

Bacteriophages
  Cystoviridae (f6)
Plant viruses
  Cryptoviruses
  Plant reoviruses (fujivirus and phytoreovirus groups)
Animal viruses
  Birnaviridae
  Reoviridae (e.g. human rotaviruses)

Single-stranded RNA

Bacteriophages
  Leviviridae (MS2)
Plant viruses
  Alfalfa mosaic virus group
  Bromoviruses
  Carlaviruses
  Carmoviruses
  Closteroviruses
  Comoviruses
  Cucumoviruses
  Dianthoviruses
  Fabaviruses
  Furoviruses
  Hordeiviruses
  Ilarviruses
  Luteoviruses
  Maize chlorotic dwarf virus group
  Marafiviruses
  Necroviruses
  Parsnip yellow fleck virus group
  Pea enation mosaic virus group
  Plant rhabdoviruses
  Potyviruses
  Potexviruses
  Sobemoviruses
  Tenuiviruses
  Tobamoviruses
  Tobraviruses
  Tomato spotted wilt virus group
  Tombusviruses
  Tymoviruses
Animal viruses
  Arenaviridae
  Bunyaviridae (e.g. Rift Valley fever)
  Caliciviridae (e.g. swine vesicular exanthema)
  Coronaviridae (e.g. human coronaviruses)
  Filoviridae
  Flaviviridae
  Nodaviridae
  Paramyxoviridae (e.g. influenza)
  Picornaviridae (e.g. poliovirus, foot-and-mouth disease)
  Retroviridae (e.g. HIV, animal tumour viruses)
  Rhabdoviridae (e.g. rabies)
  Togaviridae (e.g. rubella, yellow fever)
  Toroviridae

Appendix 11

ETYMOLOGICAL ORIGINS OF SOME COMMON WORD ELEMENTS

The word element as it appears in English is in bold type. The Greek (Gk) or Latin (L.) word from which it is derived is shown in italics and is followed by its original meaning.

a- *a* (Gk), not.
ab- *ab* (L.), from.
absci- *abscidere* (L.), to cut off.
abyss- *abyssos* (Gk), unfathomed.
acanac- *akanos* (Gk), thistle.
acanth- *akantha* (Gk), thorn.
acer- *acer* (L.), sharp.
acid- *acidus* (L.), sour.
acra- *akros* (Gk), tip.
actin- *aktis* (GK), ray.
ad- *ad* (L.), to, towards.
adeno- *aden* (Gk), gland.
adipo- *adeps* (L.), fat.
aeolian *Aeolus* (Gk), god of the winds.
aer- *aer* (L.), *aēr* (Gk), air.
-aesthesia, -aesthetic *aisthēsis* (Gk), sensation.
agrost- *agrōstis* (Gk), grass.
albo-, albu-, albino *albus* (L.), white.
alga-, algo- *alga* (Gk), seaweed.
allele, allelo- *allēlōn* (Gk), one another.
allo- *allos* (Gk), other.
ambi- *ambo* (L.), both.
amoeba *amoibē* (Gk), change.
amphi- *amphi* (Gk), both.
amylo- *amylum* (Gk), starch.
an- *an* (Gk), not.
ana- *ana* (Gk), up, again.
andr-, andro- *anēr* (Gk), male, *andrikos* (Gk), masculine.
anemo- *anemos* (Gk), wind.
angio-, -angium *anggeion* (Gk), vessel.
aniso- *anisos* (Gk), unequal.
ankylo- *agkylos* (Gk), crooked.
anlage *Anlage* (Ger.) predisposition.
annelid, annulate *annulus* (L.), a ring.
ano- *anus* (L.), anus.
anomalo- *anomalos* (Gk), uneven.
anomo- *anomos* (Gk), lawless, irregular.
ante- *ante* (L.), before.
antha-, antho-, -anthous, -anthy *anthos* (Gk), flower.
anthero- *anthēros* (Gk), flowering.
anthropo- *anthrōpos* (Gk), man.
anti- *anti* (Gk), against, opposite.
apo- *apo* (Gk), from.
arachni-, arachno- *arachnē* (Gk), spider, cobweb.
arbor- *arbor* (L.), tree.
archaeo- *archaios* (Gk), primitive, ancient.
arche- *archē* (Gk), beginning.
archi- *archis* (Gk), first.
archo- *archon* (Gk), ruler.
arci- *arcus* (L.), bow.
argent- *argentum* (L.), silver.
argyro- *argyros* (Gk), silver.
arthro- *arthron* (Gk), a joint.
artio- *artios* (Gk), even (numbered).
-asci, -ascus, asco- *askos, askidion* (Gk), bag, little bag.
astra-, astro-, -aster *astra* (Gk), star.
-atomy, -otomy *tomē* (Gk), cutting.
auri-, auricul- *auris, auricula* (L.), ear, small ear.
auto- *autos* (Gk), self.
auxi-, auxo- *auxein* (Gk), to increase.
avi- *avis* (L.), bird.
axill- *axilla* (L.), armpit.
axis, axial *axis* (L.), axle.
axo- *axōn* (Gk), axis.
barb-, barba *barba* (L.), beard.
baro- *baros* (Gk), pressure, weight.
bathy- *bathys* (Gk), deep.
batrach- *batrachos* (Gk), frog.
benthos, benthic *benthos* (Gk), depths of sea.
bi- *bis* (L.), twice.
bio-, -biotic *bios* (Gk), life, *biosis*, living, *biōtikos*, pert. life
blast- *blastos* (Gk), bud.
botany *botanē* (Gk), pasture.
bothr- *bothros* (Gk), pit.
botry- *botrys* (Gk), bunch of grapes.
brachia- *brachium* (L.), arm.
brachy- *brachys* (Gk), short.
brady- *bradys* (Gk), slow.

branchi- *branchiae* (L.), gills, or *brangchia* (Gk), gills.

brevi- *brevis* (L.), short.

bryo- *bryon* (Gk), moss.

bucco- *bucca* (L.), cheek.

caeno- *kainos* (Gk), recent.

calci- *calx* (Gk), lime.

calyptr- *kalyptra* (Gk), covering.

cambium, cambio- *cambium* (L.), change.

capit- *caput* (L.), head, *capitellum*, small head.

capsid, capso-, capsul- *capsa* (L.), box, *capsula*, little box.

carbo- *carbo* (L.), coal.

carcino- *karkinos* (Gk), crab.

cardia-, cardio- *kardia* (Gk), heart, stomach.

-carp, -carpous *karpos* (Gk), fruit.

carpa- *carpal* (L.), wrist.

cata- *katalysis* (Gk), dissolving.

cata- *kata* (Gk), down.

cauda- *cauda* (L.), tail.

caul-, cauli- *caulis* (L.), stalk or *kaulos* (Gk), stalk.

cell, cellular *cellula* (L.), small room.

centro-, -centric *kentron* (Gk), centre.

cephal- *kephalē* (Gk), head.

-ceptor, -ceptive *capere* (L.), to take.

cerca-, cerco- *kerkos* (Gk), tail.

cerebr- *cerebrum* (L.), brain.

-cerous *keras* (Gk), horn.

cervic- *cerix* (L.), neck.

ceta-, ceto- *cetus* (L.), whale.

-chaene-, -chene *chainein* (Gk), to gape.

chaet- *chaitē* (Gk), hair.

chela- *chēlē* (Gk), claw.

chemi-, chemistry *chēmeia* (Gk), trans-mutation.

chiasm- *chiasma* (Gk), cross.

chitin *chiton* (Gk), tunic.

chlamy- *chlamys* (Gk), cloak, mantle.

chloro- *chlōros* (Gk), yellow, green, pale.

chondro- *chondros* (Gk), cartilage.

-chord, chorda-, chordo- *chordē* (Gk), string.

-chore *chōros* (Gk), place.

chroma-, chromo-, -chrome *chrōma* (Gk), colour.

chrono- *chronos* (Gk), time.

-cidal *caedere* (L.), to kill.

clade, -cladous *klados* (Gk), branch.

clav- *clava* (L.), club.

cleisto- *kleistos* (Gk), closed.

clino-, -cline, -clinous *klinē* (Gk), bed.

-clinous *klinein* (Gk), to bend.

clype- *clypeus* (L.), shield.

cocc-, cocco- *koccos* (Gk), berry.

cochli- *kochlias* (Gk), a snail.

coel- *koilos* (Gk), hollow.

coen-, -coenosis *koinos* (Gk), shared in common.

coleo- *koleos* (Gk), sheath.

conch- *concha* (L.), shell, *kongchē* (Gk), shell.

cono-, -cone *kōnos* (Gk), cone.

copro- *kopros* (Gk), dung.

-corn, corne-, corni- *cornus* (L.), horn.

corona- *corona* (L.), crown.

corp-, corpor- *corpus* (L.), body.

cortex, cortic- *cortex* (L.), bark.

costa- *costa* (L.), rib.

cotyl- *kotylē* (Gk), cup.

coxa-, coxo- *coxa* (L.), hip.

crani-, crania- *kranion* (Gk), skull.

-crine *krinein* (Gk), to separate.

cruci- *crux* (L.), cross.

cryo- *kryos* (Gk), frost.

crypto- *kryptos* (Gk), hidden.

cyano- *kyanos* (Gk), dark blue.

cyath- *kyathus* (Gk), cup.

cyclo-, -cyclic *kyklos* (Gk), circle.

cyst- *kystis* (Gk), bladder.

-cyte, cyto- *kytos* (Gk), hollow.

-dactyl *daktylos* (Gk), finger.

de- *de* (L.), away.

deme, demo- *dēmos* (Gk), people.

demi- *dimidius* (L.), half.

dendr- *dendron* (Gk), tree.

dent- *dens* (L.), tooth.

derma-, dermo-, -derm *derma* (Gk), skin.

desm- *desmos* (Gk), bond.

di- *dis* (Gk), twice.

dia- *dia* (Gk), asunder.

dicho- *dicha* (Gk), in two.

dictyo- *dictyon* (Gk), net.

digito- *digitus* (L.), finger.

dino- *dinos* (Gk), rotation.

dino- *deinos* (Gk), terrible.

diplo- *diploos* (Gk), double.

dors- *dorsum* (L.), back.

-drome, -dromic, -dromous *dramein* (Gk), to run, *drōmos* (Gk), running.

-duct *ducere* (L.), to lead.

duplico- *duplex* (L.), double.

dynamo- *dynamis* (Gk), power.

dys- *dys* (Gk), mis-.

e- *ex* (L.), out of.

ec- *ek* (Gk), out of.

echino- *echinos* (Gk), spine.

eco- *oikos* (Gk), house, household.

ect- *ektos* (Gk), without, outside.

-ectomy *ektomē* (Gk), a cutting out.

elaeo-, elaio- *elaion* (Gk), oil.

electro- *elektron* (Gk) amber (a fossil resin which produces static electricity when rubbed).

embryo- *embryon* (Gk), embryo.

endo- *endon* (Gk), within, inside.

-ennial *annus* (L.), year.

entero- *enteron* (Gk), gut.

entomo- *entomon* (Gk), insect.

epi- *epi* (Gk), upon.

equi- *aequus* (L.), equal.

erg-, -ergic, -ergy *ergon* (Gk), activity, work.

erythro- *erythros* (Gk), red.

eu- *eu* (Gk), well.

eury- *eurys* (Gk), wide.

exo- *exō* (Gk), outside.

extra- *extra* (L.), beyond.

-farious *fariam* (L.), in rows.

fauna- *faunus* (L.), god of woods.

ferre-, ferri-, ferro- *ferrum* (L.), iron.

fibrino- *fibra* (L.), band.

-fid *findere* (L.), to split.

fili- *filum* (L.), thread.

flavo- *flavus* (L.), yellow.

flor- *flos* (L.), flower.

folia- *folium* (L.), leaf.

fronto- *frons* (L.), forehead.

fuco- *fucus* (L.), seaweed.

-fugal, -fuge *fugere* (L.), to flee.

galacto- *gala* (Gk), milk.

gamete, gameto- *gametes* (Gk), spouse.

gamo-, -gamy, -gamous *gamos* (Gk), marriage.

ganglio- *ganglion* (Gk), swelling.

gastro- *gaster* (Gk), stomach.

-geminal *geminus* (L.), double.

-gen, -genous *genos* (Gk), descent.

gene-, -genetic *genesis* (Gk), birth, descent, origin.

-genic, -genous *gennaein* (Gk), to produce.

genito- *gignere* (L.), to beget.

geno- *genos* (Gk), race.

geo- *gē*, or *gaia* (Gk), Earth.

germ-, germin- *germen* (L.), bud.

geronto- *gerōn* (Gk), old man.

glia-, -gloea *gloia* (Gk), glue.

-globin, -globulin *globus* (L.), sphere.

glosso- *glossa* (L.), tongue.

gluco-, glyco- *glykys* (Gk), sweet.

gnatho-, -gnath, -gnathous *gnathos* (Gk), jaw.

gonad-, -gone, -gonic *gonē* (Gk), seed.

-gone, goni- *gonos* (Gk), offspring.

-grade *gradus* (L.), step.

-gram, -graphy *graphein* (Gk), to write.

gyn-, -gynous *gynē* (Gk), female.

haem-, haema- *haima* (Gk), blood.

halo- *hals* (Gk), sea, salt.

haplo- *haploos* (Gk), simple.

hapto- *haptos* (Gk), touch.

helio- *helios* (Gk), sun.

hemi- *hēmi* (Gk), half.

hepa-, hepatico- *hepar* (Gk), liver.

hepta- *hepta* (Gk), seven.

hetero- *heteros* (Gk), other.

hex- *hex* (Gk), six.

histio- *histion* (Gk), tissue.

histo- *histos* (Gk), tissue.

holo- *holos* (Gk), whole.

homeo- *homoios* (Gk), alike.

homo- *homos* (Gk), the same.

hormone *hormaein* (Gk), to excite.

hyalo- *hyalos* (Gk), glass.

hydr-, hydro- *hydor* (Gk), water.

hygro- *hygros* (Gk), wet.

hyper- *hyper* (Gk), above.

hypo- *hypo* (Gk), under.

ichthy- *ichthys* (Gk), fish.

-icole, -icolous *colere* (L.), to dwell.

-iferous *ferre* (L.), to carry.

-ific, -ification *facere* (L.), to make.

-igen *generare* (L.), to beget.

-igerous *gerere* (L.), to bear.

im-, in- *in* (L.), not.

immuno- *immunis* (Gk), free.

in- *in* (L.), into.

infero- *inferus* (L.), beneath.

infra- *infra* (L.), below.

inter- *inter* (L.), between.

intra- *intra* (L.), within.

iso- *isos* (Gk), equal.

-jugate *jugare* (L.), to join.

jugo- *jugum* (L.), yoke.

juxta- *juxta* (L.), close to.

kary- *karyon* (Gk), nucleus, nut.

kera- *keras* (Gk), horn.

-kinesis, -kinetic *kinesis* (Gk), movement, *kinein* (Gk), to move.
labia-, labio- *labium* (L.), lip.
lacto- *lac* (L.), milk.
lati-, latero- *latus* (L.), wide.
lepido- *lepidotos* (Gk), scaly.
lepto- *leptos* (Gk), slender.
leuco-, leuko- *leukos* (Gk), white.
limn- *limne* (Gk), marsh.
lipo- *lipos* (Gk), fat.
litho-, -lith *lithos* (Gk), stone.
lopho- *lophos* (Gk), crest.
luci- *lux* (L.), light.
luteo- *luteus* (L.), orange-yellow.
-lysin, -lysis, lyso-, -lytic *lysis* (Gk), loosing, *lyein* (Gk), to dissolve.
macro- *makros* (Gk), large.
masto- *mastos* (Gk), breast.
matro- *mater* (L.), mother.
medi- *medius* (L.), middle.
mega- *megas* (Gk), large.
megalo- *megalon* (Gk), great.
meio- *meion* (Gk), smaller.
-mere, mero- *meros* (Gk), a part.
meso- *mesos* (Gk), middle.
meta- *meta* (Gk), after.
metabolism, metabolic *metabole* (Gk), change.
-metric, -metry *metron* (Gk), measure, *metreo* (Gk), to count.
micro- *mikros* (Gk), small.
mito-, mitosis *miton* (GK), thread.
mono- *monos* (Gk), alone, single.
morpho-, -morph, -morphism, -morphy *morphe* (Gk), shape, form, *morphosis* (Gk), form.
multi- *multus* (L.), many.
mutate, muta- *mutare* (L.), to change.
myco-, -mycin *mykes* (Gk), fungus.
myelo- *myelos* (Gk), marrow.
myo- *mys* (Gk), muscle.
myrme- *myrmēx* (Gk), ant.
myxo- *myxa* (Gk), slime.
nano- *nanos* (Gk), dwarf.
necro- *nekros* (Gk), dead.
nema-, nemato-, -neme *nēma* (Gk), a thread.
neo- *neos* (Gk), new.
nephr- *nephros* (Gk), kidney.
neuro- *neuron* (Gk), nerve.
nexus, -nexed *nectare* (L.), to bind.
nigro- *niger* (L.), black.
nitro- *nitron* (Gk), soda.

noci- *nocere* (L.), to hurt.
-nomics, -nomy *nomos* (Gk), law.
nomin-, -nomial *nomen* (L.), name.
noto- *noton* (Gk), back.
nucleus *nucleus* (L.), kernel.
ob- *ob* (L.), against, reversely.
occipi- *occiput* (L.), back of head.
octa-, octo- *okta* (Gk), *octo* (L.), eight.
-odont, odonto- *odous* (Gk), tooth.
-oecious, -oecium *oikos* (Gk), house.
oestro-, oestrus *oistros* (Gk), gadfly.
-ogen *genos* (Gk), birth.
-ogony *gonos* (Gk), generation.
oi-, -oo- *ōon* (Gk), egg.
-oid *eidos* (Gk), form.
olei-, oleo- *oleum* (L.), oil.
oligo- *oligos* (Gk), few.
-ology *logos* (Gk), discourse.
onco- *onkos* (Gk), bulk, mass.
onto- *on* (Gk), being.
-oo- *ōon* (Gk), egg.
ophthal- *ophthalmos* (Gk), eye.
opsi-, opsin, -opsy, opto- *opsis* (Gk), eye.
ora-, oro- *os, oris* (L.), mouth.
organ *organon* (Gk), instrument.
orni- *ornis* (Gk), bird.
ortho- *orthos* (Gk), straight.
osmo- *ōsmos* (Gk), impulse.
ost-, osteo- *osteon* (Gk), bone.
ostraco- *ostrakon* (Gk), shell.
oto- *ous* (Gk), ear.
-otomy *temnein* (Gk), to cut.
ova-, ovi-, ovo- *ovum* (L.), egg.
oxy- *oxys* (Gk), sharp.
pachy- *pachys* (Gk), thick.
palaeo- *palaios* (Gk), ancient.
palpi- *palpare* (L.), to stroke.
pan-, panto- *pan* (Gk), all.
para- *para* (Gk), beside.
-parous *parere* (L.), to produce.
patho-, -pathy *pathos* (Gk), suffering.
patri- *pater* (L.), father.
-patric *patria* (L.), native land.
ped-, -pedal *pes* (L.), foot.
-pelagic *pelagos* (Gk), sea.
penta-, pento- *pente* (Gk), five.
per- *per* (L.), through.
peri- *peri* (Gk), around.
perisso- *perissos* (Gk), odd (numbered).
petalo- *petalon* (Gk), leaf.
petro- *petros* (Gk), stone.
-phage, phago, -phagous *phagein* (Gk), to eat.

-phase *phasis* (Gk), aspect, appearance.
-phene, pheno- *phainein* (Gk), to appear.
-phil, -phile, -phily *philein* (Gk), to love.
philo- *philos* (Gk), loving.
-phobe, -phobic *phobos* (Gk), fear.
phono- *phōnē* (Gk), sound.
-phore *phorein* (Gk), to carry.
phospho- *phosphoros* (Gk), bringing light.
photo- *phos* (Gk), light.
phragmo- *phragmos* (Gk), fence.
-phylactic *phylaktikos* (Gk), fit for preserving.
-phyll, -phyllous *phyllon* (Gk), leaf.
physi-, -physis *physis* (Gk), growth.
physics *physis* (Gk), nature.
phyt-, -phyte *phyton* (Gk), plant.
pinna-, penna- *penna* (L.), feather.
pisci- *piscis* (L.), fish.
placo- *plax* (Gk), plate.
plana- *planatus* (L.), flattened.
-planetic, planeto- *planetes* (Gk), wanderer.
plano- *planos* (Gk), wandering.
-plasia *plasis* (Gk), a moulding, *plassein* (Gk), to form.
plasma-, -plasm, *plasma* (Gk), form.
-plast, -plastic, plastid *plastos* (Gk), formed.
plasti-, plasto- *plastos* (Gk), formed.
platy- *platys* (Gk), flat.
pleio- *pleion* (Gk), more.
plero- *plērēs* (Gk), full.
plesio- *plēsios* (Gk), near.
pleura-, -pleurite *pleuros* (Gk), side.
-plicate *plicare* (L.), to fold.
-ploid *aploos* (Gk), one-fold, and *eidos* (Gk), form.
pluri- *plus* (L.), more.
pneu- *pnein* (Gk), to breathe.
-pod, -podite *pous* (Gk), foot.
-poiesis, -poietic *poiesis* (Gk), making.
poly- *polys* (Gk), many.
poro-, -pore *poros* (Gk), channel.
porphyr- *porphyra* (Gk), purple.
post- *post* (L.), after.
pre- *prae* (L.), before.
primo- *primus* (L.), first.
pro- *pro* (L.), before.
proto- *prōtos* (Gk), first.
pseudo- *pseudes* (Gk), false.
psycho- *psychē* (Gk), mind.
-pter- *pteron* (Gk), wing.
pterido- *pteris* (Gk), fern.
ptero- *pteron* (Gk), wing.

pteryg- *pterygion* (Gk), little wing.
pteryg- *pterygion* (Gk), fin.
-ptile *ptilon* (Gk), feather.
pulmo- *pulmo* (L.), lung.
pycno- *pyknos* (Gk), dense.
pygo- *pygē* (Gk), rump.
-pyle *pyle* (Gk), gate.
pyreno- *pyrēn* (Gk), fruit stone.
pyri- *pyrum* (L.), pear.
pyrro- *pyrrhos* (Gk), tawny-red.
quadrato- *quadratus* (L.), squared.
quadri-, quadru- *quattuor* (L.), four.
quin-, quinque- *quinque* (L.), five.
racem- *racemus* (L.), bunch.
rachi- *rachis* (Gk), spine.
radic- *radix* (L.), a root.
radio- *radius* (L.), a ray.
-ramous *ramus* (L.), branch.
rani- *rana* (L.), frog.
re- *re* (L.), back.
rena-, reni- *renes* (Gk), kidney.
reti- *rete* (L.), net.
reticulo- *reticulum* (L.), small net.
retro- *retro* (L.), backwards.
rhabdo- *rhabdos* (Gk), rod.
rheo- *rheein* (Gk), to flow.
rhin- *rhis* (Gk), nose.
rhiza-, rhizo- *rhiza* (Gk), root.
rhodo- *rhodon* (Gk), rose.
rhyncho- *rhyngchos* (Gk), snout.
ribo- *ribes* (L.), currant.
rostra- *rostrum* (L.), beak.
rubi- *ruber* (L.), red.
rubro- *ruber* (L.), red.
sacchar- *sakchar* (Gk), sugar.
salpingo- *salpingx* (Gk), trumpet.
sangui- *sanguis* (L.), blood.
sapo- *sapo* (L.), soap.
sapro- *sapros* (Gk), decayed.
-sarc, sarco- *sarx, sarkōdēs* (Gk), flesh, fleshy.
-saur- *sauros* (Gk), lizard.
schisto-, -schist *schistos* (Gk), split.
schizo- *schizein* (Gk), to cleave.
-scopic *skopein* (Gk), to view.
scoto- *skotos* (Gk), dark.
scute, scutum *scutum* (L.), shield.
seismo- *seismos* (Gk), a shaking.
seleno- *selēnē* (Gk), moon.
-sematic *sema* (Gk), signal.
semi- *semi* (L.), half.
septa-, septi-, septo- *septum* (L.), partition.

septi- *septum* (L.), seven.
-sere *serere* (L.), to put in a row.
serum *serum* (L.), whey.
seta-, seti-, seto- *seta* (L.), bristle.
siali-, sialo- *sialon* (Gk), saliva.
sidero- *sidēros* (Gk), iron.
soma-, somato-, -some *sōma* (Gk), body.
sora-, sori-, soro- *sōros* (Gk), heap.
speleo- *spelaion* (Gk), cave.
sperma-, -sperm *sperma* (Gk), seed.
sphaero- *sphaira* (Gk), globe.
spheno- *sphen* (Gk), a wedge.
sphero- *sphaira* (Gk), globe.
spondyl- *sphondylos* (Gk), vertebra.
spor-, -spore *sporos* (Gk), seed.
squame, squama- *squama* (L.), scale.
stachy- *stachys* (Gk), ear of corn.
stamin-, -stemonous *stemon* (Gk), spun thread.
-stat, -static *stare* (L.), to stand.
stato- *statos* (Gk), stationary , standing.
stega-, stegi- *stega* (Gk), roof.
-stelic, -stely *stele* (Gk), pillar.
stereo-, -steric *stereos* (Gk), solid.
stern- *sternum* (L.), breast.
steroid, -sterone *stear* (Gk), suet.
stoma-, stomato-, -stome *stoma* (Gk), mouth.
-strate *stratum* (L.), layer.
strepto- *streptos* (Gk), twisted, pliant.
strobilus *strobilos* (Gk), fir-cone.
strom-, stroma *stroma* (Gk), bedding.
-stylic *stylos* (Gk), pillar.
sub- *sub* (L.), under.
super- *super* (L.), over.
supra- *supra* (L.), above.
sylv- *sylva* (L.), forest.
sym- *syn* (Gk), with.
syn- *syn* (Gk), with.
synaps-, synapto- *synapsis, synaptos* (Gk), union, joined.
synthesis *synthesis* (Gk), composition.
tachy- *tachys* (Gk), quick.
talo- *talus* (L.), ankle.
tarso- *tarsos* (Gk), sole of foot.
tauto- *tautos* (Gk), the same.
-taxis, -taxy *taxis* (Gk), arrangement.
taxo- *taxis* (Gk), arrangement.
tect- *tectum* (L.), roof.
tele- *tēle* (Gk) far.
teleo- *teleos* (Gk), complete.
telo-, telio- *telos* (Gk), end.
tempor- *tempora* (L.), temples.

-tene *tainia* (Gk), band.
tensin- *tonos* (Gk), tension.
terga- *tergum* (L.), back.
ternato- *terni* (L.), three.
terr- *terra* (L.), Earth.
tetra- *tetras* (Gk), four.
thallo- *thallos* (Gk), branch.
-theca, -thecium *theke* (Gk), box.
-theria *therion* (Gk), small animal.
thermo- *thermē* (Gk), heat.
thero- *theros* (Gk), summer.
thigmo- *thigēma* (Gk), touch.
thio- *theion* (Gk), sulphur.
thrombo- *thrombos* (Gk), clot.
thylako- *thylakos* (Gk), pouch.
thyro- *thyra* (Gk), door.
-tocin *tokos* (Gk), birth.
-tope, -topic, topo- *topos* (Gk), place.
toti- *totus* (L.), all.
toxico-, *-toxin* toxikon *(Gk), poison.*
tracheo- *trachia* (L.), windpipe.
trachy- *trachys* (Gk), rough.
trans- *trans* (L.), across.
trauma- *trauma* (Gk), wound.
tri- *tria* (Gk), three, *tres* (L.), three.
trich- *thrix* (Gk), hair.
trocho- *trochos* (Gk), hoop.
-troph, -trophic, tropho-, -trophy *trophē* (Gk), maintenance, nourishment.
-tropic, *-tropism* trope *(Gk), turn.*
tropo- *tropos* (Gk), turn.
tubi-, tubo- *tubus* (L.), pipe.
tympano- *tympanon* (Gk), drum.
-type *typos* (Gk), pattern.
ulna-, ulno- *ulna* (L.), elbow.
ultra- *ultra* (L.), beyond.
umbell- *umbella* (L.), sunshade.
umbona- *umbo* (L.), shield boss.
unci- *uncus* (L.), hook.
uni- *unus* (L.), one.
uredo- *urēdo* (L.), blight.
uro- *ouron* (Gk), urine or *oura* (Gk), tail.
vacuol- *vacuus* (L.), empty.
vagini- *vagina* (L.), sheath.
valv- *valvae* (L.), folding doors.
vasa-, vaso- *vas* (L.), vessel.
ventr- *venter* (L.), belly.
vermi- *vermis* (L.), worm.
versi- *versare* (L.), to turn.
vesicul- *vesicula* (L.), small bladder.
viro- *virus* (L.), poison.
vitello- *vitellus* (L.), yolk.
vitreo- *vitreus* (L.), glassy.

vivi- *vivus* (L.), living.
-vorous *vorare* (L.), to devour.
xantho- *xanthos* (Gk), yellow.
xeno-, -xenous *xenos* (Gk), host or strange.
xero- *xeros* (Gk), dry.

zo-, zoo- *zōon* (Gk), animal.
-zoic *zoikos* (Gk), pertaining to life.
zygo-, zygote *zygon* (Gk), yoke, *zygotos* (Gk), yoked.
zymo-, -zyme *zymē* (Gk), leaven.